Modern Aryne Chemistry

Modern Aryne Chemistry

Edited by
Akkattu T. Biju

WILEY-VCH

Editor

Prof. Akkattu T. Biju
Department of Organic Chemistry
Indian Institute of Science
Bangalore – 560012
India

Cover
Cover Image: (background)
PublicDomainPictures/Pixabay
(inset) Courtesy of Akkattu T. Biju

Library of Congress Card No.:
applied for

British Library Cataloguing-in-Publication Data
A catalogue record for this book is available from the British Library.

Bibliographic information published by the Deutsche Nationalbibliothek
The Deutsche Nationalbibliothek lists this publication in the Deutsche Nationalbibliografie; detailed bibliographic data are available on the Internet at <http://dnb.d-nb.de>.

Print ISBN: 978-3-527-34646-2
ePDF ISBN: 978-3-527-82307-9
ePub ISBN: 978-3-527-82309-3
oBook ISBN: 978-3-527-82308-6

Typesetting SPi Global, Chennai, India
Printing and binding CPI Group (UK) Ltd, Croydon, CR0 4YY

Printed on acid-free paper

C095372_290421

Contents

Foreword

Arynes as fleeting intermediates in organic reactions were first recognized by Stoermer and Kahlert in 1902, but the solid experimental evidence of their (benzyne) involvement came through the seminal ^{14}C tracer studies by J. D. et al. (1953) on the amination of haloarenes, and very recently Pavlićek et al. (2015) have directly observed surface-generated benzyne by atomic force microscopy (AFM). Concurrently, many preparatively useful methods for generating arynes have been devised and their diverse reactivity landscape has been extensively explored and profiled. A breakthrough in the exploitation of arynes in wide range of productive applications in organic synthesis, including total synthesis of natural products, came about through Kobayashi's discovery (1983) that stable 2-(trimethylsilyl)aryl triflates (now commercially available) undergo facile fluoride-mediated 1,2-elimination to generate arynes safely and efficiently. This convenient access to arynes gave a major fillip to mapping the diverse and potentially rich synthetic utility of these highly reactive intermediates in the syntheses of arenes harboring structural complexity and fostering molecular diversity. The next steps in the advance of aryne arena came from the hexadehydro-Diels–Alder reaction developed by Hoye, domino generation of arynes by Li, and the development of heteroarynes by Garg among many other tactical innovations that have immensely enriched the field and amplified the expanse of aryne chemistry. Intramolecular variants, domino and tandem reactions with embellished arynes still offer innovative chemical spaces that await unraveling. While chemical explorations of aryne reactivity cover a vast and varied arena, they can be broadly categorized under lead headings like pericyclic reactions, insertion reactions, multicomponent reactions, transition-metal-catalyzed transformations, and a variety of unforeseen and interesting molecular rearrangements.

The landscape of aryne chemistry has grown by leaps and bounds since the publication of R. W. Hoffmann's seminal monograph *Dehydrobenzene and Cycloalkynes* over half a century ago in 1967. A recent SciFinder search on "benzyne" and "aryne" led to −7500 and −4500 hits, respectively – clearly indicative of a fertile field and the traction it has drawn in the recent decades. These developments in aryne chemistry have been periodically captured in several timely and authoritative accounts and reviews covering diverse facets of this growing field. Nevertheless, there is a much-felt need among students and researchers in the field for an authoritative book that provides a broad, up-to-date coverage of multifaceted advances in aryne chemistry with a nuanced lens on developments that are of topical interest and likely to dominate future directions and activities and also provide a fuller flavor of the field.

Against this background and to fill a widely felt void, Akkattu T. Biju, an accomplished contributor in the arena, has put together an authoritative and timely collection of contributions from leading practitioners in the field of aryne chemistry under the title *Modern Aryne Chemistry*. This book is primarily aimed at highlighting some of the recent advances in carbon–carbon and carbon–heteroatom bond-forming reactions of arynes, as highly reactive and versatile electrophilic species, with numerous, contextual synthetic applications.

The first chapter of the book introduces the chemistry of arynes. This overarching contribution by Akkattu T. Biju (IISc Bangalore) describes the history of arynes, various methods for their generation, characterization techniques, and possible modes of reactivity. A detailed account of the application of arynes in cycloaddition reactions has been presented in the second chapter by E. Guitián from Universidade de Santiago de Compostela, Spain. The synthetic potential of arynes for accessing various polycyclic aromatic hydrocarbons (PAHs) and nanographenes has been highlighted in this chapter. The focal theme of the third chapter of the book compiled by F. Shi (WuXi AppTec Co., Ltd. Wuhan), and P. Li (Henan University) is on dipolar cycloaddition reactions of arynes and provides the interception of arynes with various dipoles for the synthesis of benzo-fused heterocycles. An overview "Recent advances in the insertion reactions of arynes" forms the fourth chapter by S. Yoshida and T. Hosoya from Tokyo Medical and Dental University, Japan. This chapter provides an interesting feature of aryne intermediates to insert into various element-to-element bonds to form 1,2-disubstituted arenes. A variety of nucleophiles can add to arynes and the generated aryl anion can be intercepted with electrophiles to orchestrate multicomponent reactions. A complete coverage of transition-metal-free aryne multicomponent reactions for the synthesis of complex 1,2-disubstituted arenes has been presented by H. Yoshida from Hiroshima University in the fifth chapter.

The sixth chapter of the book by C.-H. Cheng (National Tsing Hua University, Taiwan), M. Jeganmohan (IIT Madras), and K. Parthasarathy (University of Madras) provides a detailed account of the transition-metal-catalyzed reactions involving arynes. Transition-metal-catalyzed cycloisomerization reactions, C—H and N—H bond activations involving arynes leading to annulation reactions, three-component coupling reactions, etc. are described in this chapter. Arynes serve as versatile precursors for a number of molecular rearrangements, resulting in the synthesis of diverse and structurally attractive organic compounds that are otherwise difficult to access. The recent developments in molecular rearrangements triggered through aryne intermediates are summarized in the seventh chapter by S.-K. Tian from University of Science and Technology of China, Hefei. The eighth chapter of the book is dedicated to the new strategies and latest developments in this field and deals with new methods of aryne generation, addresses the regioselectivity and multifunctionalization issues in aryne chemistry, and is authored by Y. Li from Chongqing University. A brief history of hetarynes, cycloalkynes and related potential intermediates, different methods of their generation, various types of reactions and applications in total synthesis are briefly discussed in the ninth chapter by Akkattu T. Biju (IISc Bangalore). The penultimate chapter contributed by T. R Hoye (University of Minnesota) offers the hexadehydro Diels–Alder (HDDA) route to arynes and related chemistry. Although this reaction was first revealed in 1997, it received an expansive growth after a report in 2012 demonstrating its potential as a general strategy for generating reactive benzyne intermediates from simple triyne precursors. The potential applications of arynes in the synthesis of natural products and biologically active molecules have been highlighted by K. Suzuki (Tokyo Institute of Technology) and H. Takikawaa (Kyoto University) in the last chapter of this monograph.

It is expected that in addition to the pedagogic value of the book for the students and the general reader, the simplicity and sophistication of the synthetic strategies using aryne chemistry will also inspire and entice a wide range of organic chemists to explore new reactivity patterns and imaginative applications in total synthesis, material science, and chemical biology. It is reasonable to believe that aryne chemistry will continue to flourish and lead to many interesting/serendipitous observations in the future to enrich organic chemistry. Thus, from a wider perspective, this timely book is expected to serve many purposes to augment and advance organic chemistry and its interfaces.

Hyderabad-500046, India
4 December 2020

Goverdhan Mehta,
Dr. Kallam Anji Reddy Chair
School of Chemistry
University of Hyderabad

Preface

Carbon–carbon and carbon–heteroatom bond-forming reactions constitute the backbone of synthetic organic chemistry. When it comes to the utilization of the ring strain in these bond-forming reactions, arynes have always been at the forefront (having low-lying lowest unoccupied molecular orbital [LUMO] with a strain energy of 63 kcal mol^{-1}). Arynes are highly reactive intermediates, which are generated in situ due to their high reactivity. The chemistry of this century-old intermediate, which marked its birth in history with the vague evidence of its existence provided by Stoermer and Kahlert in 1902, has gained outstanding acceleration toward the end of the twentieth century. This unstable intermediate has been characterized using various spectroscopic methods by different research groups. In the course of time, various research groups have developed several methods for the mild generation of arynes. However, the procedure became much simpler since 1983 when Kobayashi uncovered a facile and mild method for generation of arynes from 2-(trimethylsilyl) aryl triflates using simple fluoride sources. Moreover, the reagent-free and metal-free generation and reactivity of arynes generated utilizing the concept of intramolecular hexadehydro Diels–Alder reaction (HDDA) of triynes at elevated temperature has also been studied in detail.

Arynes, being a polarizable intermediate, have a diverse reactivity profile due to their affinity toward charged and uncharged electron donors. Arynes serve as excellent dienophile and dipolarophile in pericyclic reactions such as Diels–Alder reactions, [2+2] cycloadditions and dipolar cycloaddition reactions. Arynes hold the potential to arylate a number of molecules like alcohols, amines, and thiols and to insert into various element–element σ-bonds and π-bonds. Transition-metal-free multicomponent couplings (MCCs) and molecular rearrangements are the emerging class of reactivity of arynes. Moreover, arynes undergo a variety of transition-metal catalyzed reactions. A consecutive double nucleophilic addition realizing the concept of multifunctionalization of aryne using a novel domino aryne precursor was uncovered recently by Li and coworkers. Cycloaddition reactions involving arynes give access to the synthesis of large polycyclic aromatic hydrocarbons (PAHs) that are analogous to nanosized graphene substructures. The applications of arynes are not only limited to developing novel bond-forming reactions but also in natural product synthesis. The synthesis and reactivities of several five- and six-membered hetarynes, and strained cycloalkynes have also been a subject of interest for chemists. A book on arynes is a need of the hour as the area has crossed several milestones ever since the first book on arynes *Dehydrobenzene and Cycloalkynes* by R. W. Hoffmann was published in 1967. The lack of an update to this book incorporating all the developments in the past five decades made us realize this collection of information on arynes. The focus of the present book is on the history, diverse reactivity, and application of arynes in organic synthesis. Moreover, details on hetarynes, domino generation of arynes, and HDDA method of aryne generation have also been included in the book. As the chemistry of arynes has achieved considerable growth and continues expanding further, with the strong support of Wiley-VCH, we decided to bring out a new book under the title *Modern Aryne Chemistry* to highlight the developments occurred in this interesting area. Eleven chapters highlighting the history, different modes of reactivity, and the application of arynes are presented in this book.

The foundation of this book is based on the excellent contributions of all the colleagues working in aryne chemistry, and I am thankful to them. Moreover, I would like to thank all the authors, who have contributed enormously to this project, for their valuable time, efforts as well as the expertise to make this book a source of encouragement for beginners as well as advanced chemists practicing synthetic organic chemistry. It is anticipated that the diverse reactivity and application of arynes will inspire a broad range of organic chemists to explore new opportunities and creative applications of this concept and thereby unraveling some of the remaining challenges in this field. I am also thankful to Dr Lifen Yang (*Program Manager, Books & References*) and Ms Katherine Wong (*Senior Managing Editor*) at Wiley-VCH for their unconditional support and valuable advices in organizing/developing this book.

India
30 November 2020

Akkattu T. Biju
Associate Professor
Department of Organic Chemistry
Indian Institute of Science,
Bangalore 560012

1

Introduction to the Chemistry of Arynes

Tony Roy, Avishek Guin, and Akkattu T. Biju

Indian Institute of Science, Department of Organic Chemistry, Bangalore 560012, India

1.1 Introduction

Arynes are highly reactive electrophilic intermediates proposed more than a century ago and have encountered an extraordinary resurgence of interest in the past decades. Chemists have exploited this reactive intermediate for the synthesis of a broad range of 1,2-disubstituted benzene derivatives and also several benzo-fused carbocycles and heterocycles, which are otherwise difficult to achieve by conventional methods [1–14]. With the discovery of mild methods of generation by Kobayashi [15] and Hoye [16, 17], aryne chemistry has been significantly promoted in recent years in terms of better functional group compatibility as well as accommodation of more reaction modes. The progress in heterocyclic arynes, especially the pyridynes and indolynes, has added extra aroma to the chemistry of this reactive species [18]. The development and applications of the hexadehydro-Diels–Alder (HDDA)-based arynes went parallelly over the last decade and have contributed seminally to the diversification of this field. Moreover, 1,4-benzdiyne equivalents are one of the most dependable components for the synthesis of polycyclic aromatic functional materials at present. A recently exploited domino aryne reagent, the 2-(trimethylsilyl)-1,3-phenylene bis(trifluoromethanesulfonate) (TPBT), is one of the best precursors available for the synthesis of multifunctional aromatic derivatives [19]. A discussion on the brief history of arynes, their characterization, methods of generation, and possible modes of reactivities forms the content of this introductory chapter.

1.2 History of Arynes

Initial speculation on the existence of aryne intermediate appeared in 1902. Stoermer and Kahlert provided the first evidence for the existence of arynes when they observed the formation 2-ethoxybenzofuran in the reaction of

Modern Aryne Chemistry, First Edition. Edited by Akkattu T. Biju.

3-bromobenzofuran under basic conditions. They postulated the possible intermediacy of a 2,3-didehydrobenzofuran intermediate in this reaction [20]. The unexpected product formation laid the foundation stone for the development of an interesting area based on a highly transient intermediate (Scheme 1.1).

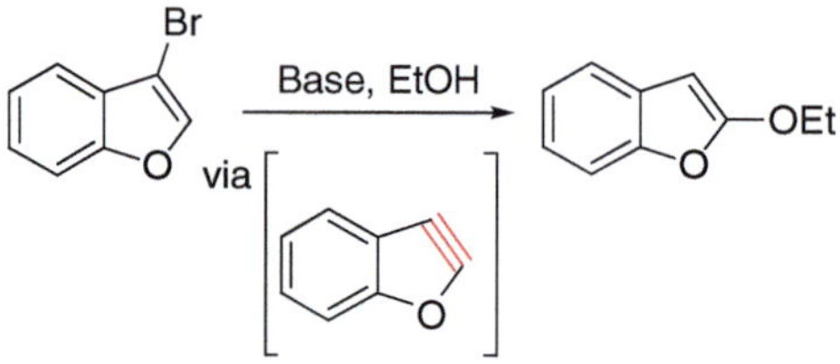

Scheme 1.1 Initial observations by Stoermer and Kahlert. Source: Based on Stoermer and Kahlert [20].

Later in 1927, Bachmann and Clarke at the Eastman Kodak Co. proposed benzyne as a reactive intermediate to explain the formation of triphenylene in a reaction, and it was taken as "decisive in favour of the free radical explanation" (Scheme 1.2) [21]. It was thought that structure **1** was predominant among the three possible structures, where the ylide structure **2** as well as the structure **3** with carbon–carbon triple bond in a six-membered ring were also considered.

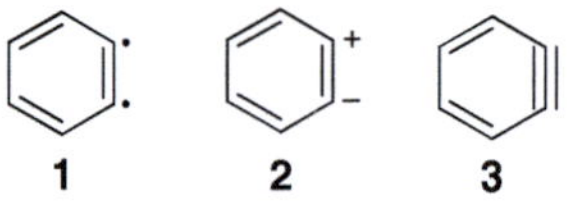

Scheme 1.2 Proposed structures of benzyne. Source: Based on Bachmann and Clarke [21].

Later, Wittig found that the formation of biphenyl in the reaction of phenyllithium with halobenzenes was fastest with fluorobenzene through nucleophilic substitution reaction via the displacement of fluoride, which was considered as a complicated task to perform [22–25]. Thus, the ylide structure **2** was proposed as a reactive intermediate for the formation of biphenyl. The trimerization of this zwitterion **2** to triphenylene was also noted under certain conditions. At the same time, Morton had also postulated the intermediate **2**, which "cannot be stabilized by double bond formation" in their study of the Wurtz reaction of pentylsodium with chlorobenzene [26]. But the zwitterionic structure failed to explain the observed regioselectivity in the reaction of substituted arynes, and hence its existence was questioned (Scheme 1.3). For instance, the reaction of aryne bearing an OMe group at 3-position with $NaNH_2$ was regioselective affording a single product, whereas the reaction of arynes having methyl substitution provided a 1 : 1 mixture of regioisomers.

It was also observed that the substitution only occurred in ipso or ortho position to that of halide substitution. Halides having no ortho hydrogen did not undergo any substitution reaction. Robert brought the official introduction of benzyne concept

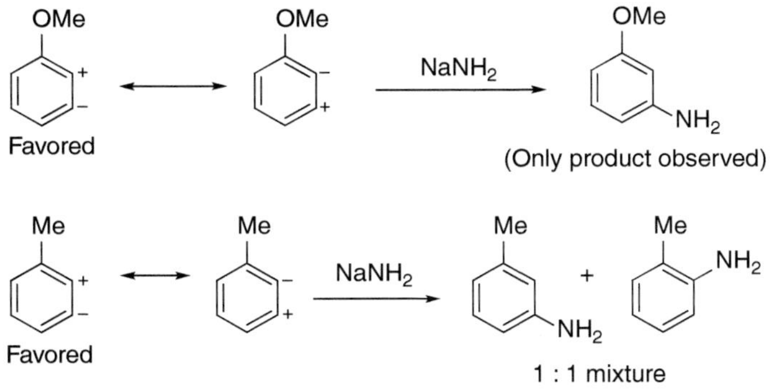

Scheme 1.3 Regioselectivity in aryne reactions.

[27, 28]. In 1953 at MIT, Robert performed a classical ^{14}C labeling experiment, which confirmed the involvement of a symmetrical, electronically neutral intermediate, benzyne **3** (Scheme 1.4). Later, Wittig performed the Diels–Alder reaction of benzyne with furan to give 76% yield of the [4+2] adduct [29].

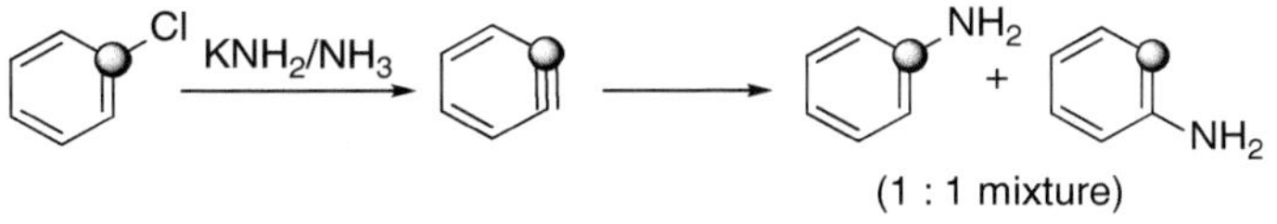

Scheme 1.4 Robert's ^{14}C-labeling experiment. Source: Roberts et al. [27]; Roberts et al. [28].

1.3 Characterization of the Aryne Intermediates

In 1963, Fisher and Lossing provided further insight to the structure **3** using mass spectrometry. They performed the pyrolysis of diiodobenzenes for all three isomers and identified **3** based on the measured ionization potential [30]. Confirmation for structure **3** also came from mass spectra. Berry et al. performed a photoinitiated benzenediazonium carboxylates decomposition in gas phase and identified the mass 76 along with other masses [31]. The same group further explained the structure **3** using UV spectra also. Radziszewsk and coworkers recorded the IR spectra providing a solid proof for the existence of structure **3**. The vibration to absorption peak emerged at 1846 cm^{-1} corresponding to *o*-benzyne, which was confirmed from different isotopomers of phthalic anhydride (Scheme 1.5) [32]. So from the IR data, it is understandable that unlike unstrained alkyne, benzyne triple bond is much weaker, as it has the stretching vibrations usually occurring in the range about 2150 cm^{-1}. IR data resemble cumulene-type structure. However, *o*-benzyne can be better explained by strained alkyne rather than biradical, which can be confirmed by alkyne-like reactivity and large singlet–triplet splitting [33, 34]. Wenthold and Squires determined the enthalpy of formation of structure **3** as

103.6–109.6 kcal mol^{-1}. The C—C triple bond length in acetylene falls at 120.3 pm and C—C double bond in ethylene at 133.9 pm. Experimental C—C triple bond length for benzyne is 122–126 pm, which is closer to that of alkyne triple bond length value, which aims at a cyclic alkyne-like structure rather than a cumulene-type structure [35–38]. Adding strength to all the above evidences, Warmuth was able to measure the nuclear magnetic resonance spectrum in solution in a hemicarcer and as a "molecular container" [39].

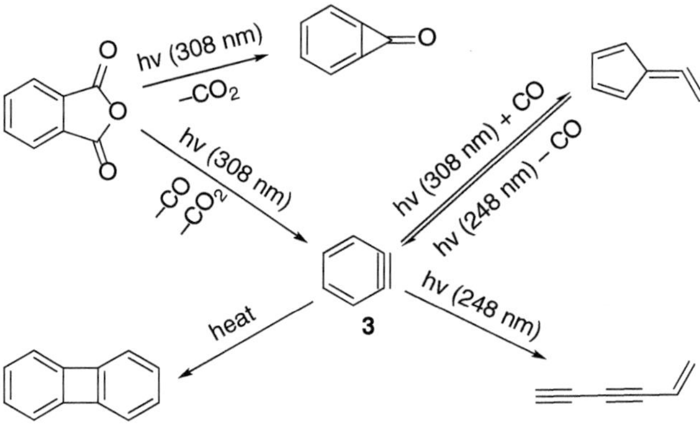

Scheme 1.5 Photochemistry of arynes. Source: Based on Berry et al. [32].

The reported ^{13}C value of 182 ppm for *o*-benzyne explains the strained alkyne character. Hoffman's demonstration of extended Hückel theory throws light on the electrophilicity of aryne. The LUMO of aryne is significantly lowered compared to dimethyl acetylene (5.1 eV) and the HOMO is also higher (0.1 eV) in energy, which make the aryne triple bond much more accessible toward different nucleophiles (Figure 1.1) [40].

Because of the high electrophilicity and reactivity of arynes, these intermediates cannot be isolated. However, stable transition metal complexes of benzynes have been prepared and analyzed by X-ray crystallography for further proof. Crystal structure of metal-bound benzyne shows that C_1—C_2 bond is more ethene-like

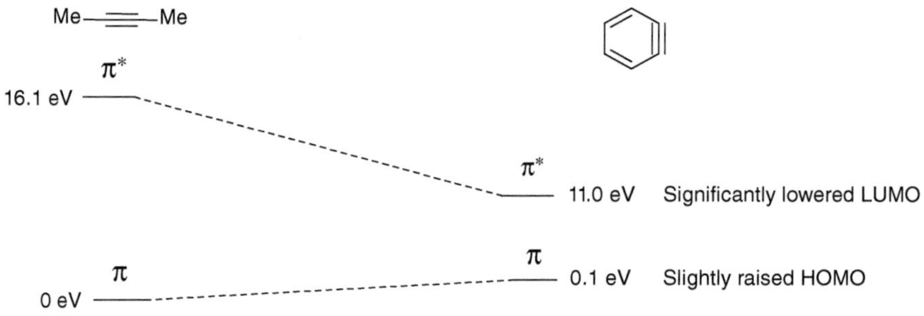

Figure 1.1 Reason for the enhanced electrophilicity of arynes. Source: Based on Hoffmann et al. [40].

(133–136 pm) and all other bond is normal benzene-like (138–140 pm). Thus, these metallocenes can be better described by metallacyclopropenes [41–43] (Scheme 1.6).

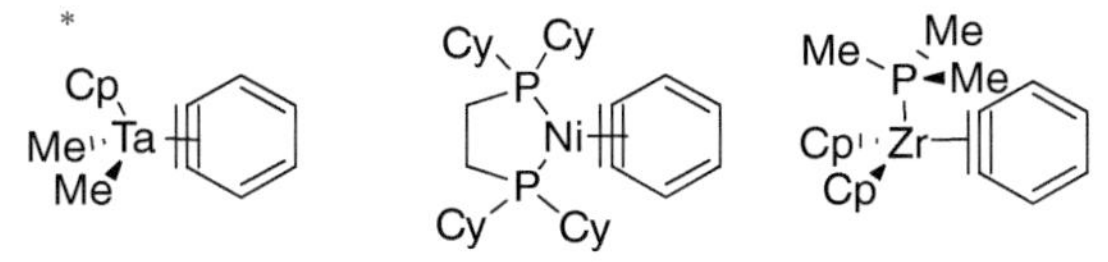

Scheme 1.6 Structures of metal-bound benzyne. Source: Buchwald et al. [41]; Bennett et al. [42]; Mclain et al. [43].

Due to the presence of a carbon–carbon triple bond in a six-membered ring, arynes are highly reactive and this also leads to the strained nature of the ring (~63 kcal mol^{-1}), and consequently, these species have low-lying LUMO, and hence the energy gap between the HOMO and LUMO is smaller than expected. In addition, arynes react as highly reactive alkynes in cycloaddition reactions. Moreover, the low-lying LUMO makes arynes a powerful electrophile for facile addition of nucleophiles.

1.4 *Ortho*-Arynes with Substitution

The effect of substitution on benzyne has game-changing effect on the reactivity of arynes. Numerous reactions go via the formation of benzyne intermediate. Substitution effect can help to understand the observed regioselectivity. Sometimes, experimental reactivity cannot be predicted due to the fact that the attack of the nucleophile is not always charge controlled. Introduction of a polar group at 3-positon influences the selectivity to a greater extent. But due to orthogonal nature of bond, classical electron-donating groups are withdrawing. Strain also can induce regioselectivity in benzyne intermediate, demonstrated by Suzuki [44]. Calculation of charge and LUMO coefficient matches prediction of bond angle strain with selectivity trend (Scheme 1.7) [45, 46].

Squires and Cramer theoretically studied naphthalynes [47]. Many naphthalyne syntheses were also reported in early 1970s. Lohmann studied the photochemistry of the two isomeric naphthalene dicarboxylic anhydrides **4** and **5**, using laserflash photolysis (LFP) and found that the dimerization of **7** is much faster than **3**; however, intermediate **6** dimerizes rarely (Scheme 1.8) [48]. Intermediates **6** and **7** are accessible when the photolysis conditions are chosen carefully [49]. Intermediate **6** can stay as alkyne but for **7**, it is more like cumulene. Intermediate **6** can be converted easily to **8** via photocarbonylation, but the analogous reaction of intermediate **7** to **9** is not observed experimentally.

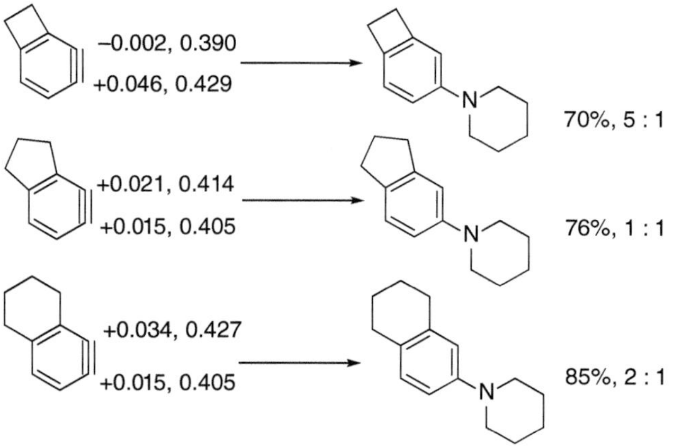

Scheme 1.7 Natural atomic charges (above) and atomic populations of LUMO coefficients (below) at C_1 and C_2 calculated with the B3LYP/6-311G(d,p) method. Source: Langenaeker et al. [45]; Johnson and Cramer [46].

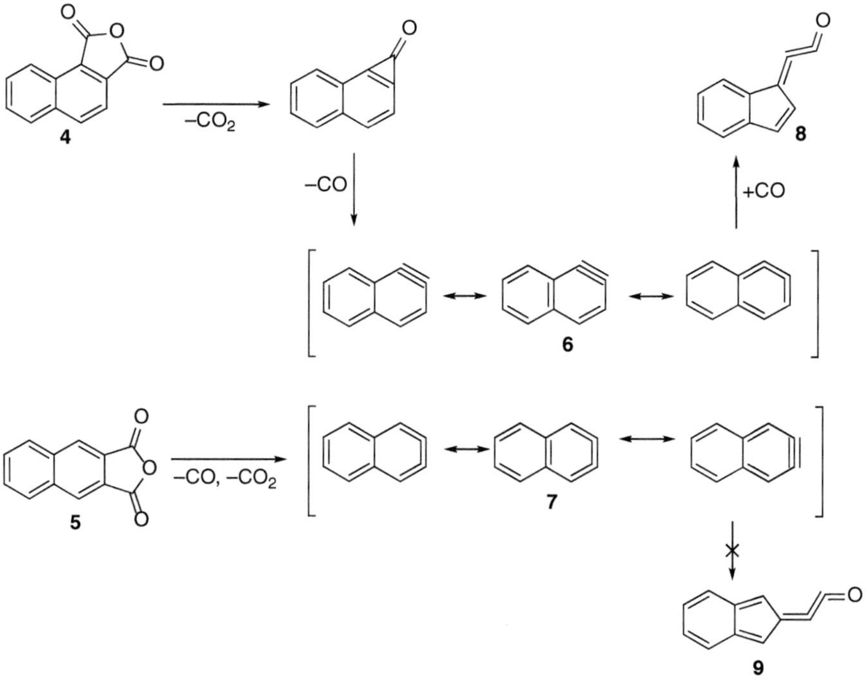

Scheme 1.8 Studies on naphthynes. Source: Lohmann [48].

1.5 *Ortho*-Arynes of Heterocycles

Although hetarynes are much older than benzynes, the physical data on hetarynes were negligible. Different groups reported the generation of five-membered biradicaloid intermediate, but direct spectroscopic evidence for these intermediates was not known due to the elevated ring strain, which helped these five-membered intermediates go through ring opening [50–53]. Thus, detecting lifetime for five-membered hetarynes by spectroscopic methods is quite less.

Among the six-membered hetarynes, the main focus was on didehydropyridines. Among didehydropyridines, the 3,4-pyridyne **11** generated by the photolysis of the precursor **10** is significantly more stable than the 2,3-isomer **13**. The bond length for the 3,4-pyridyne **11** is comparable with that of benzyne **3** [54]. Berry and coworkers detected 3,4-pyridyne **11** using mass spectrometry [55]. The trapping of the intermediate **11** in a Diels–Alder reaction attempted previously was not successful [56]. In 1988, Leroi and coworkers successfully trapped the pyridyne in nitrogen matrix and characterized it by IR spectroscopy [57]. However, the 2,3-didehydropyridine **13** generated from the precursor **12** by heating is less explored. It was assumed that compound **13** was formed as an intermediate in the gas-phase pyrolysis of 2,3-pyridine dicarboxylic anhydride **12** at 600 °C (Scheme 1.9) [58, 59].

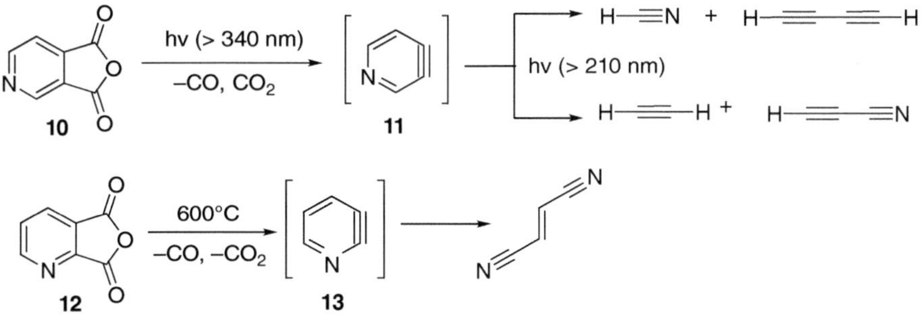

Scheme 1.9 Generation and dissociation of 3,4-pyridyne and 2,3-pyridyne. Source: Cava et al. [58]; Dunkin and MacDonald [59].

Recently, much effort has been put on the development and reactivity studies, including efficient computational model to get insight into the synthetic utility of heterocyclic arynes [60]. Moreover, different types of hetarynes can be generated at desired position of the heterocycles, thus leading to a variety of indolynes, pyridynes, benzofuranynes, and so on, which will be discussed in detail in Chapter 9 of this book.

1.6 Other Arynes

Similar to *ortho*-benzyne, arynes can also be generated in other positions of the benzene ring as well. *Meta*- and *para*-benzynes **14** and **15**, tridehydrobenzenes **16-18**, and tetradehydrobenzenes **19** are reported (Scheme 1.10) [61–64]. Tetradehydrobenzene (benzdiynes) can stay in different resonance form as illustrated in Scheme 1.10. Multireference methods are preferable for a proper quantum mechanical description [65, 66]. Due to high reactivity of 1,4-benzdiynes, it can form complex with transition metals and these complexes can be isolated in many cases [67–69].

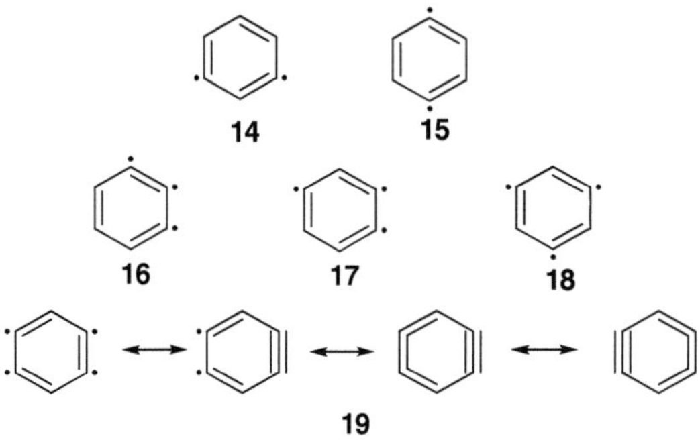

Scheme 1.10 Possibility of uncommon benzynes. Source: Wenthold [61].

The existence of *m*-benzyne has been a matter of discussion, partly due to experimental proof for the existence of an isomer, bicycle[3.1.0]hexatriene **21**, as a reactive intermediate formed by the base treatment of the bicyclohexene **20** (Scheme 1.11) [72]. Likewise, butalene **22** also denotes the existence of *p*-benzyne (Scheme 1.11) [70, 71]. However, both reactions are not fully understood due to complex nature of the reactions.

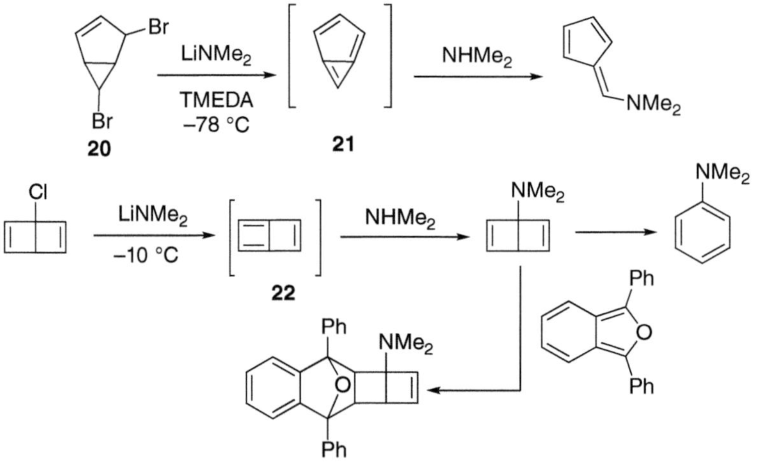

Scheme 1.11 Possible formation of *m*-benzyne and *p*-benzyne intermediates. Source: Breslow et al. [70]; Breslow and Khanna [71].

Early Hückel theory indicates that **21**, being a nonalternant "azulenoid" 6π electron system, maintains benzenoid resonance energy [73, 74]. Although it is a high-energy intermediate, it is resonance stabilized. On the other hand, iscyclobutadiene derivative **22** possesses antiaromatic character, and hence it is a high-energy species. General valence bond calculations estimated that the lowest energy forms of the 1,3- and 1,4-dehydrobenzenes are monocyclic structures having considerable biradical character [75]. In general, *m*-benzyne has much less biradical character and a larger singlet–triplet splitting than *p*-benzyne [76]. Gas-phase experimental studies and matrix isolation IR spectra are also consistent with monocyclic structure, and they do not correspond with **21** and **22**. Thus, the intermediate generated from the reactions shown in Scheme 1.11 is not clarified completely.

The Bergman cyclization (Scheme 1.12) is biologically a very important reaction because various natural products with antitumor and/or antibiotic properties contain such enediyne moieties **23** and their biological activity can be rationalized in terms of formation of transient *p*-benzynes. If *p*-benzyne abstracts hydrogen, it will generate phenyl radical, which can abstract hydrogen from DNA leading to the DNA cleavage [77–81]. The feasible interconversion of the *o*-, *m*-, and *p*-benzynes has been investigated theoretically. The experiments also indicated that the *o*- and *m*-isomers may rearrange to the *p*-isomer by H-atom tunneling and then the *p*-isomer undergoes the Bergman ring opening.

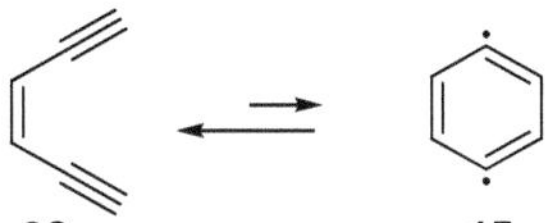

Scheme 1.12 The Bergman cyclization.

1.7 Methods of Aryne Generation

Due to the highly elevated reactivity, arynes are generated in situ in solution and cannot be isolated. However, ever since realizing the potential of this intermediate, different research groups have developed a number of methods for the mild aryne generation in solution. Selected methods are discussed as follows.

1.7.1 Selected Methods of Aryne Generation

1.7.1.1 Deprotonation of Aryl Halides

Deprotonation of aryl halides **24** followed by the dehalogenation of the anionic intermediate generated arynes in solution. However, the use of strong bases such as sodamide or *n*-BuLi limits the practical utility of this method [82]. Many base-sensitive functional groups were not compactable under the reaction conditions. Hence, this protocol is considered as a harsh method for the generation of arynes although this is traditionally important (Scheme 1.13).

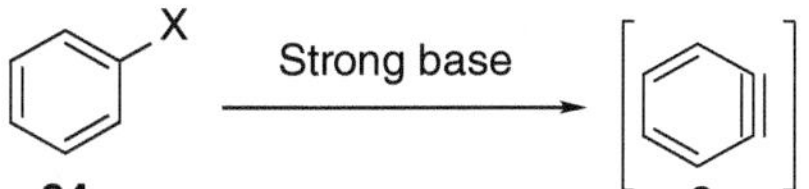

Scheme 1.13 Aryne generation from halobenzene. Source: Based on Kitamura [82].

1.7.1.2 Metal–Halogen Exchange/Elimination

Another approach involves the metal–halogen exchange/elimination of 1,2-disubstituted haloarenes **25** or haloaryl triflates mediated by metals (Mg or Li) or organometallic reagents derived from Cu, Li, and Mg [83, 84]. Many side products formed via the nucleophilic addition of the organometallic reagents itself made this route less practical (Scheme 1.14).

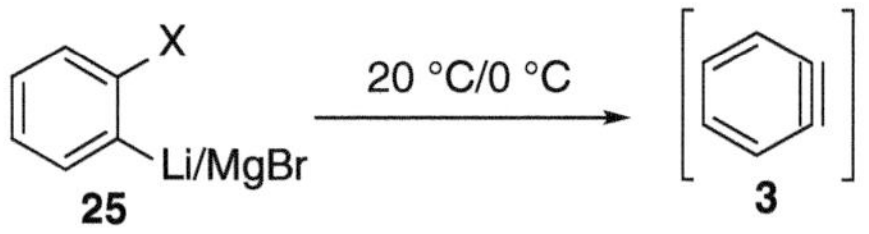

Scheme 1.14 Metal–halogen exchange/elimination route to arynes. Source: Ebert et al. [83]; Matsumoto et al. [84].

1.7.1.3 From Anthranilic Acids

The zwitterionic benzenediazonium 2-carboxylates **26** generated from anthranilic acid derivatives form arynes in the reaction course. Benzenediazonium 2-carboxylate decomposes upon heating, generating aryne with the liberation of nitrogen and carbon dioxide [85, 86]. Although the method has advantages, e.g. the side products are gases, the explosive nature of diazonium compounds limits the practical application of the method (Scheme 1.15).

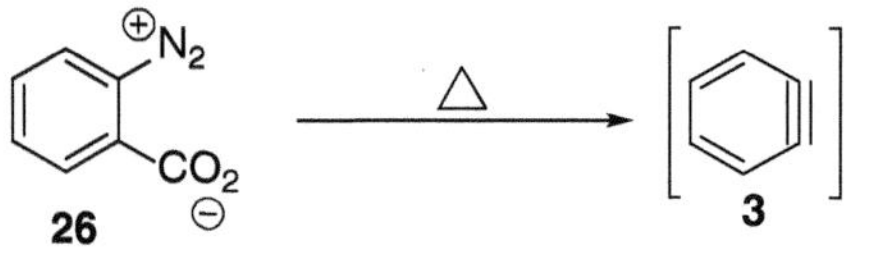

Scheme 1.15 Arynes from anthranilic acid derivatives. Source: Stiles and Miller [85]; Friedman and Logullo [86].

1.7.1.4 Fragmentation of Amino Benzotriazoles

Benzo[*d*][1,2,3]thiadiazole 1,1-dioxide **27** and amino benzotriazoles **28**, upon fragmentation, produce arynes with the evolution of nitrogen gas. The use of explosive precursor and the requirement of lead tetraacetate as oxidant appear to be the demerits of this method. Moreover, the method suffers from less functional group tolerance and hence is not widely used (Scheme 1.16) [87].

1.7.1.5 From Phenyl(2-(trimethylsilyl)phenyl)iodonium Triflate

Fluoride-induced elimination of the aryne precursor phenyl(2-(trimethylsilyl) phenyl)iodonium triflate **29** is an additional process for the generation of aryne. However, the preparation of starting material involves a complex process (Scheme 1.17) [88].

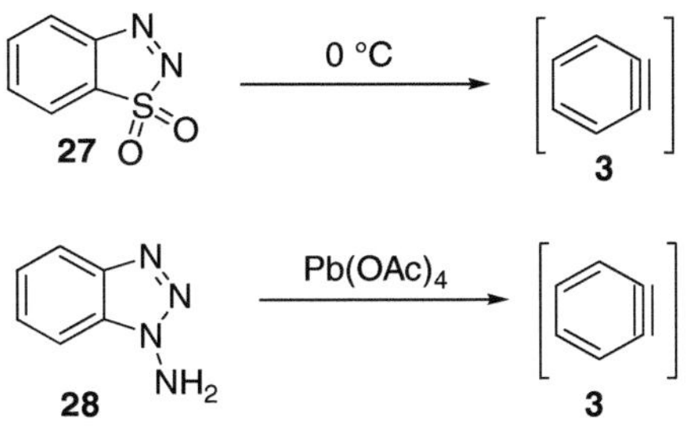

Scheme 1.16 Arynes from benzothiadiazole dioxide and amino benzotriazole. Source: Based on Campbell and Rees [87].

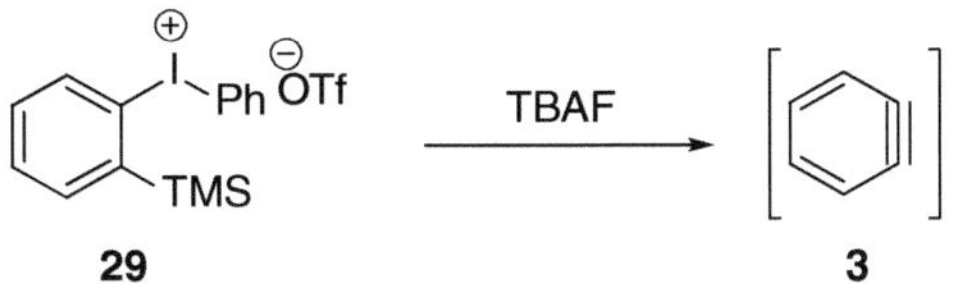

Scheme 1.17 Arynes from phenyl(2-(trimethylsilyl)phenyl)iodonium triflates. Source: Based on Kitamura and Yamane [88].

1.7.1.6 Using Hexadehydro Diels–Alder (HDDA) Reaction

Recently, Hoye and coworkers developed a new metal-free method for aryne generation using the concept of intramolecular HDDA reaction of triynes **30**, which diversified the reactivity profile of the aryne intermediate. This method allows reagent-free and metal-free thermal generation of arynes **31**; however, in some cases, it requires elevated temperature for the formation of arynes (Scheme 1.18) [16, 17].

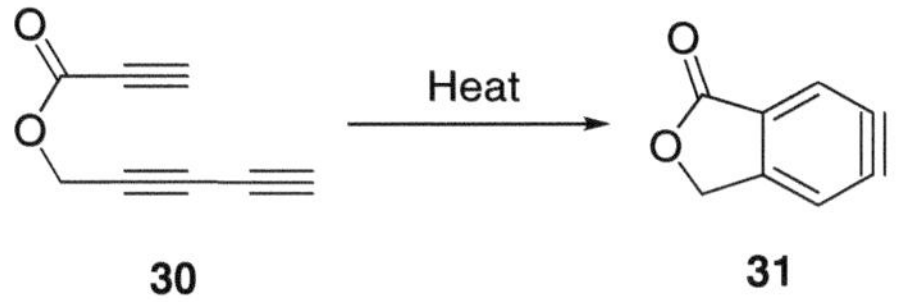

Scheme 1.18 Arynes using HDDA strategy. Source: Hoye et al. [16]; Yun et al. [17].

1.7.1.7 From *ortho*-Borylaryl Triflates

In 2013, Hosoya and coworkers developed the generation of aryne intermediates from *ortho*-borylaryl triflate **32** mediated by *sec*- or *tert*-BuLi at low temperature. However, the use of strong base limits its potential applications (Scheme 1.19) [89].

1.7.1.8 Pd(II)-Catalyzed C–H Activation Strategy Starting from Benzoic Acids

Greaney and coworkers developed a Pd(II)-catalyzed C–H activation strategy for the formation of Pd-aryne intermediate, and applied this strategy for the synthesis

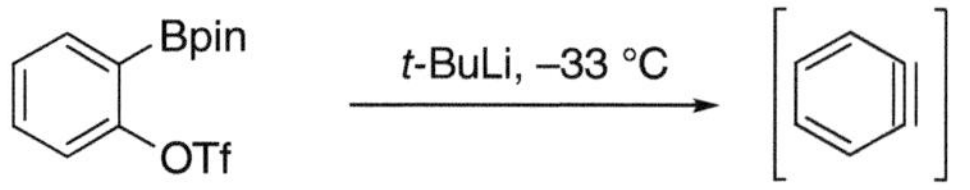

32 **3**

Scheme 1.19 Arynes from *ortho*-borylaryl triflates. Source: Based on Sumida et al. [89].

of triphenylenes starting from benzoic acids **33** (Scheme 1.20). The intermediate is formed via an *ortho*-C–H activation of benzoic acid, followed by the decarboxylation from the palladacycle [90].

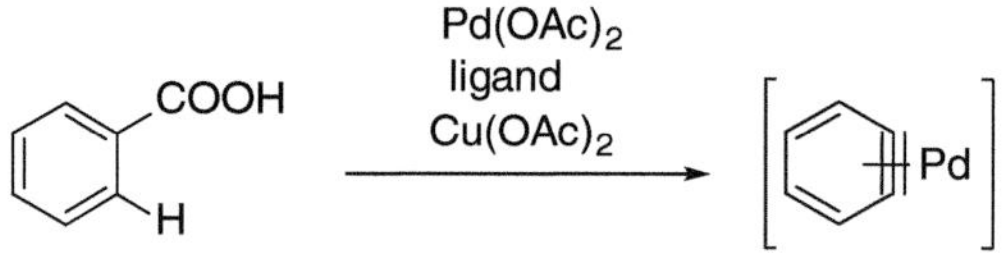

33 **3-Pd**

Scheme 1.20 Arynes via Pd(II)-catalyzed C–H activation. Source: Based on Cant et al. [90].

1.7.1.9 via Grob Fragmentation

The [2+2] cycloadducts of 3-triflyloxy arynes **34** can undergo a chemoselective ring opening to generate 2,3-aryne intermediate **35** via Grob fragmentation, which could be derivatized to trisubstituted arenes (Scheme 1.21) [91].

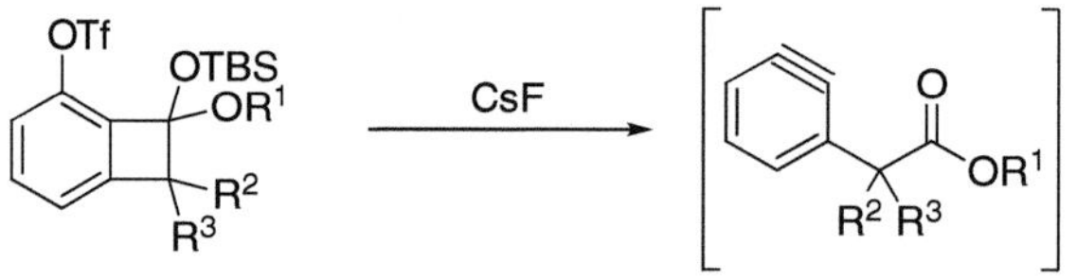

34 **35**

Scheme 1.21 Arynes via Grob fragmentation. Source: Based on Shi et al. [91].

All the methods outlined above require either metals or stoutly basic or harsh reaction conditions. The use of metals, strongly basic conditions, and high temperature are not compatible with a large number of functional groups, which considerably restricted the scope of aryne reactions in organic synthesis. The breakthrough for a mild method of aryne generation came in 1983 when Kobayashi and coworkers demonstrated a practical method for the generation of aryne intermediates.

1.7.2 Kobayashi's Fluoride-Induced Aryne Generation

In 1983, Kobayashi and coworkers, through a pioneering work, uncovered a route for the generation of arynes, which is base-free and could be carried out under mild reaction conditions. The strategy involves a fluoride-induced 1,2-elimination of 2-(trimethylsilyl)aryl triflates **36** to generate arynes in solution. This influential discovery led to a swift development in the field of aryne chemistry (Scheme 1.22)

[15]. This precursor could easily be synthesized from 2-bromophenol in three steps. Moreover, a one-pot method also could be employed for the synthesis of the precursor.

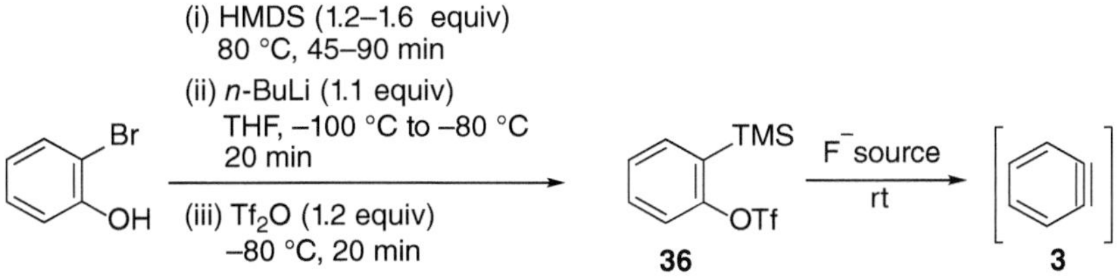

36 3

Scheme 1.22 Kobayashi's method of aryne generation. Source: Based on Himeshima et al. [15].

Kobayashi's method, unlike conventional routes, was compatible with a range of functional groups and reagents. KF (with 18-crown-6 as additive) in THF, CsF in CH_3CN, TBAT in THF, and tetrabutyl ammonium fluoride (TBAF) in THF are generally used as fluoride sources for the generation of aryne from the silyl triflate **36**. The kinetic control in the aryne reaction can be attained with cautious selection of fluoride source and solvent combination. This selection of the fluoride sources controls not only the rate of aryne generation but also the regioselectivity in the product formation. Kobayashi's method is highly preferred by synthetic chemists over traditional methods for aryne generation in recent decades. Many traditional aryne reactions were revisited to enhance the scope and yield after the establishment of mild and efficient Kobayashi's procedure of aryne generation.

1.8 Possible Reactivity Modes of Arynes

Due to their low-lying LUMO, arynes have been widely used as electrophiles in various reactions (Scheme 1.23). Arynes are well explored by synthetic organic chemists as they found that the intermediate is capable for unparallel transformations. The diverse reactivity profile of this transient intermediate is noteworthy. Aryne reactivity can be classified into:

Pericyclic reactions
Arylation reactions
Insertion reactions
Transition-metal-catalyzed reactions
Multicomponent couplings (MCCs)
Molecular rearrangements

Owing to their well-defined electrophilic nature, arynes are excellent dienophile and dipolarophile in pericyclic reactions such as Diels–Alder reactions, [2+2] cycloaddition, and dipolar cycloaddition reactions. Arynes hold the ability to arylate a number of molecules like alcohols, amines, and thiols leading to efficient

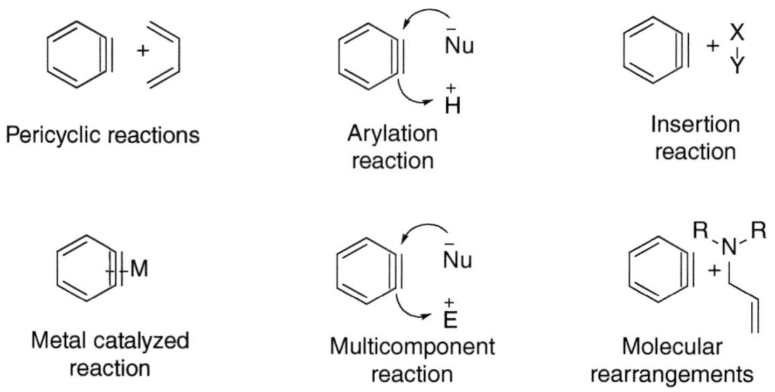

Scheme 1.23 Possible modes of reactivity of arynes.

X–H (X = O, S, and NH) insertion reactions. Arynes are well known to insert into various element–element σ-bonds and π-bonds, resulting in the formation of a library of functionalized 1,2-disubstituted arenes. Arynes have also been efficiently utilized as the electrophilic component in various transition-metal-free multicomponent couplings (MCCs). Molecular rearrangements are the emerging class of reactivity of arynes, where initial aryne addition is followed by unique molecular rearrangements (Scheme 1.23).

1.8.1 Pericyclic Reactions

Due to high electrophilicity of the carbon–carbon triple bond in arynes, they are excellent dienolphiles in pericyclic reactions. Many times, pericyclic reactions of arynes are utilized for the detection of the aryne generation in solution. Moreover, these reactions could result in the formation of complex arenes in one step under mild conditions. The Diels–Alder reaction of arynes began with the successful trapping of the in situ–generated aryne intermediate **3** by Wittig and Pohmer with furan **37** via a [4+2] cycloaddition reaction furnishing epoxynaphthalene derivative in good yields (Scheme 1.24) [29]. Independently, Huisgen and Knorr

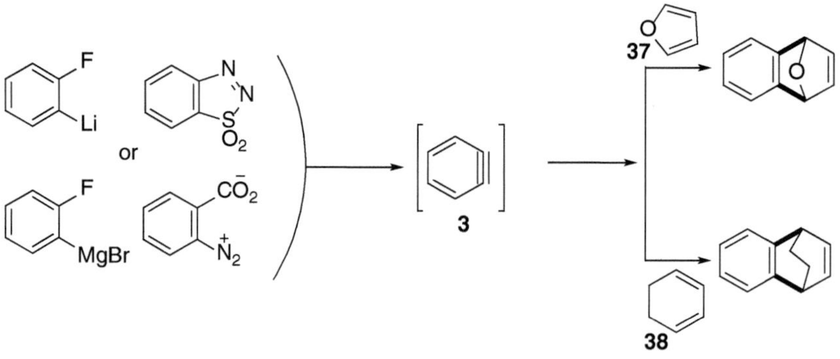

Scheme 1.24 Wittig and Huisgen's aryne cycloaddition experiment. Source: Wittig and Pohmer [29]; Huisgen and Knorr [92].

have utilized aryne as an electrophilic dienophile when they demonstrated the reaction of aryne generated from different precursors with furan **37** or cyclohexadiene **38** [92].

The Diels–Alder reaction of tropones **39** with arynes generated from the Kobayashi precursor **36** using CsF resulted in the convenient method for the synthesis of benzobicyclo[3.2.2]nonatrienones **40** in good yield and broad scope (Scheme 1.25) [93]. Although tropones can react as 4π, 6π, or 8π components in cycloaddition reactions, the reactivity in this case as diene (4π component) is noteworthy.

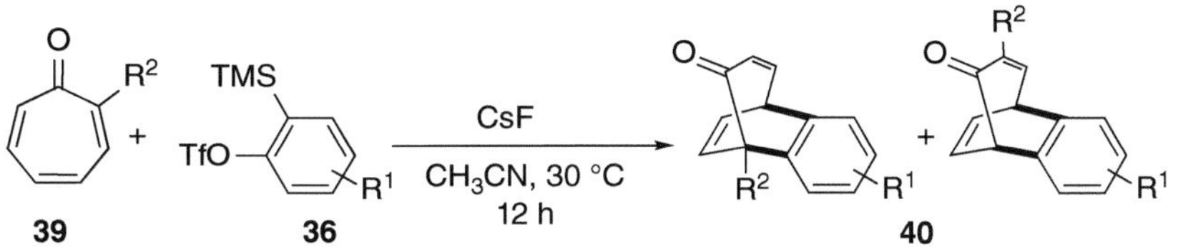

Scheme 1.25 Diels–Alder reactions of tropones with arynes. Source: Thangaraj et al. [93].

The poor yield and limited scope in the Diels–Alder reaction of styrene with arynes, known as early as 1966, has been revisited recently in a reaction of styrenes **41** with arynes generated from **36** using CsF in CH_3CN affording the 9-aryl dihydrophenanthrene derivatives **42** (Scheme 1.26) [94]. Electronics decided the products in this transformation as electron-rich and neutral styrenes reacted with arynes to form 9-aryldihydrophenanthrenes, whereas electron-poor styrenes afforded the dihydrophenanthrenes (1,1 adducts).

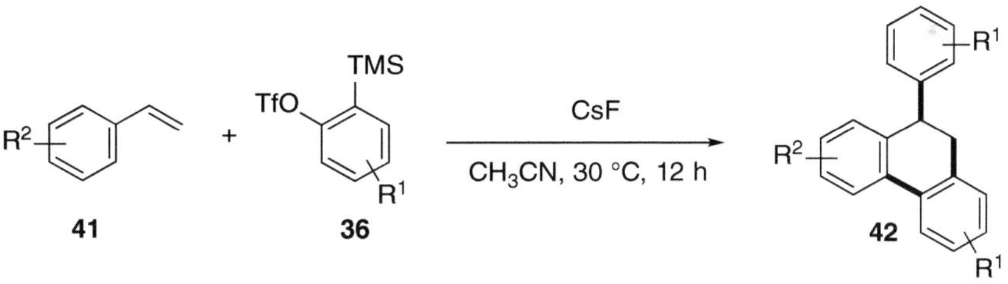

Scheme 1.26 Reaction of styrenes with arynes. Source: Bhojgude et al. [94].

However, similar reactivity was not observed when indene/benzofuran was treated with arynes under identical conditions. Using indene **43** as the diene component, reaction afforded the dihydrobenzocyclobutaphenanthrenes **44** in moderate-to-good yields (Scheme 1.27) [95]. The reaction features a unique [4+2] cycloaddition followed by a diastereoselective [2+2] cycloaddition to afford the desired product.

Due to their high electrophilicity and strong dienophilic nature, arynes are well known to participate in [2+2] cycloaddition reactions with electron-rich carbon–carbon double bonds. For instance, a tandem stereoselective aryne [2+2]

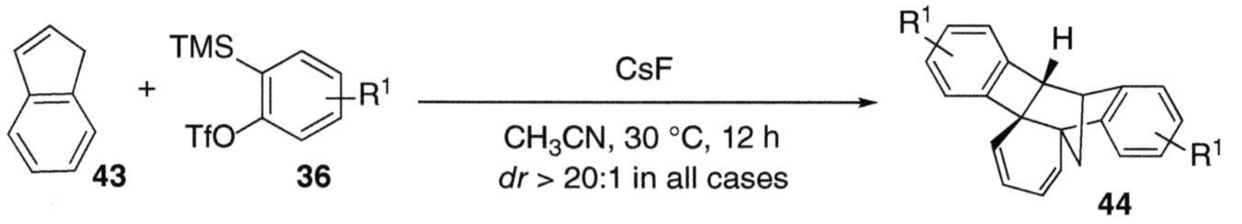

Scheme 1.27 Tandem [4+2]/[2+2] cycloaddition involving arynes and indene. Source: Bhojgude et al. [95].

cycloaddition reaction with enamides **45** for the synthesis of benzocyclobutanes **46** has been uncovered by Hsung and coworkers (Scheme 1.28) [96].

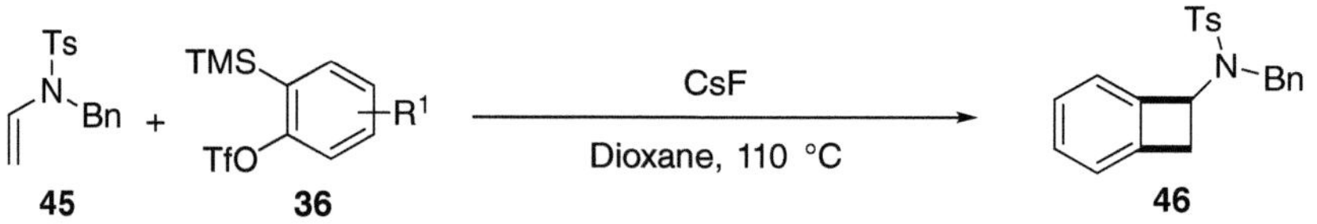

Scheme 1.28 [2+2] Cycloaddition involving arynes and enamides. Source: Based on Feltenberger et al. [96].

Apart from [4+2] and [2+2] cycloaddition reactions, arynes are well known to participate in dipolar cycloaddition reactions. Arynes are excellent dipolarophiles and can add to various 1,3-dipoles for the synthesis of benzo-fused five-membered rings. The commonly used 1,3-dipoles that can conveniently add to arynes are nitrones, nitrile oxides, nitrile imines, azomethine imines, azides, diazo compounds, etc. (Scheme 1.29).

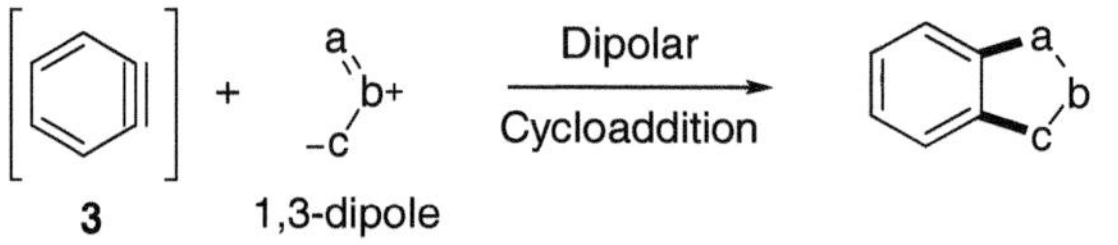

Scheme 1.29 1,3-Dipolar cycloaddition of arynes.

For example, the 1,3-dipolar cycloaddition of arynes with diazomethane derivatives for the synthesis of *N*-unsubstituted indazoles **47** has been reported by Larock and coworkers (Scheme 1.30) [97].

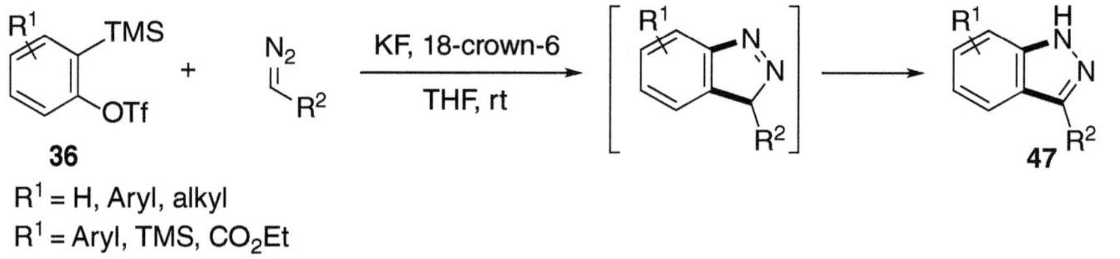

Scheme 1.30 [3+2] Cycloaddition reaction of diazo compounds with arynes. Source: Based on Liu et al. [97].

1.8.2 Arylation Reactions

Arynes are highly electrophilic species; hence, they react with a number of charged as well as uncharged nucleophiles, resulting in the in situ generation of aryl anions, which can be protonated, thus leading to efficient arylation reactions. Thus, arynes can be used as the aryl source for the arylation of OH, SH, and NH bonds under transition-metal-free conditions. In 2003, Larock and coworkers developed a mild strategy for the *N*-arylation of primary and secondary amines **48** by making use of the arylating property of arynes generated from the precursor **36** (Scheme 1.31) [98].

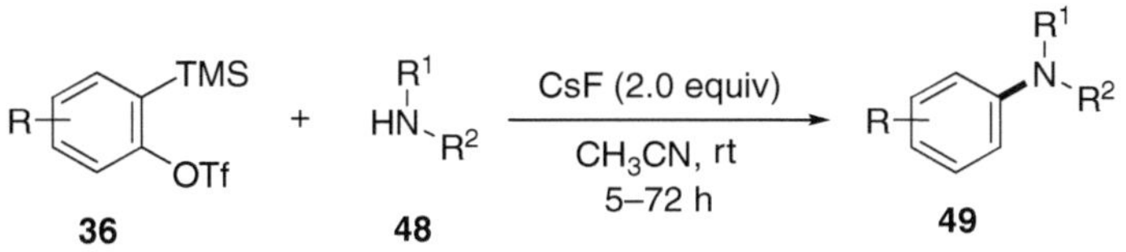

R = H, Me, OMe
R^1 = alkyl, aryl, sulfonyl
R^2 = H, alkyl, aryl

Scheme 1.31 *N*-Arylation of amines. Source: Based on Liu and Larock [98].

1.8.3 Insertion Reactions

Arynes have the exceptional ability to insert into various element–element σ-bonds and π-bonds. This insertion phenomenon of arynes has been extensively employed for the synthesis of functionalized 1,2-disubstituted arenes. For example, Shirakawa, Hiyama and coworkers utilized the benzyne insertion strategy for the synthesis of 2-aminobenzamides **50** via a N—CO bond insertion of ureas (Scheme 1.32) [99]. Also, the insertion of arynes into a carbon–carbon σ-bond of acyclic and cyclic β-ketoesters led to the one-step synthesis of 1,2-disubstituted arenes **51** as demonstrated by Stoltz and coworkers in 2005 (Scheme 1.32) [100].

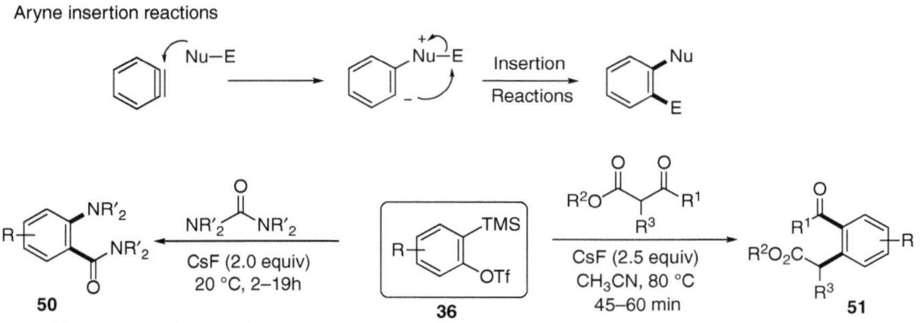

R = H, Me, OMe, –$(CH)_4$–, Ph

R = H, Me, OMe, $O(CH_2)O$, R^1 = Me, Et, *i*-Pr, Bn, Ph
R^2 = Me, Et, R^3 = Me, $(CH_2)_3$, $(CH_2)_5$

Scheme 1.32 Insertion of arynes to amides and β-keto esters. Source: Yoshida et al. [99]; Tamber and Stoltz [100].

1.8.4 Transition-Metal-Catalyzed Reactions

The development of *o*-(trimethylsilyl)aryl triflate by Kobayashi and coworkers helped to discover many novel transition-metal-catalyzed reactions involving arynes [101]. Metal-catalyzed aryne-based insertion, annulations, cycloadditions, and multicomponent reactions are well known. In 2009, Biehl and coworkers reported the coupling of a terminal alkyne **52** and aryne in the presence of CuI for the synthesis of phenyl acetylene derivatives **53** (Scheme 1.33) [102].

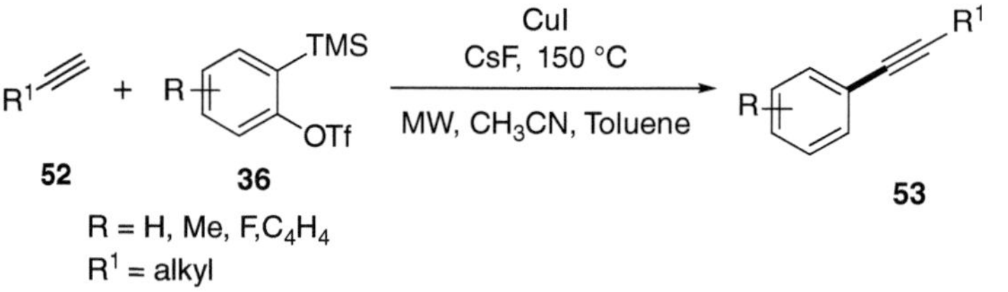

Scheme 1.33 Metal-catalyzed coupling reaction of arynes. Source: Based on Akubathini and Biehl [102].

1.8.5 Multicomponent Couplings (MCCs)

Transition-metal-free MCCs involving arynes have received considerable attention in the past two decades. A number of 1,2-disubstituted arenes are synthesized using the MCCs involving arynes in one step. In a typical aryne MCC, the nucleophiles having no acidic proton are added to the in situ–generated arynes forming the aryl anion intermediate, which is subsequently intercepted by an electrophilic third component to form complex 1,2-disubstituted arenes. If the nucleophile and the electrophile do not belong to the same molecule, these reactions result in unique MCCs. For instance, Yoshida, Kunai, and coworkers reported a facile three-component reaction of arynes with isocyanides **54** and aldehydes **55** under mild reaction conditions for the efficient synthesis of iminoisobenzofurans **56** in good yields (Scheme 1.34) [103]. The variation on all the three components is well tolerated, and the use of other third components such as activated imines and electron-poor olefins is also possible.

1.8.6 Molecular Rearrangements

Molecular rearrangements involving arynes are of high interest in recent years. These rearrangement reactions have provided a number of structurally important motifs, which is otherwise complicated to synthesize. Many well-known conventional molecular rearrangements were applied in aryne chemistry to make the aryne version of the same. Using arynes as the aryl source, many of these rearrangements proceed under mild conditions in broad scope. For example, Greaney and coworkers developed an aza-Claisen rearrangement to synthesize functionalized anilines **57** using various piperidine and morpholine-derived tertiary allylic amines **58** (Scheme 1.35) [104].

Multicomponent coupling employing arynes

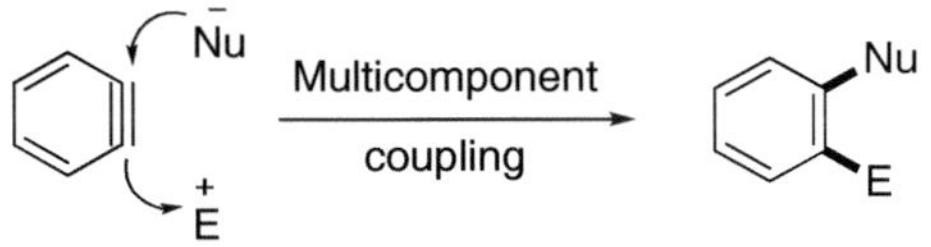

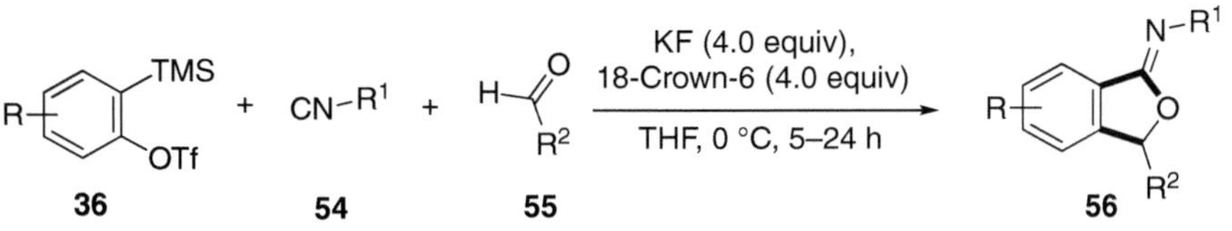

R = H, Me, $(CH_2)_3$, OMe, F
R^1 = CMe_2-CH_2-CMe_3, *t*-Bu, 1-Ad
R^2 = Et, *t*-Bu, Ph, Ar

Scheme 1.34 MCCs involving arynes, isocyanides, and aldehydes or activated imines. Source: Based on Yoshida et al. [103].

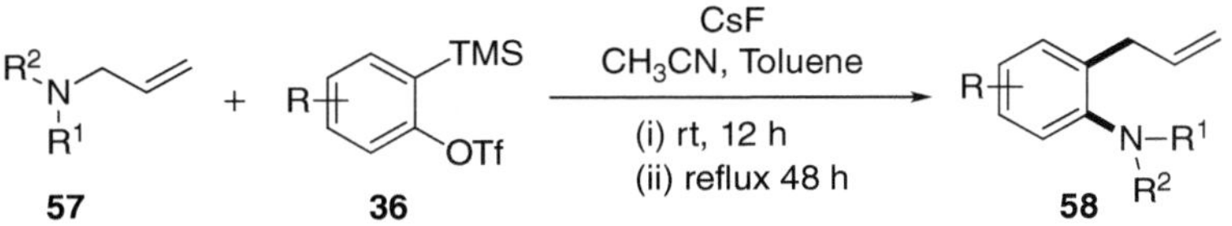

R = H, Me, $(CH_2)_3$, F
R^1 = alkyl

Scheme 1.35 Aryne aza-Claisen rearrangement. Source: Based on Cant et al. [104].

1.9 Domino Aryne Generation

The concept of multifunctionalization of aryne materialized with the realization of the concept of domino aryne generation by Li and coworkers [105]. A novel domino aryne precursor **59** has been developed (Scheme 1.36), which has vicinal electrophilic centers, capable of receiving consecutive double nucleophilic addition onto it. Utilizing this concept, a methodology for the nucleophilic addition-ene reaction end for the construction of indoline scaffolds **60** was developed by the same group [106]. A detailed discussion on the domino aryne strategy is provided in Chapter 8 of this book.

1.10 Arynes for the Synthesis of Large Polycyclic Aromatic Compounds

Cycloaddition reactions involving arynes give access to the synthesis of large polycyclic aromatic hydrocarbons (PAHs) containing five or more fused benzene rings. This strategy is of high value since the products find applications in the

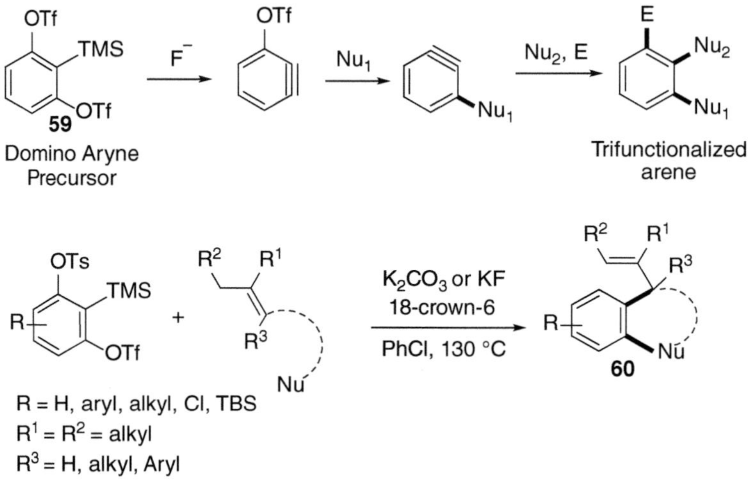

Scheme 1.36 Domino aryne strategy and the indoline synthesis. Source: Shi et al. [105]; Xu et al. [106].

field of materials science. Moreover, several PAHs are analogous to nanosized graphene substructures. The benzodiyne precursor has been efficiently utilized for the synthesis of a number of symmetrical and unsymmetrical PAHs. For example, Wudl and coworkers reported the synthesis of the stable large PAH **63**, in a reaction of cyclopentadienone **61** with benzodiyne precursor **62** (Scheme 1.37) [107]. The two-fold Diels–Alder reaction of aryne generated from **62** with the diene **61** is followed by the elimination of two molecules of CO resulting in the formation of the PAH **63** in one step.

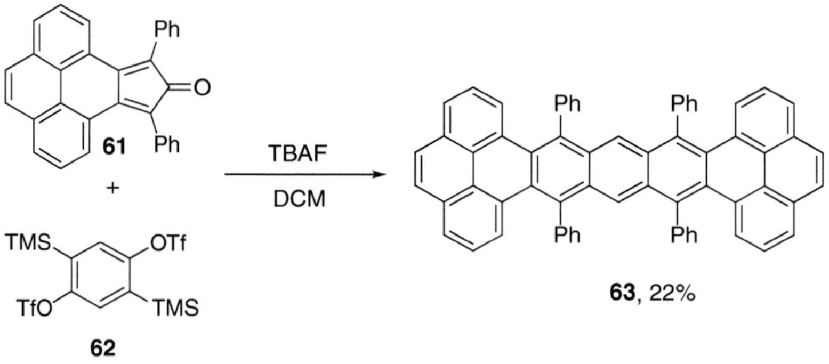

Scheme 1.37 Polycyclic aromatic hydrocarbons (PAHs) synthesis using arynes. Source: Based on Duong et al. [107].

1.11 Arynes in Natural Product Synthesis

Arynes are not only exploited in developing novel carbon–carbon and carbon–heteroatom bond-forming reactions but also in natural product synthesis

(Scheme 1.38). The motifs with new carbon–carbon or carbon–heteroatom bonds formed on the aromatic rings in the aryne reactions form the backbone for several biologically important molecules. For example, biologically important molecules such as trisphaeridine, dacarine, eupolauramine, and neuvamine could be easily synthesized using the aryne strategy. More details on application of arynes in natural products synthesis have been provided in Chapter 11 of the book.

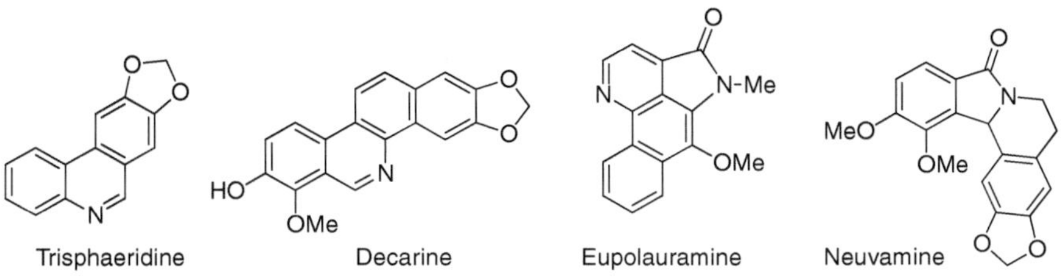

Scheme 1.38 Selected examples of natural product synthesized using the aryne concept.

1.12 Concluding Remarks

Arynes are century-old transient intermediates, which can now be accessed using simple precursors in mild and convenient ways. This has considerably broadened the diverse range of applications using this reactive intermediate. Ever since the development of Kobayashi's mild protocol for aryne generation, the field has revolutionized significantly. The reactivity of arynes can be classified into cycloaddition reactions, insertion reactions, multicomponent reactions, transition-metal-catalyzed transformations, and aryne-induced molecular rearrangements. Recent developments in the area have shown rapid emergence of a new class of reactions under molecular rearrangement employing arynes as aryl source. The potential synthetic utility of a variety of heterocyclic arynes has also been uncovered recently by synthesizing various heterocycle-fused ring systems. The generation of benzynes through the HDDA reaction has allowed this intermediate to explore new modes of intrinsic reactivities. The trifunctionalization of arynes and the domino aryne strategies are new concepts emerging recently in this field. Moreover, aryne methodologies have widely been employed for the synthesis of several natural products. It is expected that the mild and transition-metal-free method for the generation of arynes will inspire synthetic chemists to explore new reactions and potential applications using the concept of arynes.

References

1 Tadross, P.M. and Stoltz, B.M. (2012). *Chem. Rev.* 112: 3550–3577.
2 Gampe, C.M. and Carreira, E.M. (2012). *Angew. Chem. Int. Ed.* 51: 3766–3778.
3 Bhunia, A., Yetra, S.R., and Biju, A.T. (2012). *Chem. Soc. Rev.* 41: 3140–3152.

4 Dubrovskiy, A.V., Markina, N.A., and Larock, R.C. (2013). *Org. Biomol. Chem.* 11: 191–218.

5 Wu, C. and Shi, F. (2013). *Asian J. Org. Chem.* 2: 116–125.

6 Pérez, D., Peña, D., and Guitián, E. (2013). *Eur. J. Org. Chem.*: 5981–6013.

7 Roy, T. and Biju, A.T. (2018). *Chem. Commun.* 54: 2580–2594.

8 Sanz, R. (2008). *Org. Prep. Proced. Int.* 40: 215–291.

9 Takikawa, H., Nishii, A., Sakai, T., and Suzuki, K. (2018). *Chem. Soc. Rev.* 47: 8030–8056.

10 Wenk, H.H., Winkler, M., and Sander, W. (2003). *Angew. Chem. Int. Ed.* 42: 502–528.

11 Pellissier, H. and Santelli, M. *Tetrahedron* 9: 701–730.

12 Chen, Y. and Larock, R.C. (2009). Arylation reactions involving the formation of arynes. In: *Modern Arylation Methods* (ed. L. Ackermann), 401, 401–473, 473. Weinheim, Germany: Wiley-VCH Verlag GmbH & Co. KGaA.

13 Karmakar, R. and Lee, D. (2016). *Chem. Soc. Rev.* 45: 4459–4470.

14 Bhunia, A. and Biju, A.T. (2014). *Synlett* 25: 608–614.

15 Himeshima, Y., Sonoda, T., and Kobayashi, H. (1983). *Chem. Lett.* 12: 1211–1214.

16 Hoye, T.R., Baire, B., Niu, D. et al. (2012). *Nature* 490: 208–212.

17 Yun, S.Y., Wang, K.-P., Lee, N.-K. et al. (2013). *J. Am. Chem. Soc.* 135: 4668–4671.

18 Goetz, A.E., Shah, T.K., and Garg, N.K. (2015). *Chem. Commun.* 51: 34–45.

19 Shi, J., Li, Y., and Li, Y. (2017). *Chem. Soc. Rev.* 46: 1707–1719.

20 Stoermer, R. and Kahlert, B. (1902). *Dtsch. Chem. Ges.* 35: 1633–1640.

21 Bachmann, W.E. and Clarke, H.T. (1927). *J. Am. Chem. Soc.* 49: 2089–2098.

22 Wittig, G. (1942). *Naturwiss.* 30: 696–703.

23 Wittig, G. and Pohmer, L. (1956). *Chem. Ber.* 89: 1334–1351.

24 Huisgen, R. and Rist, H. (1955). *Ann. Chem.* 594: 137–158.

25 Huisgen, R. and Rist, H. (1954). *Naturwiss.* 41: 358–359.

26 Morton, A.A., Davidson, J.B., and Hakan, B.L. (1942). *J. Am. Chem. Soc.* 64: 2242–2247.

27 Roberts, J.D., Simmons, H.E., Carlsmith, L.A., and Vaughan, C.W. (1953). *J. Am. Chem. Soc.* 75: 3290–3591.

28 Roberts, J.D., Semenow, D.A., Simmons, H.E., and Carlsmith, L.A. (1956). *J. Am. Chem. Soc.* 78: 601–611.

29 Wittig, G. and Pohmer, L. (1955). *Angew. Chem. Int. Ed.* 67: 348.

30 Fisher, I.P. and Lossing, F.P. (1963). *J. Am. Chem. Soc.* 85: 1018–1019.

31 Berry, R.S., Clardy, J., and Schafer, M.E. (1964). *J. Am. Chem. Soc.* 86: 2738–2739.

32 Berry, R.S., Spokes, G.N., and Stiles, M. (1962). *J. Am. Chem. Soc.* 84: 3570–3577.

33 Radziszewski, J.G., Hess, B.A.J., and Zahradnik, R. (1992). *J. Am. Chem. Soc.* 114: 52–57.

34 Leopold, D.G., Stevens-Miller, A.E., and Lineberger, W.C. (1986). *J. Am. Chem. Soc.* 108: 1379–1384.

35 Wenthold, P.G., Squires, R.R., and Lineberger, W.C. (1998). *J. Am. Chem. Soc.* 120: 5279–5290.
36 Wenthold, P.G. and Squires, R.R. (1994). *J. Am. Chem. Soc.* 116: 6401–6412.
37 Wenthold, P.G., Paulino, J.A., and Squires, R.R. (1991). *J. Am. Chem. Soc.* 113: 7414–7415.
38 Riveros, J.M., Ingemann, S., and Nibbering, N.M.M. (1991). *J. Am. Chem. Soc.* 113: 1053.
39 Warmuth, R. (1997). *Angew. Chem. Int. Ed.* 36: 1347–1350.
40 Hoffmann, R., Imamura, A., and Hehre, W.J. (1968). *J. Am. Chem. Soc.* 90: 1499–1509.
41 Buchwald, S.L., Watson, B.T., and Huffman, J.C. (1986). *J. Am. Chem. Soc.* 108: 7411–7413.
42 Bennett, M.A., Hambley, T.W., Roberts, N.K., and Robertson, G.B. (1985). *Organomet.* 4: 1992–2000.
43 Mclain, S.J., Schrock, R.R., Sharp, P.R. et al. (1979). *J. Am. Chem. Soc.* 101: 263–265.
44 Hamura, T., Ibusuki, Y., Sato, K. et al. (2003). *Org. Lett.* 5: 3551–3554.
45 Langenaeker, W., De Proft, F., and Geerlings, P. (1998). *J. Phys. Chem. A* 102: 5944–5950.
46 Johnson, W.T.G. and Cramer, C.J. (2001). *J. Am. Chem. Soc.* 123: 923–928.
47 Squires, R.R. and Cramer, C.J. (1998). *J. Phys. Chem. A* 102: 9072–9081.
48 Lohmann, J. (1972). *J. Chem. Soc., Perkin Trans.* 168: 814.
49 Sato, T., Niino, H., and Yabe, A. (2001). *J. Phys. Chem. A* 105: 7790–7798.
50 Kauffmann, T. (1965). *Angew. Chem. Int. Ed.* 4: 543–557.
51 Kaufmann, T. and Wirthwein, R. (1971). *Angew. Chem. Int. Ed.* 10: 20–33.
52 Reinecke, M.G. (1982). Five membered hetarynes. In: *Reactive Intermediates*, vol. 2 (ed. R.A. Abramovitch), 367–526. New York, Chapter 5: Plenum.
53 van der Plas, H.C. (1982). The chemistry of triple bonded groups. In: *Supplement C of "The Chemistry of Functional Groups"* (eds. S. Patai and Z. Rappoport). New York: Wiley-Interscience.
54 Cramer, C.J. and Debbert, S. (1998). *Chem. Phys. Lett.* 287: 320–326.
55 Kramer, J. and Berry, R.S. (1972). *J. Am. Chem. Soc.* 94: 8336–8347.
56 Sasaki, T., Kanematsu, K., and Uchide, M. (1971). *Chem. Soc. Jpn.* 44: 858–859.
57 Nam, H.-H. and Leroi, G.E. (1988). *J. Am. Chem. Soc.* 110: 4096–4097.
58 Cava, M.P., Mitchell, M.J., DeJongh, D.C., and van Fossen, R.Y. (1966). *Tetrahedron Lett.* 4: 2947–2951.
59 Dunkin, I.R. and MacDonald, J.G. (1982). *Tetrahedron Lett.* 23: 4839–4842.
60 Goetz, A.E., Bronner, S.M., Cisneros, J.D. et al. (2012). *Angew. Chem. Int. Ed.* 51: 2758–2762.
61 Wenthold, P.G. (2010). *Aust. J. Chem.* 63: 1091–1098.
62 Winkler, H. and Sander, W. (2010). *Aust. J. Chem.* 63: 1013–1047.
63 Sato, T. and Niino, H. (2010). *Aust. J. Chem.* 63: 1048–1065.
64 Sander, W. (1999). *Acc. Chem. Res.* 32: 669–676.
65 Diau, E.W.-G., Casanova, J., Roberts, J.D., and Zewail, A.H. (2000). *Proc. Natl. Acad. Sci. U.S.A* 97: 1376–1379.

66 Moskaleva, L.V., Madden, L.K., and Lin, M.C. (1999). *Phys. Chem. Chem. Phys.* 1: 3967–3972.
67 Buchwald, S.L. and Nielsen, R.B. (1988). *Chem. Rev.* 88: 1047–1058.
68 Bennett, M.A. and Schwemlein, H.P. (1989). *Angew. Chem. Int. Ed.* 28: 1296–1320.
69 Frid, M., Pérez, D., Peat, A.J., and Buchwald, S.L. (1999). *J. Am. Chem. Soc.* 121: 9469–9470.
70 Breslow, R., Napierski, J., and Clarke, T.C. (1975). *J. Am. Chem. Soc.* 97: 6275–6276.
71 Breslow, R. and Khanna, P.L. (1977). *Tetrahedron Lett.* 18: 3429–3432.
72 Washburn, W.N., Zahler, R., and Chen, I. (1978). *J. Am. Chem. Soc.* 100: 5863–5874.
73 Roberts, J.D., Streitwieser, A. Jr., and Regan, C.M. (1952). *J. Am. Chem. Soc.* 74: 4579–4582.
74 Evleth, E.M. (1967). *Tetrahedron Lett.* 8: 3625–3628.
75 Dewar, M.J.S. and Li, W.-K. (1974). *J. Am. Chem. Soc.* 96: 5569–5571.
76 Jagau, J.-C., Prochnow, E., Evangelista, F.A., and Gauss, J. (2010). *J. Chem. Phys.* 132: 144110–144119.
77 Nicolaou, K.C. and Dai, W.-M. (1991). *Angew. Chem. Int. Ed.* 30: 1387–1416.
78 Nicolaou, K.C. and Smith, A.L. (1995). *Modern Acetylene Chemistry* (eds. P.J. Stang and F. Diederich). VCH: Weinheim.
79 Polukhtine, A., Karpov, A.G., Pandithavidana, D.R. et al. (2010). *Aust. J. Chem.* 63: 1099–1107.
80 Guo, X.-F., Zhu, X.-F., Shang, Y. et al. (2010). *Clin. Cancer Res.* 16: 2085–2094.
81 Roy, S. and Basak, A. (2010). *Chem. Commun.* 46: 2283–2285.
82 Kitamura, T. (2010). *Aust. J. Chem.* 63: 987–1001.
83 Ebert, G.W., Pfennig, D.R., Suchan, S.D., and Donovan, T.A. Jr., (1993). *Tetrahedron Lett.* 34: 2279–2282.
84 Matsumoto, T., Hosoya, T., Katsuki, M., and Suzuki, K. (1991). *Tetrahedron Lett.* 32: 6735–6736.
85 Stiles, M. and Miller, R.G. (1960). *J. Am. Chem. Soc.* 82: 3802.
86 Friedman, L. and Logullo, F.M. (1963). *J. Am. Chem. Soc.* 85: 1549.
87 Campbell, C.D. and Rees, C.W. (1969). *J. Chem. Soc. C* 1969: 742–747.
88 Kitamura, T. and Yamane, M. (1995). *J. Chem. Soc., Chem. Commun.* 1995: 983–984.
89 Sumida, Y., Kato, T., and Hosoya, T. (2013). *Org. Lett.* 15: 2806–2809.
90 Cant, A.A., Roberts, L., and Greaney, M.F. (2010). *Chem. Commun.* 46: 8671–8673.
91 Shi, J., Xu, H., Qiu, D. et al. (2017). *J. Am. Chem. Soc.* 139: 623–626.
92 Huisgen, R. and Knorr, R. (1963). *Tetrahedron Lett.* 16: 1017–1021.
93 Thangaraj, M., Bhojgude, S.S., Bisht, R.H. et al. (2014). *J. Org. Chem.* 79: 4757–4762.
94 Bhojgude, S.S., Bhunia, A., Gonnade, R.G., and Biju, A.T. (2014). *Org. Lett.* 16: 676–679.

95 Bhojgude, S.S., Thangaraj, M., Suresh, E., and Biju, A.T. (2014). *Org. Lett.* 16: 3576–3579.

96 Feltenberger, J.B., Hayashi, R., Tang, Y. et al. (2009). *Org. Lett.* 11: 3666–3669.

97 Liu, Z., Shi, F., Martinez, P.D.G. et al. (2008). *J. Org. Chem.* 73: 219–226.

98 Liu, Z. and Larock, R.C. (2003). *Org. Lett.* 5: 4673–4675.

99 Yoshida, H., Shirakawa, E., Honda, Y., and Hiyama, T. (2002). *Angew. Chem. Int. Ed.* 114: 3381–3383.

100 Tamber, U.K. and Stoltz, B.M. (2005). *J. Am. Chem. Soc.* 127: 5340–5341.

101 Dhokale, R.A. and Mhaske, S.B. (2018). *Synthesis* 50: 1–16.

102 Akubathini, S. and Biehl, E. (2009). *Tetrahedron Lett.* 50: 1809–1811.

103 Yoshida, H., Fukushima, H., Ohshita, J., and Kunai, A. (2004). *Angew. Chem. Int. Ed.* 43: 3935–3938.

104 Cant, A.A., Bertrand, G.H.V., Henderson, J.L. et al. (2009). *Angew. Chem. Int. Ed.* 48: 5199–5202.

105 Shi, J., Qiu, D., Wang, J. et al. (2015). *J. Am. Chem. Soc.* 137: 5670–5673.

106 Xu, H., He, J., Shi, J. et al. (2018). *J. Am. Chem. Soc.* 140: 3555–3559.

107 Duong, H.M., Bendikov, M., Steiger, D. et al. (2003). *Org. Lett.* 5: 4433–4436.

2

Aryne Cycloadditions for the Synthesis of Functional Polyarenes

Fátima García, Diego Peña, Dolores Pérez, and Enrique Guitián

Universidade de Santiago de Compostela, Departamento de Química Orgánica and CiQUS, Jenaro de la Fuente s/n, 15782, Santiago de Compostela, Spain

2.1 Introduction

The origin of aryne chemistry can be traced back to 1902, when 2,3-didehydrobenzofuran was formulated by Stoermer and Kahlert as a reaction intermediate [1]. However, arynes did not reappear in the literature as such until the middle of last century. In 1953, Roberts et al. established the structure **1** for benzyne based on their famous isotopic labeling experiments [2]. A further period of 60 years had to elapse before an image of an aryne, generated on a surface under ultra-high vacuum (UHV) conditions at cryogenic temperature, was obtained by atomic force microscopy (AFM, Figure 2.1), which curiously suggests a dominant contribution of the cumulene resonance structure (**2a**) for this particular aryne under these conditions [3, 4].

Soon after the experiments by Roberts, several groups initiated studies on the generation and reactivity of arynes. In the context of this chapter, the development of aryne Diels–Alder (D–A) reactions by Wittig and Pohmer is particularly relevant [5]. They reported the first aryne cycloaddition, the reaction between benzyne (**1**) and furan (**4**) to yield adduct **5**, which represents a fundamental milestone in the development of aryne chemistry. The formation of epoxynaphthalene **5** constitutes the first example of an aryne cycloaddition reaction or, more specifically, a [4+2] cycloaddition or D–A reaction (Scheme 2.1). Over the following decades, cycloaddition reactions involving arynes have experienced impressive progress, including a great variety of processes, such as [2+2], [4+2], [2+2+2] cycloadditions, etc. [6–11].

The increasing availability of generation methods is of utmost importance in the progress of aryne chemistry and its synthetic applications. Scheme 2.2 outlines the most commonly used methods in the 1950–1990 period for the generation of benzyne and derivatives for synthetic purposes: dehydrohalogenation of aryl halides **6** with strong base [5, 12]; treatment of *o*-dihaloaryls **7** with metals [13, 14] or organolithium compounds [15], or *o*-haloaryltosylates **8** with organolithium compounds [16]; thermal decomposition of an arenediazonium carboxylate **10** previously isolated [17] or

Modern Aryne Chemistry, First Edition. Edited by Akkattu T. Biju.

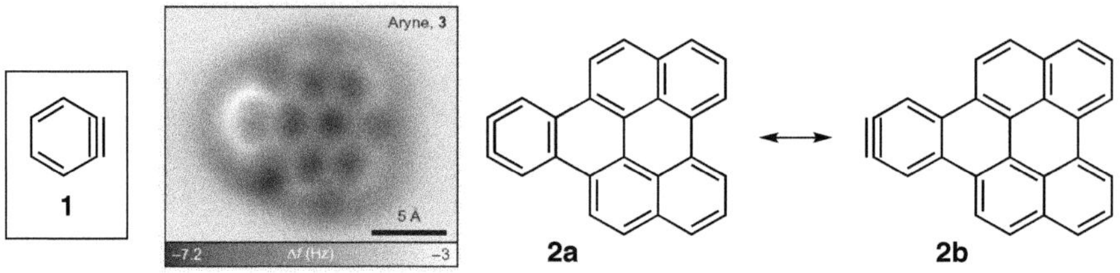

Figure 2.1 Structure of benzyne (**1**) and AFM image of *o*-aryne **2**. Source: Image adapted with permission from Pavliček et al. [3]. Copyright 2015, Springer Nature.

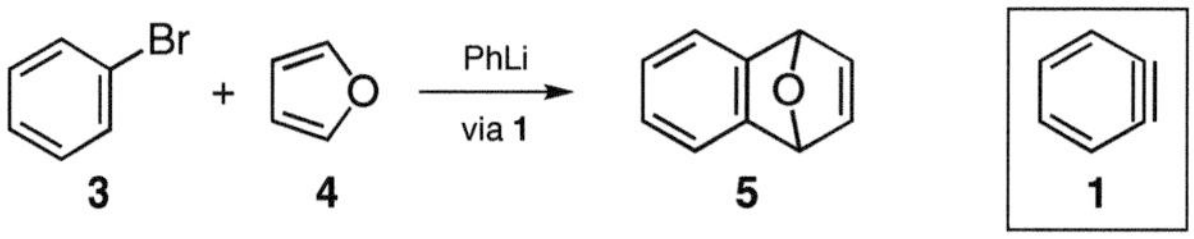

Scheme 2.1 Wittig's first aryne cycloaddition. Source: Based on Wittig and Pohmer [5].

generated in situ by diazotization of the corresponding anthranilic acid **9** [18]; thermal decomposition of a 2-diphenyliodonium carboxylate **11** [19] and oxidation of a 1-aminobenzotriazole **12** with $Pb(OAc)_4$ [20].

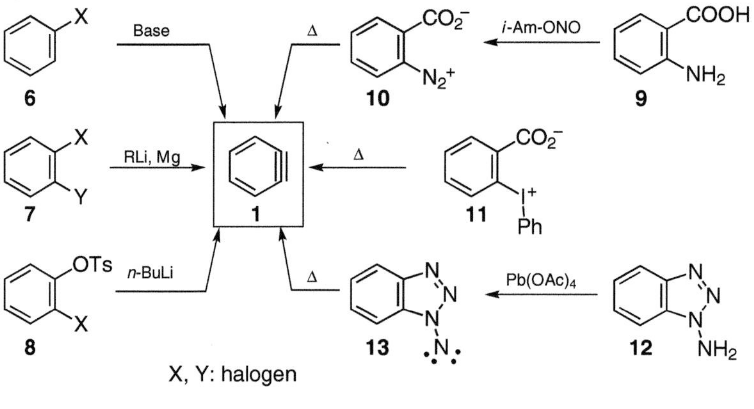

Scheme 2.2 Classical methods for aryne generation.

In 1983, Kobayashi and coworkers described a new synthetic methodology for the preparation of arynes, based on the fluoride-induced decomposition of *o*-(trimethylsilyl)phenyl triflate (**14**) (Scheme 2.3) [21]. This is a mild, efficient, and safe procedure for the generation of arynes. It does not require the use of oxidants or strong bases, and it is compatible with most substrates and reaction conditions. Despite these clear advantages, this method went unnoticed for 15 years, being reported only in three publications between 1983 and 1998.

In 1998, Guitián and Pérez's research group described the first example in the participation of benzyne, generated from triflate **14**, in a metal-catalyzed reaction, specifically in a [2+2+2] cycloaddition catalyzed by palladium(0) [22]. This, together with the establishment of a general approach for the synthesis of functionalized and

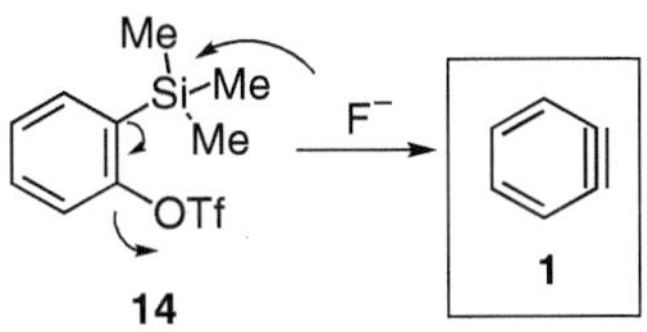

14

1

Scheme 2.3 Kobayashi's method for aryne generation. Source: Based on Himeshima et al. [21].

polycyclic *o*-(trimethylsilyl)aryl triflates developed by the same group [23], demonstrated the versatility of Kobayashi's method and, since then, aryne chemistry has experienced a reawakening.

2.2 Aryne Cycloaddition Reactions: General Considerations

This section provides an overview of the utility of [2+2], [4+2], and [2+2+2] aryne cycloaddition reactions for the preparation of extended polycyclic aromatic hydrocarbons (PAHs), which are relevant as functional organic compounds and appear in examples selected from the last 40 years' literature. The 1,3-dipolar cycloaddition reactions will be discussed in Chapter 3. General aspects of cycloaddition reactions involving arynes are described in Sections 2.2.1–2.2.3.

2.2.1 [4+2] Aryne Cycloadditions

Concerted D–A reactions involving arynes are symmetry-allowed [π4s+π2s] cycloadditions, usually controlled by the interaction between the lowest unoccupied molecular orbital (LUMO) of aryne, i.e. the C_{sp}–C_{sp} antibonding orbital, and the highest occupied molecular orbital (HOMO) of the diene. Therefore, aryne and diene must approach in almost perpendicular planes, as shown in Scheme 2.4. In this classic example, benzyne (**1**) reacts with tetraphenylcyclopentadienone (**15**), presumably through transition state **18**, to afford adduct **16**, which undergoes a cheletropic extrusion of CO to give **17** [14].

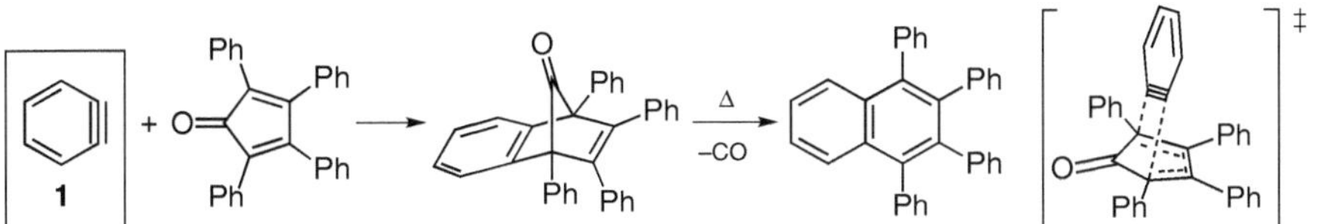

1 15 16 17 18

Scheme 2.4 [4+2] Cycloaddition of benzyne (**1**) and cyclopentadienone **15**. Source: Based on Wittig and Knauss [14].

The regioselectivity of [4+2] cycloadditions with nonsymmetric partners depends on the position and nature of the substituents on aryne and diene and has been attributed either to electronic [24], steric [25, 26], or distortion effects [27].

D–A reactions of arynes have been extensively used for the synthesis of PAHs [28, 29], heterocycles [30–33], natural products [34–37], and molecular materials (see references for Section 2.3).

2.2.2 [2+2] Aryne Cycloadditions

[π2s+π2s] Aryne cycloadditions are forbidden processes according to the Woodward–Hoffmann rules. However, there are numerous examples of [2+2] cycloadditions of arynes described in the literature, particularly with ketene acetals [6–12, 38–40].

When substituted olefins were used as reaction partners, mixtures of stereoisomers were usually observed. For example, the reaction of benzyne (**1**) and 1,2-dichloroethylene led to mixtures of stereoisomers **19** and **20** (Scheme 2.5) [41].

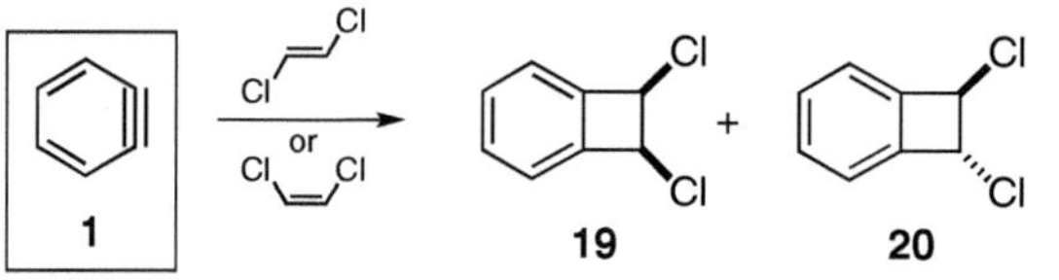

Scheme 2.5 Reaction of benzyne (**1**) and 1,2-dichloroethylene. Source: Jones and Levin [41].

These findings led to the conclusion that nonconcerted radical mechanisms are in operation, and this was supported by density functional theory (DFT) calculations [42, 43]. However, some examples of stereospecific [2+2] cycloadditions have been reported (Scheme 2.6) [39].

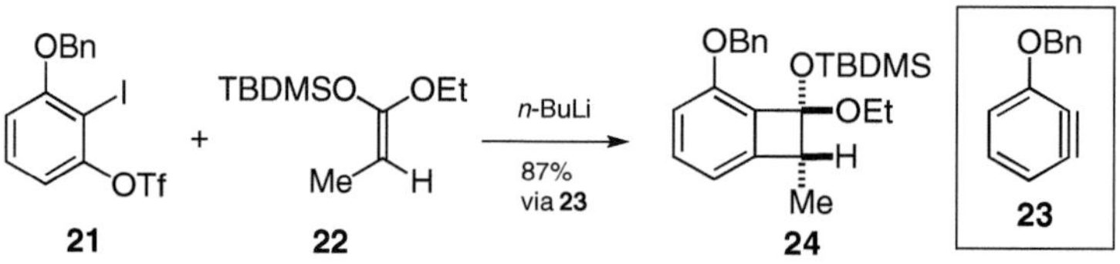

Scheme 2.6 Suzuki's stereospecific [2+2] cycloaddition. Source: Based on Hosoya et al. [39].

[2+2] Aryne cycloadditions have been used for the synthesis of natural products [37, 44] and to obtain rather complex structures, such as **25** (Figure 2.2), prepared through an iterative process of aryne generation/[2+2] cycloaddition [40]. For [2+2] cycloadditions leading to insertion reactions, see Chapter 4.

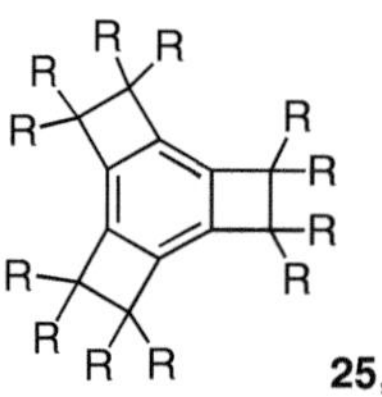

25, R = OMe

Figure 2.2 Compound **25** obtained by three consecutive [2+2] cycloadditions. Source: Based on Hamura et al. [40].

A particular case of [2+2] cycloaddition is the formation of biphenylenes, such as **26**, by dimerization of arynes (Scheme 2.7) [17]. Examples of [2+2] cycloadditions between arynes and carbon-rich structures will be discussed in Section 2.3.6.

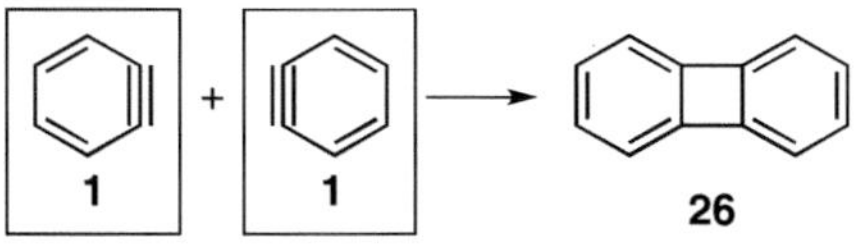

Scheme 2.7 Dimerization of benzyne to yield biphenylene (**26**). Source: Based on Logullo et al. [17].

2.2.3 [2+2+2] Aryne Cycloadditions

Despite [π2s+π2s+π2s] cycloadditions being allowed according to Woodward–Hoffmann rules, the participation of three reactants in a concerted reaction is entropically disfavored. The examples on the formation of triphenylene (**27**) by formal cyclotrimerization of benzyne described until the late 1990s are supposed to occur stepwise rather than in a concerted manner.

In their pioneering work, Guitián, Pérez, and coworkers explained in 1998 that benzyne is able to react in the presence of catalytic amounts of palladium(0) complexes, affording triphenylene (**27**) (Scheme 2.8) [22]. Experimental evidences supported a mechanism for this reaction similar to that of the well-known metal-catalyzed [2+2+2] cycloaddition of alkynes. This work represented the first example of the participation of arynes in metal-catalyzed reactions, broadening the scope and utility of these short-lived species for their application in organic synthesis.

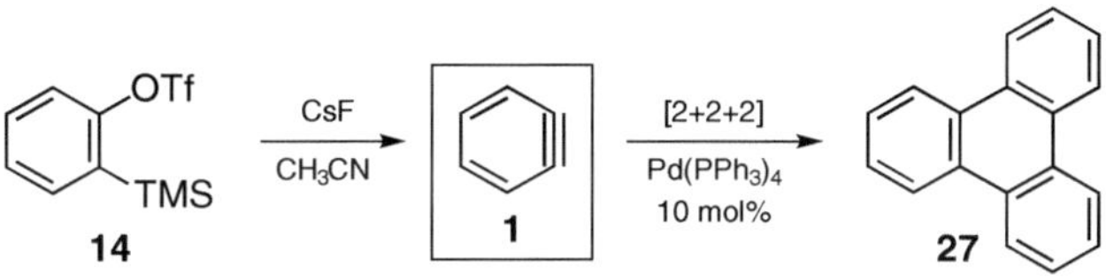

Scheme 2.8 First palladium-catalyzed [2+2+2] cycloaddition of arynes. Source: Peña et al. [22].

The authors found that the slow, controlled generation of benzyne was a crucial factor for the success of the reaction. This can be achieved by treating **14** with a fluoride source (typically, CsF) in acetonitrile or in an alternative solvent (and additives, if necessary). The reaction allows the efficient synthesis of triphenylenes, which constitutes the core of an important family of discotic liquid crystals [45], as well as the fundamental structure to access extended PAHs with trigonal symmetry, which are discussed in Section 2.3.4.

Guitián, Pérez, and coworkers also explained that the [2+2+2] cycloaddition reactions of arynes could be extended to include alkynes as reaction partners. Thus, the

Pd(0)-catalyzed cocycloaddition of benzynes with electron-deficient alkynes, such as dimethyl acetylenedicarboxylate (**28**, DMAD), was reported [46]. Surprisingly, these reactions are highly chemoselective and the product distribution depends on the choice of the catalyst. Thus, the use of $Pd(PPh_3)_4$ as catalyst allows to obtain the product **29**, bearing two units of aryne and one of alkyne, while the use of $Pd_2(dba)_3$ affords **30**, resulting from the reaction of one aryne moiety and two molecules of alkyne (Scheme 2.9).

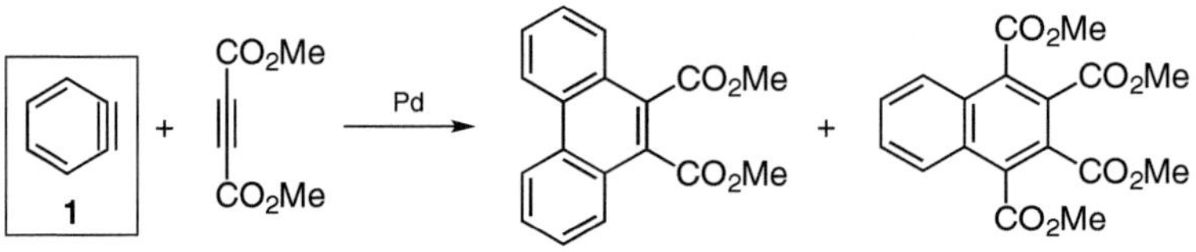

	28	29	30
	$Pd(PPh_3)_4$	84	7
	$Pd_2(dba)_3$	10	83

Scheme 2.9 First palladium-catalyzed [2+2+2] cocyclotrimerization of arynes and alkynes. Source: Based on Peña et al. [46].

Later, Yamamoto and coworkers developed an alternative catalytic system, $Pd(OAc)_2/P(o\text{-tol})_3$, which allows the [2+2+2] cocycloaddition with electron-rich alkynes [47]. Subsequently, it was demonstrated that alkenes and allenes could also participate in [2+2+2] cocycloaddition with arynes [48–50].

2.3 Aryne-Mediated Synthesis of Functional Polyarenes

Cycloaddition reactions of arynes have been used for the preparation of π-functional materials following a bottom-up approach as well as for the functionalization of materials obtained by other aryne-free methodologies. In Sections 2.3.1–2.3.6, aryne-based methodologies for the synthesis of families of compounds with potential application in materials science will be described.

2.3.1 Synthesis of Acenes

Acenes, such as rubrene and pentacene, are characterized by their low band gap and high charge carrier mobility and, therefore, are potentially useful as organic semiconductors, although their practical application is hampered by their instability and the difficulties in preparing them. To improve their properties, intensive research has been done for preparing and studying new substituted and heterocyclic acenes. In this context, [4+2] cycloadditions involving arynes as dienophiles are used for increasing the length of the acene skeleton [29]. On the other hand, it was established that the introduction of substituents or nonlinearly fused benzene rings can improve the stability of acenes [51].

Seminal work in this field was reported by Thummel et al. on the synthesis of tetracene (**35**) [52]. The [4+2] cycloaddition reactions between benzyne (**1**), generated from benzenediazonium 2-carboxylate (**10**), and 1,2-dimethylenecyclobutane (**31**) led to the formation of adduct **32** (Scheme 2.10). Furthermore, they demonstrated that **32** could be transformed into tetracene (**35**) through three reaction steps: a first electrocyclic opening at 300 °C providing the exocyclic diene **33**, followed by reaction with benzyne (**1**), and a final dehydrogenation with 2,3-dichloro-5,6-dicyano-1,4-benzoquinone (DDQ).

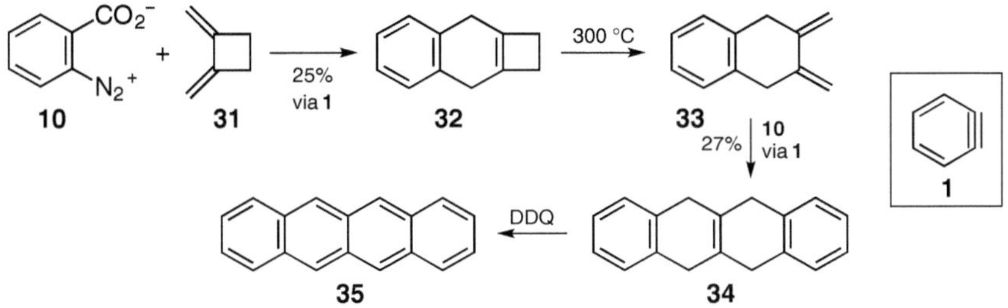

Scheme 2.10 Thummel's synthesis of tetracene (**35**). Source: Based on Thummel et al. [52].

In a related but more sophisticated approach, Herwig and Müllen described the synthesis of stable pentacene precursors **41a,b**, which can be converted in the solid state into pentacene (**42**) through heating (Scheme 2.11) [53]. The synthetic pathway started with the reaction of bisdiene **37** with two benzyne units, generated from *o*-dibromobenzene (**36**) in the presence of *n*-BuLi, to form bisadduct **38** in 54% yield. Dehydrogenation of **38** with chloranil afforded **39**. Then, **39** was reacted with tetrachlorothiophenedioxide (**40a**) or its bromo-analogue **40b** to yield adducts **41a,b**, which are soluble and easily processable as films. Finally, heating up to 180 °C, films of **41a,b** afforded pentacene (**42**) quantitatively by a retro-D–A reaction. This is a practical approach to pentacene films for the construction of field-effect transistors.

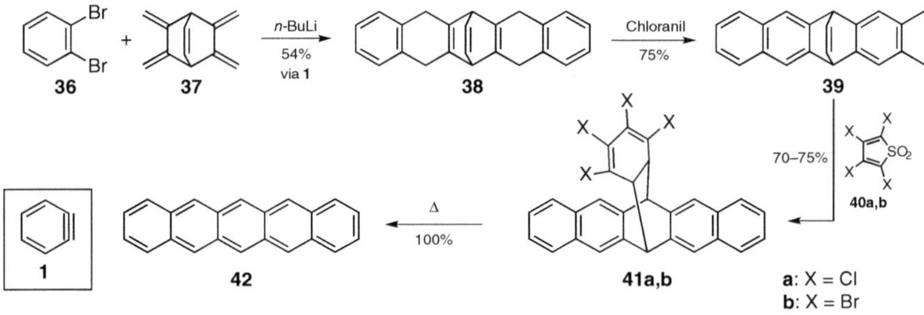

Scheme 2.11 Müllen's synthesis of pentacene (**42**). Source: Based on Herwig and Müllen [53].

A similar strategy to yield hexacene (**47**) was implemented by Neckers's group in 2007 [54]. Sequential reaction of both dienes of bisdiene **37** with benzyne (**1**) and

2,3-naphthyne (**44**) yielded bisadduct **45** in 54% yield (Scheme 2.12). Then, dehydrogenation with chloranil followed by dihydroxylation with OsO_4 and oxidation with trifluoroacetic anhydride (TFAA) and dimethyl sulfoxide (DMSO) afforded diketo-hexacene **46**, which was transformed into hexacene (**47**) by irradiation at 395 nm.

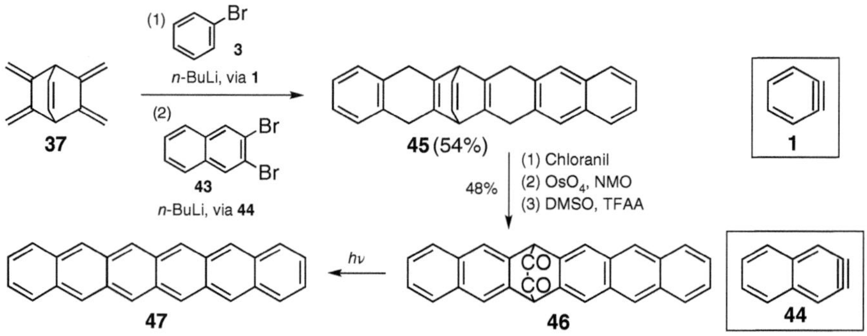

Scheme 2.12 Neckers's synthesis of hexacene (**47**). Source: Based on Mondal et al. [54].

Jancarick, Gourdon, and Levet described a general methodology for the synthesis of long acenes [55]. The key step in the synthetic route is a double D–A cycloaddition reaction between bisdiene **48** and two units of 2,3-naphthyne (**44**), generated either from 2,3-dibromonaphthalene (**43**) or from triflate **49** (Scheme 2.13). Treatment of the corresponding biscycloadduct with DDQ and deprotection of the ketal groups yielded the CO-bridged acene precursor **50**, which, upon heating in the solid state at c. 200 °C, allowed the preparation of heptacene (**51**). Other acenes, such as **52** and **53**, could be obtained when using 1,2-naphthyne (**54**) as dienophile.

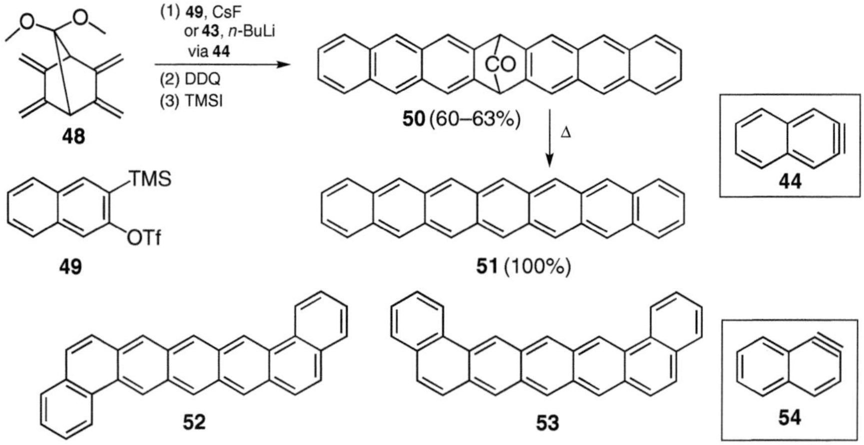

Scheme 2.13 Gourdon's synthesis of heptacene (**51**) and benzo-fused acenes **52**, **53**. Source: Based on Jancarik et al. [55].

Gribble and coworker succeeded in the generation of the tetracyclic skeleton of tetracenes in just one reaction step by a D–A reaction of 2,3-didehydronaphthalene

(**44**), generated from 2,3-dibromonaphthalene (**43**) in the presence of phenyllithium, and 2-methylisoindoles **55a,b** (Scheme 2.14) [56]. The unsubstituted and tetrafluorinated adducts **56a,b** were deaminated by treatment with *m*-chloroperbenzoic acid (*m*-CPBA) to yield tetracenes **35** and **57** in excellent yields.

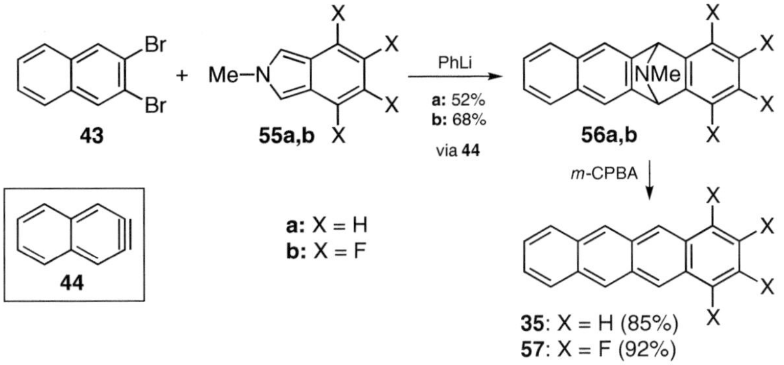

Scheme 2.14 Gribble's synthesis of tetracenes **35** and **57**. Source: Based on LeHoullier and Gribble [56].

Rickborn and coworkers also used a [4+2] cycloaddition of 2,3-didehydronaphthalene (**44**) with isobenzofuran **58** to yield adduct **59** with the tetracyclic skeleton of tetracene (Scheme 2.15) [57]. The adduct **59**, obtained in a 78% yield, was transformed into tetracene (**35**) by treatment with acetic acid, to yield ketone **60** in 90% yield, followed by its reduction with $LiAlH_4$ and dehydration with HCl.

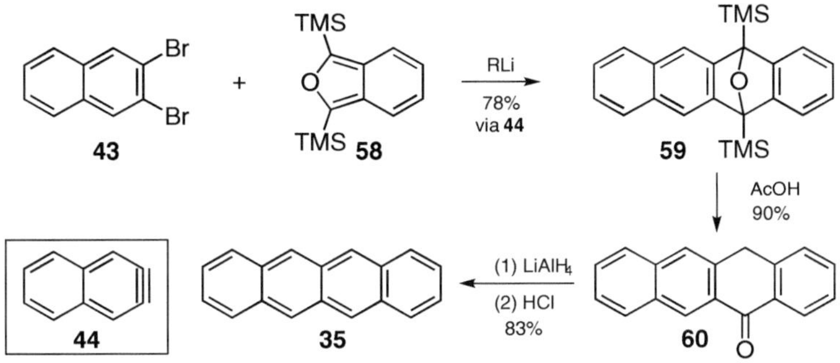

Scheme 2.15 Rickborn's synthesis of tetracene (**35**). Source: Based on Netka et al. [57].

Hamura and coworkers obtained a substituted pentacene **69** by successive D–A cycloadditions (Scheme 2.16) [58]. Epoxynaphthalene **5** was reacted with dibromoisobenzofuran **61** through a [4+2] cycloaddition to afford adduct **62**, which, in the presence of *n*-BuLi, can generate aryne **63** and react with furan through another D–A to generate endoxide **64** in 76% yield. Endoxide **64** was reacted with 3,6-di-2-pyridyl-1,2,4,5-tetrazine (**65**) and fumaronitrile to afford adduct **67**. This one-pot transformation involves a sequence of reactions, starting with a D–A reaction between **64** and tetrazine **65**, followed by a N_2 extrusion, and a final retro-D–A reaction to yield isobenzofuran **66**, which is reacted in situ with fumaronitrile. Adduct **67** by treatment with LiI and 1,8-diazabicyclo(5.4.0)undec-7-ene (DBU) was converted into diepoxypentacene **68**, which, by epoxide opening and Lewis acid–promoted aromatization, afforded pentacene **69** in 49% yield.

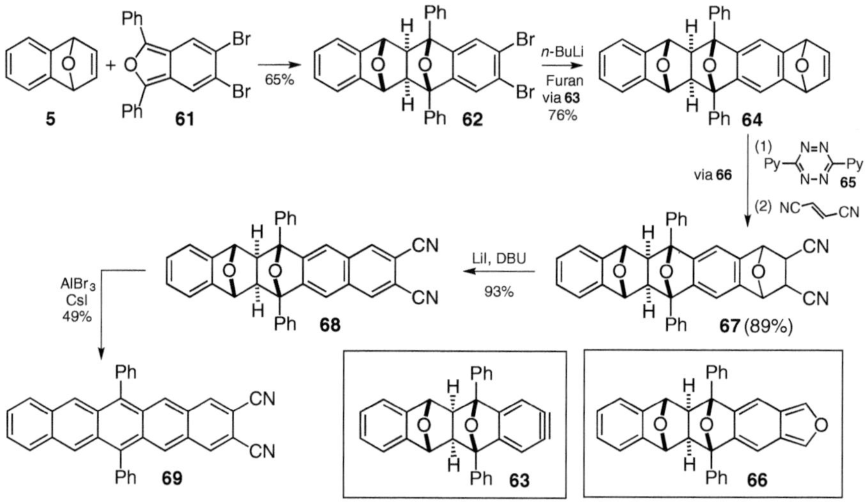

Scheme 2.16 Hamura's synthesis of substituted pentacene **69**. Source: Based on Eda et al. [58].

In a seminal work published in 1979, Hart's group introduced a convergent methodology based on the use of bisarynes (Scheme 2.17) [59]. These species formally contain two arynic triple bonds in the same molecule, but both triple bonds most probably do not coexist. In Hart's work, the transformation of **70** into bisadduct **75** can be formally represented as a double D–A over the bisaryne synthon **72**, but this process presumably involves the generation of a first monoaryne **73**, its D–A reaction with tetramethylpyrrole **71**, the generation of a second aryne **74** and its reaction with **71** to yield bisadduct **75** in 57% yield. Finally, dodecamethyltetracene (**76**) was obtained by oxidation of bisadduct **75** with *m*-CPBA and thermal elimination of the *N*-oxide in 15% yield.

A slightly modified procedure for the generation of a bisnaphthyne was implemented by Gribble et al., based on the use of tosylates as leaving groups (Scheme 2.18) [60]. For a similar synthesis using Kobayashi's method, see [61, 62].

Rickborn and coworkers followed a similar methodology for the synthesis of pentacene (**42**) (Scheme 2.19) [57]. Starting from 1,4-dibromobenzene (**82**), a formal double D–A reaction of the bisaryne synthon **83** with furan **58** yielded bisadduct **84**. This reaction is highly regioselective, observing the formation of linear pentacene derivative **84** as the only reaction product, which suggests that reaction through synthon **85** does not occur. Bisadduct **84** was subjected to a treatment with trifluoroacetic acid (TFA) followed by $LiAlH_4$ to obtain pentacene (**42**).

Luo and Hart designed an original methodology for the construction of acenes based on iterative [4+2] cycloadditions (Scheme 2.20) [63]. Starting from

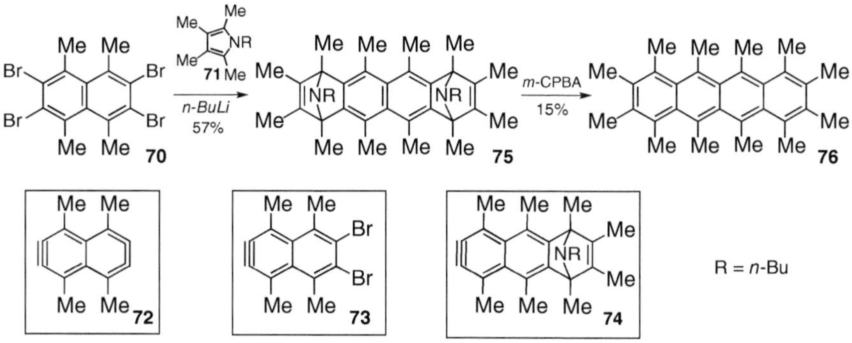

Scheme 2.17 Hart's bisaryne synthesis of dodecamethyltetracene (**76**). Source: Based on Sy and Hart [59].

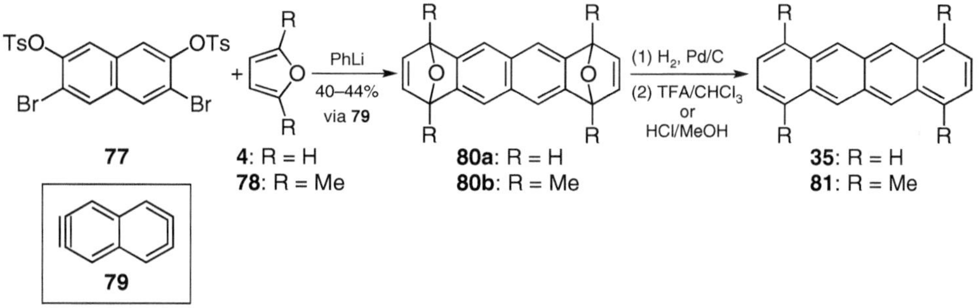

Scheme 2.18 Gribble's bisaryne synthesis of tetracenes **35** and **81**. Source: Based on Gribble et al. [60].

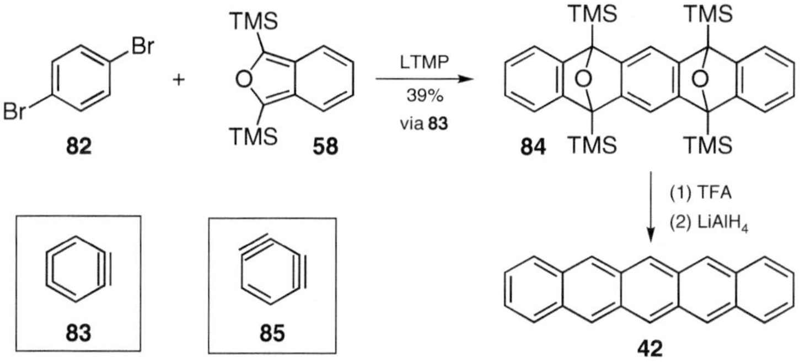

Scheme 2.19 Rickborn's synthesis of pentacene (**42**). Source: Based on Netka et al. [57].

1,2,4,5-tetrabromobenzene (**86a**) in the presence of butyllithium, bisaryne **87a** was generated and reacted with excess of furan through two D–A reactions, yielding bisadduct **88a** in a 71% yield. Then, **88a** was reacted with tetraphenylcyclopentadienone (**15**) through another [4+2] cycloaddition, affording bisadduct **89a** in high yield. Heating bisadduct **89a** to 190–200 °C generates bisdiene **91a**, which can be trapped by dienophiles. This process presumably involves two retro-D–A reactions to give norbornadienone **90** and bisfuran **91a**. Schlüter and coworkers extended this methodology to the synthesis of oligomers **92** [64, 65]. Generation of **91b** in the presence of **88b** afforded oligomers **92** with almost monomodal weight distributions ($M_n = 16\,500$; $M_w = 29\,000$) [64]. Reaction of polymer **92** with trimethylsilyliodide gave a deoxygenated material, which was not fully characterized [65].

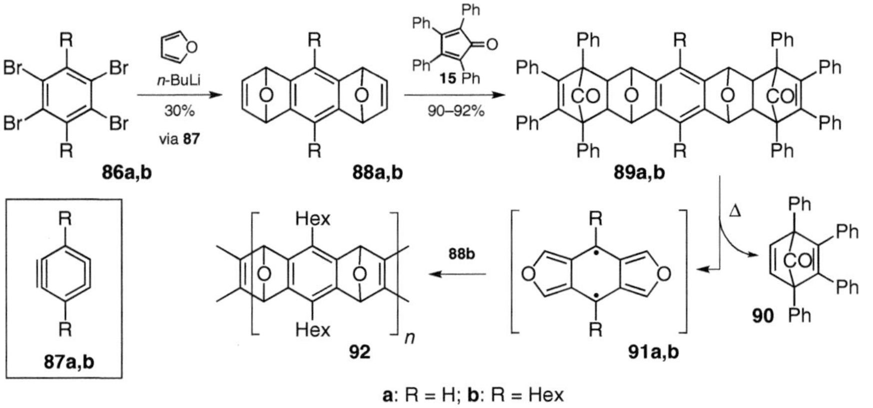

Scheme 2.20 Hart's and Schlüter's synthesis of epoxyacenes. Source: Luo and Hart [63]; Blatter and Schlüter [64]; Vogel et al. [65].

Wudl and coworkers synthesized the stable, substituted heptacene **98** [66]. Dibromoanthracene **94**, prepared by standard procedures from quinone **93**, was treated with lithium tetramethylpiperidide (LTMP), in the presence of diphenylisobenzofuran **95** to afford bisendoxide **97**. Reduction of **97** with Zn provided the expected functionalized heptacene **98** (Scheme 2.21).

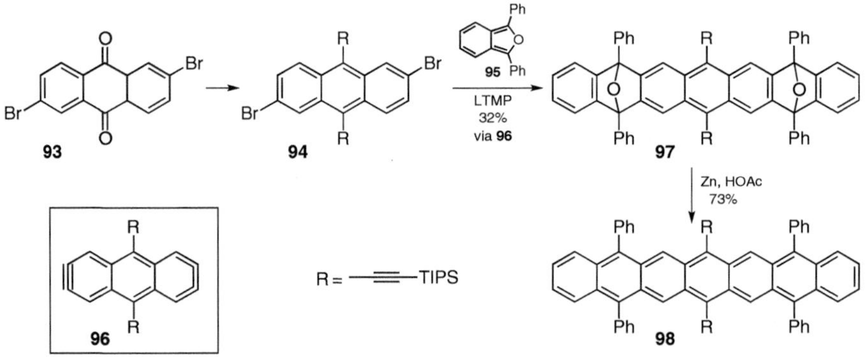

Scheme 2.21 Wudl's synthesis of heptacene **98**. Source: Based on Chun et al. [66].

Hamura's group developed a second methodology based on bisaryne and furan [4+2] cycloadditions to obtain pentacene **105** (Scheme 2.22) [67]. Selective generation of aryne **99** from 1,2,4,5-tetrabromobenzene (**86a**) by addition of *n*-BuLi and trapping with **61** yielded adduct **100**, which can also act as a formal bisbenzyne precursor (synthon **101**) in the presence of *n*-BuLi and can be trapped again with two furan units to afford triepoxypentacene **102** in 41% yield. Triepoxypentacene **102** was reacted with tetrazine **65** to give bisisobenzofuran **103**, which, in the presence of dimethyl maleate, afforded bisadduct **104**. This one-pot transformation involves a sequence of reactions, starting with a D–A reaction between terminal alkenes present in **102** and tetrazine **65**, followed by a N_2 extrusion, and a final retro-D–A reaction to yield bisisobenzofuran **103**, which is reacted in situ with dimethyl maleate. In the last step of the synthesis, triepoxypentacene **104** was transformed into pentacene **105** by treatment with H_2SO_4.

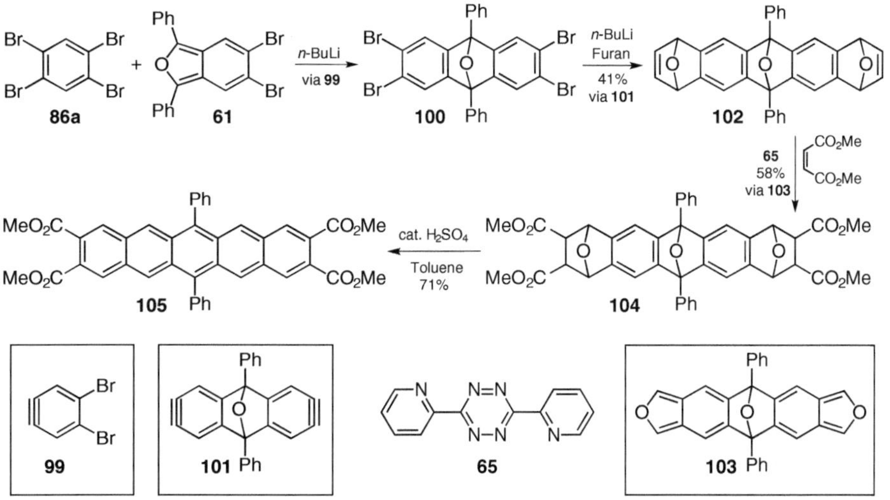

Scheme 2.22 Hamura's synthesis of substituted pentacene **105**. Source: Based on Haneda et al. [67].

Peña and coworkers modified Hamura's methodology by using Kobayashi's method for the generation of arynes and developed a methodology for the generation of acenes by combining solution chemistry with on-surface synthesis. As a proof of concept, stable bisepoxytetracene and trisepoxyhexacene were synthesized through successive D–A cycloaddition reactions between arynes and furans. Then, these epoxyacenes were deposited on Cu(111) under UHV conditions and, upon annealing at 393 K or by tip-induced manipulation with a scanning probe microscope, tetracene [68] and hexacene [69], respectively, were generated. With these precedents, the authors decided to try the generation and characterization of unknown decacene [70]. Tetraepoxydecacene (**114**) was selected as a stable precursor for decacene (**115**) (Scheme 2.23). The synthesis of **114** involved a three-step iterative sequence of four aryne cycloadditions. Bistriflate **106** by treatment with stoichiometric CsF presumably led to the formation of monoaryne **108**, which was trapped by means of a first D–A cycloaddition with highly reactive substituted isobenzofuran **107** to give a mixture of regioisomers **109a,b** in 42% yield. Further reaction of this mixture with more CsF generated aryne **110**, which, in the presence of isobenzofuran **107**, afforded regioisomers **111a,b** in 34% yield by means of a second [4+2] cycloaddition. Then, the mixture **111a,b** was treated with an excess of CsF and in situ-formed isobenzofuran **112** to obtain decacene precursor **114** in 33% yield by two consecutive aryne D–A cycloadditions. Finally, tetraepoxydecacene **114** was deposited on Au(111) under UHV conditions and transformed into decacene (**115**) by annealing at 220 °C or by voltage pulses from the scanning probe microscope tip.

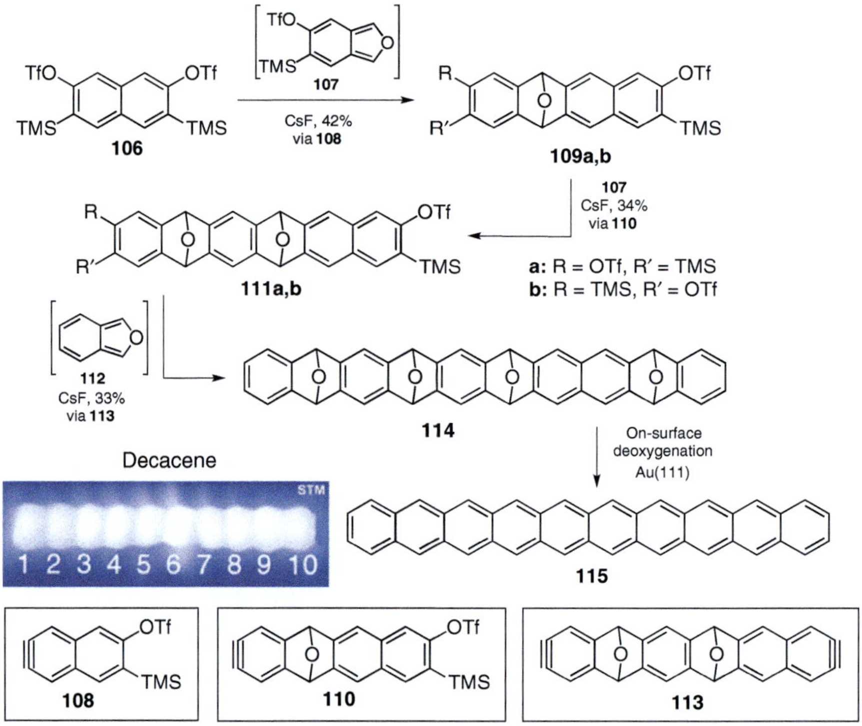

Scheme 2.23 Peña's synthesis and STM image of decacene (**115**). Source: Image adapted with permission from Krüger et al. [70]. Copyright 2017, John Wiley and Sons.

Aryne intermediates have also been used for the synthesis of undecacene – the longest unsubstituted acene prepared to date – by Bettinger and coworkers [71]. The synthesis started with the trapping of one unit of 2,3-naphthyne (**44**), generated from 2,3-dibromonaphthalene **43** in the presence of *n*-BuLi, with bisdiene **37**, generating monoadduct **116** in 47% yield (Scheme 2.24). Sequential generation of arynes from 1,2,4,5-tetrabromobenzene (**86a**) with methyllithium and trapping with two units of diene **116** afforded bisadduct **117**, bearing the undecacene skeleton in 19% yield. Bisadduct **117** was dehydrogenated with chloranil, dihydroxylated with OsO_4, oxidized with 2,2,6,6-tetramethyl-1-piperidinyloxy (TEMPO) and, finally, photoirradiated ($\lambda > 395$ nm) on a polystyrene matrix under cryogenic conditions (8 K) to generate undecacene (**118**).

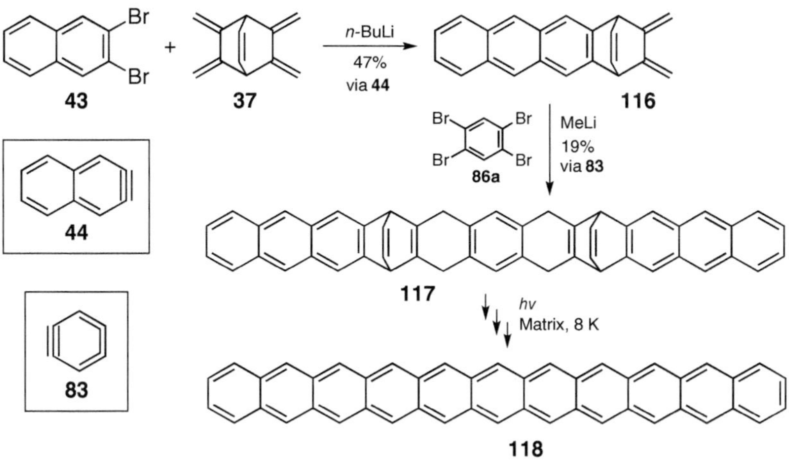

Scheme 2.24 Bettinger's synthesis of undecacene (**118**). Source: Based on Shen et al. [71].

Acenes that consist of more than four units are unstable and very reactive molecules with respect to photo-oxidation. Two main strategies have been implemented to improve their stability. The first of them entails the expansion of the conjugated system of the acene by introducing fused rings, such as phenanthrene or pyrene units. The second strategy is to introduce substituents, blocking the more reactive positions. Additionally, substituents and fused rings can create a distortion of the planarity (twistacene), improving the solubility and processability of the corresponding acenes. In this context, in 2004, Pascal and coworkers synthesized hexaphenyltetrabenzopentacene **123**, which, due to the steric constraint, is not planar, exhibiting an end-to-end twist of the pentacene skeleton of 144° [72]. The synthesis of this pentacene **123** starts with two D–A cycloadditions between *p*-diphenylbisbenzyne **121** synthon, generated from tetrabromoterphenyl **119** in the presence of *n*-BuLi, and 1,3-diphenylphenanthrofuran **120**, yielding bisendoxypentacene **122**. This compound was converted into hexaphenyltetrabenzopentacene **123** by treatment with low-valent titanium ($TiCl_3$/*n*-BuLi) in 27% isolated yield (Scheme 2.25). Moreover, the authors were able to resolve both enantiomers of hexaphenyltetrabenzopentacene **123** by preparative chiral high performance liquid chromatography (HPLC).

Very recently, Zhang, Liu, and coworkers described the synthesis of dodecatwistacene **127** by the reaction of aryne **126**, generated in situ from anthranilic acid **124** upon treatment with *i*-Am-ONO, and biscyclopentadienone **125**, followed by thermolysis of the bisadduct to force its decarbonylation (Scheme 2.26) [73]. However, **127**, although having six benzene rings fused to the dodecacene scaffold and six pendant phenyl rings, was found to be unstable on air, presumably due to the high energy of its HOMO orbital.

Wudl and coworkers used a bisbenzyne approach for the synthesis of tetraphenyltetrabenzoheptacene **130**, another twistacene [74]. Sequential aryne generation from bistriflate **128**, triggered by fluoride, followed by trapping with two units

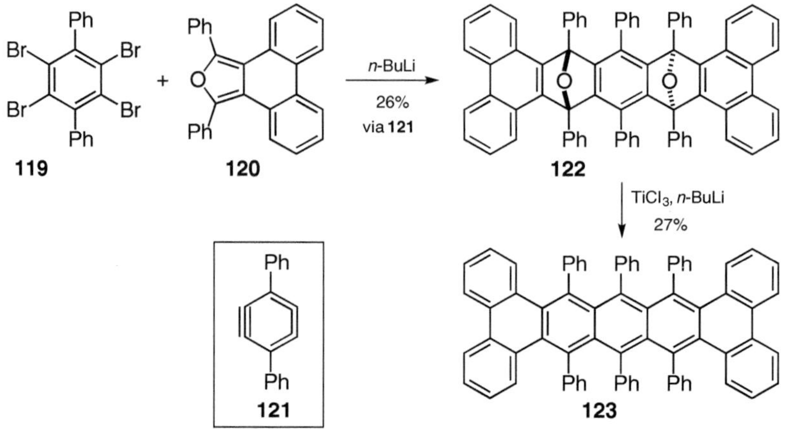

Scheme 2.25 Pascal's synthesis of twisted substituted pentacene **123**. Source: Based on Lu et al. [72].

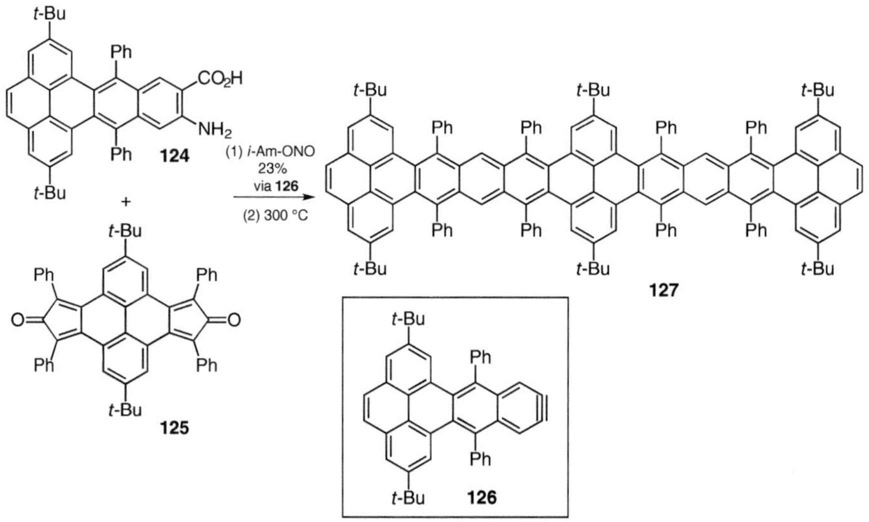

Scheme 2.26 Zhang's synthesis of dodecatwistacene **127**. Source: Based on Chen et al. [73].

of cyclopentadienone **129**, afforded tetraphenyltetrabenzoheptacene **130** as stable orange solid in 22% reaction yield (Scheme 2.27). In contrast with **127**, this relatively short twistacene is stabilized by four fused benzene rings and four phenyl substituents.

In 2015, Pérez, Peña, and coworkers described the synthesis of a series of large acenes by a sequence of two D–A cycloadditions of the 2,6-naphthodiyne synthon **79** with dienones (Scheme 2.28) [75]. Treatment of bistriflate **106** with tetrabutylammonium fluoride (TBAF) in the presence of cyclopentadienone **129** afforded acene **131**. These authors also showed that, using a controlled amount of TBAF, the

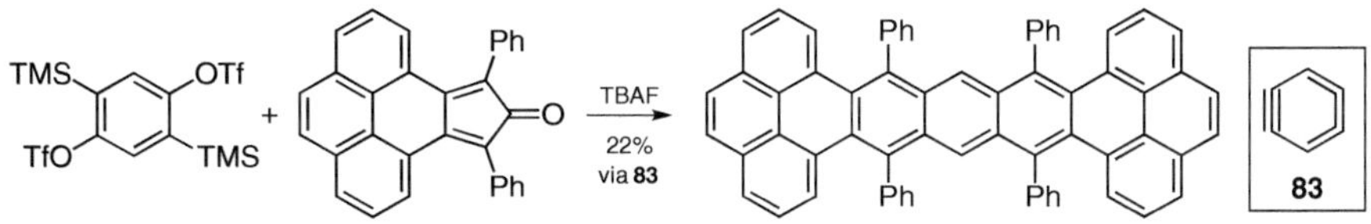

Scheme 2.27 Wudl's synthesis of benzo-fused substituted heptacene **130**. Source: Based on Duong et al. [74].

reaction of bistriflate **106** with cyclopentadienone **15** can be stopped after only one cycloaddition to isolate **132**. Then, the generation of the second aryne **134** in the presence of a different cyclopentadienone **133** afforded the asymmetric acene **135**.

Another approach to yield substituted pentacenes involving three [4+2] aryne cycloadditions was used by Nuckolls's group (Scheme 2.29) [76]. Aryne **138** was generated from the iodonium triflate **136** in the presence of TBAF, and reacted with cyclopentadienones **137** or **15** to yield bis(trimethylsilyl)naphthalenes **139a** or **139b**, respectively. These naphthalenes **139a,b** were converted into iodonium triflates **140a,b** by treatment with $PhI(OAc)_2$ and TfOH. The reaction of **140a,b** with TBAF generates the corresponding arynes **142a,b**, which, in the presence of electron-deficient diene 2,5-diaryl-6*H*-1,3,4-oxadiazine-6-one (**141**), afforded phenylated pentacenes **144a,b**. This transformation involves a D–A aryne cycloaddition followed by N_2 extrusion and another D–A cycloaddition to the resulting α-pyrone **143a,b**, followed by CO_2 extrusion.

Palladium-catalyzed [2+2+2] cocyclizations have also been used for the synthesis of functionalized acenes, as shown in Scheme 2.30 [77]. The reaction of bisaryne precursor **106** and dialkylacetylene dicarboxylates **145a–c** in the presence of $Pd_2(dba)_3$ afforded tetracene octaesters **146a–c** in one pot in 16% yield.

2.3.2 Synthesis of Perylene Derivatives

[4+2] Cycloaddition of arynes to the bay region of perylenes is an interesting one-pot synthetic strategy for extending the π-conjugation of such compounds. This transformation involves a D–A reaction followed by a spontaneous dehydrogenation. Fort and Scott reported the cycloaddition of benzyne to perylene (**147**) in the gas phase (Scheme 2.31) [78]. They used phthalic anhydride (**148**), which can generate benzyne (**1**) in the gas phase after eliminating CO_2 and CO above 650 °C. By mixing solid phthalic anhydride with perylene and, then, cosublimating the mixture at 1000 °C under reduced pressure in a quartz tube, naphtho[1,2,3,4-*ghi*]perylene (**149**) is obtained in 5% yield, compared with the 45% yield when the reaction is performed in solution and using triflate as aryne precursor.

Peña and coworkers reported the addition to perylene of substituted arynes **150** and **153**, generated from triflates **128**, and **152**, respectively, to afford the corresponding adducts **151** [79] and **154** [3] in moderate yields (Scheme 2.32).

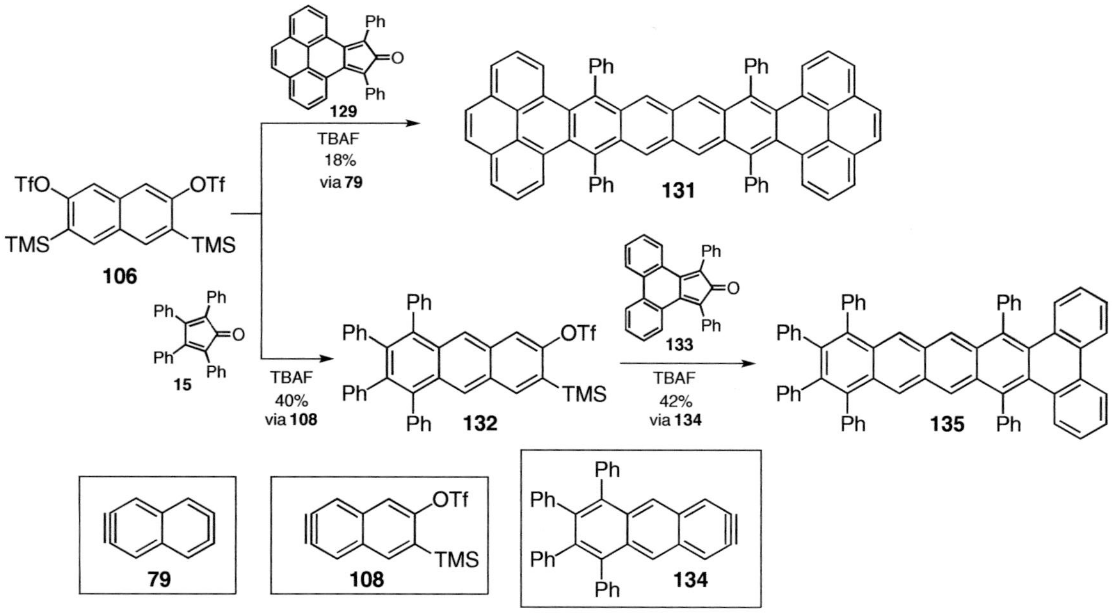

Scheme 2.28 Pérez–Peña's synthesis of benzo-fused substituted acenes. Source: Based on Rodríguez-Lojo et al. [75].

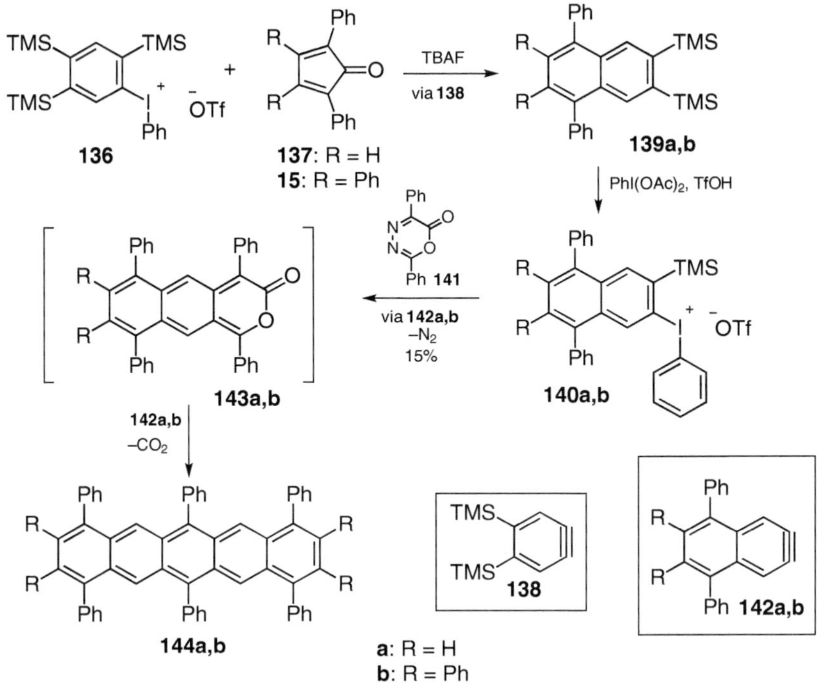

Scheme 2.29 Nuckolls's synthesis of substituted pentacenes **144**. Source: Based on Miao et al. [76].

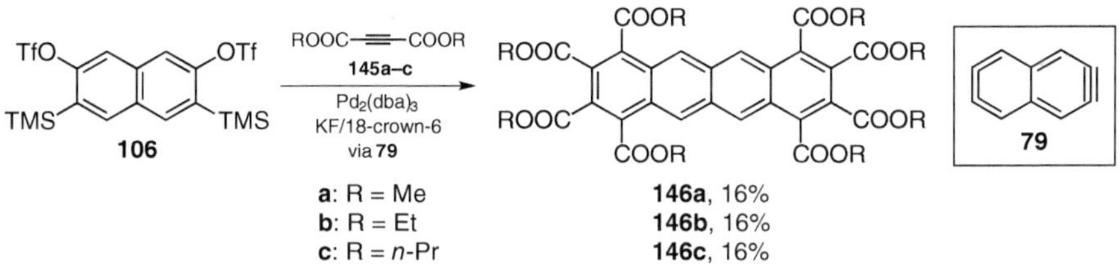

Scheme 2.30 Kitamura's synthesis of substituted tetracenes **146**. Source: Based on Kitamura et al. [77].

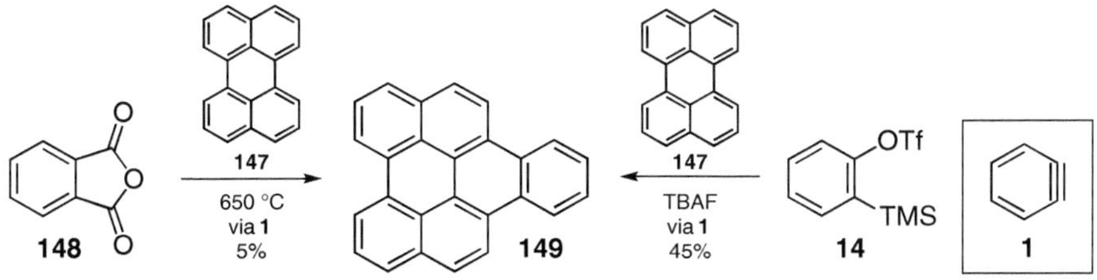

Scheme 2.31 Cycloadditions of benzyne to the bay region of perylene (**147**). Source: Based on Fort and Scott [78].

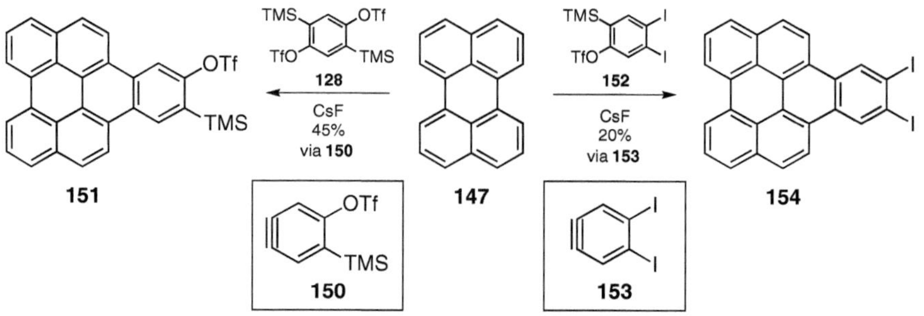

Scheme 2.32 Cycloadditions of arynes to the bay region of perylene (**147**). Source: Pavliček et al. [3]; Schuler et al. [79].

In 2013, Kubo and coworkers reported the [4+2] cycloaddition of arynes, such as **157** and **99**, to bisanthene **155**, leading to the formation of extended derivatives **158** and **159** in moderate yields (Scheme 2.33) [80].

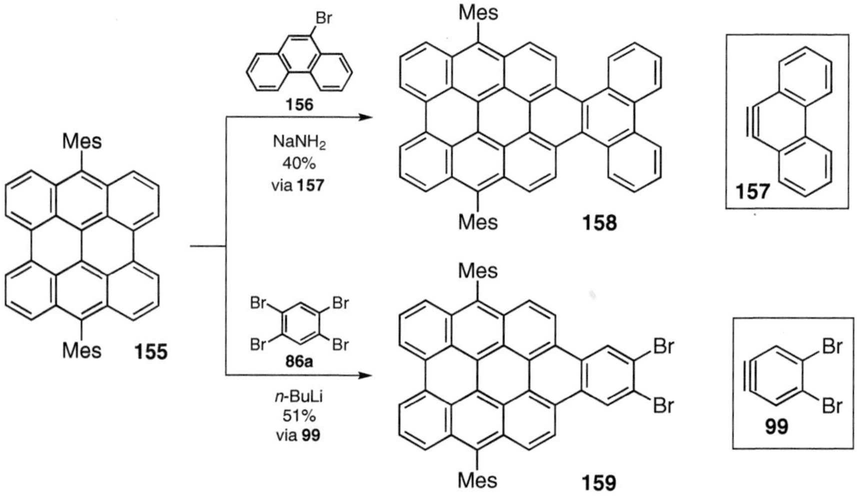

Scheme 2.33 Cycloadditions of arynes to the bay region of bisanthene **155**. Source: Based on Konishi et al. [80].

Very recently, Itami and coworkers implemented this methodology for generating π-extended perylenebisimides (PBIs) **161**, **162**, or **164** in only one reaction step from PBIs **160**, in contrast to the classical approach consisting of three reaction steps involving bromination, Suzuki–Miyaura C–C coupling, and a final cyclization (Scheme 2.34) [81].

Peña, Guitián, and coworkers described a strategy to obtain extended perylenes on a two-step synthetic route involving tandem [4+2] cycloadditions (Scheme 2.35) [82, 83]. The reaction of 1,8-difurylnaphthalene (**165**) with benzyne (**1**), generated by Kobayashi's method, afforded only the *exo,exo* diastereomer **166** in 99% yield. This key transformation involves two tandem cascade D–A reactions in a domino mode,

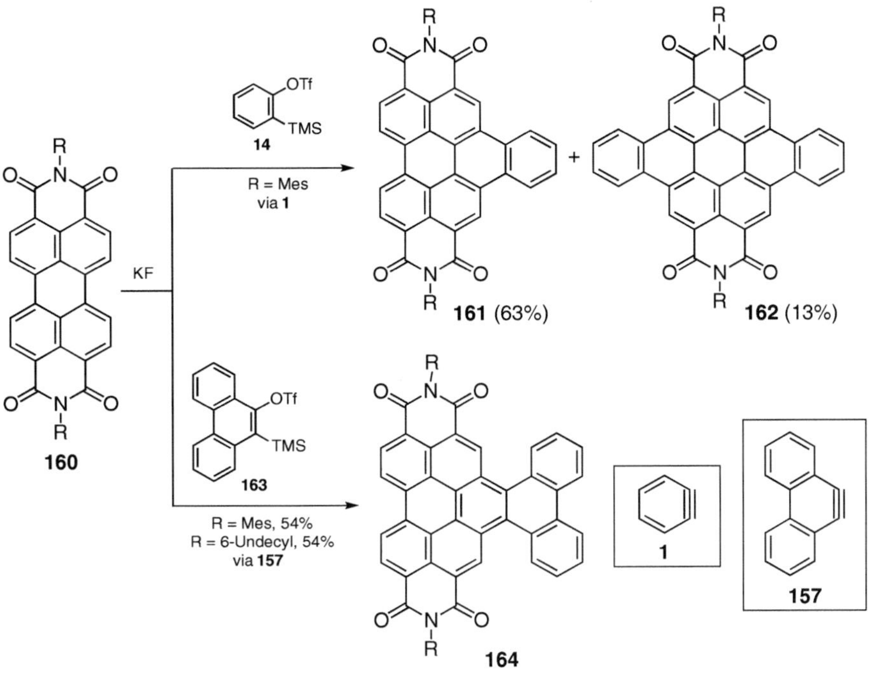

Scheme 2.34 Cycloadditions of arynes to the bay region of perylenebisimides. Source: Based on Nakamuro et al. [81].

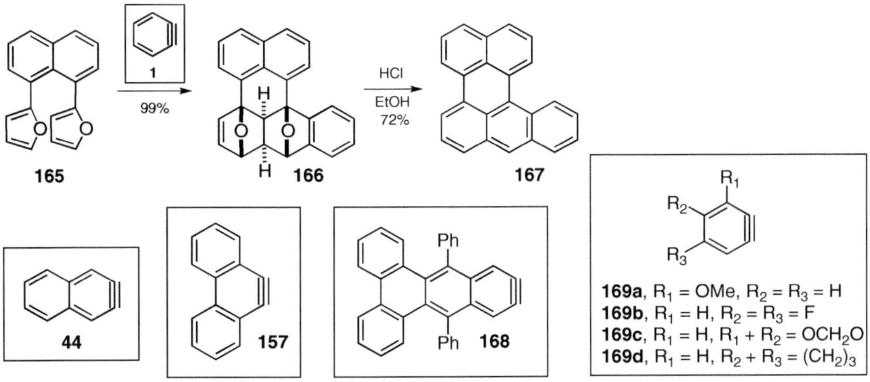

Scheme 2.35 Synthesis of perylene derivatives by tandem aryne cycloadditions. Source: Criado et al. [82]; Criado et al. [83].

with the creation of four new C—C bonds and six new stereogenic centers. Final treatment of adduct **166** with concentrated aqueous HCl in refluxing EtOH led to perylene derivative **167** in 72% yield. This methodology has been extended to more complex arynes, such as **44**, **157**, **168**, and **169**, allowing the synthesis of extended perylene derivatives.

2.3.3 Synthesis of Triptycenes

Aryne intermediates have also been used in [4+2] cycloaddition reactions with positions 9 and 10 of anthracene derivatives to yield triptycenes. Triptycenes are not planar, exhibiting a paddle-wheel configuration. Kelly et al. conceived triptycenes endowed with helicenes as molecular ratchets, where the triptycene acts as the wheel and the helicenes serve as pawls and springs [84–86]. With this goal in mind, they synthesized triptycene **172** endowed with a [4]helicene (Scheme 2.36). To obtain triptycene **172**, they reacted benzyne (**1**), generated from anthranilic acid **9** in the presence of *i*-Am-ONO, with anthracene derivative **170**. Afterwards, triptycene **171** was derivatized to the functionalized triptycene **172a** endorsed with amino and butanol groups. To yield unidirectional rotary motion, chemical energy was used for activating and biasing a thermally induced isomerization reaction, which yielded rotamer **172b** after several steps.

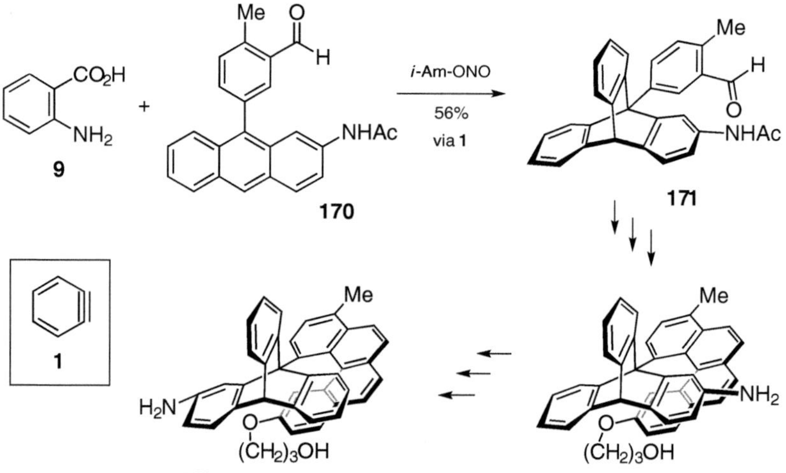

Scheme 2.36 Kelly's synthesis of molecular ratchet **172**. Source: Kelly et al. [84]; Kelly et al. [85]; Kelly et al. [86].

In 2012, iptycene-based molecules **176** were described as fluorescent sensors for explosives such as nitroaromatics and trinitrotoluene (TNT) [87, 88]. To synthesize the iptycene sensors **176**, cycloaddition reactions between bisaryne **175**, generated from haloarene **173** and *t*-BuOK, and anthracene (**174**) were employed (Scheme 2.37).

Other examples of the use of arynes for building triptycenes with specific functions include a spur gear [89], a molecular gyroscope [90], porous triptycene-based molecules [91], and microporous polymers [92].

2.3.4 Synthesis of π-Extended Starphenes or Angular PAHs

Starphenes, star-shaped PAHs having a central aromatic core with a variable number of branches, have attracted the attention of synthetic chemists due to

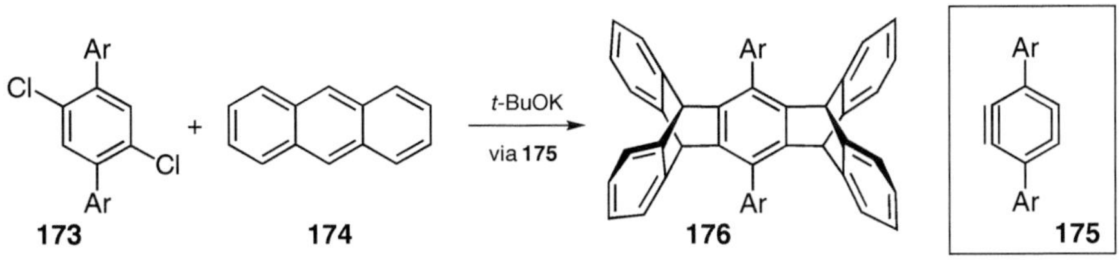

Scheme 2.37 Synthesis of triptycene sensors for explosives. Source: Based on Anzenbacher et al. [87].

their interesting properties and potential uses in electronic devices [93, 94]. Aryne cycloadditions, particularly the palladium-catalyzed [2+2+2] cycloaddition, have proved to be a useful tool for the synthesis of three-branched starphenes, also known as cloverphenes, due to their clover-like structure [95] or supertriphenylenes [96]. In 2006, our group described the synthesis of [13]cloverphenes **179a,b** by a palladium-catalyzed [2+2+2] cyclotrimerization of arynes **178a,b**. This was accomplished by the successful synthesis of novel triflates **177a,b** in four reaction steps, which can serve as precursors of arynes **178a,b** (Scheme 2.38) [97, 98]. It has been reported that [13]cloverphene **179b** exhibit columnar mesophases [96].

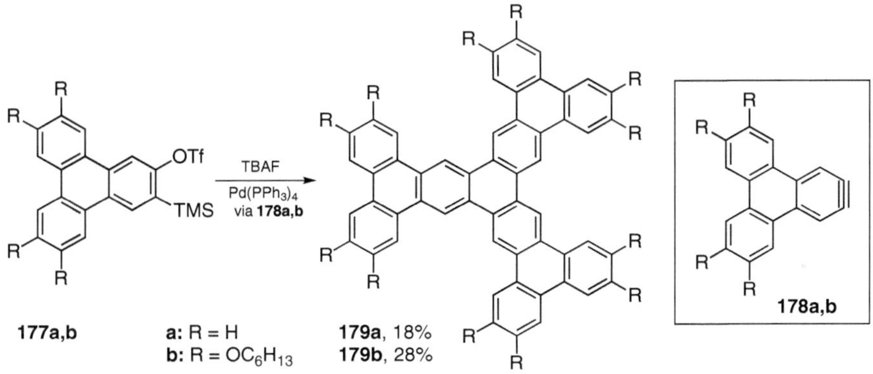

Scheme 2.38 Synthesis of starphenes **179**. Source: Romero et al. [97]; Romero et al. [98].

Moreover, the palladium-catalyzed cocyclotrimerization of arynes **178b,c**, equipped with alkoxy chains, with acetylenedicarboxylates **145a,d,e**, gave rise to tetrabenzopentaphenes **180** equipped with alkyl chains that exhibit liquid crystal behavior (Scheme 2.39) [98].

Lynett and Maly reported the synthesis of a series of substituted trinaphthylenes **183a–c** through the Pd-catalyzed [2+2+2] cyclotrimerization of substituted 2,3-naphthynes **182a–c**, generated from triflates **181a–c** (Scheme 2.40) [99]. Although no mesophases were observed for alkyl-substituted trinaphthylenes **183a–c**, in the case of **183b**, its aggregation in solution was inferred from the concentration-dependent ^{1}H NMR (nuclear magnetic resonance).

Recently, Bunz and coworkers showed that the synthesis of trinaphthylenes **183b,d–h** from dibromonaphthalenes **184b,d–h** is also possible by using Ni(0)-promoted Yamamoto coupling (Scheme 2.40) [100].

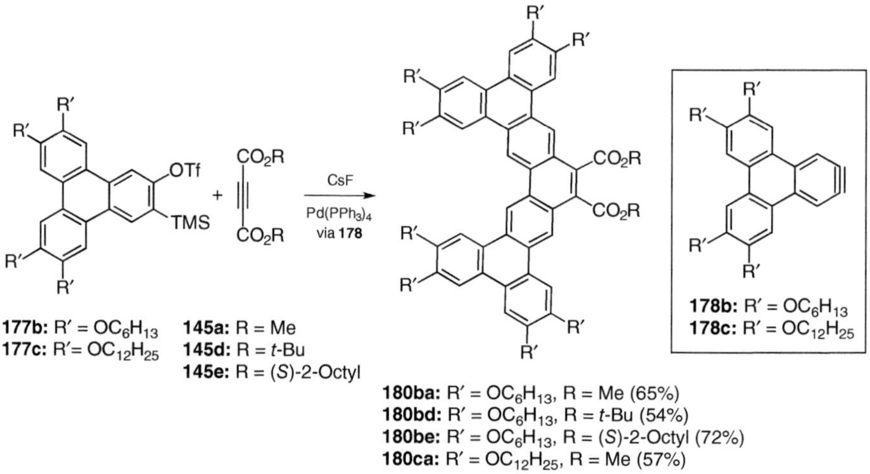

Scheme 2.39 Synthesis of mesogenic tetrabenzopentaphenes **180**. Source: Romero et al. [98].

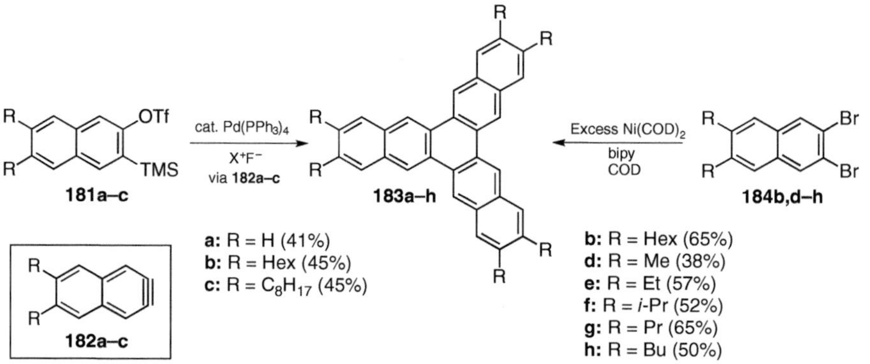

Scheme 2.40 Synthesis of trinaphthylenes **183** by Maly and coworker. Source: Lynett and Maly [99]; Rüdiger et al. [100].

Later on, Soe et al. studied the trinaphthylene **183a** deposited on Au surface by scanning tunneling microscopy (STM). They found that this molecule, when connected to Au atoms, behaves as a molecular logic gate [101].

Peña and coworkers expanded the limits of aryne cycloaddition reactions to build a clover-shaped *cata*-condensed nanographene **189**, known as [16]cloverphene, with 102 sp2 carbons (Scheme 2.41) [95]. The synthesis of [16]cloverphene started with a first D–A cycloaddition between aryne **186**, generated from triflate **185** by reaction with TBAF, and dienone **133**, followed by cheletropic extrusion of CO to afford **187**. *o*-Bromophenol **187** was then converted into the corresponding *o*-silylaryltriflate **188**, precursor of aryne **168**. The generation of this aryne in the presence of $Pd(PPh_3)_4$ led to the formation of cloverphene **189** in a 22% yield.

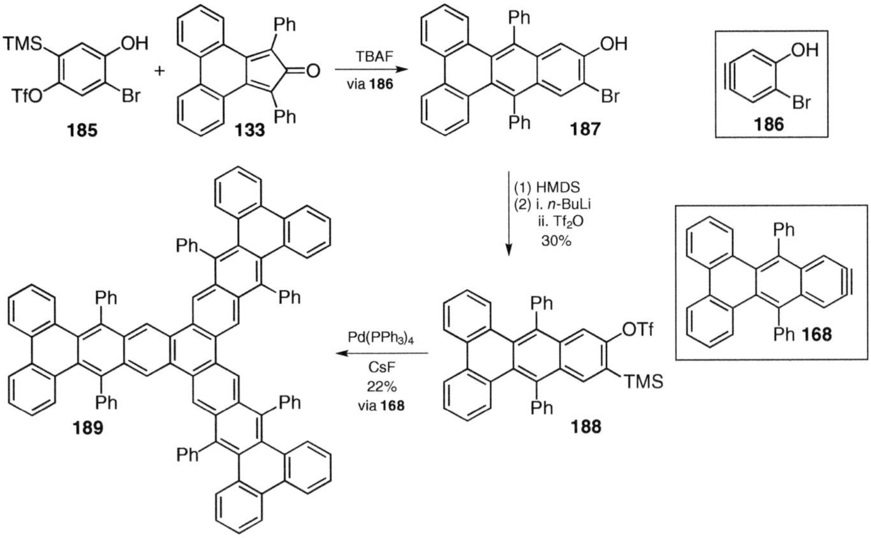

Scheme 2.41 Synthesis of twisted [16]cloverphene **189**. Source: Based on Alonso et al. [95].

A couple of years later, our group described the synthesis of a threefold symmetric $C_{78}H_{36}$ PAH with 22 fused benzene rings within its structure in only two reaction steps starting from commercially available compounds (Scheme 2.42) [79]. The synthesis of such clover-shaped nanographene **190** began with the preparation of triflate **151** (see Scheme 2.32). Then, aryne **2**, generated from triflate **151**, was transformed into cloverphene **190** by a palladium-catalyzed [2+2+2] cycloaddition. Due to its insolubility, the structure of cloverphene **190** was unambiguously elucidated by AFM on an ultrathin insulating film (Scheme 2.42b).

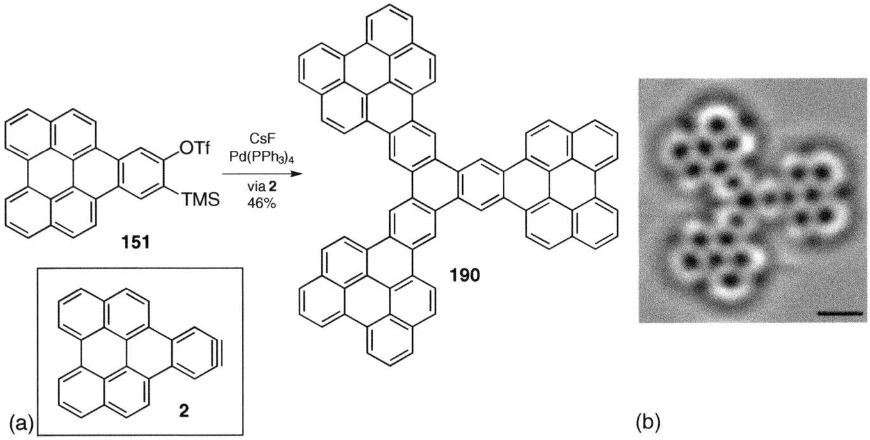

Scheme 2.42 (a) Synthesis and (b) AFM image of threefold symmetric nanographene **190**. Source: Image adapted with permission from Schuler et al. [79]. Copyright 2014, John Wiley and Sons.

More recently, our group described the synthesis of a dendritic starphene **194** with 19 *cata*-fused benzene rings distributed in six branches. This [19]dendriphene **194** was prepared in solution by a palladium-catalyzed cyclotrimerization of aryne **193**, being the largest unsubstituted *cata*-condensed PAH synthesized to date (Scheme 2.43) [102]. Triflate **192** was prepared in 19% yield by means of a palladium-catalyzed annulation of aryne **150** and iodobinaphthalene **191**.

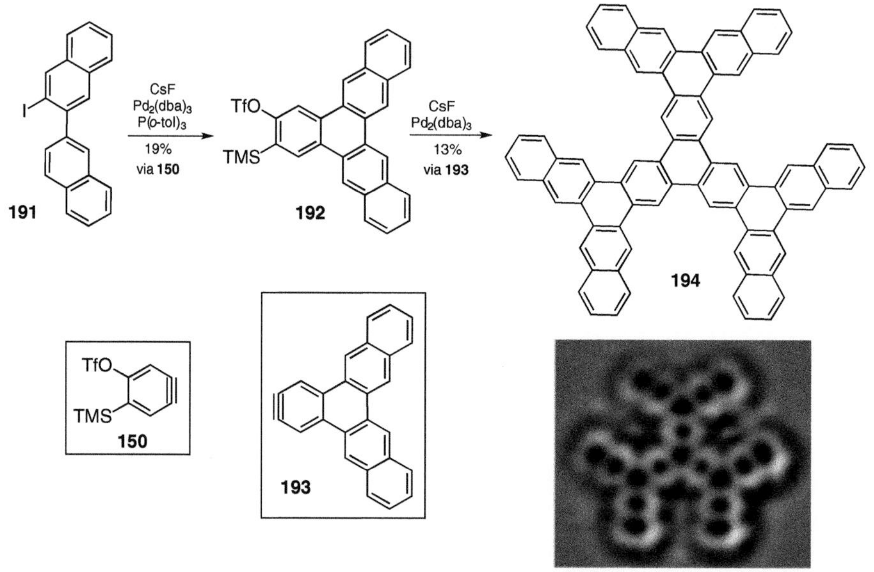

Scheme 2.43 Synthesis and AFM image of dendriphene **194**. Source: Image adapted with permission from Vilas-Varela et al. [102]. Copyright 2018, John Wiley and Sons.

The scope of the palladium-catalyzed [2+2+2] cycloaddition methodology is evidenced by the large number and diversity of molecular structures obtained. In this regard, Pérez and coworkers described the synthesis of tris(benzocyclobutadiene)triphenylenes **197a,b**, molecules consisting of three biphenylene units (bearing antiaromatic cyclobutadiene rings) fused to a central benzene ring, distorting the central triphenylene unit (Scheme 2.44) [103].

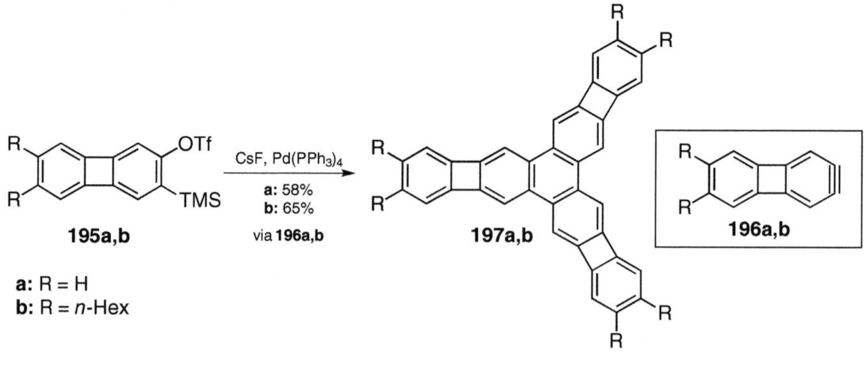

Scheme 2.44 Synthesis of biphenylene-based starphenes **197**. Source: Iglesias et al. [103].

Trimers **197a,b** were synthesized through Pd-catalyzed [2+2+2] cycloadditions of didehydrobiphenylenes **196a,b**, which were generated in good yields from their corresponding triflate precursors **195a,b** by addition of CsF. Recently, trimer **197a** was used to selectively functionalize hydrogen-passivated Ge(001):H surfaces through reversible [4+2] cycloaddition between trimers and dangling-bond Ge dimers [104].

Hamura and coworkers reported the synthesis of a star-shaped polycyclic aromatic ketone **206** involving a Pd-catalyzed [2+2+2] cyclotrimerization of aryne **202** to give **203** (Scheme 2.45) [105]. What is important about this trimer **203** is that it acts as a synthetic equivalent of isobenzofuran trimer **205**, which can be transformed into the star-shaped polycyclic ketone **206** by cycloaddition with dienophiles such as naphthoquinones **204**.

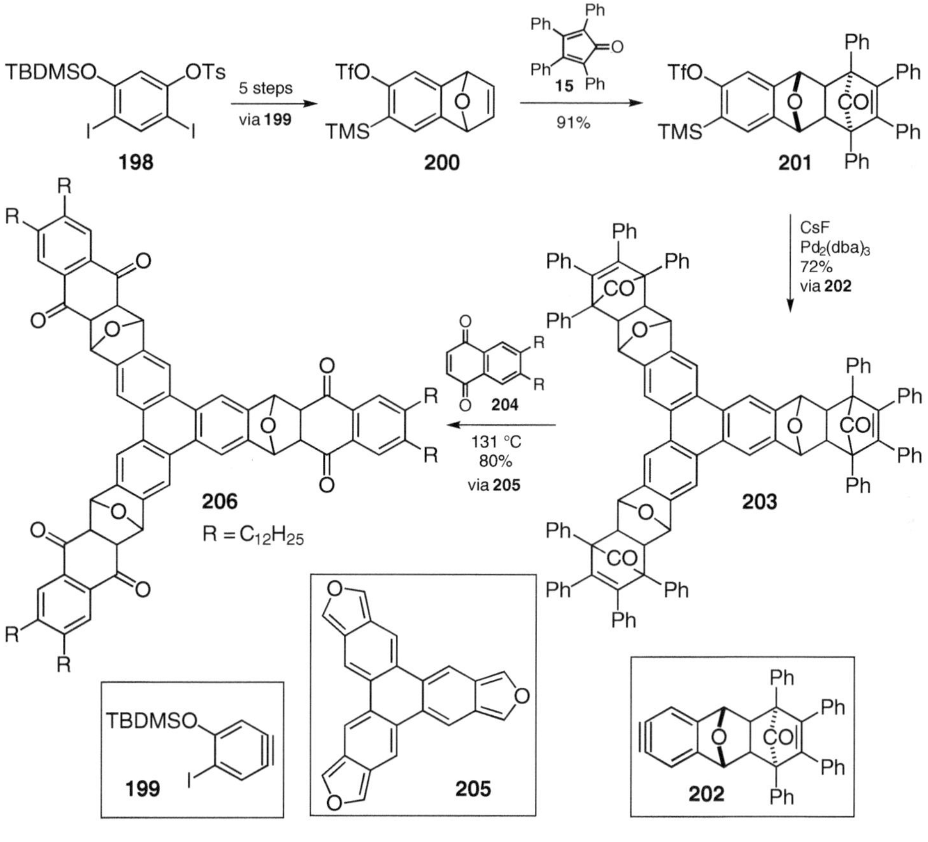

Scheme 2.45 Hamura's synthesis of starphenes **206**. Source: Based on Tozawa et al. [105].

In recent years, many organic reactions traditionally run in solution have been performed on metallic surfaces under UHV conditions [106, 107]. On-surface reactions can be promoted either by thermal annealing or by the application of voltage pulses from STM tips. For example, aryl radicals have been generated from aryl halides by using these procedures. The related on-surface generation of arynes from *o*-dihalides has already been cited in the introduction [3, 4]. With these precedents, Fasel, Meunier, and coworkers described the formation of oligomeric structures **209–211**, presumably through aryne **208**, by heating 2,3-dibromotetracene (**207**) on a Ag(111) surface (Scheme 2.46) [108].

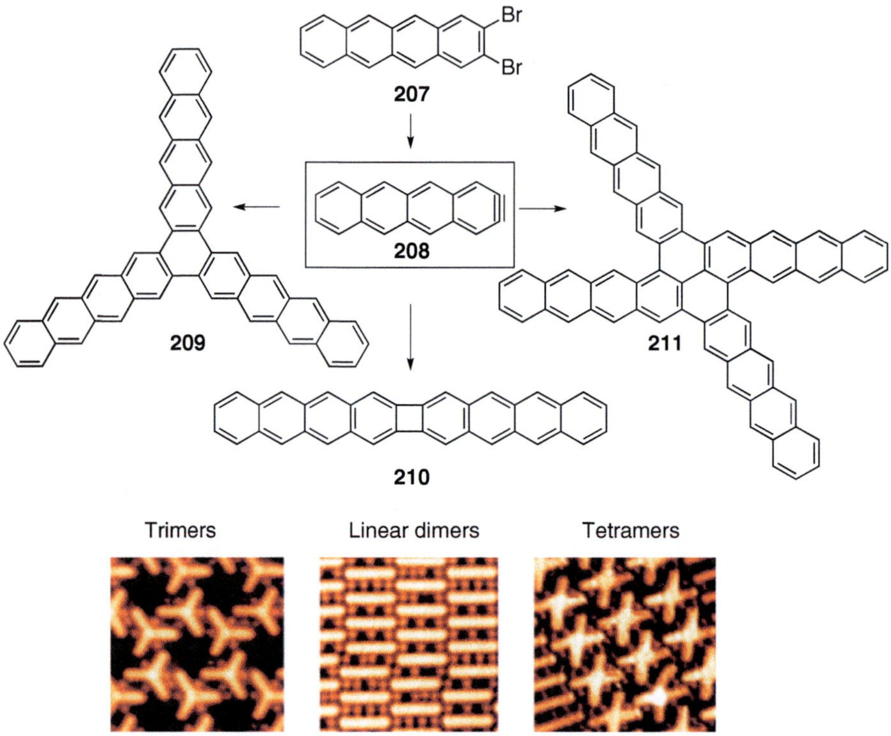

Scheme 2.46 Generation and coupling of aryne **208** on Ag(111). Source: Images adapted with permission from Sánchez-Sánchez et al. [108]. Copyright 2017 American Chemical Society.

2.3.5 Synthesis of Helicenes

Helicenes are polycyclic aromatic compounds with nonplanar helically shaped skeletons formed by *ortho*-fused aromatic rings. As helical chiral structures, they can adopt two enantiomeric configurations. Among the several methods developed for the synthesis of helicenes, organometallic catalysis plays a growing role [109]. Soon after the discovery of Pd-catalyzed [2+2+2] aryne cycloadditions, the first example of the application of this methodology to the synthesis of an helicene was reported. Guitián, Pérez, and coworkers described the preparation of the highly twisted triple-helicene **212** in 68% yield by cyclotrimerization of 9,10-didehydrophenantrene (**157**) catalyzed by $Pd_2(dba)_3$ (Scheme 2.47) [110]. The synthesis of hexabenzotriphenylene (**212**) had been previously reported but in very low yields (5–13%) [112, 113]. The authors reported experimental and computational studies on the conformational stability of this interesting triple-helicene **212**, observing the formation of the C_2-conformers, which, upon heating, evolved to the D_3 molecular propeller, which is the thermodynamic isomer [111].

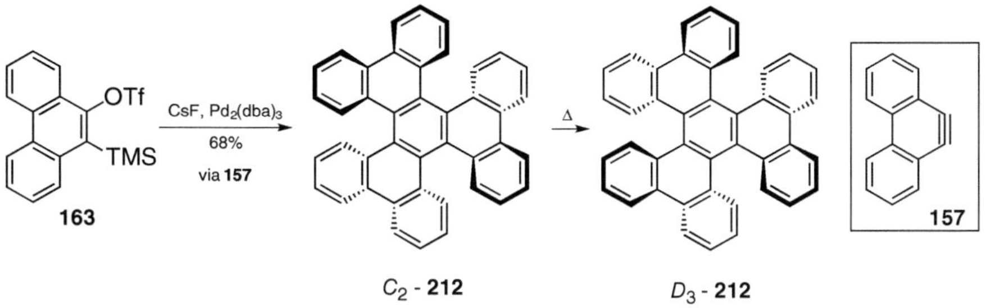

Scheme 2.47 Synthesis of hexabenzotriphenylene (**212**) and their conformers. Source: Peña et al. [110]; Peña et al. [111]; Barnett et al. [112]; Hacker et al. [113].

Our group described the preparation of triflate **213** as precursor of 3,4-didehydrophenantrene **214** (Scheme 2.48) [114]. Pd-catalyzed reaction of triflate **213** under aryne-forming conditions led to the formation of trimer **215** in 26% yield, with the other cyclization product, the C_3-symmetric trimer **216** – a triple [5]helicene – not being detected. Compound **216** was later synthesized by Watanabe and coworkers, but using oxidative photocyclization reactions, not involving arynes [115]. Pérez, Peña, and coworkers also described the cocyclizations of 3,4-didehydrophenantrene **214** and 1,2-didehydrotriphenylene with alkynes through Pd-catalyzed [2+2+2] reactions to yield other sterically congested PAHs [116].

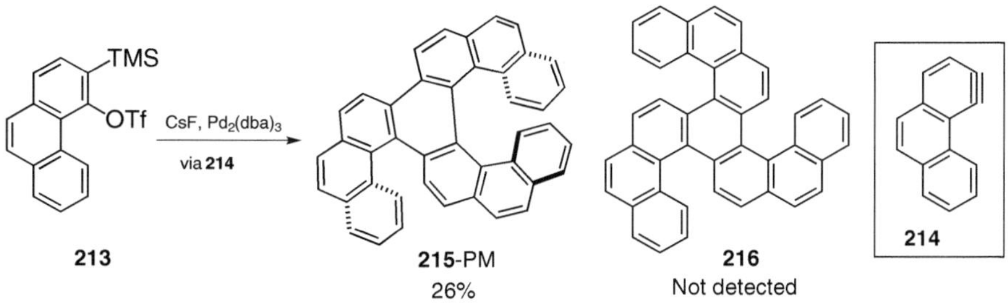

Scheme 2.48 Synthesis of double helicene **215**. Source: Based on Peña et al. [114].

The introduction of chiral ligands can lead to the formation of nonracemic [2+2+2] palladium-catalyzed cycloaddition products, as exemplified by Guitián, Pérez, and coworkers [117]. In this contribution, naphthalene triflate **217** is reacted with $Pd(PPh_3)_4$ in the presence of CsF, yielding the three possible products of the cycloaddition of two units of aryne **218** and one of DMAD (**28**) (Scheme 2.49). Even though [5]helicene **219** is not the major product from the reaction, the isolated yield (25%) was found to be satisfactory considering the limited availability of synthetic approaches to pentahelicenes. When using (2,2′-bis(diphenylphosphino)-1,1′-binaphthyl) (BINAP) as a ligand for the Pd catalyst in the reaction, they found enantioselectivity in the formation of [5]helicene **219** with reasonable enantiomeric excess. This contribution represented the first example of an enantioselective, metal-catalyzed cycloaddition involving arynes.

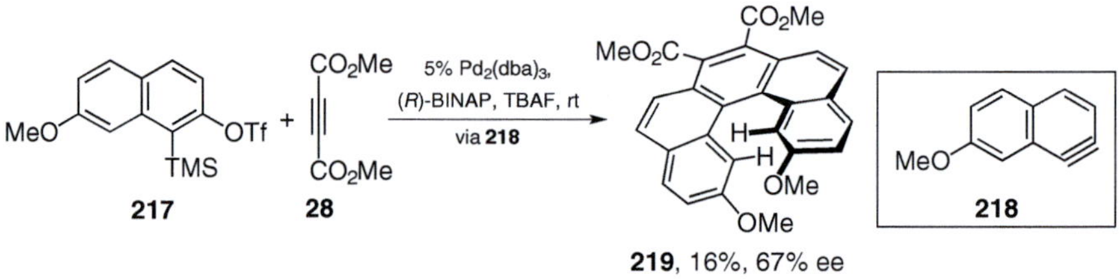

Scheme 2.49 Enantioselective synthesis of pentahelicene **219**. Source: Based on Caeiro et al. [117].

Three groups independently described the synthesis of hexapole pentahelicene **222** containing six [5]helicenes (Scheme 2.50). Kamikawa, Tsurusaki, and coworkers synthesized this hexapole helicene **222** in 54% yield by a $Pd_2(dba)_3$-catalyzed [2+2+2] cycloaddition of aryne **221**, generated from triflate **220** by treatment with CsF [118]. X-ray diffraction of **222** reveals the highest degree of twisting angle per benzene unit (35.7°) reported so far. Gingras, Coquerel, and coworkers used

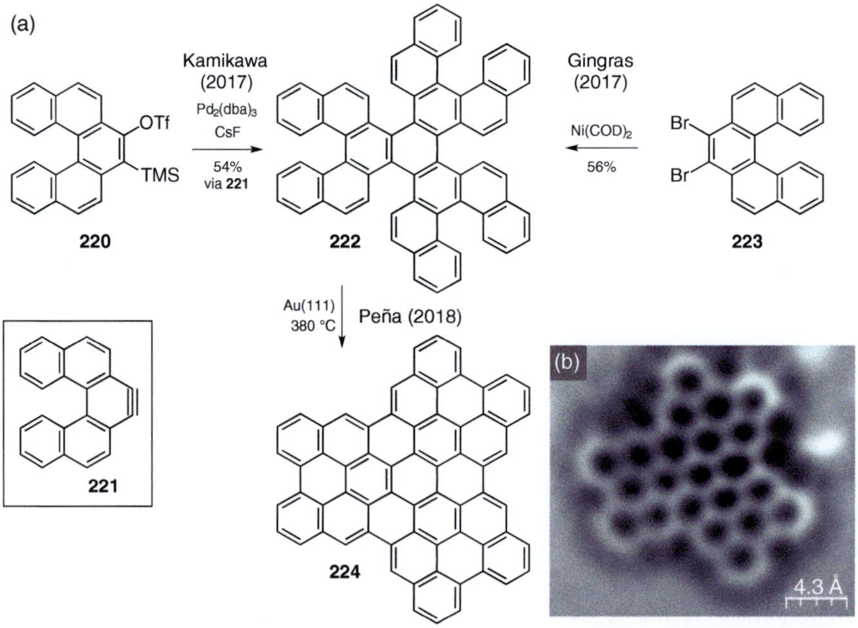

Scheme 2.50 Synthesis of hexapole pentahelicene **222**. Source: Hosokawa et al. [118]; Berezhnaia et al. [119], and AFM image of **224**. Image adapted from Zuzak et al. [120]. Copyright 2018, The Royal Society of Chemistry.

a Ni(0)-mediated trimerization of 7,8-dibromo[5]helicene (**223**) to yield hexapole **222** in one reaction step in 56% yield [119]. Our group independently prepared compound **222** by Pd-catalyzed cyclotrimerization of aryne **221**, and then **222** was deposited on Au(111) and transformed into the threefold symmetric $C_{66}H_{24}$ nanographene (**224**) by on-surface cyclodehydrogenation. This compound was characterized with submolecular resolution by AFM with functionalized tip (inset in Scheme 2.50) [120].

Sygula and coworkers synthesized helical corannulene trimers by trimerization of corannulyne **226**. Treatment of triflate **225** with CsF in the presence of $Pd_2(dba)_3$ led to hydrocarbon $C_{60}H_{24}$ **227** in 40% yield (Scheme 2.51) [121].

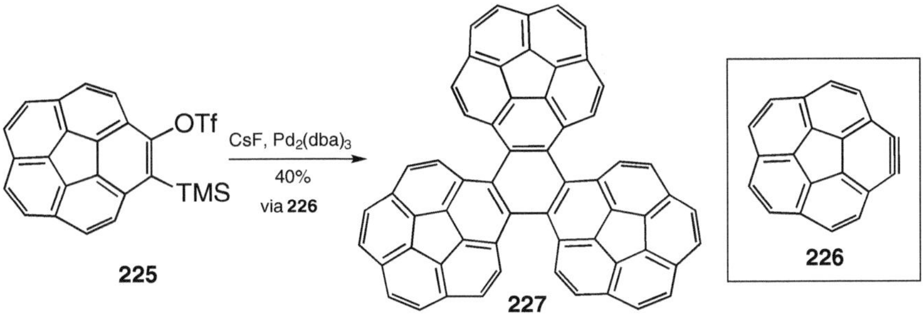

Scheme 2.51 Synthesis of triscoranylene **227**. Source: Based on Yanney et al. [121].

Garg, Houk, and coworkers reported the synthesis of helicenes containing indole moieties by palladium-catalyzed cyclotrimerization of indolynes (Scheme 2.52) [122]. For example, 3,4-indolyne **229**, generated from **228**, was trimerized using $Pd_2(dba)_3$ and BINAP to form a 1 : 2 mixture of regioisomers **230** and **231** in 93% yield. Similarly, 5,6-indolyne undergoes cyclotrimerization to give a mixture of regioisomeric cyclotrimers in 80% yield. However, the cyclotrimerization of 6,7-indoline only works in 9% yield, presumably due to steric hindrance.

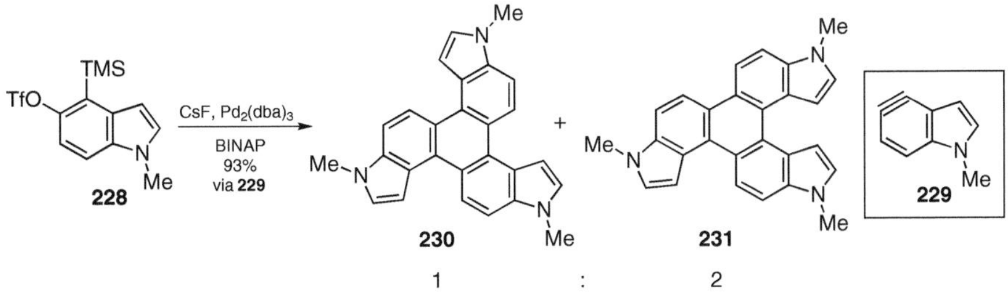

Scheme 2.52 Synthesis of heterocyclic helicenes. Source: Based on Lin et al. [122].

Finally, Pérez and coworkers reported the synthesis of [4]helicenes and other sterically congested PAHs by means of a double [4+2] cycloaddition reaction of the novel 1,7-naphthodiyne synthon **233** with different dienes (Scheme 2.53) [123]. Bistriflate **232** was reacted with furan **95** in the presence of CsF to give bisadduct **234** in 69%

yield. This bisendoxide **234** was reduced with Zn in AcOH to afford [4]helicene **235** in 45% yield. Bistriflate **232** was also reacted with cyclopentadienone **15** and CsF to afford highly congested **237** after spontaneous decarbonylation of bisadduct **236** in 23% yield.

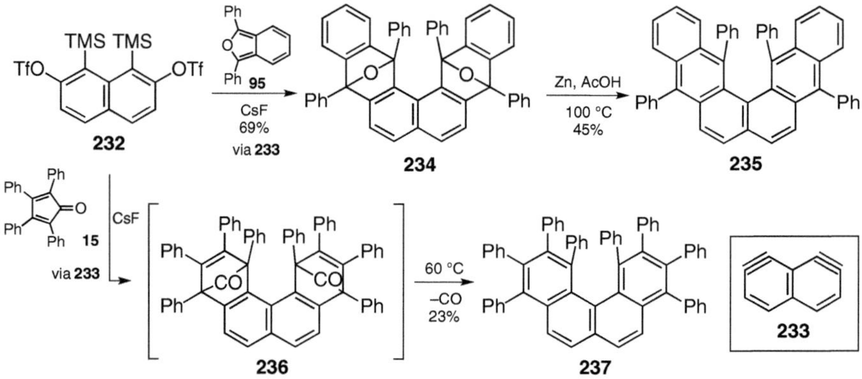

Scheme 2.53 Pérez's synthesis of angular and congested PAHs **235** and **237**. Source: Based on Pozo et al. [123].

2.3.6 Functionalization of Carbon Nanostructures

The first reports on the cycloaddition of benzyne to fullerene were published in 1992 [124, 125]. In both cases, a series of cycloadducts of benzyne, generated from anthranilic acid, and C_{60}, $[C_{60} + (C_6H_4)_n]$, were detected by mass spectrometry and NMR. Cooks, Kahr, and coworkers proposed structure **238** for the isolated monoadduct, which results from the [2+2] cycloaddition (Scheme 2.54) [124]. But it was not until 1995 that the X-ray structure of the [2+2] monoadduct was elucidated, proving that the cycloaddition process takes place at the (6,6) double bond of the fullerene sphere, generating a benzocyclobutene structure attached to a closed (6,6) bond of C_{60} [126].

In 2001, Nishimura and coworkers reported the isolation of all eight possible regioisomers (**241**) of [2+2] bisadducts from cycloaddition to C_{60} of 4,5-dimethoxybenzyne (**240a**), generated from the 4,5-dimethoxyanthranilic acid (**239**) in the presence of *i*-Am-ONO (Scheme 2.54) [127].

In 2013, Yang and coworkers claimed the formation of an open cage (5,6) monoadduct (**243**) by addition of aryne **240b** to C_{60} based on NMR, UV–vis spectroscopy, and cyclic voltammetry (Scheme 2.54) [128]. They suggested the formation of this open (5,6) monoadduct stemming from the rearrangement of the previously formed (6,6) closed-cage monoadduct. However, a recent reinvestigation of this reaction led to the conclusion that the reaction product is the (6,6) closed C_{60}-fused δ-lactone **244** [129].

Additions of up to 10 units of benzyne to C_{70} were firstly detected in 1994 by Walton and coworkers by means of mass spectrometry [130]. Shortly after, the same

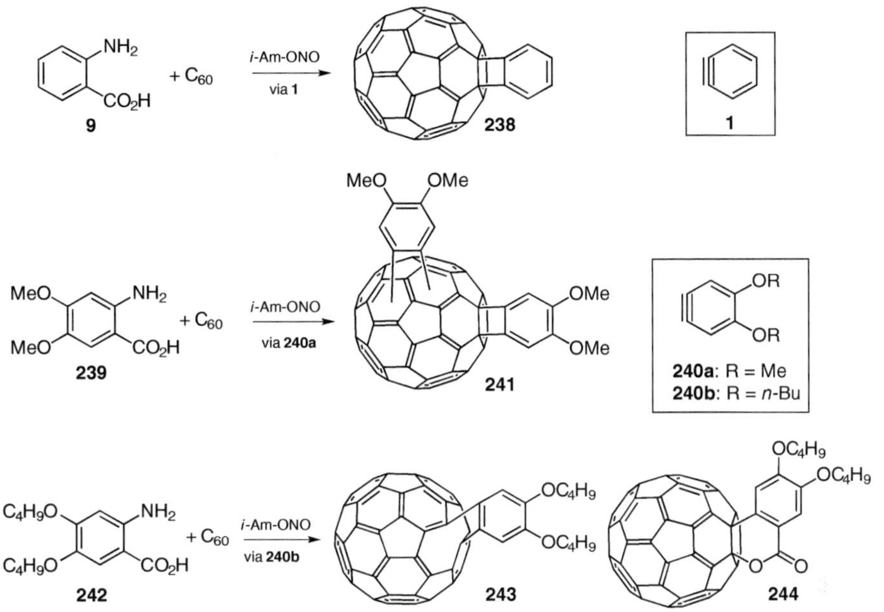

Scheme 2.54 Functionalization of C_{60} by aryne cycloadditions. Source: Hoke et al. [124]; Ishida et al. [126]; Nakamura et al. [127]; Kim et al. [128]; Mizunuma et al. [129].

group isolated the monoadducts as a mixture of four isomers according to their ^{1}H NMR spectra [131]. Three of them corresponded to [2+2] additions to (6,6) bonds and one was the result of a [4+2] cycloaddition. In 1998, Meier et al. were able to separate the four isomers and confirmed that one of them was a close monoadduct resulting from the addition of benzyne to a (6,6) bond in C_{70}, according to the X-ray crystal structure [132]. The other two were considered as other (6,6) cycloadditions, and one of them as a (5,6) according to their NMR spectra.

Arynes react with endohedral fullerenes to give [2+2] adducts, and the regioselectivity of the process depends on the metal inside the cage. While La@C_{82} gives (5,6) adducts [133], $Sc_3N@I_h$-C_{80} gives both (5,6) and (6,6) adducts [134].

Computational studies on the cycloaddition of benzyne to fullerenes [135] and endohedral fullerenes [136] support nonconcerted radical mechanisms.

In 2016, Martín, Pérez, and coworkers described the synthesis of the first aryne precursor comprising a C_{60} (**245**) (Scheme 2.55) [137]. They reacted C_{60} with monoaryne **150**, generated from bistriflate **128** and TBAF, to afford the [2+2] monoadduct **245** in 21% yield. Furthermore, upon subjecting monoadduct **245** to the presence of TBAF and a diene, such as furan, fullerobenzyne **246** was generated and trapped through a [4+2] cycloaddition to give **247** in 43% yield.

In 1998, it was suggested that addition of arynes to single-walled carbon nanotubes (SWCNTs) could produce paddle-wheel nanostructures, in which the stiff benzene fragments produced after multiple aryne cycloaddition to the SWCNTs act as teeth (Figure 2.3) [138].

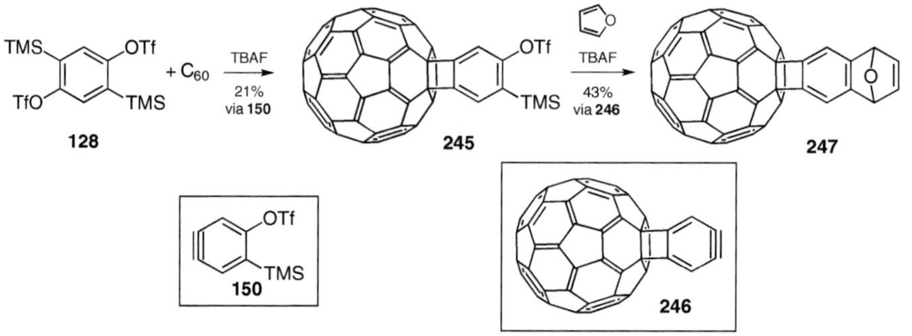

Scheme 2.55 Synthesis of a fullerobenzyne precursor. Source: Based on García et al. [137].

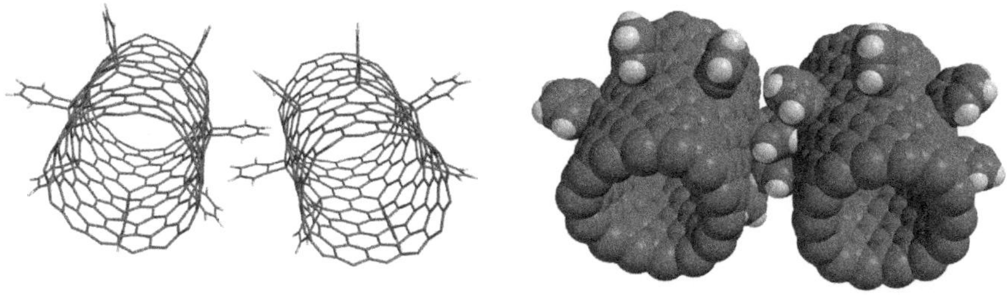

Figure 2.3 Molecular models of paddle-wheel nanostructures. Source: Globus et al. [138]. © 1998 IOP Publishing Ltd.

Langa, Guitián, and coworkers described the first functionalization of SWCNTs by aryne cycloadditions. Arynes **1** and **249b,c** were generated either by thermal decomposition of benzenediazonium 2-carboxylate (**10**) or by fluoride-induced decomposition of triflates **14**, **248**, or **163** in the presence of SWCNTs to give polyadducts, represented as monoadducts **250** and **251** for the sake of clarity (Scheme 2.56) [139]. Later on, they extended the functionalization strategy by using different arynes with microwave (MW) irradiation, increasing the functionalization up to 1 functional group per 64–120 carbon atoms (Scheme 2.56) [140].

Due to the lack of crystalline structures and NMR data caused by the high insolubility of carbon nanotubes (CNTs), it is difficult to determine whether the aryne reacts with CNT through a [2+2] or a [4+2] cycloaddition. Therefore, computational studies are essential for establishing the chemoselectivity of the cycloaddition reaction. Nagase, Zhao, and coworker reported DFT-based calculations describing the energy profiles for the [2+2] and [4+2] cycloaddition of benzyne to armchair SWCNTs [141]. They concluded that the [2+2] product is always more thermodynamically stable than the [4+2] cycloadduct, although the latter becomes kinetically favored as the diameter of the SWCNT increases. Solá, Osuna, and coworkers investigated the effect of the curvature of zigzag SWCNTs on the cycloadditions of benzyne and concluded that, as far as zigzag SWCNTs are concerned, the sidewall [4+2] cycloaddition of benzyne to SWCNTs in parallel position is the preferred reaction pathway, both kinetically and thermodynamically [135]. If zigzag SWCNT diameter increases, the energy barrier increases and the

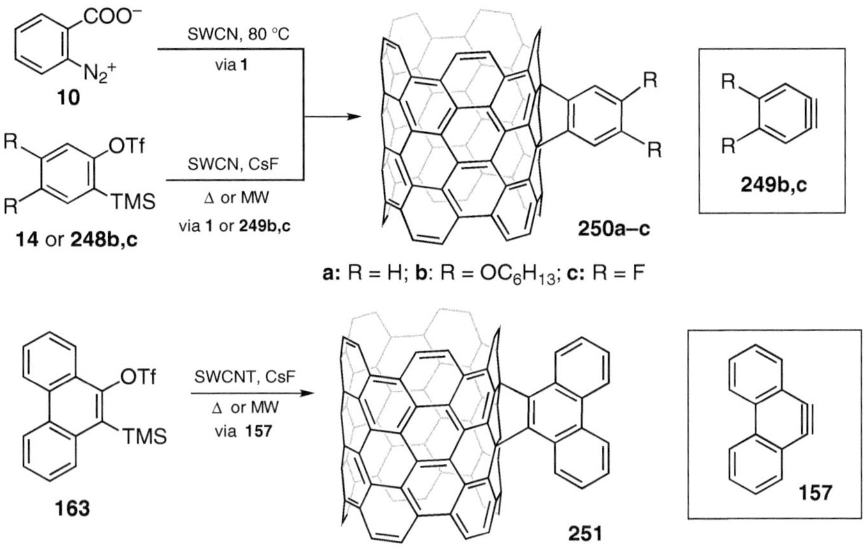

Scheme 2.56 Functionalization of carbon nanotubes by aryne cycloadditions. Source: Criado et al. [139]; Criado et al. [140].

exothermicity of the reaction decreases. Therefore, it is clear that cycloadditions of benzyne can produce both [2+2] and [4+2] adducts depending on the shape and curvature of the nanotube, as well as on their intrinsic electronic nature.

Reaction with arynes has also been used to functionalize novel carbon nanostructures such as carbon nanohorns. Thus, in 2013, Tagmatarchis and coworkers used cycloaddition of benzynes to functionalize this nanostructure (Scheme 2.57) [142]. Two different precursors were used for the generation of benzyne: anthranilic

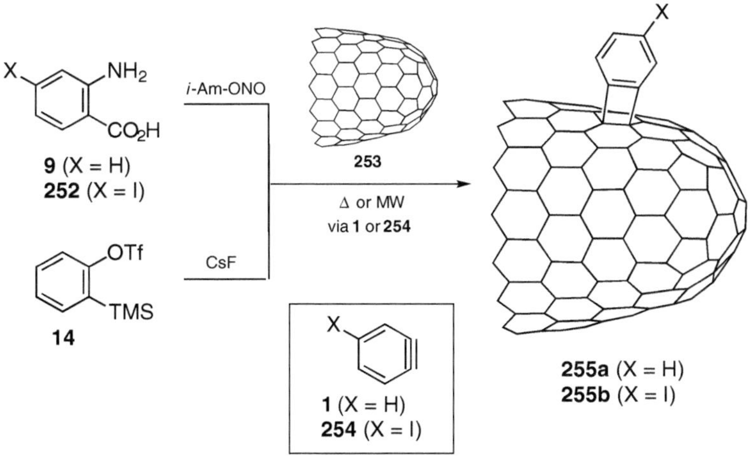

Scheme 2.57 Functionalization of carbon nanohorns by aryne cycloadditions. Source: Based on Chronopoulos et al. [142].

acid (**9**) or triflate **14**, respectively, either by normal heating or MW. Aryne cycloaddition was proved by Raman spectroscopy (increase in the D/G band intensity ratio). By using iodo anthranilic acid **252** as starting material for aryne **254**, iodine-functionalized carbon nanohorns **255b** were obtained and further derivatized to endocyclic disulfides that can immobilize gold nanoparticles.

Graphene is currently the most widely studied carbon-based material due to its outstanding properties, such as lightness, strength, stiffness, and electrical and heat conductivity. Therefore, functionalization of this material is nowadays a key issue in materials science [143].

The first aryne cycloaddition to graphene was reported in 2010 by Ma and coworkers (Scheme 2.58) [144]. Benzyne (**1**), generated by Kobayashi's procedure, in the presence of multilayered graphene afforded highly functionalized and aryne-modified graphene sheets that could be dispersed in a variety of organic solvents. Similar results were obtained by addition of some substituted arynes such as 4,5-difluorobenzyne (**249c**).

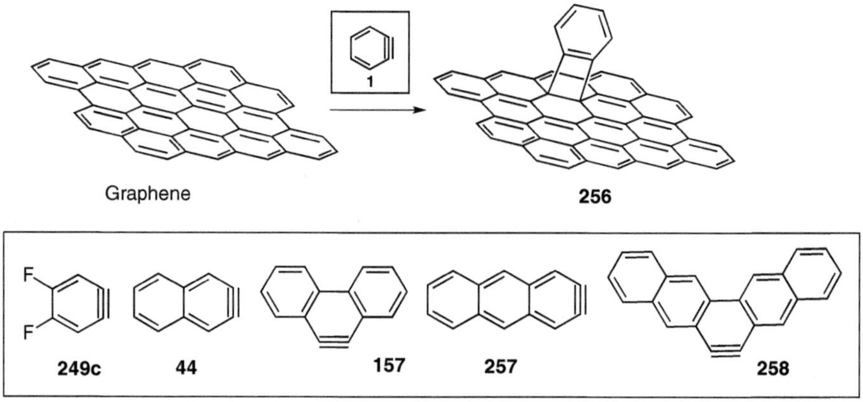

Scheme 2.58 Functionalization of graphene by aryne cycloadditions. Source: Zhong et al. [144]; Sulleiro et al. [145].

Very recently, Prato, Criado, and coworkers described the functionalization of few-layer graphene (FLG) by cycloaddition with arynes **44**, **157**, **257**, or **258** (Scheme 2.58), generated by MW-promoted decomposition of the corresponding phthalic anhydrides. Simply by MW irradiation of homogeneous powder mixtures of phthalic anhydrides and FLG, functionalization was achieved, yielding dispersions of the functionalized graphene, characterized by a number of techniques (X-ray photoelectron spectroscopy [XPS], Raman, thermogravimetric analysis [TGA], and transmission electron microscopy [TEM]) [145].

In 2016, Martín, Pérez, and coworkers described the preparation of [60]fullerene-benzyne building block **246**, which they used for functionalizing exfoliated FLG (Scheme 2.59) [137]. Triflate **245** was reacted with TBAF in the presence of FLG to yield an all-carbon hybrid material **259** comprising C_{60} and graphene. The authors carried out computational studies, which indicated that both [2+2] and [4+2] cycloadditions were feasible. Denis [146] and Zhao's [147] groups conducted other theoretical studies on cycloadditions of arynes to graphene.

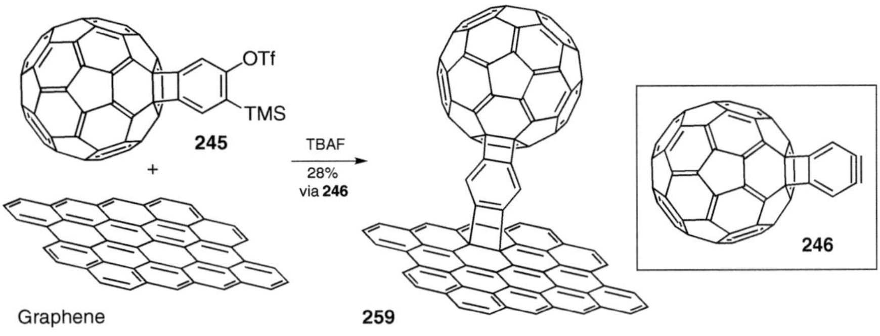

Scheme 2.59 Synthesis of C_{60}-graphene hybrids **259**. Source: Based on García et al. [137].

References

1 Stoermer, R. and Kahlert, B. (1902). *Ber. Dtsch. Chem. Ges.* 35 (2): 1633–1640.
2 Roberts, J.D., Simmons, H.E., Carlsmith, L.A., and Vaughan, C.W. (1953). *J. Am. Chem. Soc.* 75 (13): 3290–3291.
3 Pavliček, N., Schuler, B., Collazos, S. et al. (2015). *Nat. Chem.* 7 (8): 623–628.
4 Pavliček, N., Majzik, Z., Collazos, S. et al. (2017). *ACS Nano* 11 (11): 10768–10773.
5 Wittig, G. and Pohmer, L. (1955). *Angew. Chem.* 67 (13): 348.
6 Hoffmann, R.W. (1967). *Dehydrobenzene and Cycloalkynes*. New York: Academic Press.
7 Gilchrist, T.L. (1983). Arynes. In: *The Chemistry of Triple-Bonded Functional Groups* (eds. S. Patai and Z. Rappoport), 383–419. Wiley.
8 Winkler, M., Wenk, H.H., and Sander, W. (2003). Arynes. In: *Reactive Intermediate Chemistry* (eds. R.A. Moss, M.S. Platz and J.M. Jones), 741–794. Wiley.
9 Pellissier, H. and Santelli, M. (2003). *Tetrahedron* 59 (6): 701–730.
10 Sanz, R. (2008). *Org. Prep. Proced. Int.* 40 (3): 215–291.
11 Yoshida, S. and Hosoya, T. (2015). *Chem. Lett.* 44 (11): 1450–1460.
12 Caubere, P. (1993). *Chem. Rev.* 93 (6): 2317–2334.
13 Wittig, G. and Pohmer, L. (1956). *Chem. Ber.* 89 (5): 1334–1351.
14 Wittig, G. and Knauss, E. (1958). *Chem. Ber.* 91 (5): 895–907.
15 Gilman, H. and Gorsich, R.D. (1957). *J. Am. Chem. Soc.* 79 (10): 2625–2629.
16 Tochtermann, W., Stubenrauch, G., Reiff, K., and Schumacher, U. (1974). *Chem. Ber.* 107 (10): 3340–3352.
17 Logullo, F.M., Seitz, A.H., and Lester, F. (1968). *Org. Synth.* 48: 12–16.
18 Friedman, L. and Logullo, F.M. (1969). *J. Am. Chem. Soc.* 85 (10): 3089–3092.
19 Le Goff, E. (1962). *J. Am. Chem. Soc.* 84 (19): 3786.

20 Campbell, C.D. and Rees, C.W. (1969). *J. Chem. Soc. C* (5): 752–756.

21 Himeshima, Y., Sonoda, T., and Kobayashi, H. (1983). *Chem. Lett.* 12 (8): 1211–1214.

22 Peña, D., Escudero, S., Pérez, D. et al. (1998). *Angew. Chem. Int. Ed.* 37 (19): 2659–2661.

23 Peña, D., Cobas, A., Pérez, D., and Guitián, E. (2002). *Synthesis* (10): 1454–1458.

24 Díaz, M.T., Cobas, A., Guitián, E., and Castedo, L. (1998). *Synlett* 1998 (02): 157–158.

25 Akai, S., Ikawa, T., Takayanagi, S. et al. (2008). *Angew. Chem. Int. Ed.* 47 (40): 7673–7676.

26 Masson, E. and Schlosser, M. (2005). *Eur. J. Org. Chem.* 2005 (20): 4401–4405.

27 Medina, J.M., Mackey, J.L., Garg, N.K., and Houk, K.N. (2014). *J. Am. Chem. Soc.* 136 (44): 15798–15805.

28 Pérez, D., Peña, D., and Guitián, E. (2013). *Eur. J. Org. Chem.* 2013 (27): 5981–6013.

29 Wu, D., Ge, H., Liu, S.H., and Yin, J. (2013). *RSC Adv.* 3 (45): 22727–22738.

30 Alam, M.A., Shimada, K., Jahan, A. et al. (2017). *Am. J. Heterocycl. Chem.* 3 (5): 47–54.

31 Dubrovskiy, A.V., Markina, N.A., and Larock, R.C. (2013). *Org. Biomol. Chem.* 11 (2): 191–218.

32 Kovalev, I.S., Kopchuk, D.S., Zyryanov, G.V. et al. (2012). *Chem. Heterocycl. Compd.* 48 (4): 536–547.

33 Peña, D., Pérez, D., and Guitián, E. (2007). *Heterocycles* 74 (1): 89–100.

34 Gampe, C.M. and Carreira, E.M. (2012). *Angew. Chem. Int. Ed.* 51 (16): 3766–3778.

35 Tadross, P.M. and Stoltz, B.M. (2012). *Chem. Rev.* 112 (6): 3550–3577.

36 Zamiraei, Z. (2017). *Chem. Biol. Interface* 7 (4): 217–223.

37 Takikawa, H., Nishii, A., Sakai, T., and Suzuki, K. (2018). *Chem. Soc. Rev.* 47 (21): 8030–8056.

38 Stevens, R.V. and Bisacchi, G.S. (1982). *J. Org. Chem.* 47 (12): 2393–2396.

39 Hosoya, T., Hasegawa, T., Kuriyama, Y., and Suzuki, K. (1995). *Tetrahedron Lett.* 36 (19): 3377–3380.

40 Hamura, T., Ibusuki, Y., Uekusa, H. et al. (2006). *J. Am. Chem. Soc.* 128 (31): 10032–10033.

41 Jones, M. and Levin, R.H. (1969). *J. Am. Chem. Soc.* 91 (23): 6411–6415.

42 Ozkan, I. and Kinal, A. (2004). *J. Org. Chem.* 69 (16): 5390–5394.

43 Maurin, P., Ibrahim-Ouali, M., Parrain, J.-L., and Santelli, M. (2003). *J. Mol. Struct. THEOCHEM* 637 (1): 91–100.

44 Ma, Z.-X., Feltenberger, J.B., and Hsung, R.P. (2012). *Org. Lett.* 14 (11): 2742–2745.

45 Wöhrle, T., Wurzbach, I., Kirres, J. et al. (2016). *Chem. Rev.* 116 (3): 1139–1241.

46 Peña, D., Pérez, D., Guitián, E., and Castedo, L. (1999). *J. Am. Chem. Soc.* 121 (24): 5827–5828.

47 Radhakrishnan, K.V., Yoshikawa, E., and Yamamoto, Y. (1999). *Tetrahedron Lett.* 40 (42): 7533–7535.

48 Jayanth, T.T., Jeganmohan, M., and Cheng, C.-H. (2004). *J. Org. Chem.* 69 (24): 8445–8450.

49 Quintana, I., Boersma, A.J., Peña, D. et al. (2006). *Org. Lett.* 8 (15): 3347–3349.

50 Hsieh, J.-C., Rayabarapu, D.K., and Cheng, C.-H. (2004). *Chem. Commun.* (5): 532–533.

51 Li, J. and Zhang, Q. (2013). *Synlett* 24 (06): 686–696.

52 Thummel, R.P., Cravey, W.E., and Nutakul, W. (1978). *J. Org. Chem.* 43 (12): 2473–2477.

53 Herwig, P.T. and Müllen, K. (1999). *Adv. Mater.* 11 (6): 480–483.

54 Mondal, R., Adhikari, R.M., Shah, B.K., and Neckers, D.C. (2007). *Org. Lett.* 9 (13): 2505–2508.

55 Jancarik, A., Levet, G., and Gourdon, A. (2019). *Chem. Eur. J.* 25 (9): 2366–2374.

56 LeHoullier, C.S. and Gribble, G.W. (1983). *J. Org. Chem.* 48 (14): 2364–2366.

57 Netka, J., Crump, S.L., and Rickborn, B. (1986). *J. Org. Chem.* 51 (8): 1189–1199.

58 Eda, S., Eguchi, F., Haneda, H., and Hamura, T. (2015). *Chem. Commun.* 51 (27): 5963–5966.

59 Sy, A. and Hart, H. (1979). *J. Org. Chem.* 44 (1): 7–9.

60 Gribble, G.W., Perni, R.B., and Onan, K.D. (1985). *J. Org. Chem.* 50 (16): 2934–2939.

61 Kitamura, C., Abe, Y., Ohara, T. et al. (2010). *Chem. Eur. J.* 16 (3): 890–898.

62 Kitamura, C., Tsukuda, H., Yoneda, A. et al. (2010). *Eur. J. Org. Chem.* 2010 (16): 3033–3040.

63 Luo, J. and Hart, H. (1988). *J. Org. Chem.* 53 (6): 1341–1343.

64 Blatter, K. and Schlüter, A.-D. (1989). *Chem. Ber.* 122 (7): 1351–1356.

65 Vogel, T., Blatter, K., and Schlüter, A.-D. (1989). *Makromol. Chem. Rapid Commun.* 10 (8): 427–430.

66 Chun, D., Cheng, Y., and Wudl, F. (2008). *Angew. Chem. Int. Ed.* 47 (44): 8380–8385.

67 Haneda, H., Eda, S., Aratani, M., and Hamura, T. (2014). *Org. Lett.* 16 (1): 286–289.

68 Krüger, J., Pavliček, N., Alonso, J.M. et al. (2016). *ACS Nano* 10 (4): 4538–4542.

69 Krüger, J., Eisenhut, F., Alonso, J.M. et al. (2017). *Chem. Commun.* 53 (10): 1583–1586.

70 Krüger, J., García, F., Eisenhut, F. et al. (2017). *Angew. Chem. Int. Ed.* 56 (39): 11945–11948.

71 Shen, B., Tatchen, J., Sánchez García, E., and Bettinger, H. (2018). *Angew. Chem. Int. Ed.* 57 (33): 10506–10509.

72 Lu, J., Ho, D.M., Vogelaar, N.J. et al. (2004). *J. Am. Chem. Soc.* 126 (36): 11168–11169.

73 Chen, W., Li, X., Long, G. et al. (2018). *Angew. Chem. Int. Ed.* 57 (41): 13555–13559.

74 Duong, H.M., Bendikov, M., Steiger, D. et al. (2003). *Org. Lett.* 5 (23): 4433–4436.

75 Rodríguez-Lojo, D., Pérez, D., Peña, D., and Guitián, E. (2015). *Chem. Commun.* 51 (25): 5418–5420.

76 Miao, Q., Chi, X., Xiao, S. et al. (2006). *J. Am. Chem. Soc.* 128 (4): 1340–1345.

77 Kitamura, C., Takenaka, A., Kawase, T. et al. (2011). *Chem. Commun.* 47 (23): 6653–6655.

78 Fort, E.H. and Scott, L.T. (2011). *Tetrahedron Lett.* 52 (17): 2051–2053.

79 Schuler, B., Collazos, S., Gross, L. et al. (2014). *Angew. Chem. Int. Ed.* 53 (34): 9004–9006.

80 Konishi, A., Hirao, Y., Matsumoto, K. et al. (2013). *Chem. Lett.* 42 (6): 592–594.

81 Nakamuro, T., Kumazawa, K., Ito, H., and Itami, K. (2019). *Synlett* 30 (04): 423–428.

82 Criado, A., Peña, D., Cobas, A., and Guitián, E. (2010). *Chem. Eur. J.* 16 (32): 9736–9740.

83 Criado, A., Vilas-Varela, M., Cobas, A. et al. (2013). *J. Org. Chem.* 78 (24): 12637–12649.

84 Kelly, T.R., Sestelo, J.P., and Tellitu, I. (1998). *J. Org. Chem.* 63 (11): 3655–3665.

85 Kelly, T.R., Silva, R.A., De Silva, H. et al. (2000). *J. Am. Chem. Soc.* 122 (29): 6935–6949.

86 Kelly, T.R., De Silva, H., and Silva, R.A. (1999). *Nature* 401: 150–152.

87 Anzenbacher, P. Jr.,, Mosca, L., Palacios, M.A. et al. (2012). *Chem. Eur. J.* 18 (40): 12712–12718.

88 See also McQuade, D.T., Pullen, A.E., and Swager, T.M. (2000). *Chem. Rev.* 100: 2537–2574.

89 Frantz, D.K., Linden, A., Baldridge, K.K., and Siegel, J.S. (2012). *J. Am. Chem. Soc.* 134 (3): 1528–1535.

90 Godinez, C.E., Zepeda, G., Mortko, C.J. et al. (2004). *J. Org. Chem.* 69 (5): 1652–1662.

91 Taylor, R.G.D., Carta, M., Bezzu, C.G. et al. (2014). *Org. Lett.* 16 (7): 1848–1851.

92 Ghanem, B.S., Hashem, M., Harris, K.D.M. et al. (2010). *Macromolecules* 43 (12): 5287–5294.

93 Kanibolotsky, A.L., Perepichka, I.F., and Skabara, P.J. (2010). *Chem. Soc. Rev.* 39 (7): 2695–2728.

94 Zhang, H., Wu, D., Hua, L., and Yin, J. (2012). *Curr. Org. Chem.* 16 (18): 2124–2158.

95 Alonso, J.M., Díaz-Álvarez, A.E., Criado, A. et al. (2012). *Angew. Chem. Int. Ed.* 51 (1): 173–177.

96 Yatabe, T., Harbison, M.A., Brand, J.D. et al. (2000). *J. Mater. Chem.* 10 (7): 1519–1525.

97 Romero, C., Peña, D., Pérez, D., and Guitián, E. (2006). *Chem. Eur. J.* 12 (22): 5677–5684.

98 Romero, C., Peña, D., Pérez, D. et al. (2009). *J. Mater. Chem.* 19 (27): 4725–4731.

99 Lynett, P.T. and Maly, K.E. (2009). *Org. Lett.* 11 (16): 3726–3729.

100 Rüdiger, E.C., Rominger, F., Steuer, L., and Bunz, U.H.F. (2016). *J. Org. Chem.* 81 (1): 193–196.

101 Soe, W.-H., Manzano, C., Renaud, N. et al. (2011). *ACS Nano* 5 (2): 1436–1440.

102 Vilas-Varela, M., Fatayer, S., Majzik, Z. et al. (2018). *Chem. Eur. J.* 24 (67): 17697–17700.

103 Iglesias, B., Cobas, A., Pérez, D., and Guitián, E. (2004). *Org. Lett.* 6 (20): 3557–3560.

104 Godlewski, S., Engelund, M., Peña, D. et al. (2018). *Phys. Chem. Chem. Phys.* 20 (16): 11037–11046.

105 Tozawa, H., Kakuda, T., Adachi, K., and Hamura, T. (2017). *Org. Lett.* 19 (15): 4118–4121.

106 Lindner, R. and Kühnle, A. (2015). *ChemPhysChem* 16 (8): 1582–1592.

107 Shen, Q., Gao, H.-Y., and Fuchs, H. (2017). *Nano Today* 13: 77–96.

108 Sánchez-Sánchez, C., Nicolaï, A., Rossel, F. et al. (2017). *J. Am. Chem. Soc.* 139 (48): 17617–17623.

109 Shen, Y. and Chen, C.-F. (2012). *Chem. Rev.* 112 (3): 1463–1535.

110 Peña, D., Pérez, D., Guitián, E., and Castedo, L. (1999). *Org. Lett.* 1 (10): 1555–1557.

111 Peña, D., Cobas, A., Pérez, D. et al. (2000). *Org. Lett.* 2 (11): 1629–1632.

112 Barnett, L., Ho, D.M., Baldridge, K.K., and Pascal, R.A. (1999). *J. Am. Chem. Soc.* 121 (4): 727–733.

113 Hacker, N.P., McOmie, J.F.W., Meunier-Piret, J., and Van Meerssche, M. (1982). *J. Chem. Soc., Perkin Trans. 1*: 19–23.

114 Peña, D., Cobas, A., Pérez, D. et al. (2003). *Org. Lett.* 5 (11): 1863–1866.

115 Saito, H., Uchida, A., and Watanabe, S. (2017). *J. Org. Chem.* 82 (11): 5663–5668.

116 Romero, C., Peña, D., Pérez, D., and Guitián, E. (2008). *J. Org. Chem.* 73 (20): 7996–8000.

117 Caeiro, J., Peña, D., Cobas, A. et al. (2006). *Adv. Synth. Catal.* 348 (16–17): 2466–2474.

118 Hosokawa, T., Takahashi, Y., Matsushima, T. et al. (2017). *J. Am. Chem. Soc.* 139 (51): 18512–18521.

119 Berezhnaia, V., Roy, M., Vanthuyne, N. et al. (2017). *J. Am. Chem. Soc.* 139 (51): 18508–18511.

120 Zuzak, R., Castro-Esteban, J., Brandimarte, P. et al. (2018). *Chem. Commun.* 54 (73): 10256–10259.

121 Yanney, M., Fronczek, F.R., Henry, W.P. et al. (2011). *Eur. J. Org. Chem.* 2011 (33): 6636–6639.

122 Lin, J.B., Shah, T.K., Goetz, A.E. et al. (2017). *J. Am. Chem. Soc.* 139 (30): 10447–10455.

123 Pozo, I., Cobas, A., Peña, D. et al. (2016). *Chem. Commun.* 52 (32): 5534–5537.

124 Hoke, S.H., Molstad, J., Dilettato, D. et al. (1992). *J. Org. Chem.* 57 (19): 5069–5071.

125 Tsuda, M., Ishida, T., Nogami, T. et al. (1992). *Chem. Lett.* 21 (12): 2333–2334.

126 Ishida, T., Shinozuka, K., Nogami, T. et al. (1995). *Chem. Lett.* 24 (4): 317–318.

127 Nakamura, Y., Takano, N., Nishimura, T. et al. (2001). *Org. Lett.* 3 (8): 1193–1196.

128 Kim, G., Lee, K.C., Kim, J. et al. (2013). *Tetrahedron* 69 (35): 7354–7359.

129 Mizunuma, R., Tanaka, T., Nakamura, Y. et al. (2018). *Tetrahedron* 74 (5): 544–548.

130 Darwish, A.D., Abdul-Sada, A.K., Langley, G.J. et al. (1994). *J. Chem. Soc., Chem. Commun.* (18): 2133–2134.

131 Darwish, A.D., Avent, A.G., Taylor, R., and Walton, D.R.M. (1996). *J. Chem. Soc., Perkin Trans.* 2 (10): 2079–2084.

132 Meier, M.S., Wang, G.-W., Haddon, R.C. et al. (1998). *J. Am. Chem. Soc.* 120 (10): 2337–2342.

133 Lu, X., Nikawa, H., Tsuchiya, T. et al. (2010). *Angew. Chem. Int. Ed.* 49 (3): 594–597.

134 Li, F.-F., Pinzón, J.R., Mercado, B.Q. et al. (2011). *J. Am. Chem. Soc.* 133 (5): 1563–1571.

135 Martínez, J.P., Langa, F., Bickelhaupt, F.M. et al. (2016). *J. Phys. Chem. C* 120 (3): 1716–1726.

136 Yang, T., Nagase, S., Akasaka, T. et al. (2015). *J. Am. Chem. Soc.* 137 (21): 6820–6828.

137 García, D., Rodríguez-Pérez, L., Herranz, M.A. et al. (2016). *Chem. Commun.* 52 (40): 6677–6680.

138 Globus, A., Bauschlicher, C.W., Han, J. et al. (1998). *Nanotechnology* 9 (3): 192–199.

139 Criado, A., Gómez-Escalonilla, M.J., Fierro, J.L.G. et al. (2010). *Chem. Commun.* 46 (37): 7028–7030.

140 Criado, A., Vizuete, M., Gómez-Escalonilla, M.J. et al. (2013). *Carbon* 63: 140–148.

141 Yang, T., Zhao, X., and Nagase, S. (2013). *Org. Lett.* 15 (23): 5960–5963.

142 Chronopoulos, D., Karousis, N., Ichihashi, T. et al. (2013). *Nanoscale* 5 (14): 6388–6394.

143 Georgakilas, V., Otyepka, M., Bourlinos, A.B. et al. (2012). *Chem. Rev.* 112 (11): 6156–6214.

144 Zhong, X., Jin, J., Li, S. et al. (2010). *Chem. Commun.* 46 (39): 7340–7342.

145 Sulleiro, M.V., Quiroga, S., Peña, D. et al. (2018). *Chem. Commun.* 54 (17): 2086–2089.

146 Denis, P.A. and Iribarne, F. (2012). *J. Mater. Chem.* 22 (12): 5470–5477.

147 Zhao, J., Wang, H., Gao, B. et al. (2012). *J. Mol. Model.* 18 (6): 2861–2868.

3

Dipolar Cycloaddition Reactions of Arynes and Related Chemistry

Pan Li[1], *Jingjing Zhao*[1], *and Feng Shi*[2]

[1] *Henan University, Institute of Functional Organic Molecular Engineering, College of Chemistry and Chemical Engineering, Kaifeng 475004, China*
[2] *WuXi AppTec (Wuhan) Co., Ltd., International Discovery Service Unit, Research Service Division, 666 Gaoxin Rd, Wuhan East-Lake High-Tech Development Zone, Wuhan 430075, People's Republic of China*

3.1 Introduction

1,*n*-Dipoles are reactive species bearing separate positive and negative charges, and therefore exhibit both electrophilic and nucleophilic reactivity, respectively [1]. They easily react with olefins, alkynes, allenes, etc., commonly referred to as dipolarophiles, to form various rings. Such strategy has been widely used to construct polycyclic systems. Among the different dipolarophiles, aryne constitutes an important class owing to its poorly overlapped C—C triple bond that is highly reactive. Consequently, while many classical dipolar cycloadditions work better or only with polarized or activated olefins or alkynes, the strain itself of the "triple bond" of the aryne provides enough driving energy so that plain aryne is reactive enough toward various dipoles under milder conditions. Hence, aryne cycloaddition reactions have become a powerful and widely used synthetic method for constructing benzannulated heterocyclic scaffolds, which are widely found in various biologically active compounds (Scheme 3.1).

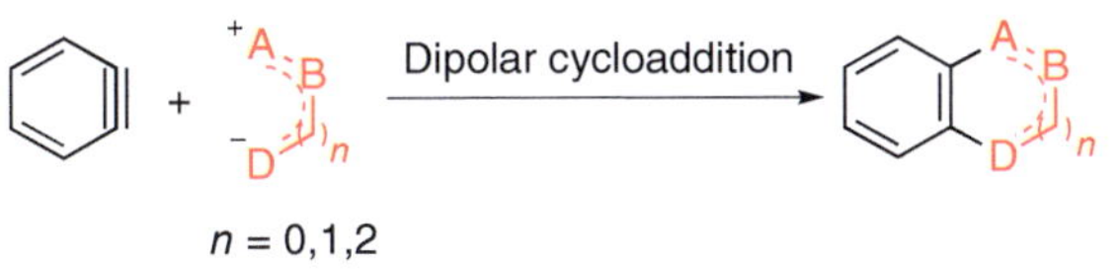

Scheme 3.1 General scheme of 1,*n*-dipolar cycloaddition of arynes.

Aryne dipolar cycloaddition takes place in either a concerted mechanism or a stepwise mechanism, the exact nature and the distinction of which have not been much studied. In an ideal situation, the concerted mechanism involves a preorganized cyclic transition state with aryne and the dipole orientated perpendicular to each other. On the other side, the stepwise mechanism involves first

Modern Aryne Chemistry, First Edition. Edited by Akkattu T. Biju.

the nucleophilic attack of the nucleophilic site of the formal "1,*n*-dipole" to aryne, followed by the back attack of the generated aryl anion to the electrophilic site of the formal "dipole." In such a situation, the formal "1,*n*-dipole" may not be zwitterionic, and any amphiphilic species where the nucleophilic site and the electrophilic site are tethered but separated by n–2 atoms would suffice. It is worth mentioning that the nonzwitterionic amphiphiles may still react with arynes in a concerted mechanism.

To date, aryne dipolar cycloadditions mainly involve 1,3-, 1,5-, and 1,7-dipoles [2]. This chapter will also follow the same classification. In addition, this chapter will primarily cover the general conceptual knowledge of aryne dipolar cycloadditions, with the emphasis on demonstrating the scope of dipoles, subsequent events and mechanism after the cycloaddition, and applications in a broader picture. The synthetic utility will be briefly mentioned, but the scope and limitation of each individual reaction type, as well as the reaction conditions and other technical details, will not be mentioned. Since most recent development of aryne chemistry uses the Kobayashi aryne precursor (2-silylaryl triflate) [3], this chapter will also focus on the application of this precursor in dipolar cycloadditions. Other versions of aryne generation will also be mentioned if necessary.[1]

3.2 1,3-Dipolar Cycloaddition Reactions of Arynes

1,3-Dipolar cycloadditions (1,3-DC) provide an efficient and straightforward route to synthesize five-membered heterocycles [4]. 1,3-DC of arynes, quickly developed in the past decade, have been recognized as a powerful tool for the synthesis of benzannulated five-membered heterocycles.

The typical classification of 1,3-dipoles is based on whether or not the three atoms are embedded into a ring. Thus, 1,3-dipoles such as nitrones, diazo compounds, azides, azomethine imines, azomethine ylides, nitrile oxides, and nitrile imines are all classified as linear dipoles (Figure 3.1). They can be further classified to straight linear (180°, such as azides or diazo compounds) where the central atoms adopt sp-hybridization, and V-shaped (120°, such as nitrones) where the central atoms adopt sp^2-hybridization. Explicitly the former dipoles would react with arynes to generate new five-membered rings that are unsaturated, while the latter dipoles would generate new five-membered rings that are saturated (*vide infra*). In both cases, the newly generated five-membered ring can undergo subsequent events, most commonly including rearrangement, elimination, and oxidation, as driven by gaining aromaticity or formation of stronger bonds after breaking weaker X–X ones (X being N or O).

Cyclic 1,3-dipoles, so-called mesoionic rings, are generally less known and utilized due to the instability and difficulty to access (Figure 3.2). Those known to react with arynes include Sydnones and Münchnones [5]. These two are azomethine imines and ylides in nature, but embedded within lactone rings. Such lactone moiety

1 This chapter will mainly cover literature from year 2000 to August 2019.

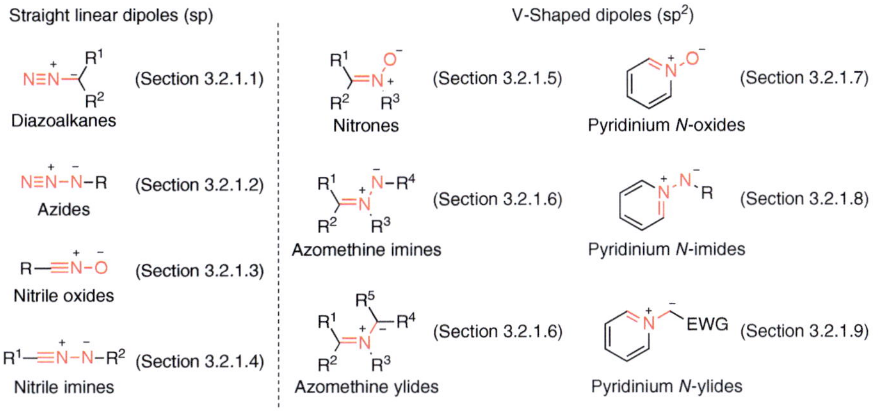

Figure 3.1 Classification of typical linear 1,3-dipoles.

Figure 3.2 Cyclic 1,3-dipoles: Sydnones and Münchnones.

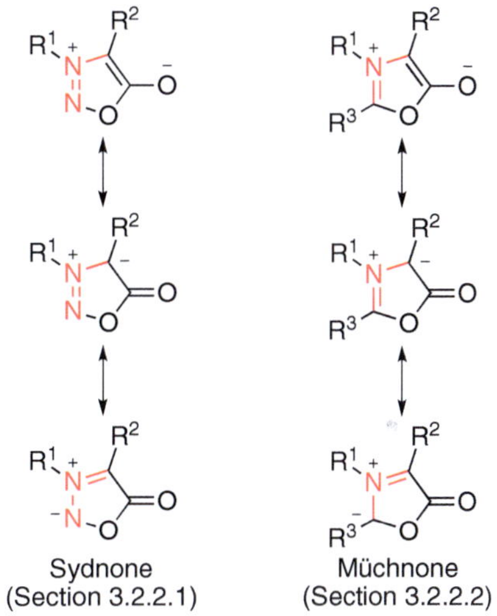

provides added stability as well as additional driving force of decarboxylation after the 1,3-DC, and thus can produce skeletons similar to those obtained by 1,3-DC of nitrile imines and ylides but with different electron organization.

Pyridine can be derivatized into a variety of inner salts, constituting a special class of dipoles. Pyridinium *N*-oxides, *N*-imides, and *N*-ylides should structurally be classified as equivalents of nitrones, azomethine imines, and ylides (see Figure 3.1). Despite that *parts of* the dipole structure are embedded into the pyridine, these dipoles fall into linear dipoles and are different from cyclic dipoles shown in Figure 3.2, where *all of* the dipole structure is embedded into rings. Pyridine can also be derived to 3-oxido- and 3-imido inner salts (see Section 3.3.1), which can react as either 1,3-dipoles or 1,7-dipoles (Figure 3.3). Thus, in this chapter, we discuss pyridinium *N*-oxides, *N*-imides, and *N*-ylides in linear dipole sections, and

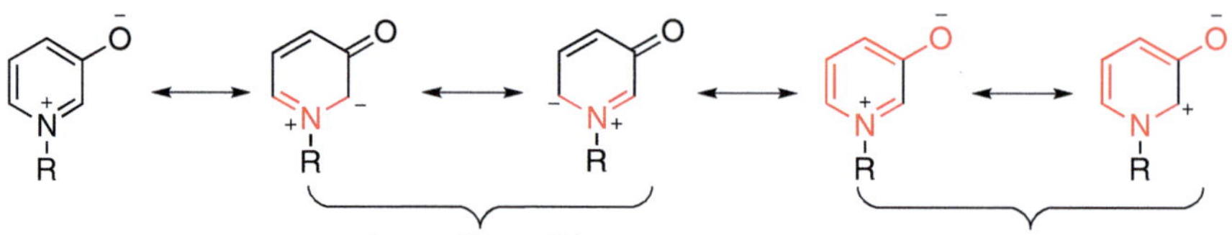

Figure 3.3 3-Oxidopyridinium dipole.

put the entire content of 3-oxido- and 3-imidopyridinium species (whatever the periselectivity is) into the 1,7-dipolar cycloaddition section.

3.2.1 [3+2] Dipolar Cycloaddition Reactions of Arynes with Linear 1,3-Dipoles

3.2.1.1 Reactions with Diazo Compounds

Diazo compounds are versatile building blocks in chemical synthesis [6], and their [3+2] cycloaddition with arynes provides a very attractive and effective strategy to construct indazoles. In this aspect, Yamamoto and Larock groups [7] have independently reported [3+2] cycloaddition of arynes and diazomethane derivatives (Scheme 3.2). The reaction, at the first stage, affords a primary [3+2] cycloadduct 3*H*-indazole **1**. Clearly spotted, the newly formed five-membered ring in **1** is not aromatic and thus prone to subsequent events, which heavily depends on the substitutions of the diazo compounds. When both R^1 and R^2 are aryl groups, the reaction stops at the stage of **1**, which can be isolated. When R^2 is H, hydrogen shift from C^3 to N^1 takes place to transform this 3*H*-indazole intermediate to a more stable and aromatic 1*H*-indazole product **2**. By controlling the reaction condition and stoichiometry, reaction can stop at this stage. When excessive aryne exists and reaction conditions permit, **2** will further react with another equivalent of aryne to yield 1-arylated 1*H*-indazole **3**.

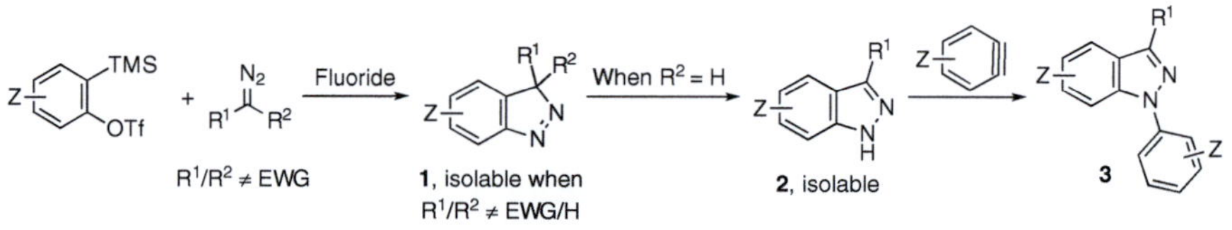

Scheme 3.2 [3+2] Cycloaddition of arynes with diazomethane derivatives.

The rearrangement of **1** to **2** takes place not only when R^2 is H. It will also occur when R^2 is an acyl group (Scheme 3.3), although acyl group shifts less readily than H. Such a process affords a 1-acylated indazole **4**. Larock and coworkers have observed that when both R^1 and R^2 are acyl groups (as in Scheme 3.2), the ketone moiety migrates more readily than the ester moiety or amide moiety [7c]. It is important to note that the apparent C^3-to-N^1 migration may result from two sequential

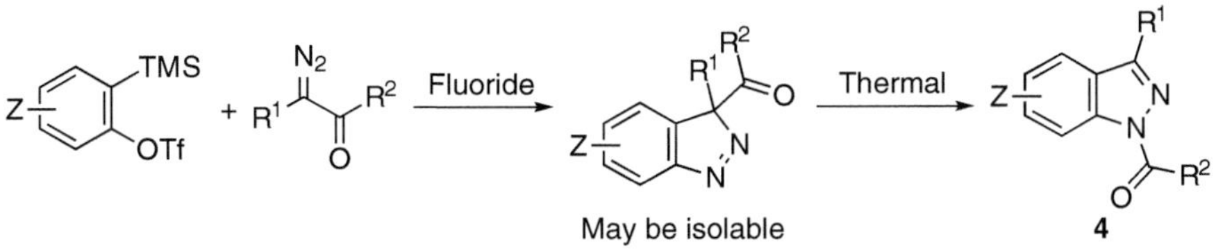

Scheme 3.3 [3+2] Cycloaddition of arynes with acylated diazomethane derivatives.

1,5-migrations from C^3 to N^2 first, and then N^2 to N^1. This rearrangement has been exploited for cyclic diazo compounds [8], where the intermediate 3*H*-indazoles **1A** and **1A′** could be stable and isolable, and further acyl migration under thermal or acidic conditions led to ring-expanded products (Scheme 3.4a). Note that here the acyl migration stopped at the N^2-position of the indazole, presumably due to the unachievable reach of the acyl group to the N^1-position given the cyclic nature of the diazo compounds.

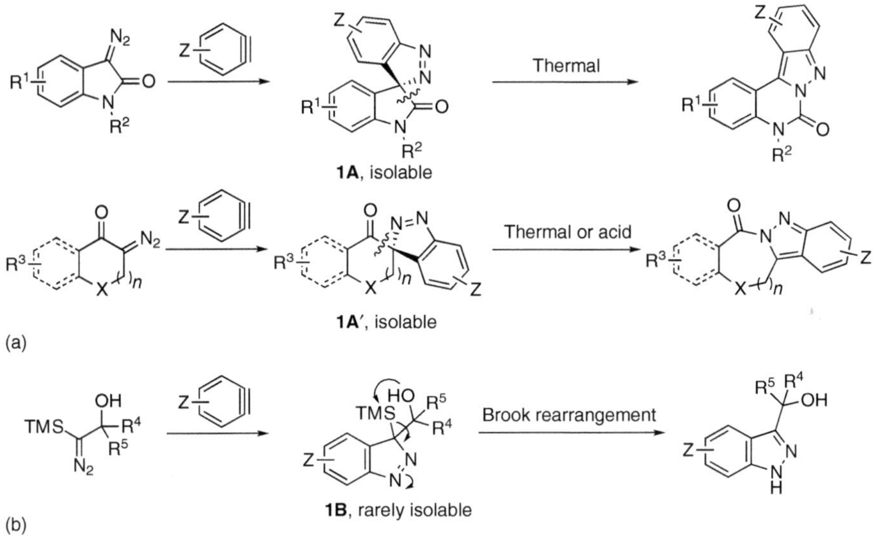

Scheme 3.4 [3+2] Cycloaddition of arynes with subsequent acyl/silyl migration. (a) Acyl shift in cyclic diazo compounds and (b) silyl shift via Brook rearrangement. Source: Based on Hari et al. [8d].

A silyl group can also shift to transform **1** to **2**. For example, trimethylsilyl (TMS)-diazomethane reacted with benzyne in a mixed solvent containing methanol to afford plain unsubstituted 1*H*-indazole [7c]. The TMS group was lost likely to the methoxy group in the solvent. A more complicated silyldiazomethane substrate has been studied in this context (Scheme 3.4b) [8d]. With an intrinsic hydroxyl group to accommodate the silyl group, the [3+2] cycloadduct **1B** readily underwent a Brook rearrangement to afford 1*H*-indazole after hydrolytic workup.

There has been no direct evidence that phosphoryl groups can migrate. When α-substituted diisopropyl α-diazophosphonates reacted with arynes, 3*H*-3-phosphoryl indazole **1C** could be obtained (Scheme 3.5, path A) [9a]. Under slightly different conditions, dimethyl α-diazophosphonates directly afforded 1*H*-indazoles **2** as the final product (Scheme 3.5, path B). Formation of **2** in this case could be explained by an in situ cleavage of the phosphate by fluoride to generate a diazomethane derivative prior to its reaction with aryne. Alternatively, it is also possible that dimethyl α-diazophosphonates reacted to afford the 3*H*-indazole adduct (as **1C**) first, followed by phosphate migration to form 1-phosphoryl-1*H*-indazole, and subsequent hydrolysis to cleave the phosphate group. A three-component variant of the latter outcome has been developed to afford 1-alkyl-1*H*-indazoles [9b].

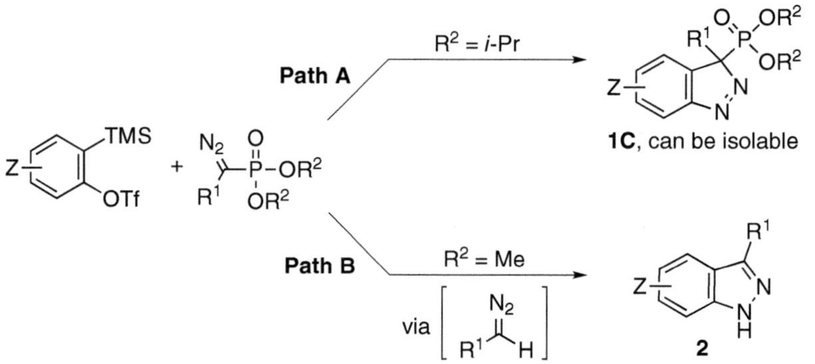

Scheme 3.5 [3+2] Cycloaddition of arynes with α-substituted α-diazophosphonate. Source: Based on Chen et al. [9a].

Apart from the aforementioned cases where 1*H*-indazoles and 3*H*-indazoles could be selectively formed, the regio-complementary 2-aryl-2*H*-indazole species **5** could also be selectively obtained under silver-catalyzed conditions (Scheme 3.6). The Ag(I) ion could intercept the conjugate base of intermediate **1D** to generate

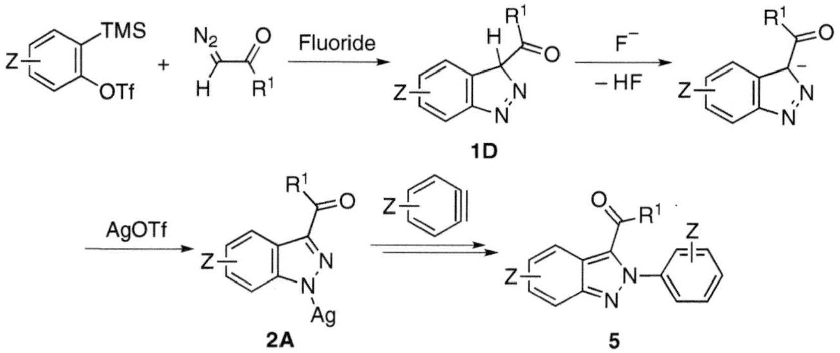

Scheme 3.6 Silver-catalyzed [3+2] cycloaddition of arynes with diazo compounds.

a possible (1*H*-indazol-1-yl)silver intermediate **2A** [10], leaving only the N^2-position of the indazole open for further arylation with another equivalent of aryne.

Diazo compounds without electron-withdrawing groups (EWGs) are unstable and barely isolable. They can be generated in situ from *N*-tosylhydrazones under basic conditions. Such substrates have been indeed employed in aryne cycloaddition to afford 3-substituted 1*H*-indazoles with aid of a phase-transfer catalyst [11]. The reaction proceeds, at least in part, through in situ formation of diazo compounds, although a stepwise annulation mechanism may also operate (Scheme 3.7, see also Section 3.4.2).

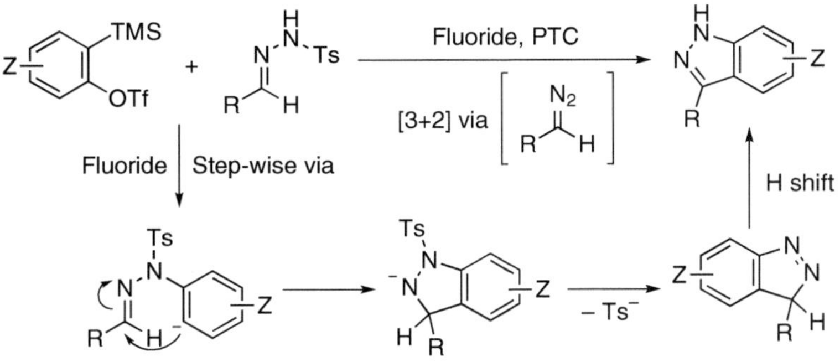

Scheme 3.7 [3+2] Cycloaddition of arynes with *N*-tosylhydrazones.

3.2.1.2 Reactions of Arynes with Azides

Azides have found wide applications in synthetic organic chemistry, polymer chemistry, material sciences, bioconjugation chemistry, and medicinal chemistry owing to their reactivity in [3+2] cycloaddition with alkynes to afford triazoles, commonly referred to as the click reaction [12]. While being generally schematic and straightforward, most click chemistry involving terminal alkynes is slow and requires either Cu or Ru to facilitate the reaction and offer regioselectivity. It has been known that strained alkynes can undergo Cu-free [3+2] cycloaddition with azides under much milder conditions [13]. Aryne, as an extreme example of "strained" alkyne, was anticipated to do the same thing to generate benzotriazoles.

Pioneering work in this field was reported simultaneously by several groups [14] using isolated azides, and indeed such [3+2] dipolar cycloaddition worked very well under mild conditions without Cu (Scheme 3.8). Quite a few studies quickly followed, including using microwave to promote the reaction [15], different aryne precursors (see Figure 3.4) [16], and different azides (including sugar-derived

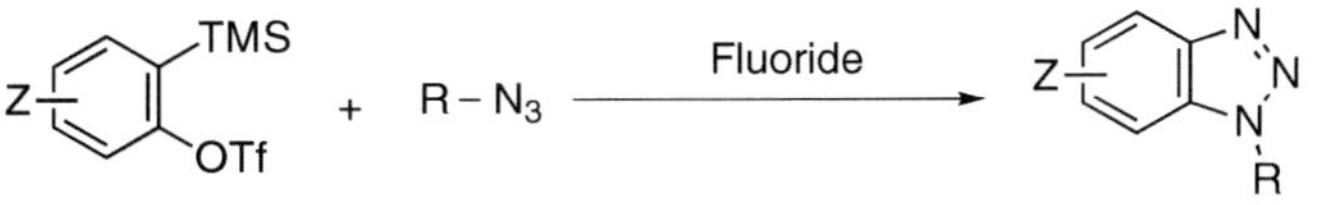

Scheme 3.8 [3+2] Cycloaddition of arynes with azides in general.

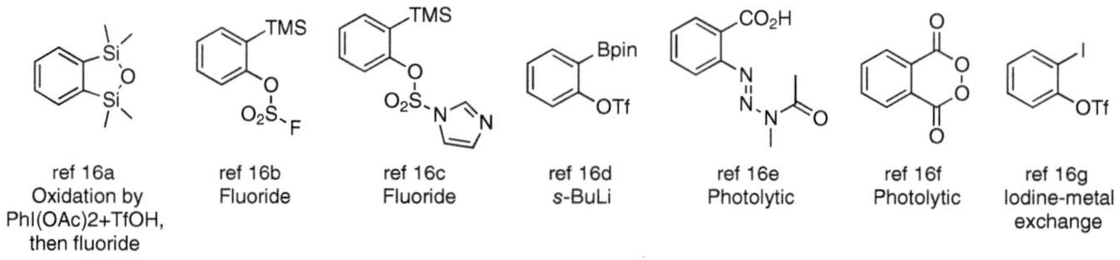

Figure 3.4 Some selected aryne precursors applied to [3+2] aryne cycloaddition with azides. Source: Lin et al. [16a]; Chen et al.[16b]; Kovács et al. [16c]; Sumida et al. [16d]; Gann et al. [16e]; Chang et al. [16f]; Yoshida et al. [16g].

azides) [17]. One interesting case worth mentioning is the one developed by Moses and coworkers [17c], where both arynes and aromatic azides were generated in situ via diazotization reactions (although sequentially) from anthranilic acids and anilines, respectively (Scheme 3.9).

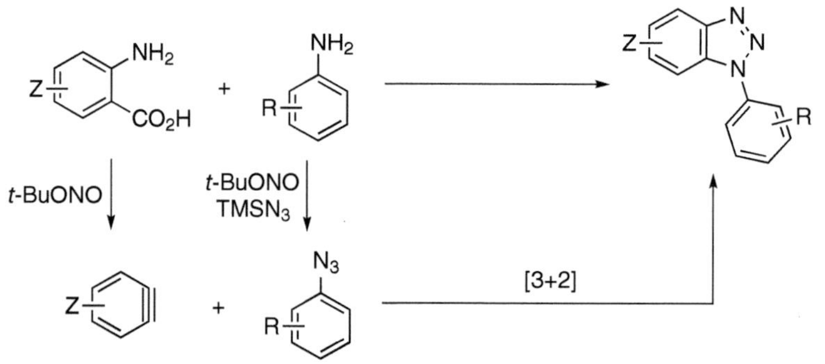

Scheme 3.9 [3+2] Cycloaddition of in situ-generated arynes with in situ-generated azides.

In the above studies, regioselectivity with respect to unsymmetrical benzynes had been observed. It is worth the effort here to summarize the regioselectivity of aryne–azide cycloaddition as part of the general, global picture of aryne cycloaddition, and to discuss the scope and regioselectivity of heteroarynes (such as indolyne and pyridyne) as well in this section. Even though similar observations have been obtained with other dipoles, data and the rationale for the cycloaddition with azides are the richest to facilitate our understanding.

Scheme 3.10 below summarizes the regioselectivity in azide cycloaddition of benzynes bearing different substitutions [18]. There are many factors that could influence the regioselectivity. Garg group has previously advanced a computational model to explain certain aryne reactions using a distortion model [19]. Thus, the *intrinsic* selectivity would involve the more nucleophilic atom of the azide, namely the internal nitrogen (Figure 3.5a), to attack the aryne position bearing a larger internal angle. Tokiwa and Akai in their report also calculated the natural bond orbital (NBO) to reveal the difference in the electron density of the two "yne" carbons (Figure 3.5b). Thus, the nucleophile, again the internal nitrogen of the

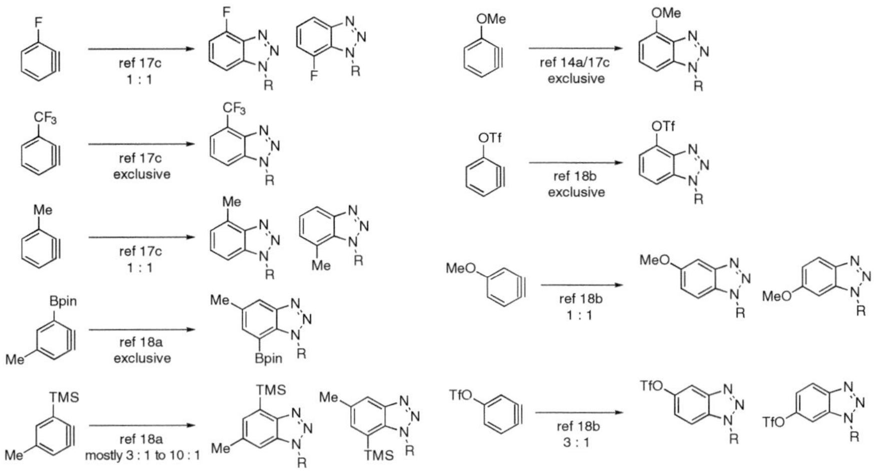

Scheme 3.10 Regioselectivity of unsymmetrical benzyne in [3+2] cycloaddition with azides. Source: Ikawa et al. [18a]; Ikawa et al. [18b].

azide, would also prefer to attack the position with smaller value in electron density. The bigger the difference in the electron density, the better the selectivity. Both rationales provide the same conclusion, and they are in agreement with empirical rationale that such EWGs (on σ-skeleton, such as OMe) on benzyne polarize the "triple" bond so that more electron density resides on the "yne" atom closer to the EWG. They are also in agreement with *most of* the observed regioselectivity, both in trend and in magnitude, and can explain similar regioselectivity in aryne cycloaddition with nitrile oxide (*vide infra*). The one that *cannot* be explained is the 3-silylbenzyne (Scheme 3.10) [18], where calculation results show that sterics induced by the big silyl group, which neither of the two rationales shown in Figure 3.5 considers, are the dictating factor.

Apart from benzynes, pyridynes were also examined in cycloaddition with azides (Scheme 3.11) [20]. Two observations should be mentioned beforehand. Firstly, pyridynes are much more reactive and short-lived than benzynes, and therefore the yields for pyridynes are generally lower. Second, 2,3-pyridyne itself is distorted and nucleophilic attack is regioselective toward its 2-position [20, 21]. Therefore, despite low yields, plain 2,3-pyridyne reacts with azide to afford 1-substituted 7-azabenzotriazole. A methoxy substituent at the 4-position reinforces this regioselectivity. In contrast, 3,4-pyridyne is less distorted and reacts to afford mixture of two regioisomers. A methoxy group positioned neighboring to the triple bond can favor one regioisomer, but a distal methoxy group lacks such capability. 4,5-Pyrimidyne was presumably too short-lived and failed to react with azides [20b]. Indolyne and 4,5-benzofuranyne have been successfully applied in cycloaddition with azides (Scheme 3.12) [22]. Although poor-to-moderate selectivity was observed for 4,5- and 5,6-indolynes due to the less extent of distortion of the ring, excellent regioselectivity was observed for 6,7-indolyne where the ring was distorted to a greater extent.

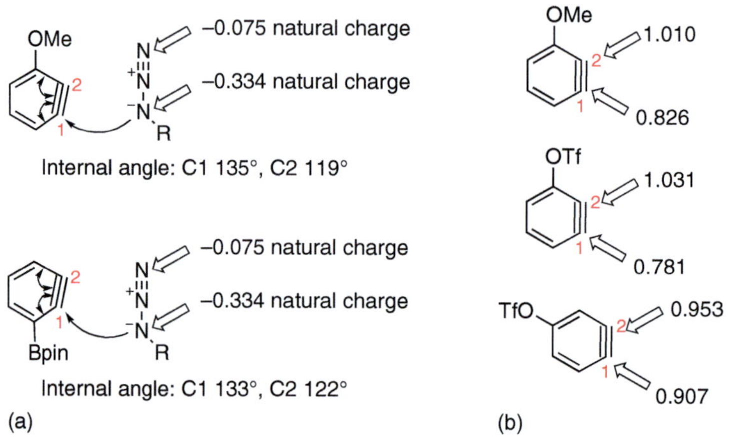

Figure 3.5 Regioselectivity rationales: (a) Aryne distortion and (b) natural bond orbital (NBO).

4,5-Benzofuranyne was distorted slightly more than 4,5-indolyne and therefore the regioselectivity was consequently better.

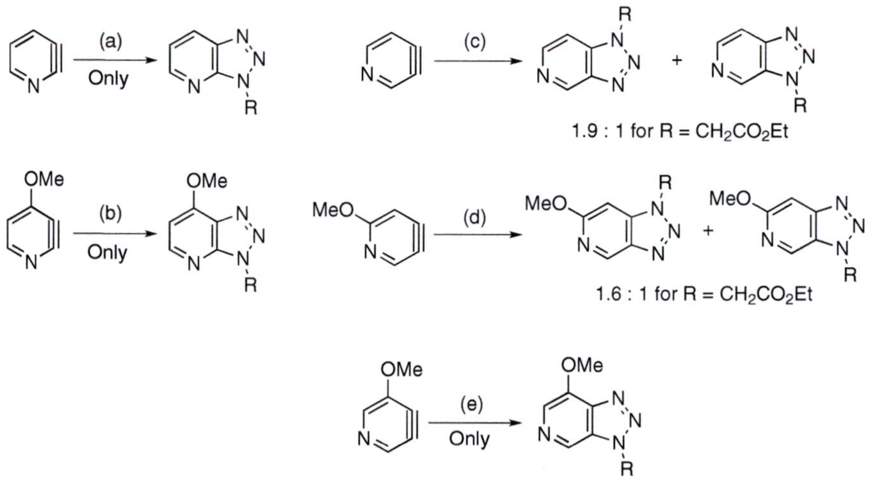

Scheme 3.11 [3+2] Cycloaddition of pyridynes with azides. Source: Saito et al. [20a]; Medina et al.[20b].

This type of heteroarynes will be more systematically discussed in Chapter 9. Their cycloaddition scope with dipoles other than azides, however, is much scattered and not systematic. Known successful examples include the cycloaddition of 4,5-benzofuranyne with diazo compound, nitrone, azomethine imine, and Sydnone [22b], as well as the cycloaddition of 2,3-pyridyne with nitrone and azomethine imine only (diazo compound, Sydnone, and nitrile oxide do not afford desired products) [20b]. Cycloaddition of 3,4-pyridyne with nitrone has also been reported

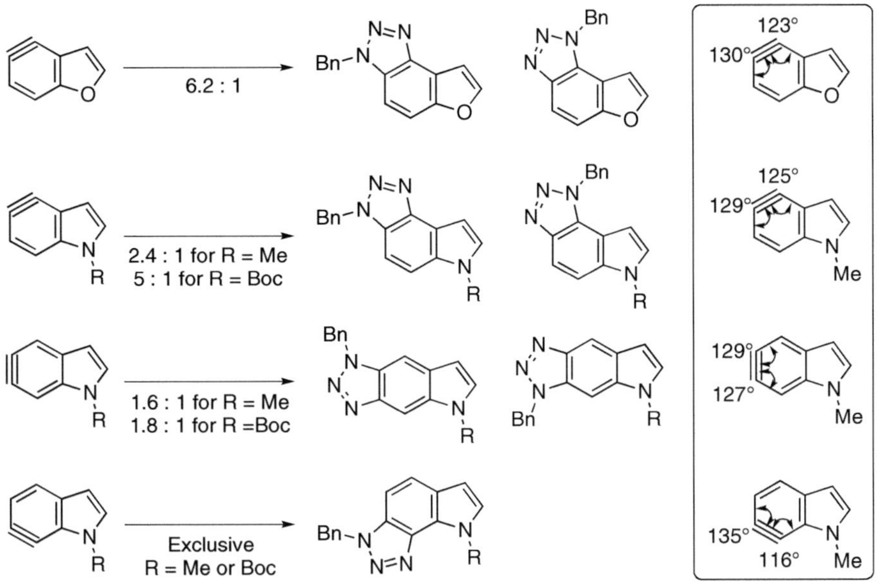

Scheme 3.12 [3+2] Cycloaddition of indolynes and benzofuranyne with azides. Source: Im et al. [22a]; Shah et al. [22b].

[19a]. These reaction scopes will not be mentioned again in the following individual sections unless necessary.

3.2.1.3 Reactions of Arynes with Nitrile Oxides

Unlike azides or diazo compounds, nitrile oxides are typically unstable and thus not isolated and used in pure form. Rather, they are generated in situ from either the corresponding hydroximoyl chlorides under basic conditions, or the nitroalkanes with a dehydration agent.

The biggest challenge for the [3+2] dipolar cycloaddition of arynes with nitrile oxides lies in the fact that both reacting partners are unstable and generated in situ. Therefore, a key factor is to match the rates at which arynes and nitrile oxides are generated so that neither of them is accumulated to undergo side reactions. In this aspect, Larock group developed a slow-vs.-slow approach using insoluble CsF as a promoter, which serves both as a base to transform hydroximoyl chlorides to nitrile oxides and a desilylation agent to generate aryne [23]. The chlorooximes had to be added slowly via syringe pump to further control the concentration of nitrile oxides (Scheme 3.13). Moses and Browne groups used the same precursors but applied a fast-vs.-fast approach, where soluble tetrabutylammonium fluoride (TBAF) was used and no syringe pump addition was required [24]. The reaction could be completed in half a minute.

Moses group also reported another 1,3-dipolar cycloaddition variant using arynes generated from diazotized anthranilic acids (Scheme 3.14) [25]. This is presumably also a fast-vs.-fast approach, and *t*-BuONO and K_2CO_3 were used to generate arynes and nitrile oxides, respectively.

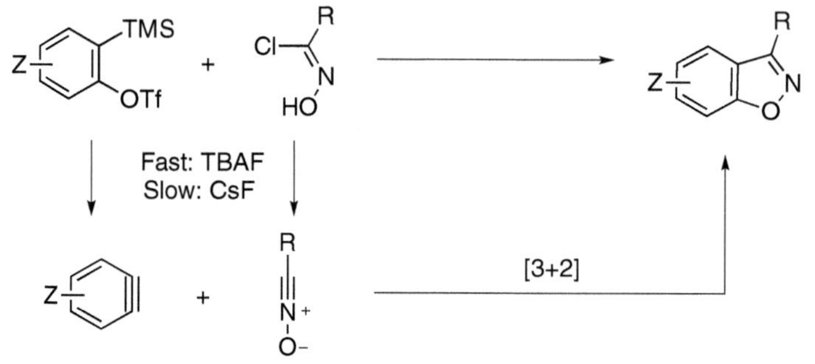

Scheme 3.13 [3+2] Cycloaddition of arynes with in situ-generated nitrile oxides.

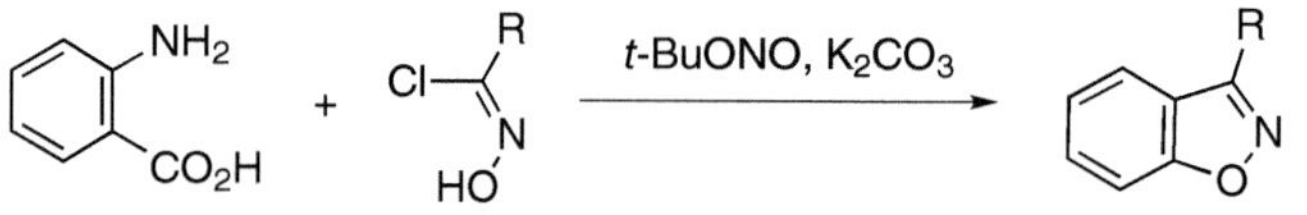

Scheme 3.14 [3+2] Cycloaddition of arynes with nitrile oxides: another variant. Source: Spiteri et al. [25].

3.2.1.4 Reactions of Arynes with Nitrile Imines

Nitrile imines are even less stable than nitrile oxides. Their reaction with aryne also requires a similar precursor and a suitable condition for their generation at a rate matching that of aryne formation so as to restrain the self-dimerization. This work was demonstrated using hydrazonyl chlorides as nitrile imine precursors, similar to that of nitrile oxides [26]. That said, due to the greater instability nature, nitrile imine cycloaddition had to follow a fast-vs.-fast approach, and either TBAF or large excess of CsF in combination of 18-C-6 would work (Scheme 3.15).

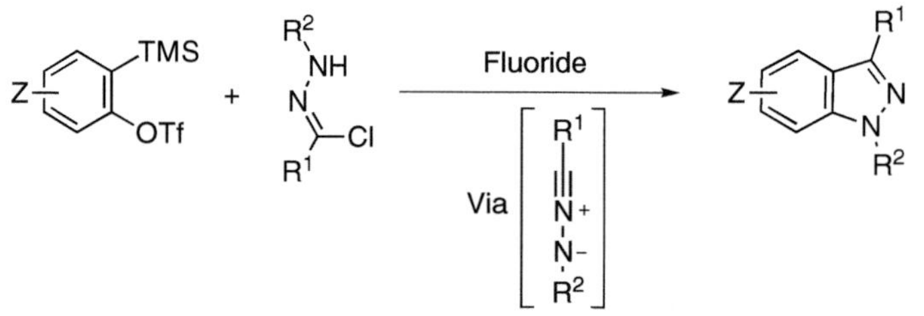

Scheme 3.15 [3+2] Cycloaddition of arynes with in situ-generated nitrile imines.

3.2.1.5 Reactions of Arynes with Nitrones

Nitrones, also known as imine oxides, are stable and isolable dipoles and have been widely used to react with various dipolarophiles to synthesize five-membered heterocycles [27]. The [3+2] cycloaddition of arynes with nitrones provides a direct and simple route to benzisoxazolines. Even before the systematic investigation of aryne–nitrone cycloaddition, Danishefsky and coworkers have applied such strategy to the synthesis of cortistatin A [28], where 2-bromophenyl triflate was treated with *n*-BuLi to generate aryne (Scheme 3.16).

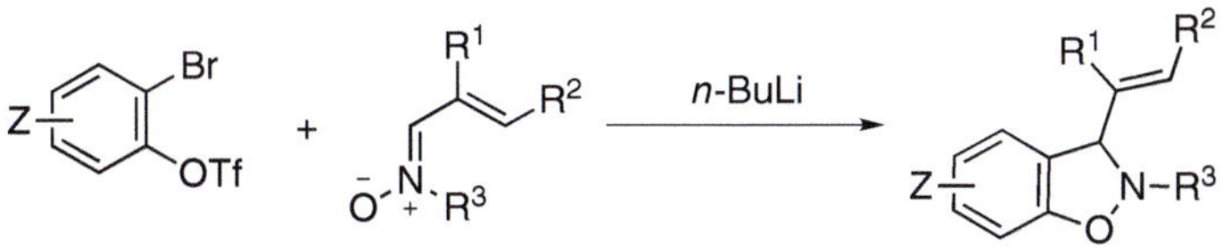

Scheme 3.16 [3+2] Dipolar cycloaddition of aryne with an α, β-unsaturated nitrone.

Systematic studies quickly followed, commencing from isolated nitrones in a general sense (Scheme 3.17) [29]. Kaliappan group expanded the scope to sugar-derived nitrones [30], and demonstrated a mild reductive cleavage of the N—O bond to open the isoxazoline ring (Scheme 3.18). Mo and coworkers reported the use of *N*-vinyl-α,β-unsaturated nitrones, whose primary [3+2] cycloadduct with arynes could undergo a stereospecific 3,3-sigmatropic rearrangement to afford a benzannulated nine-membered ring (Scheme 3.19) [31]. Subsequent N–O cleavage under different conditions could afford either pyrrole or polyketide scaffolds.

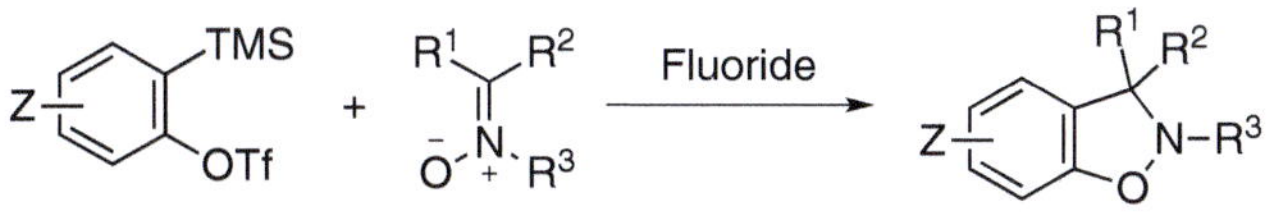

Scheme 3.17 [3+2] Dipolar cycloaddition of arynes with isolated nitrones. Source: Wu et al. [29a]; Wu et al. [29b]; Lu et al. [29c].

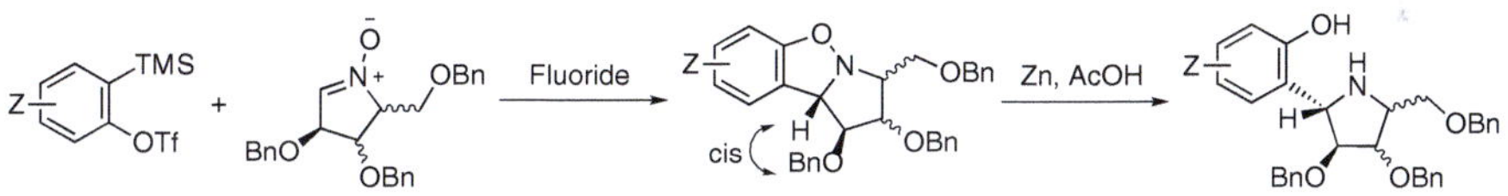

Scheme 3.18 [3+2] Dipolar cycloaddition of arynes with sugar-derived nitrones.

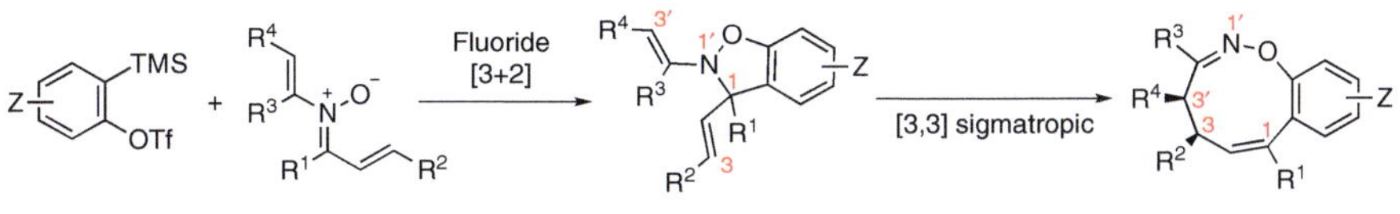

Scheme 3.19 [3+2] Dipolar cycloaddition with *N*-vinyl-α,β-unsaturated nitrones: ring enlargement by sigmatropic rearrangement. Source: Based on Ma et al. [31].

Despite being stable and isolable, nitrones exhibit high polarity and often cause difficulty in isolation and purification. Hence, much efforts were devoted to bypass this isolation and purification process and use in situ-generated nitrones or nitrone surrogates. Oxaziridines were found to react with arynes to give the same product as that of nitrones [32]. The reaction *could* possibly proceed through in situ ring

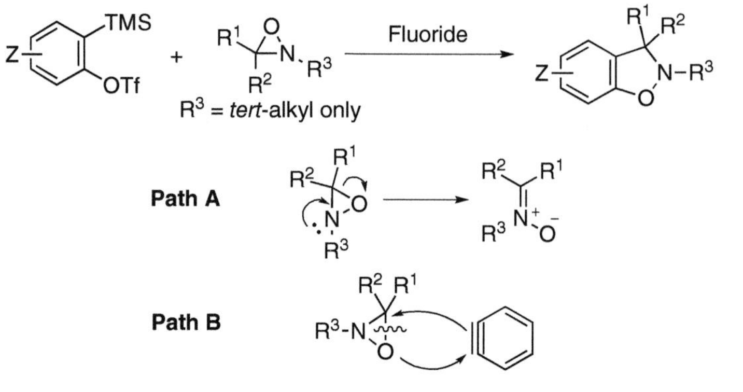

Scheme 3.20 Cycloaddition of arynes with oxaziridines.

opening of oxaziridines to nitrones, followed by [3+2] cycloaddition with arynes (Scheme 3.20, path A). However, it is also possible that arynes directly insert into the C—O bond of the oxaziridines (Scheme 3.20, path B), affording the same product. In addition, hydroxylamines could react with acetylenedicarboxylates to form nitrones in situ and this type of nitrones, bearing active methylene units, could react with arynes preferentially in a [3+2] cycloaddition manner (Scheme 3.21) [33]. Another interesting way to form nitrones is to react ketoximes with aryne in an *N*-arylation fashion, and subsequent cycloaddition with another equivalent of aryne could take place smoothly (Scheme 3.22) [34]. It is interesting to note that only ketoximes worked in this way and aldoximes simply underwent *O*-arylation with arynes (as seen for OH groups in phenols and aromatic carboxylic acids). It is not clear why ketoximes and aldoximes behave differently and why isoelectronic hydrazones behave differently from the oxime counterparts (cf. Scheme 3.7 and *vide infra* Section 3.4.2).

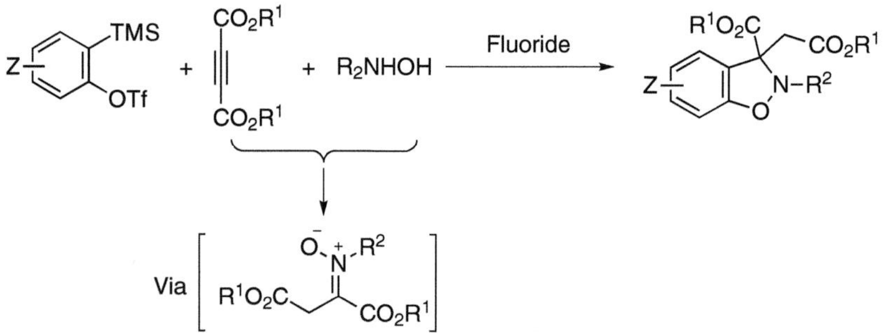

Scheme 3.21 [3+2] Cycloaddition of arynes with nitrones in situ generated from hydroxylamines and acetylenedicarboxylates. Source: Based on Li et al. [33].

In the latter report, it has been observed that the afforded cycloadduct dihydrobenzisoxazole **6** could readily undergo thermal rearrangement to a more stable dihydrobenzoxazole **7**. This rearrangement seems intrinsic and can occur

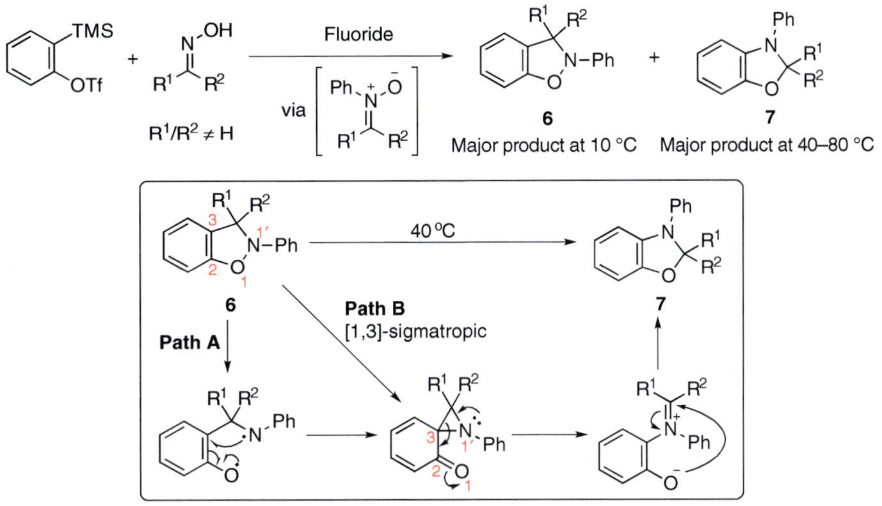

Scheme 3.22 [3+2] Cycloaddition of arynes with ketoxime-derived nitrones and subsequent rearrangement. Source: Based on Yao et al. [34].

at temperature as low as 40 °C, yet it was not mentioned in the previous reports, where even higher temperature (such as 65 °C [29c]) has been used to facilitate the cycloaddition. Structurally, nitrones in this report are ketone-derived *and* bear aryl groups at the nitrogen. However, it is not clear whether these factors are necessary or sufficient to trigger this rearrangement. Putting that aside, the rearrangement was proposed to take place via a radical dissociation-rearrangement-cyclization pathway (Scheme 3.22, path A), yet theoretically a 1,3-sigmatropic rearrangement could also explain the same phenomenon (Scheme 3.22, path B). This type of rearrangements is commonly seen for analogous structures derived from other aryne cycloadditions, and will be discussed more in Section 3.2.1.7, where another type of post-transformation of aryne–nitrone cycloadduct will also be mentioned.

3.2.1.6 Reactions of Arynes with Azomethine Imines and Ylides

Azomethine imines are the nitrogen analogue of nitrones but much less stable. Despite their reactivity with different dipolarophiles [35], azomethine imines are more often only transient intermediates and not isolated. However, some azomethine imine with part of the dipole structure incorporated in a ring and with an EWG to stabilize the negative charge (Scheme 3.23) can be stable and isolable.

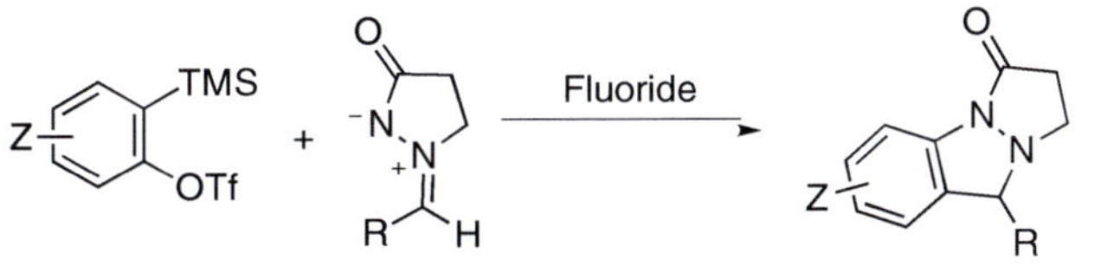

Scheme 3.23 [3+2] Dipolar cycloaddition of arynes with azomethine imines.

Thus, Yamamoto and Larock groups simultaneously reported [3+2] cycloaddition of arynes with such dipole [7, 36], where the cycloadduct can be obtained in moderate-to-good yields.

Azomethine ylides are the carbon analogue and even less stable. Using the aryne activation approach as shown in Scheme 3.22, imine **8** could be successfully transformed into an azomethine ylide, but this ylide was apparently so short-lived that it reacted with another molecule of **8** as opposed to aryne (Scheme 3.24) [37].

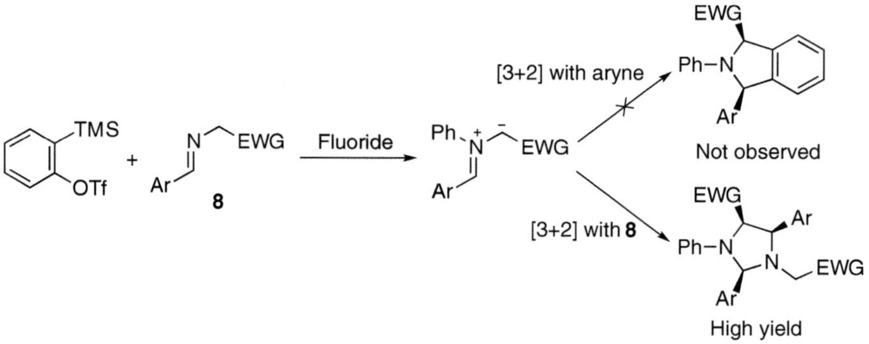

Scheme 3.24 Failed [3+2] cycloaddition of arynes with azomethine ylide. Source: Swain et al. [37a]; Jia et al. [37b].

A successful example of azomethine ylide cycloaddition was indeed reported (Scheme 3.25) [38] for an isatin imine substrate. This ylide was presumably stabilized both by the strong electron-withdrawing CF_3 group and the extra conjugation into the oxyindole system. The lifetime of the ylide was long enough to react with arynes, presumably through an extra stabilization of an internal hydrogen bonding, as proposed.

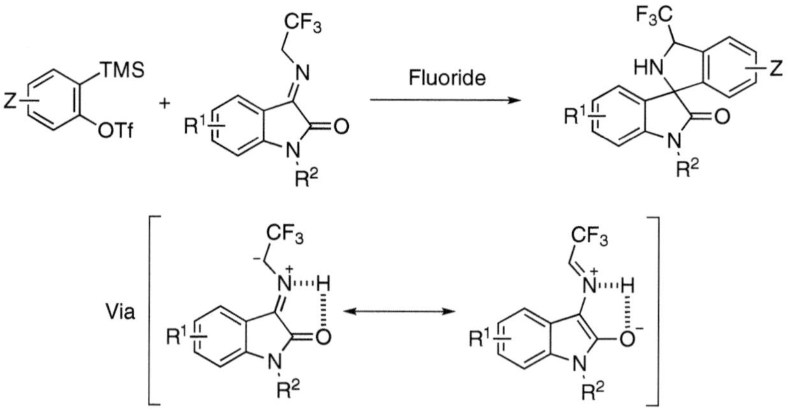

Scheme 3.25 [3+2] Dipolar cycloaddition of arynes with azomethine ylides. Source: Based on Ryu et al. [38].

3.2.1.7 Reactions of Arynes with Pyridinium *N*-Oxides

Pyridinium *N*-oxides can be classified as a cyclic version of nitrones, with part of the dipole structure embedded into a pyridine ring. However, one should keep in mind that [3+2] cycloaddition with arynes breaks the aromaticity of the pyridine and forms a rather weak N—O single bond, rendering the intrinsic driving force of the cycloadducts to undergo a variety of post-transformations to form more stable products.

Larock and Liu groups sequentially disclosed two comprehensive studies of aryne cycloaddition with pyridinium *N*-oxides under different conditions [39]. In both cases, the reaction involved a first [3+2] dipolar cycloaddition to afford an aromaticity-broken intermediate **9** (Scheme 3.26). This intermediate can then undergo either a deprotonation/elimination to cleave the N—O bond to afford 2-arylpyridine **10** (path A), or a 3,5-sigmatropic-like rearrangement to afford **11** (path B) followed by internal rearrangement to finally afford 3-arylpyridine **12**. The exact choice is complicated and depends on the substitution of the pyridine, the reaction conditions, and even the aryne precursor and the conditions under which aryne is generated. Thus, Larock's conditions involving excess aryne with CsF/MeCN preferred **12**, while Liu's conditions with reversed stoichiometry involving TBAF/dichloromethane (DCM)-tetrahydrofuran (THF) preferred **10**.

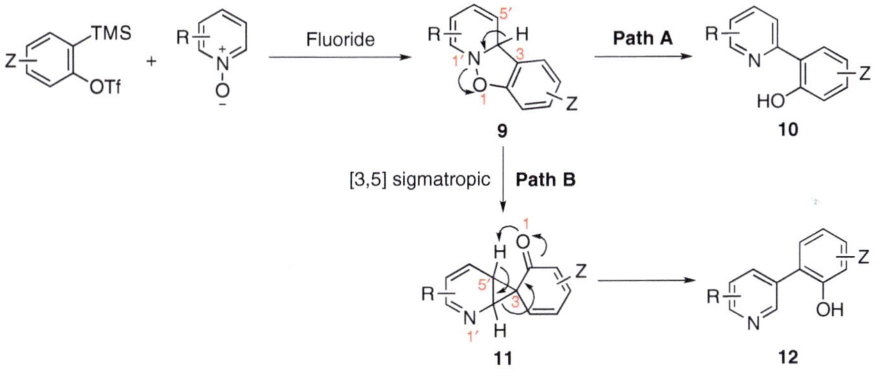

Scheme 3.26 [3+2] Dipolar cycloaddition of arynes with pyridinium *N*-oxides.

Quinolinium *N*-oxides behave similarly, but are more sensitive to the substitution and reaction conditions. 2-Unsubstituted quinolinium *N*-oxides reacted with arynes under CsF conditions to afford two products **10A** and **13** (Scheme 3.27) [40]. Formation of **10A** came from elimination (path A) and formation of **13**, preferred with excessive arynes under harsher conditions, presumably arose from an additional aryne insertion to the N—O single bond at the stage of the cycloadduct **9A** (path C). 2-Substiuted quinolinium *N*-oxides under KF/18-C-6 conditions could afford **12A** preferably (path B) [41]. Acridinium *N*-oxides reacted smoothly as well and the cycloadduct **9B** underwent a 3,9-sigmatropic-like rearrangement to finally afford 4-aryl acridines **12B** (Scheme 3.28).

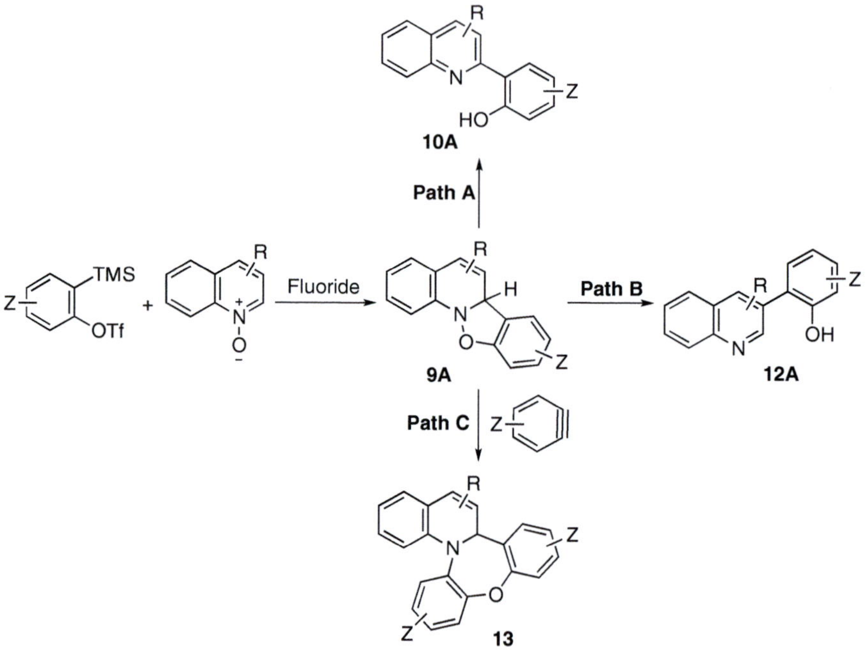

Scheme 3.27 [3+2] Dipolar cycloaddition of arynes with quinolinium *N*-oxides.

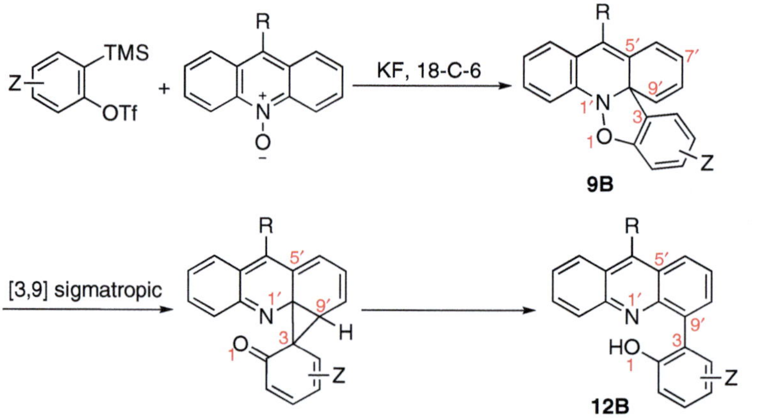

Scheme 3.28 [3+2] Dipolar cycloaddition of arynes with acridinium *N*-oxides.

The [3+2] cycloadduct of arynes with regioisomeric isoquinolinium *N*-oxides, generated from Ag-catalyzed cyclization of 2-alkynylbenzaldoximes, underwent a different post-transformation [42]. Here, a bicyclic [3.2.2] skeleton **14** was formed via either a radical dissociation/recombination or an equivalent 1,3-sigmatropic-like rearrangement (Scheme 3.29). In this case, the 5′-position belongs to another aromatic system and is not accessible for 3,5-rearrangement as seen in Scheme 3.26. In contrast, however, the cycloadduct of dihydroisoquinolinium *N*-oxide, a real nitrone rather than a pyridinium *N*-oxide analogue, could undergo CsF-promoted deprotonative elimination to give rise to an imine product **10B** (Scheme 3.30) [40].

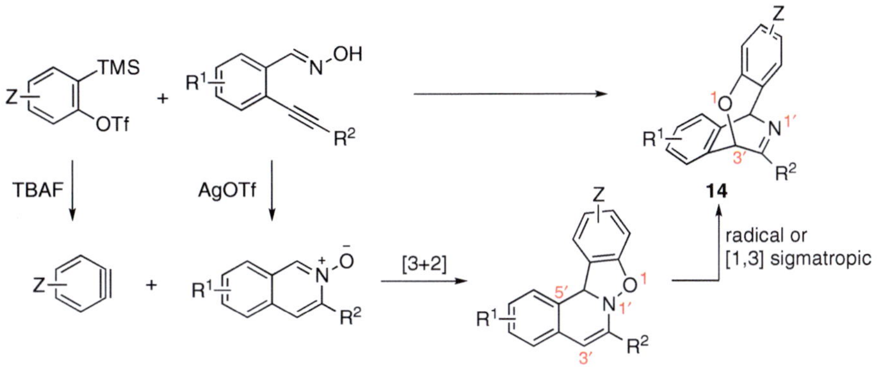

Scheme 3.29 [3+2] Dipolar cycloaddition of arynes with isoquinolinium *N*-oxides.

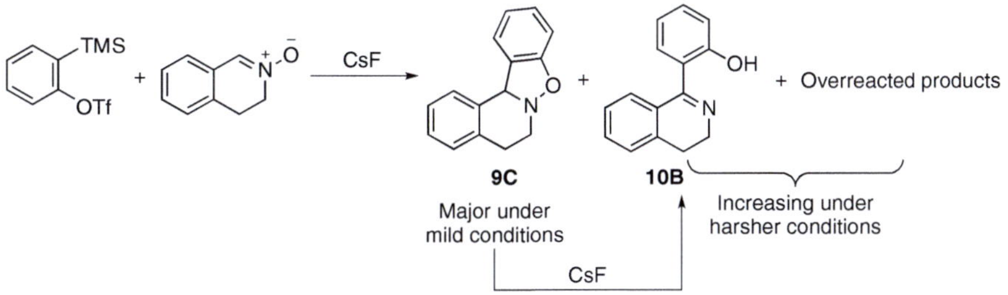

Scheme 3.30 [3+2] Dipolar cycloaddition of aryne with a dihydroisoquinolinium *N*-oxide (a nitrone). Source: Based on Okuma et al. [40].

3.2.1.8 Reactions of Arynes with Pyridinium *N*-Imides

Pyridinium *N*-imide is isoelectronic to pyridinium *N*-oxide and can be considered as an azomethine imine with part of the dipole structure incorporated in a pyridine ring. Pyridinium *N*-imide is less stable than the corresponding oxide but more stable than regular acyclic azomethine imines. If the exocyclic nitrogen is equipped with an EWG, the pyridinium *N*-imide can be isolated. An early study in this regard demonstrated a feasible cycloaddition with subsequent elimination at the intermediate that afforded a final product resembling **10** (cf. Scheme 3.26, path A) [43]. This is likely facilitated by the weak N—N bond, as well as the fact that EWG-equipped nitrogen could serve as a leaving group. An alternative solution has been recently disclosed [44], where the subsequent event can be tuned mechanistically by replacing the EWG (at the exocyclic nitrogen) by a tosyl group. Tosyl group itself could function as a leaving group, and therefore could strategically switch the mechanism at the stage of **9D** to a 1,4-elimination to afford pyrido[1,2-*b*]indazoles **15** (Scheme 3.31). This alternative outcome exploited the tosyl group as a traceless group and provided another mechanistic solution to the instability of the cycloadduct. Under the same conditions, quinolinium and isoquinolinium imides behaved similarly, and the latter could be prepared in situ from the corresponding *N′*-(2-alkynylbenzylidene)-tosylhydrazides in the presence of a silver catalyst [44b].

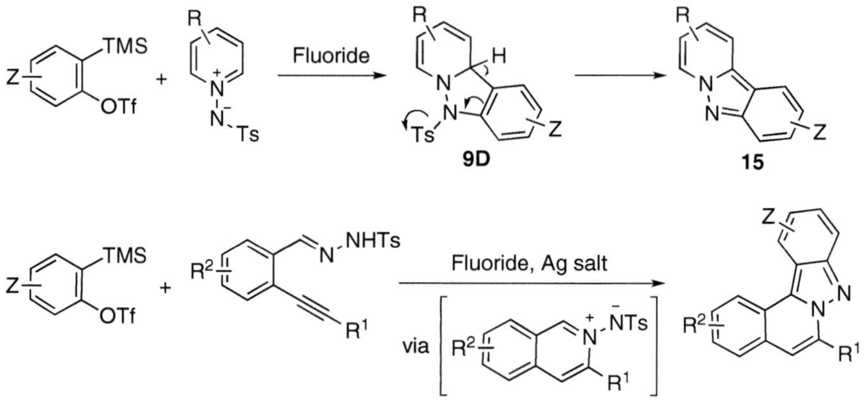

Scheme 3.31 [3+2] Dipolar cycloaddition of arynes with *N*-tosylpyridinium and isoquinolinium imides.

A recent study has also revealed, somewhat strikingly, that the direct cycloadduct **9E** could be stable enough to be isolated for quinolinium *N*-benzoylimides and would not experience 1,2-elimination to cleave the N—N bond [45]. Scheme 3.32 shows this reaction and the similar one with the *N*-tosyl variant for a comparison [44b].

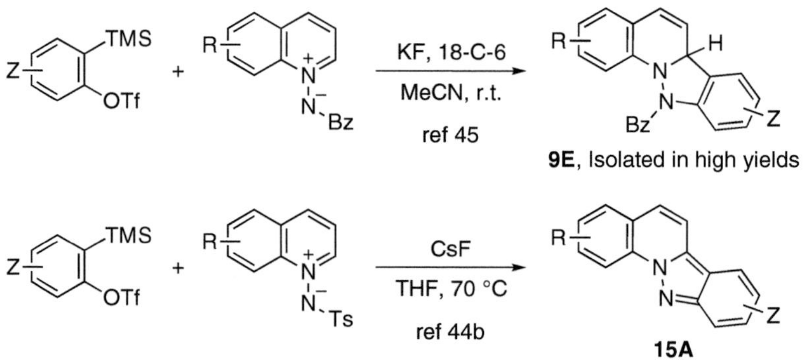

Scheme 3.32 Divergent outcome from [3+2] dipolar cycloaddition of arynes with *N*-benzoyl and *N*-tosylquinolinium imides. Source: Based on Zhao et al. [44b].

Another interesting event observed during this study was the pyridinium imide bearing two leaving groups at 3,5-positions [44b]. In this case, the intermediates **9F** underwent a 1,5-sigmatropic-like rearrangement to afford intermediates **11A**, which, upon further elimination, afforded 5*H*-pyrido[3,2-*b*]indoles **16** in good combined yield (Scheme 3.33).

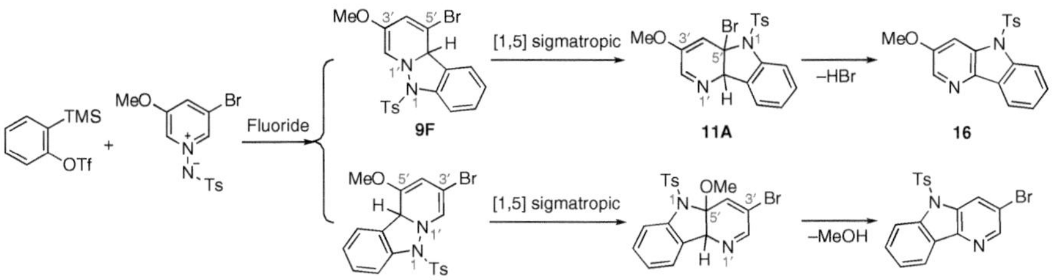

Scheme 3.33 [3+2] Dipolar cycloaddition of aryne with a special *N*-tosylpyridinium imide.

The cycloaddition with pyridynes is also worth special mentioning [46]. While the parent 3,4-pyridyne reacted with in situ-generated *N*-tosylisoquinolinium imides uneventfully to afford product **15B** via elimination of tosyl anion, it was found that 2-chloro-3,4-pyridyne afforded **14A** instead via apparent 1,3-sigmatropic rearrangement (or a radical dissociation/recombination) (Scheme 3.34).

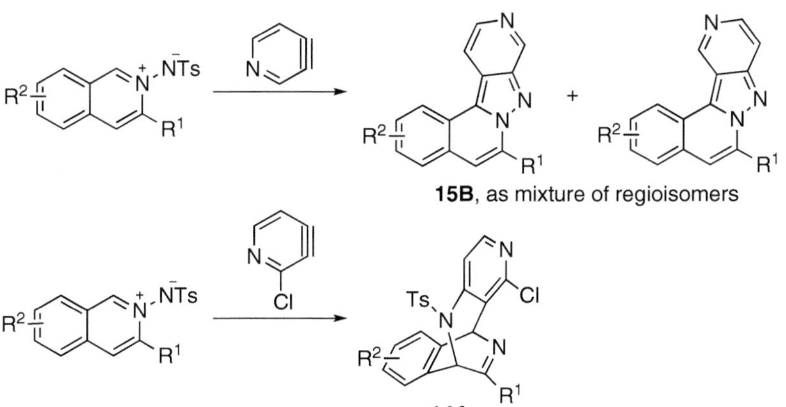

Scheme 3.34 Divergent outcome from [3+2] cycloaddition of *N*-tosylisoquinolinium imides with different pyridynes.

3.2.1.9 Reactions of Arynes with Pyridinium Ylides

Pyridinium *N*-ylides are more stable than the acyclic azomethine ylides, particularly if bearing EWGs to stabilize the carbanion. Such ylides could readily undergo [3+2] cycloaddition with arynes. The direct cycloadduct in this reaction could undergo a 1,4-elimination if a leaving group is positioned at the ylide carbon [47]. Alternatively, the cycloadduct could be dehydrogenated by air to furnish pyrido[2,1-*a*]-isoindoles as the final product (Scheme 3.35). The latter reaction could be easily performed in a one-pot fashion, where pyridinium *N*-ylides were generated from pyridines and α-haloketones/esters in situ [48].

Another very interesting reaction in this type exploited the dipole nature of indolizine that electronically resembles an azomethine ylide. Thus, indolizines successfully react with arynes, followed by a dehydrogenation step to afford indolizino[3,4,5-*ab*]isoindoles (Scheme 3.36) [49].

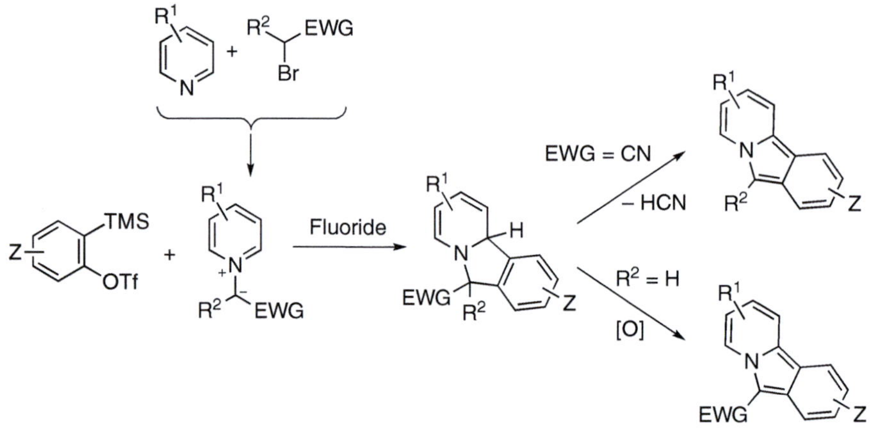

Scheme 3.35 [3+2] Dipolar cycloaddition of arynes with in situ-generated pyridinium *N*-ylides.

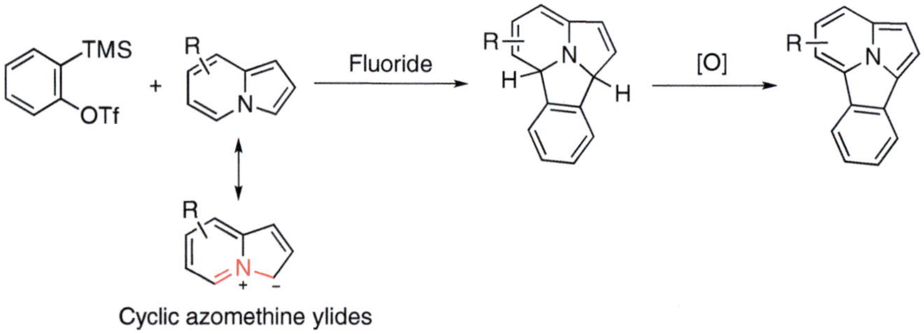

Scheme 3.36 [3 + 2] Dipolar cycloaddition of arynes with indolizines. Source: Based on Shen et al. [49].

3.2.2 [3+2] Dipolar Cycloaddition Reactions of Arynes with Cyclic 1,3-Dipoles

3.2.2.1 Reactions of Arynes with Sydnones

Sydnone is a representative stable and isolable cyclic 1,3-dipole. Structurally, it can be either viewed as an aromatic system in the form of 1,2,3-oxadiazolium inner salt, or as an azomethine imine possessing an internal lactone unit. Sydnones have been widely used in dipolar cycloadditions with various dipolarophiles to synthesize corresponding indazole derivatives and analogues, and their application in aryne [3+2] dipolar cycloaddition was advanced as a versatile and efficient way to synthesize 2*H*-indazole derivatives [50]. Different from the aforementioned [3+2] dipolar cycloaddition of arynes with linear 1,3-dipoles, the cycloaddition with Sydnones afforded a bicyclic [2.2.1] system, which would readily undergo a [4+2] cycloreversion to re-establish a planar structure along with the extrusion of CO_2 (Scheme 3.37).

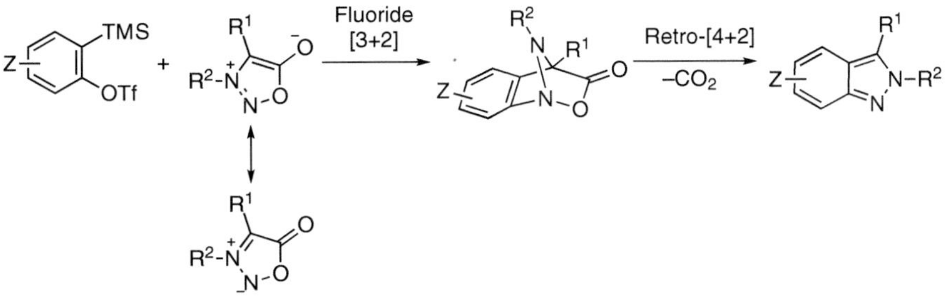

Scheme 3.37 [3+2] Dipolar cycloaddition of arynes with Sydnones.

This reaction has been expanded in two interesting directions. In one direction, the scope of Sydnone was extended to a thiazolidine derivative [51]. The product obtained, after oxidation, could extrude SO_2 and form another dipole **17**, which can be exploited in quite a handful of subsequent transformations (Scheme 3.38). The substrate here functions a dual role: the Sydnone functions as a dipole, and the thiazolidine functions as a masked secondary dipole precursor.

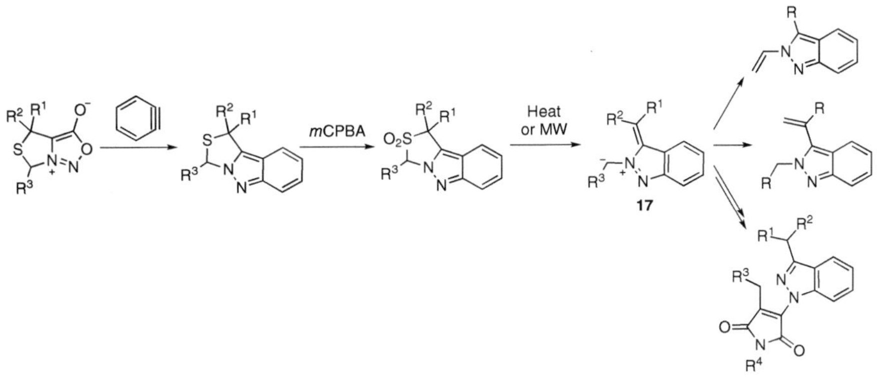

Scheme 3.38 [3+2] Dipolar cycloaddition of arynes with thiazolidine-derived Sydnones.

The other direction was aimed to helicene synthesis by employing 1,2-naphthalyne and 3,4-phenanthryne as arynes, and densely fused Sydnones derived from at least an isoquinoline system [52]. An interesting regioselectivity was observed in favor of the formation of heterohelicenes (as opposed to the zig-zag orientation) when 3,4-phenanthryne reacted with Sydnones bearing fused aromatic system positioned in a suitable region (Scheme 3.39). A comprehensive computational study revealed that a C–H…π interaction (an aromatic C—H bond of the aryne to the π system of the Sydnone shown in red in Scheme 3.39) during the early transition state contributed to this observed regioselectivity. It is interesting to note that the product could also be obtained theoretically from isoquinolinium imides (cf. Scheme 3.31) and benzo[*h*]isoquinolinium imides, but no study has been performed to confirm so.

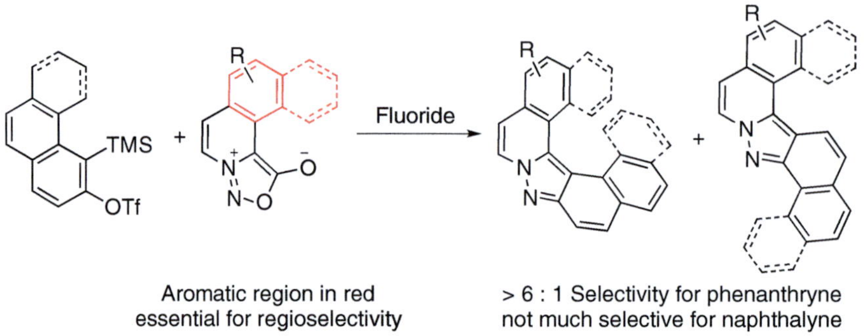

Scheme 3.39 [3+2] Dipolar cycloaddition of fused arynes with Sydnones.

3.2.2.2 Reactions of Arynes with Münchnones

As another representative cyclic 1,3-dipole, Münchnone is isoelectronic to Sydnone, and can be viewed either as oxazolium inner salt or as an azomethine ylide with an internal lactone unit. Münchnones are much less stable than Sydnones, and only those with an EWG at the 4-position can be stable enough to be isolated. The cycloaddition of arynes with Münchnones follows that of Sydnones in a [3+2] cycloaddition/[4+2] cycloreversion pathway (Scheme 3.40). That being said, the afforded 2*H*-indole product from Münchnones is much more reactive and will readily react with another molecule of aryne in a Diels–Alder manner to furnish a bicyclic [2.2.1] product. Different conditions have been developed and with different combinations of fluoride sources, solvents, and stoichiometry, one can often obtain either the 2*H*-indole or the bicyclic product as the major product [53]. Münchnones can be prepared from cyclodehydration of amino acid derivatives, and the less stable ones without EWGs at 4-position can be used in a one-pot fashion without isolation.

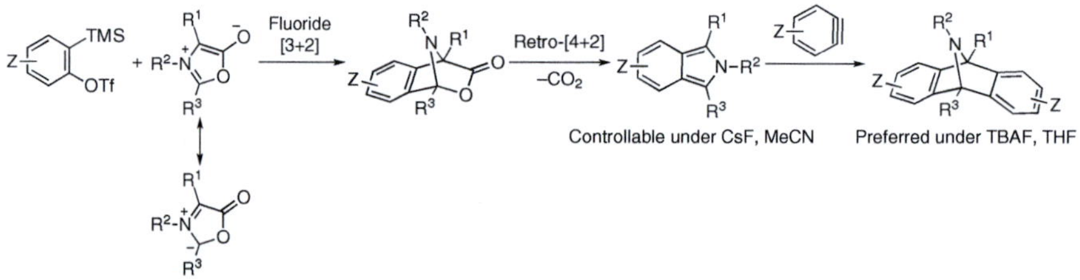

Scheme 3.40 [3+2] Dipolar cycloaddition of arynes with Münchnones.

3.3 Other [n+2] Dipolar Cycloaddition Reactions of Arynes

3.3.1 Cycloaddition with Other Dipoles

Apart from 1,3-dipoles, other dipoles such as 1,5-dipoles and 1,7-dipoles have also been known to react with arynes to yield corresponding heterocyclic products.

A [4+2] cycloaddition of Au-bearing pyrylium cation with aryne has been reported [54]. The aurated pyrylium species was generated in situ from Au-catalyzed cyclization of 2-alkynylphenyl ketones. The [4+2] cycloaddition was followed by an elimination to afford formal *O*-transpositioned ketones. Whether this [4+2] cycloaddition should be called a hetero-Diels–Alder reaction or a dipolar variant, considering the formal charges, is probably in a vague area, yet it remains necessary to be mentioned (Scheme 3.41).

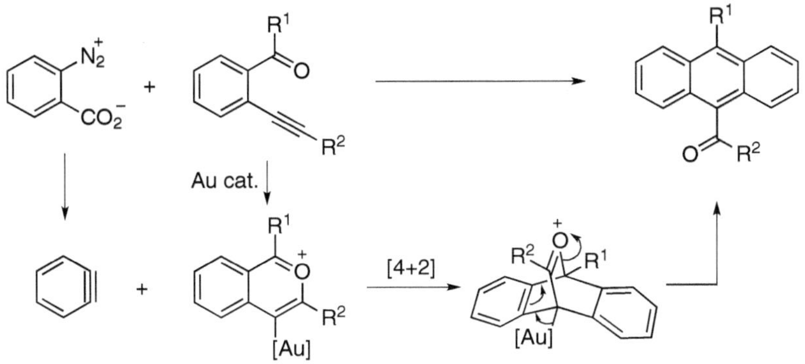

Scheme 3.41 [4+2] Cycloaddition of arynes with aurated pyrylium species.

Vinylogous pyridinium *N*-imide, a pyridinium *N*-imide bearing an olefin unit inserted into the N—N bond, has recently been developed as an air-stable 1,5-dipole. This dipole has been successfully utilized in the [5+2] cycloaddition with arynes to construct a 1,4-benzodiazepine scaffold [55]. Despite the broken aromaticity, the primary cycloadduct **18** here cannot undergo any of the processes known for that of pyridinium oxides/imides (cf. Sections 3.2.1.7 and 3.2.1.8). Instead, it was observed that the $C^3{=}C^4$ double bond of the original pyridine moiety quickly reacted with another equivalent of aryne in a [2+2] cycloaddition fashion (Scheme 3.42). Studies demonstrated that the subsequent [2+2] cycloaddition was faster than the initial [5+2] and **18** could be hardly isolated. Vinylogous quinolinium imides followed the same mechanism. It is worth mentioning that the [2+2] cycloaddition was not observed in the cycloadducts of *N*-benzoylquinolinium imides despite a similar aromaticity-broken structure (cf. Scheme 3.32).

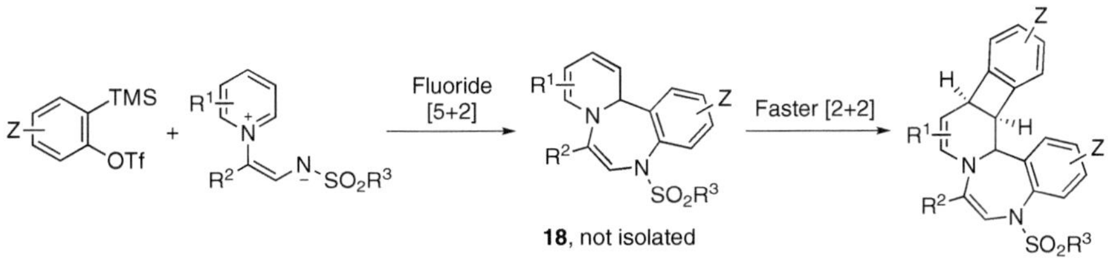

Scheme 3.42 [5+2]/[2+2] Cycloaddition of arynes with vinylogous pyridinium imides.

3-Oxidopyridinium species is another latent dipole buried in a pyridine scaffold [56]. This kind of dipole is stable and isolable and has been applied to a variety of dipolar cycloaddition reactions exhibiting various periselectivity. A systematic study of its cycloaddition with arynes has been performed and confirmed a [3+2] cycloaddition mode, where the reaction took place at the C^2,C^6-positions of the pyridine [57]. This [3+2] periselectivity (although sometimes called [5+2] in early literature) can be viewed as involving C^2–N–C^6 moiety as a cyclic azomethine ylide, and afforded a bicyclic [3.2.1] scaffold in moderate-to-good yield (Scheme 3.43).

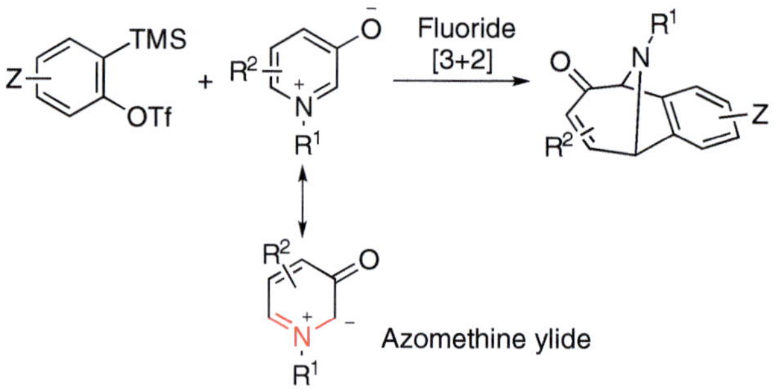

Scheme 3.43 [3+2] Dipolar cycloaddition of arynes with 3-oxidopyridinium species.

The isoelectronic analogue of 3-imidopyridinium species, however, underwent a totally different pathway. For this substrate, reaction took place at C^2,*N*(exocyclic)-positions. This can be viewed either as a formal [3+2] cycloaddition (if one counts only C^2–C^3–*N*(exocyclic) atoms) or a [7+2] cycloaddition, where the "7" counts the *N*(exocyclic)-C^3–C^4–C^5–C^6–*N*(endocyclic)-C^2 atoms in a full conjugation system carrying 8 electrons in total (cf. Figure 3.3). This [7+2] cycloaddition features an aromaticity-broken intermediate and, interestingly, this intermediate underwent a retro-6π electrocyclization to afford a two-substituted indole structure eventually in low yields (Scheme 3.44).

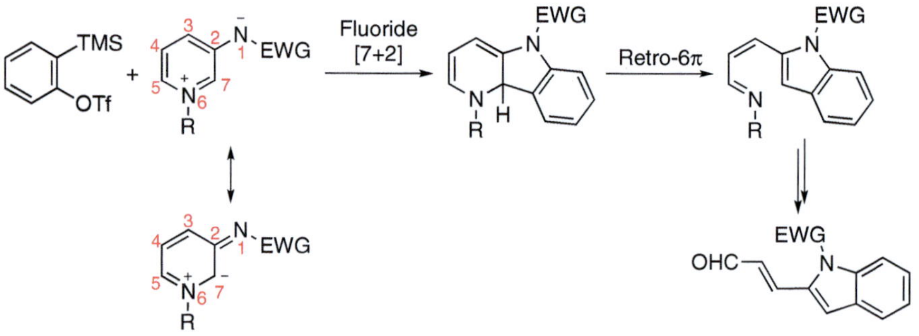

Scheme 3.44 [7+2] Dipolar cycloaddition of arynes with 3-imidopyridinium species.

3-Oxidopyridinium species can also follow this pathway when benzynes bearing 3-methoxy groups were used (Scheme 3.45). For unknown reasons, such benzynes delivered both the two regioisomeric [3+2] cycloadducts and a benzofuran product that came out of the [7+2] cycloaddition. In contrast, 4-methoxybenzyne only delivered [3+2] products. What role the 3-methoxy substitution plays in twisting the periselectivity remains to be elucidated.

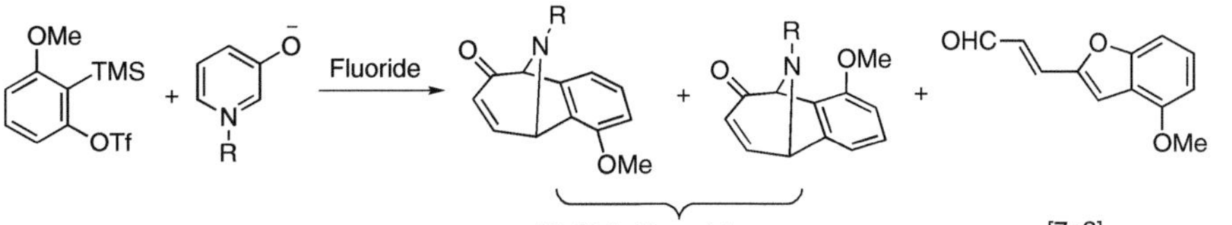

Scheme 3.45 Dipolar cycloaddition of 3-methoxybenzyne with 3-oxidopyridinium species.

3.3.2 Cycloaddition of Extended Scope of Arynes

As can be seen, most of the recent advances in aryne cycloaddition reactions have relied on the utilization of the Kobayashi aryne precursor, which can be counted as one of the most important stepstones in recent aryne chemistry. Despite the wide application and successful commercialization, Kobayashi aryne precursors used so far have limited functionality partly due to the necessity to use BuLi during the preparation. Although a few individual functionalized precursors have been used, limited availability of functionalized aryne precursors remains a key issue for further expansion.

In this regard, a recent study has shed light on broadening the scope of functionalized aryne precursors by demonstrating that Ir-catalyzed C–H borylation [58] can be applied to the assembled Kobayashi precursors [59]. Such last-stage functionalization affords a handful of borylated aryne precursors (Scheme 3.46a), which have been successfully applied to several dipolar cycloaddition reactions (Scheme 3.46b). The regioselectivity of the borylation generally installs the boryl group at a distal position and the tolerance of the boryl group in cycloaddition reactions allows an orthogonal functionality remaining at the aryne moiety. Needless to mention, the boryl group can be oxidized by peroxide, aminated via Chan–Lam reaction, iodinated under Cu-mediated conditions, and can undergo Suzuki coupling to form biaryl units, or Rh-catalyzed conjugate addition to Michael acceptors, thus providing a versatile handle for manipulation.

Another interesting direction to expand the scope is benzdiynes. In general, the concept of "benzdiyne" refers to generation of the two "yne" units sequentially, not simultaneously. The general aspects of 1,2-benzdiyne will be discussed in Chapter 8, and two cases for 1,3- and 1,4-benzdiyne are shown here.

A Kobayashi-type benzdiyne precursor bearing two differentiated silyl groups was prepared (Scheme 3.47). Activation of the more labile TMS group could trigger the formation of the first "yne" under controlled reaction time. Subsequent activation of the less labile *t*-butyldimethylsilyl (TBS) group over elongated period triggered the second "yne" formation [60]. Thus, as a proof of concept, two sequential dipolar

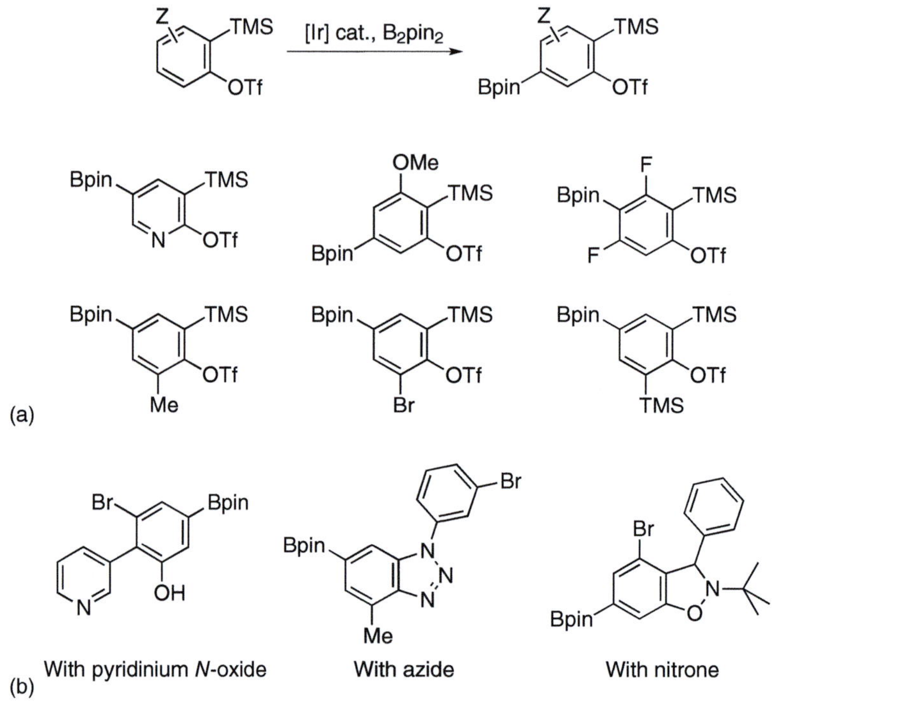

Scheme 3.46 Borylation of aryne precursors: more arynes and more dipolar cycloadducts. (a) Scope of more aryne precursors and (b) scope of dipolar cycloaddition products.

cycloadditions, or one dipolar cycloaddition with one reaction of another type, have been successfully applied to the two differentiated "yne" units. The challenges for the synthetic applications are the moderate overall yield and the moderate regio-selectivity (see Scheme 3.10 for that of 3-sillyarynes).

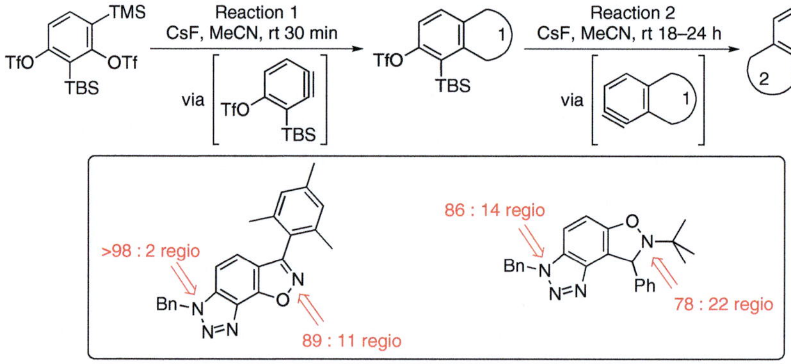

Scheme 3.47 Sequential cycloaddition of 1,3-benzdiyne from disilylaryl ditriflate.

Another benzdiyne precursor was the benzobis(oxadisilole) precursor (see Figure 3.4). By controlling the reaction conditions and stoichiometry, activation of only one oxadisilole could be achieved. Formation of the first "yne" and subsequent trapping readily took place and the sequence could be repeated to activate and trap the second "yne" unit (Scheme 3.48) [61]. This precursor is quite versatile and can offer more diversity of the polyynes. However, preparation and activation of this precursor is a little step intensive.

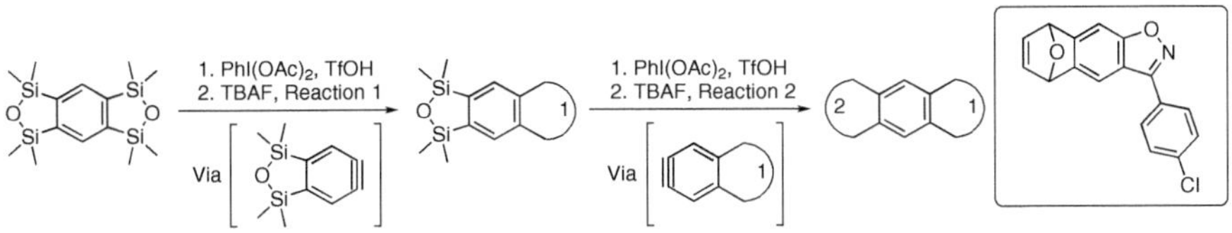

Scheme 3.48 Sequential cycloaddition of 1,4-benzdiyne from benzobis(oxadisilole). Source: Based on Ma et al. [61].

3.4 Formal Cycloaddition Reactions of Arynes

Formal cycloaddition could refer to as widely as reactions of arynes with amphiphilic reagents that could result in a ring formation. There are indeed many smartly designed amphiphilic substrates with tethered nucleophiles and electrophiles, leading to interesting and useful benzannulated heterocycles [62]. For the scope of the discussion, we wish to narrow it down to substrates that at least involve π systems or resemble dipole-like structures. Herein, three cherry-picked systems are summarized and discussed.

3.4.1 Formal Cycloaddition with N–C–C Systems Forming Indole/Indoline/Oxyindole Scaffolds

Enamines are nucleophiles, and those with strategically positioned EWGs can act as amphiphilic reagents.

N-Boc-2-methyleneglycine derivatives stand as an interesting substrate of this class [63a]. The enamine nitrogen could serve as the nucleophile, and the Michael acceptor functioned as the electrophile for the aryl anion to back attack (Scheme 3.49). The reaction afforded indoline products and the enamine substrate exhibited overall reactivity equivalent to an N(δ−)–C=C(δ+) system in a formal [3+2] cycloaddition reaction.

Somewhat counterintuitively, it was demonstrated that replacing the Boc group on the enamine nitrogen to an acyl group could shift the polarity of the enamine system. The nucleophilic site changed from the nitrogen to the enamine carbon (which functioned as the electrophilic site in the former case). Thus, the formal [3+2] cycloaddition was changed to a polarity-reversed formal [4+2] cycloaddition that final afforded isoquinolines after a dehydration (Scheme 3.50) [63].

A more ambiguous example was the reaction of aryne with iminophosphorane **19** (Scheme 3.51), generated in situ from a reaction of 2-azidoacrylates with stoichiometric phosphine [64]. Intermediate **19** could undergo formal [3+2] cycloaddition

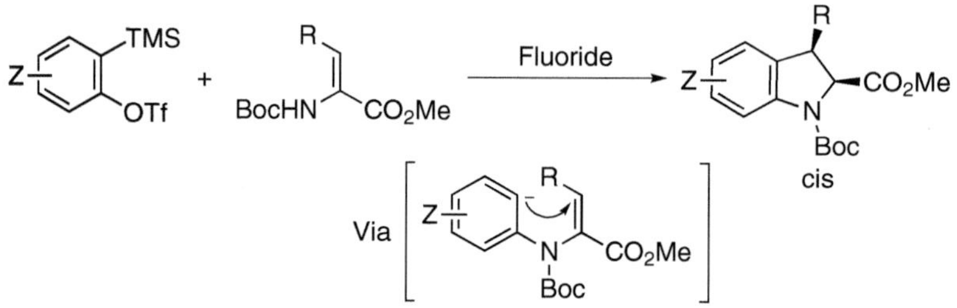

Scheme 3.49 Formal [3+2] cycloaddition of arynes with *N*-Boc-2-methyleneglycine derivatives.

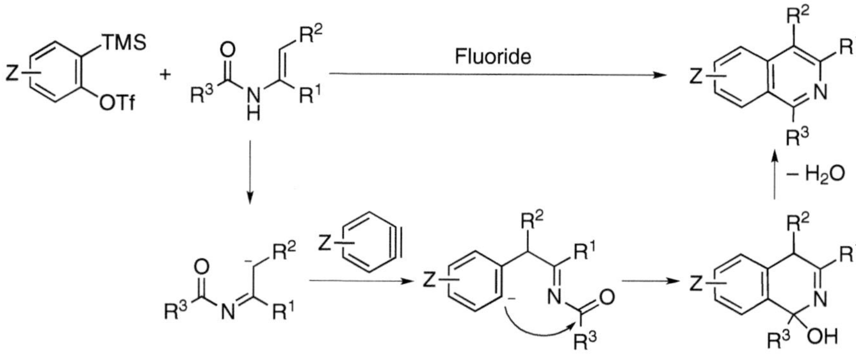

Scheme 3.50 Formal [4+2] cycloaddition of arynes with *N*-acylenamines. Source: Gilmore et al. [63a]; Zhao et al. [63b].

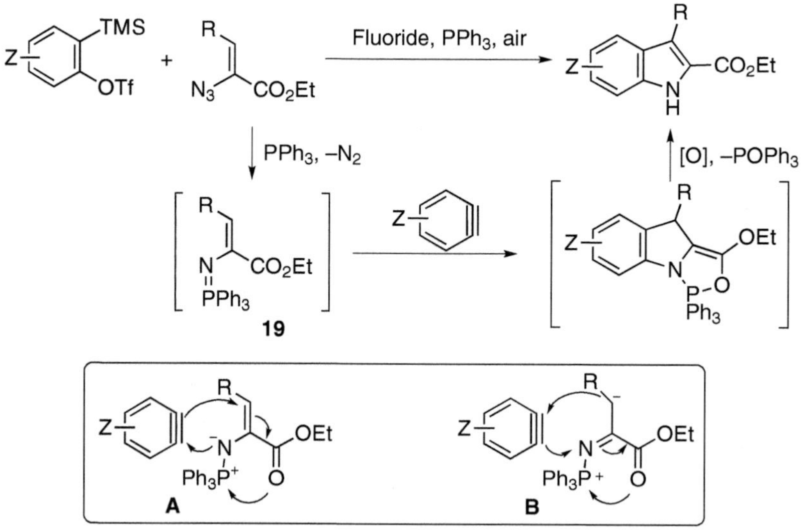

Scheme 3.51 Formal [3+2] cycloaddition of arynes with 2-azidoacrylates.

with arynes, and subsequent extrusion of phosphine oxide through hydrolysis and air oxidation/dehydrogenation could afford an indole product. Despite the success of the reaction, the electronics of **19** remain in the vague area. The arrow-pushing mechanism **A**, proposed by the authors, mimics the electron flow of that in Scheme 3.49 and a similar N(δ−)–C=C(δ+) system as well. However, reaction with 3-methoxybenzyne afforded a regioselectivity opposite to this electron flow. This regioselectivity could, however, be accounted for by an alternative arrow-push mechanism **B**, exhibiting a polarity reversed N(δ+)=C–C(δ−) system. The real scenario is elusive and a computational study could potentially help to answer the question.

Apart from the N(δ−)–C=C(δ+) and the N(δ+)=C–C(δ−) systems described above, a third class, namely α-halo-*N*-alkoxycarboxyamides, was reported to undergo formal [3+2] cycloaddition with arynes to afford *N*-alkoxy oxyindoles in good yields (Scheme 3.52) [65]. Here, the reactive partner exhibited a conceptually different N(δ−)–C–C(δ+) system. Unlike the two aforementioned systems that involve π bonds and are equivalent to four-electron units, this new system does not have a π bond and is perhaps more suitably considered as a two-electron unit. Although multiple mechanistic pathways could explain the formation of the described product, experiment using a chiral α-bromoamide led to complete erosion of stereochemistry (Scheme 3.53). This evidence is contradictory to the stepwise mechanism where at least some chiral memory should retain, and points to an in situ elimination of HBr first to afford a system that looks like a hetero-trimethylenemethane (hetero-TMM) unit bearing an N(δ−)–C–C(δ+) typed electron distribution, followed by its reaction with arynes.

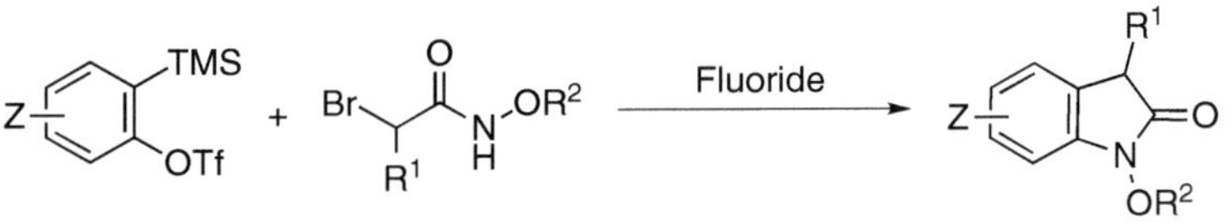

Scheme 3.52 Formal [3+2] cycloaddition of arynes with α-halo-*N*-alkoxyamides. Source: Singh et al. [65].

3.4.2 Formal Cycloaddition with Hydrazone-Derived N–N–C Systems

Formal [3+2] cycloaddition of arynes with different hydrazones has been extensively studied. The mechanism and outcomes are heavily influenced by the hydrazone substitution and the reaction conditions.

As mentioned before in Section 3.2.1.4, *N*-monosubstituted hydrazonyl chlorides can generate nitrile imines by in situ elimination of HCl under basic conditions (cf. Scheme 3.15). It was later found that *N,N*-disubstituted hydrazonyl chlorides could react with arynes to afford very similar products as monosubstituted hydrazonyl chlorides do. Clearly, the disubstituted ones cannot generate nitrile imines and the reaction has to proceed through a presumable stepwise manner to afford a formal [3+2] cycloadduct **20** (Scheme 3.54) [66a]. This intermediate could subsequently

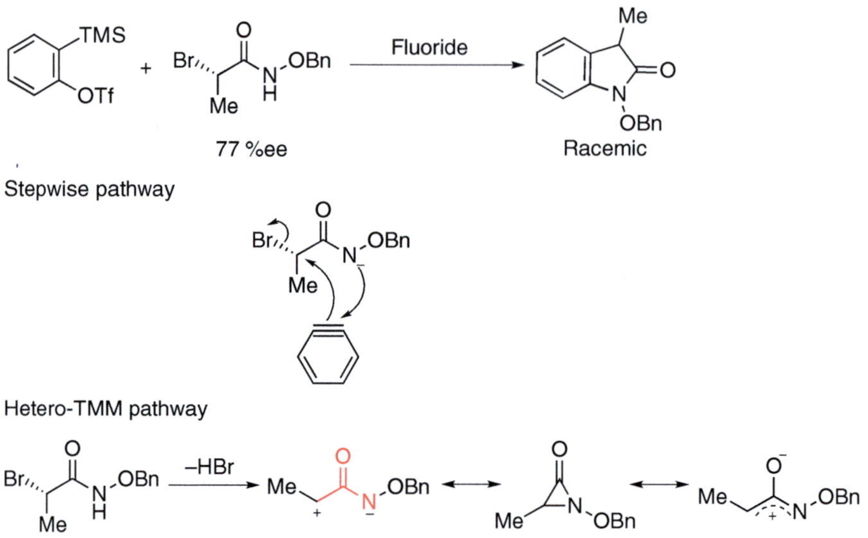

Scheme 3.53 Evidence supporting involvement of a 2-electron hetero-TMM.

eliminate the chloride, and one of the methyl groups was then scavenged by an external nucleophile to afford indazoles.

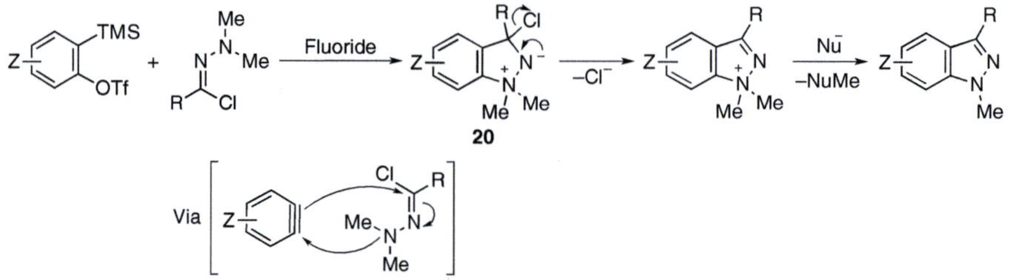

Scheme 3.54 Formal [3+2] cycloaddition of arynes with *N,N*-dimethylhydrazonyl chlorides. Source: Based on Markina et al. [66a].

Without the chloride at the imidoyl position, aldehyde-derived *N,N*-dialkyl hydrazones could still react with arynes to afford a similar formal [3+2] adduct **20A** (Scheme 3.55). As there is no chloride to eliminate, a proton shift took place to form a carbanion, which triggered N–N cleavage to open the newly formed five-membered ring to afford an imine derivative [66b]. This imine can be manipulated in different workup procedures to afford different products [66c].

In addition to this pathway, cycloadduct **20A** can be trapped by electrophiles such as Ac_2O. Treatment of the trapped adduct **21**, without isolation, with hydrazine under high temperature could afford indazoles (Scheme 3.56). This reaction is quite similar to the overall transformation shown in Scheme 3.54 without being necessary to form the chloride. Although the authors did not mention, the hydrazine workup theoretically requires an oxidant to give rise to the aromatized structure.

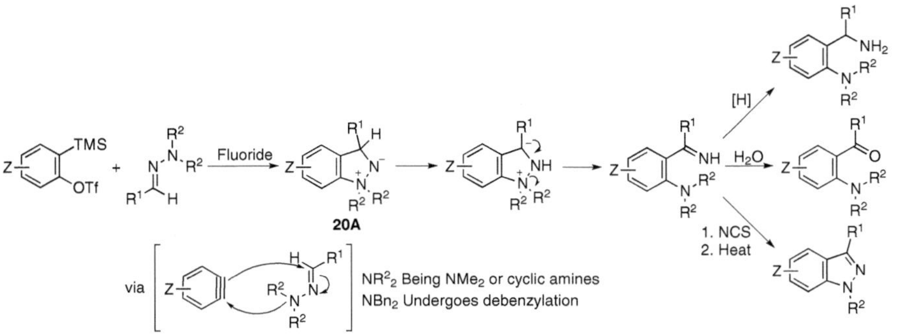

Scheme 3.55 Formal [3+2] cycloaddition of arynes with *N,N*-dialkylhydrazones.

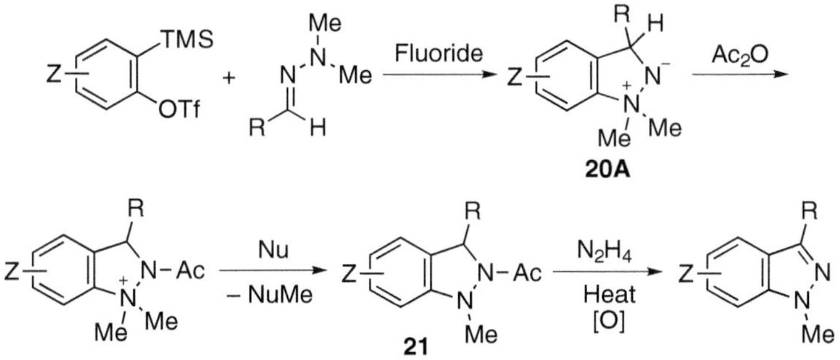

Scheme 3.56 Formal [3+2] cycloaddition of arynes with *N,N*-dimethylhydrazones.

Parallel to this work, another report disclosed a similar oxidative strategy to form indazoles from aryne formal [3+2] cycloaddition. In this case, aldehyde-derived *N*-monosubstituted hydrazones were used as the substrate and the reaction was carried out under aerobic oxidation conditions with a desired volume of air serving as the oxidant (Scheme 3.57) [67]. This reaction can be additionally viewed in comparison with the aforementioned examples, as well as the reaction involving *N*-tosylhydrazones shown in Scheme 3.7. The oxidation could occur on the

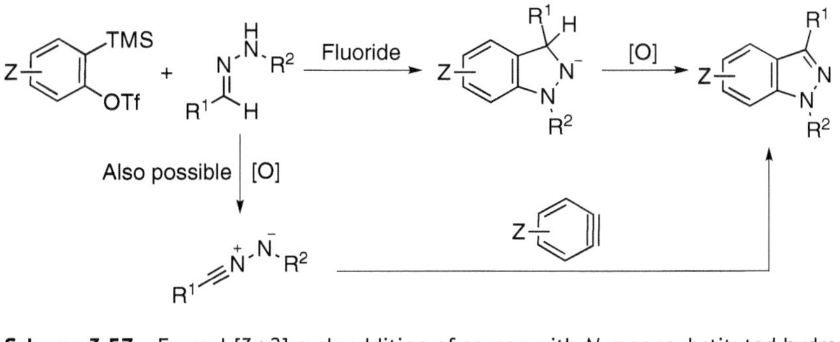

Scheme 3.57 Formal [3+2] cycloaddition of arynes with *N*-monosubstituted hydrazones under oxidative conditions. Source: Based on Li et al. [67].

cycloadduct to convert it to the final indazole. However, attempts in trapping this cycloadduct by Ac_2O were unsuccessful, which brings the additional likelihood that the role of air was perhaps to convert the hydrazones to nitrile imines, followed by a real [3+2] cycloaddition.

Shown above are several examples, where different substitution patterns could be exploited in manipulation of the primary cycloadduct in different subsequent reactions to give rise to different products. The overall events could be complicated and multiple mechanistic pathways are available for the primary cycloadducts, and subtle changes in substrate structure and reaction conditions may alter the mechanism and consequently the final outcome. It is worth mentioning that despite the already-existing complexity, to date, studies still have not covered all substitution patterns in this series.

3.4.3 Formal Cycloaddition and with Sulfur-Containing Substrates

Sulfur is a soft nucleophile and can extend its valence. Thus, even sulfur atoms in thioethers and thiocarbonyls remain nucleophilic. Several sulfur-containing substrates have been known to react with arynes in formal [3+2] cycloaddition manner. Interestingly, despite their reactivity, many substrates of this type may not appear amphiphilic.

Aryl vinyl sulfides bearing EWGs at the distal vinylic position (i.e. β-(arylthio) acrylates or analogues) have been found to react with arynes in a [3+2] dipolar cycloaddition manner [68]. The sulfur atom serves as the nucleophile to generate a primary cycloadduct **22** (Scheme 3.58), which is essentially a sulfur ylide. Computational studies showed that formation of **22** is actually a concerted cycloaddition process despite the lack of dipole-like structural feature of the substrate. Subsequent proton shift at the stage of **22** took place via a water-facilitated protonation–deprotonation process [69]. A final elimination afforded a diaryl sulfide. It is noteworthy that the overall reaction looks identical to an aryne insertion into the S—C bond of the substrate, and Chapter 4 will focus on aryne insertion chemistry of this type. However, mechanistically, this reaction does not follow the typical aryne insertion and has more cycloaddition feature.

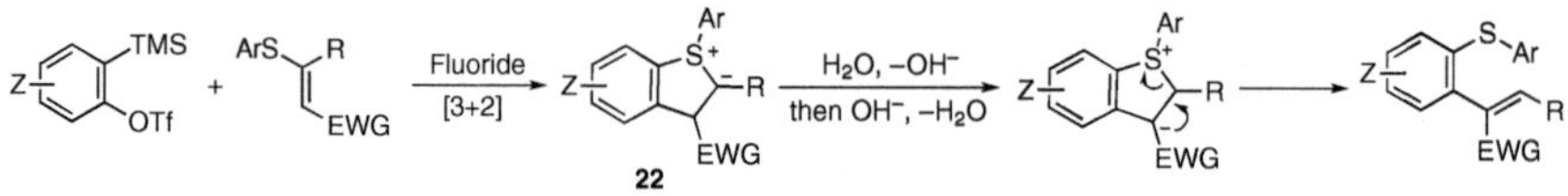

Scheme 3.58 [3+2] Cycloaddition of arynes with vinyl sulfides.

When a second EWG was present and in the absence of water, **22A** underwent a further arylation with aryne (Scheme 3.59). Subsequent deprotonation triggered a similar elimination to afford a stilbene product.

Similar to vinyl sulfides, ketene dithioacetals have been known to act as a formal 1,3-dipole, and have been used to construct five-membered carbocycles through formal [3+2] cycloaddition with electrophiles [70]. A special ketene dithioacetal

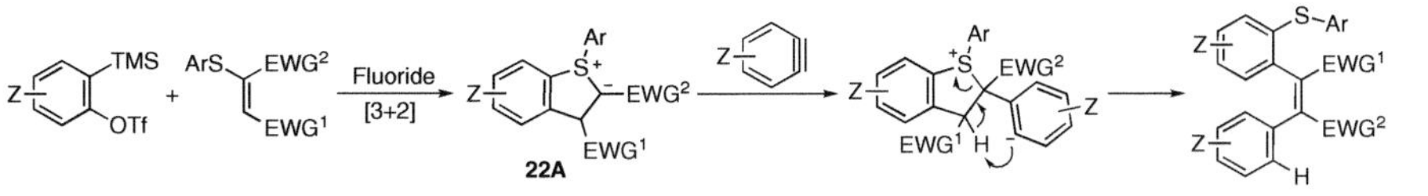

Scheme 3.59 [3+2] Cycloaddition of arynes with vinyl sulfides: further arylation variant.

substrate derived from ethanedithiol has been studied in aryne cycloaddition. The reaction presumably took place via the zwitterionic resonance structure of the ketene dithioacetal, and the ethanedithiol unit allowed for an either subsequent or simultaneous extrusion of ethylene to afford 2-mercaptobenzothiophenes [71], which reacted further with aryne in an S-arylation event (Scheme 3.60).

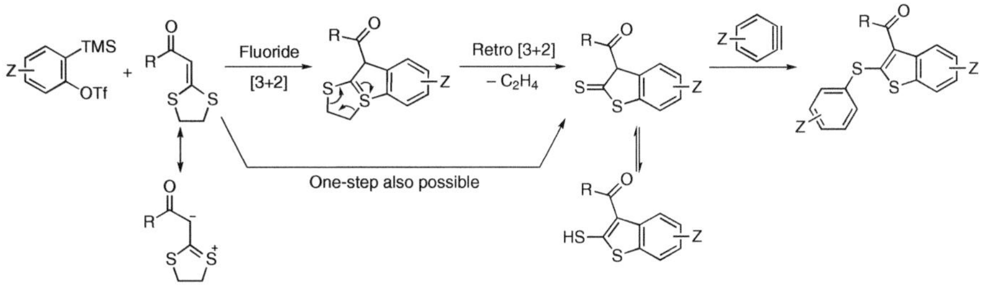

Scheme 3.60 Formal [3+2] cycloaddition of arynes with ketene dithioacetals.

Apart from the thiol ether category, 1,2,5-thiadiazoles represent another class of formal dipole that could react with arynes. Symmetrical 1,2,5-thiadiazoles were known long time ago to undergo aryne [3+2] cycloaddition, followed by a subsequent [3+2] cycloreversion to afford benzoisothiazoles with extrusion of a cyanide [72a]. This chemistry has been recently revisited to incorporate both symmetrical and unsymmetrical 1,2,5-thiadiazoles [72b]. For the unsymmetrical substrates, it was found that when R^1 and R^2 are hydroxyl and amino groups, respectively, the highest level of discrimination was achieved, and extrusion of cyanic acid occurred selectively (Scheme 3.61).

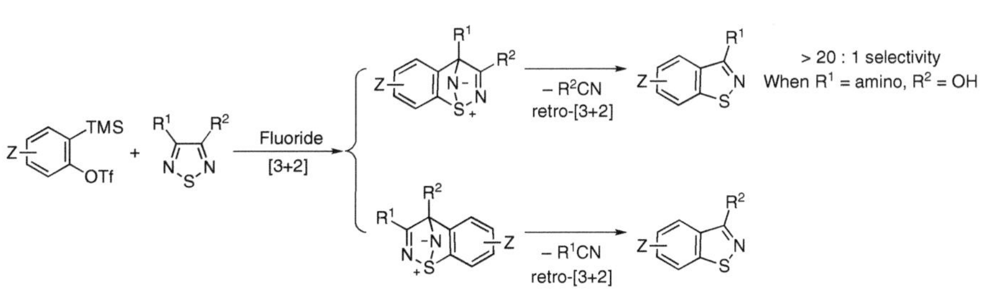

Scheme 3.61 Formal [3+2] cycloaddition of arynes with 1,2,5-thiadiazoles.

Last but not least, electron-deficient thioamides have also been known to react with a special class of benzyne [73]. This will be mentioned in Chapter 10.

3.5 Summary

Aryne dipolar cycloaddition reactions have received considerable attention. Major efforts have been devoted to the synthetic applications and reactivity exploration. Mechanistic studies have also been widely conducted to answer questions or provide rationale to scientific curiosity. In the past decade, a handful of heterocyclic scaffolds have been made attainable through aryne dipolar cycloadditions, and many more will surely come in the future. Yet more importantly, these studies have brought us not only more methods in the synthetic toolbox, but also more and deeper understanding of the reactivity of arynes, which, in turn, would promote the next round of reaction design and mechanistic study. Computational studies, more (hetero)arynes scaffolds, more dipoles, and perhaps more kinetic studies will advance the frontline of our understanding in aryne dipolar cycloadditions and enrich the aryne chemistry in a broader scope of view.

List of Abbreviations

18-C-6	18-crown-6
Ac	acetyl
Ar	aryl
Bn	benzyl
Boc	*t*-butoxycarbonyl
Bpin	pinacolboryl
Bz	benzoyl
DCM	dichloromethane
EWG	electron-withdrawing group
*m*CPBA	*meta*-chloroperbenzoic acid
MW	microwave irradiation
NBO	natural bond orbital
NCS	*N*-chlorosuccinimide
Nu	nucleophile
PTC	phase transfer catalyst
TBAF	tetrabutylammonium fluoride
TBS	*t*-butyldimethylsilyl
Tf	trifluoromethanesulfonyl
THF	tetrahydrofuran
TMM	trimethylenemethane
TMS	trimethylsilyl
Ts	*p*-toluenesulfonyl

References

1 (a) De, N. and Yoo, E.J. (2018). *ACS Catal.* 8: 48–58. (b) Hashimoto, T. and Maruoka, K. (2015). *Chem. Rev.* 115: 5366–5412.

2 (a) Dubrovskiy, A.V., Markina, N.A., and Larock, R.C. (2013). *Org. Biomol. Chem.* 11: 191–218. (b) Wu, C. and Shi, F. (2013). *Asian J. Org. Chem.* 2: 116–125. (c) Tadross, P.M. and Stoltz, B.M. (2012). *Chem. Rev.* 112: 3550–3577. (d) Roy, T. and Biju, A.T. (2018). *Chem. Commun.* 54: 2580–2594. (e) Bhojgude, S.S., Bhunia, A., and Biju, A.T. (2016). *Acc. Chem. Res.* 49: 1658–1670. (f) Kaicharla, T. and Biju, A.T. (2014). Transition-metal-free synthesis of benzo-fused five- and six-membered heterocycles employing arynes. In: *Green Synthetic Approaches for Biologically Relevant Heterocycles* (ed. G. Brahmachari), 45–76. Elsevier Science. (g) Pérez, D., Peña, D., and Guitián, E. (2013). *Eur. J. Org. Chem.*: 5981–6013. (h) Gampe, C.M. and Carreira, E.M. (2012). *Angew. Chem. Int. Ed.* 51: 3766–3778. (i) Bhunia, A., Yetra, S.R., and Biju, A.T. (2012). *Chem. Soc. Rev.* 41: 3140–3152. (j) Okuma, K. (2012). *Heterocycles* 85: 515–544. (k) Chen, Y. and Larock, R.C. (2009). Arylation reactions involving the formation of arynes. In: *Modern Arylation Methods* (ed. L. Ackermann), 401–473. Weinheim (Germany): Wiley-VCH. (l) Sanz, R. (2008). *Org. Prep. Proced. Int.* 40: 215–291. (m) Wenk, H.H., Winkler, M., and Sander, W. (2003). *Angew. Chem. Int. Ed.* 42: 502–528. (n) Pellissier, H. and Santelli, M. (2003). *Tetrahedron* 9: 701–730.

3 Himeshima, Y., Sonoda, T., and Kobayashi, H. (1983). *Chem. Lett.* 12: 1211–1214.

4 (a) Huisgen, R. (1984). 1,3-Dipolar cycloadditions – introduction, survey, mechanism. In: *1,3-Dipolar Cycloaddition Chemistry*, vol. 1 (ed. A. Padwa), 3–6. New York: Wiley. (b) Gothelf, K.V. and Jørgensen, K.A. (1998). *Chem. Rev.* 98: 863–909.

5 (a) Decuypère, E., Plougastel, L., Audisio, D., and Taran, F. (2017). *Chem. Commun.* 53: 11515–11527. (b) Stewart, F.H.C. (1964). *Chem. Rev.* 64: 129–147. (c) Lopchuk, J.L. (2012). *Top. Heterocycl. Chem.* 29: 381–413. (d) Kawase, M., Sakagami, H., and Motohashi, N. (2009). *Top. Heterocycl. Chem.* 16: 135–152. (e) Ollis, W.D. and Ramsden, C.A. (1976). *Adv. Heterocycl. Chem.* 19: 1–122. (f) Gribble, G. (2002). *Synthetic Applications of 1,3-Dipolar Cycloaddition Chemistry Toward Heterocycles and Natural Products*, vol. 59 (eds. A. Padwa and W.H. Pearson), 681–753. Wiley.

6 Ford, A., Miel, H., Ring, A. et al. (2015). *Chem. Rev.* 115: 9981–10080.

7 (a) Jin, T. and Yamamoto, Y. (2007). *Angew. Chem. Int. Ed.* 46: 3323–3325. (b) Jin, T., Yang, F., and Yamamoto, Y. (2009). *Collect. Czech. Chem. Commun.* 74: 957–972. (c) Liu, Z., Shi, F., Martinez, P.D.G. et al. (2008). *J. Org. Chem.* 73: 219–226.

8 (a) Reddy, B.V.S., Reddy, R.R.G., Thummaluru, V.R., and Sridhar, B. (2017). *ChemistrySelect* 2: 4290–4293. (b) Cheng, B., Zu, B., Bao, B. et al. (2017). *J. Org. Chem.* 82: 8228–8233. (c) Cheng, B., Bao, B., Zu, B. et al. (2017). *RSC Adv.* 7: 54087–54090. (d) Hari, Y., Sone, R., and Aoyama, T. (2009). *Org. Biomol. Chem.* 7: 2804–2808.

9 (a) Chen, G., Hu, M., and Peng, Y. (2018). *J. Org. Chem.* 83: 1591–1597. (b) Phatake, R.S., Mullapudi, V., Wakchaure, V.C., and Ramana, C.V. (2017). *Org. Lett.* 19: 372–375.

10 Wang, C.-D. and Liu, R.-S. (2012). *Org. Biomol. Chem.* 10: 8948–8952.

11 (a) Li, P., Zhao, J., Wu, C. et al. (2011). *Org. Lett.* 13: 3340–3343. (b) This reaction is only applicable to hydrazones derived from aldehydes or non-enolizable ketones. For those derived from enolizable ketones, a Fischer indole type reaction sequence can take place, see, McAusland, D., Seo, S. et al. (2011). *Org. Lett.* 13: 3667–-3669.

12 (a) Moses, J.E. and Moorhouse, A.D. (2007). *Chem. Soc. Rev.* 36: 1249–1262. (b) Gil, M.V., Arevalo, M.J., and Lopez, O. (2007). *Synthesis*: 1589–1620.

13 (a) Agard, N.J., Prescher, J.A., and Bertozzi, C.R. (2004). *J. Am. Chem. Soc.* 126: 15046–15047. (b) Jewett, J.C. and Bertozzi, C.R. (2010). *Chem. Soc. Rev.* 39: 1272–1279. (c) Müller, M., Maichle-Mossmer, C., and Bettinger, H.F. (2014). *J. Org. Chem.* 79: 5478–5483. (d) Gold, B., Batsomboon, P., Dudley, G.B., and Alabugin, I.V. (2014). *J. Org. Chem.* 79: 6221–6232.

14 (a) Shi, F., Waldo, J.P., Chen, Y., and Larock, R.C. (2008). *Org. Lett.* 10: 2409–2412. (b) Campbell-Verduyn, L., Elsinga, P.H., Mirfeizi, L. et al. (2008). *Org. Biomol. Chem.* 6: 3461–3463. (c) Chandrasekhar, S., Seenaiah, M., Rao, L., and Reddy, C.R. (2008). *Tetrahedron* 64: 11325–11327.

15 Ankati, H. and Biehl, E. (2009). *Tetrahedron Lett.* 50: 4677–4682.

16 (a) Lin, Y., Chen, Y., Ma, X. et al. (2011). *Tetrahedron* 67: 856–859. (b) Chen, X., Yu, H., Xu, Z. et al. (2015). *J. Org. Chem.* 80: 6890–6896. (c) Kovács, S., Csincsi, Á.I., Nagy, T.Z. et al. (2012). *Org. Lett.* 14: 2022–2025. (d) Sumida, Y., Kato, T., and Hosoya, T. (2013). *Org. Lett.* 15: 2806–2809. (e) Gann, A.W., Amoroso, J.W., Einck, V.J. et al. (2014). *Org. Lett.* 16: 2003–2005. (f) Chang, D., Zhu, D., and Shi, L. (2015). *J. Org. Chem.* 80: 5928–5933. (g) Yoshida, S., Morita, T., and Hosoya, T. (2016). *Chem. Lett.* 45: 726–728.

17 (a) Subba, R., Basi, V., Praneeth, K., and Yadav, J.S. (2011). *Carbohydr. Res.* 346: 995–998. (b) Singh, G., Kumar, R., Swett, J., and Zajc, B. (2013). *Org. Lett.* 15: 4086–4089. (c) Zhang, F. and Moses, J.E. (2009). *Org. Lett.* 11: 1587–1590.

18 (a) Ikawa, T., Takagi, A., Goto, M. et al. (2013). *J. Org. Chem.* 78: 2965–2983. (b) Ikawa, T., Kaneko, H., Masuda, S. et al. (2015). *Org. Biomol. Chem.* 13: 520–526.

19 (a) Goetz, A.E. and Garg, N.K. (2013). *Nat. Chem.* 5: 54–60. (b) Goetz, A.E. and Garg, N.K. (2014). *J. Org. Chem.* 79: 846–851.

20 (a) Saito, N., Nakamura, K., and Sato, Y. (2014). *Heterocycles* 88: 929–937. (b) Medina, J.M., Jackl, M.K., Susick, R.B., and Garg, N.K. (2016). *Tetrahedron* 72: 3629–3634.

21 Fang, Y. and Larock, R.C. (2012). *Tetrahedron* 68: 2819–2826.

22 (a) Im, G.-Y.J., Bronner, S.M., Goetz, A.E. et al. (2010). *J. Am. Chem. Soc.* 132: 17933–17944. (b) Shah, T.K., Medina, J.M., and Garg, N.K. (2016). *J. Am. Chem. Soc.* 138: 4948–4954.

23 (a) Dubrovskiy, A.V. and Larock, R.C. (2010). *Org. Lett.* 12: 1180–1183. (b) Dubrovskiy, A.V., Jain, P., Shi, F. et al. (2013). *ACS Comb. Sci.* 15: 193–201.

24 (a) Spiteri, C., Sharma, P., Zhang, F. et al. (2010). *Chem. Commun.* 46: 1272–1274. (b) Crossley, J.A. and Browne, D.L. (2010). *Tetrahedron Lett.* 51: 2271–2273.

25 Spiteri, C., Mason, C., Zhang, F. et al. (2010). *Org. Biomol. Chem.* 8: 2537–2542.

26 Spiteri, C., Keeling, S., and Moses, J.E. (2010). *Org. Lett.* 12: 3368–3371.

27 Murahashi, S.-I. and Imada, Y. (2019). *Chem. Rev.* 119: 4684–4716.

28 Dai, M., Wang, Z., and Danishefsky, S.J. (2008). *Tetrahedron Lett.* 49: 6613–6616.

29 (a) Wu, K., Chen, Y., Lin, Y. et al. (2010). *Tetrahedron* 66: 578–582. (b) Wu, Q.-C., Li, B.-S., Lin, W.-Q. et al. (2007). *Chin. J. Synth. Chem.* 15: 292–295. (c) Lu, C., Dubrovskiy, A.V., and Larock, R.C. (2012). *J. Org. Chem.* 77: 2279–2284.

30 Khangarot, R.K. and Kaliapan, K.P. (2012). *Eur. J. Org. Chem.*: 5844–5854.

31 Ma, X.-P., Li, L.-G., Zhao, H.-P. et al. (2018). *Org. Lett.* 20: 4571–4574.

32 Kivrak, A. and Larock, R.C. (2010). *J. Org. Chem.* 75: 7381–7387.

33 Li, P., Wu, C., Zhao, J. et al. (2013). *Can. J. Chem.* 91: 43–50.

34 Yao, T., Ren, B., Wang, B., and Zhao, Y. (2017). *Org. Lett.* 19: 3135–3138.

35 (a) Nájera, C., Sansano, J.M., and Yus, M. (2015). *Org. Biomol. Chem.* 13: 8596–8636. (b) Coldham, I. and Hufton, R. (2005). *Chem. Rev.* 105: 2765–2809.

36 Shi, F., Mancuso, R., and Larock, R.C. (2009). *Tetrahedron Lett.* 50: 4067–4070.

37 (a) Swain, S.P., Shi, Y.-C., Tsay, S.-C. et al. (2015). *Angew. Chem. Int. Ed.* 54: 9926–9930. (b) Jia, H., Guo, Z., Liu, H. et al. (2018). *Chem. Commun.* 54: 7050–7053.

38 Ryu, H., Seo, J., and Ko, H.M. (2018). *J. Org. Chem.* 83: 14102–14109.

39 (a) Raminelli, C., Liu, Z., and Larock, R.C. (2006). *J. Org. Chem.* 71: 4689–4691. (b) Shaibu, B.S., Kawade, R.K., and Liu, R.-S. (2012). *Org. Biomol. Chem.* 10: 6834–6839.

40 Okuma, K., Hirano, K., Shioga, C. et al. (2013). *Bull. Chem. Soc. Jpn.* 86: 615–619.

41 Dhiman, A.K., Kumar, R., Kumar, R., and Sharma, U. (2017). *J. Org. Chem.* 82: 12307–12317.

42 Ren, H., Luo, Y., Ye, S., and Wu, J. (2011). *Org. Lett.* 13: 2552–2555.

43 Yamashita, Y., Hayashi, T., and Masumura, M. (1980). *Chem. Lett.* 9: 1133–1136.

44 (a) Zhao, J., Wu, C., Li, P. et al. (2011). *J. Org. Chem.* 76: 6837–6843. (b) Zhao, J., Li, P., Wu, C. et al. (2012). *Org. Biomol. Chem.* 10: 1922–1930.

45 Kumar, R., Chaudhary, S., Kumar, R. et al. (2018). *J. Org. Chem.* 83: 11552–11570.

46 Jiang, L., Yu, X., Fang, B., and Wu, J. (2012). *Org. Biomol. Chem.* 10: 8102–8107.

47 (a) Matsumoto, K., Katsura, H., Uchida, T. et al. (1996). *J. Chem. Soc., Perkin Trans. 1*: 2599–2602. (b) Tominaga, Y., Shiroshita, Y., Matsuda, Y., and Hosomi, A. (1987). *Heterocycles* 26: 2073–2075.

48 (a) Xie, C., Zhang, Y., and Xu, P. (2008). *Synlett*: 3115–3120. (b) Huang, X. and Zhang, T. (2009). *Tetrahedron Lett.* 50: 208–211.

49 Shen, Y.-M., Grampp, G., Leesakul, N. et al. (2007). *Eur. J. Org. Chem.*: 3718–3726.

50 (a) Wu, C., Fang, Y., Larock, R.C., and Shi, F. (2010). *Org. Lett.* 12: 2234–2237. (b) Fang, Y., Wu, C., Larock, R.C., and Shi, F. (2011). *J. Org. Chem.* 76: 8840–8851.

51 Soares, M.I.L., Nunes, C.M., Gomes, C.S.B., and Pinho e Melo, T.M.V.D. (2013). *J. Org. Chem.* 78: 628––637.

52 Yen-Pon, E., Champagne, P.A., Plougastel, L. et al. (2019). *J. Am. Chem. Soc.* 141: 1435–1440.

53 (a) Fang, Y., Larock, R.C., and Shi, F. (2014). *Asian J. Org. Chem.* 3: 55–57. (b) Lopchuk, J.M. and Gribble, G.W. (2014). *Tetrahedron Lett.* 55: 2809–2812.

54 Asao, N. and Sato, K. (2006). *Org. Lett.* 8: 5361–5363.

55 Shin, J., Lee, J., Ko, D. et al. (2017). *Org. Lett.* 19: 2901–2904.

56 Katritzky, A.R. and Dennis, N. (1989). *Chem. Rev.* 89: 827–861.

57 Ren, H., Wu, C., Ding, X. et al. (2012). *Org. Biomol. Chem.* 10: 8975–8984.

58 (a) Preshlock, S.M., Ghaffari, B., Maligres, P.E. et al. (2013). *J. Am. Chem. Soc.* 135: 7572–7582. (b) Ishiyama, T., Takagi, J., Hartwig, J.F., and Miyaura, N. (2002). *Angew. Chem. Int. Ed.* 41: 3056–3058.

59 Demory, E., Devaraj, K., Orthaber, A. et al. (2015). *Angew. Chem. Int. Ed.* 54: 11765–11769.

60 Ikawa, T., Masuda, S., Takagi, A., and Akai, S. (2016). *Chem. Sci.* 7: 5206–5211.

61 Ma, X., Chen, Y., Zhang, Y. et al. (2012). *Eur. J. Org. Chem.*: 1388–1393.

62 For selected examples on formal cycloadditions of arynes with substrates including nucleophilic and electrophilic sites, see:(a)Huang, X. and Zhang, T. (2010). *J. Org. Chem.* 75: 506–509. (b) Lu, C., Dubrovskiy, A.V., and Larock, R.C. (2012). *Tetrahedron Lett.* 53: 2202–2205. (c) Zhao, J. and Larock, R.C. (2005). *Org. Lett.* 7: 4273–4275. (d) Zhao, J. and Larock, R.C. (2007). *J. Org. Chem.* 72: 583–588. (e) Okuma, K., Nojima, A., Matsunaga, N., and Shioji, K. (2009). *Org. Lett.* 11: 169–171. (f) Rogness, D.C. and Larock, R.C. (2010). *J. Org. Chem.* 75: 2289–2295. (g) Fang, Y., Rogness, D.C., Larock, R.C., and Shi, F. (2012). *J. Org. Chem.* 77: 6262–6270. (h) Samineni, R., Madapa, J., Srihari, P., and Mehta, G. (2017). *Org. Lett.* 19: 3119–3122. (i) Xu, D., Zhao, Y., Song, D. et al. (2017). *Org. Lett.* 19: 3600–3603. (j) Pandya, V.G. and Mhaske, S.B. (2018). *Org. Lett.* 20: 1483–1486. (k) Hu, W., Zhang, C., Huang, J. et al. (2019). *Org. Lett.* 21: 941–945.

63 (a) Gilmore, C.D., Allan, K.M., and Stoltz, B.M. (2008). *J. Am. Chem. Soc.* 130: 1558–1559. (b) Zhao, M.-N., Ren, Z.-H., Wang, Y.-Y., and Guan, Z.-H. (2012). *Chem. Commun.* 48: 8105–8107.

64 Hong, D., Chen, Z., Li, X., and Wang, Y. (2010). *Org. Lett.* 12: 4608–4611.

65 Singh, R., Nagesh, K., Yugandhar, D., and Prasanthi, A.V.G. (2018). *Org. Lett.* 20: 4848–4853.

66 (a) Markina, N.A., Dubrovskiy, A.V., and Larock, R.C. (2012). *Org. Biomol. Chem.* 10: 2409–2412. (b) Dubrovskiy, A.V. and Larock, R.C. (2011). *Org. Lett.* 13: 4136–4139. (c) Dubrovskiy, A.V. and Larock, R.C. (2012). *J. Org. Chem.* 77: 11232–11256.

67 Li, P., Wu, C., Zhao, J. et al. (2012). *J. Org. Chem.* 77: 3149–3158.

68 Li, Y., Muck-Lichtenfeld, C., and Studer, A. (2016). *Angew. Chem. Int. Ed.* 55: 14435–14438.

69 For a similar proton shift via water-facilitated protonation-deprotonation, see, Xia, Y., Liang, Y. et al. (2007). *J. Am. Chem. Soc.* 129: 3470––3471.

70 (a) Dong, J., Pan, L., Xu, X., and Liu, Q. (2014). *Chem. Commun.* 50: 14797–14800. (b) Fang, Z., Liu, J., Liu, Q., and Bi, X. (2014). *Angew. Chem. Int. Ed.* 53: 7209–7213.

71 Garg, P. and Singh, A. (2018). *Org. Lett.* 20: 1320–1323.
72 (a) Bryce, M.R., Dransfield, T.A., Kandeel, K.A., and Vernon, J.M. (1988). *J. Chem. Soc., Perkin Trans. 1*: 2141–2144. (b) Chen, Y. and Willis, M.C. (2015). *Org. Lett.* 17: 4786–4789.
73 Zhang, J., Page, A.C.S., Palani, V. et al. (2018). *Org. Lett.* 20: 5550–5553.

4

Recent Insertion Reactions of Aryne Intermediates

Suguru Yoshida and Takamitsu Hosoya

Tokyo Medical and Dental University (TMDU), Institute of Biomaterials and Bioengineering (IBB), Laboratory of Chemical Bioscience, 2-3-10 Kanda-Surugadai, Chiyoda-ku, Tokyo 101-0062, Japan

4.1 Introduction

A wide variety of insertion reactions of aryne intermediates have been reported so far [1, 2]. Such reactions have enabled the straightforward synthesis of diverse aromatic compounds that were difficult to achieve by the conventional methods. Since arynes show a remarkably high electrophilic reactivity, due to their significantly low LUMO level, various reactions involving arynes proceed via nucleophilic addition of arynophiles to arynes [3]. The resulting carbanions react intramolecularly to form a new covalent bond at the electrophilic moiety of the attacked arynophiles via substitution reactions, including protonation (Figure 4.1a) or addition reactions (Figure 4.1b). A broad range of transformations have been achieved through either of these stepwise mechanisms. This chapter overviews recently reported diverse insertion reactions of arynes by mainly categorizing the nucleophilic atoms of the arynophiles.

4.2 Amination and Related Transformations

4.2.1 Transformations Involving the Formation of C—N and C—H Bonds

The synthesis of aniline derivatives via arynes had been studied more than 50 years ago. For example, addition reactions of amines **1** to arynes, followed by protonations to afford aniline derivatives **2** (Figure 4.2a). In particular, aniline synthesis from aryl halides **3** by treatment with sodium amide in liquid ammonia is often described in basic organic chemistry textbooks as a representative aryne reaction (Figure 4.2b) [4].

Numerous studies on the aniline synthesis by aryne insertion reactions have been performed from the viewpoint of substituted aniline synthesis. For example, in 2003, Larock and coworkers reported the amination of arynes, which were generated from *o*-silylaryl triflates (Kobayashi precursors) [5] by treatment with cesium fluoride,

Modern Aryne Chemistry, First Edition. Edited by Akkattu T. Biju.

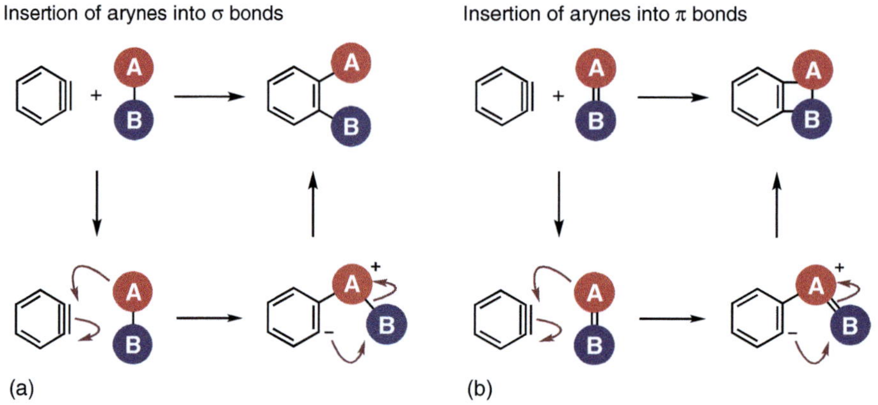

Figure 4.1 Typical reactions of benzyne.

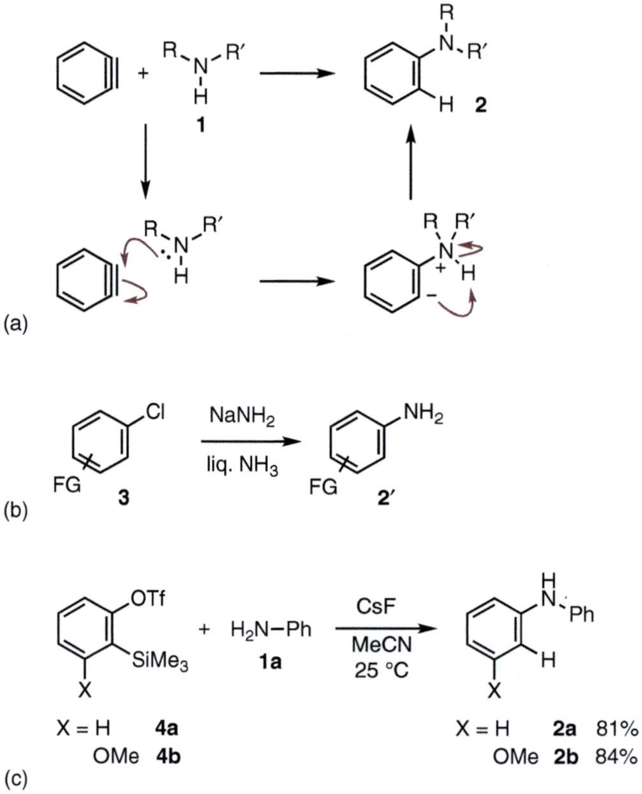

Figure 4.2 Addition reactions of amines to arynes.

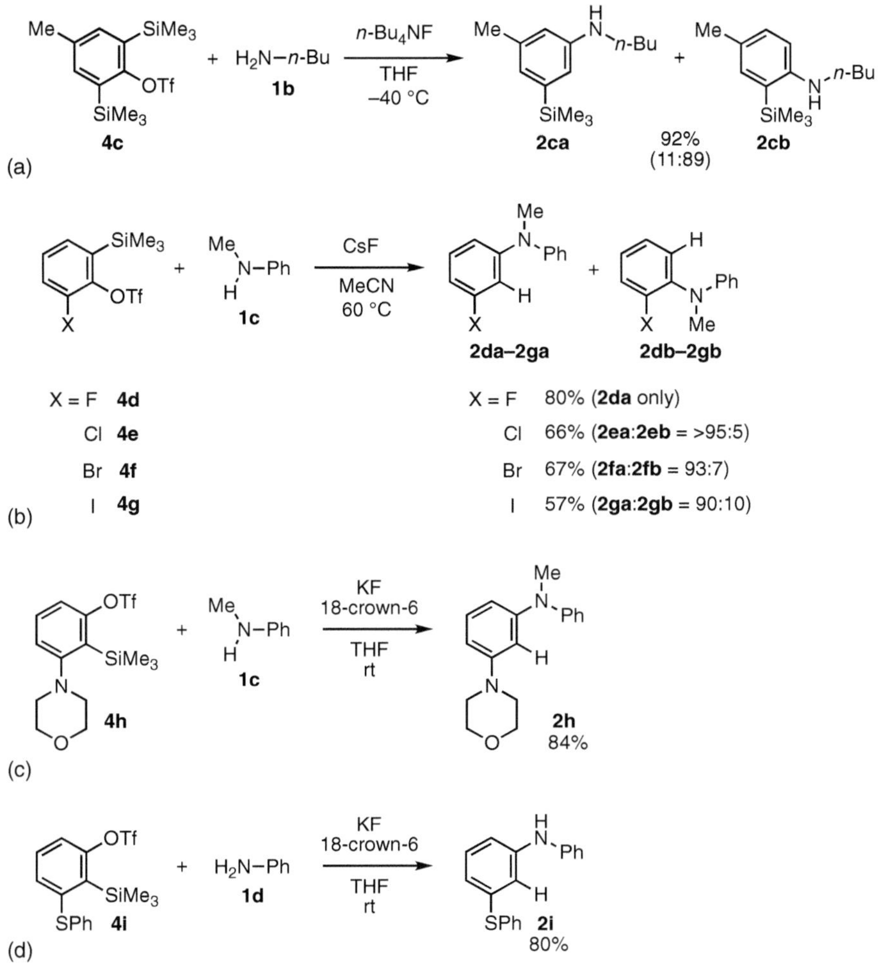

Figure 4.3 Regioselective amination reactions of arynes.

using various amines (Figure 4.2c) [6]. In particular, *meta*-methoxyaniline derivatives were obtained by the amination of 3-methoxybenzyne generated from **4b** due to the inductive and steric effects of the methoxy group. A wide range of amines, including anilines, bulky *tert*-butyl amine, amines bearing functional groups such as terminal alkyne moiety and unprotected hydroxy group, and sulfonamides, participated in this *N*-arylation reaction with arynes.

Several regioselective amination reactions of arynes have been reported, in which the selectivity depends on the substituent on the arynes (Figure 4.3). In 2011, Akai and coworkers reported the *ortho*-selective amination of 3-silylarynes based on the electron-donating effect of the silyl group (Figure 4.3a) [7]. Garg and coworkers showed *meta*-selective amination of 3-haloarynes, in which the *meta*-selectivity decreased with decreasing electronegativity of the halogen atoms

(Figure 4.3b) [8]. We also reported the *meta*-amination of 3-morpholinobenzyne [9a] and 3-(phenylthio)benzyne [9b], which were generated from the corresponding precursors **4h** and **4i**, prepared via 3-(triflyloxy)benzyne [10], as described later (Figures 4.3c,d). Several examples for regioselective amination of pyridynes and various ring-fused arynes, including hetarynes, have also been reported [11]. The regioselectivities in these aryne amination reactions were determined by the effect of the substituent on the benzyne ring, particularly by the characteristics of the substituted atom.

Various nitrogen nucleophiles have been used in aryne amination reactions; for instance, Greaney and coworkers reported in 2011 that hydrazone **5** smoothly reacted with arynes to furnish *N*-arylhydrazone **6** through the insertion of arynes into the N—H bond (Figure 4.4a) [12]. Subsequent rearrangement by treating *N*-arylhydrazone **6** with boron trifluoride under Fischer indole synthesis conditions provided indole **7** in high yield. Addition reaction to arynes by hydrazine derivative **I** generated in situ by the reaction between diethyl azodicarboxylate and triphenylphosphine was reported by Cheng et al. in 2017 (Figure 4.4b) [13]. *N*-Arylation of amidine **10** with arynes also proceeded efficiently in the presence of sodium bicarbonate (Figure 4.4c) [14].

Several reactions of arynes with tertiary amines have been reported so far; for example, in 2013, Biju and coworkers found that treatment of a mixture of *o*-silylaryl triflate **4a** and *N*,*N*-dimethylaniline (**12**) with potassium fluoride and 18-crown-6 in the presence of ammonium bicarbonate at 60 °C yielded aniline derivative **13** through demethylation in excellent yield (Figure 4.4d, upper scheme) [15]. Diesendruck and coworkers reported the efficient synthesis of ammonium salt **14** in 2016 by the reaction between benzyne and *N*,*N*-dimethylaniline (**12**) at room temperature (Figure 4.4d, lower scheme) [16]. In 2017, we also reported the reaction of *N*,*N*-dimethylaniline (**12**) with aryne **III** generated by the hexadehydro-Diels–Alder (HDDA) reaction of aryne **II** having a 1,3-diyne moiety, which was generated from *o*-iodoaryl triflate **15a** by treating it with a silylmethyl Grignard reagent as an activator to promote the iodine–magnesium exchange. In this case, carboamination product **16** was obtained through intramolecular arylation of **IV** in high yield instead of ammonium salt **17** (Figure 4.4e) [17].

Various types of organonitrogen compounds have been prepared easily by *N*-arylation reactions involving arynes. The *N*-arylation of amino sugars with arynes was demonstrated by Nilsson and coworkers in 2018, which was achieved using sugars without protection of hydroxy groups (Figure 4.5a) [18]. Jin and coworkers reported the *N*-arylation of *N*-alkoxyamides with arynes in 2016 (Figure 4.5b) [19]. In the case of the reaction of 2-(trifluoroacetylamino)pyridine (**22**) with arynes, *N*-arylation of the pyridine nitrogen took place (Figure 4.5c) [20]. In two independent studies, Singh and coworkers [21] and our group [22] reported the *N*-arylation of *N*-H sulfoximine **24a** with arynes (Figure 4.5d, upper scheme). We also found that aminosulfinylation of arynes involving migratory *N*-arylation proceeded when using diaryl sulfoximine **24b** (Figure 4.5d, lower scheme) similar to the reaction with *N*-H sulfilimines as described later [22].

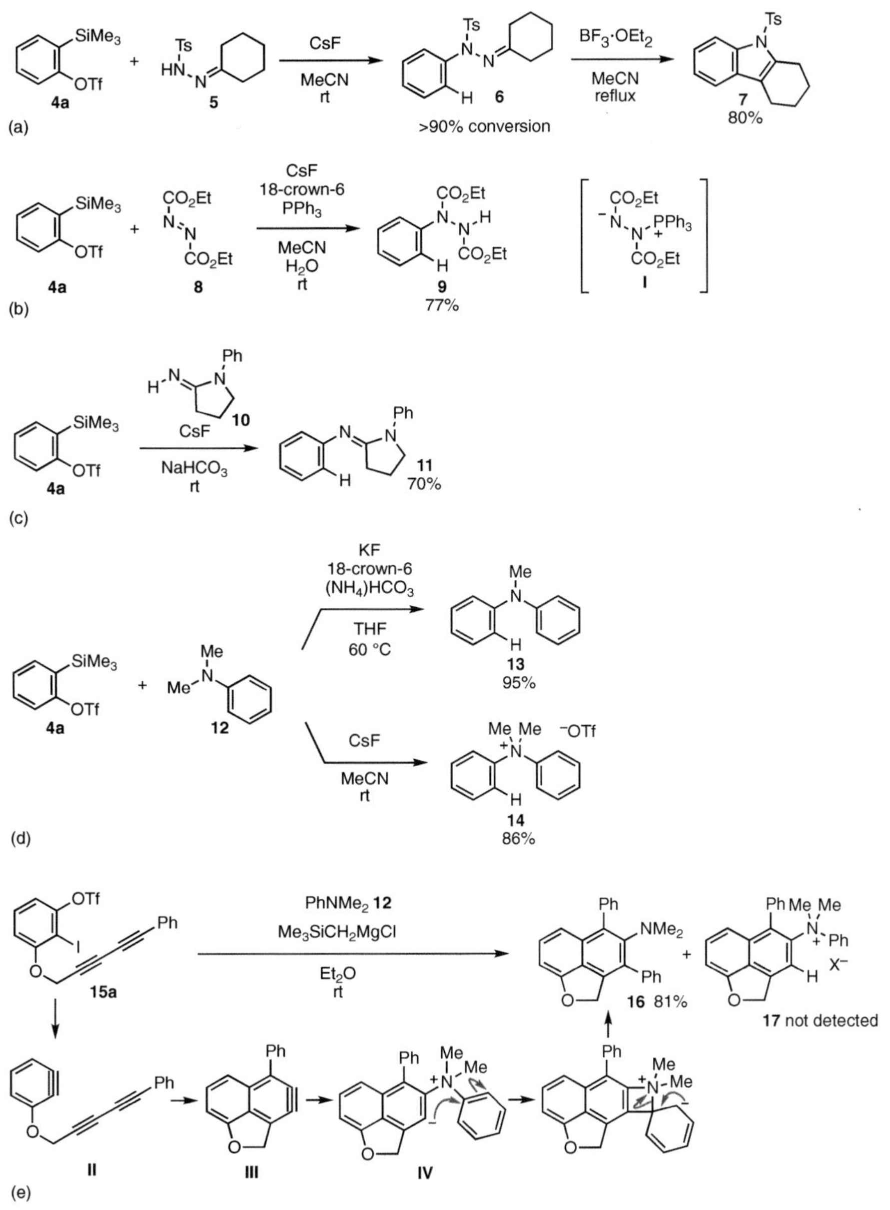

Figure 4.4 Aryne amination reactions by various nitrogen nucleophiles.

4.2.2 Transformations Involving the Formation of C—N and C—Mg Bonds

The aminometalation of arynes was reported by Knochel and coworkers, providing an efficient preparation method for various arylmagnesium reagents [23]. Based on this approach, Greaney and coworkers developed an efficient method to synthesize

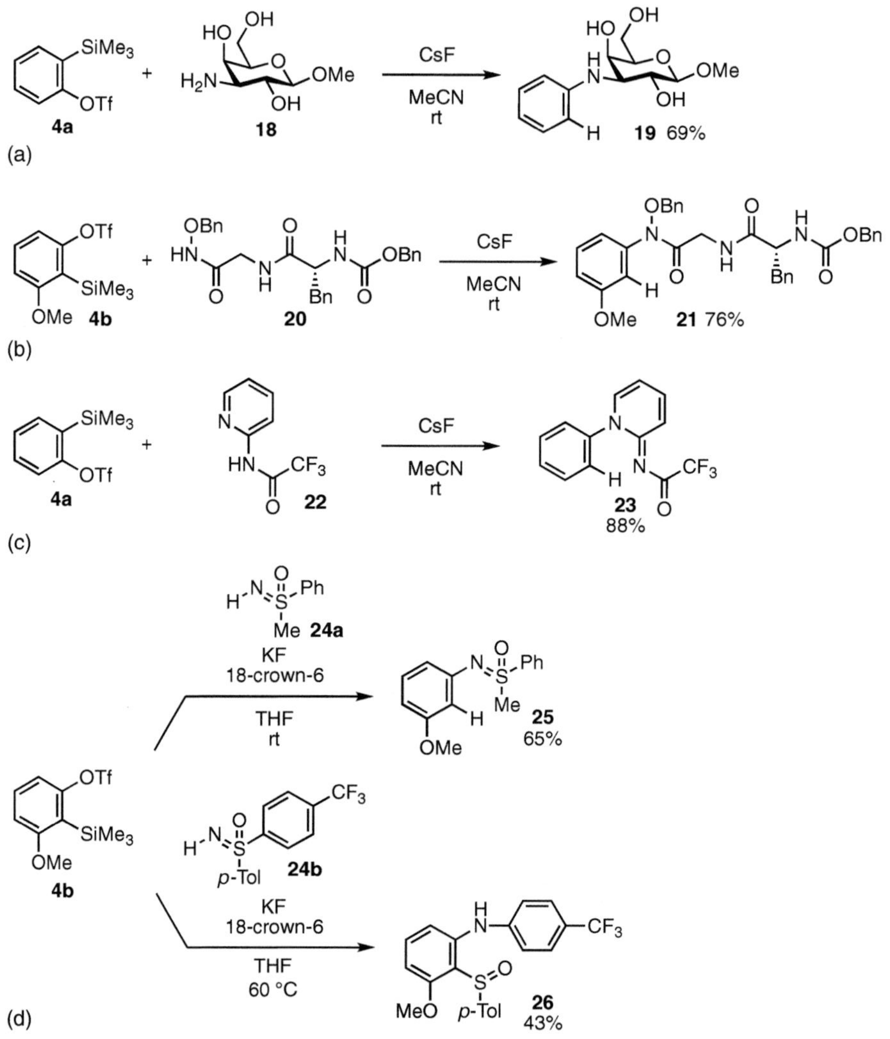

Figure 4.5 Various *N*-arylation reactions involving arynes.

aniline derivative **29** via aminomagnesiation of benzyne generated from *o*-iodoaryl sulfonate **15b** with isopropylmagnesium chloride and subsequent copper-catalyzed amination using *N*-benzoyloxymorpholine (**28**) (Scheme 4.1) [24].

4.2.3 Transformations Involving the Formation of C—N and C—C Bonds

A wide range of carboamination reactions of arynes have been reported using various amides or related organonitrogen compounds. In 2002, Hiyama and coworkers reported the synthesis of benzodiazepine **31** by the reaction of arynes with urea **30**, which proceeded through *N*-arylation followed by ring expansion of the resulting carbanion intermediate **V** (Figure 4.6a) [25]. Larock and coworkers

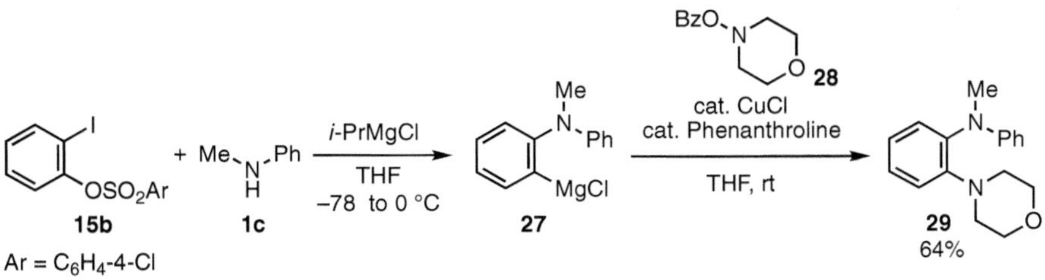

Scheme 4.1 Transformation via aminomagnesiation of benzyne [24].

reported (in 2005) that *N*-trifluoroacetylanilide **32** reacted smoothly with arynes to provide the carboaminated products through the C—N bond insertion to **VI** (Figure 4.6b) [26]. Carboamination of *N*-benzoylanilide (**34**) with arynes was reported by Greaney and coworkers in 2010, enabling the preparation of acridone **37** by further intramolecular *N*-arylation (Figure 4.6c) [27]. In 2007,

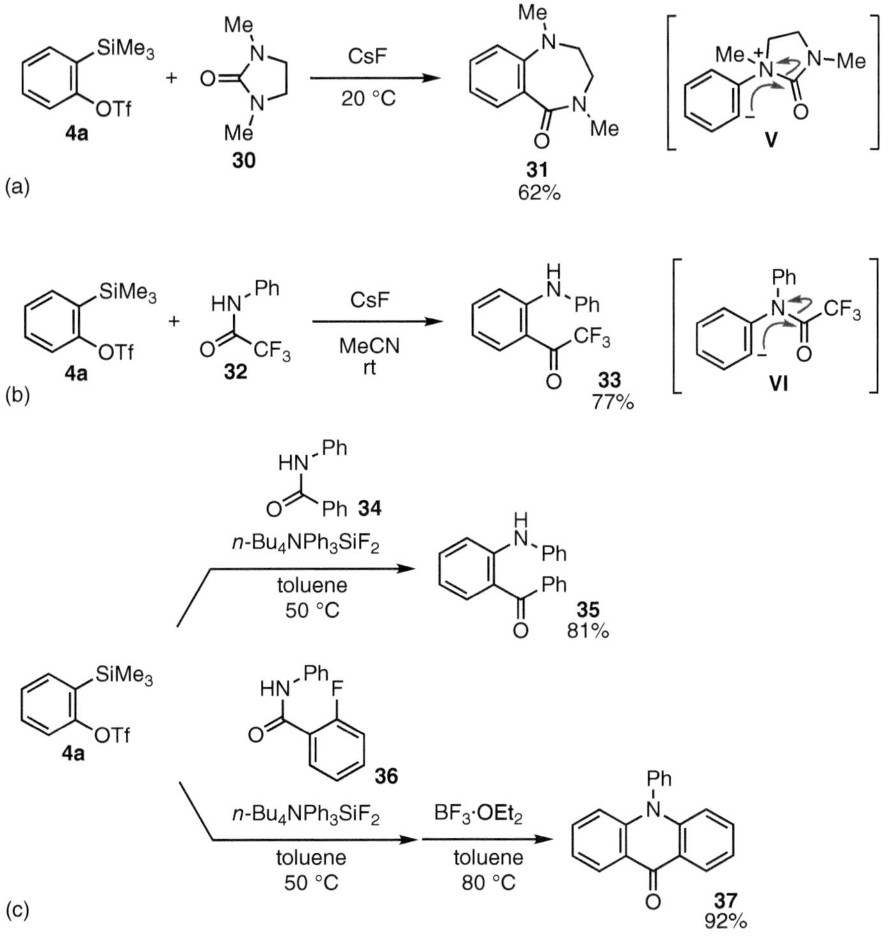

Figure 4.6 Carboamination reactions of benzyne with amides.

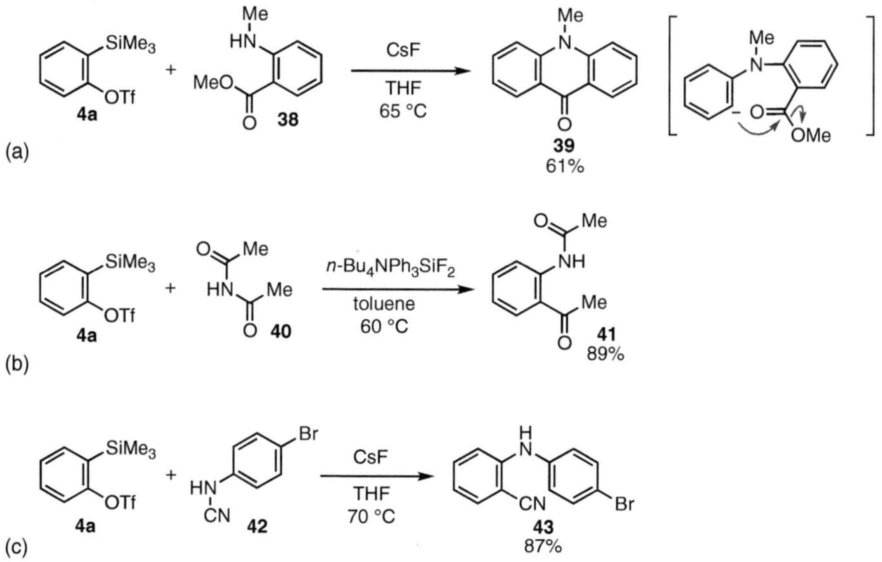

Figure 4.7 Other carboamination and aminocyanation reactions of benzyne.

Larock and coworkers reported acridone synthesis by the reaction between arynes and *o*-(methoxycarbonyl)aniline derivative **38** (Figure 4.7a) [28] based on the xanthone synthesis, as described later. The insertion reaction of arynes to the C—N bond of imide **40** was also reported by Stoltz and coworkers in 2016, providing carboaminated product **41** in high yield (Figure 4.7b) [29]. The aminocyanation of arynes with *N*-cyanoanilide **42** was also reported by Zeng and coworkers in 2014 (Figure 4.7c) [30]. In 2008, Stoltz and coworkers reported that the carboamination of arynes with α-(*tert*-butoxycarbonylamino)acrylic acid methyl ester **44** led to indoline derivative **45** (Figure 4.8a), while isoquinolines were obtained with two C—C bond formations when α-(acylamino)acrylic acid esters were used [31] as described in Section 4.3.3. Moreover, the reaction of arynes with *N*,*N*-dimethylhydrazone **46** was reported to afford carboaminated products such as **47** through the addition of an amino group to aryne, followed by intramolecular cyclization and cleavage of the N—N bond (Figure 4.8b) [32]. In 2010, Wang and coworkers achieved the indole synthesis via the reaction of arynes with azaylide **VII** prepared from alkenyl azide **48** and triphenylphosphine (Figure 4.8c) [33].

4.2.4 Transformations Involving the Formation of C—N and C—S, C—P, C—Cl, or C—Si Bonds

Several methods for the difunctionalization of arynes involving the formation of C—N and C—S bonds have been reported. Reaction of *N*-(trifluoromethylsulfinyl) anilide **50** with arynes efficiently afforded aminosulfinylated product **51** (Figure 4.9a) [26], as is the case using *N*-(trifluoroacetyl)anilides (Figure 4.6b).

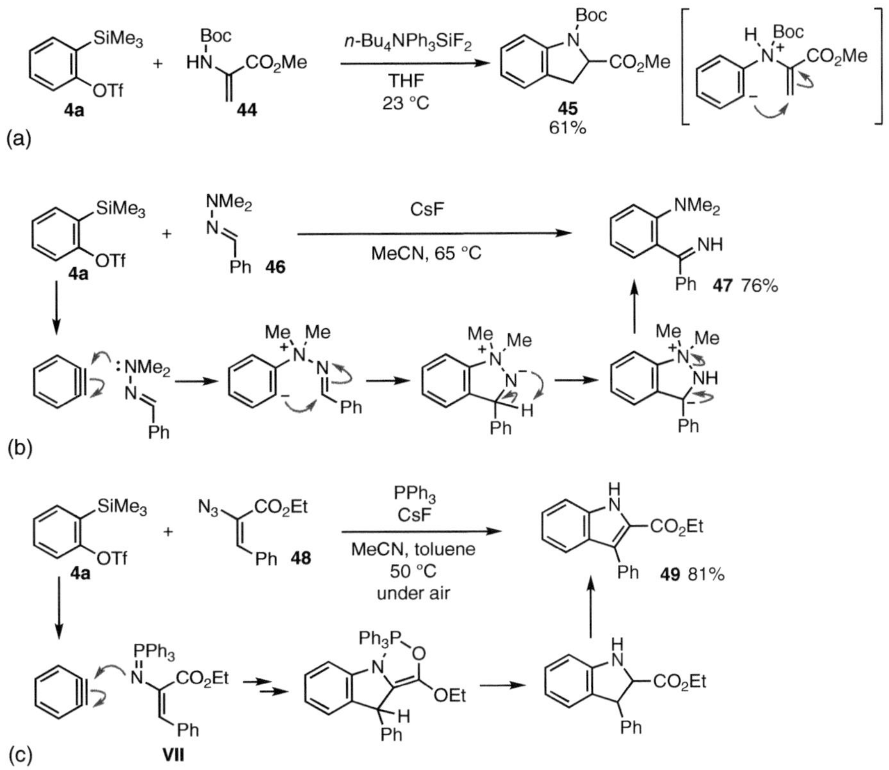

Figure 4.8 Other carboamination reactions of benzyne.

Hoye and coworkers reported the sulfonylamination of aryne **VIII** generated by the HDDA reaction (Figure 4.9b) [34]. We developed a method for *o*-thioaniline synthesis by the reaction of arynes with sulfilimines affording the products such as **55** via aminosulfanylation and migratory *N*-arylation (Figure 4.9c) [35]. Recently, Biju and coworkers reported the reaction of arynes with *N*-sulfanylamines, including **56**, affording aminosulfanylated products (Figure 4.9d) [36].

Difunctionalization of arynes involving C—N and C—P bond formations was reported by Zhang and coworkers in 2013 (Figure 4.10a) [37]. The reaction between arynes and azaylide **60** provided phosphonium salt **61** (Figure 4.10b) [38]. When *N*-chloroamine **62** was used as the arynophile, chloroamination of arynes proceeded to afford *o*-chloroaniline derivative **63** (Figure 4.10c) [39].

Silylamination of arynes produces *o*-aminoarylsilanes. The pioneering work, in which these results were presented, was reported by Hiroto Yoshida and coworkers in 2005 [40]. They used *N*-silylamines such as **64a** containing a bulky silyl group to prevent the decomposition by the fluoride ion used as the activator for generating arynes from *o*-silylaryl triflates (Figure 4.11a). We recently found that treatment of *o*-iodoaryl triflate **15c** with (trimethylsilyl)methylmagnesium chloride in the presence of *N*-silylamine **64b** efficiently afforded 3-amino-2-silylaryl triflate **4h** via silylamination of 3-(triflyloxy)aryne intermediate, owing to the mild and soft

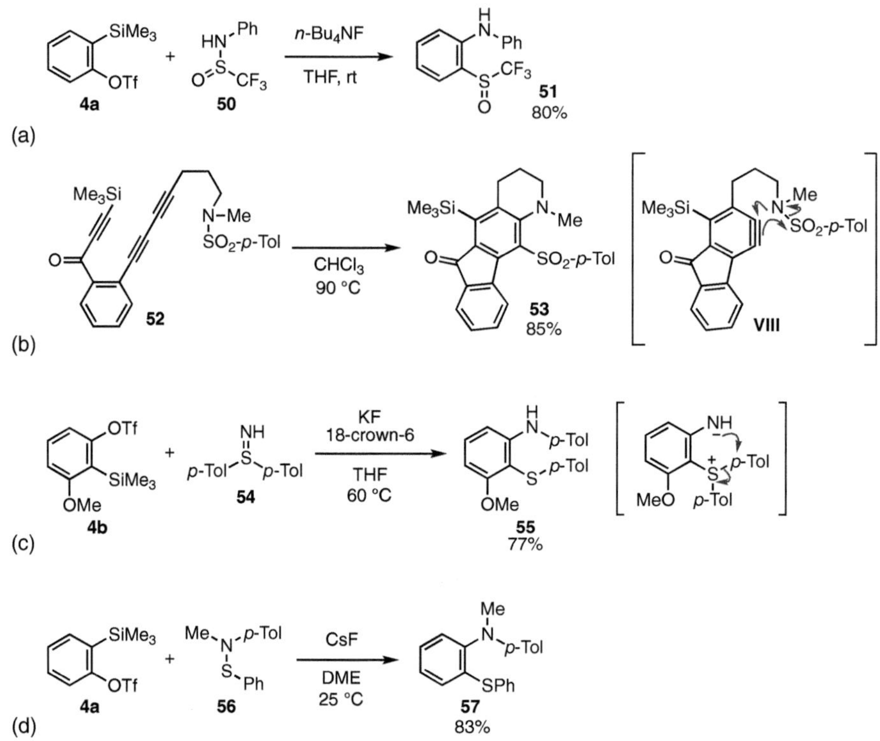

Figure 4.9 Thioamination reactions of arynes.

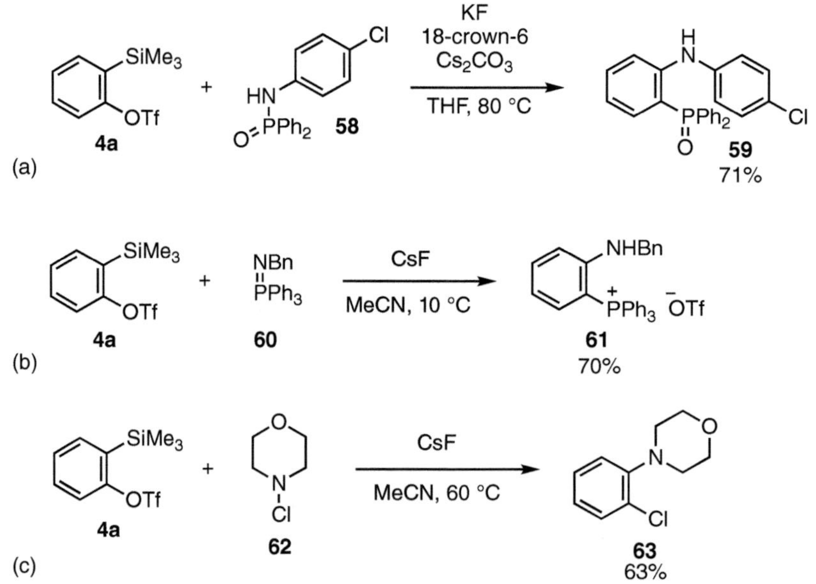

Figure 4.10 Phosphinyl-/phosphinoamination and chloroamination reactions of benzyne.

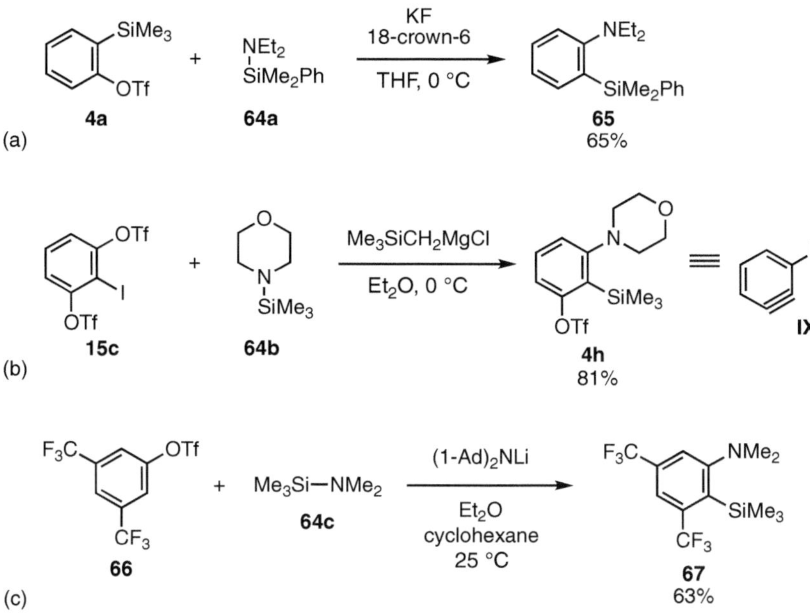

Figure 4.11 Silylamination reactions of arynes.

reactivity of the silylmethyl Grignard reagent (Figure 4.11b) [9a]. The subsequent 3-aminoaryne generation, triggered by fluoride ion in the presence of various arynophiles, allowed for the synthesis of diverse aniline derivatives. Thus, this "aryne relay" approach enabled the facile synthesis of anilines in a modular synthetic manner. In 2018, Daugulis and coworkers also reported the silylamination of arynes generated from aryl triflates or aryl halides by treatment with lithium di(1-adamantyl)amide (Figure 4.11c) [41].

4.3 Transformations Involving Bond Formation with Nucleophilic Carbons

4.3.1 Transformations Involving Carbometalation

The carbometallation of arynes is a fascinating approach for biaryl synthesis under transition metal-free conditions. In 1985, Hart et al. reported the elegant synthesis of 1,3-diarylbenzenes **69** from 1,3-dihalo-2-iodobenzenes **68** via sequential selective carbomagnesiations to consecutively generated arynes, which enabled the facile preparation of various *m*-terphenyls with functional groups (Figure 4.12a) [42]. This type of transition-metal-free approach to biaryl synthesis has been intensively studied by Leroux and coworkers. For example, they reported carbolithiation of arynes generated from 1,2-dibromobenzene (**70**) with aryllithium **71** afforded biaryl **72** via bromination of aryllithium intermediate **X** (Figure 4.12b) [43]. In 2015, the synthesis of chiral biaryls by atropo-diastereoselective aryne coupling was reported by Panossian and coworkers based on chiral sulfoxide chemistry (Figure 4.12c) [44].

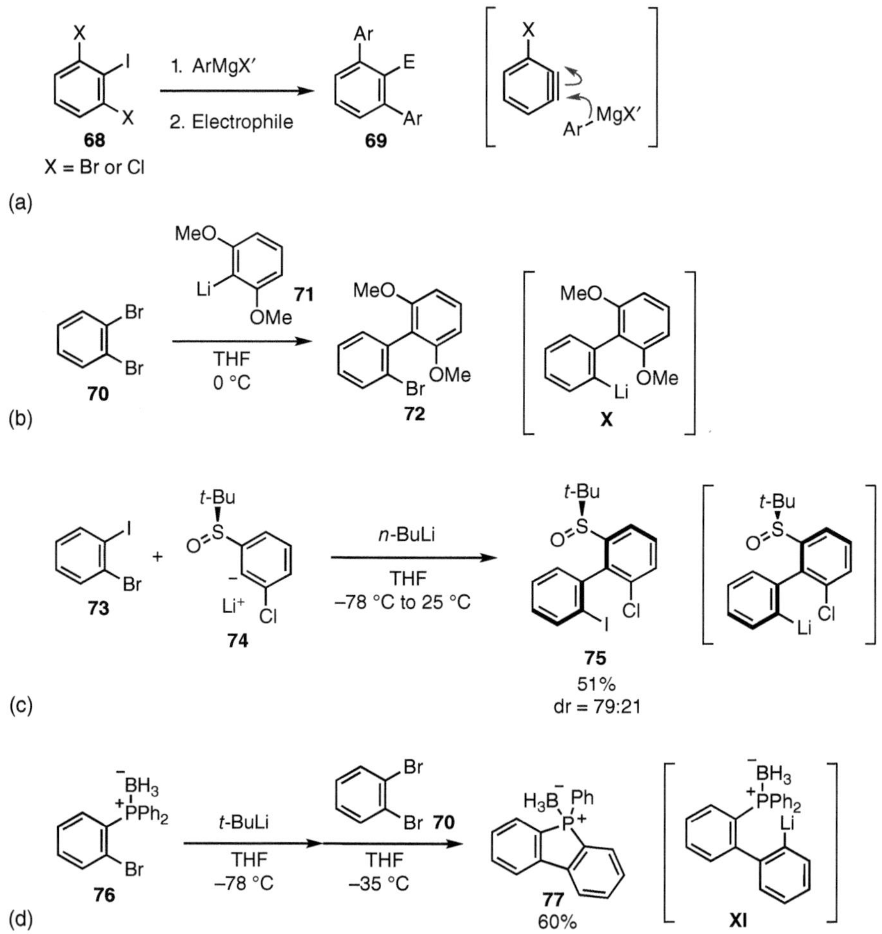

Figure 4.12 Transformations via carbolithiation of arynes.

In 2012, Jugé and coworkers reported the stereoselective synthesis of dibenzophosphole derivative **77** via aryllithium intermediate **XI** prepared in situ by carbolithiation of benzyne generated from 1,2-dibromobenzene (**70**) (Figure 4.12d) [45].

4.3.2 Benzocyclobutene Synthesis by [2+2] Cycloaddition

The synthesis of benzocyclobutenes by the [2+2] cycloaddition of arynes with electron-rich olefins, including vinyl ethers and ketene acetal derivatives, has been shown as one of the most useful aryne transformations. Although the thermal [2+2] cycloaddition is forbidden in terms of the Woodward–Hoffmann rules, nucleophilic C—C bond formation between arynes and alkenes, followed by cyclization, affords the formal [2+2] cycloadducts [46, 47]. For example, in 1965, the nucleophilic attack of alkenes **79** or **81** to benzyne followed by intramolecular *C*-arylation was reported to afford [2+2] cycloadduct **80** and acylalkylated product

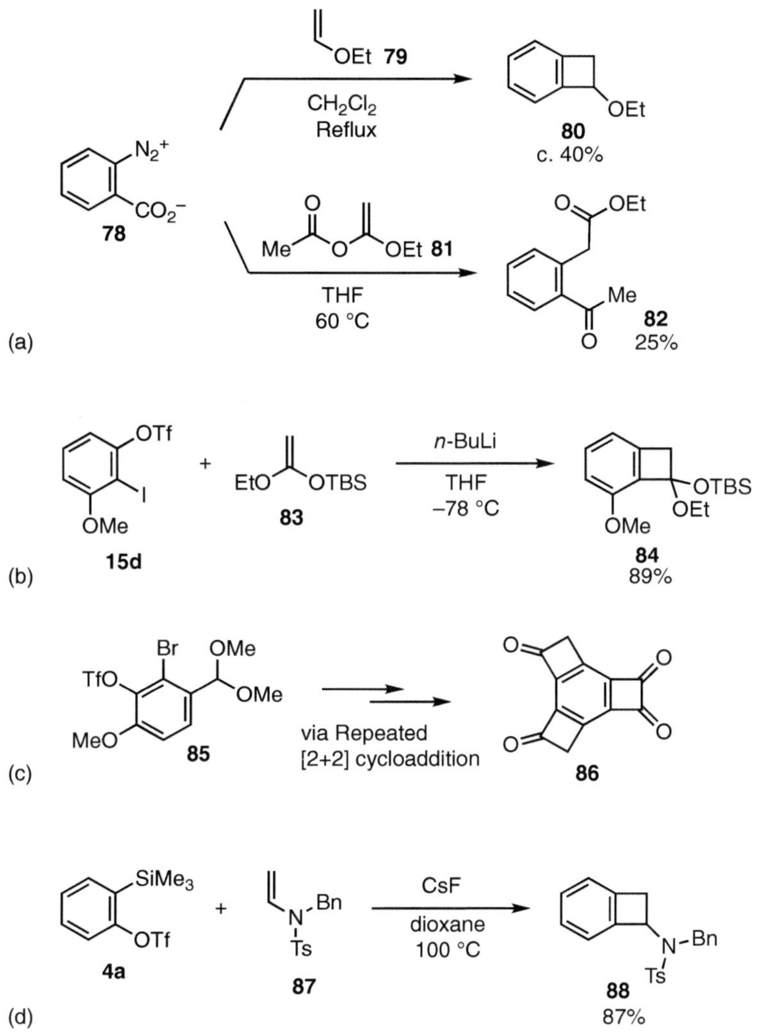

Figure 4.13 Synthesis of benzocyclobutenes via arynes.

82, respectively (Figure 4.13a) [47a]. Recent remarkable advances in synthetic aryne chemistry have enabled diverse transformations, including [2+2] cycloadditions and difunctionalizations. In particular, Suzuki and coworkers extensively studied synthetic chemistry applying the [2+2] cycloaddition of arynes. In 1995, they reported an efficient benzocyclobutene synthesis by the [2+2] cycloaddition of arynes, generated from *o*-iodoaryl triflates by treatment of butyllithium at −78 °C, with ketene silyl acetals (Figure 4.13b) [48]. The rapid iodine–lithium exchange reaction enabled aryne generation at a quite low temperature. Based on this [2+2] cycloaddition reaction, elegant syntheses of various tricylobutabenzene derivatives such as **86** were accomplished (Figure 4.13c) [49]. In addition, [2+2] cycloaddition of arynes with *N*-benzyl-*N*-tosylenamide **87** was reported by Hsung and coworkers in 2009 (Figure 4.13d) [50].

4.3.3 Acylalkylations and Related Transformations

The acylalkylation of arynes using 1,3-dicarbonyl compounds has attracted the attention of many synthetic organic chemists because the products are useful synthetic intermediates. In 2005, Stoltz and coworkers [51] and Hiroto Yoshida et al. [52] independently reported the acylalkylation of arynes with 1,3-dicarbonyl compounds (Figures 4.14a,b). These reactions enabled the facile preparation of a range of benzene-fused macrocyclic compounds **93** by ring expansion (Figure 4.14c) [52a] and multisubstituted benzenes such as **90b** (Figure 4.14d) [51c].

After these pioneering works, various reactions between arynes generated from *o*-silylaryl triflates and active methylene or methyne compounds have been reported. These include acylalkylation reactions of arynes with α-cyanoketones (Figure 4.15a) [52c], 9-acyl-9*H*-fluorenes (Figure 4.15b, lower scheme) [52d], and trifluoromethyl benzyl ketones (Figure 4.15c, upper scheme) [52f] that efficiently afforded the corresponding ketones, while acylclkylation did not undergo using ethyl α,α-diphenylacetate (Figure 4.15b, upper scheme). When trifluoromethyl phenethyl ketone (**102**) was used, benzocyclobutenol **103** was obtained (Figure 4.15c, lower scheme) [52f]. Cyclic ketones, such as β-tetralone (**104**) (Figure 4.16a) [53] and

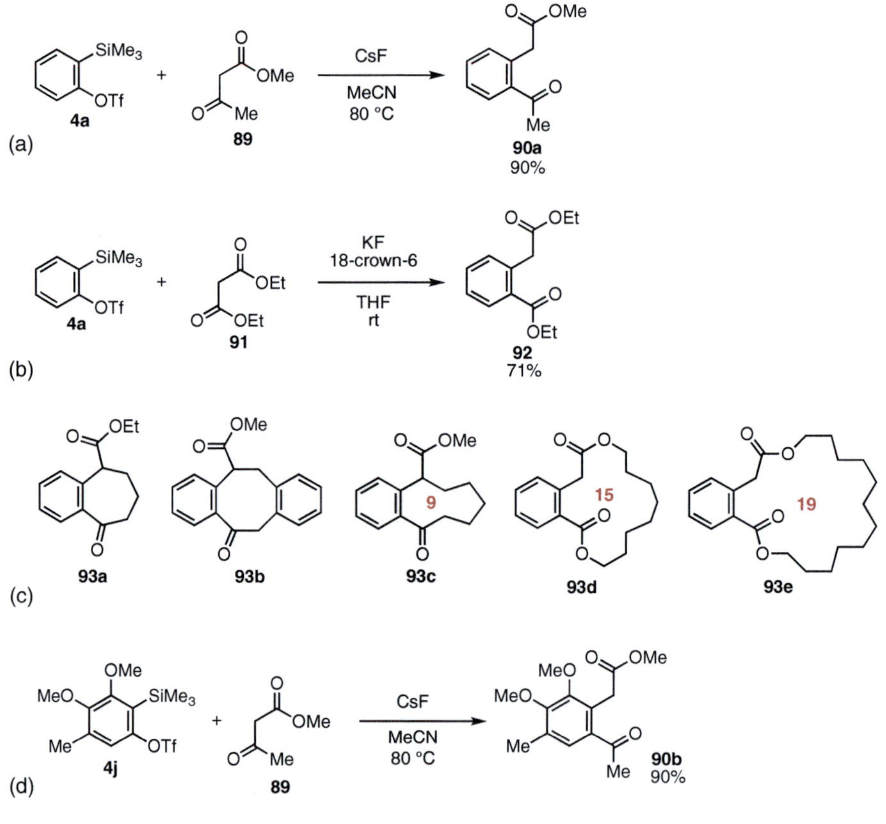

Figure 4.14 Acylalkylation reactions of arynes.

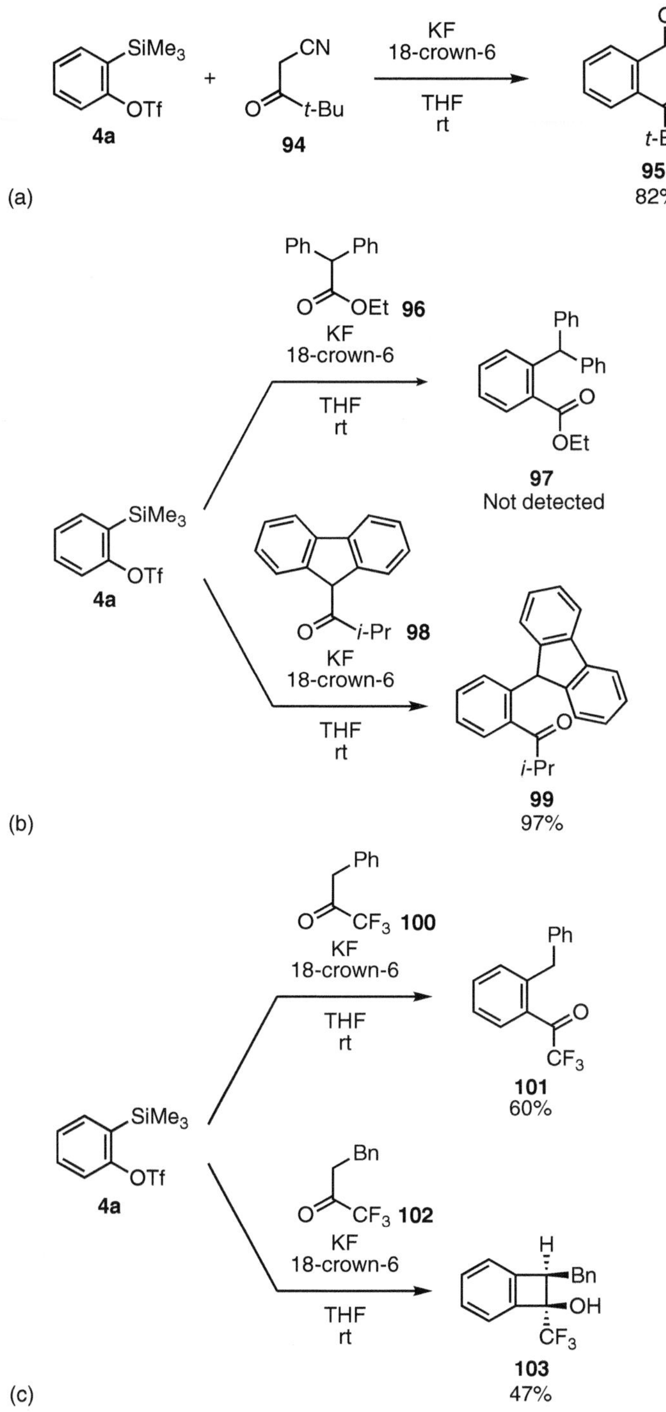

Figure 4.15 Various acylalkylation reactions of benzyne.

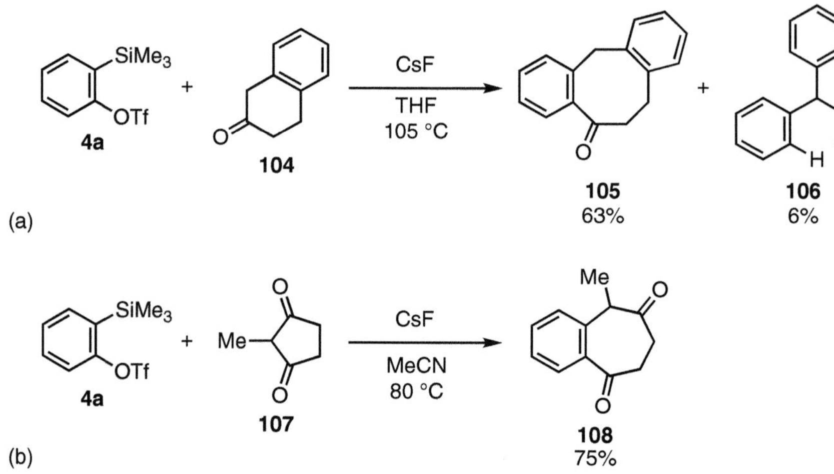

Figure 4.16 Acylalkylation reactions of benzyne with cyclic ketones.

1,3-pentanedione **107** (Figure 4.16b) [54], also participated in the reaction to afford eight- and seven-membered compounds, respectively. The reaction of benzyl phenyl ketone (**109**) (Figure 4.17a) [55] and imidoester **112** (Figure 4.17b, top scheme) [56] with arynes afforded acyl- and imidoylalkylated products **110** and **113**, respectively. In the reaction between arynes and amidine **114**, imidoylalkylated product **115** was obtained as a major product by using tetrabutylammonium fluoride as an activator (Figure 4.17b, middle scheme), while *N*-arylation with arynes proceeded to afford diarylated product **116** as a major product when cesium fluoride was used as an activator (Figure 4.17b, bottom scheme) [56].

Acylalkylation of arynes has been useful in constructing fused skeletons; for example, Okuma et al. reported the synthesis of isocoumarin **118** (Figure 4.18a) [57a] and 1-naphthol **119** (Figure 4.18b) [57b] via the acylalkylation of arynes using diketones **117a** and **117b**, respectively, followed by cyclization. When 2-benzyl-3-hydroxy-1,4-naphthoquinone (**120**) was used as an arynophile, tetracyclic compound **121** was obtained diastereoselectively (Figure 4.18c) [58]. In 2016, Suzuki and coworkers reported an efficient method for the synthesis of 3-alkoxy-2-naphthol **124** via the reaction of arynes with 1,3,5-tricarbonyl equivalent **122** (Figure 4.19a) [59]. The reaction of arynes with tricarbonyl compound **125** afforded indane derivative **126** via acylalkylation and subsequent intramolecular cyclization (Figure 4.19b) [60].

Further arylation of the acylalkylated products by arynes at the active methylene group proceeds depending on the substrates. For example, diarylmethyl sulfone **128a** and diarylacetonitrile **128b** were obtained through a second arylation of the acylalkylated products formed from the reaction between arynes and sulfonylacetonitrile **127a** and malononitrile (**127b**), respectively (Figure 4.20a) [61]. The reaction of α-sulfonyl-substituted ethyl acetate with aryne afforded **130** without further arylation (Figure 4.20b, upper scheme) [62]. Acylalkylation followed by ring formation with fumaric acid diester **131** yielded naphthol **132** (Figure 4.20b, lower scheme).

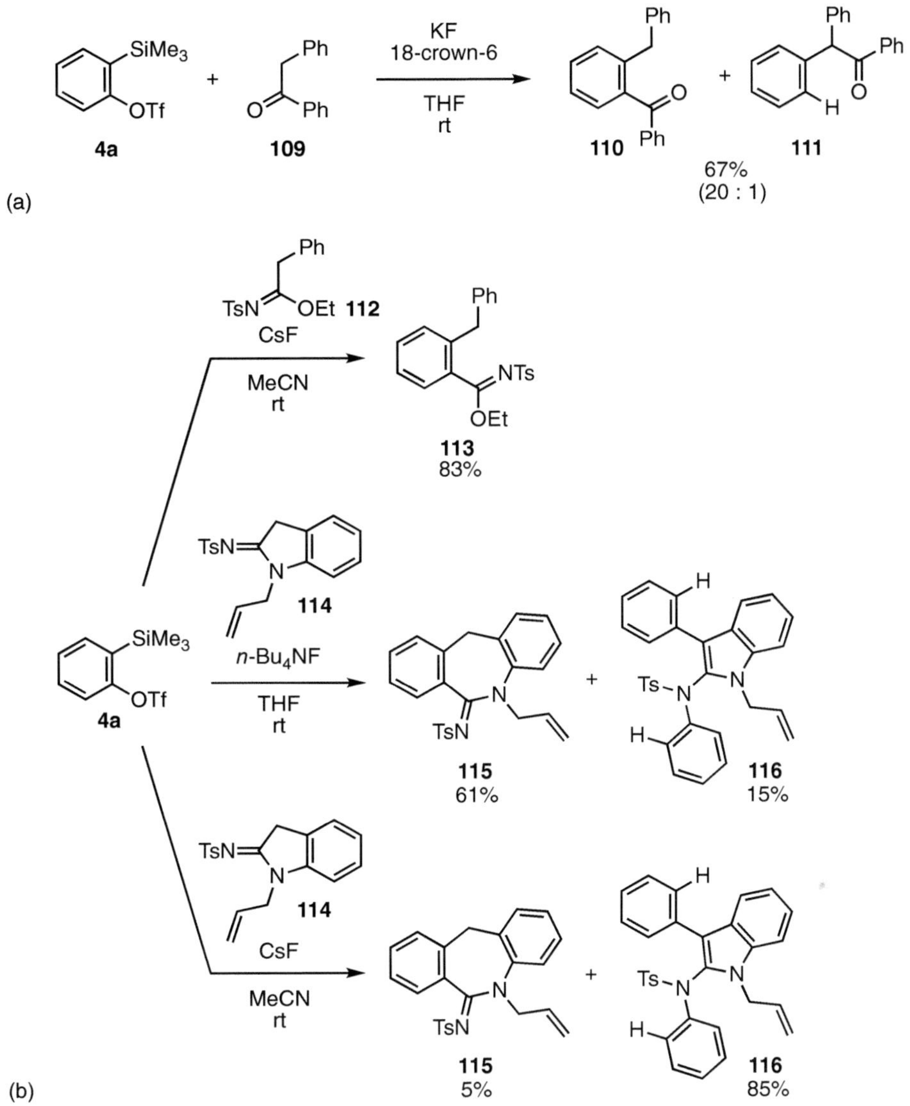

Figure 4.17 Reactions of benzyl phenyl ketone, imidoester, and amidine with benzyne.

In contrast to the reaction of arynes with α-sulfonyl esters, in which no C—S bonds were formed, C—P bond formation proceeded in the reaction of arynes with α-phosphinylacetonitrile **133** (Figure 4.21a) [52b]. On the other hand, the reaction of arynes with β-ketophosphonic acid diester **135** resulted in a selective acylalkylation (Figure 4.21b) [63]. When sulfonium salt **137** was used as an arynophile, methylthioalkylated product **138** was obtained via nucleophilic addition and subsequent intramolecular C—S bond formation, where acylalkylation did not proceed (Figure 4.21c) [64].

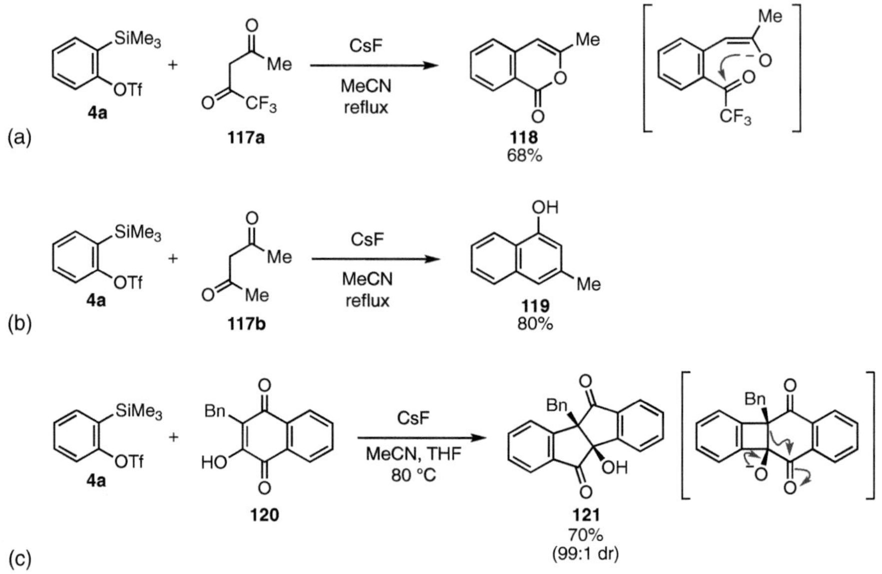

Figure 4.18 Synthesis of cyclic compounds by acylalkylation reactions of benzyne.

Isoquinoline synthesis via the reaction between arynes and α-(acylamino)acrylic acid ester **139** involving nucleophilic C—C bond formation, followed by intramolecular acylation and dehydration, was reported by Stoltz and coworkers in 2008 (Figure 4.22a) [31]. As described in Section 4.2.3, the reaction pathway depended on the substituent at the nitrogen. Formal formylalkylation of arynes was achieved using enamine **141** as an arynophile (Figure 4.22b) [65]. In addition, alkynyl ether **143** reacted with arynes to afford *o*-ethoxyphenylacetylene (**144**) via C–O insertion (Scheme 4.2) [66].

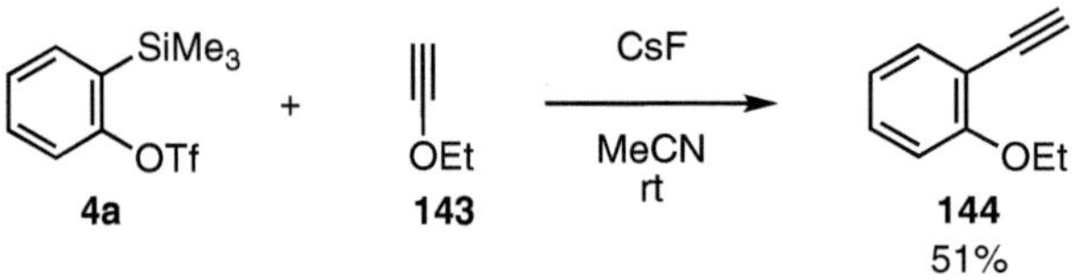

Scheme 4.2 Reaction of benzyne with an alkynyl ether [66].

4.3.4 Transformations Involving C—C and C—H Bond Formations

Aryne transformations involving C—C and C—H bond formations have also been reported. The carbene-catalyzed hydroformylation of arynes with aldehydes was reported by Biju and Glorius in 2010 (Figure 4.23a) [67]. *C*-arylations of arynes with enamide **147** [68] or amides **149a** and **149b** [69] were also reported (Figures 4.23b,c), while a diarylation took place to give **151** when the latter amide was employed

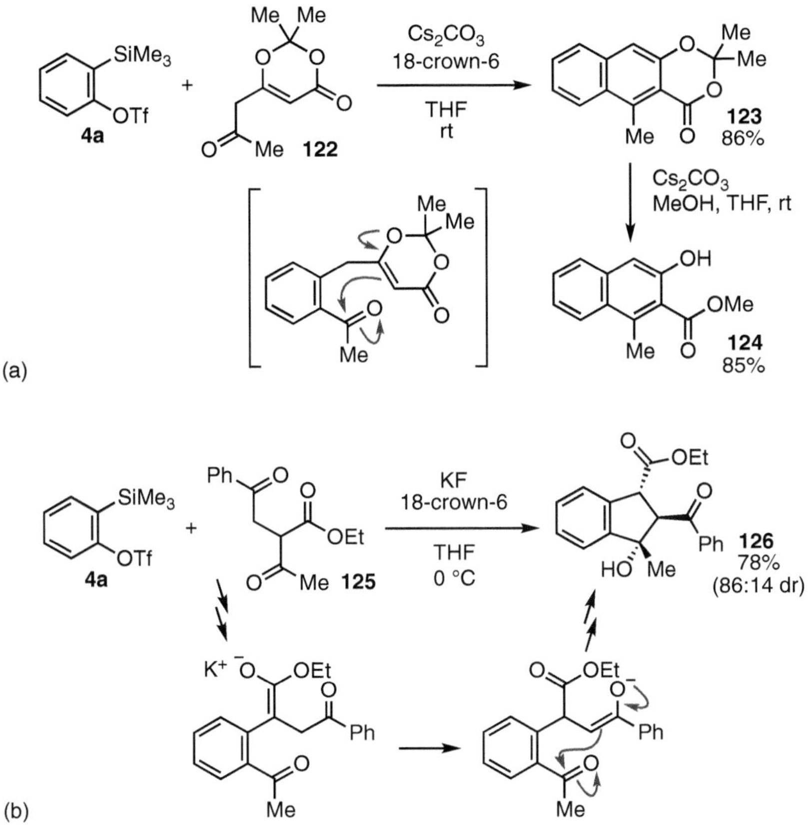

Figure 4.19 Synthesis of multisubstituted cyclic compounds by acylalkylation reactions of benzyne.

(Figure 4.23c, lower scheme). Coquerel, Rodriguez, et al. reported the arylation of arynes with α-amido-substituted cyclopentanone (Figure 4.24a), enabling the synthesis of chiral amide using a catalytic amount of thiourea as an organocatalyst, albeit in low enantiomeric excess [70]. The arylation was successfully applied to the chiral ketoester synthesis by Garg and Houk et al. using a chiral amine **155** (Figure 4.24b) [71]. Furthermore, arylation of arynes with amide **158** having a fluoro group and subsequent removal of the ethoxycarbonyl group efficiently afforded α-aryl-α-fluororactamide derivative **159** (Figure 4.24c) [72].

4.4 Etherification and Related Transformations

The addition of phenols or carboxylic acids to arynes proceeds smoothly enabling the *O*-arylation under transition metal-free conditions as reported by Larock and coworkers in 2004 (Figure 4.25a and top scheme of Figure 4.25b) [73a]. They found that oxyacylated product **164** was obtained from the reaction of arynes with aliphatic

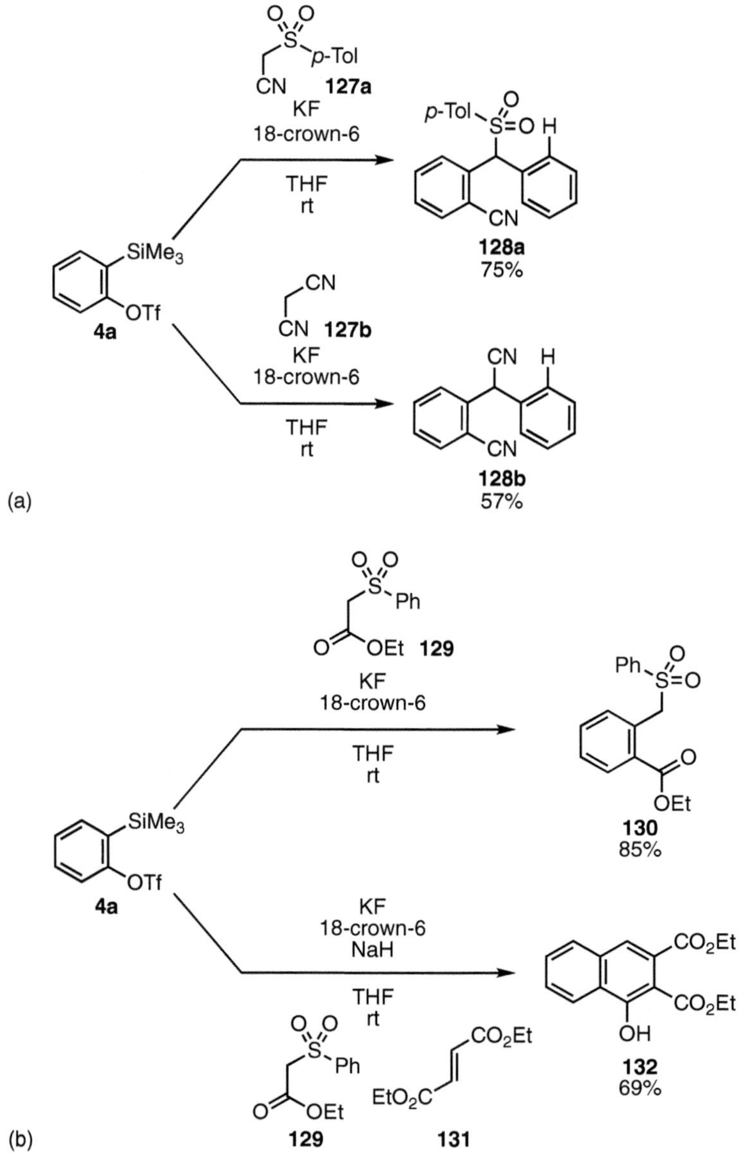

Figure 4.20 Reactions of benzyne with various compounds having an active methylene group.

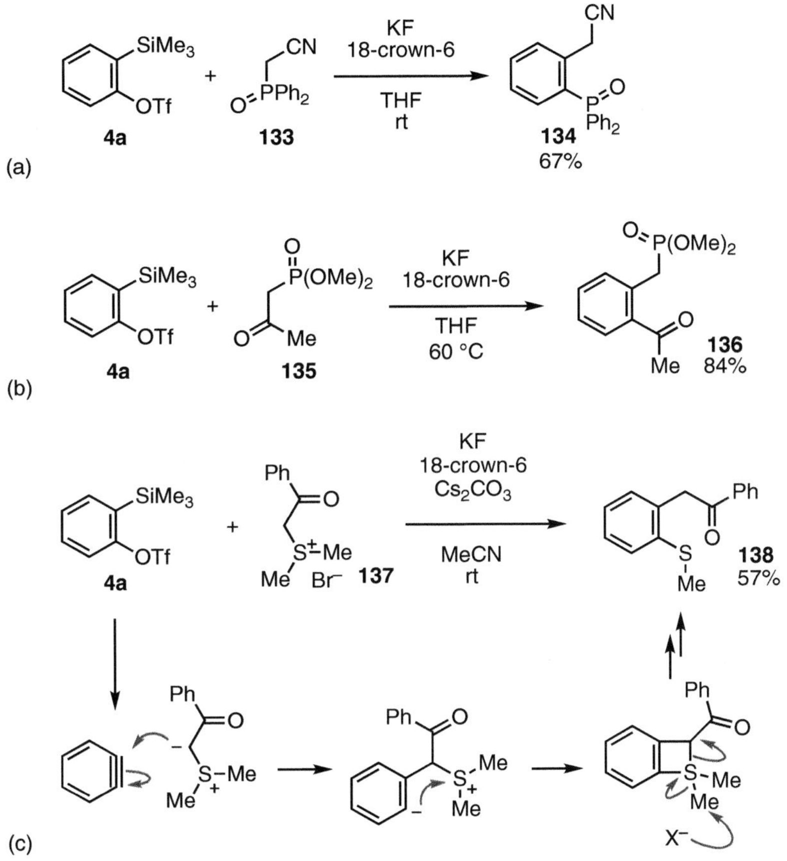

Figure 4.21 Reactions of benzyne with other compounds having an active methylene group.

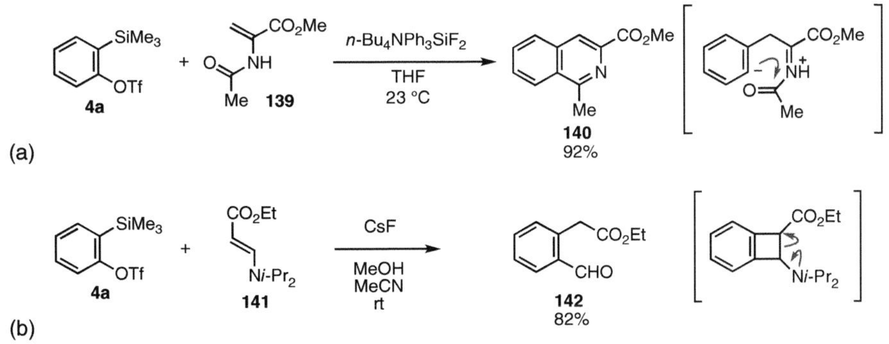

Figure 4.22 Reactions of benzyne with aminoacrylates.

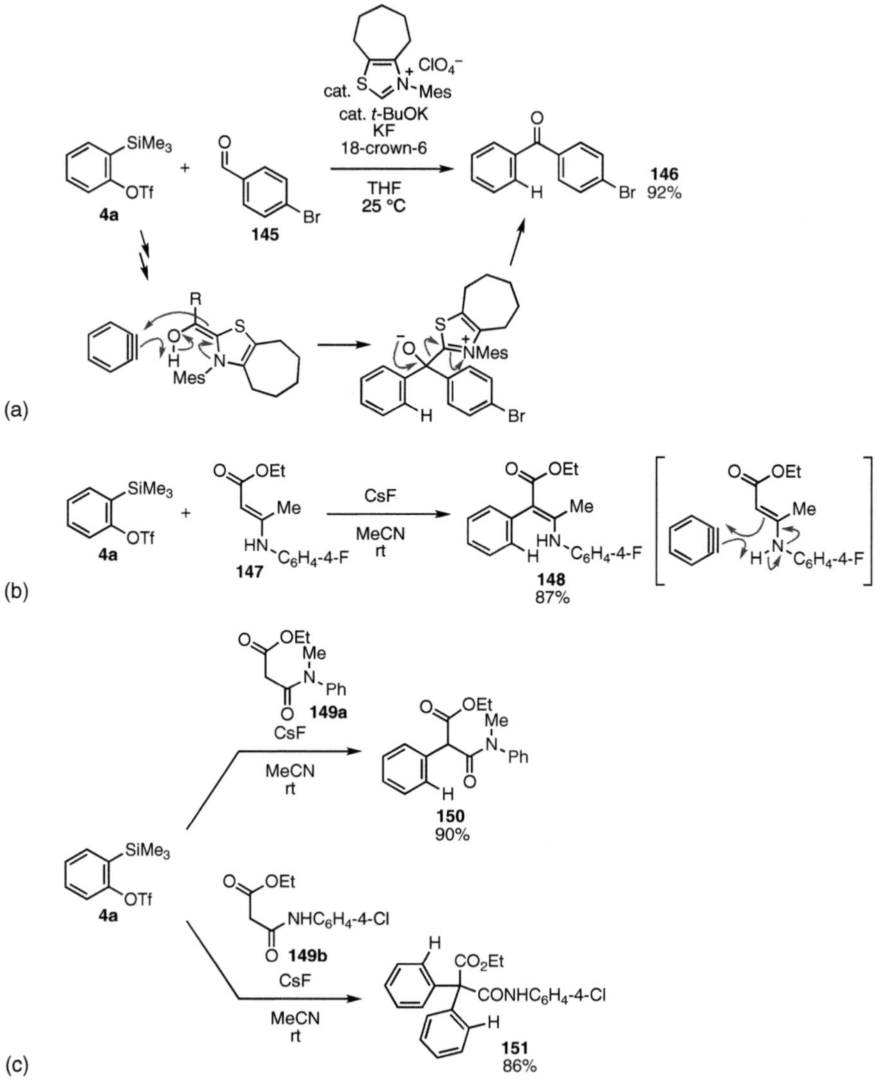

Figure 4.23 Benzyne transformations involving C—C and C—H bond formations.

carboxylic acid **162b** (Figure 4.25b, middle scheme) [73c, e],while xanthene **165** was available via further *O*-arylation with second arynes, followed by cyclization (Figure 4.25b, bottom scheme) [28, 73e]. Addition of the hydroxy group of phenols **166** and **168** to arynes and subsequent ring closure afforded xanthone (**167**) [28, 73b] and xanthene **169** [73d], respectively (Figures 4.25c,d).

Oxysulfanylations of aryne intermediates with sulfoxides bearing a nucleophilic oxygen have also been reported. In 2017, Studer and coworkers found that alkenyl sulfoxide **170** smoothly reacted with arynes to afford oxysulfanylated products such as **171** via the S—O bond cleavage (Figure 4.26a) [74]. In 2017, we achieved the difunctionalization of arynes using diaryl sulfoxides at 110 °C, while the

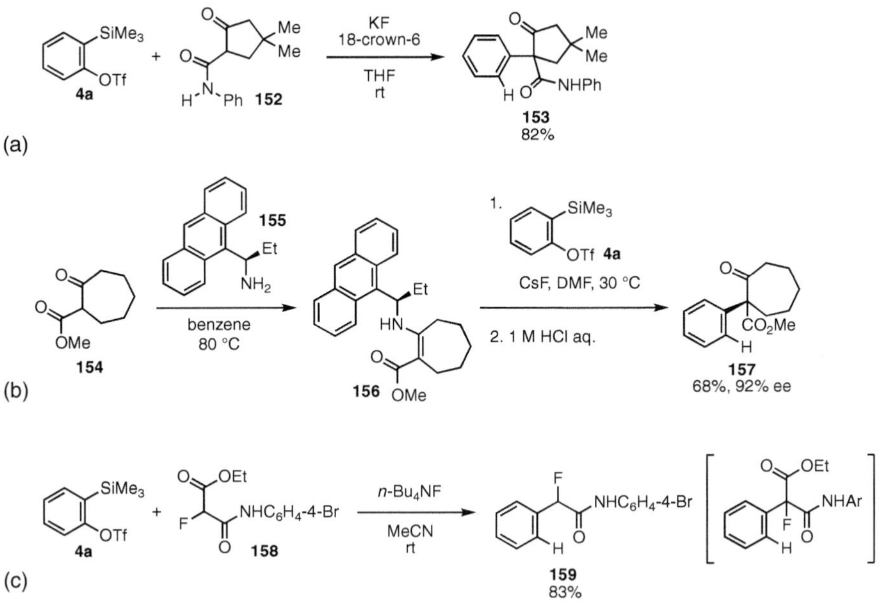

Figure 4.24 α-Phenylation of β-carbonyl amide or ester derivatives with benzyne.

reaction performed at 60 °C resulted in the formation of sulfonium salt **173** [75] (Figure 4.26b) [76].

Transformations of arynes involving C—O and C—N bond formations have also been reported. The reaction of arynes with nitrosobenzene (**175**) led to the formation of carbazole (**176**) via oxyamination, N—O bond cleavage, ring closure, and dehydration (Figure 4.27a) [77]. The oxyamination of arynes with *N*-hydroxyimide **177** afforded *N*-aryloxindole derivative **178** (Figure 4.27b) [78]. Furthermore, transformations of arynes involving C—O and C—Si bond formations were reported by Hoye et al., who demonstrated that the intramolecular oxysilylation of aryne **XII** having a silyloxy group through an appropriate linker proceeded efficiently (Figure 4.27c) [79].

4.5 Sulfanylation and Related Transformations

4.5.1 Hydrosulfanylation of Arynes

Several sulfanylation reactions of arynes with various nucleophilic organosulfur compounds have been reported so far. The efficient synthesis of aryl sulfides by sulfanylation of arynes generated from *o*-silylaryl triflates with thiols was reported by Larock and coworkers in 2006 (Figure 4.28a) [6b]. Thioxantone synthesis was achieved by the reaction of arynes with benzene thiol **183** having an ester moiety at the ortho position (Figure 4.28b) [28]. We found that the generation of 3-fluorobenzyne from 3-fluoro-2-iodophenyl triflate (**15e**) by treatment of a silylmethyl Grignard reagent in the presence of 1-dodecanethiol (**185**) at room

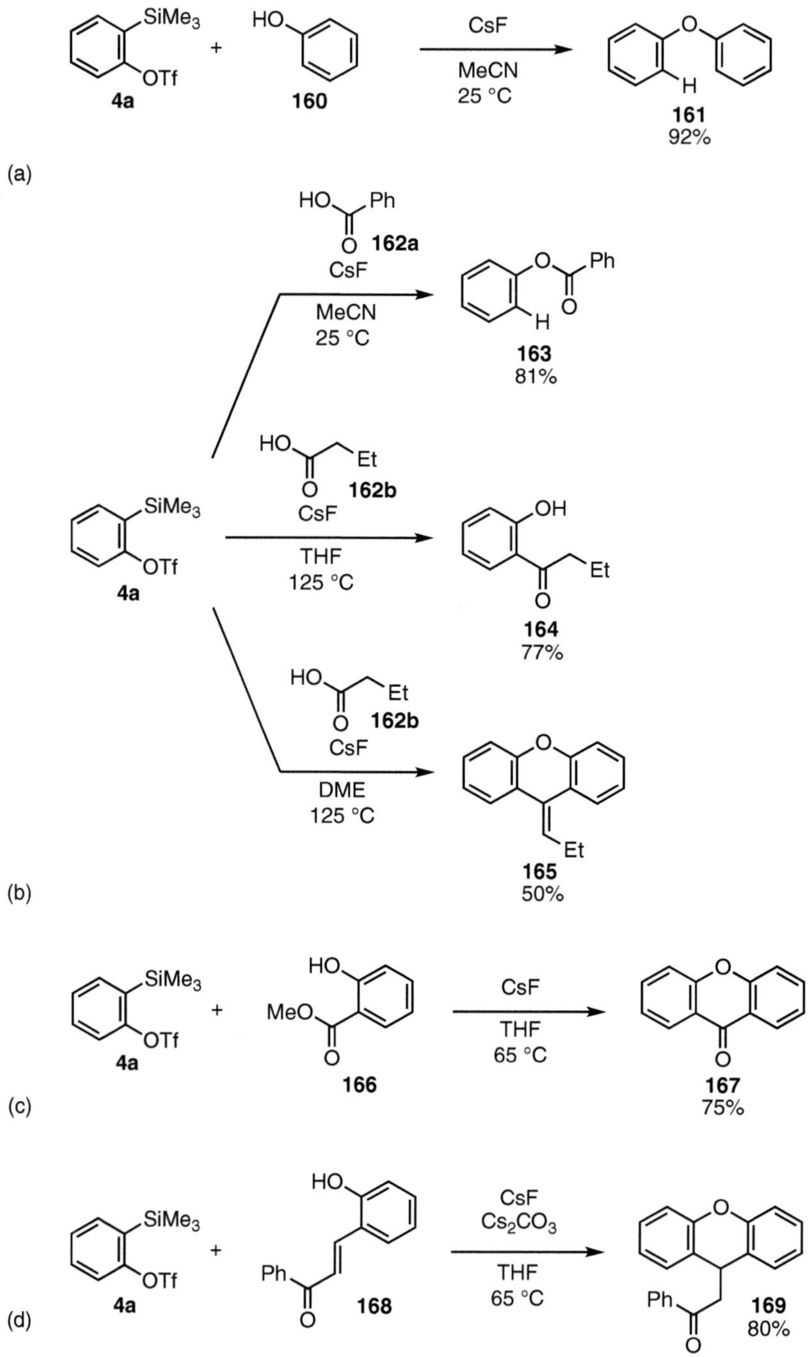

Figure 4.25 Addition reactions of phenols or carboxylic acids to benzyne.

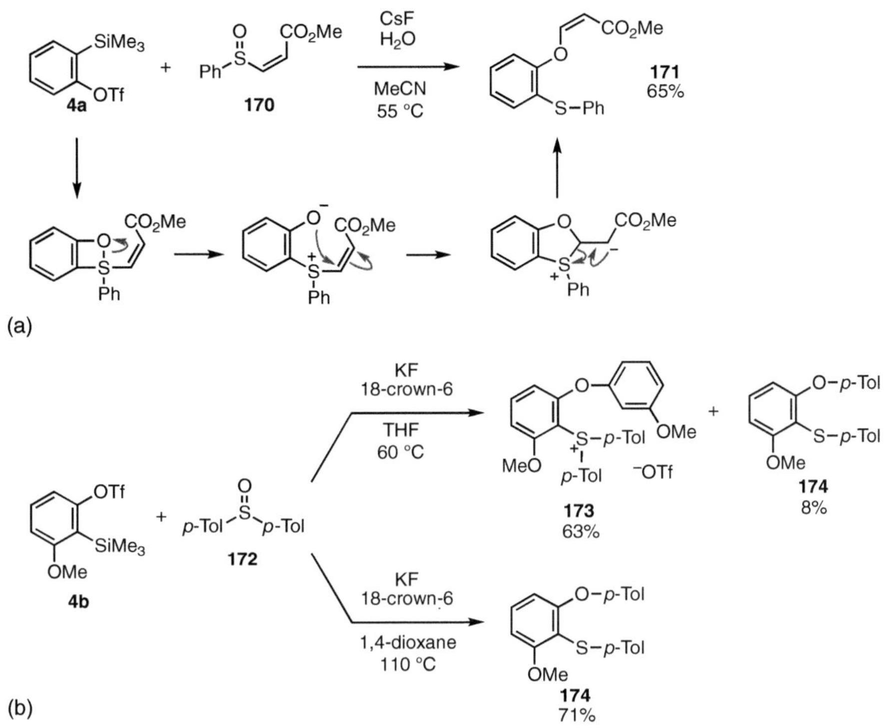

Figure 4.26 Oxysulfanylations of arynes.

temperature afforded sulfanylated arene **186**, while performing the reaction at −78 °C provided three-component coupling product **187**, which was obtained by initial formation of a C—O bond through the reaction of aryne with THF used as a solvent (Figure 4.28c) [80]. Phenylsulfonylation and triflylation of arynes with sodium sulfinates **188** and **190** were reported by Mhaske and coworkers [81] and Shibata and coworkers [82], respectively (Figures 4.28d,e).

Phenylsulfanylation of arynes was also achieved efficiently using ethyl phenyl sulfide (**192a**), from which ethylene was eliminated after the nucleophilic attack of sulfide to aryne (Figure 4.29a) [83]. In 2017, Peng and coworkers reported sulfonium salt formation by the reaction between arynes and diphenyl sulfide (**192b**) (Figure 4.29b) [84].

4.5.2 Transformations Involving C–S and C–C Bond Formations

In 2016, Studer et al. reported the carbosulfanylation of arynes by alkenyl sulfide **195**, which afforded *o*-alkenylaryl sulfide **196** (Figure 4.30a) [85]. The carbosulfanylation of arynes that resulted in the formation of a benzisothiazole skeleton was reported by Willis and coworkers in 2015, wherein 3,4-dichloro-1,2,5-thiadiazole (**197**) was used to react with arynes followed by S—N and C—C bond cleavage (Figure 4.30b) [86].

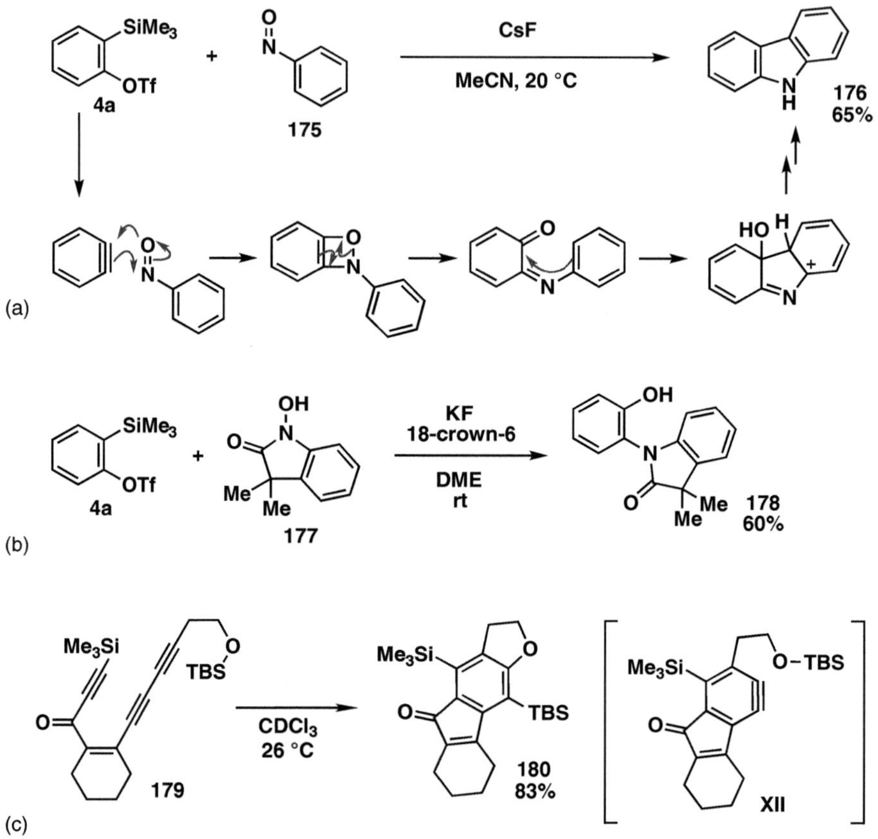

Figure 4.27 Transformations of arynes involving C—O and C—N bond formations.

4.5.3 Other Transformations Involving C—S and C—X Bond Formations

Various transformations of arynes involving C–S and C–Mg, C–S, C–P, C–Si, C–Sn, or C–Bi have been reported. Sulfanylmagnesiations of arynes were achieved by treating *o*-iodoaryl sulfonate-type precursors such as **15b** with isopropylmagnesium chloride in the presence of magnesium thiolates or thiols (Figures 4.31a,b) [23b, 24]. Thianthrene (**204**) was synthesized by disulfanylation of benzyne with 1,3-benzodithiol-2-imine (**203**) that was promoted by the liberation of a cyanide ion (Figure 4.31c) [87]. We found a reaction involving the sulfanylation of arynes by diphenyl sulfide (**192b**), followed by a thia-Fries rearrangement to afford zwitterionic sulfonium salt **205** (Figure 4.31d) [10a]. Disulfanylations of arynes with disulfides **206a** and **206b** were reported by Raminelli and coworkers [88] and Daugulis and coworkers [89], respectively (Figures 4.32a,b). Furthermore, Hiroto Yoshida et al. reported the sulfanylstannylation of arynes by *S*-stannyl sulfides in 2004 (Figure 4.33a) [90]. We recently reported the silylsulfanylation of 3-(triflyloxy)arynes using *S*-silyl sulfides, which enabled the facile preparation

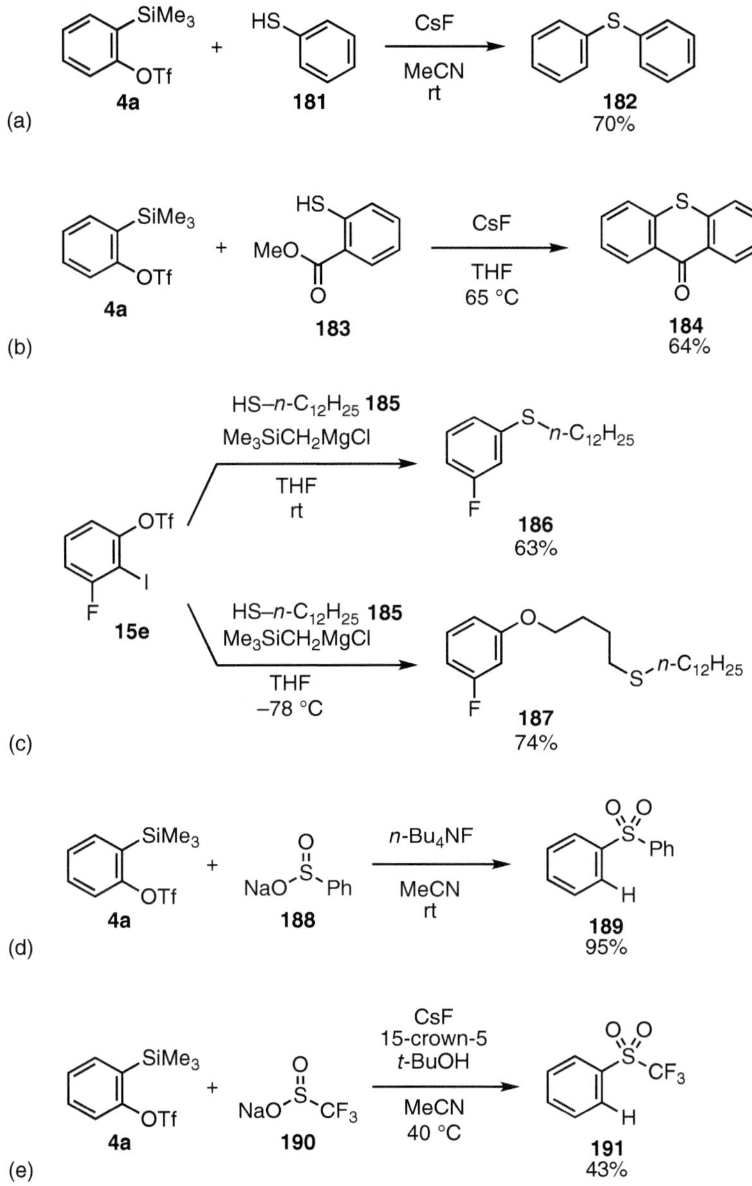

Figure 4.28 Reactions of arynes with various thiols and sulfinates.

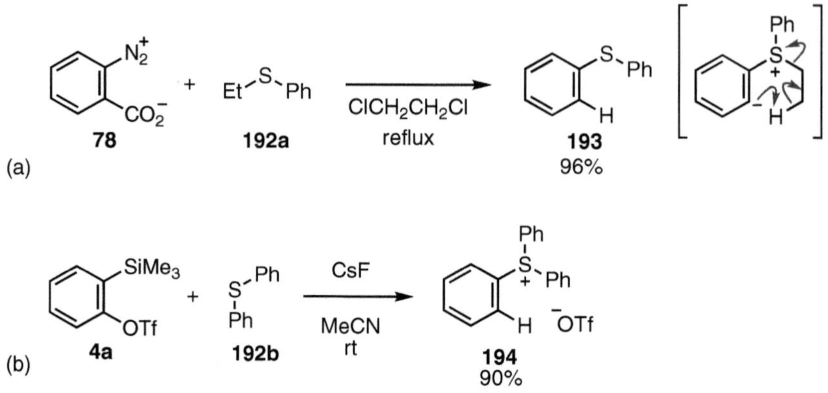

Figure 4.29 Reactions of sulfides with benzyne.

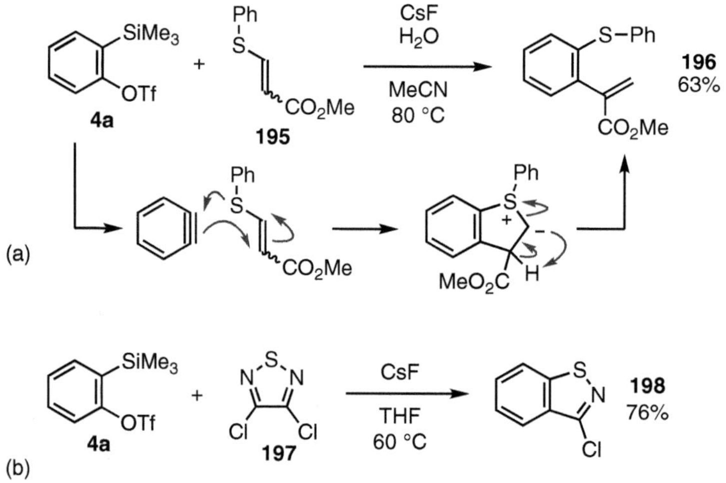

Figure 4.30 Carbosulfanylation of benzyne.

of *o*-silylaryl triflate-type 3-sulfanylaryne precursors (Figure 4.33b) [9b]. In 2018, Daugulis and coworkers achieved the silylsulfanylation of arynes generated by deprotonation of aryl halides or triflates with lithium di(1-adamantyl)amide (Figure 4.33c) [41]. Bismuthosulfanylation of arynes was reported by Murafuji and coworkers in 2011 (Figure 4.34a) [91]. In 2011, Alajarin et al. reported the synthesis of phosphonium salt **216** by the reaction of arynes with phosphine sulfide **215** via C—S and C—P bond formations followed by S—P bond cleavage and cyclization (Figure 4.34b) [92].

Reactions of arynes with various thiocarbonyl compounds have been reported to afford a range of cyclic sulfides. For example, the synthesis of pyridine-fused benzothiophene **219** through the reaction of benzyne, generated from *o*-diazoniumphenyl carboxylate, with *N*-acetoxythiopyridone (**218**) was reported by Biehl and coworkers in 2002 (Figure 4.35a) [93]. In 2016, Hoye and coworkers found

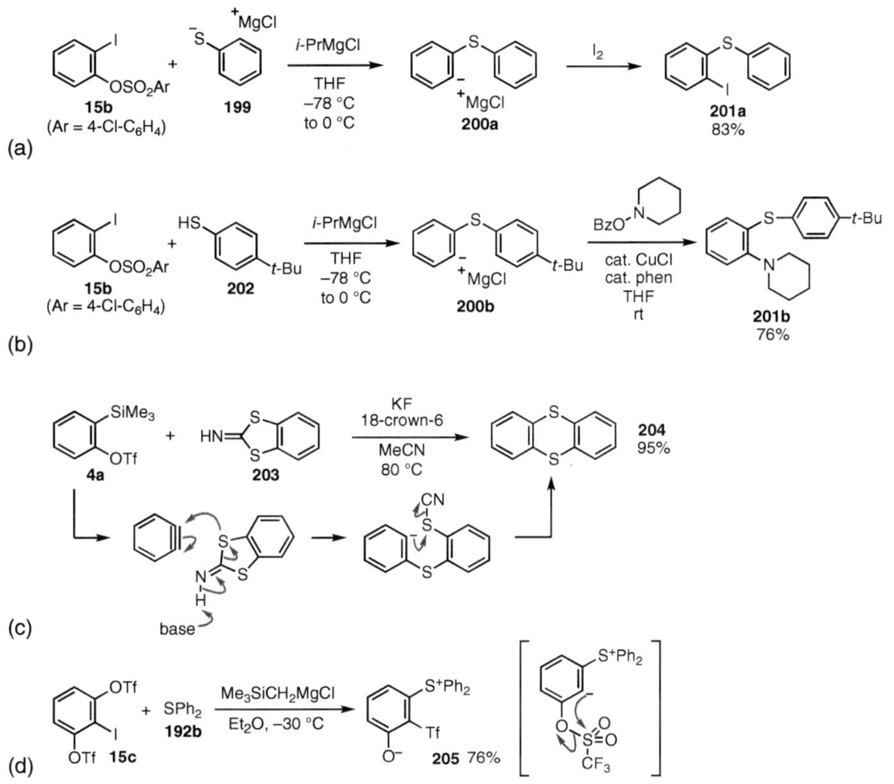

Figure 4.31 Various transformations of arynes involving C–S and C–X formations.

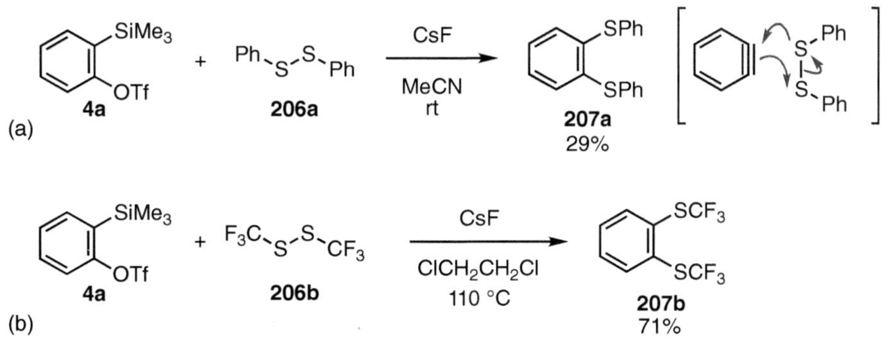

Figure 4.32 Disulfanylations of benzyne with disulfides.

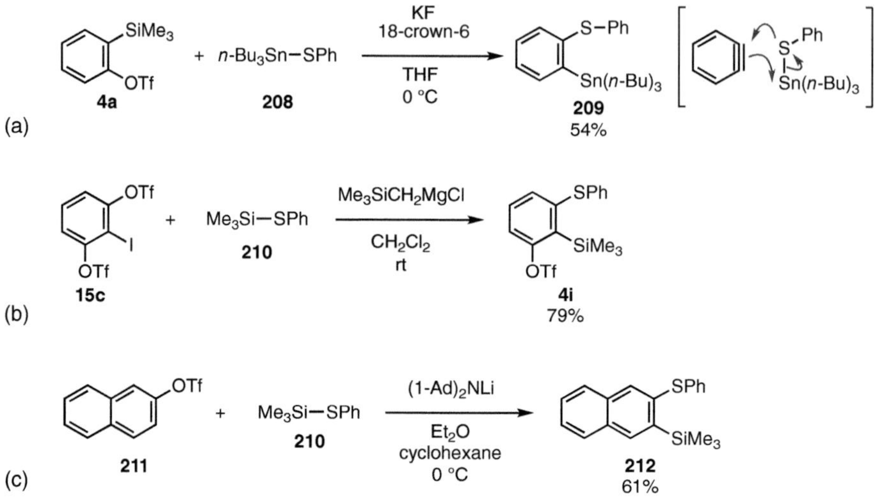

Figure 4.33 Sulfanylstannylation and silylsulfanylations of arynes.

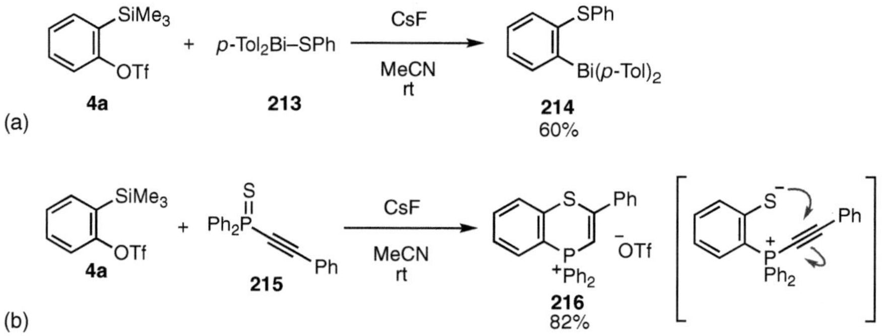

Figure 4.34 Bismuthosulfanylation and phosphinosulfanylation of benzyne.

that thioamide **221** reacted with arynes generated from triyne **220** by the HDDA reaction to form a six-membered ring via C—S bond cleavage and C—S bond formation (Figure 4.35b) [94]. In addition, [2+2] cycloaddition between benzyne and thioketone **224** was reported by Okuma et al. in 1998 (Figure 4.35c) [95].

4.6 Transformations Involving Bond Formation with Other Heteroatom Nucleophiles

4.6.1 Transformations Involving C—P Bond Formation

Reactions between arynes and nucleophilic organophosphorus compounds have also received great attention. Insertion reaction of arynes to P—Li bonds was achieved by the reaction between lithium phosphide, generated from

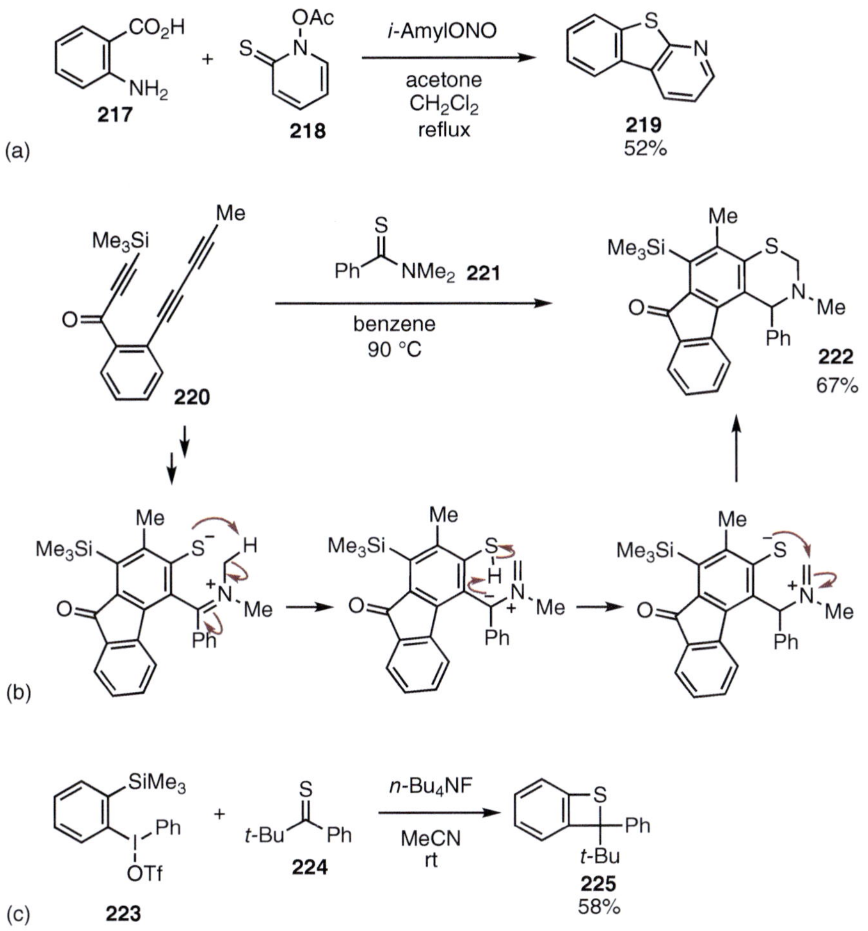

Figure 4.35 Reactions of arynes with various thiocarbonyl compounds.

phosphine–borane **226**, and 1,2-dibromobenzene (**70**) by Leroux, Jugé, et al. in 2012 (Figure 4.36a) [96]. Addition reactions of organophosphorus compounds to arynes, including Michaelis–Arbuzov-type reactions, were reported independently by our group [97] and Mhaske's group et al. [98] in 2013 (Figures 4.36b,c). In 2016, Chen et al. reported the *P*-arylation of arynes using phosphonic acid diester **232** (Figure 4.36d) [99].

The stannylphosphination of arynes generated from *o*-iodoaryl sulfonates triggered by the turbo-Grignard reagent was reported by Studer et al. in 2016 (Figure 4.37a) [100]. In addition to the silylamination and silylsulfanylation reactions, Daugulis and coworkers reported the silylphosphination of arynes (Figure 4.37b) [41]. In 2018, Miura, Hirano, et al. reported an efficient synthesis of 1,2-diphosphinobenzene derivatives such as **236** by treatment of *o*-silylaryl triflates with tetrabutylammonium difluorotriphenylsilicate (TBAT) as an activator in the presence of diphosphine **235** (Figure 4.37c) [101].

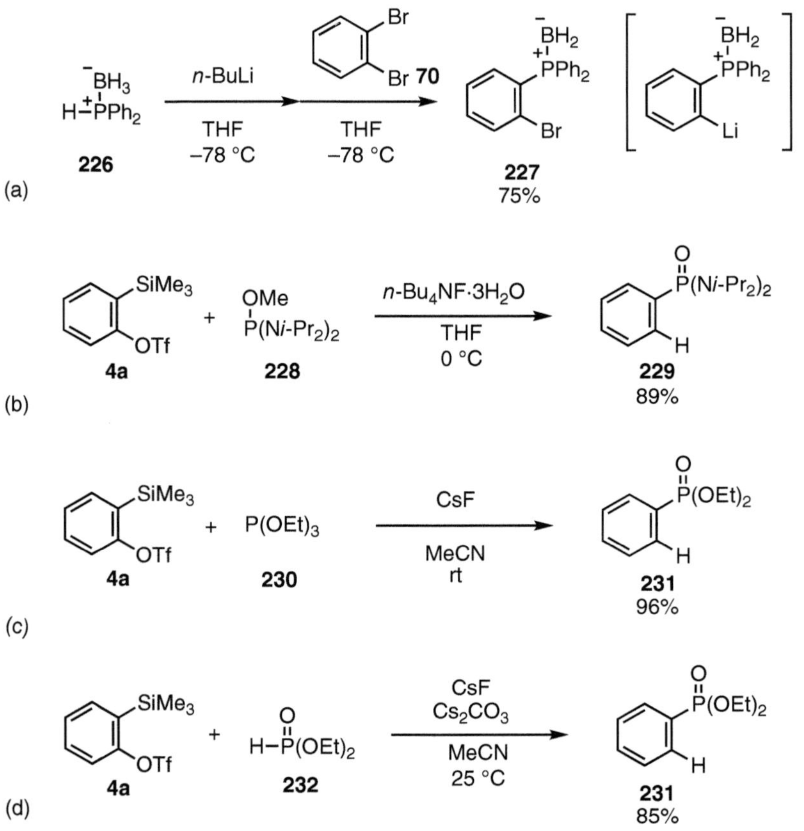

Figure 4.36 Reactions of benzyne with organophosphorus compounds.

4.6.2 Transformations Involving C—B, C—I, or C—Cl Bond Formations

The borylzincation of arynes was reported by Uchiyama and coworkers in 2013, where the key to the success of the reaction was the formation of *o*-borylatedarylzincate **XIII** via the reaction of arynes with borylzincate intermediate generated from diboron **237**, dimethyl zinc, and magnesium *tert*-butoxide (Figure 4.38a) [102]. Diiodination of arynes was reported by Guitián, Pérez, and coworkers in 2012 (Figure 4.38b) [103]. Hiroto Yoshida, Kunai, and coworkers succeeded in the chloroacylation of arynes using acid chloride **240** as an arynophile (Figure 4.39a) [104a]. This reaction was extended to the chloroarylation of arynes using cyanuric chloride derivative **242** (Figure 4.39b) [104b].

4.7 Conclusions

This chapter has summarized recent insertion reactions and related transformations of arynes. A wide range of arynophiles have become available with the increase in the number of precursors and generating conditions of arynes, greatly expanding the

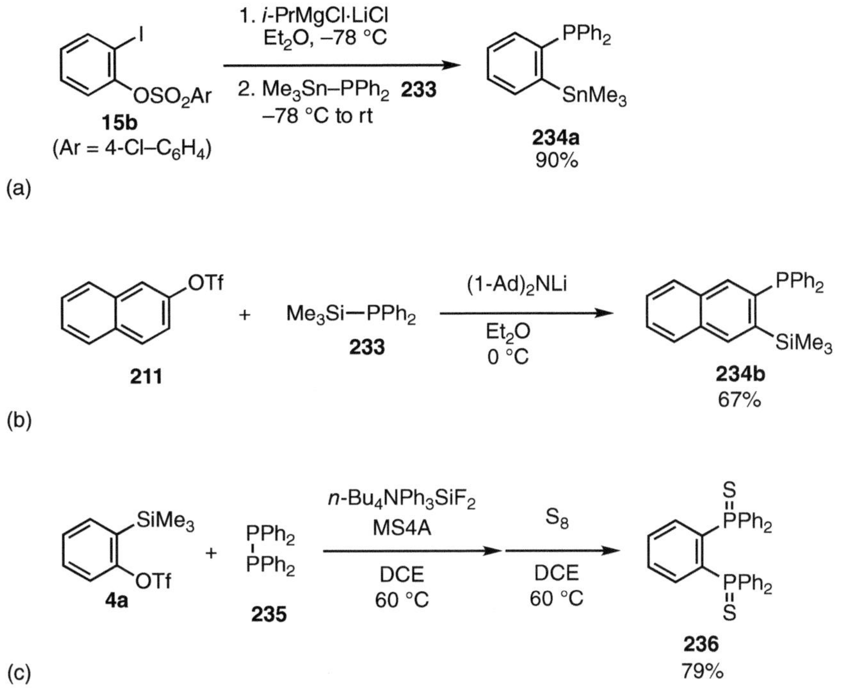

Figure 4.37 Stannylphosphination, silylphosphination, and diphosphination of arynes.

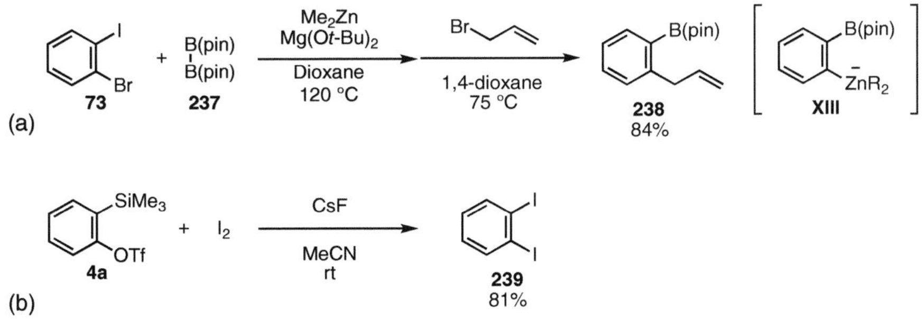

Figure 4.38 Borylzincation and diiodination of benzyne.

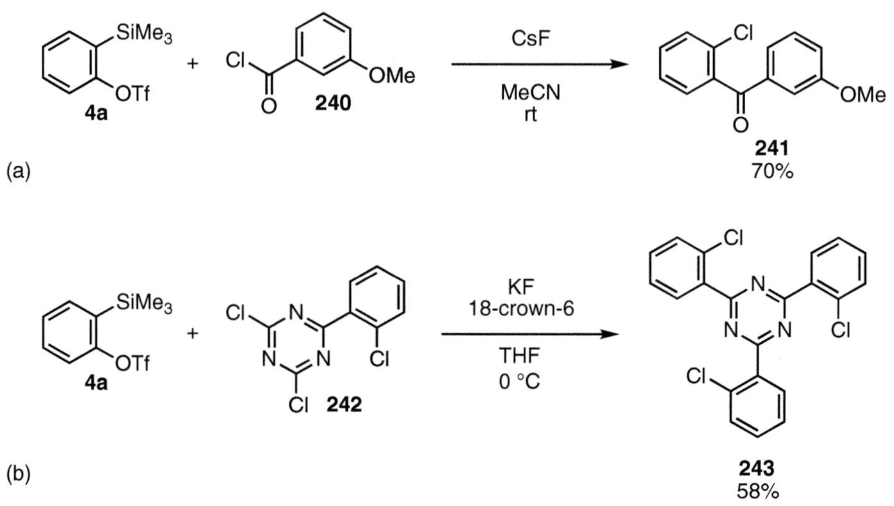

Figure 4.39 Chloroacylation and chloroarylation of benzyne.

range of synthesizable compounds. Efficient insertion reactions, including difunctionalizations of arynes, enable the facile preparation of multisubstituted benzenes, which are often difficult to synthesize by the conventional methods. Thus, continuous efforts to further expand the utility of synthetic aryne chemistry would allow the expeditious development of compounds that could be useful in a broad range of research fields, such as materials science and pharmaceutical chemistry.

References

1 Hoffmann, R.W. (ed.) (1967). *Dehydrobenzene and Cycloalkynes*. New York: Academic Press.

2 For some recent reviews on arynes, see:(a)Tadross, P.M. and Stoltz, B.M. (2012). *Chem. Rev.* 112: 3550. (b) Dubrovskiy, A.V., Markina, N.A., and Larock, R.C. (2013). *Org. Biomol. Chem.* 11: 191. (c) Yoshida, S. and Hosoya, T. (2015). *Chem. Lett.* 44: 1450. (d) Goetz, A.E., Shah, T.K., and Garg, N.K. (2015). *Chem. Commun.* 51: 34. (e) García-López, J.-A. and Greaney, M.F. (2016). *Chem. Soc. Rev.* 45: 6766. (f) Shi, J., Li, Y., and Li, Y. (2017). *Chem. Soc. Rev.* 46: 1707. (g) Idiris, F.I.M. and Jones, C.R. (2017). *Org. Biomol. Chem.* 15: 9044. (h) Roy, T. and Biju, A.T. (2018). *Chem. Commun.* 54: 2580. (i) Yoshida, S. (2018). *Bull. Chem. Soc. Jpn.* 91: 1293. (j) Matsuzawa, T., Yoshida, S., and Hosoya, T. (2018). *Tetrahedron Lett.* 59: 4197. (k) Takikawa, H., Nishii, A., Sakai, T., and Suzuki, K. (2018). *Chem. Soc. Rev.* 47: 8030.

3 (a) Kessar, S.V. (1992). Chap. 2.3: Nucleophilic coupling with arynes. In: *Comprehensive Organic Synthesis: Additions to and Substitutions at C—C π-Bonds*, vol. 4 (eds. B.M. Trost, M.F. Semmelhack and I. Fleming), 483–516. Pergamon.

(b) Hoffman, R., Imamura, A., and Hehre, W.J. (1968). *J. Am. Chem. Soc.* 90: 1499.

4 (a) Wotiz, J.H. and Huba, F. (1959). *J. Org. Chem.* 24: 595. (b) de Graaff, G.B.R., den Hertog, H.J., and Melger, W.C. (1965). *Tetrahedron Lett.* 6: 963.

5 Himeshima, Y., Sonoda, T., and Kobayashi, H. (1983). *Chem. Lett.* 12: 1211.

6 (a) Liu, Z. and Larock, R.C. (2003). *Org. Lett.* 5: 4673. (b) Liu, Z. and Larock, R.C. (2006). *J. Org. Chem.* 71: 3198.

7 Ikawa, T., Nishiyama, T., Shigeta, T. et al. (2011). *Angew. Chem. Int. Ed.* 50: 5674.

8 Medina, J.M., Mackey, J.L., Garg, N.K., and Houk, K.N. (2014). *J. Am. Chem. Soc.* 136: 15798.

9 (a) Yoshida, S., Nakamura, Y., Uchida, K. et al. (2016). *Org. Lett.* 18: 6212. (b) Nakamura, Y., Miyata, Y., Uchida, K. et al. (2019). *Org. Lett.* 21: 5252.

10 (a) Yoshida, S., Uchida, K., Igawa, K. et al. (2014). *Chem. Commun.* 50: 15059. (b) Ikawa, T., Kaneko, H., Masuda, S. et al. (2015). *Org. Biomol. Chem.* 13: 520. (c) Shi, J., Qiu, D., Wang, J. et al. (2015). *J. Am. Chem. Soc.* 137: 5670. (d) Uchida, K., Yoshida, S., and Hosoya, T. (2016). *Synthesis* 48: 4099.

11 (a) Hamura, T., Ibusuki, Y., Sato, K. et al. (2003). *Org. Lett.* 5: 3551. (b) Bronner, S.M., Bahnck, K.B., and Garg, N.K. (2009). *Org. Lett.* 11: 1007. (c) Cheong, P.H.-Y., Paton, R.S., Bronner, S.M. et al. (2010). *J. Am. Chem. Soc.* 132: 1267. (d) Im, G.-Y.J., Bronner, S.M., Goetz, A.E. et al. (2010). *J. Am. Chem. Soc.* 132: 17933. (e) Bronner, S.M., Goetz, A.E., and Garg, N.K. (2011). *J. Am. Chem. Soc.* 133: 3832. (f) Goetz, A.E., Bronner, S.M., Cisneros, J.D. et al. (2012). *Angew. Chem. Int. Ed.* 51: 2758. (g) Goetz, A.E. and Garg, N.K. (2013). *Nat. Chem.* 5: 54.

12 McAusland, D., Seo, S., Pintori, D.G. et al. (2011). *Org. Lett.* 13: 3667.

13 Cheng, B., Bao, B., Chen, Y. et al. (2017). *Org. Chem. Front.* 4: 1636.

14 Yao, T. (2015). *Tetrahedron Lett.* 56: 4623.

15 Bhojgude, S.S., Kaicharla, T., and Biju, A.T. (2013). *Org. Lett.* 15: 5452.

16 Hirsch, M., Dhara, S., and Diesendruck, C.E. (2016). *Org. Lett.* 18: 980.

17 Yoshida, S., Shimizu, K., Uchida, K. et al. (2017). *Chem. Eur. J.* 23: 15332.

18 Pal, K.B., Mahanti, M., and Nilsson, U.J. (2018). *Org. Lett.* 20: 616.

19 Zhang, L., Geng, Y., and Jin, Z. (2016). *J. Org. Chem.* 81: 3542.

20 Cheng, B., Wei, J., Zu, B. et al. (2017). *J. Org. Chem.* 82: 9410.

21 Aithagani, S.K., Dara, S., Munagala, G. et al. (2015). *Org. Lett.* 17: 5547.

22 Yoshida, S., Nakajima, H., Uchida, K. et al. (2017). *Chem. Lett.* 46: 77.

23 (a) Sapountzis, I., Lin, W., Fischer, M., and Knochel, P. (2004). *Angew. Chem. Int. Ed.* 43: 4364. (b) Lin, W., Sapountzis, I., and Knochel, P. (2005). *Angew. Chem. Int. Ed.* 44: 4258. (c) Lin, W., Ilgen, F., and Knochel, P. (2006). *Tetrahedron Lett.* 47: 1941.

24 García-López, J.-A., Çetin, M., and Greaney, M.F. (2015). *Angew. Chem. Int. Ed.* 54: 2156.

25 Yoshida, H., Shirakawa, E., Honda, Y., and Hiyama, T. (2002). *Angew. Chem. Int. Ed.* 41: 3247.

26 Liu, Z. and Larock, R.C. (2005). *J. Am. Chem. Soc.* 127: 13112.

27 Pintori, D.G. and Greaney, M.F. (2010). *Org. Lett.* 12: 168.
28 Zhao, J. and Larock, R.C. (2007). *J. Org. Chem.* 72: 583.
29 Wright, A.C., Haley, C.K., Lapointe, G., and Stoltz, B.M. (2016). *Org. Lett.* 18: 2793.
30 Rao, B. and Zeng, X. (2014). *Org. Lett.* 16: 314.
31 Gilmore, C.D., Allan, K.M., and Stoltz, B.M. (2008). *J. Am. Chem. Soc.* 130: 1558.
32 Dubrovskiy, A.V. and Larock, R.C. (2012). *J. Org. Chem.* 77: 11232.
33 Hong, D., Chen, Z., Lin, X., and Wang, Y. (2010). *Org. Lett.* 12: 4608.
34 Wang, Y., Zheng, L., and Hoye, T.R. (2018). *Org. Lett.* 20: 7145.
35 Yoshida, S., Yano, T., Misawa, Y. et al. (2015). *J. Am. Chem. Soc.* 137: 14071.
36 Gaykar, R.N., Bhattacharjee, S., and Biju, A.T. (2019). *Org. Lett.* 21: 737.
37 Shen, C., Yang, G., and Zhang, W. (2013). *Org. Lett.* 15: 5722.
38 Lopez-Leonardo, C., Raja, R., López-Ortiz, F. et al. (2014). *Eur. J. Org. Chem.*: 1084.
39 (a) Hendrick, C.E., McDonald, S.L., and Wang, Q. (2013). *Org. Lett.* 15: 3444. (b) Hendrick, C.E. and Wang, Q. (2014). *J. Org. Chem.* 80: 1059.
40 Yoshida, H., Minabe, T., Ohshita, J., and Kunai, A. (2005). *Chem. Commun.*: 3454.
41 Mesgar, M., Nguyen-Le, J., and Daugulis, O. (2018). *J. Am. Chem. Soc.* 140: 13703.
42 Hart, H., Harada, K., and Du, C.J.F. (1985). *J. Org. Chem.* 50: 3104.
43 Diemer, V., Begaud, M., Leroux, F.R., and Colobert, F. (2011). *Eur. J. Org. Chem.*: 341.
44 Berthelot-Bréhier, A., Panossian, A., Colobert, F., and Leroux, F.R. (2015). *Org. Chem. Front.* 2: 634.
45 Diemer, V., Berthelot, A., Bayardon, J. et al. (2012). *J. Org. Chem.* 77: 6117.
46 Hayes, D.M. and Hoffmann, R. (1972). *J. Phys. Chem.* 76: 656.
47 (a) Wasserman, H.H. and Solodar, J. (1965). *J. Am. Chem. Soc.* 87: 4002. (b) Stevens, R.V. and Bisacchi, G.S. (1982). *J. Org. Chem.* 47: 2396.
48 (a) Hosoya, T., Hasegawa, T., Kuriyama, Y. et al. (1995). *Synlett*: 177. (b) Hosoya, T., Hasegawa, T., Kuriyama, Y., and Suzuki, K. (1995). *Tetrahedron Lett.* 36: 3377.
49 (a) Hamura, T., Ibusuki, Y., Uekusa, H. et al. (2006). *J. Am. Chem. Soc.* 128: 3534. (b) Hamura, T., Ibusuki, Y., Uekusa, H. et al. (2006). *J. Am. Chem. Soc.* 128: 10032. (c) Shinozaki, S., Hamura, T., Ibusuki, Y. et al. (2010). *Angew. Chem. Int. Ed.* 49: 3026.
50 Feltenberger, J.B., Hayashi, R., Tang, Y. et al. (2009). *Org. Lett.* 11: 3666.
51 (a) Tambar, U.K. and Stoltz, B.M. (2005). *J. Am. Chem. Soc.* 127: 5340. (b) Tadross, P.M., Virgil, S.C., and Stoltz, B.M. (2010). *Org. Lett.* 12: 1612. (c) Tambar, U.K., Ebner, D.C., and Stoltz, B.M. (2006). *J. Am. Chem. Soc.* 128: 11752. (d) Tadross, P.M., Gilmore, C.D., Bugga, P. et al. (2010). *Org. Lett.* 12: 1224.
52 (a) Yoshida, H., Watanabe, M., Ohshita, J., and Kunai, A. (2005). *Chem. Commun.*: 3292. (b) Yoshida, H., Watanabe, M., Ohshita, J., and Kunai, A. (2005).

Chem. Lett. 34: 1538. (c) Yoshida, H., Watanabe, M., Ohshita, J., and Kinai, A. (2005). *Tetrahedron Lett.* 46: 6729. (d) Yoshida, H., Kishida, T., Watanabe, M., and Ohshita, J. (2008). *Chem. Commun.*: 5963. (e) Yoshida, H., Morishita, T., and Ohshita, J. (2010). *Chem. Lett.* 39: 508. (f) Yoshida, H., Ito, Y., Yoshikawa, Y. et al. (2011). *Chem. Commun.* 47: 8664.

53 Rao, B., Tang, J., Wei, Y., and Zeng, X. (2016). *Chem. Asian J.* 11: 991.

54 Samineni, R., Srihari, P., and Mehta, G. (2016). *Org. Lett.* 18: 2832.

55 Rao, B., Tang, J., and Zeng, X. (2016). *Org. Lett.* 18: 1678.

56 Kranthikumar, R., Chegondi, R., and Chandrasekhar, S. (2016). *J. Org. Chem.* 81: 2451.

57 (a) Okuma, K., Hirano, K., Tanabe, Y. et al. (2014). *Chem. Lett.* 43: 492. (b) Okuma, K., Itoyama, R., Sou, A. et al. (2012). *Chem. Commun.* 48: 11145.

58 Kumar, A.S., Thirupathi, G., Reddy, G.S., and Ramachary, D.B. (2019). *Chem. Eur. J.* 25: 1177.

59 Takikawa, H., Nishii, A., and Suzuki, K. (2016). *Synthesis* 48: 3331.

60 Hu, W., Zhang, C., Huang, J. et al. (2019). *Org. Lett.* 21: 941.

61 Yoshida, H., Watanabe, M., Morishita, T. et al. (2007). *Chem. Commun.*: 1505.

62 Huang, X. and Xue, J. (2007). *J. Org. Chem.* 72: 3965.

63 Liu, Y.-L., Liang, Y., Pi, S.-F., and Li, J.-H. (2009). *J. Org. Chem.* 74: 5691.

64 Ahire, M.M., Thoke, M.B., and Mhaske, S.B. (2018). *Org. Lett.* 20: 848.

65 Li, R., Wang, X., Wei, Z. et al. (2013). *Org. Lett.* 15: 4366.

66 Łączkowski, K.Z., García, D., Peña, D. et al. (2011). *Org. Lett.* 13: 960.

67 Biju, A.T. and Glorius, F. (2010). *Angew. Chem. Int. Ed.* 49: 9761.

68 Ramtohul, Y.K. and Chartrand, A. (2007). *Org. Lett.* 9: 1029.

69 Dhokale, R.A., Thakare, P.R., and Mhaske, S.B. (2012). *Org. Lett.* 14: 3994.

70 Mohanan, K., Coquerel, Y., and Rodriguez, J. (2012). *Org. Lett.* 14: 4686.

71 Picazo, E., Anthony, S.M., Giroud, M. et al. (2018). *J. Am. Chem. Soc.* 140: 7605.

72 Gupta, E., Kant, R., and Mohanan, K. (2017). *Org. Lett.* 19: 6016.

73 (a) Liu, Z. and Larock, R.C. (2004). *Org. Lett.* 6: 99. (b) Zhao, J. and Larock, R.C. (2005). *Org. Lett.* 7: 4273. (c) Dubrovskiy, A.V. and Larock, R.C. (2010). *Org. Lett.* 12: 3117. (d) Lu, C., Dubrovskiy, A.V., and Larock, R.C. (2012). *Tetrahedron Lett.* 53: 2202. (e) Dubrovskiy, A.V. and Larock, R.C. (2013). *Tetrahedron* 69: 2789.

74 Li, Y. and Studer, A. (2017). *Org. Lett.* 19: 666.

75 Li, X., Sun, Y., Huang, X. et al. (2017). *Org. Lett.* 19: 838.

76 Matsuzawa, T., Uchida, K., Yoshida, S., and Hosoya, T. (2017). *Org. Lett.* 19: 5521.

77 Chakrabarty, S., Chatterjee, I., Tebben, L., and Studer, A. (2013). *Angew. Chem. Int. Ed.* 52: 2968.

78 Chen, Z. and Wang, Q. (2015). *Org. Lett.* 17: 6130.

79 Hoye, T.R., Baire, B., Niu, D. et al. (2012). *Nature* 490: 208.

80 Yoshida, S., Nagai, A., Uchida, K., and Hosoya, T. (2017). *Chem. Lett.* 46: 733.

81 Pandya, V.G. and Mhaske, S.B. (2014). *Org. Lett.* 16: 3836.

82 Sumii, Y., Sugita, Y., Tokunaga, E., and Shibata, N. (2018). *Chemistry Open* 7: 204.

83 Nakayama, J., Fujita, T., and Hoshino, M. (1983). *Chem. Lett.* 12: 249.

84 Zhang, L., Li, X., Sun, Y. et al. (2017). *Org. Biomol. Chem.* 15: 7181.

85 Li, Y., Mück-Lichtenfeld, C., and Studer, A. (2016). *Angew. Chem. Int. Ed.* 55: 14435.

86 Chen, Y. and Willis, M.C. (2015). *Org. Lett.* 17: 4786.

87 Pawliczek, M., Garve, L.K.B., and Werz, D.B. (2015). *Chem. Commun.* 51: 9165.

88 Toledo, F.T., Marques, H., Comasseto, J.V., and Raminelli, C. (2007). *Tetrahedron Lett.* 48: 8125.

89 Mesgar, M. and Daugulis, O. (2017). *Org. Lett.* 19: 4247.

90 Yoshida, H., Terayama, T., Ohshita, J., and Kunai, A. (2004). *Chem. Commun.*: 1980.

91 Chen, J., Murafuji, T., and Tsunashima, R. (2011). *Organometallics* 30: 4532.

92 Alajarin, M., Lopez-Leonardo, C., Raja, R., and Orenes, R.-A. (2011). *Org. Lett.* 13: 5668.

93 Rao, U.N. and Biehl, E. (2002). *J. Org. Chem.* 67: 3409.

94 Palani, V., Chen, J., and Hoye, T.R. (2016). *Org. Lett.* 18: 6312.

95 Okuma, K., Shiki, K., and Shioji, K. (1998). *Chem. Lett.* 27: 79.

96 Bayardon, J., Laureano, H., Diemer, V. et al. (2012). *J. Org. Chem.* 77: 5759.

97 Yoshida, S. and Hosoya, T. (2013). *Chem. Lett.* 42: 583.

98 Dhokale, R.A. and Mhaske, S.B. (2013). *Org. Lett.* 15: 2218.

99 Chen, Q., Yan, X., Du, Z. et al. (2016). *J. Org. Chem.* 81: 276.

100 Li, Y., Chakrabarty, S., Mück-Lichtenfeld, C., and Studer, A. (2016). *Angew. Chem. Int. Ed.* 55: 802.

101 Okugawa, Y., Hayashi, Y., Kawauchi, S. et al. (2018). *Org. Lett.* 20: 3670.

102 Nagashima, Y., Takita, R., Yoshida, K. et al. (2013). *J. Am. Chem. Soc.* 135: 18730.

103 Rodríguez-Lojo, D., Cobas, A., Peña, D. et al. (2012). *Org. Lett.* 14: 1363.

104 (a) Yoshida, H., Mimura, Y., Ohshita, J., and Kunai, A. (2007). *Chem. Commun.*: 2405. (b) Yoshida, H., Mimura, Y., and Ohshita, J. (2009). *Chem. Lett.* 38: 1132.

5

Multicomponent Reactions Involving Arynes and Related Chemistry

Hiroto Yoshida

Hiroshima University, Graduate School of Advanced Science and Engineering, Higashi-Hiroshima 739-8526, Japan

5.1 Introduction

Multicomponent reactions enable three or more starting materials to be put together generally in one chemical step in one pot; a wide range of organic molecules become readily accessible, as compared with conventional two-component, multistep reactions, with higher molecular diversity, structural complexity, and synthetic convenience, verifying their outstanding significance in chemical synthesis [1]. Arynes, one of the most representative reactive intermediates in organic chemistry [2], have been demonstrated to serve as a useful and reliable component in multicomponent reactions, thus leading to direct formation of multisubstituted arenes and benzo-annulated structures in a selective manner, despite their highly reactive and transient characters. In particular, this field has experienced remarkable progress in the last 15 years, triggered mainly by the use of a combination of 2-(trimethylsilyl)aryl triflates with a fluoride ion for generating arynes (Kobayashi's method) [3]. With Kobayashi's method, arynes can be generated at a moderate temperature and pace under neutral conditions, being compatible with a wide variety of functionalized reagents, thus expanding considerably reaction patterns and accessible products in the aryne-based multicomponent reactions. This chapter aims to summarize the recent advance in the transition metal-free multicomponent reactions of arynes based upon their highly electrophilic character, focusing on those reported during the last five years; the previous results published before 2013 have been thoroughly collected in Chapter 4.9 of Ref. [4]'s book and Chapter 3 of Ref. [4]'s book and review articles [2, 4]. Because the concept of "multicomponent reaction" covers vast scale of examples, this chapter will not include transition metal-catalyzed reactions and reactions via hexadehydro Diels–Alder (HDDA) route, which will be individually discussed in Chapters 6 and 10.

Modern Aryne Chemistry, First Edition. Edited by Akkattu T. Biju.

5.2 Classification of Multicomponent Reactions

The multicomponent reactions of arynes, discussed in this chapter, are mainly triggered by addition of neutral nucleophiles, and can basically be classified into three categories depending on reaction modes (Scheme 5.1).

1) An initially formed zwitterion is directly trapped by an electrophile (*El*) to afford 1,2-disubstituted arenes. The reaction often includes cyclization depending upon the structures of nucleophile/electrophile, resulting in the formation of benzo-annulated heterocycles.
2) An initially formed zwitterion first interacts with a pronucleophile (X–Nu′); a generated nucleophile (Nu'^{-}) then reacts with an electrophilic site (Nu^{+}) to afford products.
3) A formal [2+2] cycloaddition between an aryne and a nucleophilic–electrophilic double bond (Nu=El) provides a benzocyclobutene species. It then undergoes 4π-electrocyclization (ring-opening) to give an ortho-quinoid intermediate, which is finally trapped by a third component.

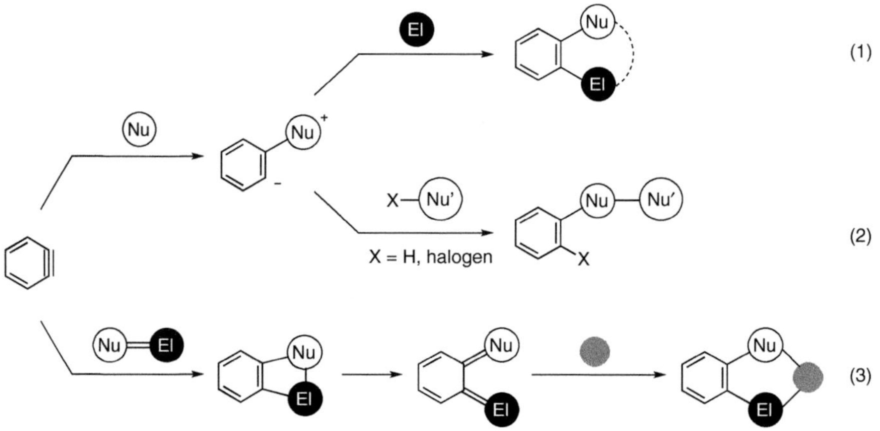

Scheme 5.1 Classification of multicomponent reactions involving arynes.

5.3 Carbon Nucleophile-Based Multicomponent Reactions

5.3.1 Isocyanide

Since Yoshida first developed the three-component reaction of arynes, isocyanides, and aldehydes [5], that produced benzo-annulated *O*-heterocycles, isocyanides have continued to be convenient and useful nucleophiles for multicomponent reactions of arynes. A zwitterion (**1**) is a common intermediate, and Biju reported that this intermediate could be trapped by atmospheric pressure of CO_2 to provide phthalimide derivatives **3** (Scheme 5.2) [6]. As shown in Scheme 5.2, CO_2 reacts with zwitterion **1** to afford a tentative product (**2**), which is facilely converted into **3** by fluoride ion–induced rearrangement. A similar three-component synthesis of phthalimides was later disclosed by Wang and Ji (Scheme 5.3) [7]; various arylisocyanides smoothly gave the respective products under their conditions (KF/18-crown-6 in THF), while the reaction with alkylisocyanides became sluggish, being in sharp contrast to Biju's results (conditions: CsF in MeCN).

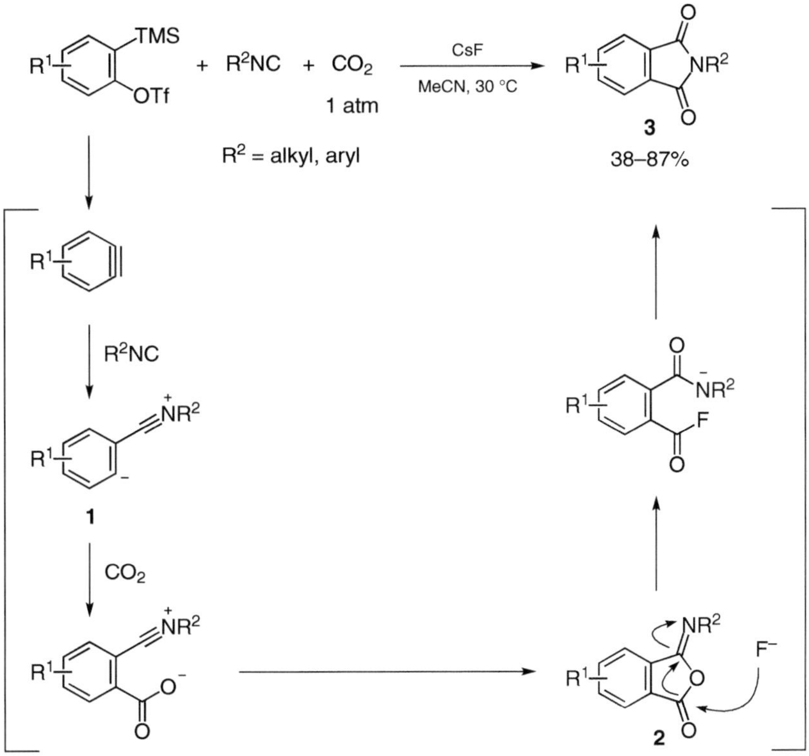

Scheme 5.2 Three-component reaction of arynes, isocyanides, and CO_2. Source: Based on Kaicharla et al. [6].

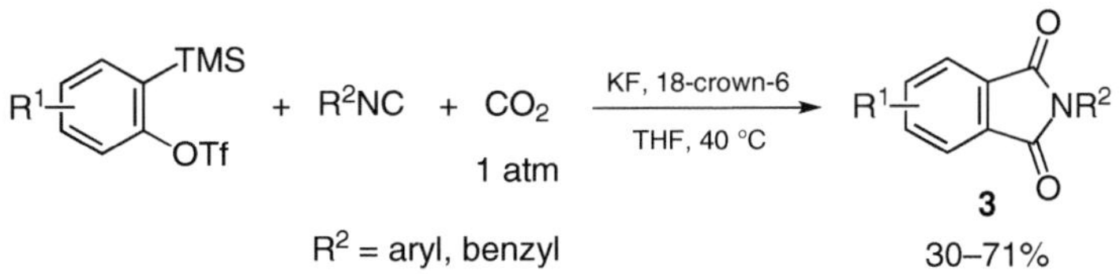

Scheme 5.3 Three-component reaction of arynes, isocyanides, and CO_2. Source: Based on Fang et al. [7].

Biju also demonstrated that water could serve as a pronucleophile for capturing zwitterion **1**, resulting in the direct formation of diverse amides (**4**) (Scheme 5.4)

[6]. The proposed reaction pathway was confirmed by the deuterium incorporation reaction using D_2O (Scheme 5.5), and Pirali applied this system to the synthesis of α-aroylamino amides (**5**) using α-isocyanoacetamides as a nucleophile (Scheme 5.6) [8]. The synthetic utility of the reaction was exemplified by the total synthesis of proglumide **6**, a cholecystokinin antagonist used in the treatment of stomach ulcers.

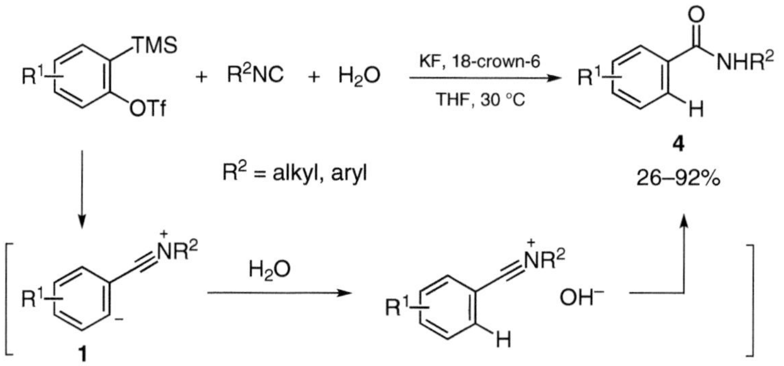

Scheme 5.4 Three-component reaction of arynes, isocyanides, and H_2O. Source: Based on Kaicharla et al. [6].

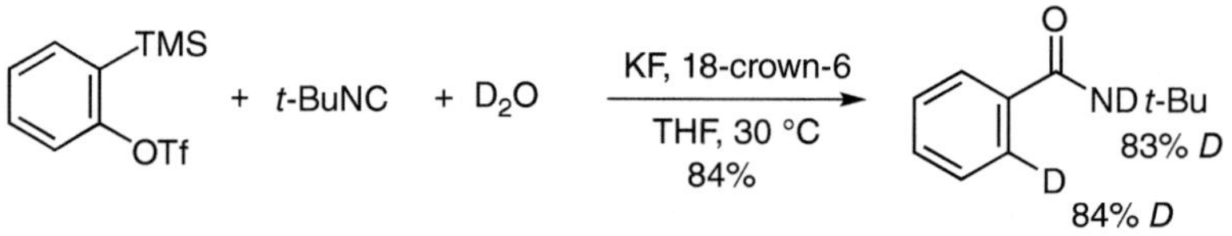

Scheme 5.5 Deuterium incorporation reaction using D_2O.

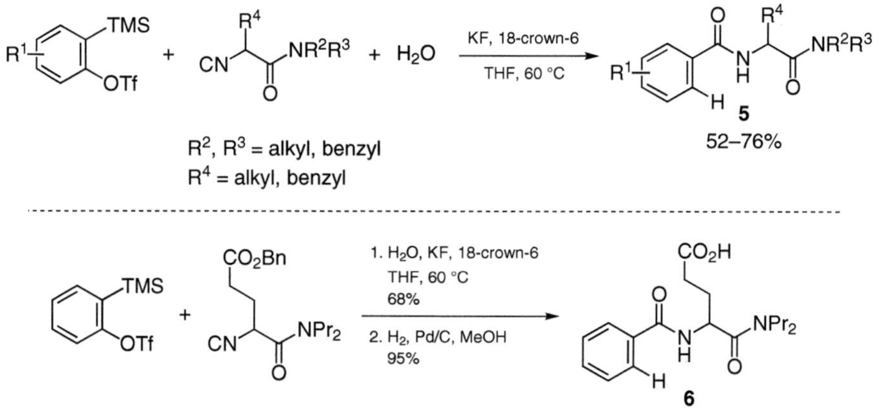

Scheme 5.6 Three-component reaction of arynes, α-isocyanoacetamides, and H_2O. Source: Based on Serafini et al. [8].

5.3.2 Active Methylene Compounds

Insertion reaction of arynes into a C(*methylene*)—C(*EWG*) bond of active methylene compounds was well established by Stoltz [9] and Yoshida [10], and the insertion followed by formal [4+2] cycloaddition by use of methyl 3-phenylpropiolates was reported by Shu and Wu (Scheme 5.7) [11]. Various multisubstituted naphthalenes (**7** and **8**) were straightforwardly accessible by the three-component reaction with aroylacetonitriles or diaroylmethanes; the aroyl group at the α position of **8**, in the latter case, was lacking via elimination ($ArCO_2^-$)-aromatization process as described in Scheme 5.8.

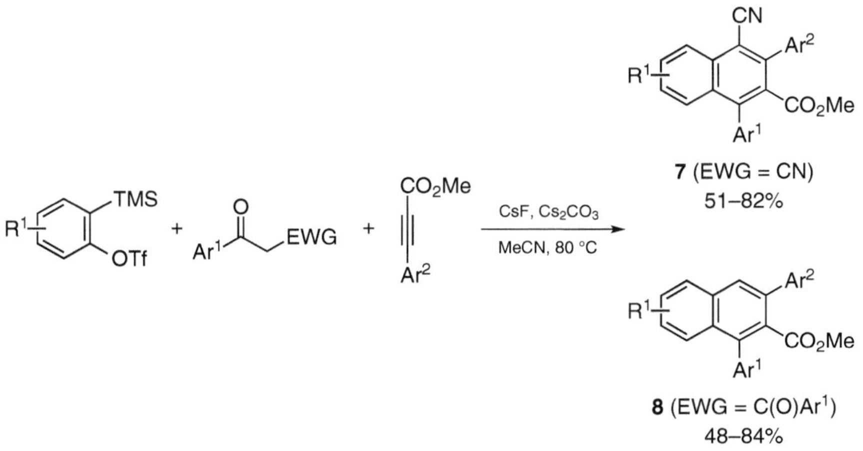

Scheme 5.7 Three-component reaction of arynes, active methylene compounds, and electron-deficient alkynes. Source: Based on Shu et al. [11].

The same group also developed formal [2+2+2] cycloaddition of arynes, active methylene compounds, and acetylene dicarboxylates, which gave multisubstituted naphthalenes (**9**) (Scheme 5.9) [12]. The difference in the reaction modes between this and the above may be attributable to the different electrophilicity of alkynoates employed, and an acetylene dicarboxylate first accepts nucleophilic attack by an anion of an active methylene compound to forma 1,4-dipole (**10**), which then undergoes formal [4+2] cycloaddition with an aryne to afford the final product (Scheme 5.10).

5.4 Nitrogen Nucleophile-Based Multicomponent Reactions

5.4.1 Amine

Aziridines bearing a carbonyl substituent at the 2-position were found to be added to arynes to generate zwitterion **11**, which were convertible into α-fluoro-β-amino acid derivatives (**12**) through ring opening with a fluoride (Scheme 5.11) [13]. A proton

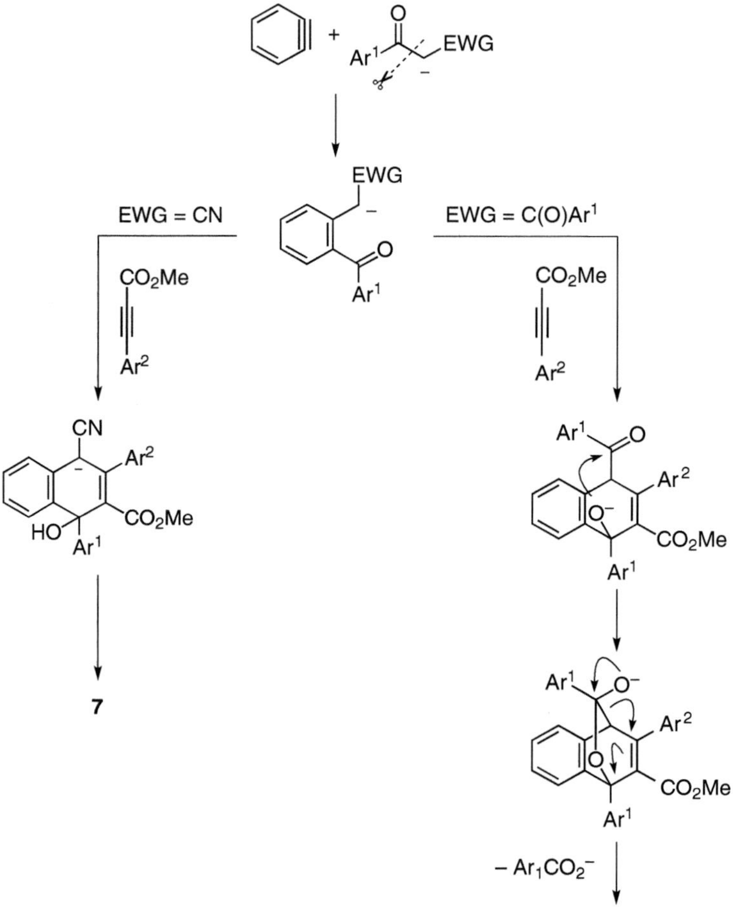

7

8

Scheme 5.8 Reaction pathways for three-component reaction of arynes, active methylene compounds, and electron-deficient alkynes.

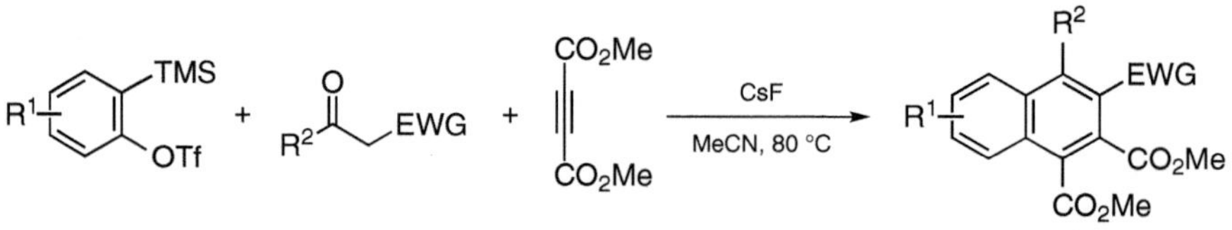

9
52–89%

R^2 = alkyl, aryl
EWG = C(O)R, CO_2R, CN

Scheme 5.9 Three-component reaction of arynes, active methylene compounds, and acetylene dicarboxylate. Source: Based on Shu et al. [12].

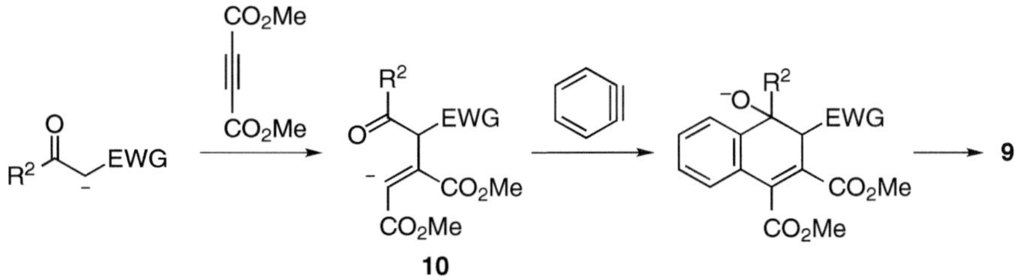

10

Scheme 5.10 A reaction pathway for three-component reaction of arynes, active methylene compounds, and acetylene dicarboxylate.

source, H_2O, which captures the aryl anion moiety in **11**, is necessary for the reaction to proceed, and thus the use of hydrated TBAF as a fluoride source is the key to the successful transformation.

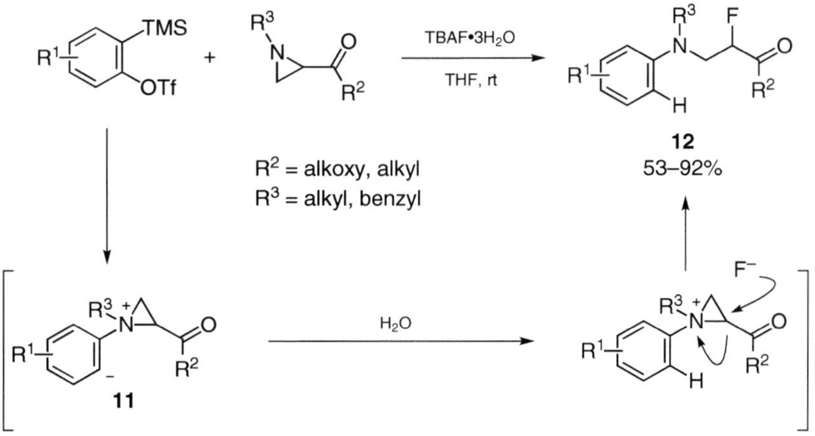

12
53–92%
11

Scheme 5.11 Four-component reaction of arynes, aziridines, H_2O, and a fluoride. Source: Based on Tang et al. [13].

Biju reported that aziridine-derived zwitterion **11** could also be transformed into ring-opened three-component products (**13** and **14**) using carboxylic acids or phenols as pronucleophiles (Scheme 5.12) [14]. The reaction was applicable to azetidines to produce γ-amino alcohol derivatives, and furthermore, the use of water in combination with trifluoroacetic acid led to a similar three-component reaction, where in situ-generated trifluoroacetate ion acts as an actual nucleophile toward aziridinium cation (Scheme 5.13) [15].

Aziridines with an electron-withdrawing functionality (CN or CO_2R) were found to undergo unique three-component reaction with arynes and aldehydes (or isatins) to provide amino epoxides (**16**) in a stereoselective manner (Scheme 5.14) [16]. An initially formed zwitterion (**11**) is not directly trapped by an aldehyde, but is converted into an aziridinium ylide (**15**) via proton transfer, which is finally trapped by an aldehyde to give epoxide **16**.

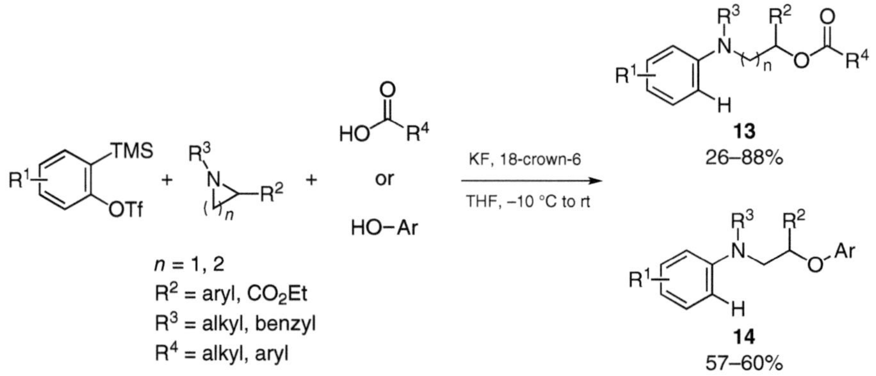

Scheme 5.12 Three-component reaction of arynes, aziridines, and carboxylic acids/phenols. Source: Based on Roy et al. [14].

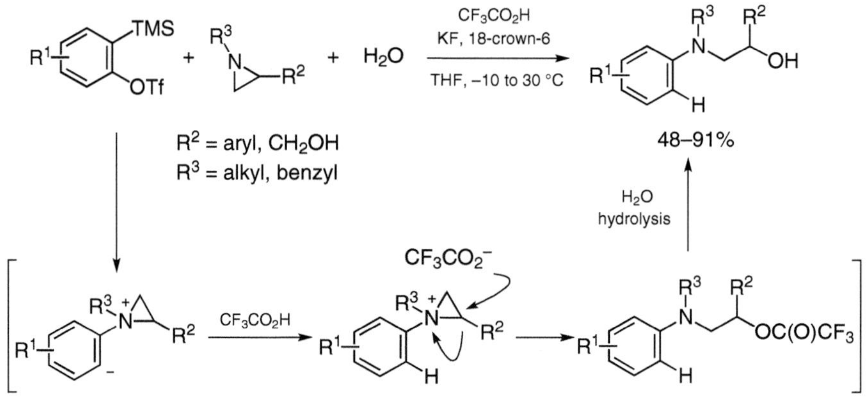

Scheme 5.13 Three-component reaction of arynes, aziridines, and trifluoroacetic acid. Source: Based on Roy et al. [15].

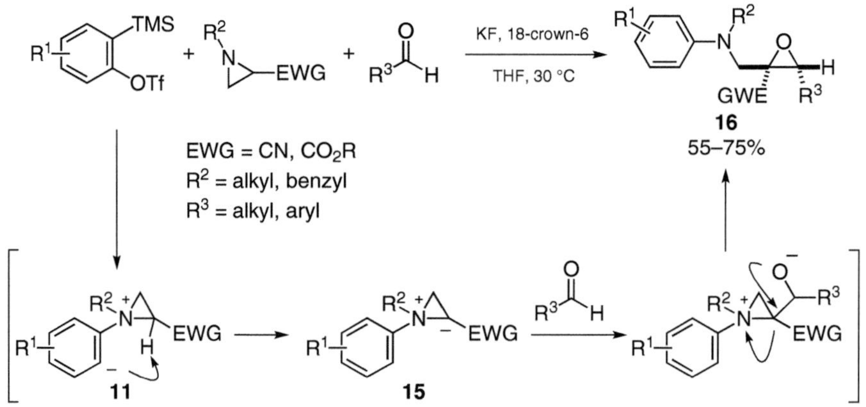

Scheme 5.14 Three-component reaction of arynes, aziridines, and aldehydes. Source: Based on Roy et al. [16].

Three-component reaction with cyclic amines accompanied by their ring opening was also achieved by using DABCO, where various thiols served as an effective pronucleophile for capturing zwitterion **17** (Scheme 5.15) [17]. A variety of *N*-alkyl-*N*′-aryl piperazines (**18**) were straightforwardly accessible in good yields, and structurally related quinuclidine could participate in the reaction.

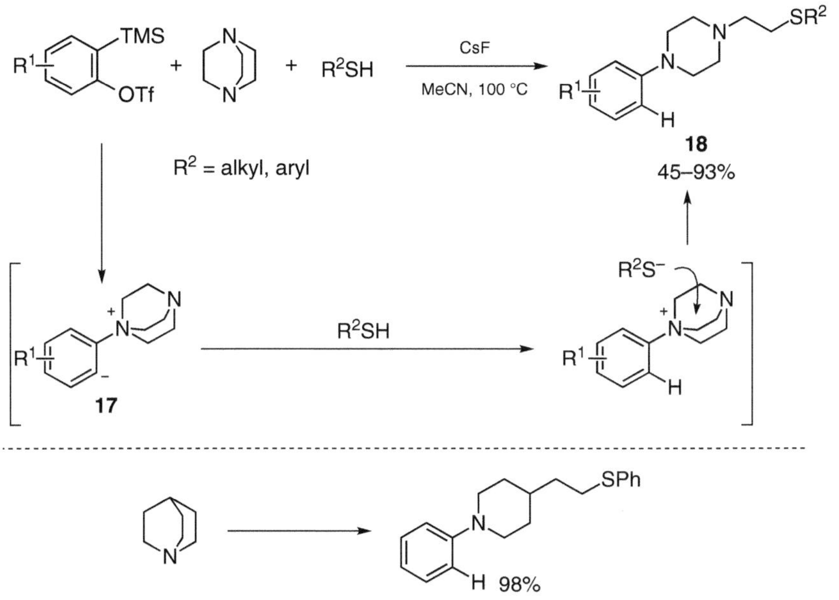

Scheme 5.15 Three-component reaction of arynes, DABCO, and thiols. Source: Based on Min et al. [17].

Biju disclosed that zwitterion **19** arising from nucleophilic attack of *N*,*N*-dialkylanilines were readily coupled with aldehydes or activated ketones (isatins or trifluoroacetophenone) to afford **21** (Scheme 5.16) [18]. The reaction proceeds through intramolecular migration of an aryl group from nitrogen to oxygen in intermediate **20**, being similar to the process of the Smiles rearrangement.

A similar three-component reaction was found to also take place by employing an atmospheric pressure of CO_2 for trapping the zwitterion (**19**) [19]; the aryl migration from nitrogen to oxygen likewise occurs to provide aryl benzoates (**22**) insofar as the aryl group possesses an electron-withdrawing substituent at the para-position (Scheme 5.17). In marked contrast, the reaction with electron-rich and -neutral *N*,*N*-dialkylanilines experienced alkyl migration, furnishing solely alkyl benzoates (**23**).

Okuma extended the above methodology to direct synthesis of nine- and ten-membered dibenzo-*N*,*O*-heterocycles (**24** and **25**) through the S_NAr pathway by use of *N*-methylindoline or *N*-methyltetrahydroquinoline as a nitrogen nucleophile and aldehydes as an electrophile (Scheme 5.18) [20].

Recently, Kim reported that an atmospheric pressure of sulfuryl fluoride serves as a good electrophile in the reaction with arynes and secondary amines, resulting

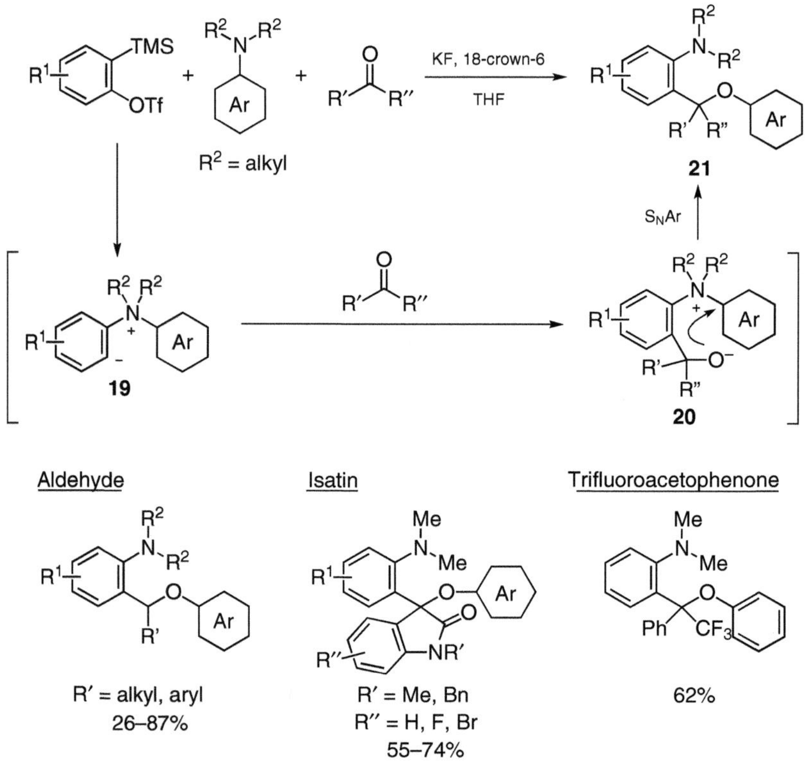

Scheme 5.16 Three-component reaction of arynes, anilines, and carbonyl compounds. Source: Based on Bhojgude et al. [18].

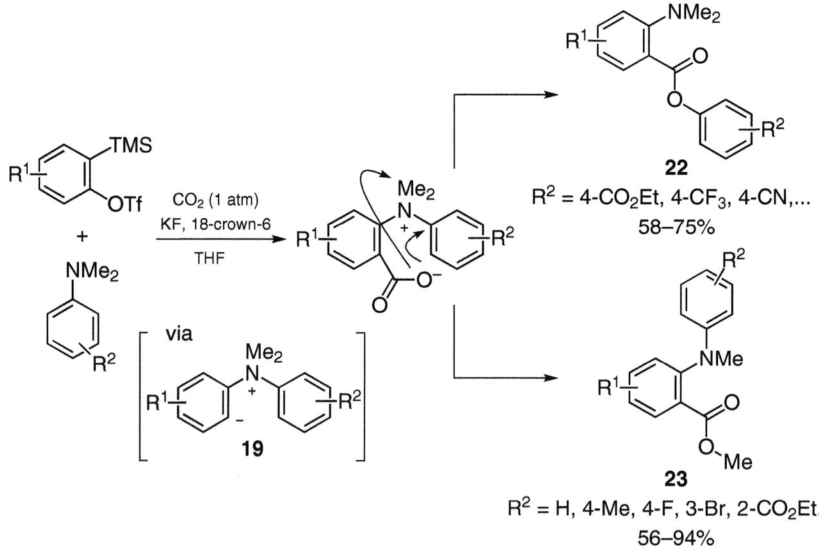

Scheme 5.17 Three-component reaction of arynes, anilines, and CO_2.

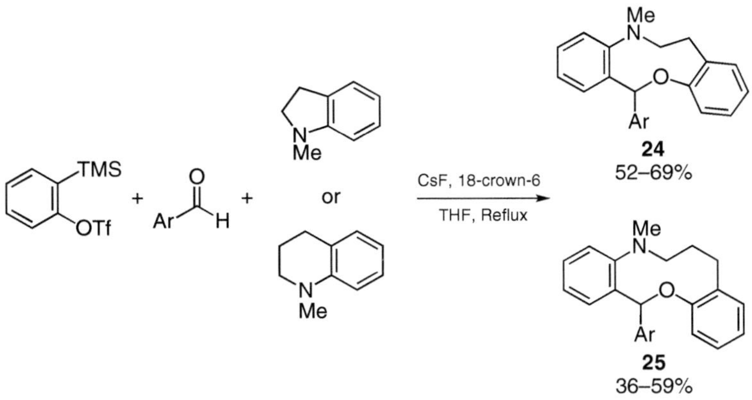

Scheme 5.18 Three-component reaction of arynes, *N*-heterocycles, and aldehydes. Source: Based on Okuma et al. [20].

in the formation of 2-amino-substituted arenesulfonyl fluorides (**26**) in good yields (Scheme 5.19) [21]. A variety of dialkyl-, alkylaryl-, and diarylamines were convertible into the respective products, and furthermore, the reaction was applicable to other nucleophiles, including *tert*-butylisocyanide and triphenylphosphine.

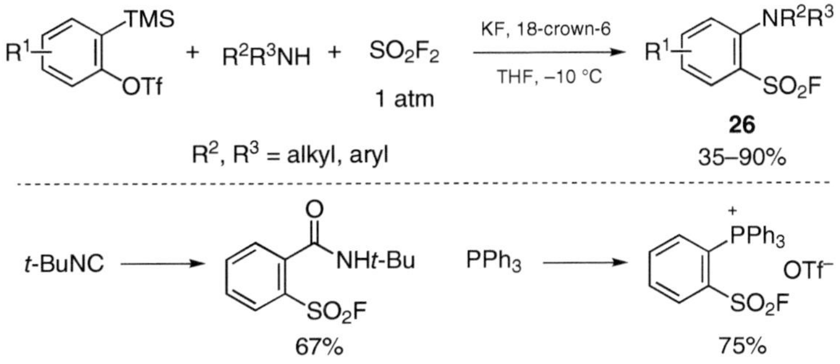

Scheme 5.19 Three-component reaction using arynes and sulfuryl fluoride. Source: Based on Kwon et al. [21].

5.4.2 Imine

A multicomponent reaction by use of imines as nucleophiles was first developed by Yoshida [22], where an initially formed zwitterion (**27**) was captured by CO_2 to give benzoxazinone derivatives. In 2017, Tian reported that **27** could smoothly react with such pronucleophiles as chloroform, acetonitrile, and methyl propiolate, affording a variety of functionalized tertiary amines (**28**) (Scheme 5.20) [23]. *N*-Aryl-2,2,2-trichloroethanamine is the common molecular structure obtained with chloroform; however, it further undergoes elimination of HCl, when the aryl moiety contains an electron-withdrawing group, leading to the formation of *N*-aryl-2,2-dichloroethenamine **29** (Scheme 5.21).

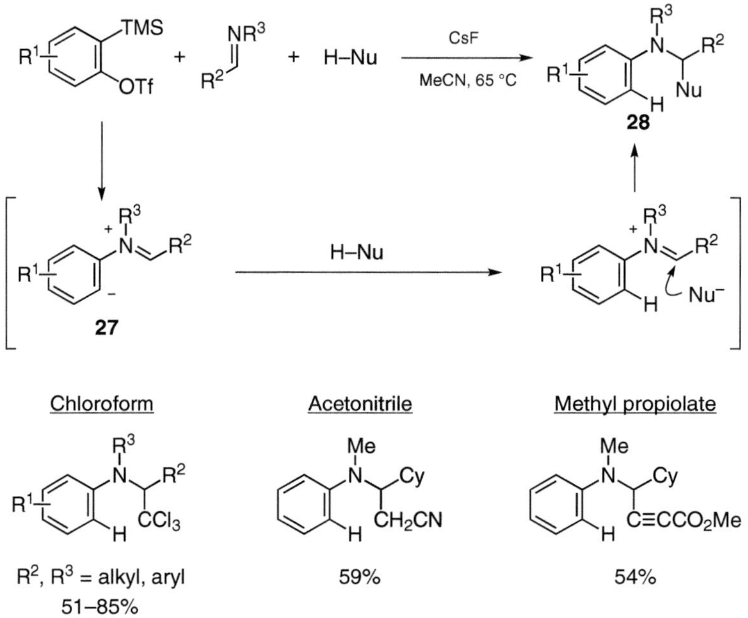

Scheme 5.20 Three-component reaction of arynes, imines, and pronucleophiles. Source: Based on Xu et al. [23].

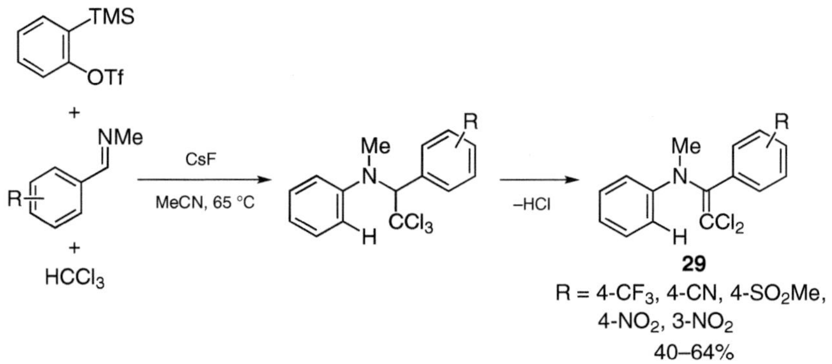

Scheme 5.21 Three-component reaction of arynes, imines, and chloroform.

A quite similar three-component reaction with oxazoline and chloroform was developed by Peng (Scheme 5.22) [24], and moreover, Tian and Yu first demonstrated that carbon tetrachloride could act as a pronucleophile for the three-component reaction with arynes and imines (or *N*-heteroarenes, see Section 5.4.3) [25]. In the latter reaction, the aryl anion moiety of zwitterion **27** attacks the chlorine atom of carbon tetrachloride to give product **30** accompanied by Ar—Cl bond-forming process (Scheme 5.23).

In contrast to the results above, imines having a $-CH_2CO_2Me$ substituent on the nitrogen atom were reported to generate readily azomethine ylides (**32**) from

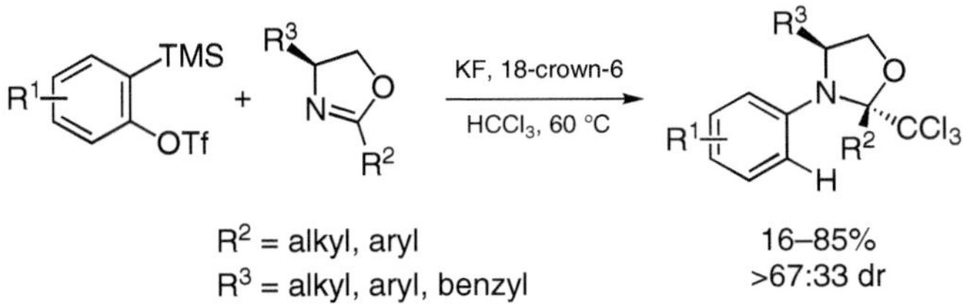

Scheme 5.22 Three-component reaction of arynes, oxazolines, and chloroform.

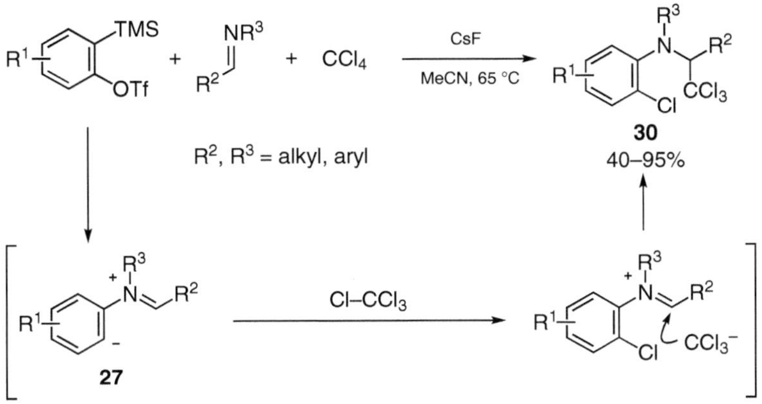

Scheme 5.23 Three-component reaction of arynes, imines, and carbon tetrachloride.

initially formed zwitterions (**31**) upon treatment with arynes via proton transfer, and the resulting azomethine ylides (**32**) were transformable to multisubstituted pyrrolidines (**33**) stereoselectively by [3+2] cycloaddition with electron-deficient alkenes, including methyl vinyl ketone, acrylonitrile, and dimethyl maleate (Scheme 5.24) [26].

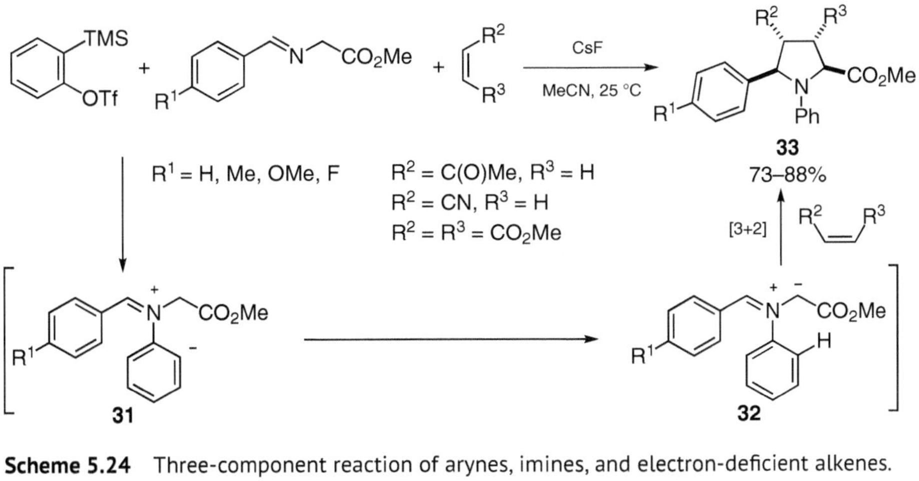

Scheme 5.24 Three-component reaction of arynes, imines, and electron-deficient alkenes. Source: Based on Swain et al. [26].

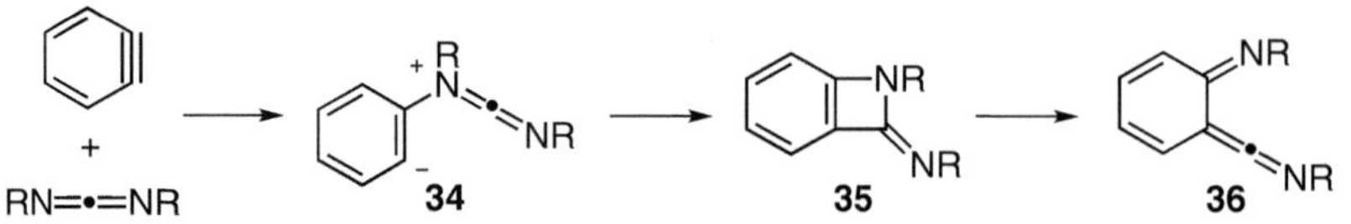

Scheme 5.25 Reaction of arynes with carbodiimides.

Carbodiimides, having cumulative C=N double bonds, were found to be converted into benzoazetine derivatives (**35**) by formal [2+2] cycloaddition with arynes probably through zwitterionic intermediates (**34**) (Scheme 5.25), as were the cases with imines [27]. The resulting nitrogen four-membered ring then underwent 4π-electrocyclic ring opening to give aza-*ortho*-quinone methides (**36**), which were demonstrated to serve as key intermediates in multicomponent reactions with pronucleophiles. In 2014, Zhang disclosed that **36** could be trapped by terminal alkynes to provide alkynylimine derivatives (**38**) with dual incorporation of arynes (Scheme 5.26) [28]. The formation of **38** can be rationally explained by *N*-arylation of the aniline moiety in primarily generated **37**, arising from the capture of **36** with terminal alkynes. The reaction was also applicable to bromo- and iodoalkynes to give similar products, although the reaction pathway is unclear.

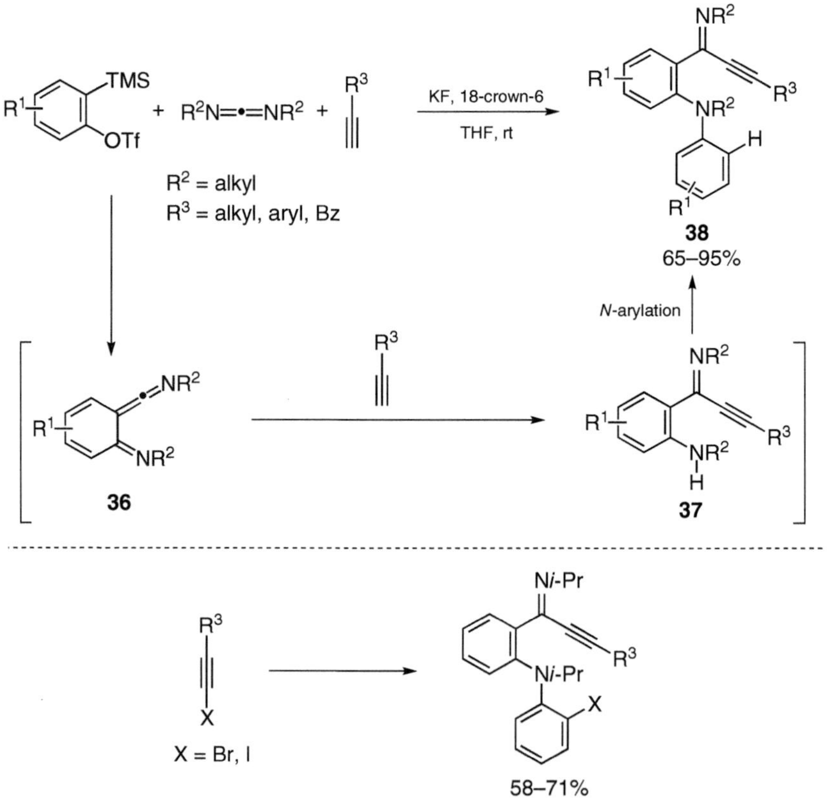

Scheme 5.26 Three-component reaction of arynes, carbodiimides, and alkynes. Source: Based on Zhou et al. [28].

Water and alcohols were effective pronucleophiles for trapping the carbodiimide-derived aza-*ortho*-quinone methide (**36**), resulting in the direct formation of amides (**39**) and imidates (**40**), respectively (Scheme 5.27) [29]; two molar amounts of arynes were incorporated into the final products also in this case.

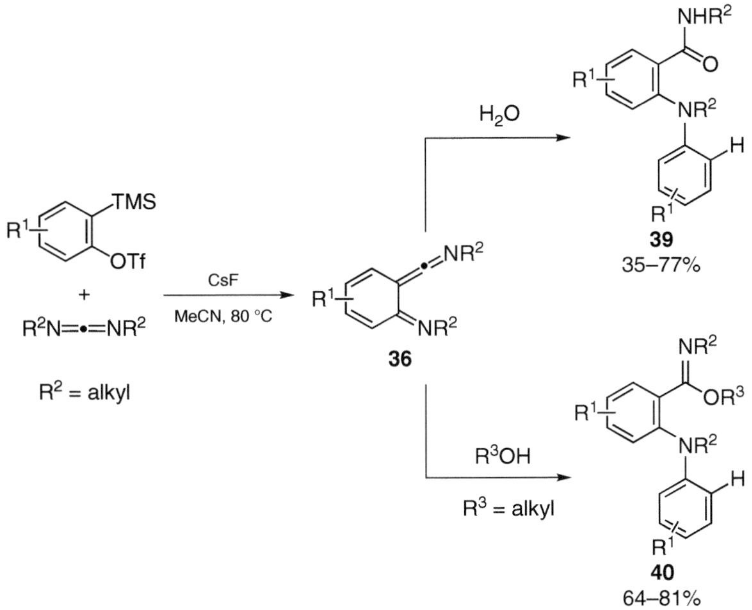

Scheme 5.27 Three-component reaction of arynes, carbodiimides, and H_2O/alcohols. Source: Based on Li et al. [29].

5.4.3 *N*-Heteroarene

Biju developed a three-component reaction with quinolines and aldehydes that produced various six-membered *N*,*O*-heterocycles (**42**) straightforwardly (Scheme 5.28) [30], after his pioneering work on the three-component reaction using isoquinolines as *N*-heteroarene nucleophiles [31]. As was the case with an isoquinoline, nucleophilic attack of a quinoline to an aryne triggers the reaction; an intermediary formed zwitterion (**41**) is then trapped by an aldehyde to give **42**. Under Biju's conditions (KF/18-crown-6 in THF, −10 °C to rt), the diastereoselectivity was high on the whole, and cis-isomers were generated preferentially. In addition, such ketones as benzophenone, *p*-benzoquinone, and ethyl benzoylformate could also participate in the reaction. Around the same time, Lei and Hu also reported the reaction by use of quinolines and aldehydes [32], although the diastereoselectivity was much lower than that of Biju's results under their conditions (CsF in THF, 60 °C).

Zwitterions (**43** and **41**) derived from isoquinolines or quinolines were demonstrated to be transformed into dearomatized trichloro-substituted products (**44** and

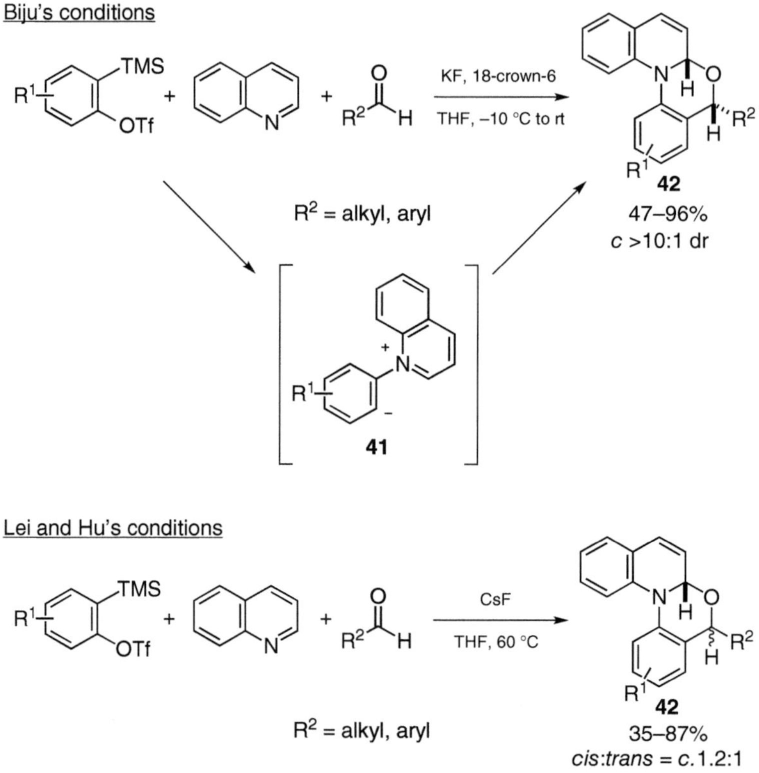

Scheme 5.28 Three-component Reaction of arynes, quinolines, and aldehydes. Source: Based on Bhunia et al. [30].

45) by employing chloroform as a pronucleophile (Scheme 5.29) [33]. Dichloro- and dibromomethane could also act as a pronucleophile for capturing **43**, and furthermore, the reaction with carbon tetrachloride proceeded in a similar fashion to that with imines (see Section 5.4.2) to afford products **46** with concurrent construction of the Ar—Cl bond (Scheme 5.30) [25]. It should be noted that acridine acted as a nucleophile in the latter case, where an aryne and carbon tetrachloride were added in 1,4-manner rather than usual 1,2-manner.

Dai and He disclosed that dialkoxyphosphites could be involved as a pronucleophile in the three-component reaction with quinolines or isoquinolines, giving rise to dearomatized phosphonylated *N*-heterocycles (**47**) in good yields (Scheme 5.31) [34]. Monophosphonylation took place exclusively with 1,10-phenanthrene, and acridine underwent 1,4-addition to provide **48** and **49**.

5.4.4 Diazene

Multicomponent access to cinnoline derivatives by the use of arynes, tosylhydrazine, and α-bromoketones was reported by Wu [35]. A variety of α-bromoacetophenones were convertible into the respective cinnolines (**50**) in a straightforward manner,

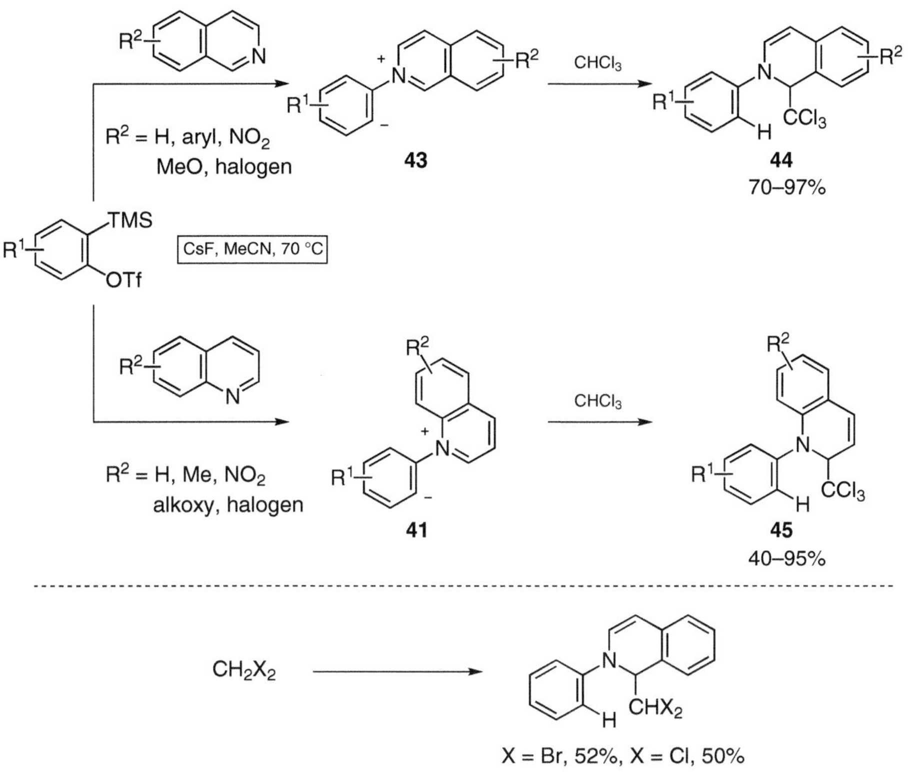

Scheme 5.29 Three-component reaction of arynes, *N*-heteroaromatics, and pronucleophiles. Source: Based on Tan et al. [33].

and the actual nucleophile in the reaction is diazene generated in situ from tosylhydrazine (Scheme 5.32).

5.4.5 Nitrite

The highly electrophilic character of arynes allows even sodium nitrite to act as a nucleophile, leading to a new method for nitrating an aromatic ring [36]. This concept was further extended to the three-component reaction; aromatic aldehydes were suitable electrophiles in the reaction to produce 2-nitrobenzhydrols (**51**), although the yields still remained moderate (Scheme 5.33).

5.5 Oxygen Nucleophile-Based Multicomponent Reactions

5.5.1 Dimethylformamide

Since Miyabe [37] and Yoshida [38] developed the multicomponent reactions for synthesizing benzo-annulated *O*-heterocycles, with a DMF-derived ortho-quinone

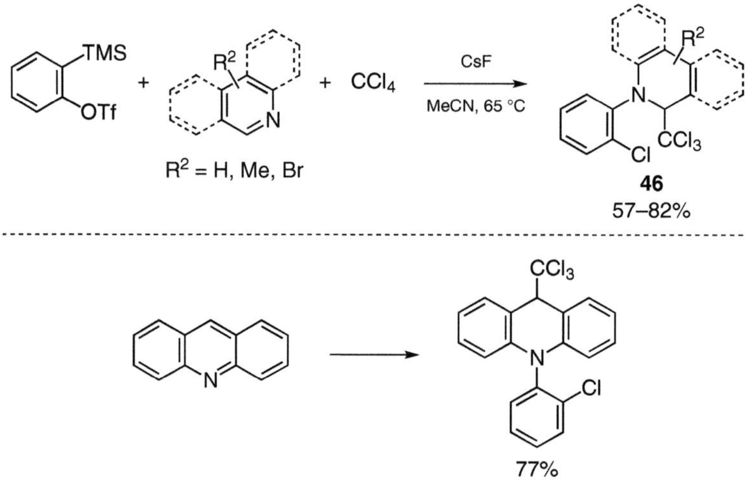

Scheme 5.30 Three-component reaction of arynes, *N*-heteroaromatics, and carbon tetrachloride. Source: Based on Li et al. [25].

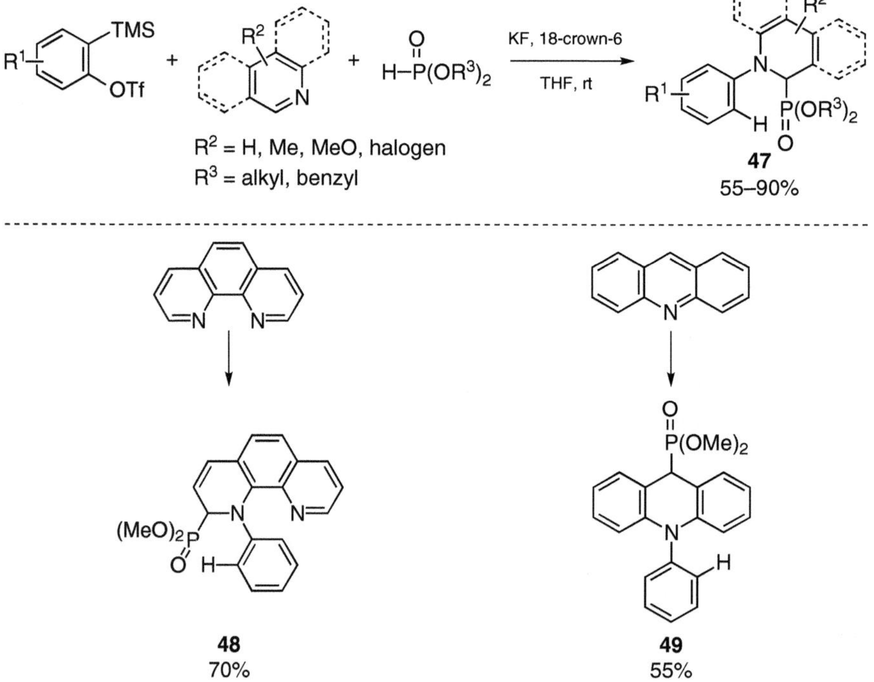

Scheme 5.31 Three-component reaction of arynes, *N*-heteroaromatics, and dialkoxy phosphites. Source: Based on Liu et al. [34].

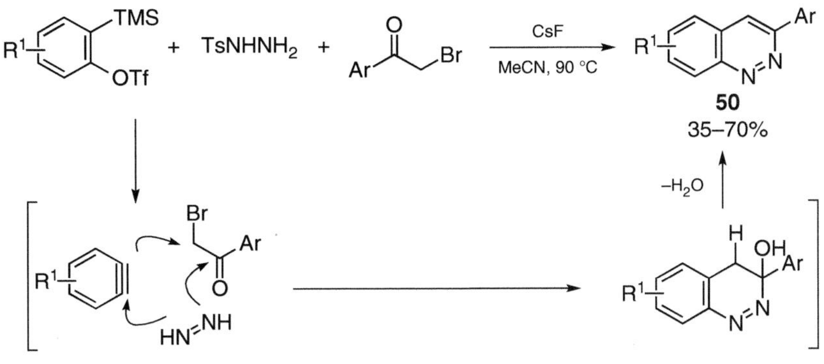

Scheme 5.32 Three-component reaction of arynes, diazene, and α-bromoketones.

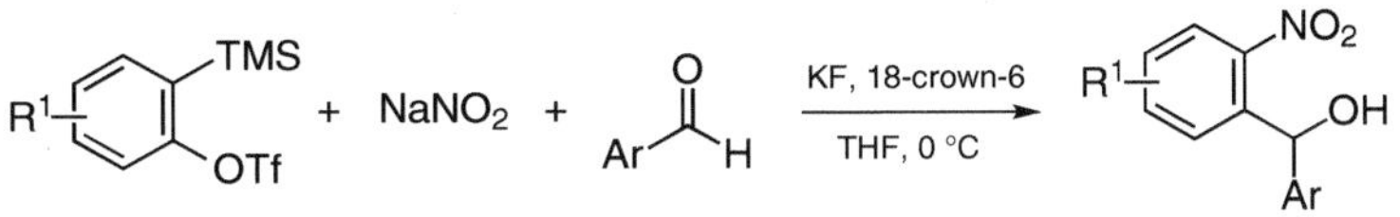

Scheme 5.33 Three-component reaction of arynes, sodium nitrite, and aldehydes.

methide, arising from formal [2+2] cycloaddition with arynes, this reactive intermediate has continued to be the key scaffold for the development of new multicomponent reactions. Similar to the above pioneering works, the ortho-quinone methide (**52**) was found to undergo facilely [4+2] cycloaddition with acetylene dicarboxylates to provide dimethylamino-4*H*-chromenes (**53**) in good yields (Scheme 5.34) [39].

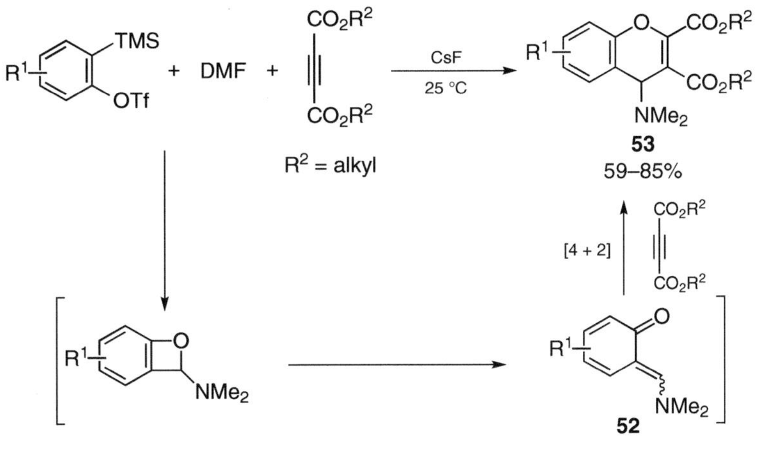

Scheme 5.34 Three-component reaction of arynes, DMF, and acetylene dicarboxylates. Source: Based on Yoshioka et al. [39].

Li reported that *N*,*S*-keteneacetals could serve as efficient dienophiles for trapping ortho-quinone methide **52** to afford diverse 2-aryliminochromenes (**54**) in high yields (Scheme 5.35) [40]. An intermediary generated [4+2] cycloadduct from **52** and an *N*,*S*-keteneacetal is transformed into the final product via elimination of dimethylamine and methyl mercaptan. A related three-component reaction with in situ-generated sulfonyl allenes from 2-bromoallyl sulfones, which directly produced 3-(arylsulfonyl)-2*H*-chromene-2-ols (**55**), was disclosed by Gogoi (Scheme 5.36) [41].

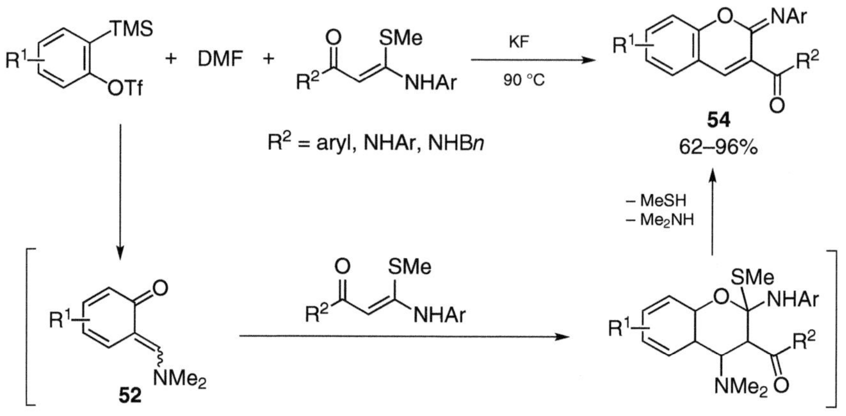

Scheme 5.35 Three-component reaction of arynes, DMF, and *N*,*S*-keteneacetals. Source: Based on Wen et al. [40].

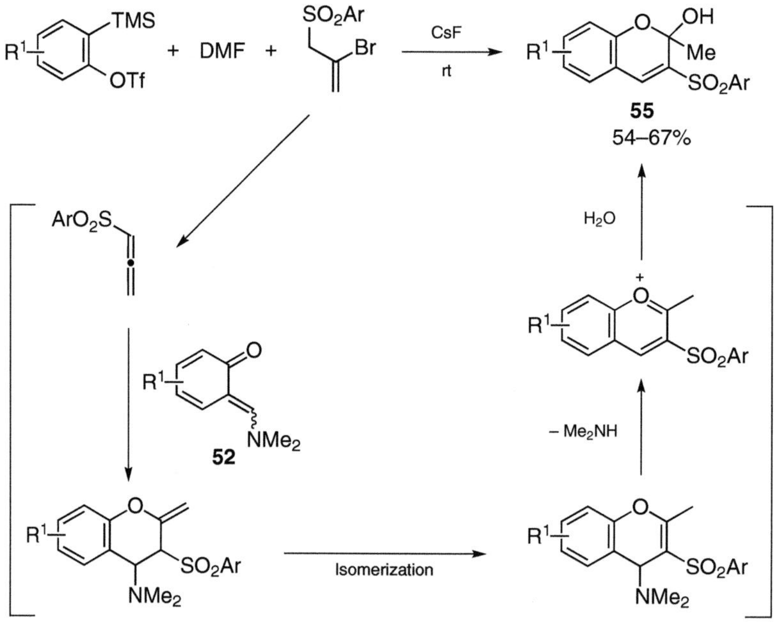

Scheme 5.36 Three-component reaction of arynes, DMF, and sulfonyl allenes. Source: Based on Sharma and Gogoi [41].

The ortho-quinone methide (**52**) was convertible into 2-aroylbenzofurans (**57**) by formal [4+1] cycloaddition followed by elimination of dimethylamine from intermediary formed five-membered ring **56** (Scheme 5.37). Acetophenone-derived sulfonium salts (*A*) [42] and α-bromoacetophenones (*B*) [43] were utilized as C1 source for the three-component reaction, and sulfoniumylides were proposed to be actual trapping agents in the former case.

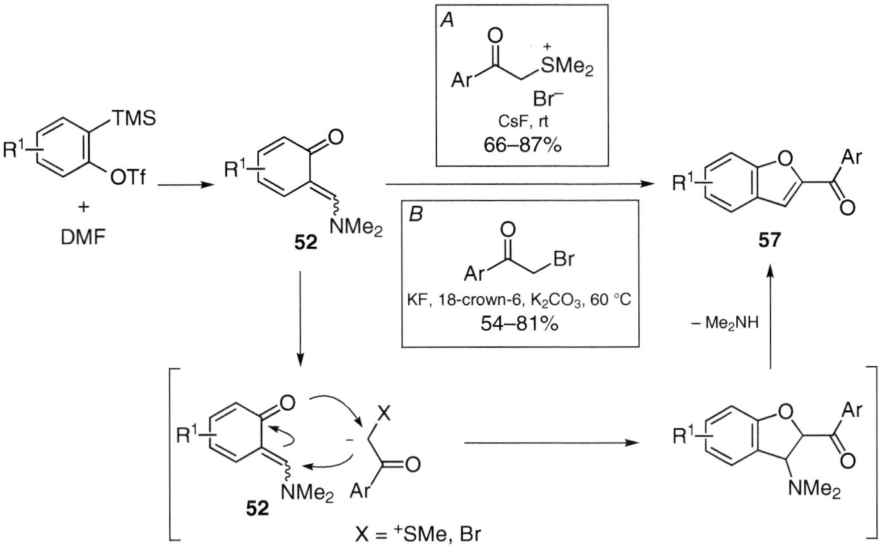

Scheme 5.37 Three-component [4+1] cycloaddition using arynes and DMF.

Jiang reported that treatment of arynes with diaryliodonium salts in DMF gave *ortho*-formyl diaryl ethers (**58**) (Scheme 5.38) [44]. *para*-Anisyl moiety in unsymmetrical diaryliodonium salts (Ar–I^+–*p*-anisyl X^-) was found to act as "dummy group," leading to the selective installation of an Ar moiety into the final product; ortho-quinone methide**52** is the key intermediate also in the reaction, whose nucleophilic carbonyl oxygen reacts with an electrophilic diaryliodonium salt to give **58**.

Gogoi disclosed that the nucleophilic carbonyl oxygen of ortho-quinone methide **52** could also be coupled with allyl bromides to give allyl *ortho*-formylaryl ethers (**59**) after hydrolytic work-up (Scheme 5.39) [41].

Lu and Wang developed a unique three-component reaction using aroyl cyanides, in which ortho-quinone methide **52** was inserted into the ArCO—CN bond to provide α-amino-α-aryl carbonitriles (**61**) in a straightforward manner [45]. As shown in Scheme 5.40, the nucleophilic carbonyl oxygen of **52** attacks the carbonyl carbon of an aroyl cyanide to give intermediate **60**, whose cyano group then migrates to the iminium carbon to form **61**.

5.5.2 Sulfoxide

In 2014, Chen and Xiao first demonstrated that a S=O bond of DMSO underwent formal [2+2] cycloaddition with arynes to generate *S*,*O*-containing

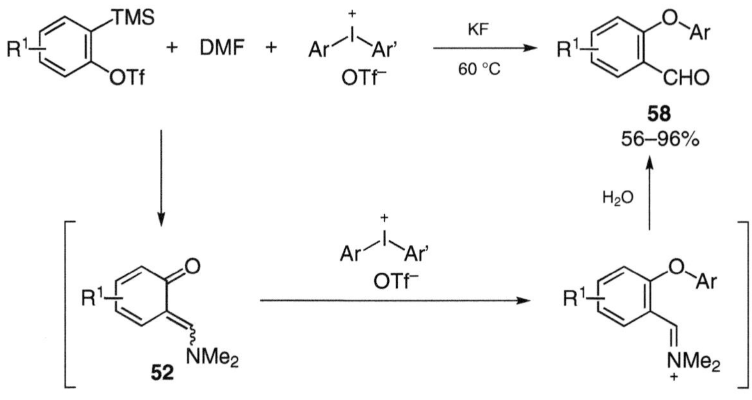

Scheme 5.38 Three-component reaction of arynes, DMF, and diaryliodonium salts. Source: Based on Liu et al. [44].

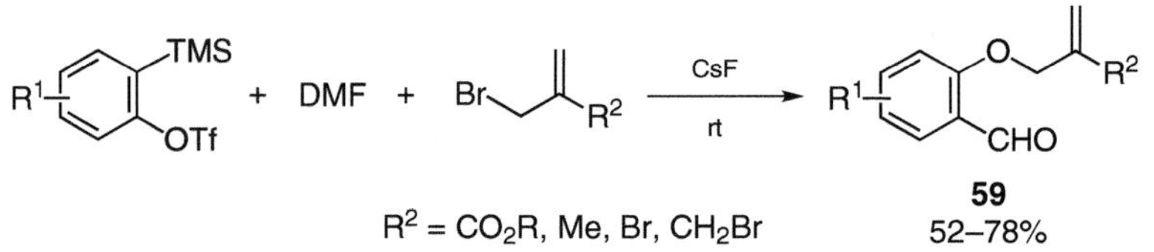

Scheme 5.39 Three-component reaction of arynes, DMF, and allyl bromides. Source: Based on Sharma and Gogoi [41].

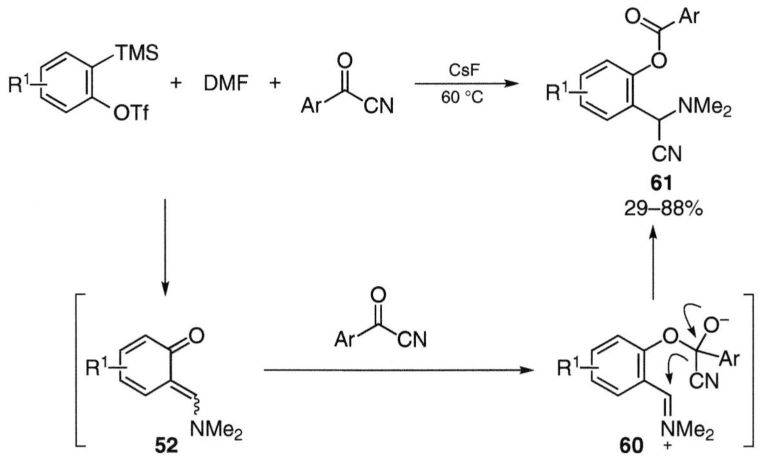

Scheme 5.40 Three-component reaction of arynes, DMF, and aroyl cyanides.

four-membered ring **62**, which was converted into the respective ortho-quinoid species (**63**) (Scheme 5.41) [46]. In the presence of activated alkyl bromides such as α-bromoketones, α-bromoesters, and a benzyl bromide, ortho-quinoid species **63** was smoothly alkylated at the oxygen atom to afford product **64** through concurrent demethylation at the sulfur atom.

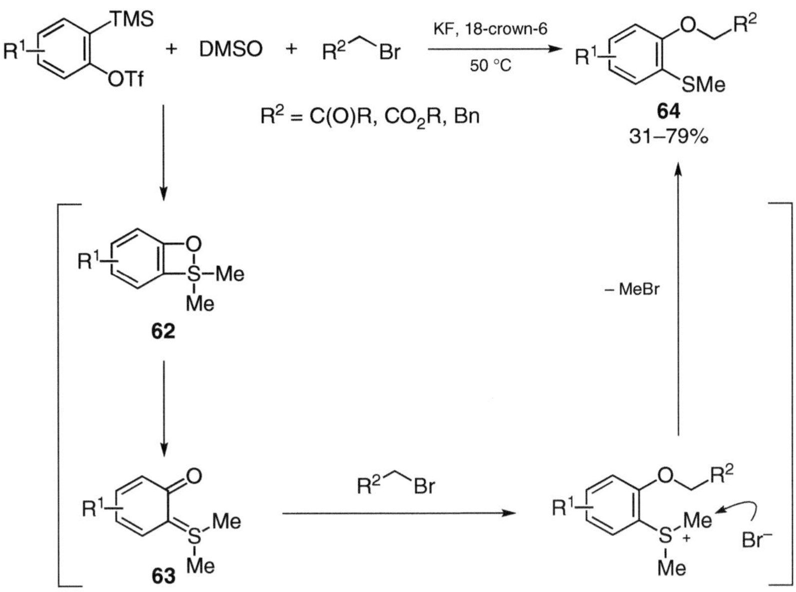

Scheme 5.41 Three-component reaction of arynes, DMSO, and alkyl bromides. Source: Based on Liu et al. [46].

The DMSO-derived ortho-quinoid species (**63**) was also found to be captured with acetylene dicarboxylates to provide maleate derivatives **65** with high stereoselectivity (Scheme 5.42) [47]. The reaction proceeds through conjugate addition of the nucleophilic carbonyl oxygen to a carbon–carbon triple bond of an acetylene dicarboxylate, accompanied by the demethylation.

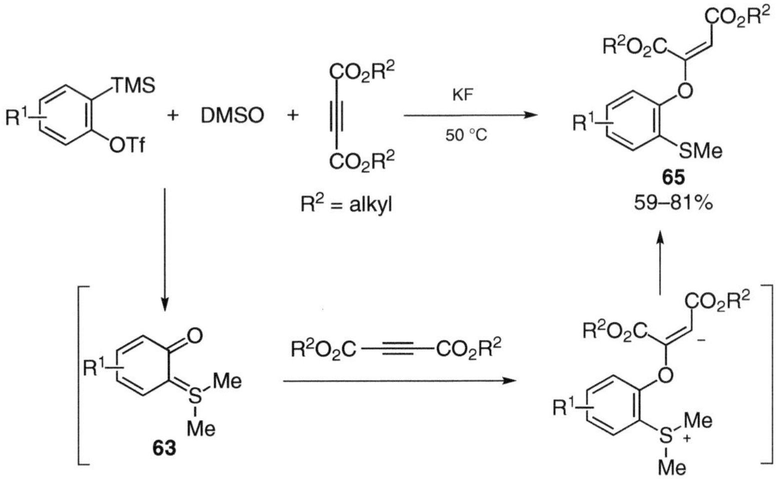

Scheme 5.42 Three-component reaction of arynes, DMSO, and acetylene dicarboxylates. Source: Based on Hazarika et al. [47].

A unique 1,2,3-trifunctionalization of an aromatic ring was developed by Li [48]; treatment of aryl allyl sulfoxides with arynes and carbon electrophiles (activated alkyl bromides and di-*tert*-butyl dicarbonate) led to consecutive construction of Ar^1—O, Ar^2—S, and Ar^3—C bonds, giving **69** in good yields (Scheme 5.43). As were the cases with DMSO, formal [2+2] cycloaddition between an aryl allyl sulfoxide and an aryne triggers the reaction to form *S*,*O*-four-membered ring **66**, which is converted into **67** via *S* to *O* allyl migration. Oxonium Claisen rearrangement followed by ring-opening then gives **68**, which is finally trapped by a carbon electrophile to provide trifunctionalized product **69**.

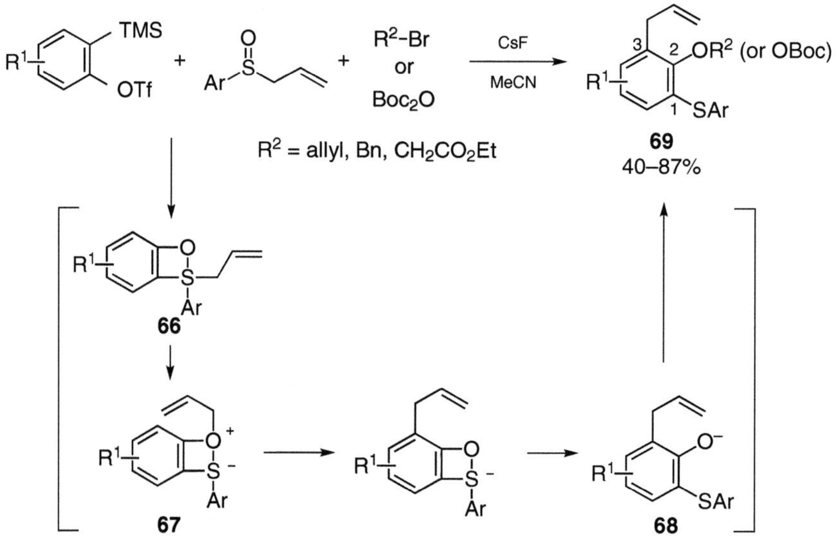

Scheme 5.43 Three-component reaction of arynes, aryl allyl sulfoxides, and carbon electrophiles.

5.5.3 Cyclic Ether

Such a cyclic ether as THF could serve as a good nucleophile toward arynes, and the resulting zwitterions (**70**) were found to react facilely with alcohol to offer three-component reaction products (**71**) (Scheme 5.44) [49]. Other cyclic ethers with different ring size could also participate in the reaction; however, the yields became much lower than those obtained with THF.

Qi and Jiang developed a four-component reaction of cyclic ethers, arynes, CO_2, and amines, that provided a variety of functionalized carbamates (**74**) (Scheme 5.45) [50]. In stark contrast to other multicomponent reactions, attachment of a triflyloxy group (OTf) at the adjacent position to the aryne triple bond was indispensable for the reaction to proceed, which may be due to the enhanced aryne electrophilicity induced by the OTf substituent. It should be noted that tetrabutylammonium iodide (TBAI), instead of commonly used fluoride, was the best promoter for the reaction.

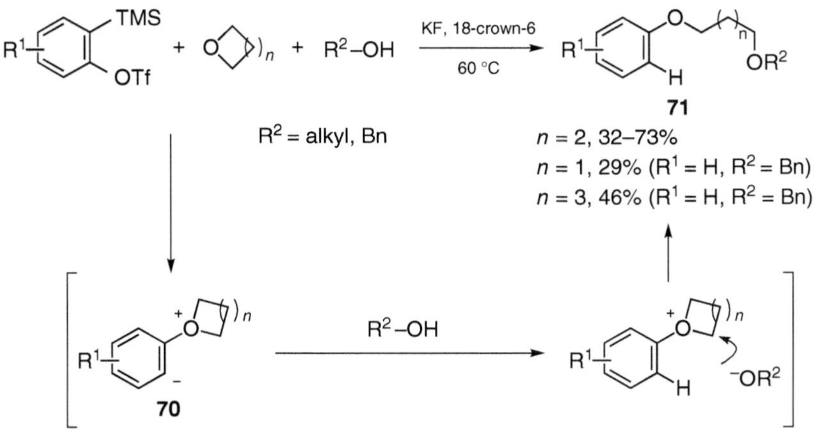

Scheme 5.44 Three-component reaction of arynes, cyclic ethers, and alcohols. Source: Based on Thangaraj et al. [49].

The regioselective formation of zwitterion **72**, where THF attacks the meta position to the OTf group, commences the reaction. The resulting zwitterion (**72**) then undergoes protonation and ring opening with in situ-generated carbamate (**73**) from CO_2 and an amine.

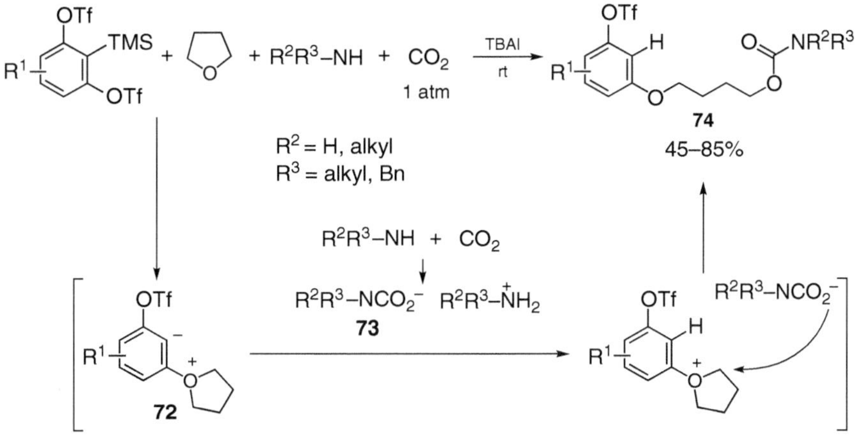

Scheme 5.45 Four-component reaction of arynes, cyclic ethers, amines, and CO_2. Source: Based on Xiong et al. [50].

5.5.4 Trifluoromethoxide

A trifluoromethoxide anion, generated efficiently from trifluoromethyl benzoate and a fluoride, was reported to be added across arynes, and the resulting *ortho*-(trifluoromethoxy)aryl anion could be trapped by a Br^+ equivalent (PhC≡Br, $C_6F_{13}Br$, and C_6F_5Br) to afford trifluoromethoxylation–bromination products (**75**) (Scheme 5.46) [51]. Iodine and chlorine could also be installed by use of C_6F_5I or

CCl_4 as an electrophile, respectively, and furthermore, this protocol was extended to other prefluoroalkoxylation–bromination reactions.

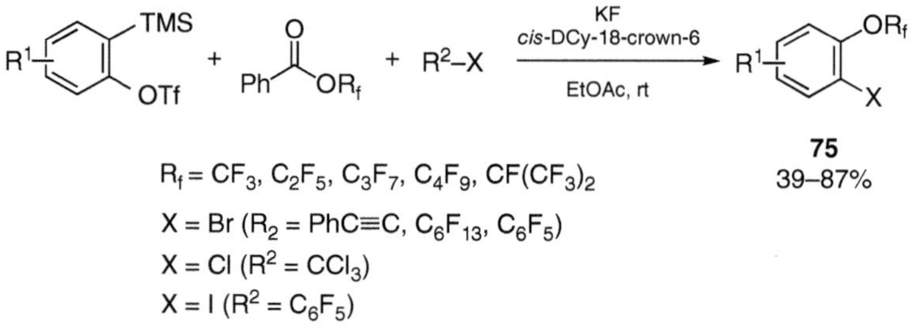

Scheme 5.46 Three-component reaction using arynes and perfluoroalkoxides. Source: Based on Zhou et al. [51].

5.6 Phosphorus Nucleophile-Based Multicomponent Reactions

In 2013, Hosoya disclosed that the phosphorus atom of methyl *N,N,N′,N′*-tetra-isopropylphosphorodiamidite [$(i\text{-}Pr_2N)_2POMe$] could act as a good nucleophile toward arynes [52]; the generated zwitterion (**76**) was then captured by an atmospheric pressure of CO_2 to give methyl esters (**77**) via methyl group migration (Scheme 5.47).

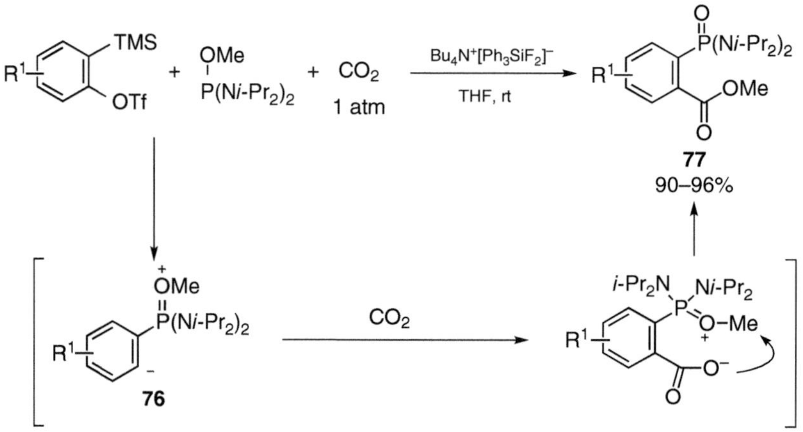

Scheme 5.47 Three-component reaction of arynes, phosphorodiamidite, and CO_2.

A three-component reaction with phosphines and activated ketones such as trifluoroacetophenones, α-ketoesters, and isatins was reported by Biju (Scheme 5.48) [53]. The reaction proceeds via the initial formation of zwitterion **78** from a phosphine and an aryne, which is intercepted by an ketone in a formal [3+2] cycloaddition fashion

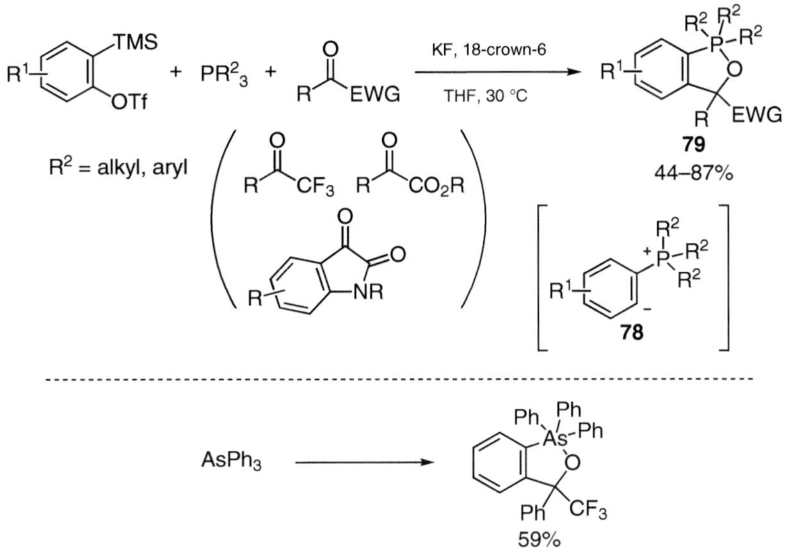

Scheme 5.48 Three-component reaction of arynes, phosphines, and ketones. Source: Based on Bhunia et al. [53].

to afford a benzoxaphosphole derivative (**79**). Triaryl- and trialkylphosphine, and even triphenylarsine, could be used as a nucleophilic trigger.

A quite similar three-component reaction also took place by use of aldehydes as an electrophile, and a variety of benzoxaphosphole derivatives (**80**) were directly available from triaryl- and trialkylphosphines (Scheme 5.49) [54]. In contrast to the above results, the reaction with triphenylarsine led to the sole formation of arsonium triflate.

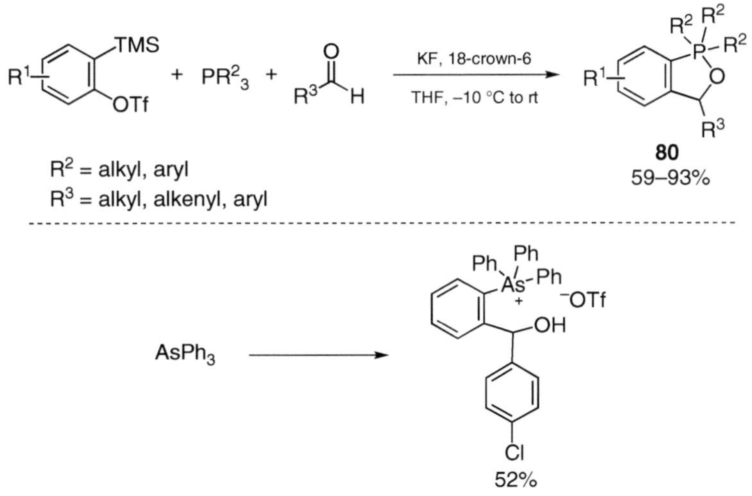

Scheme 5.49 Three-component reaction of arynes, phosphines, and aldehydes. Source: Based on Bhunia et al. [54].

5.7 Sulfur Nucleophile–Based Multicomponent Reactions

Tan and Xu reported that such cyclic thioethers with four- to six-membered ring as thietane, tetrahydrothiophene, and thiane readily formed zwitterions (**81**) by treating with arynes, which were finally transformable into the three-component products (**82**) via ring-opening capture with pronucleophiles (Scheme 5.50) [55]. A wide variety of carbon-, nitrogen-, oxygen-, and sulfur-centered pronucleophiles with pK_a range (~13–19) efficiently served as a ring-opening agent: the representative ones, which furnished high yields of the products (**82**), were bis(phenylsulfonyl)methane (*C*), 2-benzoxazolinone (*N*), 2-naphthol (*O*), and ethyl 2-mercaptoacetate (*S*).

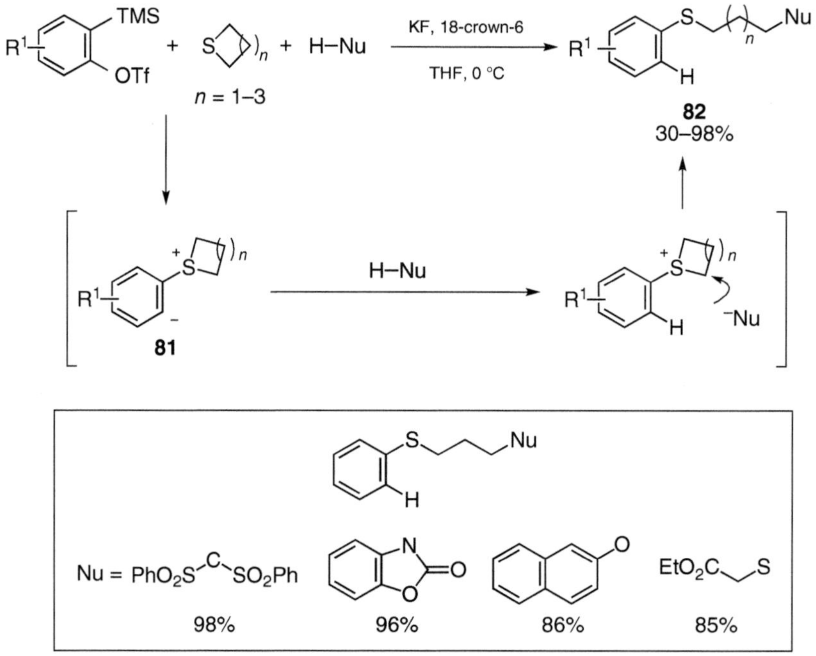

Scheme 5.50 Three-component reaction of arynes, cyclic thioethers, and pronucleophiles. Source: Based on Zheng et al. [55].

The ring-opening three-component reaction by use of cyclic thioethers was also applicable to a fluoride ion, leading to the direct formation of fluorinated alkyl aryl sulfides (**83**). A proton source for trapping the aryl anion moiety of an intermediary-generated zwitterion (**81**) was necessary for the smooth reaction: Tan and Xu utilized 2,3-dimethylindole (Scheme 5.51) [56], and He utilized H_2O for this purpose (Scheme 5.52) [57]. Under the latter reaction conditions, He disclosed that chlorine, bromine, and thiocyanate were incorporated into the ring-opened products (**84**) in the presence of KCl, KBr, or KSCN (Scheme 5.53). Moreover, trimethylsilyl cyanide, azide, and chloride were demonstrated to serve as a good ring-opening agent in the three-component reaction (Scheme 5.54).

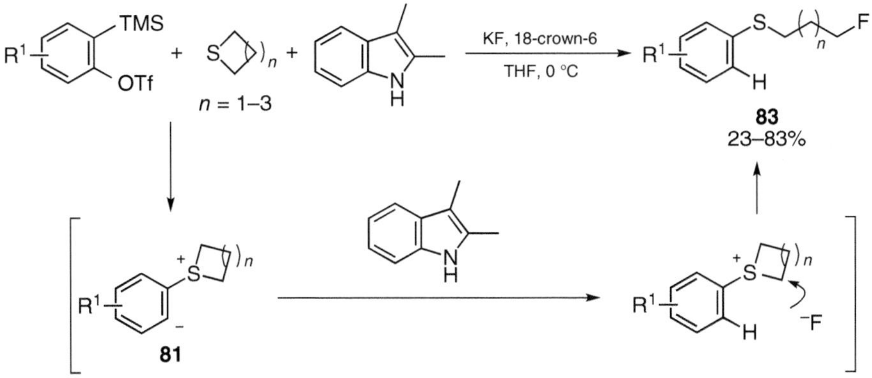

Scheme 5.51 Four-component reaction of arynes, cyclic thioethers, a proton source, and a fluoride. Source: Based on Fan et al. [56].

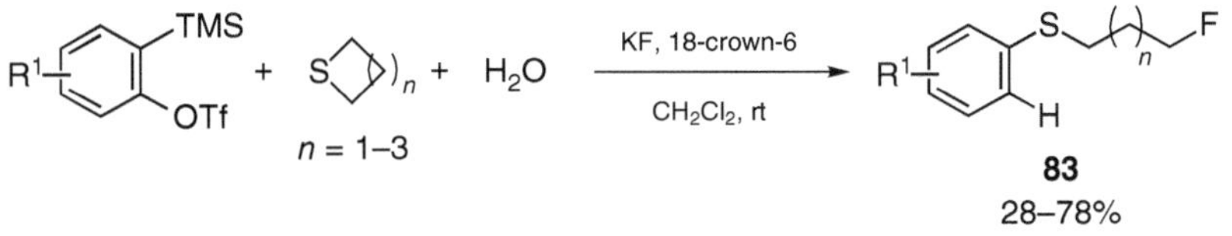

Scheme 5.52 Four-component reaction of arynes, cyclic thioethers, H_2O, and a fluoride. Source: Based on Jian et al. [57].

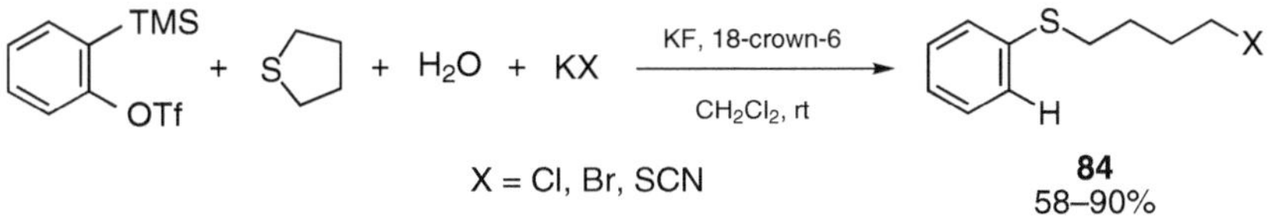

Scheme 5.53 Four-component Reaction of arynes, cyclic thioethers, H_2O, and potassium salts.

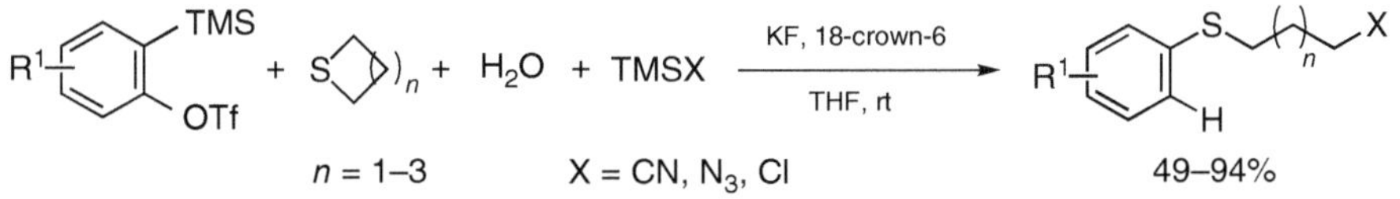

Scheme 5.54 Four-component reaction of arynes, cyclic thioethers, H_2O, and trimethylsilyl compounds.

5.8 Halogen Nucleophile-Based Multicomponent Reactions

Although a fluoride ion has rarely been utilized as a nucleophile toward arynes, probably owing to "hard (F^-)"–"soft (aryne)" mismatch, Hu reported that this could be involved in the three-component reaction with I^+ equivalent, C_4F_9I, in the presence of a catalytic amount of diphenyliodonium triflate, affording various *ortho*-fluoroaryl iodides (**86**) (Scheme 5.55) [58]. The reaction proceeds through nucleophilic attack of a fluoride ion (from CsF and/or an in situ-generated complex [**85**]), and subsequent reaction with C_4F_9I provides **86**.

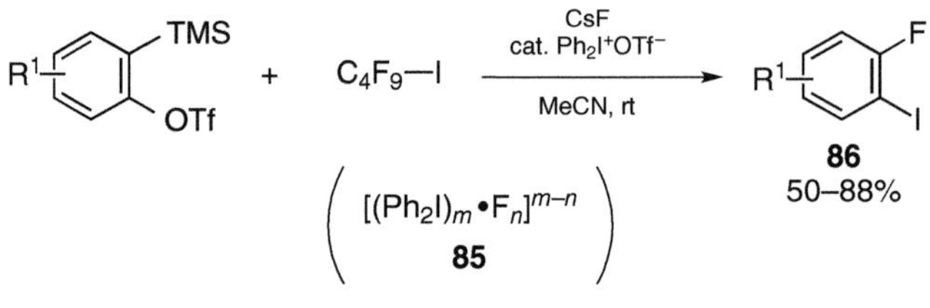

Scheme 5.55 Three-component reaction of arynes, a fluoride, and I^+ equivalent. Source: Based on Zeng et al. [58].

A four-component reaction with a chloride ion (from KCl), an atmospheric pressure of CO_2, and alkyl chlorides, where diverse alkyl *ortho*-chlorobenzoates (**87**) were straightforwardly produced, was reported by Jiang (Scheme 5.56) [59]. A chloride ion first attacks an aryne to give an *ortho*-chloroaryl anion, which is sequentially trapped by CO_2 and an alkyl chloride to lead to the final product (**87**).

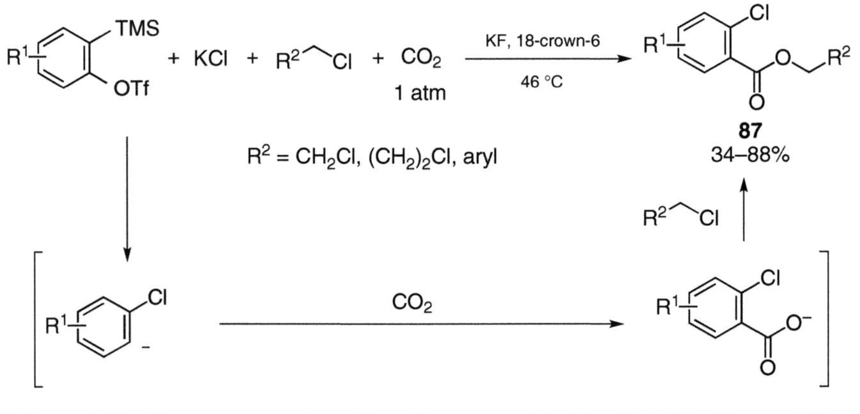

Scheme 5.56 Four-component reaction of arynes, a chloride, alkyl chlorides, and CO_2. Source: Based on Jiang et al. [59].

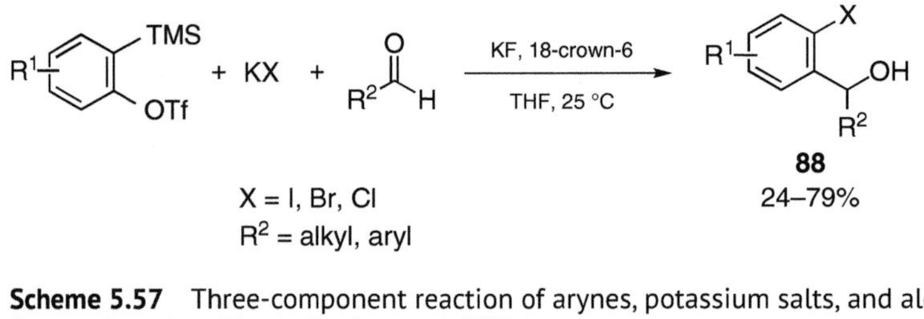

Scheme 5.57 Three-component reaction of arynes, potassium salts, and aldehydes. Source: Based on Bhattacharjee et al. [60].

Biju reported that the use of KI allowed a three-component iodination to occur with aldehydes, giving a variety of *ortho*-iodobenzhydrol derivatives **88** (Scheme 5.57) [60]. An isatin could also serve as an electrophile in the reaction, and furthermore, the use of KBr or KCl instead of KI resulted in the three-component bromination or chlorination.

5.9 Miscellaneous

Treatment of a borylzincate species (**89**), generated in situ from bis(pinacolato)diboron, Me_2Zn, and $Mg(Ot\text{-}Bu)_2$, with 2-bromoaryl iodides resulted in the formation of arynes, which further accepted nucleophilic attack by **89** to provide 2-borylarylzincate species (**90**) (Scheme 5.58) [61]. The resulting zincates (**90**) were finally transformed into three-component allylboration products (**91**) via capturing with allyl bromide.

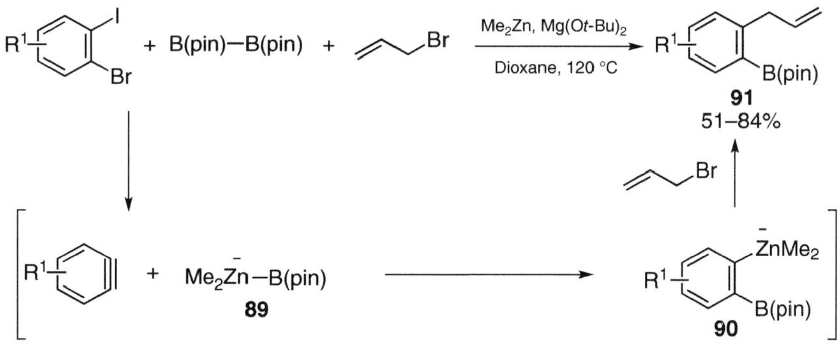

Scheme 5.58 Three-component reaction of arynes, borylzincate, and allyl bromide. Source: Based on Nagashima et al. [61].

Yoshida and Hosoya developed a three-component reaction of triflyloxy (OTf)-substituted benzocyclobutenones, organolithium reagents, and arynophiles such as furans, pyrrole, and aniline (Scheme 5.59) [62]. The reaction is triggered by nucleophilic addition of an organolithium reagent to the carbonyl moiety of a benzocyclobutenone. The resulting benzocyclobutenoxide intermediate (**92**) then undergoes carbon–carbon bond cleavage/elimination of the OTf moiety to give an aryne, which is converted into the final product (**93**) by the reaction with an arynophile.

5.10 Conclusive Remarks

This chapter depicted that a wide array of multicomponent reactions involving arynes have now become feasible depending upon their highly electrophilic character, and that this field has witnessed remarkable progress during the last five years. Various combinations of nucleophiles and electrophiles/pronucleophiles are utilizable for the multicomponent reactions under Kobayashi's conditions, which leads to convenient and direct access to diverse benzo-annulated structures and

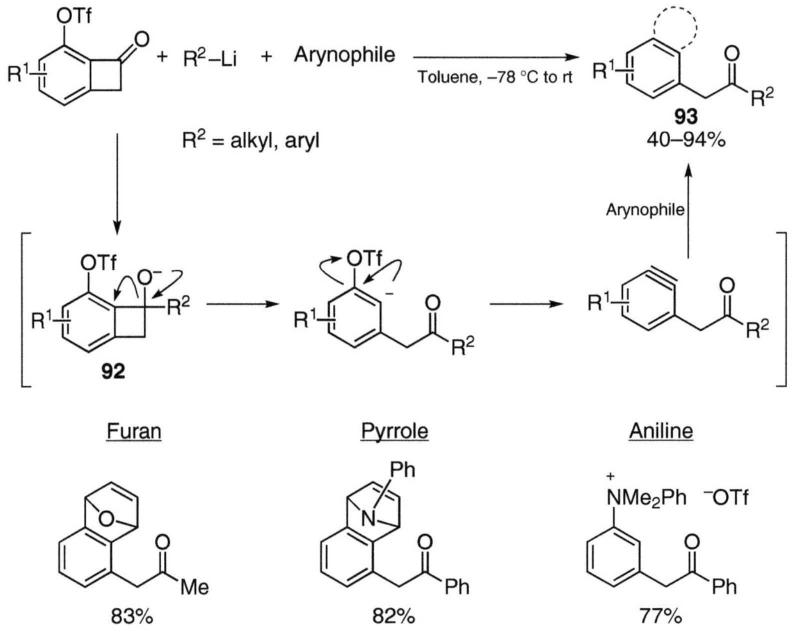

Scheme 5.59 Three-component reaction using triflyloxy-substituted benzocyclobutenones, and organolithium reagents. Source: Based on Uchida et al. [62].

multisubstituted arenes. In view of the highly reactive yet controllable features of their strained triple bond, which would enable unique bond cleavage/construction, development of new multicomponent reactions involving arynes continues to be an important subject in synthetic organic chemistry.

References

1 (a) Zhu, J., Wang, Q., and Wang, M.-X. (eds.) (2015). *Multicomponent Reactions in Organic Synthesis*. Weinheim: Wiley-VCH. (b) Brauch, S., van Berkel, S.S., and Westermann, B. (2013). *Chem. Soc. Rev.* 42: 4948. (c) Zarganes-Tzitzikas, T., Chandgude, A.L., and Dömling, A. (2015). *Chem. Rec.* 15: 981.

2 (a) Chen, Y. and Larock, R.C. (2009). Arylation reactions involving the formation of arynes. In: *Modern Arylation Methods* (ed. L. Ackermann), 401–473. Weinheim: Wiley. (b) Pellissier, H. and Santelli, M. (2003). *Tetrahedron* 59: 701. (c) Dyke, A.M., Hester, A.J., and Lloyd-Jones, G.C. (2006). *Synthesis*: 4093. (d) Sanz, R. (2008). *Org. Prep. Proced. Int.* 40: 215. (e) Kitamura, T. (2010). *Aust. J. Chem.* 63: 987. (f) Yoshida, H., Ohshita, J., and Kunai, A. (2010). *Bull. Chem. Soc. Jpn.* 83: 199. (g) Tadross, P.M. and Stoltz, B.M. (2012). *Chem. Rev.* 112: 3550. (h) Bhunia, A., Yetra, S.R., and Biju, A.T. (2012). *Chem. Soc. Rev.* 41: 3140. (i) Yoshida, H. and Takaki, K. (2012). *Synlett* 23: 1725. (j) Dubrovskiy, A.V., Markina, N.A., and Larock, R.C. (2013). *Org. Biomol. Chem.* 11: 191. (k) García-López, J.-A. and Greaney, M.F. (2016). *Chem. Soc. Rev.* 45: 6766.

3 Himeshima, Y., Sonoda, T., and Kobayashi, H. (1983). *Chem. Lett.*: 1211.

4 (a) Yoshida, H. (2014). Nucleophilic coupling with arynes. In: *Comprehensive Organic Synthesis*, vol. 4 (eds. P. Knochel and G.A. Molander), 517–579. Amsterdam: Elsevier. (b) Yoshida, H. (2015). Aryne-Based Multicomponent Reactions. In: *Multicomponent Reactions in Organic Synthesis* (eds. J. Zhu, Q. Wang and M.-X. Wang), 39–71. Weinheim: Wiley-VCH. (c) Yoshida, H. and Takaki, K. (2012). *Heterocycles* 85: 1333. (d) Miyabe, H. (2015). *Molecules* 20: 12558. (e) Bhojgude, S.S., Bhunia, A., and Biju, A.T. (2016). *Acc. Chem. Res.* 49: 1658.

5 Yoshida, H., Fukushima, H., Ohshita, J., and Kunai, A. (2004). *Angew. Chem. Int. Ed.* 43: 3935.

6 Kaicharla, T., Thangaraj, M., and Biju, A.T. (2014). *Org. Lett.* 16: 1728.

7 Fang, Y., Wang, S.-Y., and Ji, S.-J. (2015). *Tetrahedron* 71: 2768.

8 Serafini, M., Griglio, A., Viarengo, S. et al. (2017). *Org. Biomol. Chem.* 15: 6604.

9 Tambar, U.K. and Stoltz, B.M. (2005). *J. Am. Chem. Soc.* 127: 5340.

10 (a) Yoshida, H., Watanabe, M., Ohshita, J., and Kunai, A. (2005). *Chem. Commun.*: 3292. (b) Yoshida, H., Watanabe, M., Ohshita, J., and Kunai, A. (2005). *Tetrahedron Lett.* 46: 6729. (c) Yoshida, H., Watanabe, M., Morishita, T. et al. (2007). *Chem. Commun.*: 1505.

11 Shu, W.-M., Liu, S., He, J.-X. et al. (2018). *J. Org. Chem.* 83: 9156.

12 Shu, W.-M., Zheng, K.-L., Ma, J.-R., and Wu, A.-X. (2016). *Org. Lett.* 18: 3762.

13 Tang, C.-Y., Wang, G., Yang, X.-Y. et al. (2014). *Tetrahedron Lett.* 55: 6447.

14 Roy, T., Bhojgude, S.S., Kaicharla, T. et al. (2016). *Org. Chem. Front.* 3: 71.

15 Roy, T., Baviskar, D.R., and Biju, A.T. (2015). *J. Org. Chem.* 80: 11131.

16 Roy, T., Thangaraj, M., Gonnade, R.G., and Biju, A.T. (2016). *Chem. Commun.* 52: 9044.

17 Min, G., Seo, J., and Ko, H.M. (2018). *J. Org. Chem.* 83: 8417.

18 Bhojgude, S.S., Baviskar, D.R., Gonnade, R.G., and Biju, A.T. (2015). *Org. Lett.* 17: 6270.

19 Bhojgude, S.S., Roy, T., Gonnade, R.G., and Biju, A.T. (2016). *Org. Lett.* 18: 5424.

20 Okuma, K., Kinoshita, H., Nagahora, N., and Shioji, K. (2016). *Eur. J. Org. Chem.*: 2264.

21 Kwon, J. and Kim, B.M. (2019). *Org. Lett.* 21: 428.

22 Yoshida, H., Fukushima, H., Ohshita, J., and Kunai, A. (2006). *J. Am. Chem. Soc.* 128: 11040.

23 Xu, J.-K., Li, S.-J., Wang, H.-Y. et al. (2017). *Chem. Commun.* 53: 1708.

24 Huang, X., Zhao, W., Chen, D.-L. et al. (2019). *Chem. Commun.* 55: 2070.

25 Li, S.-J., Wang, Y., Xu, J.-K. et al. (2018). *Org. Lett.* 20: 4545.

26 Swain, S.P., Shih, Y.-C., Tsay, S.-C. et al. (2015). *Angew. Chem. Int. Ed.* 54: 9926.

27 (a) Aly, A.A., Mohamed, N.K., Hassan, A.A., and Mourad, A.-F.E. (1999). *Tetrahedron* 55: 1111. (b) Singal, K.K. and Kaur, J. (2001). *Synth. Commun.* 31: 2809. (c) Yoshida, H., Kuriki, H., Fujii, S. et al. (2017). *Asian J. Org. Chem.* 6: 973.

28 Zhou, Y., Chi, Y., Zhao, F. et al. (2014). *Chem. Eur. J.* 20: 2463.

29 Li, R., Tang, H., Fu, H. et al. (2014). *J. Org. Chem.* 79: 1344.

30 Bhunia, A., Porwal, D., Gonnade, R.G., and Biju, A.T. (2013). *Org. Lett.* 15: 4620.

31 Bhunia, A., Roy, T., Pachfule, P. et al. (2013). *Angew. Chem. Int. Ed.* 52: 10040.

32 Liu, P., Lei, M., and Hu, L. (2013). *Tetrahedron* 69: 10405.

33 Tan, J., Liu, B., and Su, S. (2018). *Org. Chem. Front.* 5: 3093.

34 Liu, K., Liu, L.-L., Gu, C.-Z. et al. (2016). *RSC Adv.* 6: 33606.

35 Shu, W.-M., Ma, J.-R., Zheng, K.-L., and Wu, A.-X. (2016). *Org. Lett.* 18: 196.

36 Dhokale, R.A. and Mhaske, S.B. (2016). *Org. Lett.* 18: 3010.

37 Yoshioka, E., Kohtani, S., and Miyabe, H. (2011). *Angew. Chem. Int. Ed.* 50: 6638.

38 Yoshida, H., Ito, Y., and Ohshita, J. (2011). *Chem. Commun.* 47: 8512.

39 Yoshioka, E., Tamenaga, H., and Miyabe, H. (2014). *Tetrahedron Lett.* 55: 1402.

40 Wen, L.-R., Man, N.-N., Yuan, W.-K., and Li, M. (2016). *J. Org. Chem.* 81: 5942.

41 Sharma, A. and Gogoi, P. (2019). *Org. Biomol. Chem.* 17: 333.

42 Gouthami, P., Chavan, L.N., Chegondi, R., and Chandrasekhar, S. (2018). *J. Org. Chem.* 83: 3325.

43 Neog, K., Das, B., and Gogoi, P. (2018). *Org. Biomol. Chem.* 16: 3138.

44 Liu, F., Yang, H., Hu, X., and Jiang, G. (2014). *Org. Lett.* 16: 6408.

45 Zhou, C., Wang, J., Jin, J. et al. (2014). *Eur. J. Org. Chem.*: 1832.

46 Liu, F.-L., Chen, J.-R., Zou, Y.-Q. et al. (2014). *Org. Lett.* 16: 3768.

47 Hazarika, H., Neog, K., Sharma, A. et al. (2019). *J. Org. Chem.* 84: 5846.

48 Li, Y., Qiu, D., Gu, R. et al. (2016). *J. Am. Chem. Soc.* 138: 10814.

49 Thangaraj, M., Bhojgude, S.S., Mane, M.V., and Biju, A.T. (2016). *Chem. Commun.* 52: 1665.

50 Xiong, W., Qi, C., Cheng, R. et al. (2018). *Chem. Commun.* 54: 5835.

51 Zhou, M., Ni, C., Zeng, Y., and Hu, J. (2018). *J. Am. Chem. Soc.* 140: 6801.

52 Yoshida, S. and Hosoya, T. (2013). *Chem. Lett.* 42: 583.

53 Bhunia, A., Roy, T., Gonnade, R.G., and Biju, A.T. (2014). *Org. Lett.* 16: 5132.

54 Bhunia, A., Kaicharla, T., Porwal, D. et al. (2014). *Chem. Commun.* 50: 11389.

55 Zheng, T., Tan, J., Fan, R. et al. (2018). *Chem. Commun.* 54: 1303.

56 Fan, R., Liu, B., Zheng, T. et al. (2018). *Chem. Commun.* 54: 7081.

57 Jian, H., Wang, Q., Wang, W.-H. et al. (2018). *Tetrahedron* 74: 2876.

58 Zeng, Y., Li, G., and Hu, J. (2015). *Angew. Chem. Int. Ed.* 54: 10773.

59 Jiang, H., Zhang, Y., Xiong, W. et al. (2019). *Org. Lett.* 21: 345.

60 Bhattacharjee, S., Guin, A., Gaykar, R.N., and Biju, A.T. (2019). *Org. Lett.* 21: 4383.

61 Nagashima, Y., Takita, R., Yoshida, K. et al. (2013). *J. Am. Chem. Soc.* 135: 18730.

62 Uchida, K., Yoshida, S., and Hosoya, T. (2017). *Org. Lett.* 19: 1184.

6

Transition-Metal-Catalyzed Reactions Involving Arynes and Related Chemistry

Kanniyappan Parthasarathy[1], Jayachandran Jayakumar[2], Masilamani Jeganmohan[3], and Chien-Hong Cheng[2]

[1] *University of Madras, Department of Organic Chemistry, Guindy Campus, Chennai, Tamil Nadu 600025, India*
[2] *National Tsing Hua University, Department of Chemistry, No. 101, Section 2, Guangfu Road, East District, Hsinchu 30013, Taiwan*
[3] *Indian Institute of Technology Madras, Department of Chemistry, Chennai, Tamil Nadu 600036, India*

6.1 Introduction

Aryne (or) benzyne is one of the most versatile organic electrophilic species and it has received great attention in structural aspects as well as a reactive carbon–carbon π-component in the metal-catalyzed organic reaction [1]. In general, an aryne is an extremely reactive species and it must be generated in situ. The high reactivity of aryne is due to the less energy gap between the highest occupied molecular orbital (HOMO) and the lowest unoccupied molecular orbital (LUMO) as compared with alkynes. Thus, an aryne is always called the superior electrophile in the synthetic community. Several methods are known to generate benzynes, including strongly basic conditions, reducing metals, thermally/photochemically driven, and fluoride sources (mild condition). Among them, a mild method for the in situ generation of an aryne at ambient temperatures from *ortho*-silyl aryltriflates **1** (Kobayashi precursor) has received considerable interest in the past two decades [2]. The aryne generation from *ortho*-(trimethylsilyl)phenyl triflate **1** can be controlled by varying the concentration of fluoride ion in the solution. Usually, CsF, KF, *tetra*-*n*-butylammonium fluoride, and AgF are commonly used as fluoride ion sources. From the mechanism point of view, the aryne generation takes place via 1,2-elimination of *ortho*-silyl aryltriflates by fluoride (F^- sources). As a result, Kobayashi's aryne precursor has found numerous applications in transition-metal-catalyzed reactions. The benzyne can be transformed into a wide range of new organic molecules by involving cyclo and cocyclotrimerization, annulations (via C–H and N–H activation), multicomponent coupling reactions, etc. (Scheme 6.1: see journey map). In this chapter, we have organized the metal-catalyzed reaction types according to the journey of metal-catalyzed (Pd, Pt, Rh, Ni, Cu, Ag, and Au) benzyne chemistry, including new breakthroughs,

Modern Aryne Chemistry, First Edition. Edited by Akkattu T. Biju.

substrate scopes, reaction selectivity, mechanism, limitations as well as applications toward natural products and bioactive molecules.

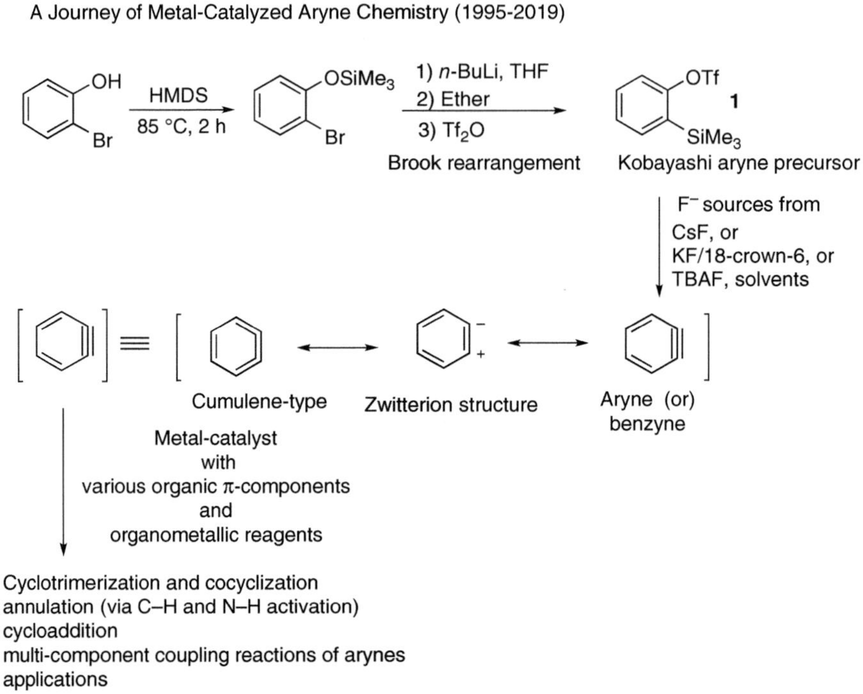

Scheme 6.1 Aryne generation and general metal-catalyzed reactions.

6.2 Metal-Catalyzed Cyclotrimerization and Cocyclization of Arynes

6.2.1 Palladium-Catalyzed Cyclotrimerization and Cocyclization with Arynes

Palladium (Pd) serves as a unique catalyst for various metal-catalyzed organic transformations. These transformations often achieved good reactivity and high functional group tolerance, and provided the expected products in excellent stereo- and regioselectivity [3]. These factors can be finetuned by varying the ligand, base, solvent, temperature as well as additives. In general, palladium-catalyzed reaction goes in Pd(0), Pd(II), and Pd(IV) oxidation states. In particular, Pd(0) and Pd(II) oxidation states are majorly involved in aryne reactions to make multiple C—C and C—X bonds in one pot. It is important to mention that most of palladium-catalyzed aryne reaction begins with a Pd(0) oxidation state. When Pd(II)-complex is used, the Pd(II) complex reduces into Pd(0) in the first step of the catalytic cycle with the aid of phosphine ligands or bases or solvents in most of the reactions. The Pd(0) and Pd(II) complexes are effectively utilized in benzyne-based reactions and also deliver various valuable new products in organic synthesis.

In general, benzyne easily undergoes trimerization with the help of a metal-giving cyclotrimerization product in good-to-excellent yields under mild reaction conditions. The cocyclization can also be done with one benzyne and two other carbon–carbon π-components or two benzynes with one carbon–carbon π-component in a highly selective manner (Scheme 6.2). By employing this strategy, symmetrical as well as unsymmetrical fused polycyclic aromatic compounds can be prepared in good-to-excellent yields.

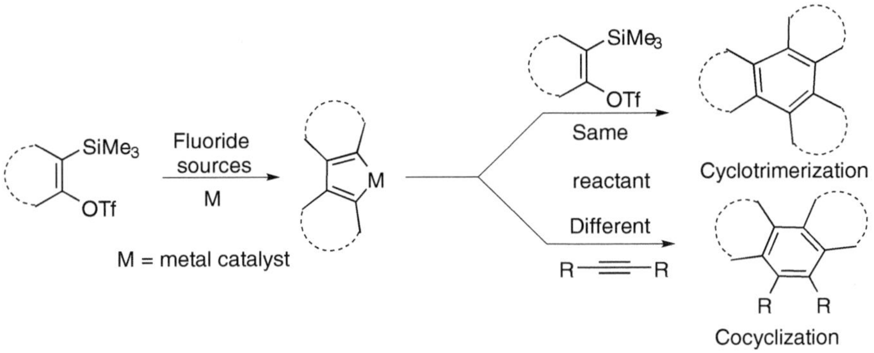

Scheme 6.2 Cyclotrimerization and cocyclization employing arynes.

In 1998, palladium-catalyzed cyclotrimerization of arynes was observed (Scheme 6.3) [4]. When 2-trimethylsilylphenyl triflate (**1**) was treated with CsF in the presence of a $Pd(PPh_3)_4$ (3 mol%) in CH_3CN, a triphenylene **2** was obtained in 83% yield. It is noteworthy that, in the absence of a Pd catalyst, the triphenylene was not detected. The starting material **1** was recovered when the **1** was treated with $[Pd(PPh_3)_4]$ in the absence of fluoride source. This observation indicates that both Pd and fluoride ion are vital for the product formation. Better regioselectivity was observed for the unsymmetrical *ortho*-methoxy-substituted aryne precursor **1a**. In the reaction, a mixture of 1,5,12- and 1,5,9- trimethoxytriphenylenes (**2a** and **2a′**, respectively) was observed in 93 : 7 ratio in 81% combined yield. Meanwhile, the phosphine-free $Pd_2(dba)_3$ catalyst was also effective for the reaction. The cyclotrimerization of unsymmetrical benzyne precursor **1b** provides a mixture of strained polycyclic hydrocarbons **2b** and **2b′** in 60% yield with a 2.7 : 1 ratio, respectively.

A year later, the same group demonstrated the catalyst-controlled synthesis of phenanthrenes and naphthalenes via cocyclization of arynes with electron-deficient alkynes (hexafluoro-2-butyne and dimethyl acetylenedicarboxylate [DMAD]). In the reaction, cocyclization products such as two benzynes and one alkyne, phenanthrene derivatives (**4**), were observed as a major product in the presence of $Pd(PPh_3)_4$. The phosphine-free $Pd_2(dba)_3$ complex provided cocyclization of one benzyne and two alkynes product, naphthalene derivatives (**5**), as a major product in good yields. Notably, the reaction of benzyne with other alkynes (3-hexyne and diphenylacetylene) afforded lower yields as well as selectivity in the formation of phenanthrenes and naphthalenes in the presence of both catalysts (Scheme 6.4) [5].

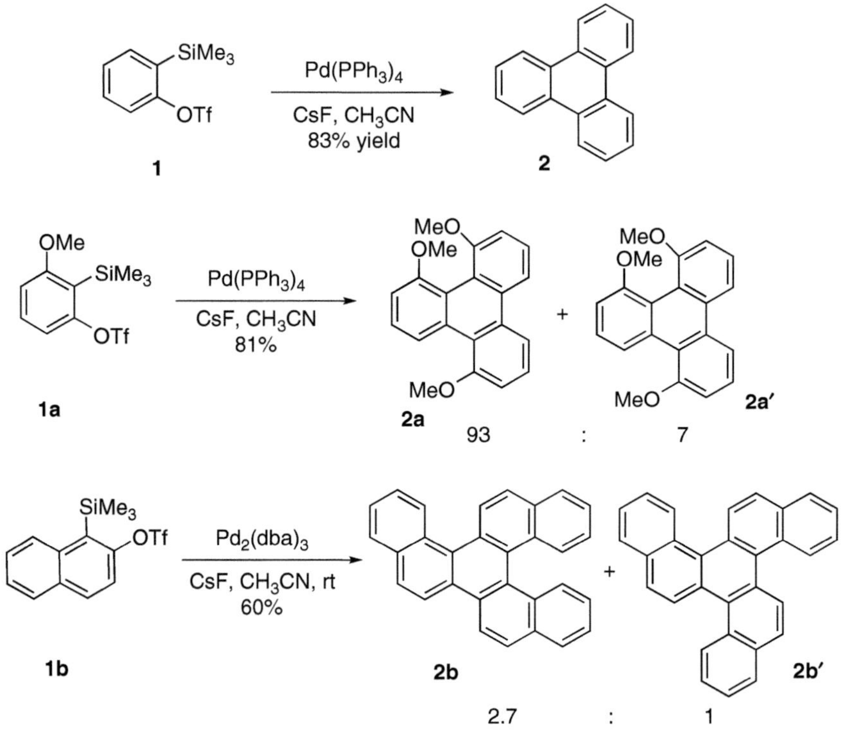

Scheme 6.3 Palladium-catalyzed [2+2+2] cocyclotrimerization of arynes. Source: Peña et al. [4a]; Peña et al. [4b]; Peña et al. [4c].

Later on, Yamamoto and coworkers observed a selective synthesis of phenanthrene derivatives in good yields in the presence of $Pd(OAc)_2$/P(*o*-tol)$_3$/CsF system in CH_3CN at 60 °C for four hours.

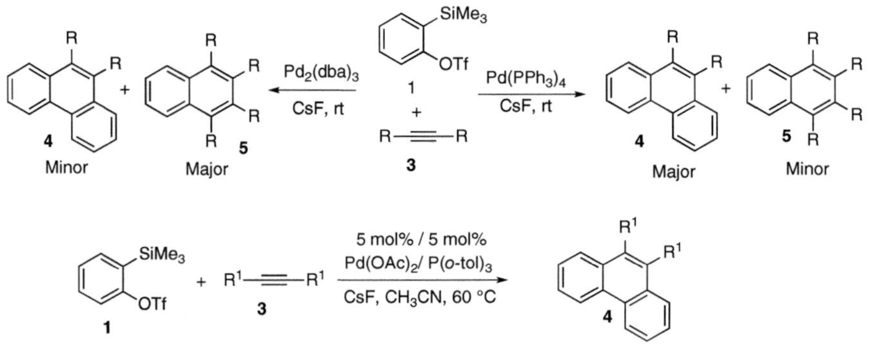

Scheme 6.4 Palladium-catalyzed [2+2+2] cycloaddition of arynes with alkynes. Source: Peña et al. [5a]; Peña et al. [5b].

In 2012, Pena and coworkers synthesized a novel hexaphenyl-substituted [16]cloverphene derivative by a Pd-catalyzed [2+2+2] cycloaddition of polycyclic aryne precursors (Scheme 6.5). In this reaction, the polycyclic aryne was slowly generated by using a mixture of dichloromethane/MeCN (1 : 6 ratio) at 40 °C with CsF [6]. Under the slow generation, benzyne undergoes cyclotrimerization in the presence of 10 mol% of [Pd(PPh_3)$_4$] to give cloverphene **2c** in 22% yield as a yellow solid. The prepared cloverphenes are well characterized by ^{1}H NMR, ^{13}C NMR, and MALDI mass spectrum, in which the ^{1}H NMR spectrum showed a singlet and a doublet in a deshielded zone at $\delta = 8.71$ and 8.30 ppm. In ^{13}C NMR spectrum, eight CH carbon and seven C aromatic signals were observed. In the HRMS, the molecular ion peak was observed with a mass of m/z 1284.4. Further, tetrabenzoheptaphenes **4a** were prepared by the cocyclotrimerization of bisarynes with DMAD under similar reaction conditions.

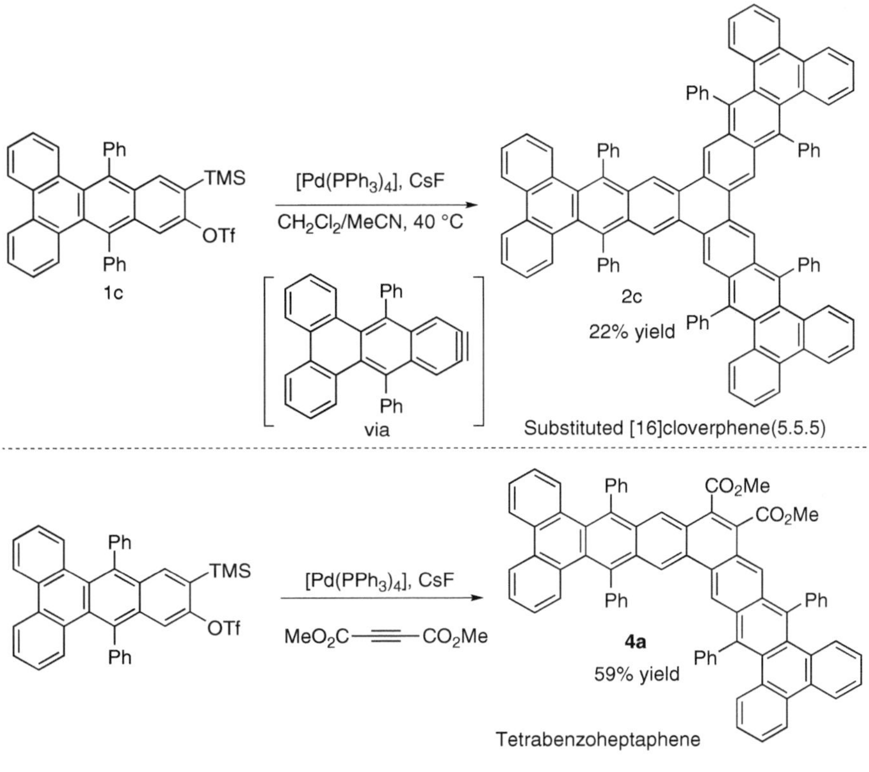

Scheme 6.5 Synthesis of cloverphenes and tetrabenzoheptaphene by using Pd-catalyzed cyclotrimerization and cocyclotrimerization strategy .

In the meantime, Yamamoto's group reported the [2+2+2] cycloaddition of benzynes with allylic chlorides (2 : 1), benzynes with alkynes (2 : 1), and benzynes with allylic chlorides and alkynes (1 : 1 : 1) in the presence of a palladium catalyst (Scheme 6.6) [7].

Cheng and coworkers reported a palladium-catalyzed [2+2+2] cocyclotrimerization of benzynes with a bicyclic alkene providing norbornane anellated 9,10-dihydrophenanthrene derivatives [8a]. The reaction of **1** with **7** was carried

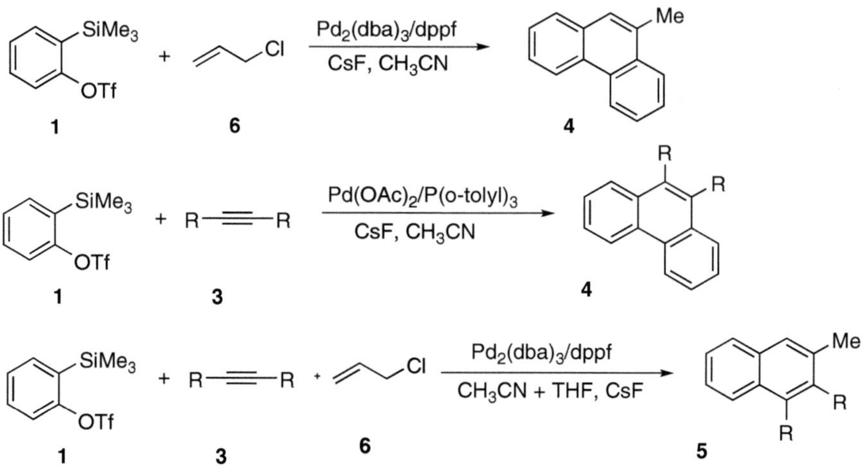

Scheme 6.6 Palladium-catalyzed reaction of arynes with allylic chlorides or alkynes. Source: Radhakrishnan et al. [7a]; Yoshikawa et al. [7b]; Yoshikawa et al. [7c].

out in the absence of metal catalyst. In the reaction, only [2+2] cycloaddition product was observed in quantitative yield [8b, c]. The palladium catalyst is necessary to achieve [2+2+2] cocyclotrimerization product. Similarly, the bicyclic alkene **7** undergoes [2+2+2] cocyclotrimerization with arynes in the presence of $PdCl_2(PPh_3)_2$ in CH_3CN at ambient temperature yielding annulated 9,10-dihydrophenanthrenes **8** in good-to-excellent yields (Scheme 6.7). The oxa- and azabicyclic alkenes also nicely participated in the reaction. Interestingly, the observed product readily undergoes deoxyaromatization in the presence of $BF_3{\cdot}OEt_2$ in DCM solvent yielding polycyclic aromatic compounds in good-to-excellent yields. Further, when cycloaddition product **8** was refluxed in toluene for 24 hours in the presence of diethyl acetylenedicarboxylate, the corresponding cycloadduct product **7a** and the substituted phenanthrene **4** were obtained in high yields (Scheme 6.7). In this transformation, the cycloaddition product **8** undergoes retro Diels–Alder reaction readily generating isobenzo[*c*]furan, which rapidly reacts with diethyl acetylenedicarboxylate giving product **7a**.

In 2003, a Pd(0)-catalyzed synthesis of benzo[*b*]fluorenones through partially intramolecular [2+2+2] cycloaddition of benzodiynes **9** with benzyne precursor **1** was reported (Scheme 6.8) [9]. This transformation proceeds via the coordination of the alkynes **9** with a palladium metal followed by oxidative cyclization giving a five-membered palladacycle **8-A**. Coordinative insertion of benzyne with the palladacycle **8-A** followed by reductive elimination provides the desired cycloaddition product **10** and regenerates the active palladium(0) species for the next catalytic cycle.

Mori's group reported a new Pd-catalyzed [2+2+2] cocyclization of substituted diynes with arynes giving naphthalene skeletons in good-to-excellent yields [10]. The cycloaddition product was used as a key intermediate for the total synthesis of Taiwanins C and E (Scheme 6.9).

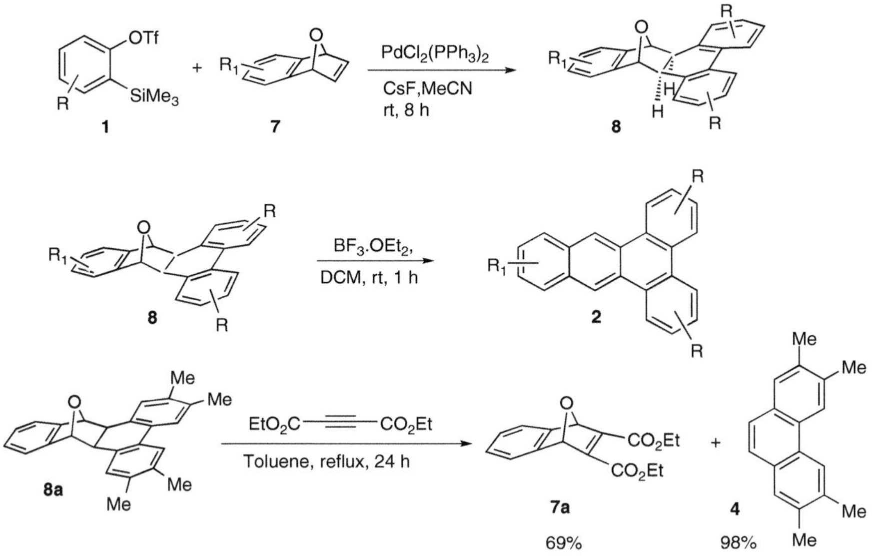

Scheme 6.7 Pd-catalyzed cocyclotrimerization of benzynes with a bicyclic alkene.

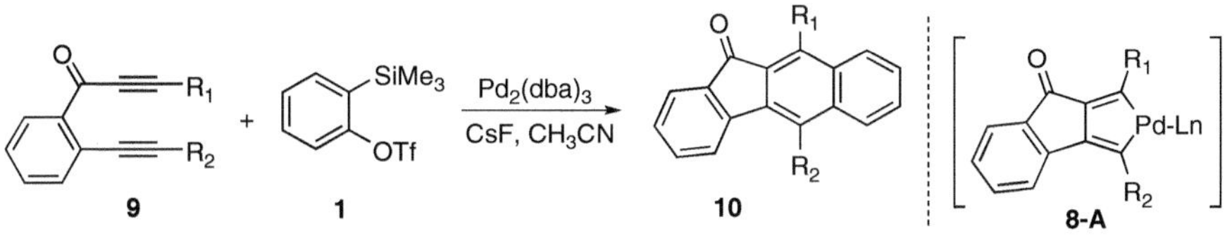

Scheme 6.8 Pd-catalyzed [2+2+2] cycloaddition of benzodiynes with arynes. Source: Based on Peña et al. [9].

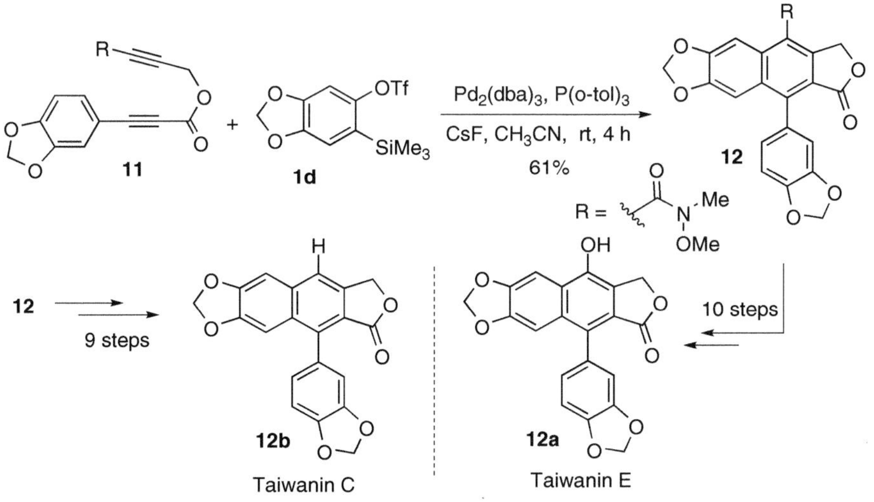

Scheme 6.9 Pd-catalyzed [2+2+2] cocyclization of substituted diynes with arynes.

Very recently, Li's group observed a palladium-catalyzed [2+2+2] cocyclotrimerization of benzynes with allenes (Scheme 6.10) [11]. In the reaction, the internal double bond of allene **13** reacted with two same moieties of benzynes giving product **4**.

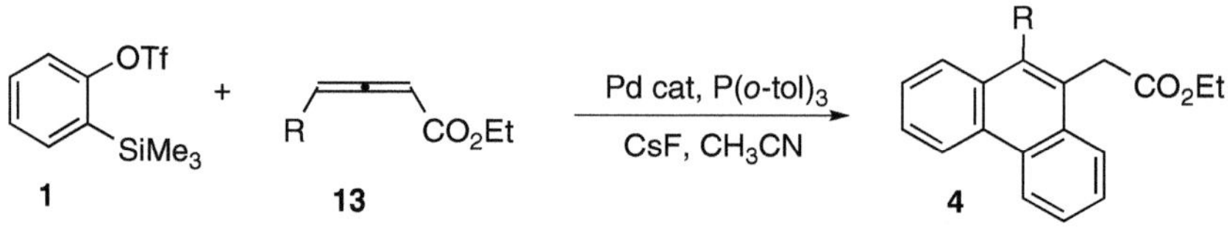

Scheme 6.10 Pd-catalyzed cocyclotrimerization of benzynes with allenes. Source: Liu et al. [11].

The selective syntheses of 9,10-dihydrophenanthrenes (**15**) as well as *ortho*-olefinated biphenyls (**16**) were achieved through cotrimerization of arynes with electron-deficient alkenes (**14**). The phosphine ligand plays a crucial role for the selectivity. A palladacycloheptadiene intermediate **11-A** was proposed as a key intermediate for the reaction. The bulkier P(*o*-tolyl)$_3$ phosphine ligand favors the β-H elimination to give *ortho*-olefinated biphenyls, whereas the PPh_3 ligand favors the reductive elimination to form 9,10-dihydrophenanthrenes (Scheme 6.11). Notably, electron-rich alkenes (e.g. ethyl vinyl ether) did not undergo cotrimerization reaction. Further, the cotrimerization of disubstituted alkenes with arynes was very efficient with a nickel catalyst as compared with a palladium catalyst. For example, when the dimethyl fumarate (**14a**) was treated with aryne in the presence

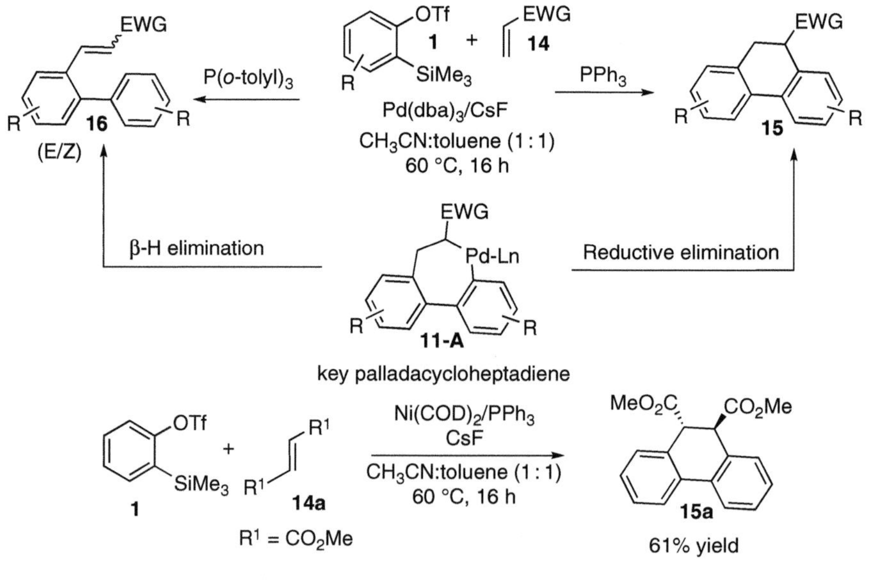

Scheme 6.11 Palladium-catalyzed coupling of alkenes with arynes. Source: Based on Quintana et al. [12].

of a $Ni(cod)_2/PPh_3$, *trans*-dihydrophenanthrene **15a** was observed in good yield without the formation of biaryls **16** (Scheme 6.11) [12].

A new Pd-mediated aryne generation was achieved via simultaneous δ-carbon elimination and decarboxylation from methyl 2-bromobenzoates **17** [13]. It undergoes Pd-mediated [2+2+2] cycloaddition to give triphenylenes in moderate yields (Scheme 6.12). Notably, methyl 2-chlorobenzoates and dimethyl 2-bromoterephthalate failed to give trimerization products under similar reaction conditions.

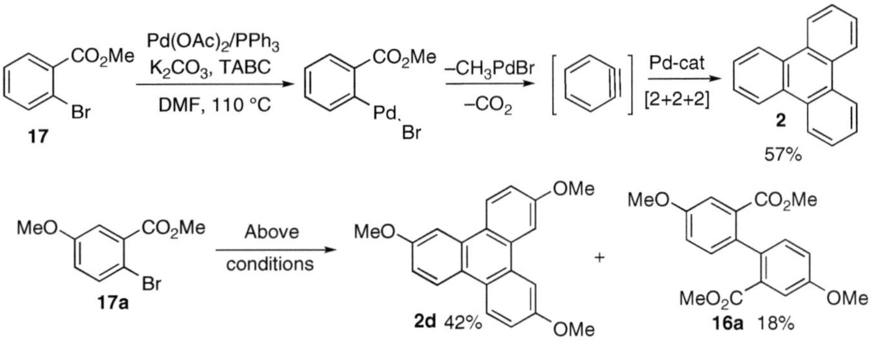

Scheme 6.12 Pd-catalyzed [2+2+2] cycloaddition of arynes from methyl 2-bromobenzoates.

Greaney and his coworkers explored the synthesis of triphenylenes (**2**) by using benzoic acids (**18**) as alternative aryne precursors (Scheme 6.13). The reaction

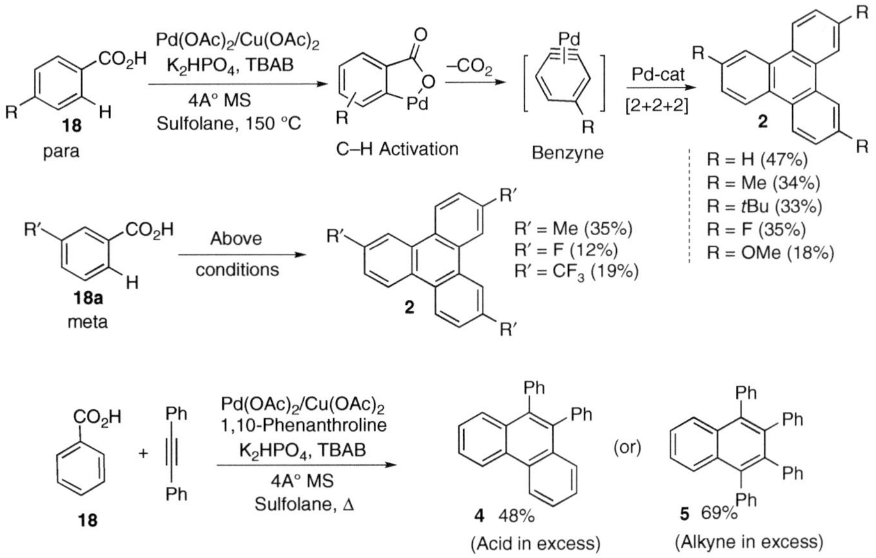

Scheme 6.13 Pd-catalyzed [2+2+2] cycloaddition of arynes from benzoic acids as an aryne precursor. Source: Based on Cant et al. [14].

is believed to proceed via a Pd-mediated ortho C—H bond activation followed by decarboxylation to give a palladium-coordinated aryne species. It undergoes [2+2+2] cycloaddition with benzynes providing triphenylenes. The [2+2+2] cycloaddition reaction of *para*- and *meta*-substituted benzoic acids (**18**) produced triphenylenes as single isomers in moderate yields. The cocyclization was also achieved with alkynes leading to phenanthrene (**4**) and naphthalene (**5**) derivatives. For example, the excess amount of benzoic acid reacted with diphenylacetylene (1 equiv) selectively producing phenanthrene in 48% yield (Scheme 6.13) [14], whereas an excess amount of alkyne provides naphthalene derivative (**5**) at a slightly lower temperature of 120 °C.

Further, the new strategy for the synthesis of triphenylenes by trimerization of arynes was reported by Greaney and his coworker [15]. In this reaction, benzyne was generated by *ortho*-halo (or) triflate arylboronates (Scheme 6.14). The $Pd(dba)_2$ catalyst, DPEphos ligand, and potassium *tert*-butoxide base in toluene as a solvent at 100 °C are the optimized reaction conditions for the cycloaddition reaction. Notably, the starting material 2-bromophenylboronic ester (**19**) is easily prepared from haloarenes within two steps. In the reaction, a series of Pd(0) and Ni(0) benzyne complexes were prepared from (2-bromophenyl)boronic esters with the aid of ^{t}BuOK base.

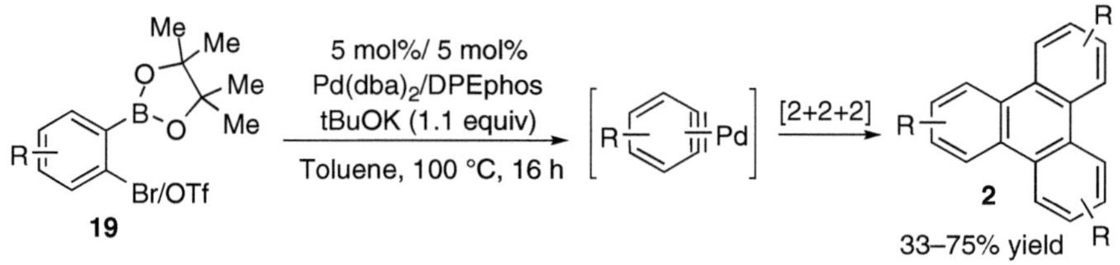

Scheme 6.14 Pd-catalyzed trimerization of (2-bromophenyl)boronic esters via arynes.

Argade and coworker reported an efficient method for the synthesis of highly substituted naphthalenes by a palladium/NHC-promoted [2+2+2] cocyclization of arynes with unsymmetrical conjugated dienes (**20**) [16]. To achieve the synthetic target, the authors studied systematic optimization of [2+2+2] cycloaddition reaction of arynes with unsymmetrical conjugated dienes. The optimization studies clearly revealed that the $Pd_2(dba)_3$ (15 mol%), IMes·HCl (30 mol%), and CsF (3 equiv) in CH_3CN reflux for 24 hours are the ideal condition for the cycloaddition reaction. Other metal complexes such as $Ni(cod)_2$, $Pd(PPh_3)_4$, $Pd(OAc)_2$, and $PdCl_2(PPh_3)_2$ failed to give the product. Particularly, one of the naphthalene derivatives was used as a key intermediate for the synthesis of anti-HIV natural products justicidin B (**12d**) and retrojusticidin B (**12e**) in three steps (Scheme 6.15).

In 2017, Houk and his coworkers prepared a novel class of indole-based conjugated trimers by a palladium-catalyzed [2+2+2] cycloaddition strategy (Scheme 6.16). The prepared indole-based trimer molecule has potential application in optoelectronic materials [17]. This reaction proceeds via a palladium-catalyzed cyclotrimerization of substituted indole-based arynes by the [2+2+2] cycloaddition strategy. In this

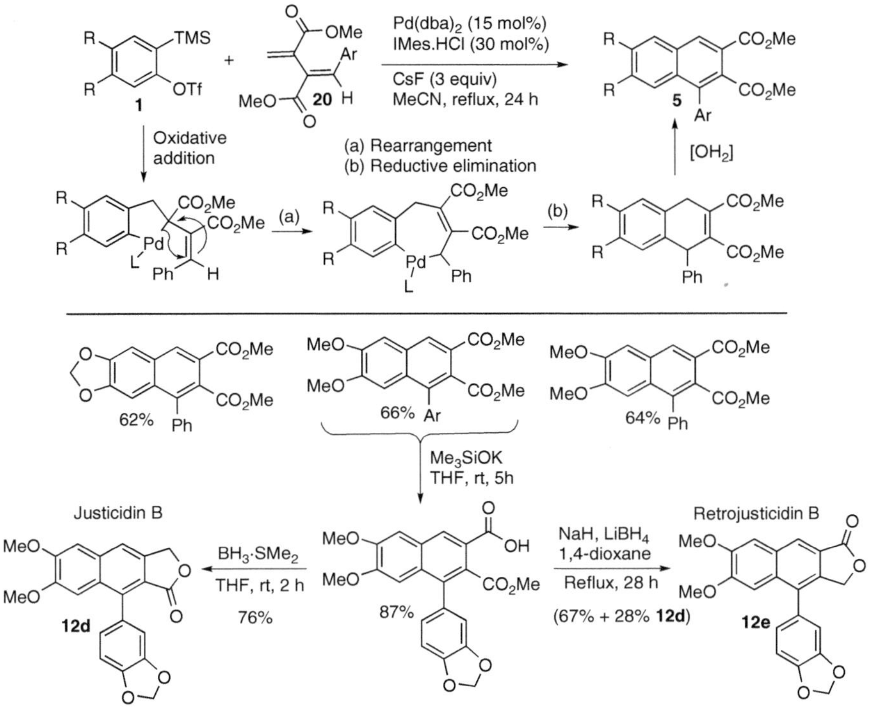

Scheme 6.15 Cocyclization of arynes with conjugated dienes by Pd/NHC ligand.

strategy, three carbon–carbon bonds were constructed in one pot. In addition, computational studies showed that most of the indole-based trimers are out of planarity to improve the steric interactions.

6.2.2 Ni-Catalyzed Cyclotrimerization and Cocyclization with Benzynes

In 2005, Cheng and Hsieh reported a Ni-catalyzed [2+2+2] cocyclotrimerization of diyenes with arynes giving naphthalenes containing five to seven-membered rings [18]. The cocyclotrimerization reaction shows excellent functional group tolerance and leads to the expected products in moderate-to-good yields. The reaction of hex-5-ynenitrile with arynes under standard conditions gave a phenanthrene (**4**) and naphthalene (**21**) derivative in 56 : 11 ratios, respectively, in a combined 67% yield (Scheme 6.17). The cycloaddition reaction proceeds via oxidative cyclization of C≡C bonds of diynes with a Ni(0)-center providing intermediate (**17-II**). Next, aryne inserted into the complex (**17-II**) affords a seven-membered nickelacycle intermediate (**17-III**). It easily undergoes reductive elimination, affords naphthalene derivative, and regenerates the Ni(0) catalyst.

In 2004, Cheng and his coworkers reported a nickel-catalyzed [2+2+2] cocyclotrimerization of allenes (**22**) with arynes (**1**) leading to 10-methylene-9,10-dihydrophenanthrenes (**23**) in moderate-to-good yields [19]. In this method, the

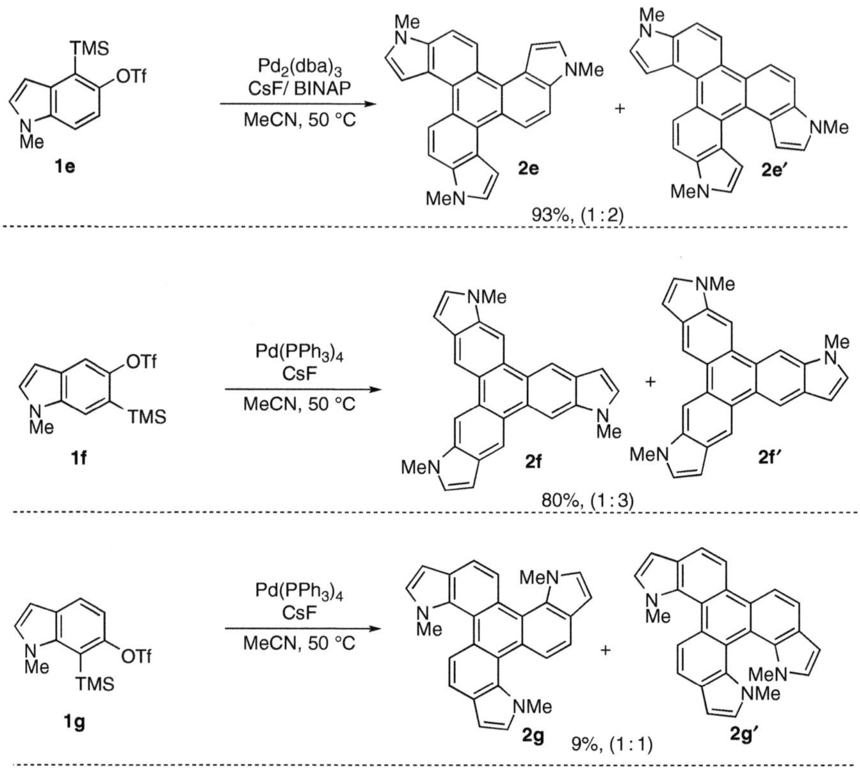

Scheme 6.16 Palladium-catalyzed cyclotrimerization of substituted indole-based aryne precursors.

internal double bond of the allene is selectively reacted with aryne providing 9,10-dihydrophenathrenes (Scheme 6.18). However, the mixture of regioisomers was formed with 1,1-disubstituted allenes. The proposed mechanism starts with coordinative oxidative cyclization of an aryne and allene with a Ni(0) to form a five-membered nickelacycle (**18-I**). Next, the insertion of another molecule of aryne into the Ni—C bond affords a seven-membered nickelacycle (**18-II**). It further undergoes reductive elimination, provides the expected product, and regenerates the Ni(0) catalyst.

In 2008, Cheng and coworker developed a Ni-catalyzed [2+2+2] cycloaddition reaction of *ortho*-dihaloarenes (**24**) as an aryne precursor with alkynes or bis-alkynes or nitriles to give substituted naphthalene and phenanthridine derivatives (Scheme 6.19) [20]. It should be noted that the $Pd(PPh_3)_4$, $Pd(dba)_2$, $PdCl_2(PPh_3)_2$, and $CoI_2(dppe)$ complexes were ineffective for this reaction. The Ni/Zn combination proved the expected octaphenylanthracene (**5A**) in 64% yield by the reaction of 1,2,4,5-tetraiodobenzene (**24a**) with 4 equiv of diphenyl acetylene (**3**). The propargylic ether (**11**) was also underwent cycloaddition with arynes under similar reaction condition affording dihydrofuronaphthales (**20**) in good yields.

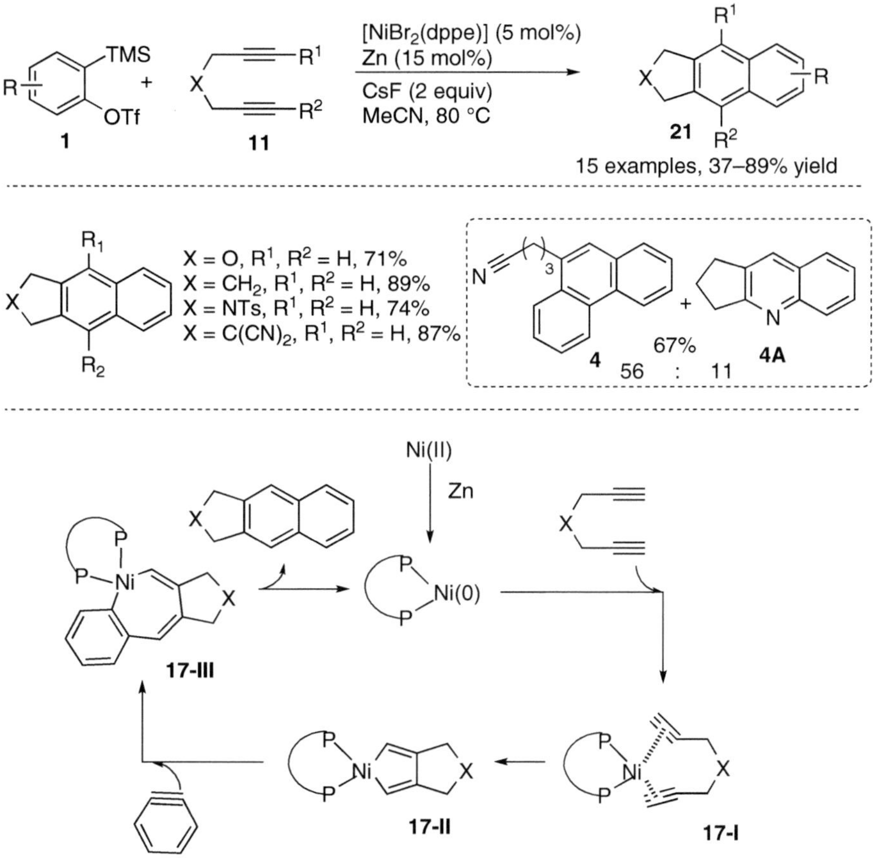

Scheme 6.17 Ni-catalyzed cocyclotrimerization of arynes with diynes.

Further, phenanthridines (**25**) were prepared in good yields via cycloaddition of two molecules of diiodobenzenes with acetonitrile as a π-component in the presence $NiBr_2$(dppe) and zinc.

In 2010, Sato and Iwayama developed an efficient methodology for the synthesis of highly substituted isoquinolines through a Ni(0)-catalyzed [2+2+2] cycloaddition of 3,4-pyridynes with 2 equiv of arynes under the mild reaction conditions (Scheme 6.20) [21]. It has been found that the 2-butyn-1,4-diol as well as substituted 1,3-diynes were also compatible for the reaction. It was found that the propargylic oxygen of an alkyne is crucial for the reactivity as well as the selectivity of the reaction. The reaction proceeds via oxidative cyclization of alkynes having propargylic oxygen substituent with a nickel species providing a nickelacyclopentadiene intermediate (**20-II**) in a highly stereoselective manner. Next, the insertion of 3,4-pyridyne into the Ni—C bond of **20-II** affords seven-membered nickelacycle intermediate **20-III** or **20-III′**. Subsequent reductive elimination of **20-III** or **20-III′** gives isoquinoline product (**26**) in a highly stereoselective manner and regenerates a Ni(0)-catalyst for the next catalytic cycle.

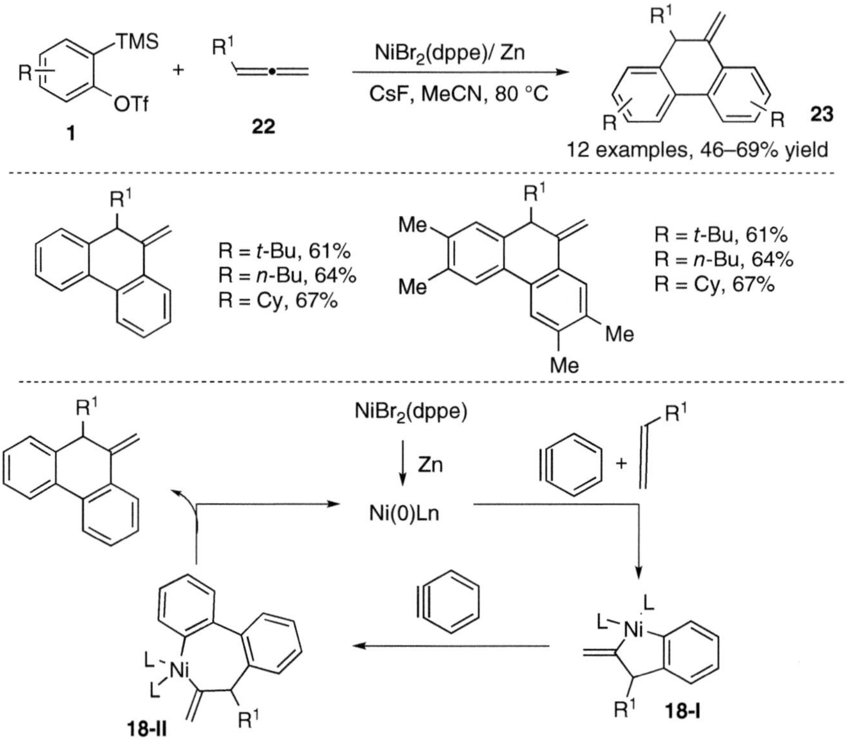

Scheme 6.18 Ni-Catalyzed cocyclotrimerization of arynes with allenes.

There are few reports on the [2+2+2] cycloaddition of unactivated alkenes with aryne moieties. In 2009, Sato and his coworkers reported a Ni(0)-catalyzed [2+2+2] cycloaddition reaction of two arynes with alkenes to afford unexpected 9,10-dihydrophenanthrene derivatives in good yields. In the reaction, a trace amount of diene, functionalized with a biaryl ring at one of the double bonds, was also observed (Scheme 6.21) [22]. However, the reaction of substituted dienes with unsubstituted arynes provided the expected products in moderate yields.

In 2011, Candito and Lautens have developed the stereoselective nickel-catalyzed [2+2+2] cycloaddition of 1,6-enynes (**28**) and arynes (**1**) [23]. Particularly, 1,6-enynes having heteroatoms such as nitrogen and oxygen efficiently participated in the reaction with arynes providing the expected cycloaddition products **29** in good yields (Scheme 6.22). Notably, bulkier substitution on the olefin or alkyne of 1,6-enynes has poor conversion or failed to undergo the reaction. The authors proposed that the reaction mechanism proceeds via five- and seven-membered nickelacycle intermediates **22-I** and **22-II**.

In 2009, Xie and coworker reported a novel Ni-catalyzed [2+2+2] carboannulation reaction of activated alkenes with alkynes and arynes (Scheme 6.23) [24]. This methodology offers an efficient route to 1,2-dihydronaphthalenes (**30**)

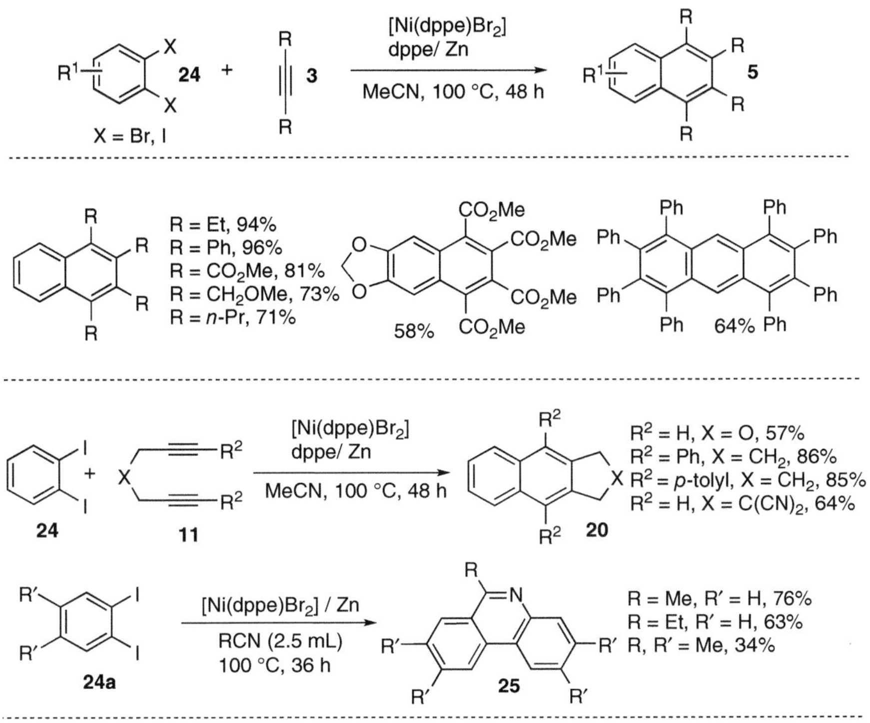

Scheme 6.19 Ni-catalyzed [2+2+2] cycloaddition of arynes with alkynes/eneynes. Source: Based on Hsieh and Cheng [20].

from commercially available starting materials. The proposed reaction mechanism initiated by oxidative coupling of alkene and benzyne with a Ni(0) forms a five-membered nickelacycle (**23-I**), which is possibly stabilized by an intramolecular coordination of the carbonyl group. Further, the insertion of an alkyne into the Ni—$C_{(aryl)}$ bond affords a seven-membered nickelacycle (**23-II**). It undergoes reductive elimination, gives the final product, and regenerates the Ni(0)-catalyst.

6.2.3 Au-Catalyzed Cyclotrimerization of Arynes

The gold(I) complex effectively catalyzes the cyclotrimerization of arynes providing triphenylenes (**2**) under the mild reaction conditions (Scheme 6.24) [25]. It is important to note that this type of trimerization can also be done by gold as well as palladium catalysts. The silver-catalyzed trimerization reaction mechanism proceeds in a similar way like gold- or palladium-catalyzed reactions. Particularly, unsymmetrical aryne precursors such as Me, OMe, and OCF_3 gave regioisomeric triphenylenes with a 1 : 3 ratio. This result has good agreement with a known Pd(0)-catalyzed cyclotrimerization of arynes [4].

A gold-catalyzed cyclization of *o*-alkynyl(oxo)benzenes (**31**) with benzene diazonium 2-carboxylates (**32**) as benzyne precursor for the construction of a variety of anthracenes (**33**) having a ketone group at the 9-position was reported by Asao and

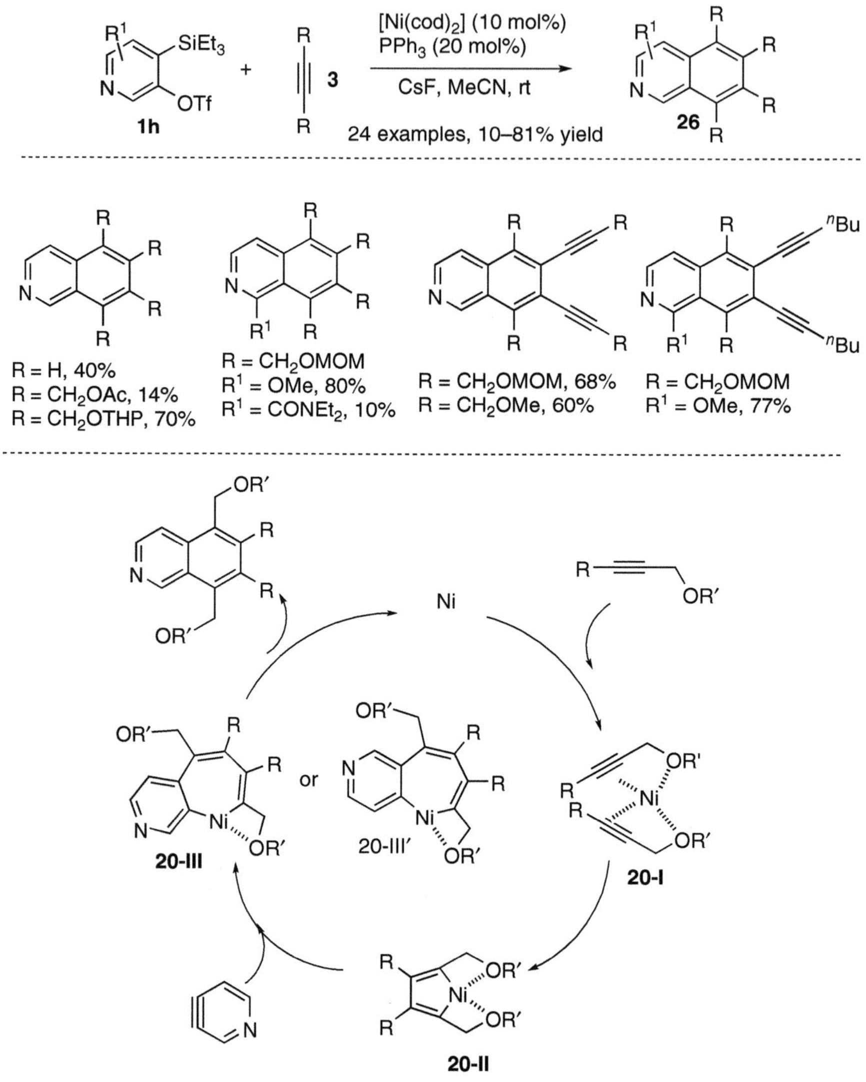

Scheme 6.20 Ni-catalyzed [2+2+2] cycloaddition of 3,4-pyridyne with 1,3-diynes. Source: Modified from Iwayama and Sato [21].

Sato [26]. This transformation proceeds through the reverse electron demand-type Diels–Alder reaction, followed by bond rearrangement, affording anthracene derivative **33** and regenerating AuCl for the next catalytic cycle (Scheme 6.25). It should be noted that, in the absence of gold catalyst, no benzannulation product was obtained.

6.2.4 Au-Catalyzed [4+2] Cycloaddition of o-Alkynyl(oxo)benzenes with Arynes

See Scheme 6.25.

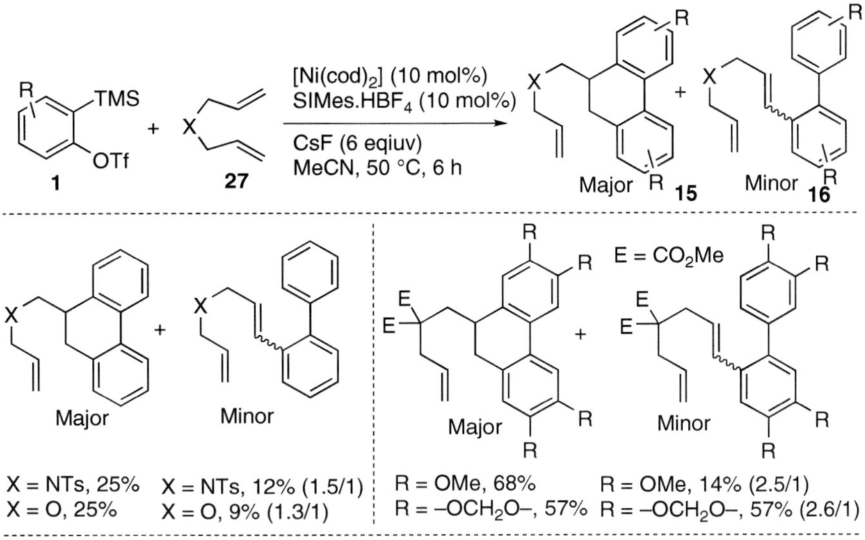

Scheme 6.21 Ni-catalyzed [2+2+2] cycloaddition of arynes with unactivated alkenes. Source: Based on Saito et al. [22].

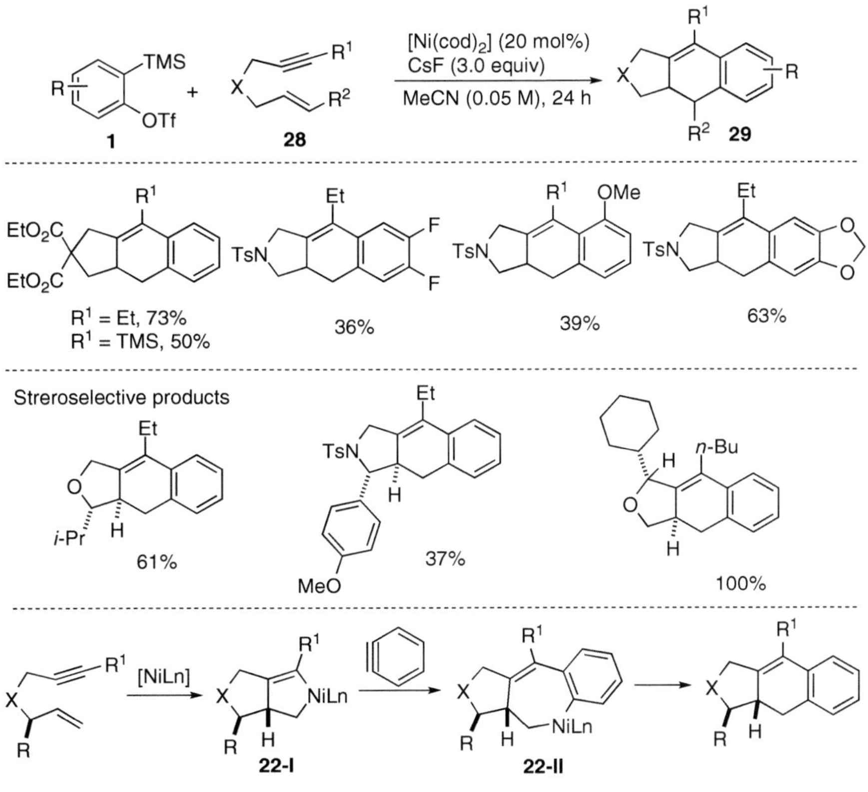

Scheme 6.22 Ni-catalyzed [2+2+2] cycloaddition of enynes and arynes.

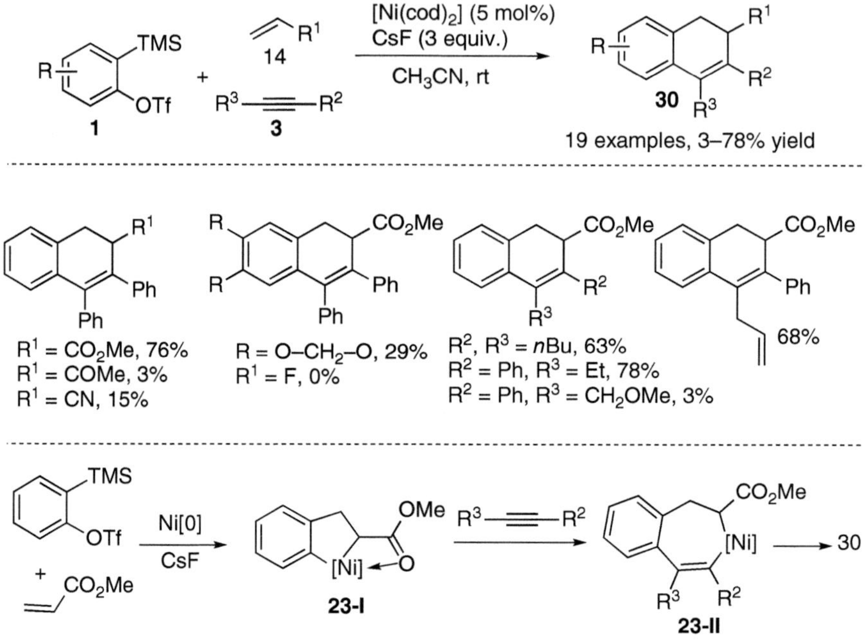

Scheme 6.23 Ni(0)-catalyzed cycloaddition of arynes, activated alkenes, and alkynes. Source: Based on Qiu and Xie [24].

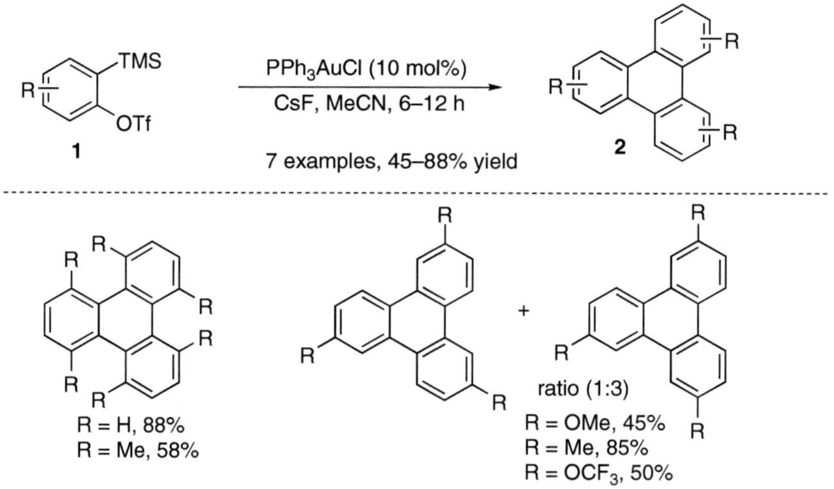

Scheme 6.24 Gold-catalyzed self-trimerization of arynes. Source: Based on Chen et al. [25].

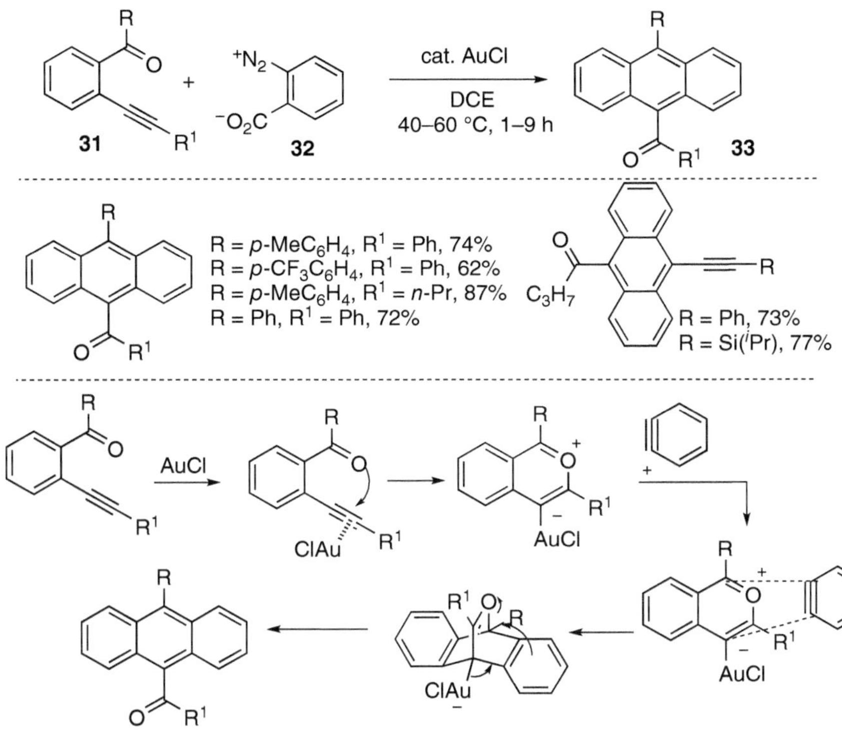

Scheme 6.25 Au(I)-catalyzed benzannulation of *o*-alkynyl(oxo)benzenes with arynes.

6.3 Metal-Catalyzed Annulation with Arynes via C—H and N—H Bond Activation

6.3.1 Palladium-Catalyzed Carbocyclization Reaction by C–H Activation

Benzynes can be successfully used for the transition-metal-catalyzed annulation reaction involving C—H bond activation as a key step.

A palladium-catalyzed carbocyclization of 2-halobiaryls **34** with benzynes giving fused polycyclic aromatic compounds was reported by Larock's group (Scheme 6.26) [27]. The slow generation of benzyne is crucial to the success of this annulation reaction. In this reaction, 5 mol% of $Pd(dba)_2$, 5 mol% of $P(o\text{-tolyl})_3$, and 3.0 equiv of CsF in 1 : 9 ratio of MeCN/toluene at 110 °C for 24 hours are needed for better transformation. This reaction was effective for the annulation of electron-donating as well as electron-withdrawing substituted 2-halobiaryls. The annulation reaction can also be applied to the synthesis of phenanthrenes by using the related vinyl halides. The annulation reaction proceeds via the oxidative addition of 2-halobiaryls with a Pd(0) and gives arylpalladium species **26-I** (Scheme 6.26). Coordinative insertion of aryne with intermediate **26-I** was followed by reductive elimination to afford the desired product. To support the proposed mechanism, palladium intermediate

26-I was prepared and allowed to react with 2.0 equiv of the benzyne precursor **1** in the presence of 3.0 equiv of CsF. In the reaction, the desired annulation product **2** was obtained in 22% yield along with the trimerization product. This observation clearly supports the proposed reaction mechanism. Further, this method can also be extended to double annulation of arynes with electron-withdrawing aryl halides (**34**). In the reaction, unsymmetrical triphenylenes were formed in good-to-moderate yields.

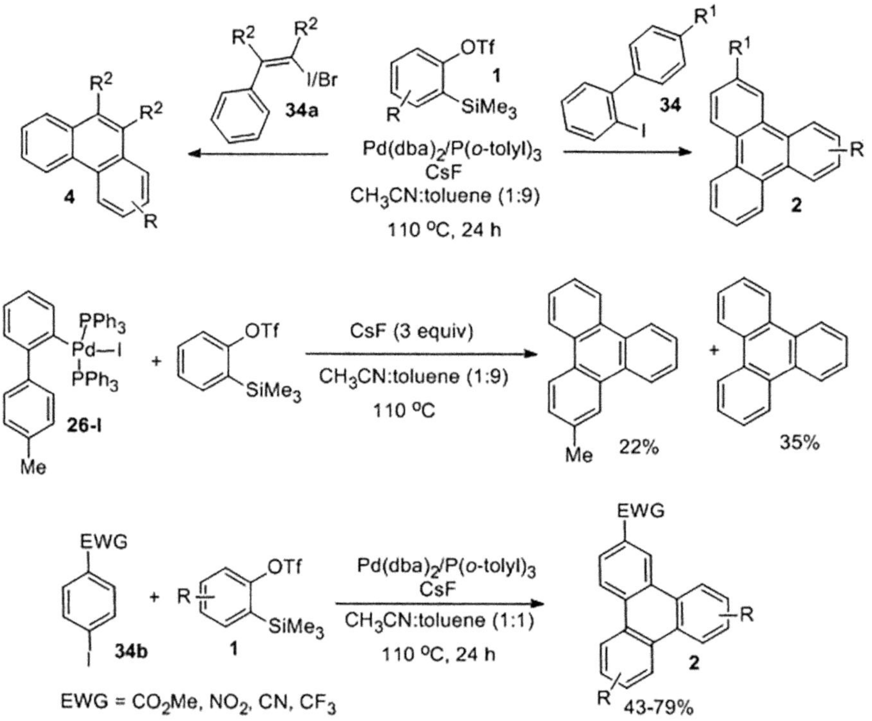

Scheme 6.26 Palladium-catalyzed carbocyclization of 2-halobiaryls with arynes. Source: Liu et al. [27a]; Liu and Larock [27b].

Cheng's group also found a palladium-catalyzed intermolecular 1 : 2 cyclization of aromatic iodides **34** with benzynes producing unsymmetrical triphenylene derivatives **2** in good yields (Scheme 6.27) [28]. TlOAc was crucial for the success of the intermolecular reaction, which is used as iodide scavenger and also accelerates the ortho metalation via C–H activation.

Further, Larock group's reported a palladium-catalyzed intermolecular annulation reaction of *o*-halobenzaldehydes (**34c**) with benzynes [29a]. This annulation strategy provides a convenient route to fluoren-9-ones **10** in good yields (Scheme 6.28). The authors proposed possible intermediates **28-I**, **28-II**, and **28-II′** for the formation of annulation product.

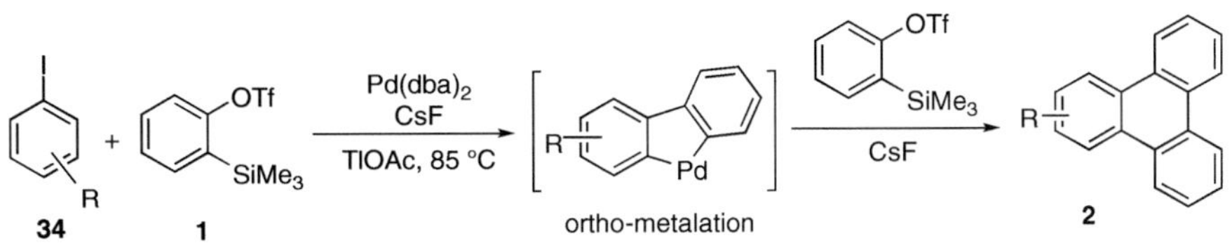

Scheme 6.27 Palladium-catalyzed carbocyclization of aromatic halides with arynes. Source: Based on Jayanth and Cheng [28].

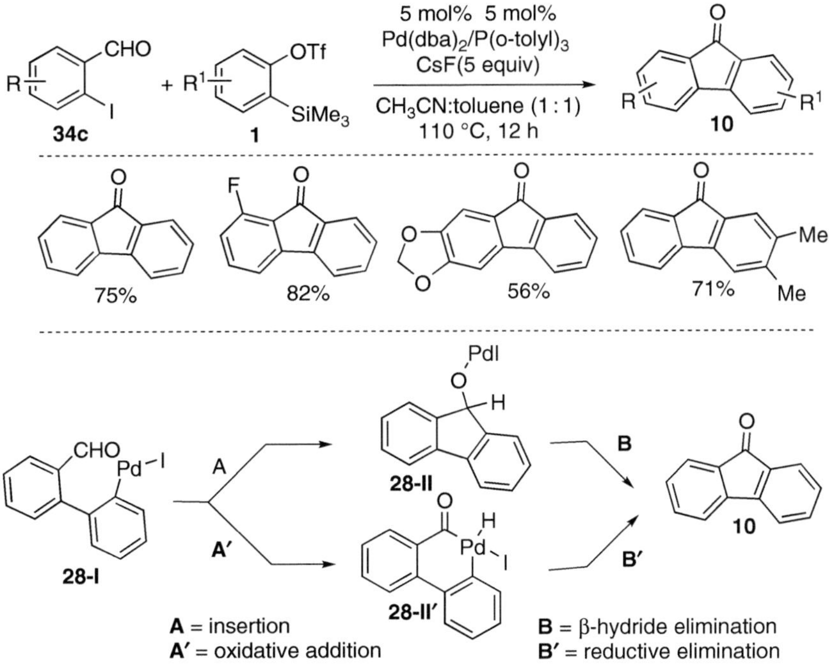

Scheme 6.28 Pd-catalyzed annulations of *o*-halobenzaldehydes with arynes.

During the mechanistic investigation, a palladium-catalyzed intermolecular sequential 1 : 1 : 1 carbocyclization of aromatic halides **34** with alkynes **3** and benzynes **1** was reported (Scheme 6.29) [29b]. In the reaction, additive Tl(OAc) was used to activate the C—H bond of aromatic iodides. The mechanism for the carbocyclization involves the stepwise regio- and chemoselective carbopalladation **29-I** of an aryl halide with internal alkyne, and subsequent insertion of aryne to give a seven-membered palladacycle **29-II**. It undergoes reductive elimination to give the cyclized product and regenerates the Pd(0) catalyst for the next catalytic cycle.

Zhang and his coworkers demonstrated a palladium-catalyzed annulation of 1-(2-bromophenyl)-1*H*-indoles (**35**) with benzynes in the presence of CsF giving indolo-[1,2-*f*]phenanthridines (**36**) in good yields (Scheme 6.30) [30]. The intramolecular C—H bond activation as a key step was proposed in the annulation reaction.

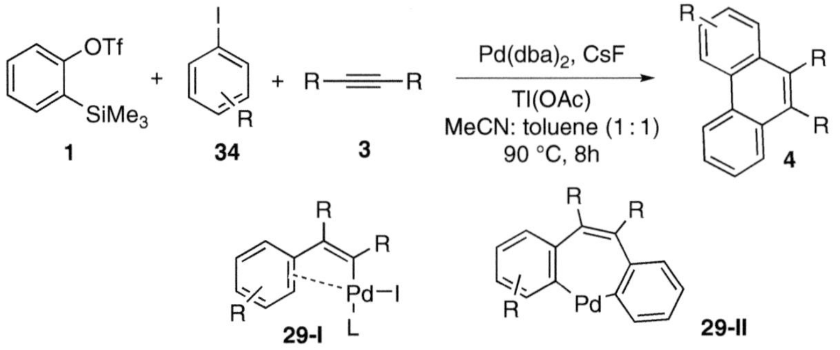

Scheme 6.29 Palladium-catalyzed three-component coupling of aromatic halides with alkynes and arynes. Source: Based on Liu and Larock [29b].

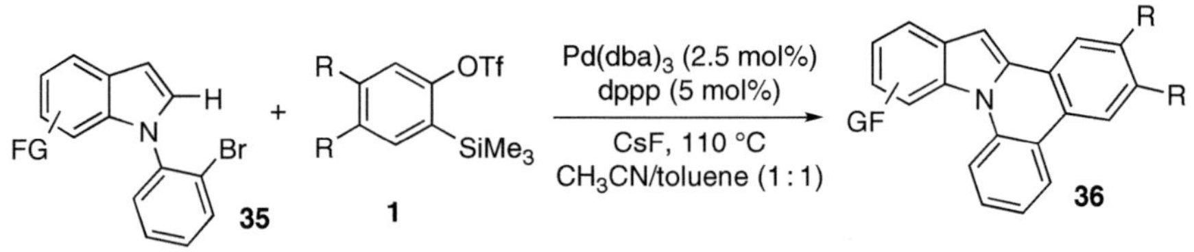

Scheme 6.30 Pd-catalyzed annulation of 1-(2-bromophenyl)-1*H*-indoles with benzynes. Source: Based on Xie et al. [30].

Cheng group observed very interesting and highly regioselective palladium-catalyzed complete intermolecular 1 : 1 : 1 carbocyclization of aromatic iodides **34** with bicyclic/heterobicyclic alkenes **7** and benzynes (Scheme 6.31) [31a]. In the reaction, external additive is not necessary to activate the C—H bond of aromatic iodides. The carbocyclization product undergoes deoxyaromatization reaction in the presence of $BF_3{\cdot}OEt_2$ (1.5 equiv) in DCM at room temperature for one hour providing polyaromatic hydrocarbons **2i** and **2j** in 85% and 90% yields, respectively. The polycyclic aromatic hydrocarbons show strong photoluminescence in the solid state as well as in solutions. This compound has potential application in material chemistry as an electroluminescent and photoluminescent material [31b, c].

By employing a similar strategy, polycyclic dibenzocoronene bis(dicarboximide) (**38**) and dinaphthocoronene tetracarboxdiimide (**38a**) were prepared in 80% and 46% yields, respectively (Scheme 6.32) [32]. Interestingly, the prepared compounds showed excellent optical properties with absorption wavelengths ranging from 380 to 600 nm, high absorption coefficients, and high fluorescence quantum yields.

A palladium-catalyzed annulation of 2-(2-iodophenoxy)-1-substituted ethanones (**34**) with arynes providing 6*H*-benzo[*c*]-chromenes (**39**) was reported (Scheme 6.33). In this reaction, two new carbon–carbon bonds (sp^2 and sp^3) are formed through an arylation/annulation strategy in one pot [33].

A palladium-catalyzed cyclization of 2-iodobenzyl-3-phenylpropiolates (**40**) or 1-iodo-2-(2-(phenylethynyl)benzyloxy)benzenes (**40a**) with arynes to give isochromen-6-ones (**41**) and phenanthro[1,10-*bc*]oxepines (**42**) in one pot was described (Scheme 6.34). This transformation involves an interesting cascade

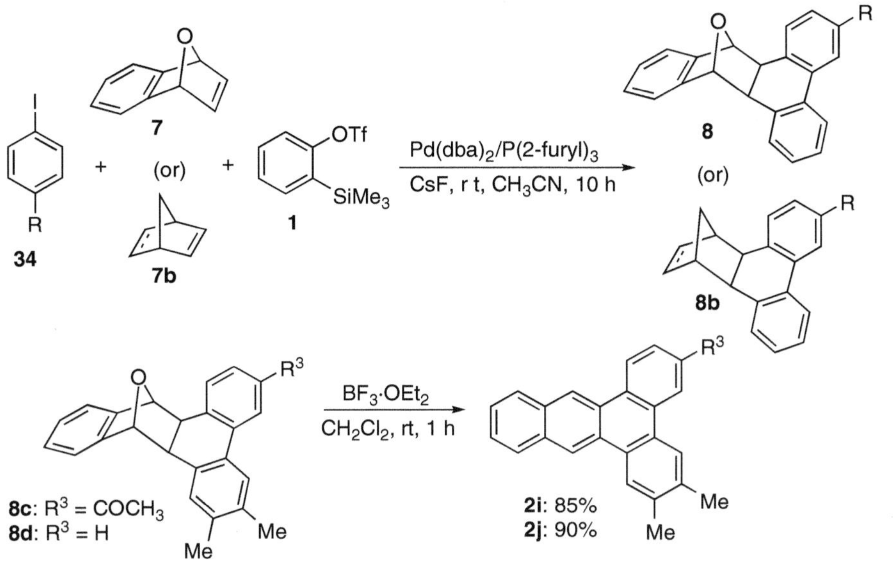

Scheme 6.31 Pd-catalyzed three-component coupling of aromatic iodides with bicyclic/heterobicyclic alkenes and arynes. Source: Modified from Bhuvaneswari et al. [31a].

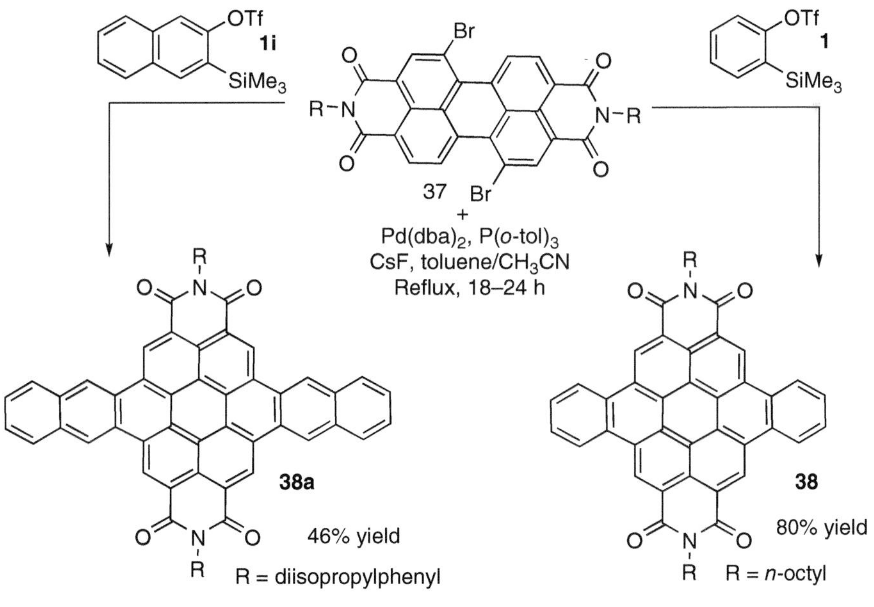

Scheme 6.32 Pd-catalyzed carbocylization strategy of arynes. Source: Avlasevich et al. [32a]; Lütke Eversloh et al. [32b].

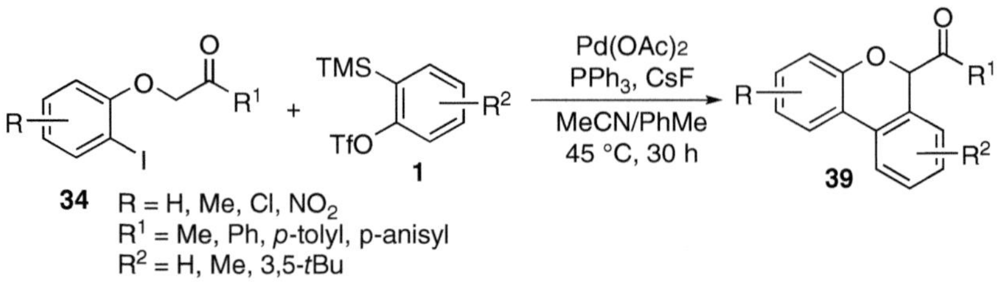

Scheme 6.33 Palladium-catalyzed annulation of 2-(2-iodophenoxy)-1-substituted ethanones with arynes.

biscarbocyclization of alkyne followed by benzyne through the intramolecular C—H bond activation [34]. The role of TlOAc was unclear in the reaction. However, it may act as a halogen scavenger on the palladium center of the key metallacycle.

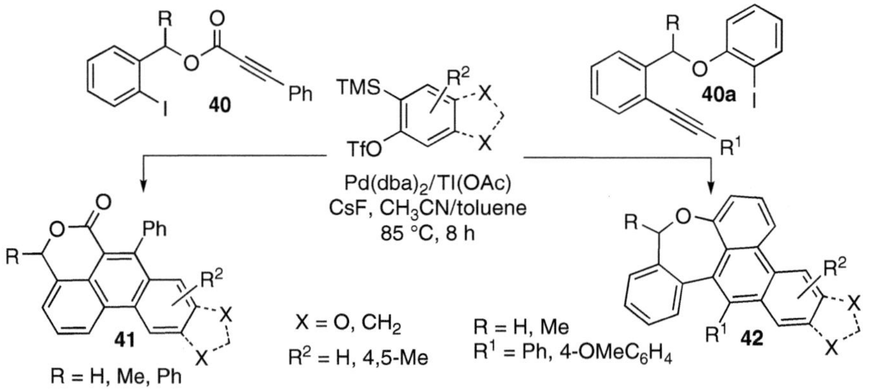

Scheme 6.34 Pd-catalyzed cascade biscarbocyclization reactions.

In 2011, Wu and coworkers reported an easy protocol to synthesize 11-(diarylmethylene)-11*H*-benzo[*b*]fluorenes (**44**) via a palladium-catalyzed [3+2] cycloaddition of (diarylmethylene)cyclopropa[*b*]naphthalenes (**43**) with arynes **1** (Scheme 6.35) [35]. The palladacyclobutene intermediate **35-I** possibly formed by the oxidative addition of palladium(0) with the highly strained three-membered ring in (diarylmethylene)cyclopropa[*b*]naphthalenes followed by aryne coordinative

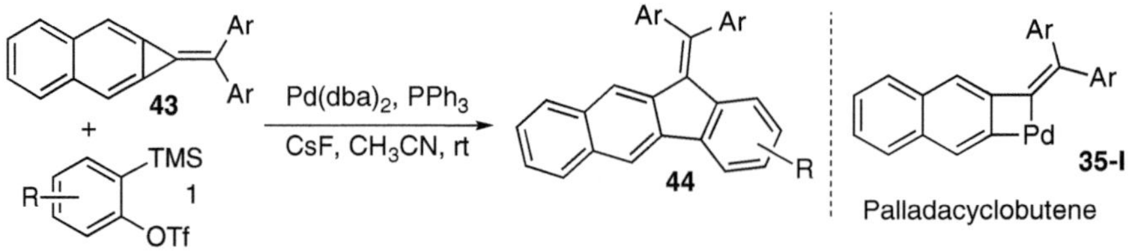

Scheme 6.35 Pd-catalyzed [3+2] cycloaddition of (diarylmethylene)cyclopropa[*b*]-naphthalenes with arynes. Source: Based on Lin et al. [35].

insertion and reductive elimination to give the desired product. It is important to note that the (diarylmethylene)cyclopropa[*b*]naphthalenes bearing electron-donating as well as electron-withdrawing groups reacted with arynes providing the expected products in good-to-excellent yields.

The intermolecular coupling of benzynes with propargylic carbonates (**3**) catalyzed by a palladium complex was reported by Ma and coworkers (Scheme 6.36). The 5 mol% of $Pd(OAc)_2$ and 10 mol% of TFP was needed for the formation of 9-butyl-10-(prop-1-en-2-yl)phenanthrenes (**4**) in good yields [36]. Other palladium complexes such as $Pd(O_2CCF_3)_2$/TFP, PdI_2/TFP, $PdCl_2(MeCN)_2$/TFP, $Pd(PPh_3)_4$, and $Pd(OAc)_2$ with different phosphine ligands (PCy_3, $(4\text{-}MeOC_6H_4)_3P$, and $(2,4,6\text{-}MeOC_6H_2)_3P$) were less effective. After the initial formation of 1,2-allenylpalladium (**36-I**) through oxidative addition of palladium(0) with **1a**, two molecules of arynes inserted into the palladium intermediate (**36-II**) to afford phenanthrenes through cyclization and β-H elimination.

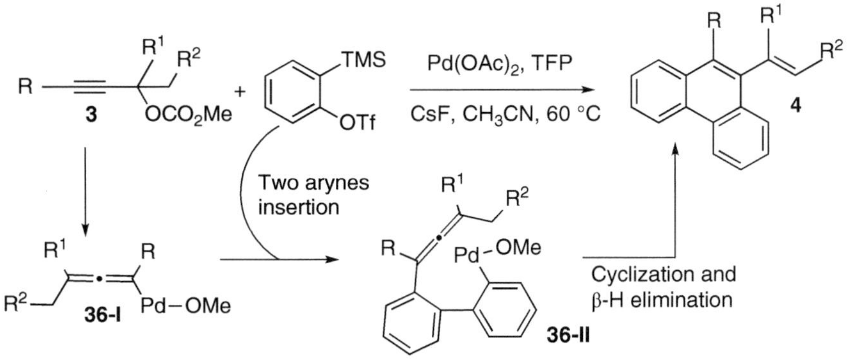

Scheme 6.36 Palladium-catalyzed reaction of arynes with propargylic carbonates.

Larock and Worlikar reported the palladium-catalyzed annulation of arynes with substituted *o*-halostyrenes (**34**) giving substituted 9-fluoroenylidenes (**44**) in good yields [37a]. This protocol tolerates various functional groups, including ester, cyano, ketone, and aldehyde on the aromatic moiety. Interestingly, in the reaction of *o*-halo allylic benzenes (**45**) with arynes, phenanthrene derivatives were observed in good yields (Scheme 6.37). The reaction proceeds via the oxidative addition of aryl iodide with a palladium complex followed by coordinative insertion of arynes gives intermediate **37-II**. The palladium–carbon bond of intermediate, inserted into the C=C bond and followed by β-hydride elimination, provides the cyclized product and regenerates the active palladium catalyst for the next cycle [37b]. Later, this methodology was extended to synthesize 9-fluorenylidene (**44**) and 9,10-phenanthrenes (**4**) by the reaction of *o*-halostyrenes or *o*-halo allylic benzenes with arynes in the presence of a palladium catalyst [37c]. The scope of the reaction was examined with various functional groups such as cyano, ester, aldehyde, and ketone-substituted aromatics. The proposed reaction mechanism was similar to that of Larock and coworker [37a].

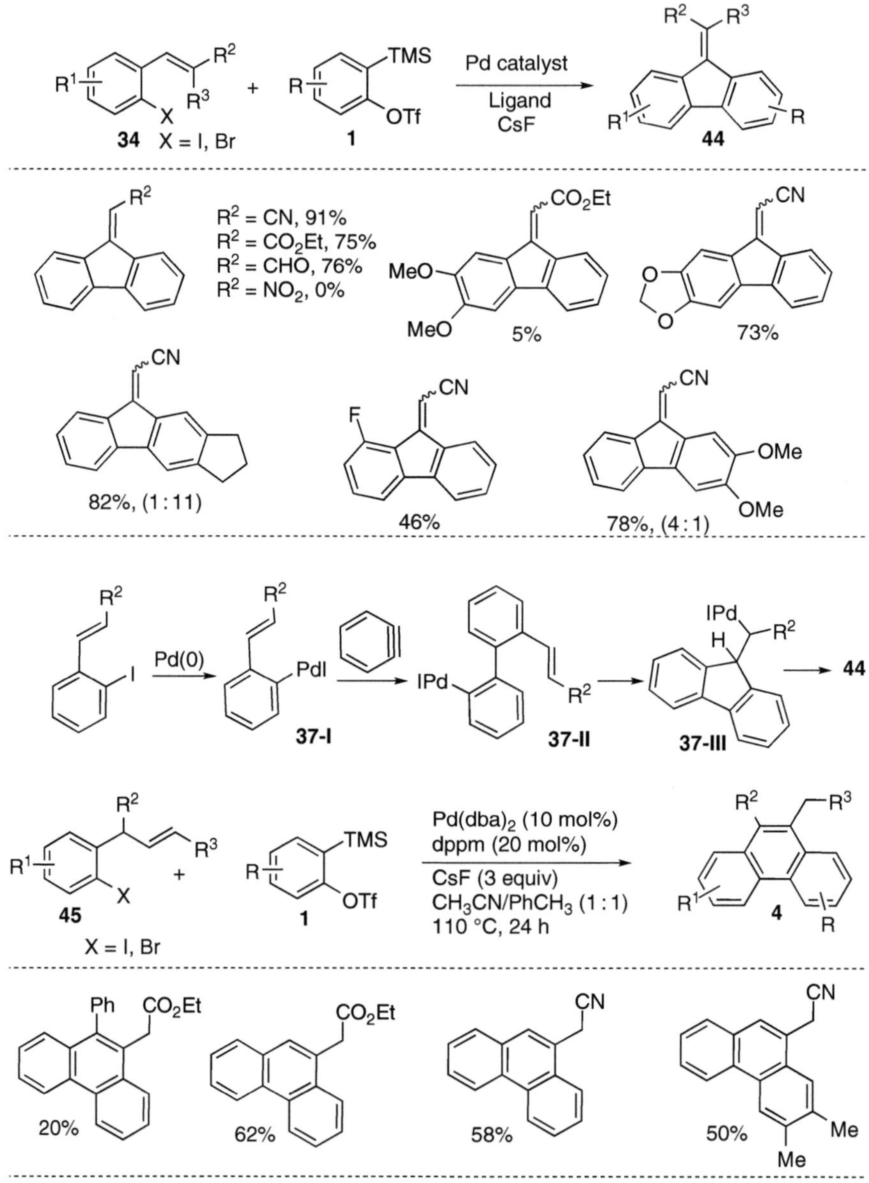

Scheme 6.37 Palladium-catalyzed annulation of substituted *o*-halostyrenes with arynes.

The Pd-catalyzed facile method for the synthesis of hydrophenanthren-1(2*H*)-ones **4a** and naphtho[2,1-*c*]furan-3(1*H*)-ones **4a′** has been explored through carboannulation of arynes with allyl-substituted iodocyclohexenones and iodofuranones [38]. The poor regioselectivity was observed in the case of unsymmetrical arynes (*p*-fluoro- or *o*-methylaryne), whereas *o*-methoxybenzyne precursor showed good regioselectivity owing to the coordination of the methoxy group with the palladium atom in the insertion step. However, the allyl moiety bearing a methyl group on compound **45c** failed to undergo β-hydride elimination with benzyne precursor **1**, rather giving a complex mixture. Notably, when the same reaction was carried out with trapping agent **14** (butyl acrylate), it underwent two benzyne insertion followed by termination with butyl acrylate and afforded product **4A** in 67% yield (Scheme 6.38).

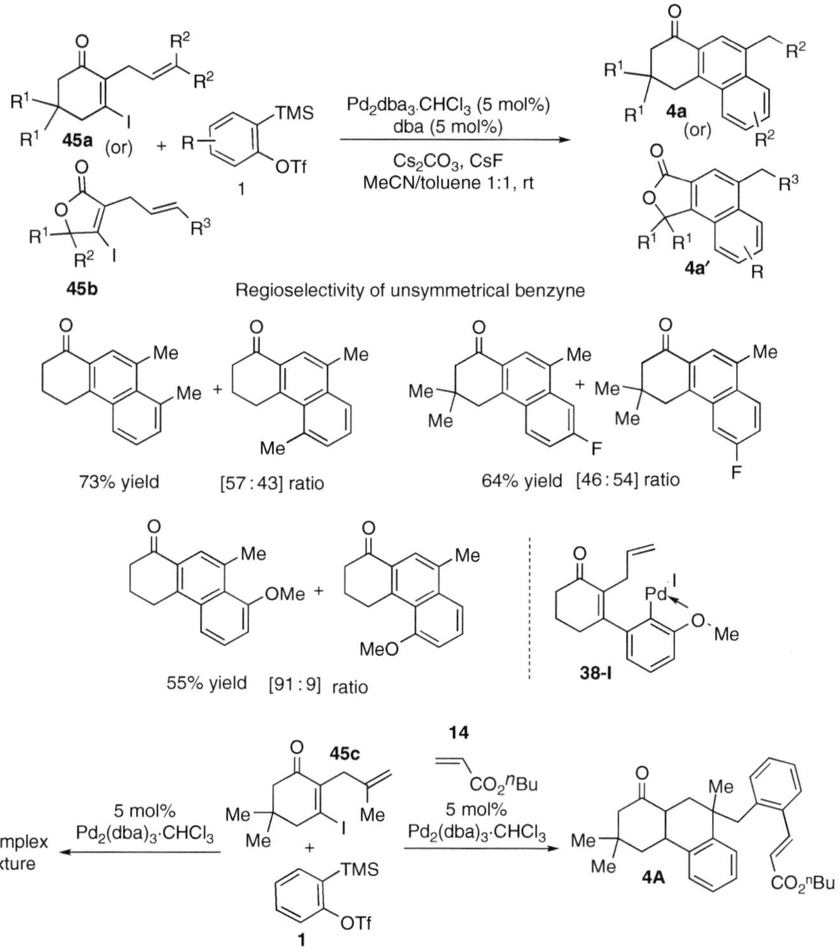

Scheme 6.38 Pd-catalyzed carboannulation of arynes with allyl-substituted iodocyclohexenones and iodofuranones. Source: Based on Yang et al. [39].

The *N*-substituted-*N*-(2-halophenyl)formamides (**46**) effectively annulated with benzynes in the presence of a $Pd(OAc)_2$ and $P(o\text{-tolyl})_3$ to give *N*-substituted phenanthridinones (**47**) in one-pot process (Scheme 6.39) [39]. The scope of reaction was very broad and various *N*-substituted phenanthridinone derivatives were prepared in good-to-excellent yields. This annulation strategy is useful to construct two new C—C bonds via an arylation/annulation sequences from easily available starting materials. The authors proposed that the acetamino coordination on a key six-membered palladacycle intermediate (**39-I**) is crucial for the success of cyclization reaction.

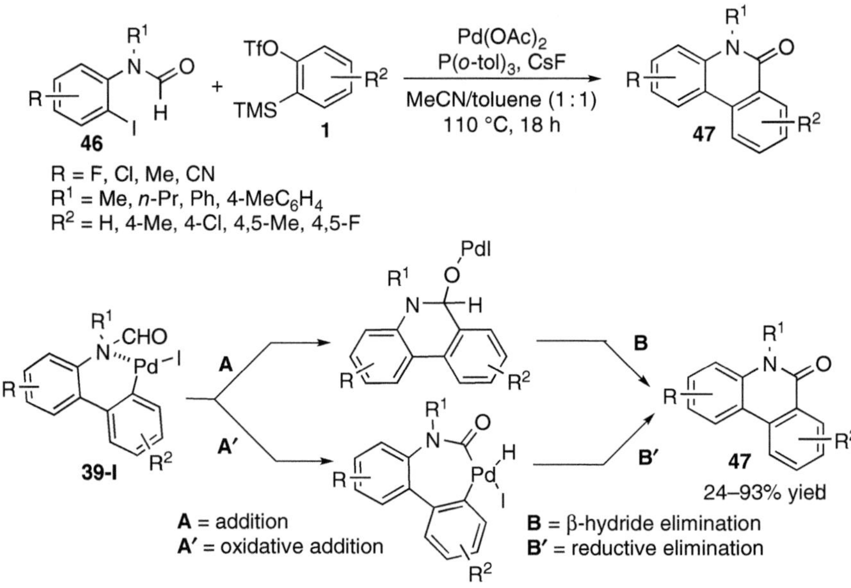

Scheme 6.39 Palladium-catalyzed annulation of *N*-(2-halophenyl)formamides with arynes.

A selective synthesis of 9,10-phenanthrenes in moderate-to-good yields was achieved through a palladium-catalyzed annulation of in situ-generated arynes with *o*-halostyrenes (**34**) [40]. The reaction proceeds through the insertion of arylpalladium(II) species into aryne followed by intramolecular *endo*-Heck reaction and isomerization of the *exo* olefin leading to the phenanthrene derivatives (Scheme 6.40). The variety of functional groups such as nitrile, ester, amide, and ketone on the aromatic moiety were well tolerated in the reaction. Further, this annulation concept was successfully applied to the formal total synthesis of a biologically active alkaloid (±)-tylophorine **48**.

An efficient one-pot palladium-catalyzed cascade reaction of *N*-(2-phenylallyl) sulfonamides (**49**) with arynes to provide spiro heterocyclic compounds in good yields was demonstrated (Scheme 6.41). Pd(OAc)$_2$ (10 mol%), PPh$_3$ (20 mol%), and CsF (3 equiv) in 1 : 1 mixture of toluene and MeCN at 90 °C is the best system for the formation of spiro heterocyclic compound [41]. This cascade transformation proceeds via two possible pathways: in path (a): the first step involves the intramolecular Heck reaction of the tethered alkene via 5-*exo-trig* cyclization to give σ-alkyl Pd(II) species (**41-I**), which undergoes C–H activation leads to spiro five-membered palladacycle (**41-II**). Then, the intermediate (**41-II**) undergoes co-ordinative insertion with aryne followed by reductive elimination to

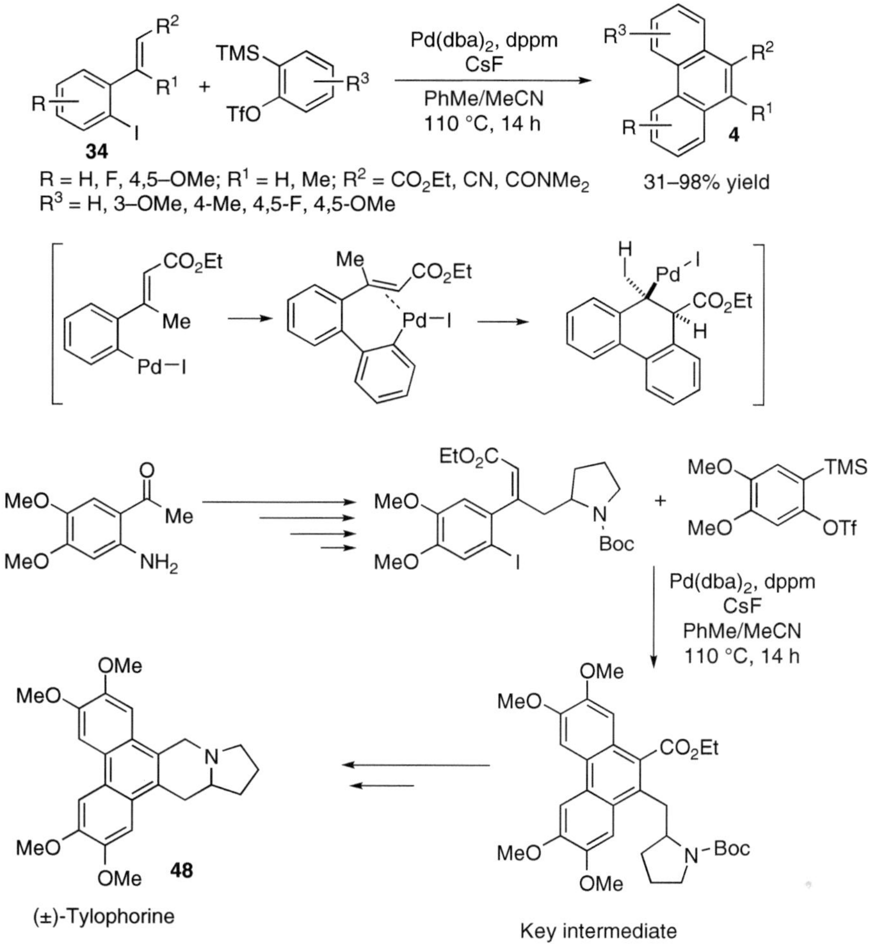

Scheme 6.40 Palladium-catalyzed cascade reaction of *ortho*-halostryrenes with arynes.

give desired product. In path (b), the σ-alkyl Pd(II) species/aryne insertion/C–H activation/reductive elimination sequences are followed for the formation of product. In the case of longer tethered alkene, regioisomeric products **50** and **50a′** were observed in 44% and 25% yield. It is believed that the palladacycle (**41-II′**) is the possible key intermediate, which formed via Heck 6-*exo-trig* cyclization for the formation of product **50a′**. As expected, unsymmetrical arynes afforded the expected product in the regioisomeric mixtures.

Lautens and coworkers synthesized a range of spirooxindoles (**50a**) and spirodihydrobenzofurans (**50b**) in good-to-excellent yield via a palladium-catalyzed intramolecular domino Heck spirocyclization (Scheme 6.42) [42]. This spiro cyclization proceeds through the oxidative addition of aromatic iodide with a palladium followed by carbopalladation to generate alkylpalladium(II) intermediate. It further undergoes domino-Heck spirocyclization via the C–H activation, benzyne insertion, and reductive elimination to give the corresponding cyclized product.

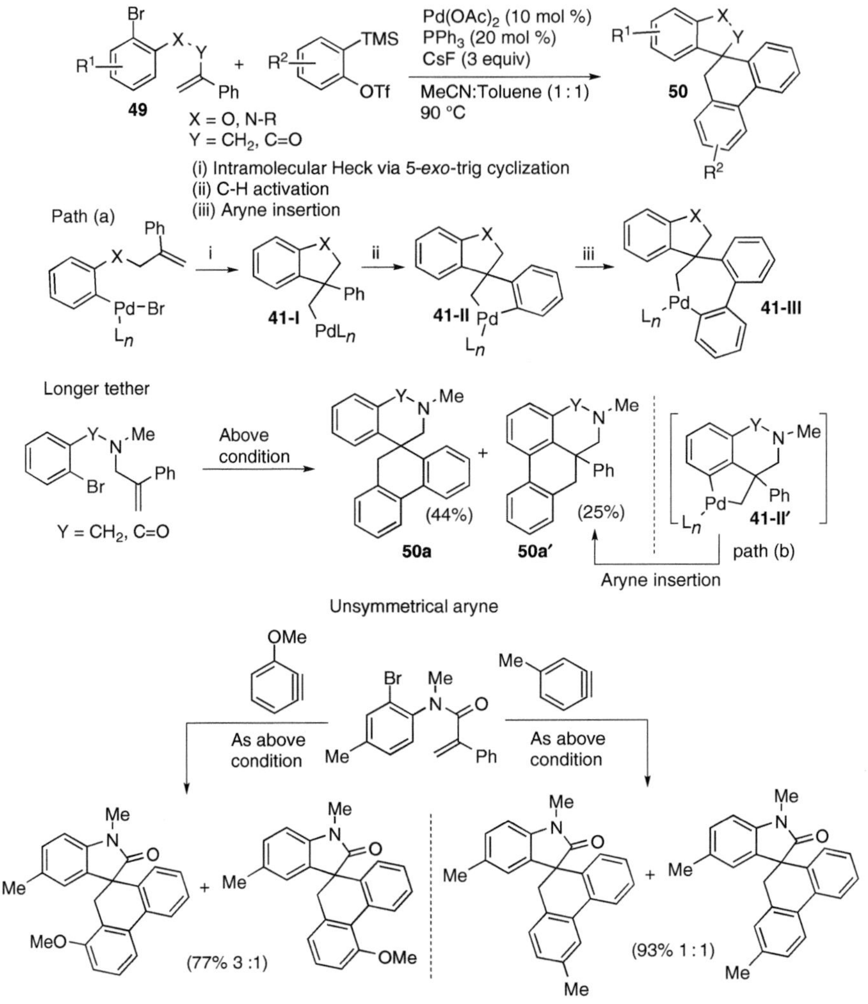

Scheme 6.41 Palladium-catalyzed cascade reaction of *N*-(2-phenylallyl)sulfonamides with arynes. Source: Based on Yoon et al. [42].

The authors have isolated the spiropalladacycle intermediate (**42-I**) to support the proposed mechanism and the structure was confirmed by a single-crystal X-ray crystallography. Further, the spiropalladacycle (**42-I**) was treated with in situ-generated benzyne to give cyclized product **50** in 83% NMR yield. This result clearly supports the proposed reaction mechanism. The observed spirocycles can be expended into the cyclic product in the presence of a catalytic amount of peroxide and NBS.

In 2017, Yao and He reported a novel Pd-catalyzed Heck/aryne carbopalladation/C–H functionalization reaction of arynes with iodo-aryl-alkene substrates (**52**) for the synthesis of 9,10-dihydrophenanthrenes (**53**) [43]. This approach offers a moderate-to-excellent yields of 9,10-dihydrophenanthrenes through a

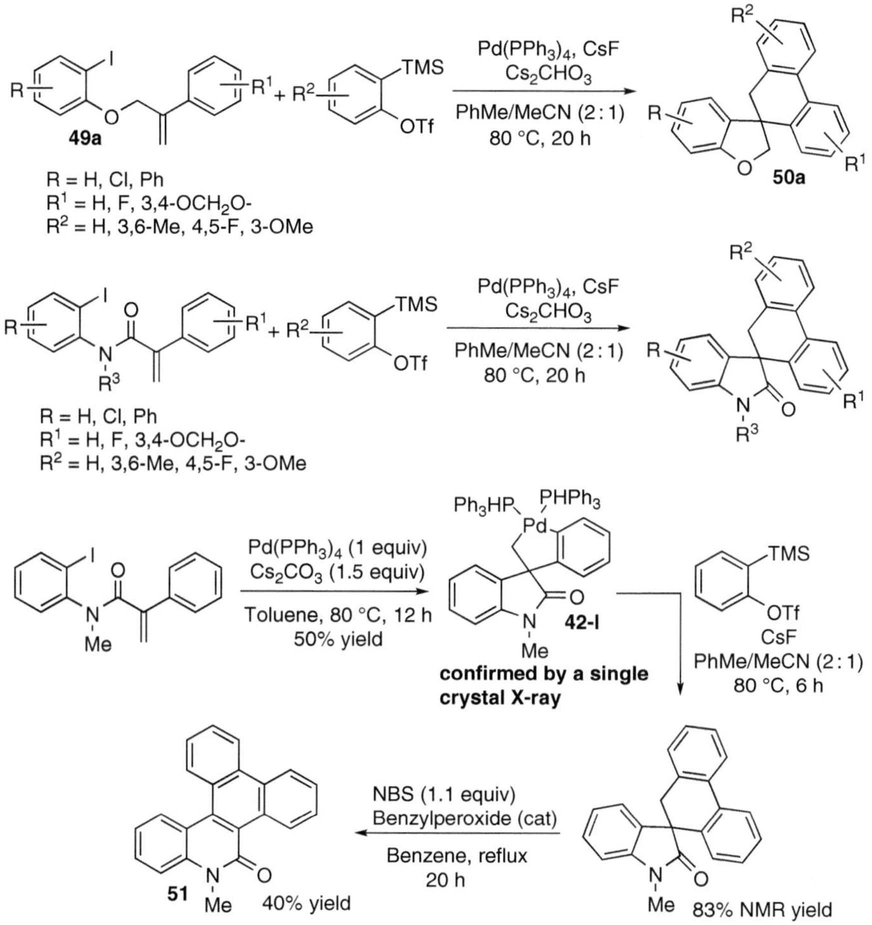

Scheme 6.42 Palladium-catalyzed domino Heck spirocyclization of arynes.

three new C—C bond along with the formation of a carbon quaternary center. The proposed reaction mechanism is shown in Scheme 6.43. The arylpalladium(II) intermediate (**43-I**) is generated from the oxidative addition of Pd(0) into the aryl-iodide. It undergoes 6-*exo-trig* ($n = 1$) or 5-*exo-trig* ($n = 0$) cyclization giving the key neopentyl-type Pd(II) σ-complex (**43-II**). In route (a), aryne can undergo carbopalladation with intermediate (**43-II**) to give arylpalladium intermediate (**43-III**). Next, the intramolecular C–H activation through base-induced palladation forms a seven-membered palladacycle (**43-IV** and **43-IV′**). Further, it undergoes reductive elimination affording product **53** and regenerates the Pd(0) catalyst. Alternatively, in route (b), initially intermediate (**43-II**) performs an intramolecular C–H activation to form five-membered palladacycle (**43-V**), followed by aryne insertion on C_{aryl}–Pd (VI) and subsequent reductive elimination to form the cyclized product **53** and regenerates a Pd(0) catalyst [43b].

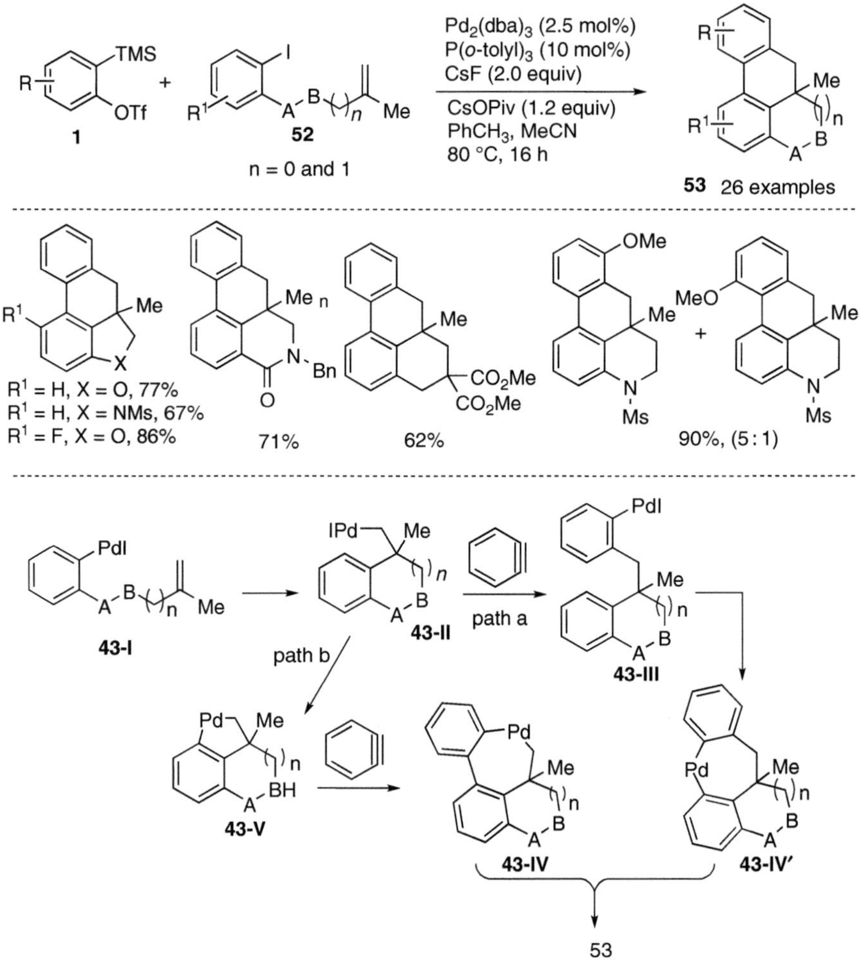

Scheme 6.43 Synthesis of 9,10-dihydrophenanthrenes by Pd-catalyzed carbopalladation of arynes with iodo-aryl-alkenes.

Luan and his coworkers developed a Pd(0)-catalyzed chemoselective [3+2] spiroannulation of 2-halobiaryls (**34**) with arynes providing spirofluorenes in good-to-excellent yields (Scheme 6.44) [44]. This protocol proceeds through the unusual [3+2] spiroannulation by dearomative 5-*exo-trig* cyclization, distal-hydride elimination (or), base-mediated dearomatization followed by reductive elimination. Particularly, the CH_2CO_2R on the naphthalene ring is crucial for the spiro-cyclization reaction. After the detailed optimization studies, the authors found that the $[Pd(cinnamyl)Cl]_2$, $P(o\text{-anisyl})_3$, in MeCN as solvent at 90 °C is the best condition for the annulation reaction. Interestingly, *p*-bromonaphthalene (**34a**) also underwent similar [2+2+1] annulation with benzyne leading to the desired product **54a** in 60% yield. Under the optimized reaction conditions, the nonacidic substrate **34b** smoothly undergoes cyclization with aryne to give a new type of spirocyclic product **55** in 56% yield.

6.3.2 Palladium-Catalyzed Arynes in C–X Annulations (X = N, O)

Larock's group described a palladium-catalyzed synthesis of *N*-substituted phenanthridinones (**47**) through annulation of arynes with *N*-substituted *o*-halobenzamides (**56**). In the reaction, C—C and C—N bonds can be constructed in one pot (Scheme 6.45). Various *N*-substituted *o*-halobenzamides worked well under the reaction conditions. However, 2-bromobenzamide, *N*-phenyl-2-bromobenzamide, and *N*-dimethyl-2-bromobenzamide failed to give the expected product [45].

A palladium-catalyzed selective synthesis of various *N*-acylcarbazoles was reported in the reaction of 2-bromo/iodoacetanilides with arynes (Scheme 6.46) [46]. To achieve this annulation, various mono- and bidentate phosphine ligands were examined. Among them, dppf was found to be an effective ligand for the reaction. The cyclization reaction was compatible with various substituted benzamides. However, *N*-(2-bromophenyl)pivalamide and methoxy and difluoro-substituted benzyne precursors did not work in the annulation reaction. The same group has also reported the carbocyclization of benzynes with 2-iodophenols providing heterocyclic molecules in good-to-excellent yields.

A palladium-catalyzed domino annulation of benzophenone *O*-perfluorobenzoyl oximes (**58**) with benzynes giving phenanthridines (**25**) was demonstrated (Scheme 6.47). The optimization studies clearly reveal that in the absence of a Pd catalyst, the reaction leads to mixture of products. Thus, the palladium catalyst is necessary for selectivity as well as high yields. The reaction proceeds via oxidative addition of Pd(0) into N—O bond of an oxime giving key aminopalladium species (**47-I**). Then, the species (**47-I**) undergoes cyclization in two possible pathways: (i) *ortho* C–H activation leads to five-membered palladacycle followed by benzyne insertion to form a seven-membered palladacycle (**47-IV**). (ii) Alternatively, palladium species (**47-I**) undergoes co-ordinative *syn* insertion with benzyne to give aryl palladium complex (**47-III**) followed by C–H activation to form intermediate (**47-IV**). Later, intermediate (**47-IV**) undergoes reductive elimination to give phenanthridine derivative **25** and regenerates a palladium catalyst for the next catalytic cycle (Scheme 6.47) [47]. However, acetophenone *O*-perfluorobenzoyl oxime reacted slowly with benzyne affording phenanthridine in the lower yield.

The aryl ketone *O*-acetyloximes also underwent the Pd-catalyzed annulation with arynes providing synthetically useful phenanthridine derivatives (**25**) via the C—H bond activation (Scheme 6.48). This conversion involved a C—H bond activation/aryne insertion/followed by cyclization and reductive elimination reaction sequences. Under similar reaction conditions, various symmetrical and unsymmetrical acetophenone and benzophenone *O*-acetyloximes underwent annulation smooth with arynes providing the expected products in good-to-excellent yields [48]. Interestingly, the reaction of 4-methoxy-4′-chlorobenzophenone *O*-acetyloxime with aryne afforded phenanthridine derivative as a sole product in 60% yield. It is important to mention that the C–H activation/annulation occurred exclusively

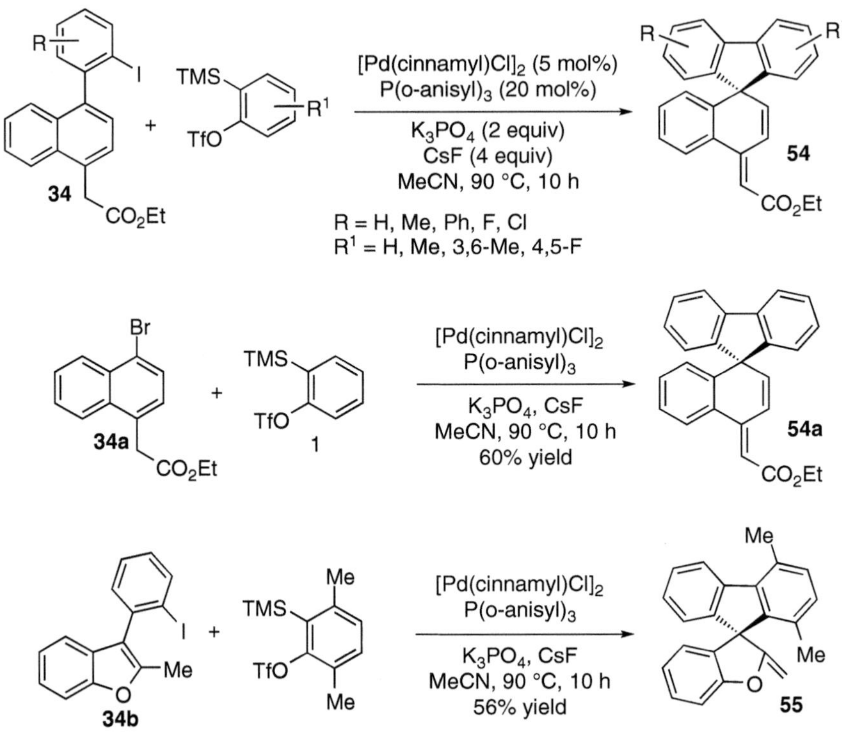

Scheme 6.44 Pd(0)-catalyzed chemoselective [3+2] spiroannulation of 2-halobiaryls with arynes. Source: Based on Zuo et al. [44].

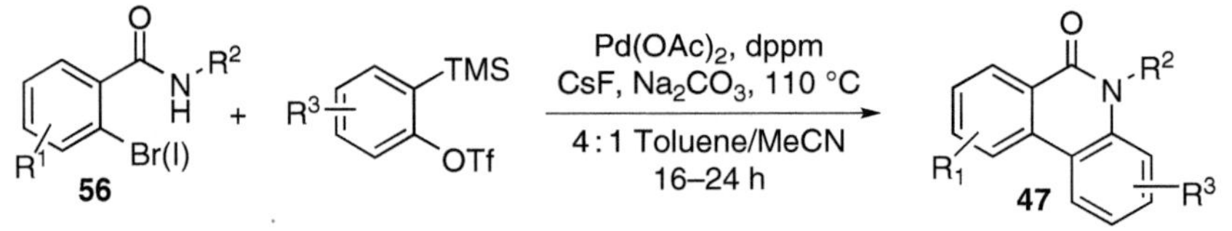

Scheme 6.45 Palladium-catalyzed annulation of arynes with *ortho*-halobenzamides.

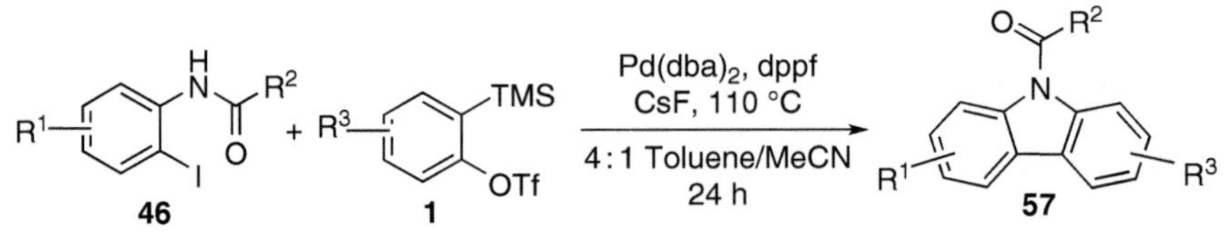

Scheme 6.46 Synthesis of *N*-acylcarbazoles by palladium-catalyzed cyclization of 2-iodoacetanilides with arynes. Source: Lu et al. [46].

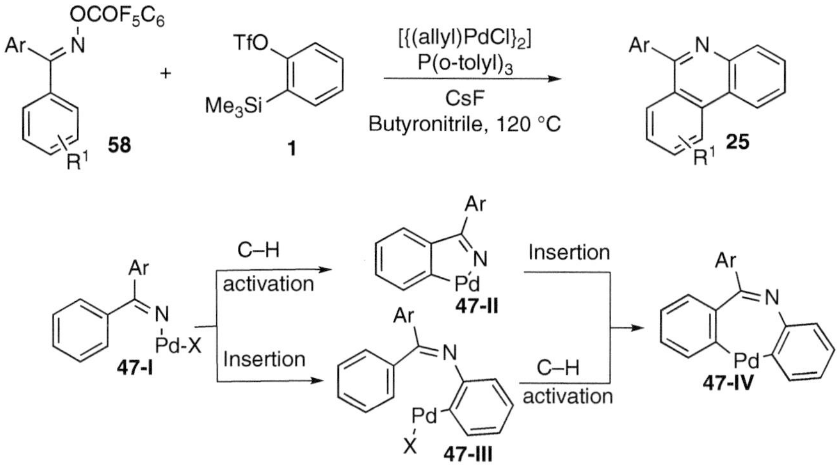

Scheme 6.47 Synthesis of phenanthridines by palladium-catalyzed domino annulation of benzophenone *O*-perfluorobenzoyl oximes with arynes. Source: Based on Gerfaud et al. [47].

on the –OMe substituted aryl ring. In the case of 3-methyl and 3-chloro acetophenone *O*-acetyloximes, mixture of regioisomeric products was observed. Similarly, unsymmetrical aryne reacted with 4-methoxy acetophenone *O*-acetyloxime giving regioisomeric products in 46% combined yield in a 1 : 1 ratio. The ratio of isomers was determined by the ^{1}H NMR analysis of the crude sample.

A one-pot tricyclic phenanthridinones was efficiently synthesized by the cyclization of *N*-methyl or methoxybenzamides (**56**) with benzynes in the presence of $Pd(OAc)_2$, 1-adamantanecarboxylic acid and $K_2S_2O_8$ in CH_3CN. In this conversion, the 1-adamantanecarboxylic acid (Adm-1-COOH) acted as an effective additive and enhanced the yield of the cyclized product (Scheme 6.49) [49]. Benzyne acts as a highly reactive organic π-component for this annulation reaction. Notably, electron-donating substituted *N*-methoxy benzamides favors the *ortho* C—H bond activation and cyclization reaction, whereas halogen and electron-deficient aromatic benzamides give only direct *N*-arylated product **56a**. The authors observed that the palladacycle of 4-chloro benzamide reacted with benzyne **1** to give expected cyclization product **47** in 45% yield. These observations clearly indicate that the *ortho* C—H bond activation is very slow in the electron-deficient benzamides, and the competitive nucleophilic N–H addition is very fast.

The palladium-catalyzed oxidative C–H annulation of *N*-methoxybenzamides (**56**) with arynes was achieved for the synthesis of phenanthridinones (**47**) in the presence of $Cu(OAc)_2$ as a co-oxidant (Scheme 6.50) [50]. In this transformation, electron-withdrawing and electron-donating substitutions were tolerated on both benzamides and aryne-coupling partners. The authors have confirmed the proposed mechanism by isolating the key five-membered palladacycle. It has been prepared by treating 1 equiv of $Pd(OAc)_2$ with benzamide in acetic acid at 120 °C. The palladacycle was further converted into the expected product under the optimized

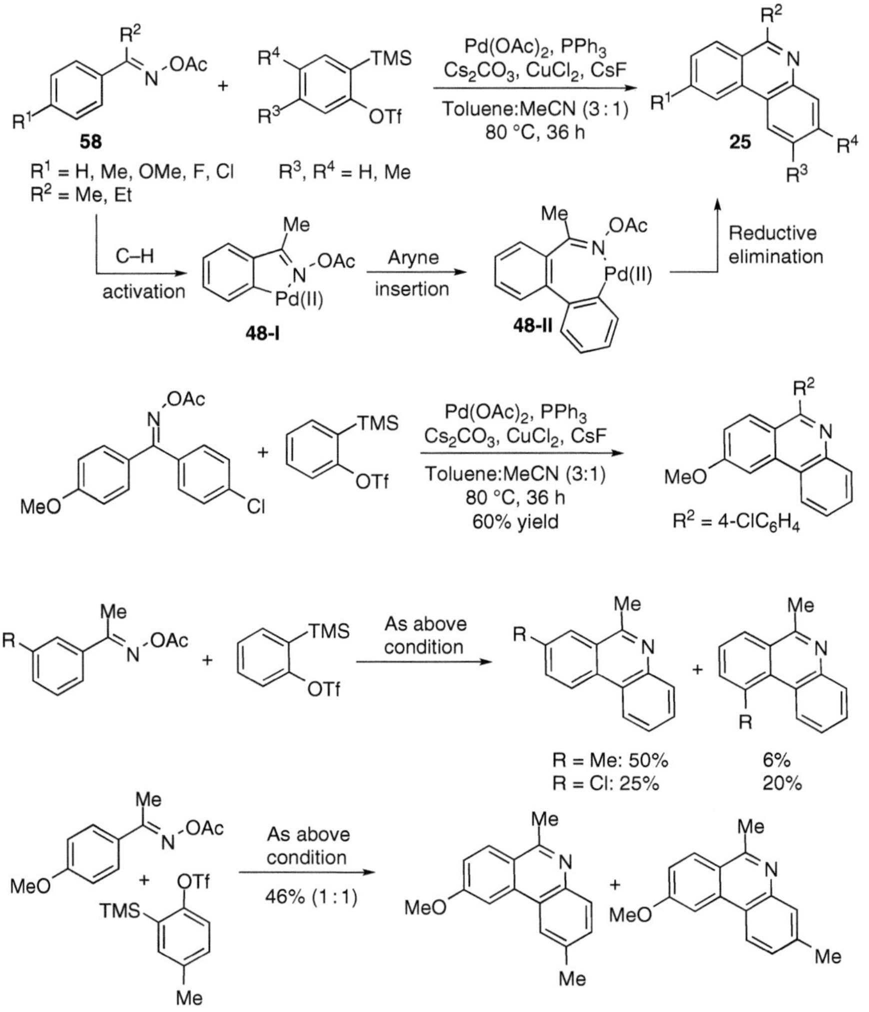

Scheme 6.48 Pd-catalyzed annulation of aryl ketone *O*-acetyloximes with arynes.

reaction conditions. Further, the observation of the KIE value of 4.7 revealed that the C–H activation is the rate-determining step in the catalytic cycle.

This protocol was also applicable to synthesize various quinolinones (**60**) by using substituted acrylamides (**59**) and benzynes (Scheme 6.51). This annulation transformation is a very challenging task due to the less reactive nature of the C—H bond of alkene [51]. The quinolinone formation was not observed in the absence of a palladium catalyst. The reaction mechanism takes place via: (i) N–H and C–H activation/benzyne insertion/reductive elimination (or) (ii) benzyne insertion/6-*endo-trig* Heck cyclization/reductive elimination sequences. In either pathway, after the reductive elimination step, the Pd(0) was released. It again

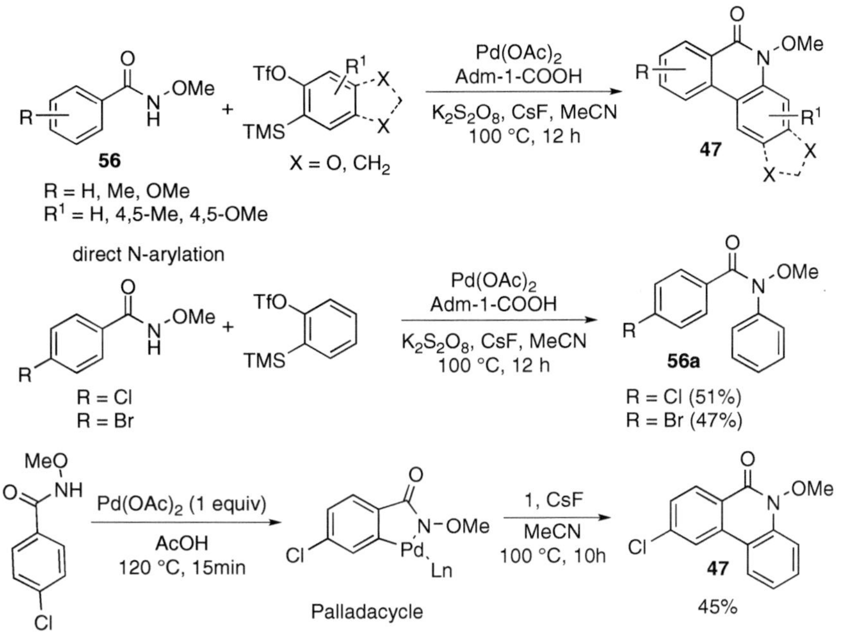

Scheme 6.49 Phenanthridinones by Pd-catalyzed C–H annulation of *N*-methoxybenzamides with arynes. Source: Based on Pimparkar and Jeganmohan [49].

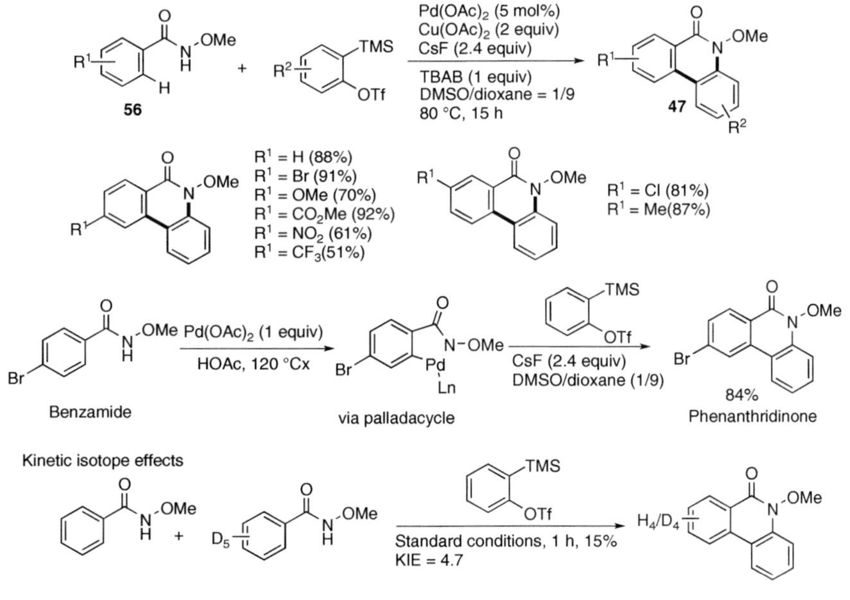

Scheme 6.50 Synthesis of phenanthridinones via palladium-catalyzed oxidative C–H annulation of *N*-methoxybenzamides with arynes. Source: Peng et al. [50].

reoxidized into the active Pd(II) species in the presence of $Cu(OAc)_2$ for the next catalytic cycle.

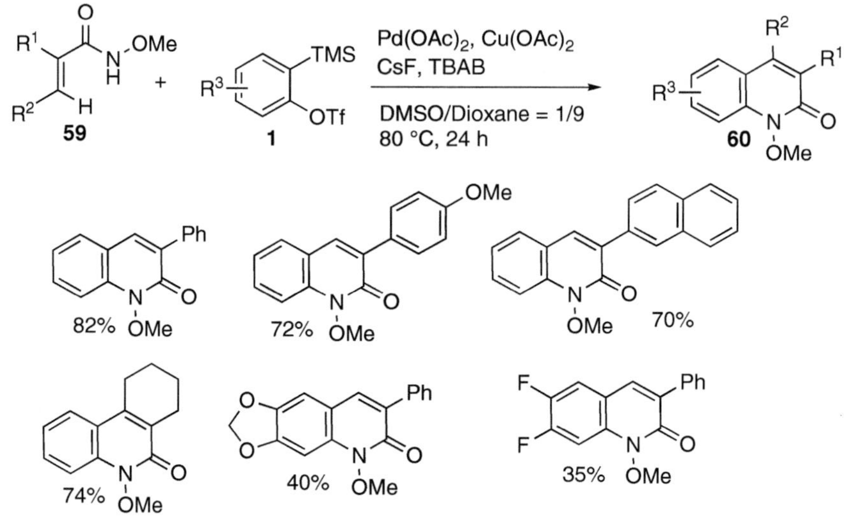

Scheme 6.51 Pd(II)-catalyzed synthesis of quinolinones by acrylamides and arynes.

An interesting and useful synthesis of coumestans **62** was reported by Gogoi and his coworkers by the cyclization of 4-hydroxycoumarins (**61**) with arynes in the presence of a palladium catalyst and $Cu(OAc)_2$ as an external oxidant [52]. In the reaction, $Cu(OAc)_2$ was used to regenerate the active Pd(II) catalyst from Pd(0). This cascade strategy proceeds via the C—H bond activation/C—O and C—C bond formations in one pot. This methodology affords the expected products in moderate-to-good yields. Further, this methodology was applied to the synthesis of natural product flemichapparin C **62a** (Scheme 6.52).

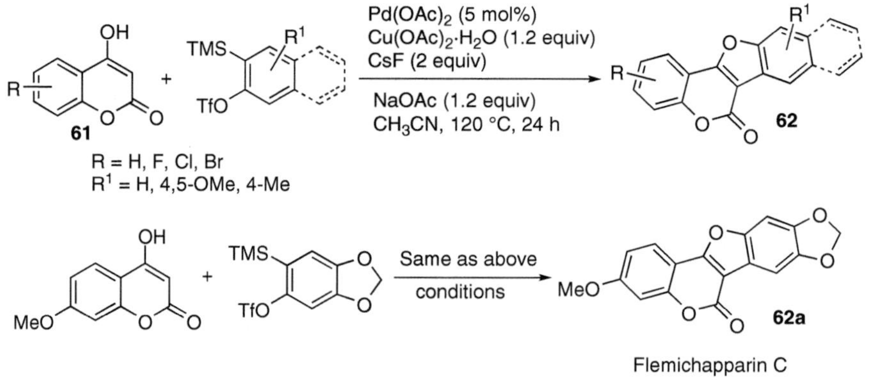

Scheme 6.52 Synthesis of coumestans and natural product flemichapparin C.

A combination of palladium and photoredox catalysis was used for the synthesis of phenanthridinones (**47**) at room temperature via annulation of benzamides (**56**) with arynes (Scheme 6.53). The Na_2-eosin Y is reacted with molecular oxygen to give superoxygen anion radical species with the blue LED [53]. Therefore, the superoxygen anion radical species regenerates the Pd(0) to Pd(II) catalyst for the next catalytic cycle. To find the annulation mechanism, the authors have carried out the kinetic isotopic effect experiment. The observed value of $K_H/K_D = 5.9$ indicates that the C–H activation is the rate-determining step in this catalytic reaction. Controlled experiments such as without eosin Y, TEMPO experiments, and tapping for superoxygen anion radical species supported the proposed annulation mechanism.

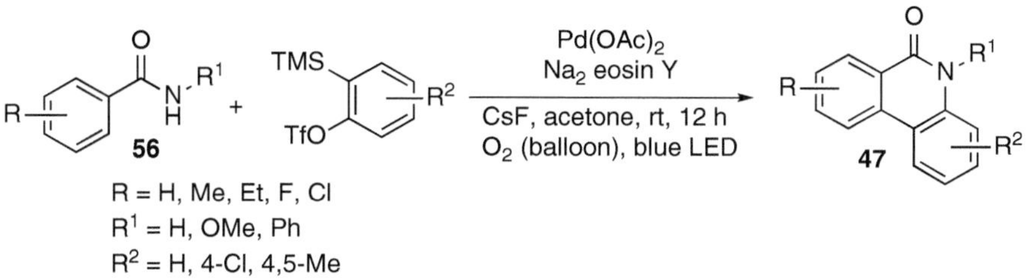

Scheme 6.53 $Pd(OAc)_2$/photoredox-catalyzed synthesis of phenanthridinones.

One-pot palladium(II)-catalyzed oxidative annulation of *N*-methyl benzylamines (**63**) with arynes (**1**) in the presence of $Cu(OAc)_2$ as a co-oxidant to give 5,6-dihydro-phenanthridine derivatives (**64**) through the C–H activation was reported (Scheme 6.54) [54]. The electron-donating group substituted *N*-methyl benzylamines provided the expected products in good yields, whereas the electron-withdrawing substituted benzylamines gave the expected products in the lower yields. This reaction proceeds through the five-(**54-I**) and seven-(**54-II**) membered metallacycles as key intermediates.

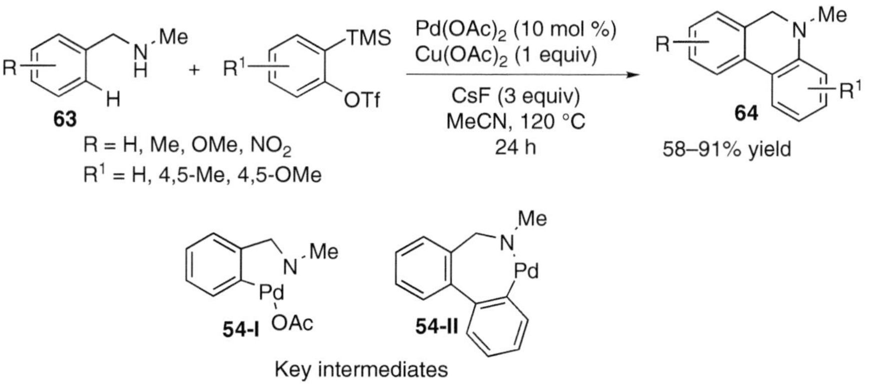

Scheme 6.54 Pd(II)-catalyzed oxidative annulation of *N*-methyl benzylamines with arynes. Source: Modified from Asamdi et al. [54].

Recently, a palladium-catalyzed cyclization of *N*-alkoxybenzsulfonamides (**65**) with highly reactive benzynes affording dibenzosultams (**66**) in one pot was reported (Scheme 6.55) [55]. The reaction proceeds via a Pd(II)-catalyzed directed C–H/N–O alkyl activation to give consecutive C—C/C—N bond formation via the unexpected N—O bond cleavage. The control experiment revealed that the use of palladium catalyst, $Cu(OAc)_2$, CsF, and $PivONa \cdot H_2O$ is essential to increase the efficiency of the reaction. The potential utility of this protocol was broad with respect to benzsulfonamides as well as arynes. A mechanistic study shows that the reaction proceeds through a rate-determining C—H bond cleavage as a key step.

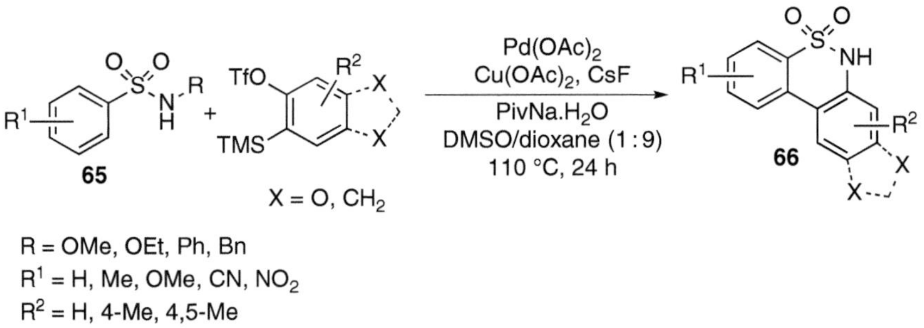

Scheme 6.55 Pd(II)-catalyzed direct synthesis of dibenzosultams by C-H activation. Source: Modified from Feng et al. [55].

A new method for the synthesis of phenanthridinones (**47**) by a palladium-catalyzed decarbonylative annulation of phthalimides (**67**) with arynes was developed (Scheme 6.56) [56]. The reaction proceeds through a palladium-catalyzed decarbonylation by the assistance of a 8-amino quinoline group. Meanwhile, the *N*-phenyl and methyl substitutions on the nitrogen atom also failed to undergo decarbonylative annulation reaction. Further, the key five-membered Pd(II) complex was isolated to support the mechanism. The structure of complex was confirmed by a single crystal X-ray crystallography. This catalytic system was compatible with various phthalimides and symmetrical benzynes. However, unsymmetrical arynes gave a mixture of regioisomeric products. Surprisingly, the reaction of 2-(quinolin-8-yl)-4,5,6,7-tetrahydro-1*H*-isoindole-1,3(2*H*)-dione (**67**) with benzyne providing cascade aryne insertion/dehydroaromatization product **67a** in 66% yield under similar reaction conditions (Scheme 6.56).

6.3.3 Ni-Catalyzed C–N Annulations by Denitrogenative Process

In 2018, Cheng and his coworkers demonstrated a Ni-catalyzed denitrogenative/annulation of 1,2,3-benzotriazin-4-(3*H*)-ones (**68**) with arynes providing structurally diverse phenanthridinones (**47**) in good-to-excellent yields (Scheme 6.57) [57a]. A wide range of functional groups on the aromatic system was well tolerated. This protocol was successfully applied to synthesize natural product *N*-methylchrinasiadine in 84% yield from the reaction of triazinone with arynes

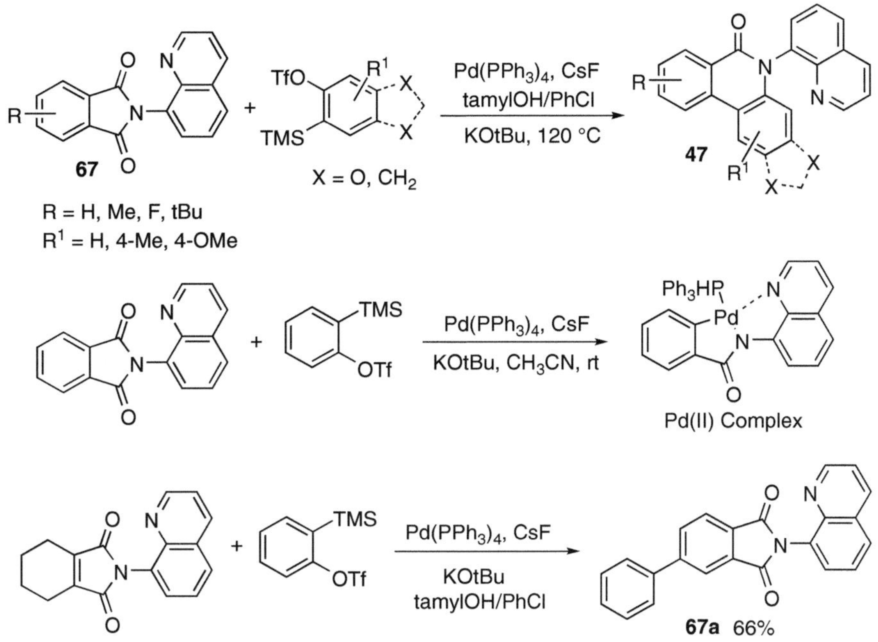

Scheme 6.56 Palladium-catalyzed 8-amino quinoline directed decarbonylative annulation of phthalimides with arynes.

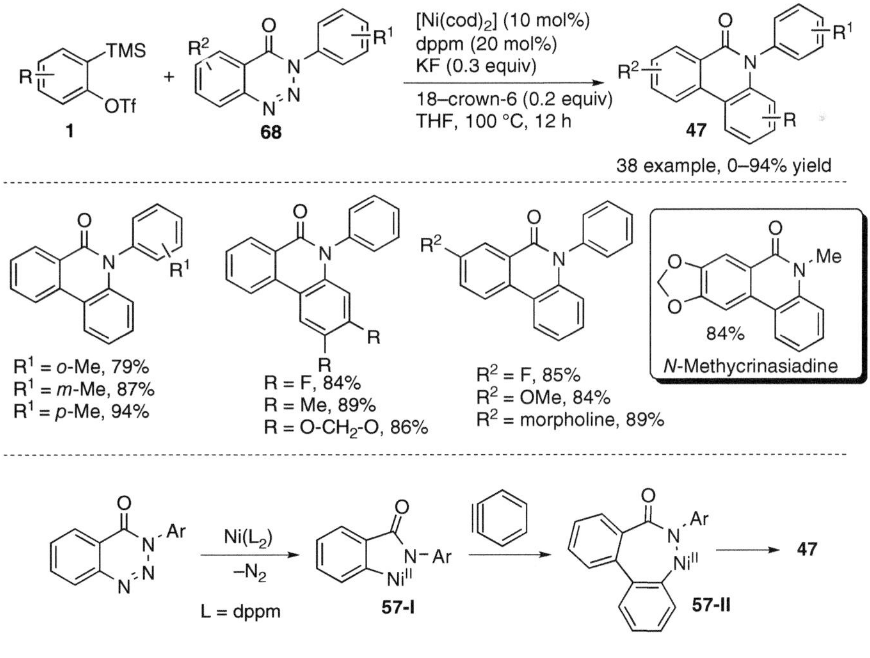

Scheme 6.57 Ni-catalyzed denitrogenative/annulation of 1,2,3-benzotriazin-4-(3*H*)-ones with arynes. Source: Based on Thorat et al. [57a].

[57b]. This reaction proceeds via the formation of a five-membered azanickelacycle (**57-I**) by the reaction of triazinone and Ni(0) through the extrusion of molecular nitrogen. Then, coordination of arynes with **57-I** was followed by the insertion of arynes into the Ni—C bond to form a seven-membered nickelacycle (**57-II**). Finally, reductive elimination of (**57-II**) affords the final product **47** and regenerates a Ni(0)-catalyst.

6.3.4 Cu-Catalyzed C–H and N–H Annulations of Arynes

A copper-mediated bidentate directed *ortho*-C–H/N–H annulation reaction of benzamides (**56**) with arynes to give the corresponding phenanthridinone (**47**) in good-to-excellent yields was reported (Scheme 6.58) [58]. The reaction was conducted by using tetrabutylammonium iodide (TBAI) in a mixture of solvent DMF:MeCN (1 : 1) under an O_2 atmosphere. This protocol was highly regioselective as well as compatible with various functional group substitutents on the benzamides. By employing this strategy, crinasiadine and phenaglydon natural products were synthesized. The 8-aminoquinoline-directing group was cleaved by using BBr_3

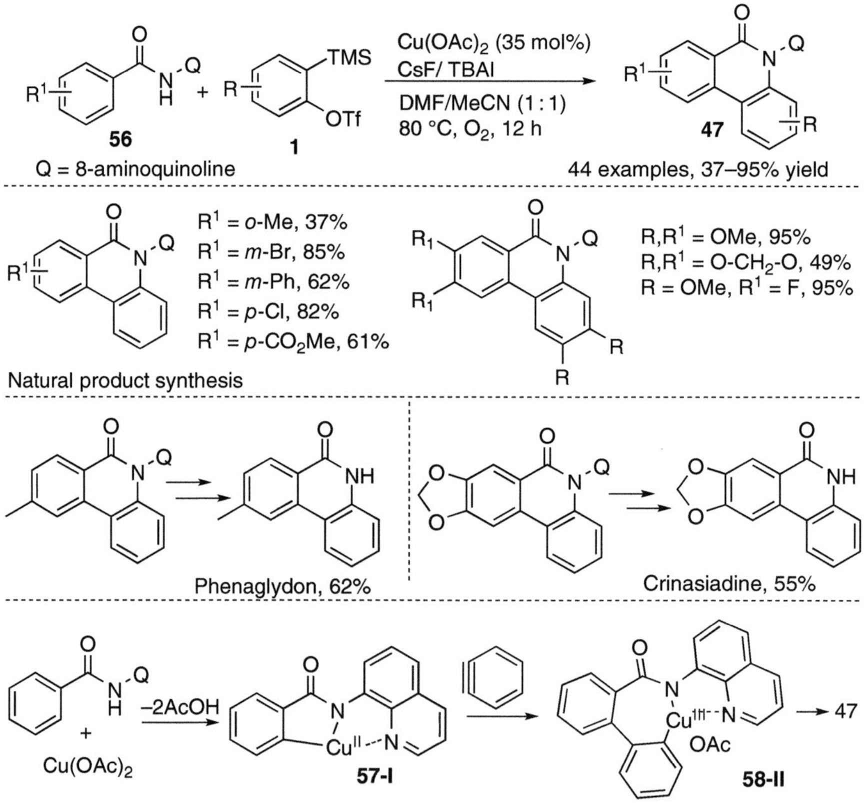

Scheme 6.58 Copper-mediated C–H/N–H annulation reaction of benzamides with arynes for the synthesis of phenanthridinones. Source: Based on Zhang et al. [58].

and $PhI(TFA)_2$ reagents. The proposed reaction mechanism was almost similar to the reported Pd-catalyzed reaction.

In 2018, the same group reported a Cu(II)-mediated *ortho*-C–H/N–H annulation reaction of indolobenzamides (**69**) with arynes providing indoloquinoline alkaloids (**70**) in good-to-excellent yields (Scheme 6.59) [59]. Notably, this protocol was utilized to synthesize isocryptolepine and 5*H*-indolo[2,3-*c*]quinolin-6(7*H*)-one containing natural products by removal of directing group using BBr_3 and $PhI(TFA)_2$ reagents.

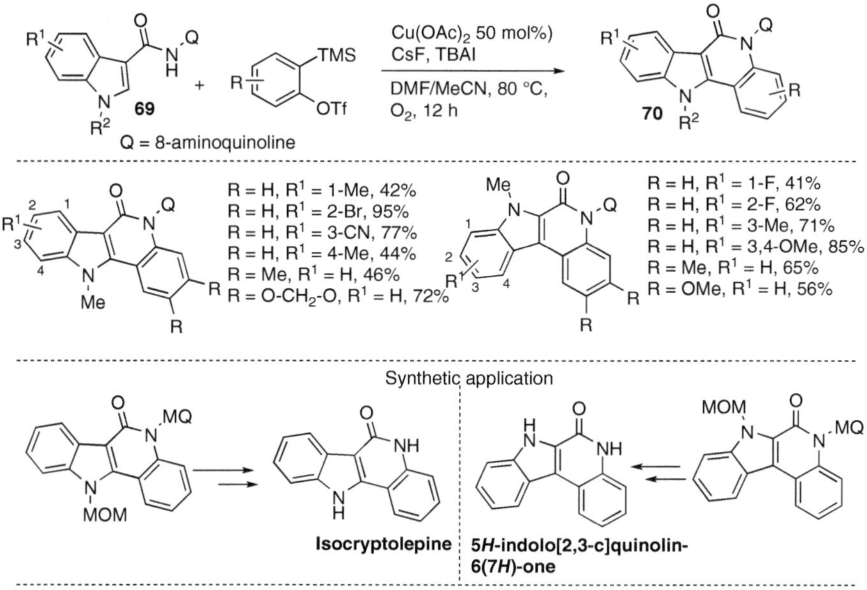

Scheme 6.59 Cu-mediated C-H/N-H annulation of indolobenzamides and arynes. Source: Based on Zhang et al. [59].

6.4 Transition-Metal-Catalyzed Three-Component Coupling Reactions

6.4.1 Palladium-Catalyzed Three-Component Coupling in Arynes

The transition-metal-catalyzed aryne involving three-component coupling is particularly interesting in organic synthesis due to the fact that it can construct two different chemical bonds at the ortho positions of an aromatic ring in one pot. The construction of two different chemical bonds at the ortho positions of an aromatic ring is extremely difficult in traditional organic synthesis. Particularly, the palladium-catalyzed three-component coupling reaction has fundamentally revolutionized the synthetic concepts, owing to the coupling of very reactive aryne species with organic electrophiles and organometallic reagents for the formation of carbon–carbon bonds exclusively at the ortho-position.

Yamamoto's group prepared 1,2-diallylated benzenes **72** via a palladium-catalyzed bis-allylation of benzynes with a bis-π-allylpalladium complex (Scheme 6.60) [60]. In this reaction, amphiphilic bis-π-allylpalladium intermediate was prepared by allylic chloride **6** and allylic stannane **71**. In the substituted allyl chlorides and allyl stannanes, very low chemoselectivity was observed (Scheme 6.60). For example, in the reaction of methallyl chloride **(6a)** with allyltributylstannane **(71)** and benzyne, three different types of bis allylated products **72**, **72′**, and **72″** were observed. This is probably due to the ready σ–π exchange of the bis-allyl intermediates (**59-I** and **59-I′**). It leads to the interchange of the nucleo- and electrophilicity of bis-allyl intermediate.

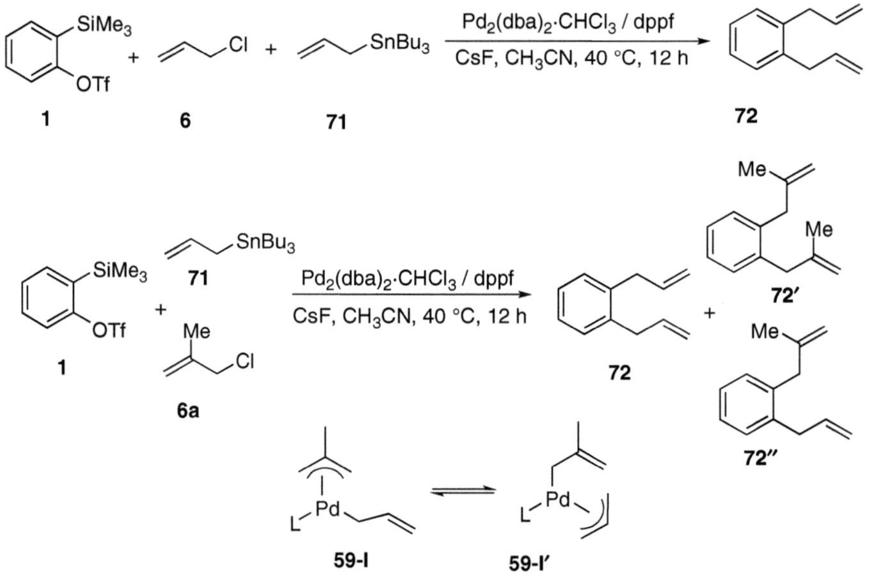

Scheme 6.60 Pd-catalyzed bis-allylation of arynes. Source: Based on Yoshikawa et al. [60].

Subsequently, Cheng's group demonstrated a series of three-component coupling reaction of benzynes with electrophiles and nucleophiles in the presence of a palladium catalyst (Scheme 6.61) [61]. Benzynes react with allylic halides **6** and alkynyl stannanes **73** in the presence of $Pd(dba)_2$, dppe, CsF in CH_3CN to give

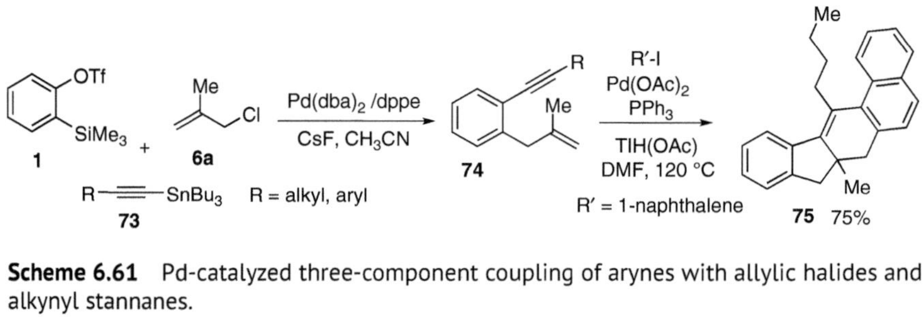

Scheme 6.61 Pd-catalyzed three-component coupling of arynes with allylic halides and alkynyl stannanes.

substituted 1-allyl-2-alkynylbenzenes **74** in good-to-excellent yields (Scheme 6.61). In the three-component assembling reaction, a variety of allylic chlorides and alkynyl stannanes are compatible with various benzyne precursors to construct two new C—C bonds in one pot. Further, the 1-allyl-2-alkynylbenzenes **74** could be converted to the multicyclic product **75** in 75% yield in the presence of $Pd(OAc)_2$ (5 mol%), PPh_3 (10 mol%), and Tl(OAc) (1.2 equiv) in DMF at 120 °C for 10 hours.

In the coupling reaction, various organometallic reagents such as allenyl stannanes **73a**, alkenyl metal reagents **73b**, and aromatic metal reagents **78** were used as nucleophilic reagents. In the reaction, organometallic reagents coupled with benzynes and allylic halides to give 1,2-disubstituted benzenes **76**, **77**, and **79** in good-to-excellent yields (Scheme 6.62) [62].

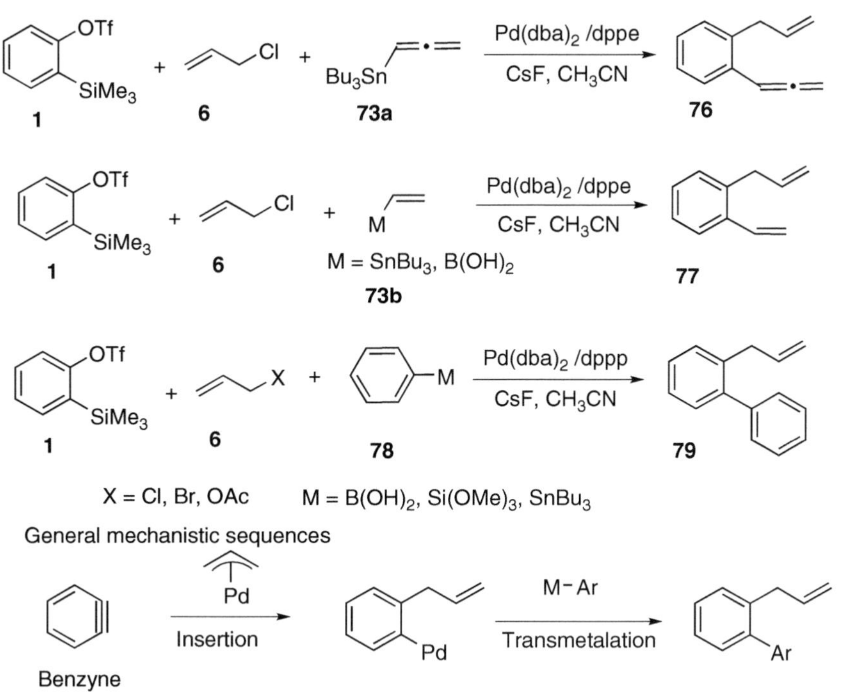

Scheme 6.62 Three-component Stille and Suzuki coupling reactions of arynes. Source: Jeganmohan and Cheng [62a]; Jayanth et al. [62b].

Further, Cheng's group reported a palladium-catalyzed Sonogashira-type coupling of benzynes with allylic acetates or halides **6** and terminal alkynes **80** promoted by CuI affording 1-allyl-2-alkynylbenzene derivatives **74** (Scheme 6.63) [63]. The coupling reaction shows several interesting features when compared with a previously reported Pd-catalyzed three-component coupling of benzynes with allylic chlorides and alkynyl stannanes.

In this type of three-component coupling reaction, allylic acetates, carbonates, and halides were used as an organic electrophile effectively. Cheng's group introduced allylic epoxide as an electrophile for this type of coupling reaction.

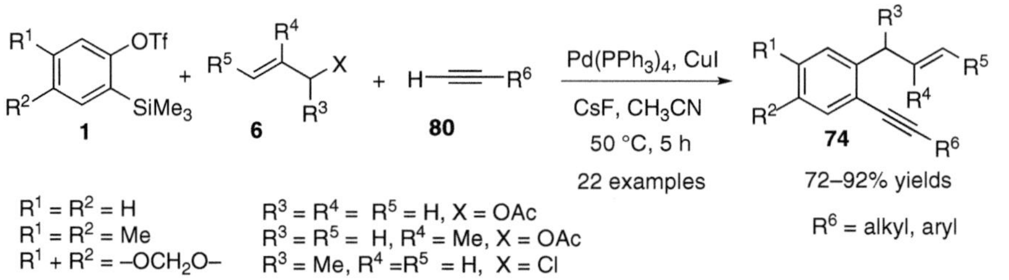

Scheme 6.63 Pd-catalyzed Sonogashira-type coupling of arynes. Source: Bhuvaneswari et al. [63].

A cooperative palladium- and copper-catalyzed highly regio- and chemoselective three-component coupling of benzynes with allylic epoxides **81** and terminal alkynes **80** to give 1,2-disubstituted benzenes **76** in good-to-excellent yields (Scheme 6.64) [64]. Further, the observed 1,2-disubstituted benzenes were converted into substituted 1,4-dioxane derivatives **82** in the presence of [AuCl(PPh_3)] (2 mol%) and $AgSbF_6$ (6 mol%), in DCM at room temperature for six hours. The gold-catalyzed reaction proceeds via intramolecular 6-*endo-dig* cyclization of the enyne group of **76** followed by a cationic intermediate (**64-I**) formation and subsequent dimerization.

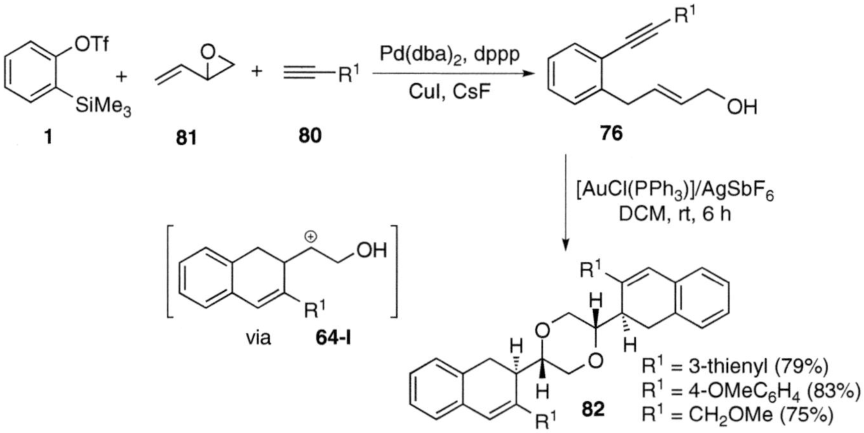

Scheme 6.64 Cooperative palladium- and copper-catalyzed three-component coupling of benzynes with allylic epoxides and terminal alkynes. Source: Based on Jeganmohan et al. [64].

Greaney and coworkers reported a palladium-catalyzed three-component Heck-type coupling of benzynes with benzyl bromide (or) methyl bromoacetate **83** and activated alkenes **14** to give 1,2-disubstituted benzenes **84** in good-to-excellent yields (Scheme 6.65) [65a]. This three-component coupling reaction worked with two distinct palladium complexes. For example, methyl bromoacetate effectively involved in the three-component coupling reaction to yield the corresponding products in good yields in the presence of 5 mol% of $PdCl_2dppf$ in DME at

50 °C. The combination of $Pd(OAc)_2$ and bidentate dppe ligand was effective for three-component coupling of benzyl bromides with benzynes and acrylates. Further, the synthetic utility of the palladium-catalyzed multicomponent reaction of benzynes with aryl halides and activated alkenes was also developed by the same group [65b]. By using this method, a wide range of biphenyls **16** were obtained in good yields. The authors have proposed possible key intermediates (**65-I** to **65-IV**) for this three-component coupling reaction.

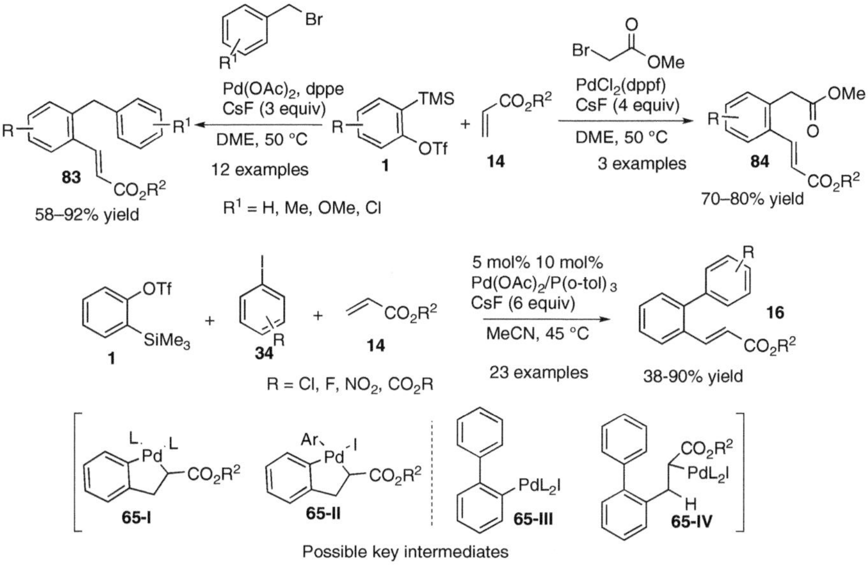

Scheme 6.65 Pd-catalyzed three-component Heck-type coupling of arynes. Source: Based on Henderson et al. [65a].

Werz and his coworker reported a three-component coupling of in situ-generated arynes with terminal alkynes **80**, and vinyl cyclopropanedicarboxylate **85** in the presence of a catalytic amount of $Pd(dba)_2$/dppp [66]. This protocol is specifically interesting and useful for the coupling of sp and sp^3 bonds in one pot under mild reaction conditions. Various electron-donating and electron-withdrawing group substituted arylacetylenes worked very well. However, the ethyl propiolate did not participate in the reaction. Meanwhile, aryne precursor **1** bearing two fluorine substituents provided the expected coupling product **86f** in only 11% yield (Scheme 6.66). The unsymmetrical aryne precursors provided a mixture of regioisomeric products.

A cationic palladium complex and KF/18-crown-6 system was found to be effective for a three-component coupling of arynes, isocyanides **87**, and cyanoformates **88** providing cyano-substituted iminoisobenzofurans **89** as well as unexpected α-iminonitriles **90** (Scheme 6.67) [67]. In general, the synthesis of α-iminonitriles is a very challenging task. In the present method, α-iminonitriles were prepared effectively. The cyano-substituted imino isobenzofurans smoothly undergo skeletal rearrangement to give α-iminonitriles with the aid of DIBAL-H or $AlMe_3$. However,

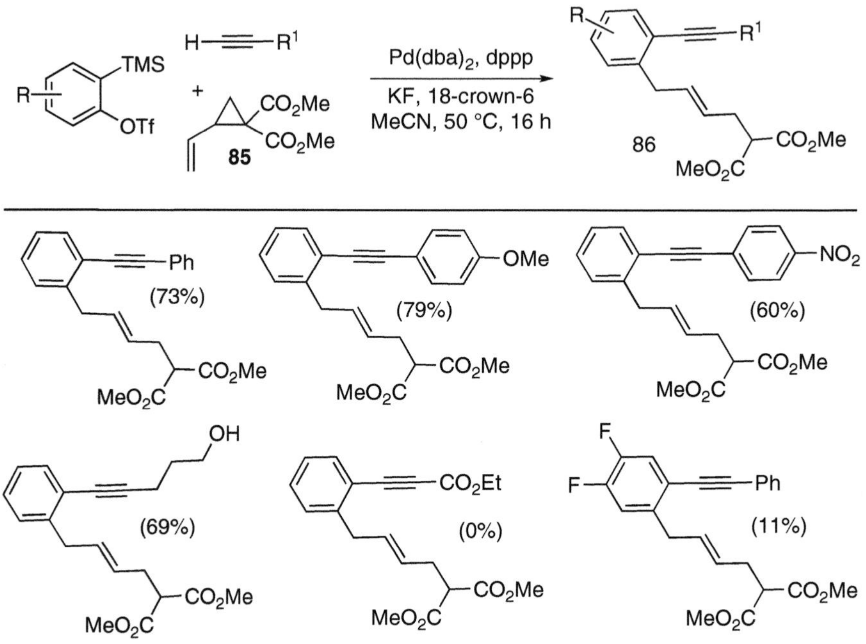

Scheme 6.66 Pd-catalyzed three-component coupling of arynes, terminal alkynes, and vinyl cyclopropanedicarboxylate.

in the unsymmetrical aryne such as 1-(trimethylsilyl)naphth-2-yl triflate (**1k**), mixtures of cyano-substituted iminoisobenzofurans and α-iminonitriles were observed.

6.4.2 Nickel-Catalyzed Three-Component Coupling in Arynes

In 2007, Cheng and Jayanth reported a Ni-catalyzed three-component coupling of enones with organoboronic acids and arynes providing biaryl frameworks **91** with an alkyl chain in the ortho-position (Scheme 6.68) [68]. The proposed mechanism initiated by the reaction Ni(0) with the aryne and enone to form a five-membered nickelacycle (**68-I**). The protonation occurs at the α-carbon atom of the ketone group with the aid of boronic acid followed by transmetalation of aryl group to the Ni-atom gives intermediate **68-III**. Subsequent reductive elimination provides the desired product and regenerates the Ni(0) catalyst.

6.4.3 Copper-Catalyzed Three-Component Coupling in Arynes

Very recently, Xiao, Chen, and coworkers reported a Cu-catalyzed three-component coupling of arynes with *o*-benzoxazolylhydroxylamines (**92**) and terminal alkynes (**80**) (or) benzoxazoles **94**. In this protocol, *o*-alkynyl anilines **93** and *o*-benzoxazolyl anilines **95** were selectively achieved by using terminal alkynes (or) benzoxazoles in one pot with two distinct reaction conditions (Scheme 6.69) [69]. The reaction

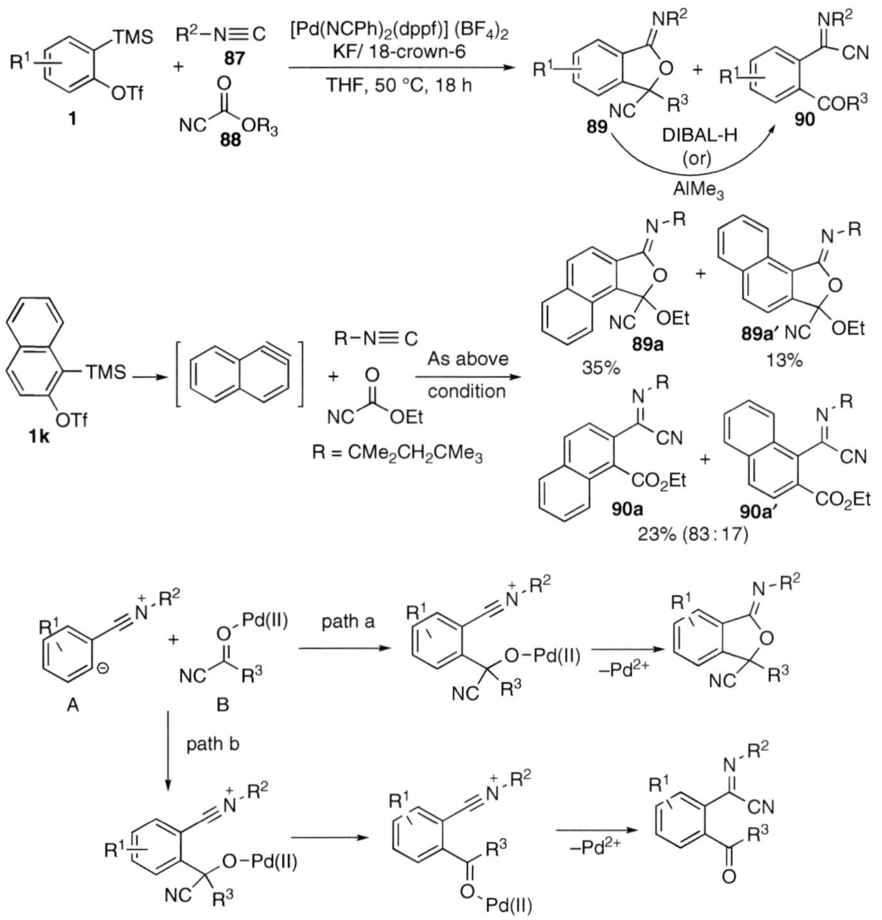

Scheme 6.67 Pd-catalyzed three-component coupling of arynes, isocyanides, and cyanoformates. Source: Modified from Li et al. [67].

has excellent functional group compatibility and the reaction can be done from readily available starting materials in one step. The reaction mechanism starts with the R^1–Cu(I) intermediate (**69-I**), which is generated from terminal alkynes (or) benzoxazoles in the presence of CuI and base. Then, R^1–Cu(I) intermediate (**69-I**) undergoes addition with aryne giving aryl C—Cu bond (**69-II**). It reacts with *o*-benzoxazolylhydroxylamine to give Cu(III) species (**69-III**). Reductive elimination of intermediate **69-III** affords coupling product and regenerates the active Cu(I).

A copper(I)-catalyzed difunctionalization of arynes through the three-component coupling reaction of terminal alkynes (**80**) with benzenesulfonothioates (**96**) and arynes (**1**) was achieved successfully. This transformation consists of an intermolecular C—C and C—S bond formation in one pot (Scheme 6.70) [70]. This three-component reaction offers an efficient method for the synthesis of *ortho*-alkynylarylsulfides (**97**) from the commercially available starting materials.

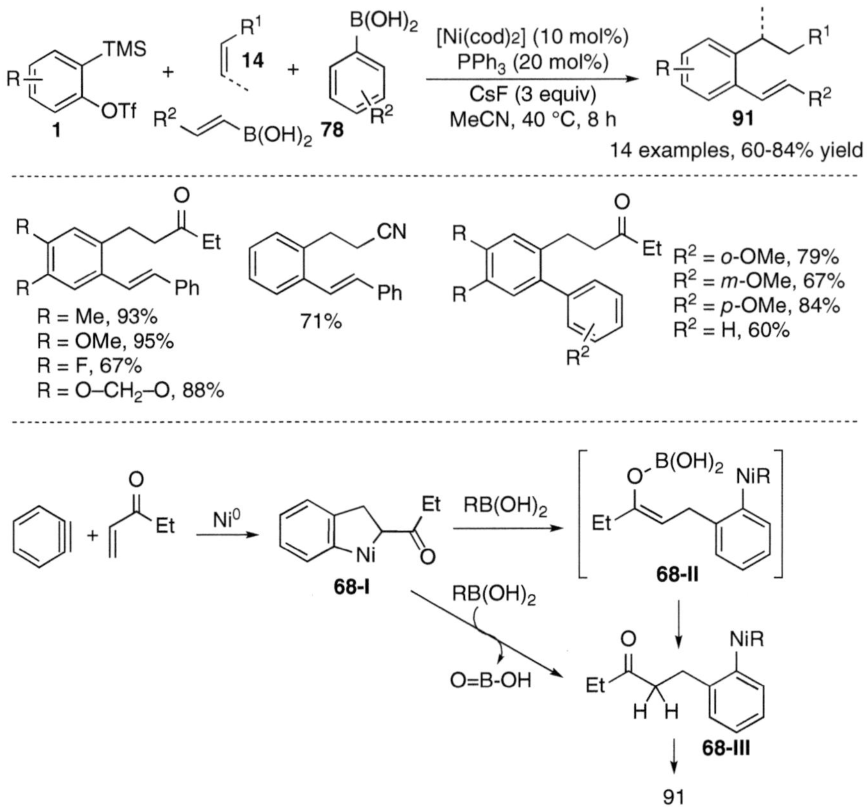

Scheme 6.68 Ni(0)-catalyzed three-component coupling of arynes with α,β-unsaturated ketones and organo boronic acids. Source: Based on Jayanth et al. [68].

The reaction was compatible with various substituted terminal alkynes, arynes, and benzenesulfonothioates. The reaction proceeds via coordinative insertion of benzyne with copper(I) acetylide (**70-I**) giving aryl copper intermediate (**70-II**) in the presence of base. The aryl–C bond undergoes nucleophilic addition with $PhSO_2SR$, affords the final product, and regenerates the active copper catalyst.

Pineschi and coworkers successfully developed three-component coupling of vinylaziridines (**98**) with terminal alkynes and arynes in the presence of CuI/PPh_3 under mild reaction conditions (Scheme 6.71) [71]. Various vinylaziridines react efficiently with benzynes and terminal alkynes to afford functionalized allylic amines **99** in moderate-to-good yields in a highly regio- and stereoselectivity. Notably, allylic aziridines undergo selective S_N2' ring-opening catalyzed by a copper–phosphine complex. On the other hand, a highly reactive ethyl propiolate reacted with benzyne and cyclic allylic aziridine affording S_N2' ring-opening product as well as tetrahydrophenanthridine **100** in 54% and 30% yields, respectively.

Zhang and his coworkers reported the efficient Cu(I)-catalyzed two-component coupling of terminal alkynes with arynes as well as three-component coupling of

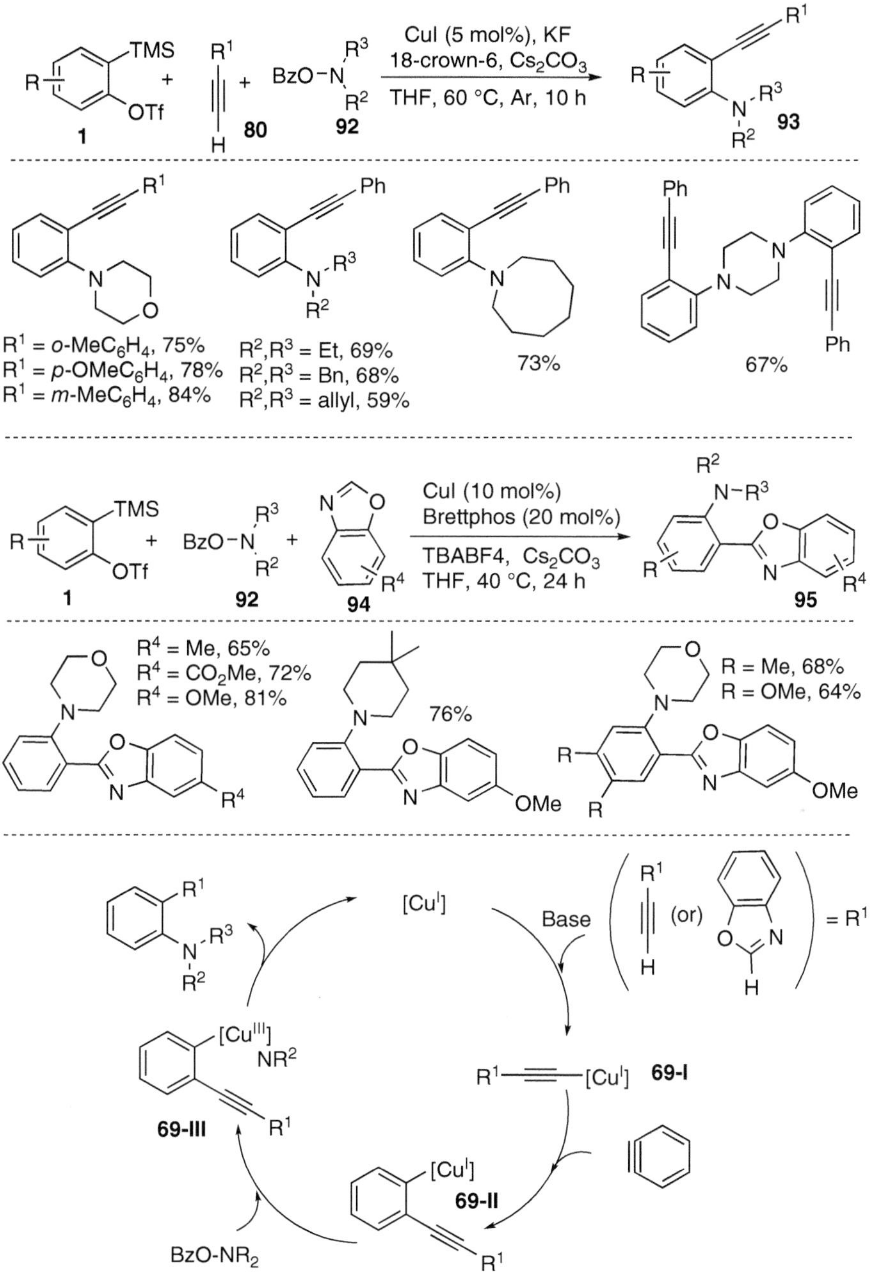

Scheme 6.69 Copper-catalyzed three-component carboamination of arynes. Source: Based on Niu et al. [69].

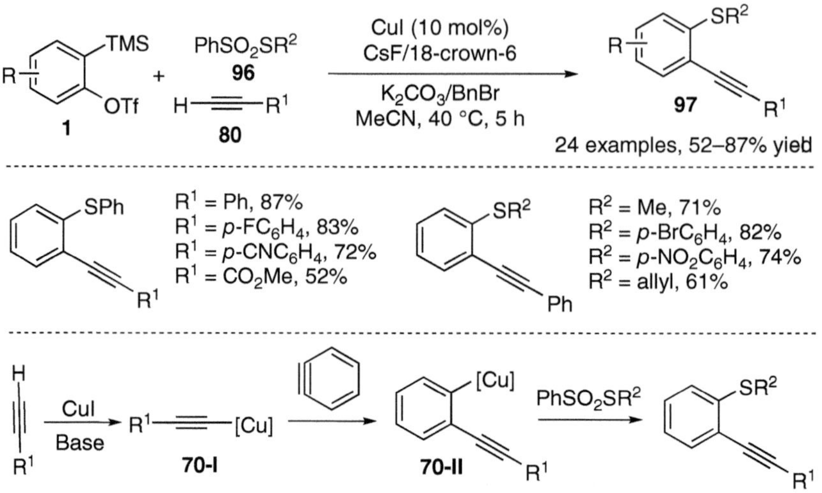

Scheme 6.70 CuI-Catalyzed three-component reaction of arynes. Source: Based on Peng et al. [70].

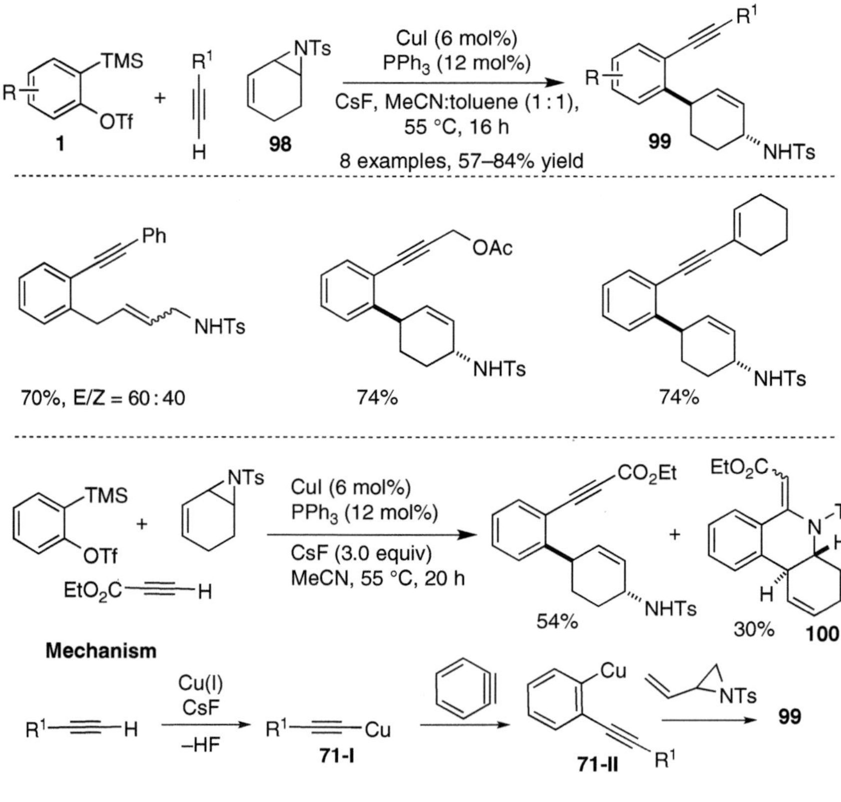

Scheme 6.71 Copper-catalyzed multicomponent coupling reaction of arynes. Source: Based on Berti et al. [71].

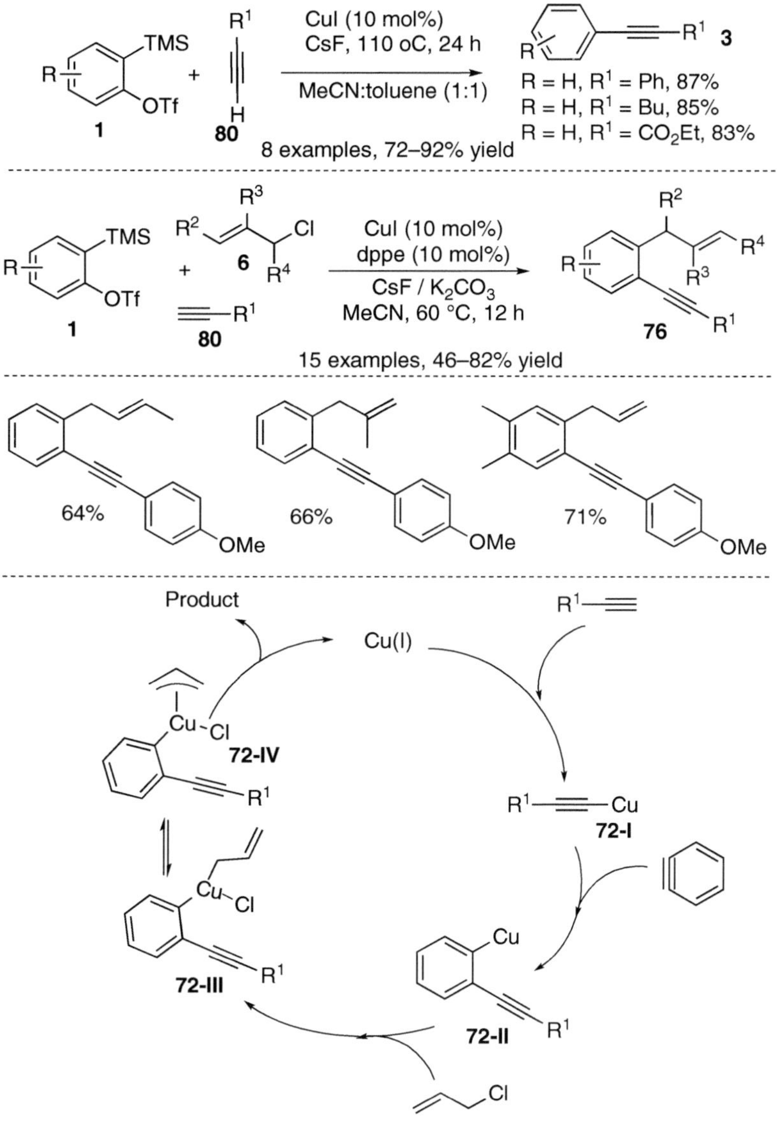

Scheme 6.72 Multicomponent reactions of terminal alkynes, arynes, and allylic chlorides by copper catalyst. Source: Modified from Xie et al. [72].

terminal alkynes with allylic chlorides and arynes under mild reaction conditions (Scheme 6.72) [72]. This method provides a facile and novel route for the synthesis of substituted acetylenes **3** and 1,6-enyne derivatives **76** in moderate-to-good yields. Initially, alkynyl-Cu(I) intermediate (**72-I**) is formed from the alkyne and Cu(I), followed by the insertion of benzyne gives an alkynylated-aryl copper species (**72-II**). Subsequently, allyl chloride undergoes oxidative addition with intermediate **72-II** providing intermediate **72-III**. Further, this intermediate rearranged into an allyl-copper complex (**72-IV**). Finally, the reductive elimination of intermediate **72-III** or **72-IV** affords the expected product. In the substituted allyl chlorides, the elimination selectively takes place at the less hindered end of the π-allylic system. For example, 3-chlorobut-1-ene gave a similar product like chlorobut-2-ene in the reaction. Notably, the phosphine ligand plays a crucial role in the reactivity/selectivity of the reaction.

In 2008, Cheng and his coworkers reported the three-component coupling of arynes with terminal alkynes (**80**) and activated alkenes (**14**) in the presence of CuI/PCy_3 system in a mixture of MeCN:THF giving 1-alkyl-2-alkynylbenzenes (**101**) in good yields [73]. A possible mechanism of this reaction was shown in Scheme 6.73. Alkynyl cuparation of arynes with copper acetylide complex (**72-I**) affords arylcuprous intermediate (**72-II**). The conjugate addition of intermediate **72-II** with alkene gives intermediate (**72-III**). Later, the protonation of intermediate **72-III** provides the expected product and regenerates the active catalyst.

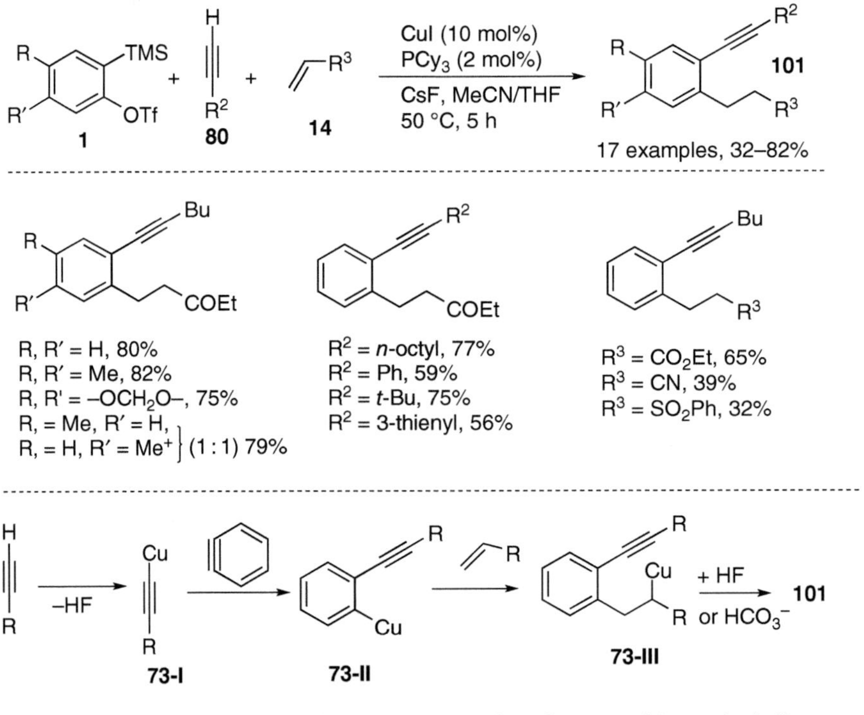

Scheme 6.73 Cu-catalyzed multicomponent coupling of arynes with terminal alkynes and activated alkenes.

Kobayashi and coworkers developed a copper-catalyzed three-component coupling of terminal alkynes with arynes and carbon dioxide providing isocoumarins in moderate-to-good yields (Scheme 6.74). Notably, the NHC-copper complex plays a crucial role in controlling the selectivity and yielding the expected isocoumarins **102** in a highly selective manner [74]. The reaction was well tolerated with substituted aryne precursors as well as aryl/alkyl-substituted terminal alkynes. In the catalytic cycle, the first copper-acetylide complex (**74-1**) was generated by NHC-copper carbonate or hydroxide with terminal alkyne. Coordinative insertion of benzyne with copper-acetylide complex (**74-I**) provides *ortho*-alkynyl copper complex (**74-II**). Next, the intermediate (**74-II**) undergoes insertion with CO_2 generates copper carboxylate intermediate (**74-III**). It further undergoes a 6-*endo-dig* cyclization and gives an endocyclic copper heterocycle (**74-IV**). Finally, the transmetalation of intermediate (**74-IV**) with a cesium salt (**73-V**) followed by protonation gives the expected isocoumarin **102** and regenerates the active catalytic for the next catalytic cycle.

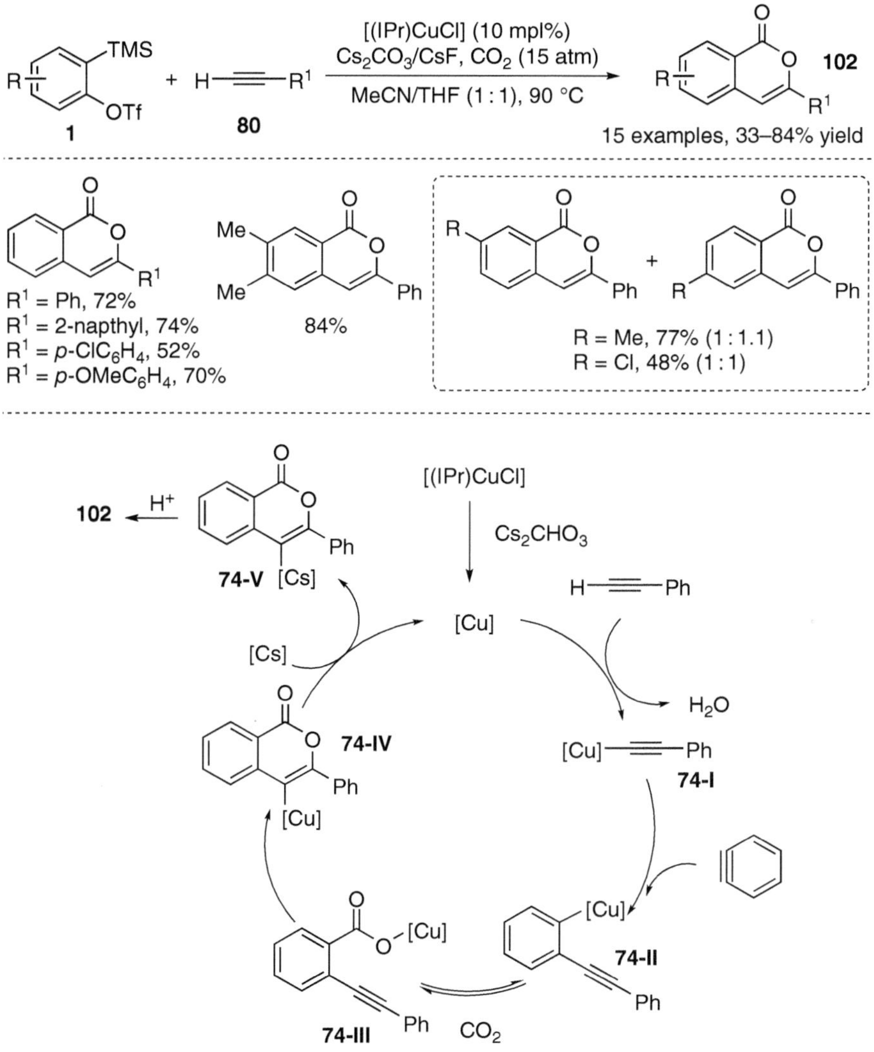

Scheme 6.74 [(IPr)CuCl]-catalyzed synthesis of isocoumarins from multicomponent reaction between alkynes, arynes, and CO_2.

A copper-mediated three-component coupling of arynes, allyl bromides, and a CF_3 source providing trifluoromethylated allylbenzenes **103** in good yields was reported (Scheme 6.75) [75a]. This protocol provides a wide range of structurally diverse trifluoromethylated allylarenes in good-to-excellent yields. Unsymmetrical benzynes afforded a mixture of regioisomeric products with different ratios. In addition, this method was successfully applied to the synthesis of CF_3-containing analogue of the antispasmodic drug papaverine [75b]. The [$CuCF_3$] reagent is generated from the cheap industrial byproduct fluoroform (CF_3H). Initially, the insertion of the Cu—CF_3 bond to aryne affords the *ortho*-trifluoromethyl arylcopper intermediate **75-I**. Similar types of metal–CF_3 bond [75c] or Cu—C bond [70] to arynes have been known in the literature. The intermediate **75-I** is transformed into a Cu-π-allyl type (**75-II**) through oxidative addition with allyl bromide. Finally, the reductive elimination of **75-II** affords the corresponding trifluoromethylated allylarenes **103**.

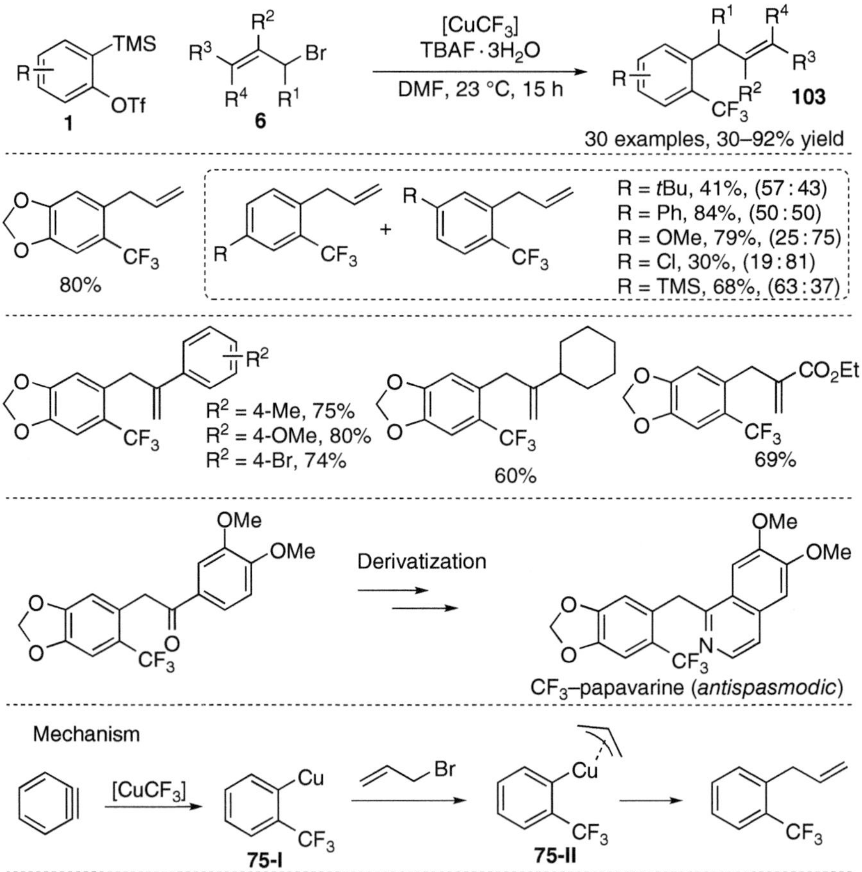

Scheme 6.75 Trifluoromethylation–allylation of aryne with [$CuCF_3$]. Source: Based on Yang and Tsui [75a].

A copper-catalyzed versatile synthesis of *o*-alkynyl aryl iodides (**105**) has been developed via three-component coupling of arynes with terminal alkynes and NIS

[76]. Phenyl acetylenes having an electron-donating or electron-withdrawing substituents on benzene ring smoothly participated in the three-component reaction in good (Scheme 6.76). Moreover, this catalytic reaction is highly compatible with amino, thioether, halides, ester, carbonyl, trifluoromethyl, cyano, and nitro functional group on the aromatic moiety.

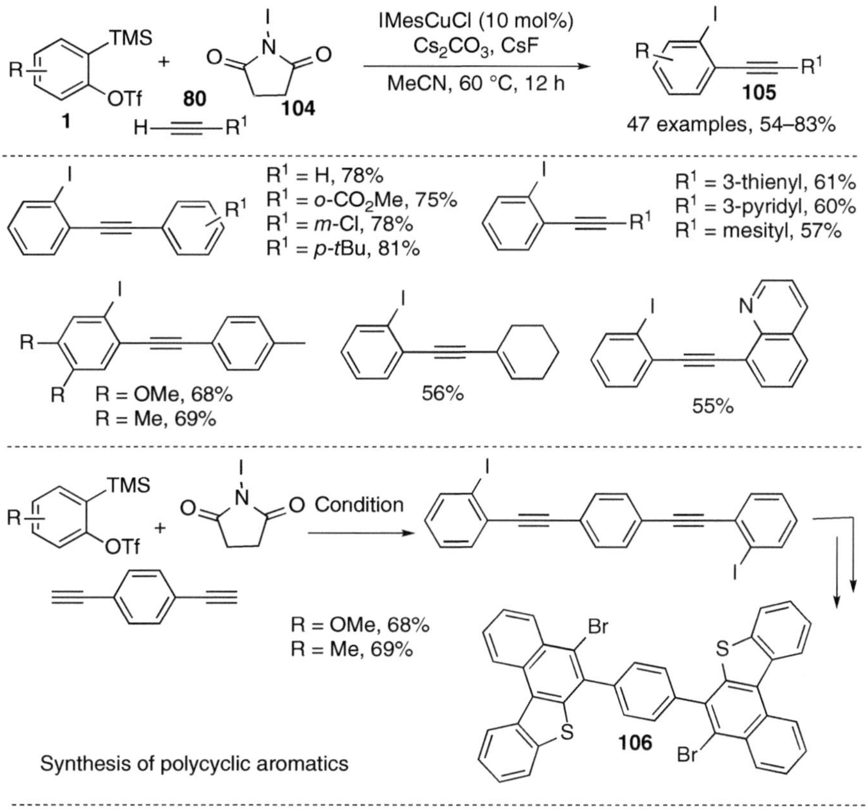

Scheme 6.76 Copper-catalyzed iodoalkynylation reaction of arynes, terminal alkynes, and NIS.

6.4.4 Silver-Catalyzed Three-Component Coupling in Arynes

Wu and coworkers reported the multicomponent reaction of alkynylbenzaldehyde (**31**) with tosylhydrazine (**107**), and benzyne **1** in the presence of a silver catalyst providing substituted pyrazolo[5,1-*a*]isoquinolines **108** in good-to-excellent yields (Scheme 6.77) [77]. The catalytic reaction proceeds via the formation of isoquinolinium-2-ylamide intermediate by the reaction of tosylhydrazine with alkynylbenzaldehyde in the presence of Ag. It further undergoes [3+2] cycloaddition with the benzyne unit that affords pyrazolo[5,1-*a*]isoquinoline **108** and regenerates the silver catalyst for the next catalytic cycle.

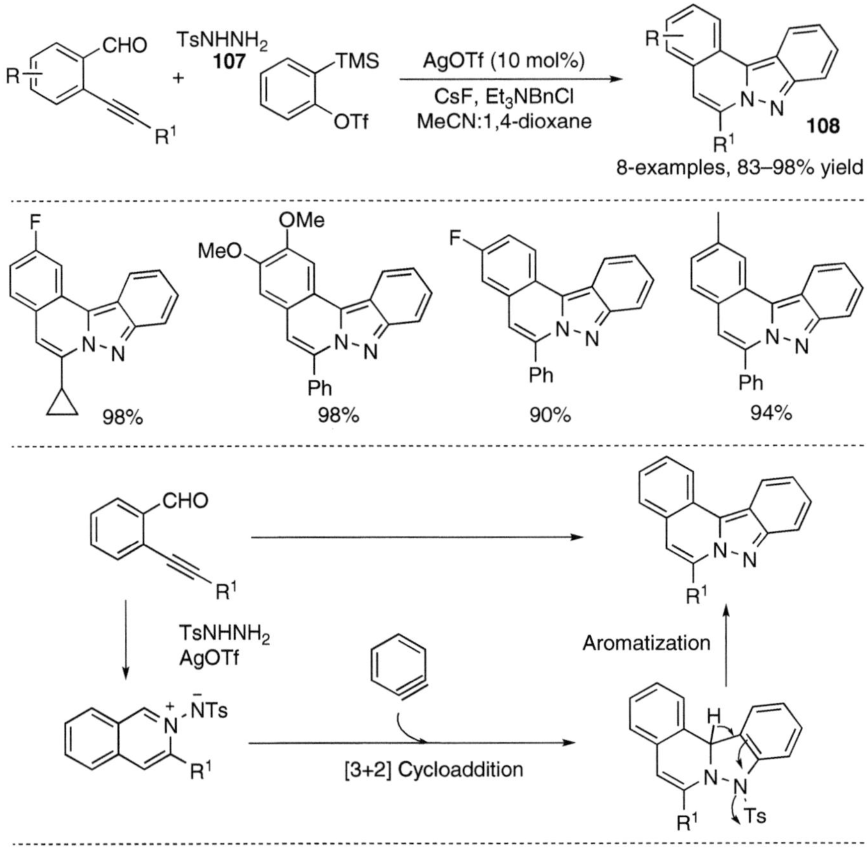

Scheme 6.77 Synthesis of *H*-pyrazolo[5,1-*a*]isoquinolines via three-component reaction of 2-alkynylbenzaldehyde, tosylhydrazine, and benzyne. Source: Based on Yang et al. [77].

6.5 Metal-Catalyzed Addition of Metal–Metal (or) Metal–Carbon and C—X bonds into Arynes

The direct addition of metal–metal (or) metal–carbon bond to the organic π-components remains the challenging task in metal-catalyzed organic synthesis. This method provides for the direct synthesis of C—X (X = B, Sn, and Si) bonds from various E—E (E = B, Sn, and Si) interelemental bonds. This type of product has widespread synthetic application across the chemical field, including the synthesis of pharmaceuticals, natural products, agrochemicals, etc.

6.5.1 Palladium-Catalyzed C—Sn Bond Addition to Arynes

Hiyama and his coworkers reported a palladium-catalyzed σ-bond addition of C–Sn of alkynylstannane with benzynes (Scheme 6.78) [78]. Treatment of benzyne precursors **1** with tributyl(phenylethynyl)tins (**75**) in the presence of a palladium catalyst and CsF at 50 °C for three hours in CH_3CN produced carbostannylation products

109 in 30–54% yields. In this reaction, a palladium-catalyzed direct crosscoupling of C—OTf bond of **1** with alkynylstannane **75** followed by fluoride ion–induced protodesilylation was also observed. In the reaction, unsymmetrical aryne precursor provided a mixture of two regioisomeric products **109** and **109′**.

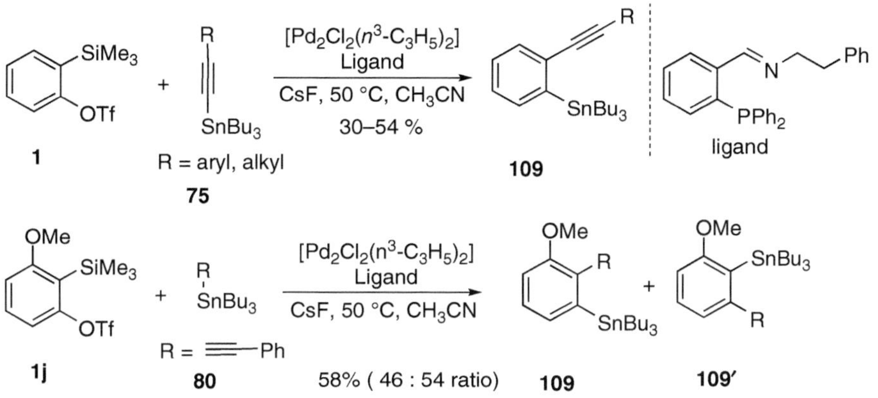

Scheme 6.78 Palladium-catalyzed carbon–Sn bond addition to arynes. Source: Based on Yoshida et al. [78].

6.5.2 Palladium-Catalyzed Sn–Sn/Si–Si Bond Addition to Arynes

Kunai and coworkers described a palladium-catalyzed addition reaction of σ-bonded Si–Si with benzynes (Scheme 6.79) [79a]. When benzyne precursor **1** was treated with disilane **110** in the presence of $Pd(OAc)_2$, *t*-OcNC, KF, and 18-crown-6 in tetrahydrofuran (THF), substituted disilane **111** was observed in good-to-moderate yields. The reaction proceeds via oxidative addition of silicon–silicon bond with a metal to give disilyl-palladium species. Coordinative insertion of aryne with disilyl-palladium species followed by reductive elimination provides bis-silylation product **111**. The bis-stannylation reaction was also achieved with arynes as well as cyclohexynes in moderate yields under similar reaction conditions. Interestingly, the bisaryne precursor **1** converted into bis[3,4-bis(tributylstannyl)phenyl] ether **113a**, in which four carbon–Sn bonds are formed in one pot. When the palladium-catalyzed reaction was carried in the presence of ETPO (4-ethyl-2,6,7-trioxa-1-phosphabicyclo-[2.2.2]octane) ligand **79A**, dimerization–distannylation product **113** was observed along with a minor amount of **114** [79b].

6.5.3 Palladium-Catalyzed Ar–SCN Bond Addition to Arynes

The direct ArS–CN addition to aryne C—C multiple bond with the aid of a palladium catalyst has been reported by Werz and coworkers (Scheme 6.80) [80]. In this conversion, the new C—SAr and C—CN bonds are formed in one step with the combination of $Pd(OAc)_2$ and a large bite angle having Xantphos ligand. Interestingly, under oxygen atmosphere, the yields increased besides minimizing the formation

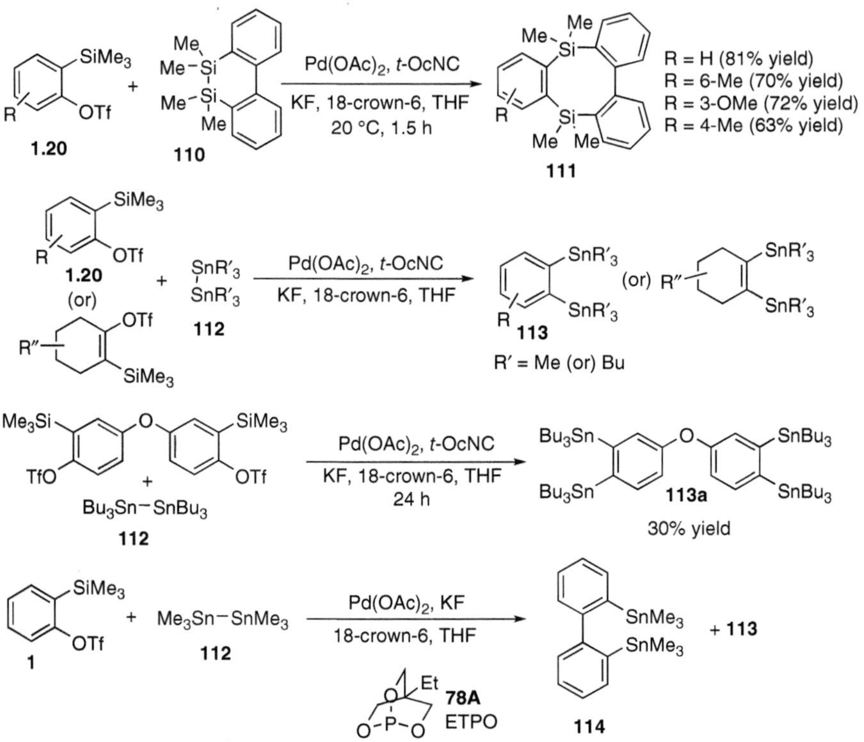

Scheme 6.79 Palladium-catalyzed bis-silylation/bis-stannylation of arynes. Source: Based on Yoshida et al. [79a].

of side products. The electron-withdrawing 4-CF_3-thiobenzonitrile was less reactive for the cyclization with aryne. Meanwhile, two aryne-bearing fluorine substituents gave the desired thio-cyanation product in less than 5% yield, which was measured by GC-MS.

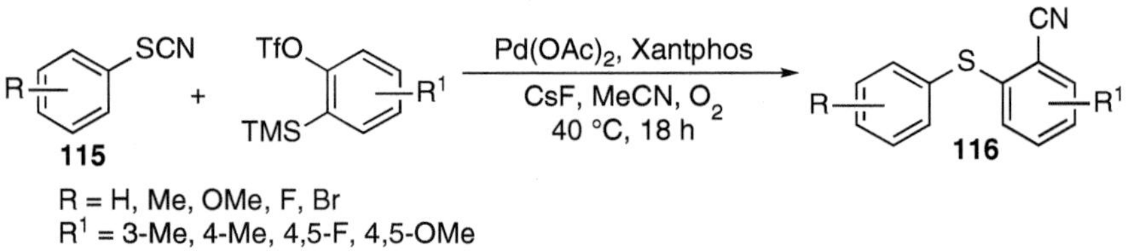

Scheme 6.80 Direct ArS−CN addition to aryne C−C multiple bonds by Pd catalyst. Source: Pawliczek et al. [80].

In 2014, Ma and Yuan developed a cascade reaction for the synthesis of 2,3-disubstituted benzofuran derivatives in moderate-to-good yields from the alkynyl-substituted arynes with aryl halides in the presence of a Pd(0) catalyst (Scheme 6.81) [81]. This protocol offers an efficient and different approach to benzofurans. This heterocyclic compound has potential application in pharmacological and biological chemistry. The reaction proceeds via intramolecular ene reaction of benzyne-yne intermediate (**81-I**) giving allene intermediate (**81-II**). Further, the ArPdI undergoes insertion with (**81-II**) forming a π-allylic palladium intermediate (**81-III**). Subsequent, β-H elimination of intermediate (**81-III**) affords unexpected isomeric benzofuran derivatives [81].

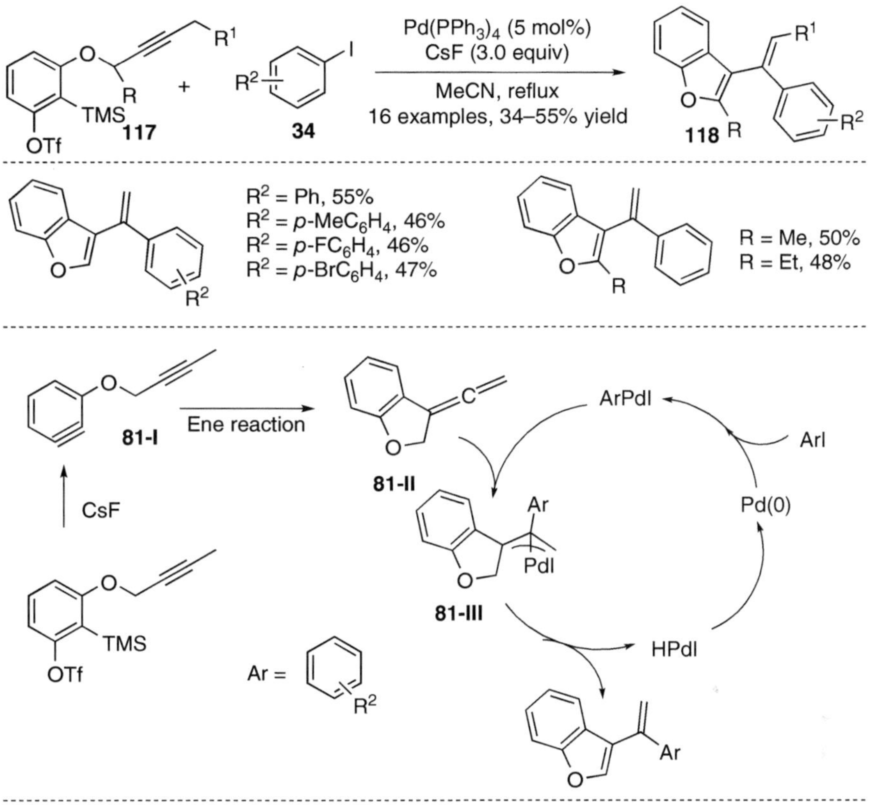

Scheme 6.81 Synthesis of 2,3-disubstituted benzofurans by palladium catalyst. Source: Modified from Yuan and Ma [81].

6.5.4 Platinum-Catalyzed Boron–Boron Bond Addition to Arynes

It is known that the Pt(0) complex was effective for the addition of E—E interelement bonds into the C—C multiple bonds. With this idea, Yoshida et al. developed a platinum-catalyzed addition of B_2pin_2 to arynes giving 1,2-diborylarenes in good-to-excellent yields [82]. This transformation is effective only with the platinum-isocyanide complex. Various other arynes were found to undergo the same transformation with bis(pinacolato)-diboron. It is significant to note that the 1,2-diborylarenes are relatively stable to air, moisture, and column chromatography and thus offer various synthetic advantages. For example, 1,2-diborylarene was converted into symmetrical and unsymmetrical *ortho*-terphenyls via a Pd-catalyzed Suzuki–Miyaura crosscoupling reaction (Scheme 6.82).

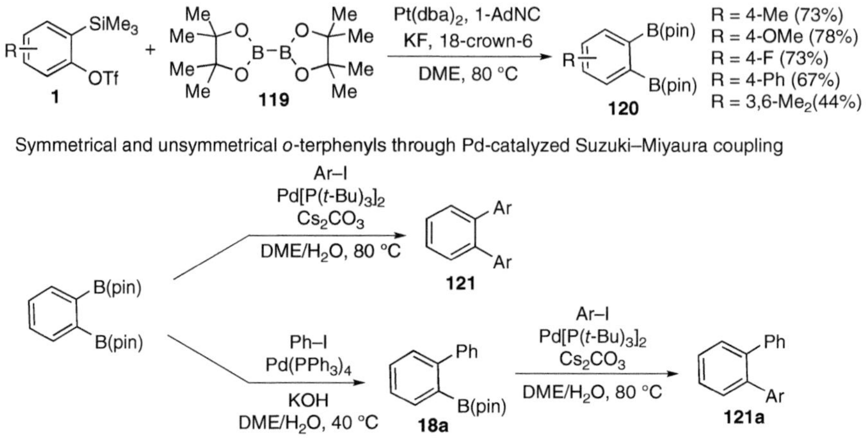

Scheme 6.82 Platinum-catalyzed addition of B_2pin_2 to arynes.

Similarly, a platinum-catalyzed diborylation reaction of indolynes with bis(pinacolato)diboron in the presence of *t*BuNC ligand was reported (Scheme 6.83) [83]. This protocol is useful for the synthesis of unprecedented indole-building blocks. The reactivity of two boron sites in the 6,7-bis[(pinacolato)boryl]indole **120a** moiety can be differentiated by a stepwise Suzuki–Miyaura crosscoupling reaction. In the reaction, the arylation was done site-selectively at the C7 position of indole without disturbing the C6 position. Later, the C6 position was also subjected to Suzuki–Miyaura crosscoupling in the presence of a palladium catalyst and Cs_2CO_3 base. On the other hand, the authors have observed that the 4,5-bis[(pinacolato)boryl]indole **120b** (or) 4,5-bis[(pinacolato)boryl]indole **120c** provided mixtures of crosscoupling products under similar reaction conditions.

6.5.5 Copper-Catalyzed B—B Bond Addition to Arynes

In 2012, Yoshida et al. developed an efficient copper-catalyzed vicinal-diborylation of arynes with B–B reagent [84]. This transformation is highly selective with a different mono- and di-substituted arynes. In this transformation, the expected *vic*-diborylarenes were observed in good yields. The borylcopper species **84A** is a key intermediate for the diborylation of benzyne was proposed. It has been generated by the reaction of copper acetate with the diboron reagent (Scheme 6.84).

6.5.6 Copper-Catalyzed Ar—Sn Bond Addition to Arynes

Yoshida and his coworkers developed a Cu-catalyzed coupling of arylstannes with benzynes giving *ortho*-stannylbiaryls and teraryls in good yields [85]. In this reaction, electron-deficient (fluoro (or) trifluoro arenes) substituted tin played a vital role for the efficient conversion. Notably, no product was formed with phenylstannane under similar reaction conditions. To prove this, the authors estimated the tin electron deficiency by ^{119}Sn NMR chemical shift, which indicates upfield shift

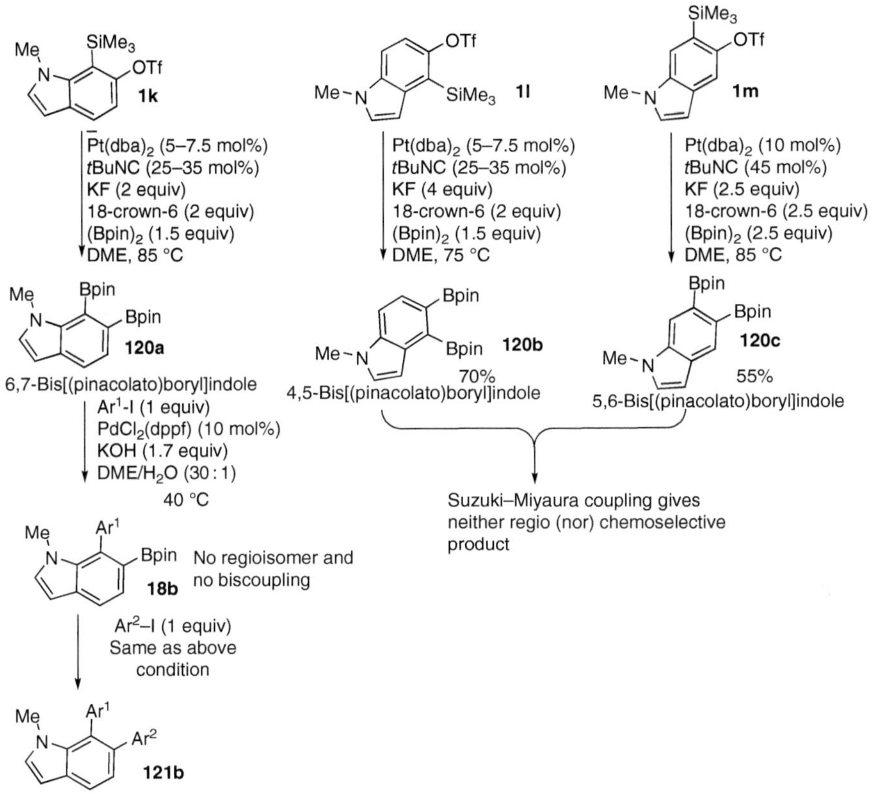

Scheme 6.83 Platinum-catalyzed diborylation reaction of indolynes with bis(pinacolato)diboron. Source: Based on Pareek et al. [83].

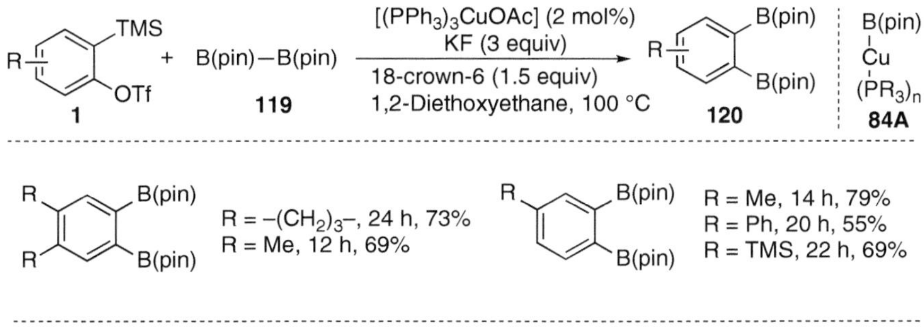

Scheme 6.84 Cu-catalyzed vicinal-diborylation of arynes.

of a number of fluorine decreases. It should be noted that other electron-deficient substitution on the arylstannanes showed less reactive (or) no reaction with arynes. For example, 2-cyanophenylstannane gave the expected product in 24% yield. However, 4-pyridylstannane failed to give the expected product. The selective formation of *ortho*-stannylbiaryls and teraryls can be achieved by changing the equivalent of aryne precursor (Scheme 6.85).

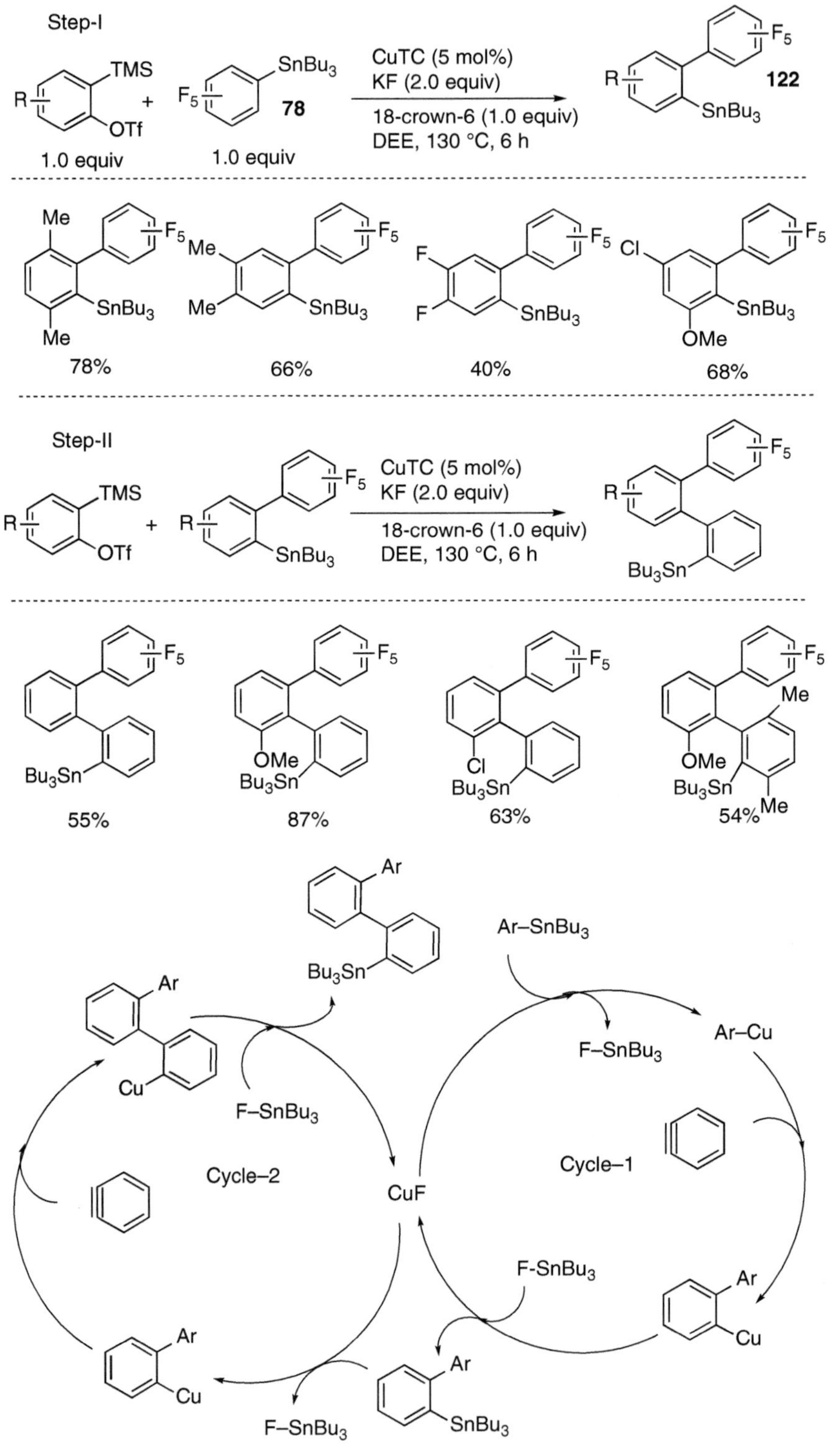

Scheme 6.85 Copper-catalyzed synthesis of *ortho*-stannylbiaryls from arynes and arylstannanes.

6.5.7 Copper-Catalyzed sp C–H Bond Addition to Arynes

In 2009, Biehl and his coworker reported a copper(I)-catalyzed coupling of terminal alkynes with arynes in a mixture of acetonitrile:toluene (1 : 1) solvent under the microwave-assisted heating for 30 minutes at 150 °C (Scheme 6.86) [86]. This method offers symmetrical and unsymmetrical alkynes in a short reaction time in moderate-to-excellent yields as compared with the conventional methods.

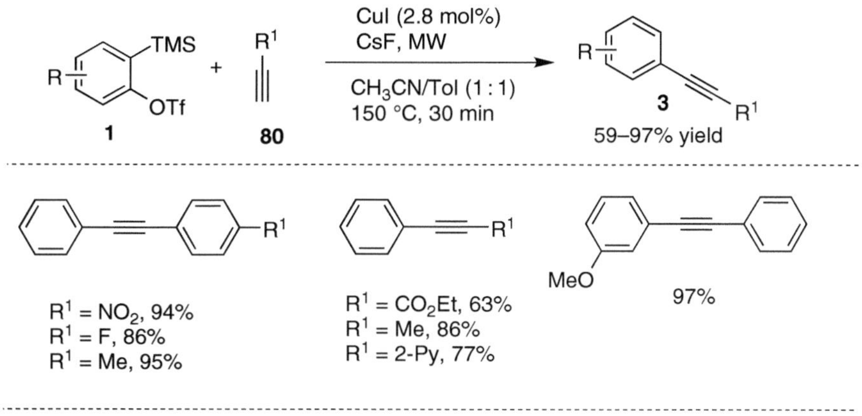

Scheme 6.86 Copper-catalyzed terminal alkynes sp C–H bond addition to arynes. Source: Based on Akubathini and Biehl et al. [86].

6.5.8 Gold/Copper-Catalyzed sp C–H Bond Addition to Arynes

In 2008, Zhang and coworkers developed a gold(I)- and copper(I)-catalyzed coupling of terminal alkynes (**80**) with arynes leading to 2-alkynylated biaryl frameworks (**123**) in good-to-excellent yields (Scheme 6.87) [87]. The catalytic cycle involves the initial coordination of an aryne to the Au(I)-catalyst gives aryne-gold complex. Meanwhile, an alkynyl-Cu(I) intermediate (**87-I**) that reacted with benzyne gives a alkynylated-aryl copper species (**87-II**). Finally, the nucleophilic addition of alkynylated-aryl copper to the aryne-Au(I) complex (**87-III**) affords intermediate (**87-IV**). Later, intermediate (**87-IV**) undergoes protonation giving the 2-alkynylated biaryl compounds.

Similarly, Yoshida et al. reported a copper(I)-catalyzed coupling of terminal alkynes with two arynes under the modified reaction conditions (Scheme 6.88) [88]. Terminal alkynes bearing a *m*-tolyl, methyl propargyl ether, and *t*-butyl substituents selectively gave 2 : 1 coupling products **123** in reasonable yield. 1-Naphthylacetylene afforded 2 : 1 coupling product along with 13% of direct addition product **3**. Interestingly, ethyl propiolate with benzyne gave ethyl 2-(9*H*-fluoren-9-ylidene)-2-phenylacetate (**44**) in 26% yield under standard reaction conditions. The product **44** is constructed through arylcopper intermediates (**88-I** and **88-II**) and arynes.

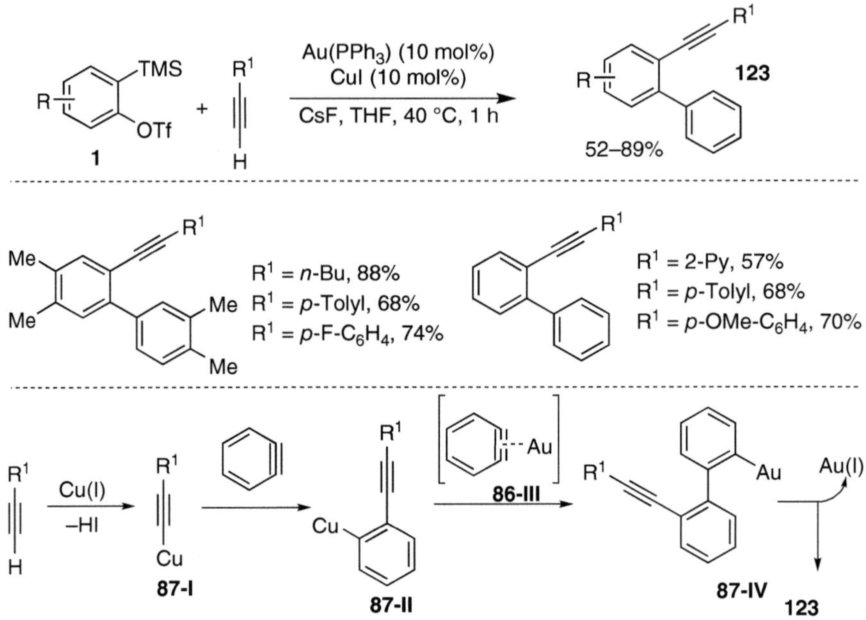

Scheme 6.87 Gold and copper cocatalyzed coupling reactions of terminal alkynes with arynes. Source: Based on Xie et al. [87].

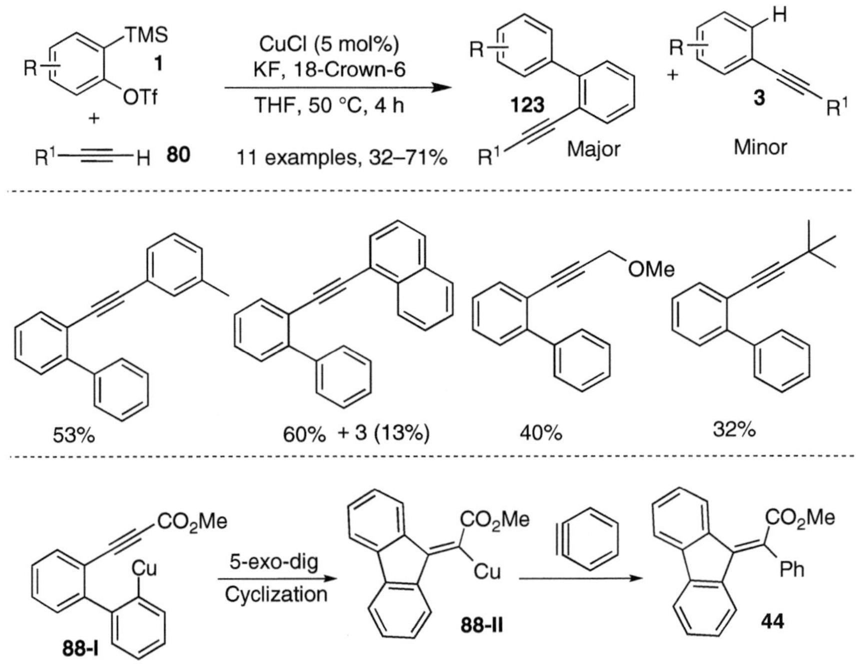

Scheme 6.88 Copper-catalyzed two-component coupling reactions. Source: Based on Yoshida et al. [88].

6.5.9 Copper-Catalyzed C–Br Bond Addition to Arynes

Subsequently, Yoshida and his coworkers reported a copper-catalyzed bromo alkynylation of arynes with bromoalkynes giving mixtures of 2-bromo-alkynyl-biphenyls **123** (major) and 2-bromo-alkynylarene **105** (minor) in good yields with high selectivity (Scheme 6.89) [89]. The catalytic reaction proceeds through the bromocuparation (**89-I**) of an aryne with $CuBr_2$. The second insertion of aryne moiety into the Cu—C bond leads to the intermediate (**89-II**). It undergoes further coupling with the bromo-alkyne and provides the substituted-biphenyl product. In the reaction, 2-bromo-alkynylarene was also formed as a side product. In the reaction of aryl-substituted allyl bromide as well as aryl-substituted propargyl bromide, mixture of regioisomeric products, 2-bromo aromatics, were observed.

6.5.10 Copper-Mediated 1,2-Bis(trifluoromethylation) of Arynes

A novel copper-mediated 1,2-bis(trifluoromethylation) of arynes with $[CuCF_3]$ was reported (Scheme 6.90) [90]. This protocol provides structurally diverse 1,2-bis(trifluoromethylation) in one pot under mild and safe conditions. Furthermore, estrone-based aryne was also successfully involved in the reaction giving 1,2-bis(trifluoromethylated) derivative in good yields. In the reaction, CuCl reacts with CF_3H in the presence of base giving fluoroform $[Cu^{I}CF_3]$. It is highly reactive and quickly oxidized by DDQ to $[Cu^{II}CF_3]$. The $[Cu^{II}CF_3]$ reacts with aryne providing an aryl radical intermediate **90-I**. The formation of radical intermediate was supported by the radical clock experiment [90b, c]. Intermediate (**90-I**) combined with a second equivalent of $[Cu^{II}CF_3]$ provides a $[Cu^{III}CF_3]$ species (**90-II**) [90d]. Finally, reductive elimination of intermediate (**90-II**) affords 1,2-bis(trifluoromethylation) product.

In general, the introduction of the CF_3 group into an adjacent aromatic ring is a challenging and important task for the synthetic chemist. In 2013, Hu and his coworkers developed a method for the synthesis of *ortho*-trifluoromethyl iodoarenes **129** by the three-component assembling of arynes with 1-iodophenylacetylene **80c** and $AgCF_3$ in the presence of 2,2,6,6-tetramethylpiperidine (TMP) [75c]. This method offers a facile and efficient route for several *ortho*-trifluoromethyl iodoarenes in excellent yields, which is difficult to achieve with other methods (Scheme 6.91) [91]. It was found that the TMP has played an important role in the vicinal difunctionalization of the aryne. The possible catalytic cycle involves the formation of intermediate **91-I** by the reaction of aryne with $AgCF_3$ and TMP. The intermolecular hydrogen bond of TPM of intermediate **91-I** with another TMP unit gives intermediate **91-II**. Next, the abstraction of H^+ from **91-II** followed by coordination of **91-I** group of 1-iodophenylacetylene forms intermediate **91-III**. Finally, the "ate" complex **91-IV** was formed by intramolecular nucleophilic attack, which reshuffles to afford the expected *ortho*-trifluoromethyl iodoarenes. It is noteworthy that the *m*-phenyl- and methyl group substituted aryne precursors provided mixtures of regioisomers in 1 : 1 ratio.

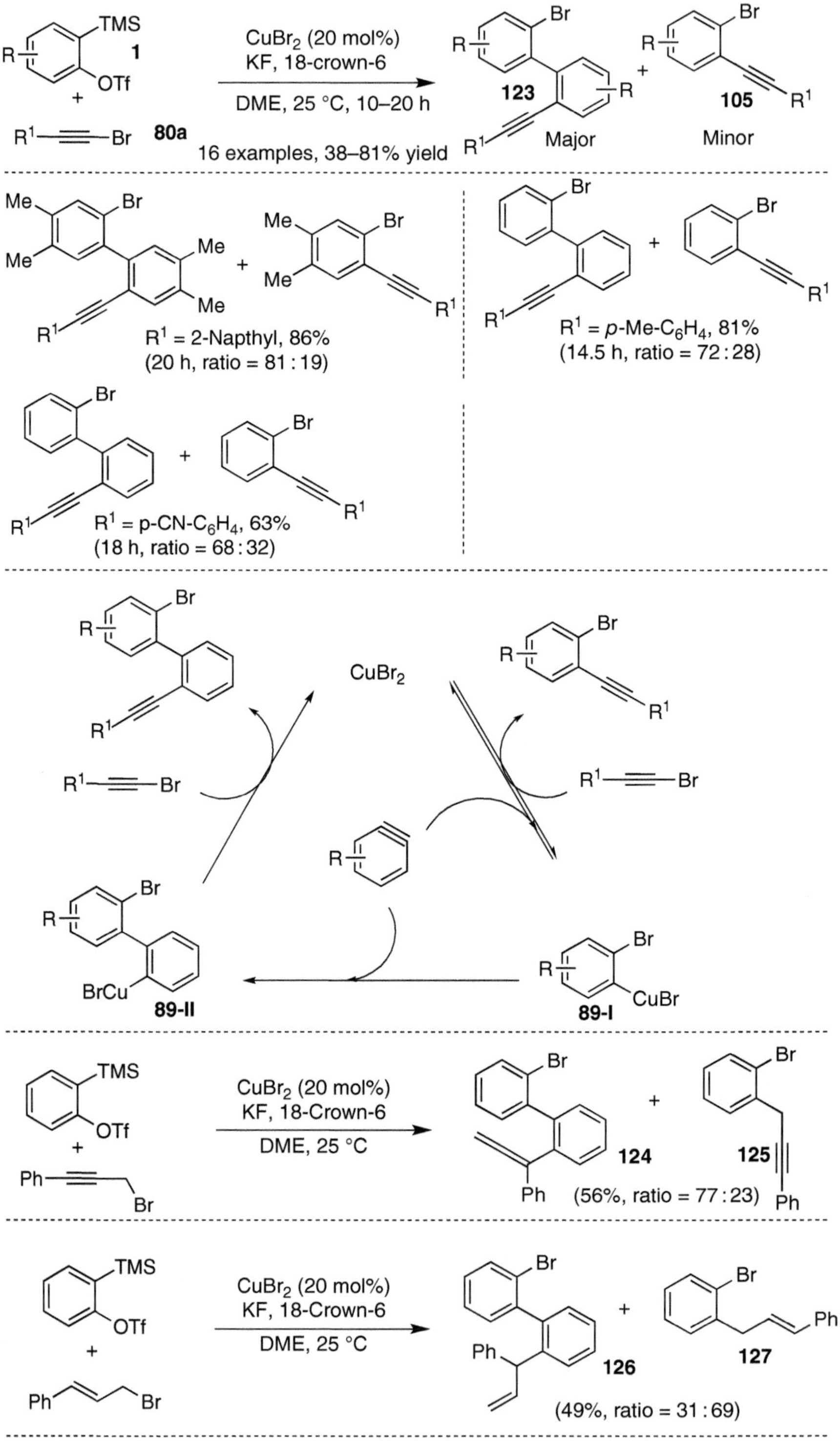

Scheme 6.89 Copper-catalyzed bromo alkynylation of arynes. Source: Based on Morishita et al. [89].

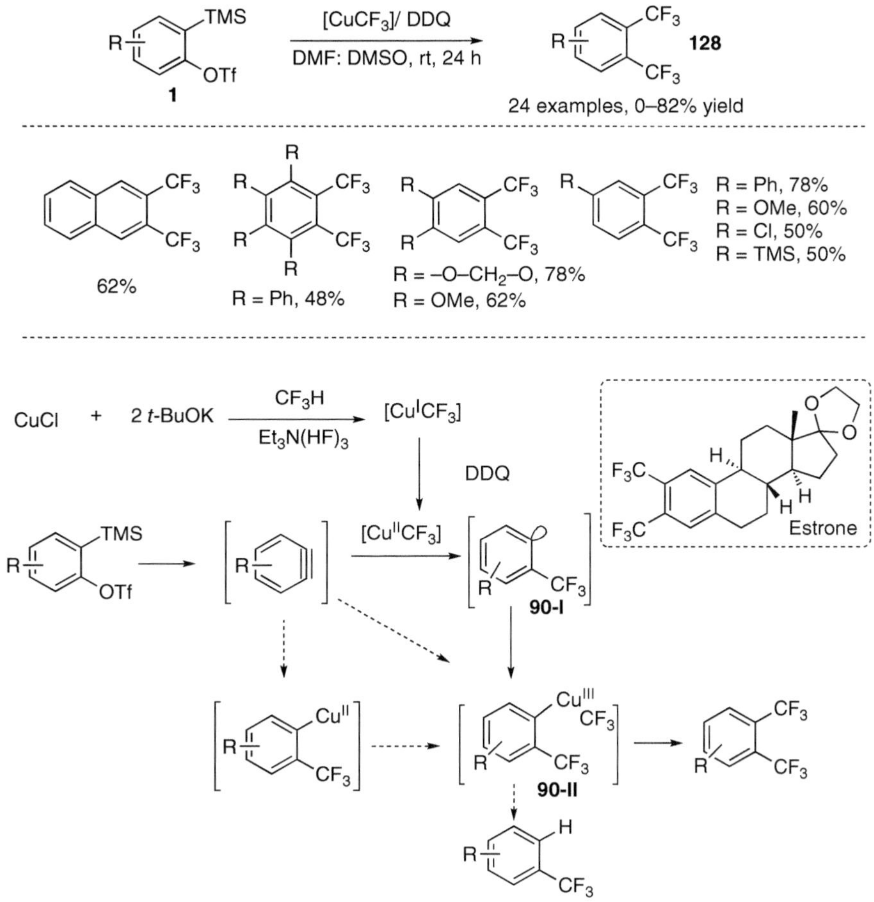

Scheme 6.90 Copper-mediated 1,2-bis(trifluoromethylation) of arynes. Source: Yang and Tsui [90a]; Okuma et al. [90b]; Yang et al. [90c]; Zhang et al. [90d].

Efficient silver-catalyzed insertion of perfluoroalkene-halide (R_f—I) bond with arynes was achieved. This methodology provides an easy access to *ortho*-perfluoroalkyl iodoarenes as well as I atom that has also transferred efficiently (Scheme 6.92) [92]. The reaction mechanism proceeds via the insertion of arynes into perfluoroalkyl-metal (MR_f) forming arylmetal intermediate. It readily undergoes s-bond metathesis with R_fI affording the expected product and regenerates the MR_f. Notably, unsymmetrical benzyne gave mixture of regioisomers in 8 : 1 ratio.

6.5.11 Copper- and Silver-Catalyzed Hexadehydro-Diels–Alder-Cycloaddition of a Triyne (or) Tetrayne (HDDA Arynes) with Terminal Alkynes

Interesting HDDA benzyne generation has been achieved by hexadehydro-Diels–Alder-cycloaddition of a suitable triyne (or) tetrayne **131** by heating conditions

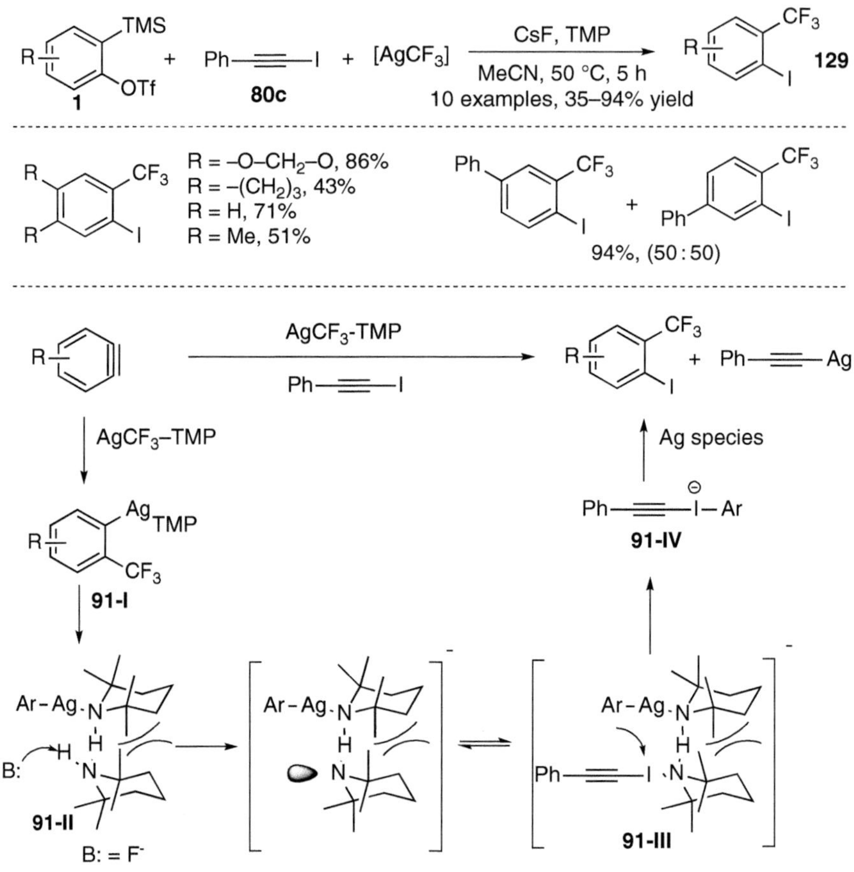

Scheme 6.91 Synthesis of *ortho*-trifluoromethyl iodoarenes from the multicomponent reactions of arynes, with 1-iodophenylacetylene, TMP, and $AgCF_3$. Source: Based on Uchiyama et al. [91].

[93]. This benzyne is efficiently converted into bromoalkynylation **132** and hydroalkynylation **133** by using Cu(I) catalyst (Scheme 6.93). 1-Bromo-1-alkynes reacted with benzynes in the presence of CuBr (10 mol%), CH_3CN at 80 °C giving bromoalkynylation products in good-to-excellent yields, whereas CuCl (5 mol%) induces the addition of terminal alkyne to HDDA benzyne to give hydroalkynylation products in good yields. However, the regiochemistry of both formed products is entirely different from one another. In the mechanism, the aryl copper(I) species was formed by nucleophilic addition of CuBr with benzyne. Then, it reacts with a bromoalkyne to generate the Cu(III) species followed by reductive elimination giving desired product with a regeneration of Cu(I) source for the next catalytic cycle. On the other hand, hydroalkynylation reaction initiated through the formation of Cu-alkyne from arylcopper species of one benzyne with CuCl. The Cu-acetylide undergoes addition with HDDA benzyne to

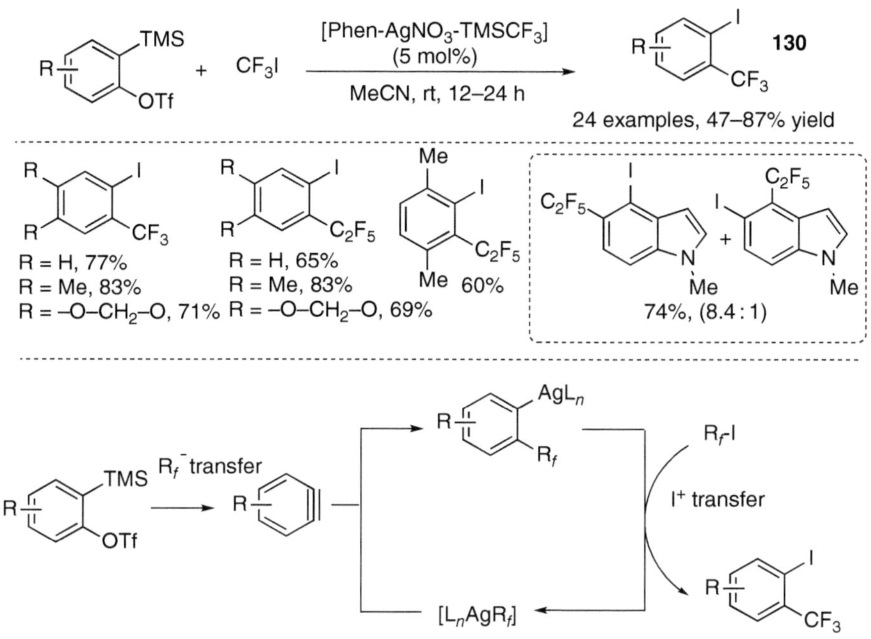

Scheme 6.92 Silver-catalyzed formal insertion of arynes into R_f—I bonds. Source: Based on Zeng and Hu [92].

produce aryl copper(I) species, which abstracts the proton from terminal alkyne, leading to the product with the regeneration of Cu-acetylide for the next catalytic cycle.

An interesting approach for fluorinated, trifluoromethylated, and trifluoromethylthiolated indoline and isoindoline derivatives **134** has been efficiently achieved by silver catalysts or silver-containing stoichiometric reagents (Scheme 6.94) [94]. In this transformation, stoichiometric amount of $AgBF_4$ in toluene afforded fluoroarenes in good yields. The trifluoromethylation was achieved by in situ generation of $AgCF_3$ with the combination of AgF and $TMSCF_3$ in acetonitrile solvent. The trifluoromethylthiolation reaction worked very efficiently in toluene solvent with $AgSCF_3$. The reaction proceeds via the formation of arylcation intermediate **94-II** from Ag^+ co-ordinated aryne species **94-I** through thermal hexadehydro Diels–Alder reaction (or) vinyl cation **94A**. Then, the addition of fluoride ion or trifluoromethyl or trifluoromethyl-thiol group onto the aryl cation **94-II** leads to organosilver species **94-III**. It undergoes subsequent protodeargentation giving the final product and regenerates the active Ag(I)-catalyst.

6.5.12 Copper-Catalyzed P—H Bond Addition to arynes

A P-arylation of phosphine oxides via a ligand-free copper-catalyzed addition of H—P(O) bond to aryne at room temperature was reported by Chen et al.

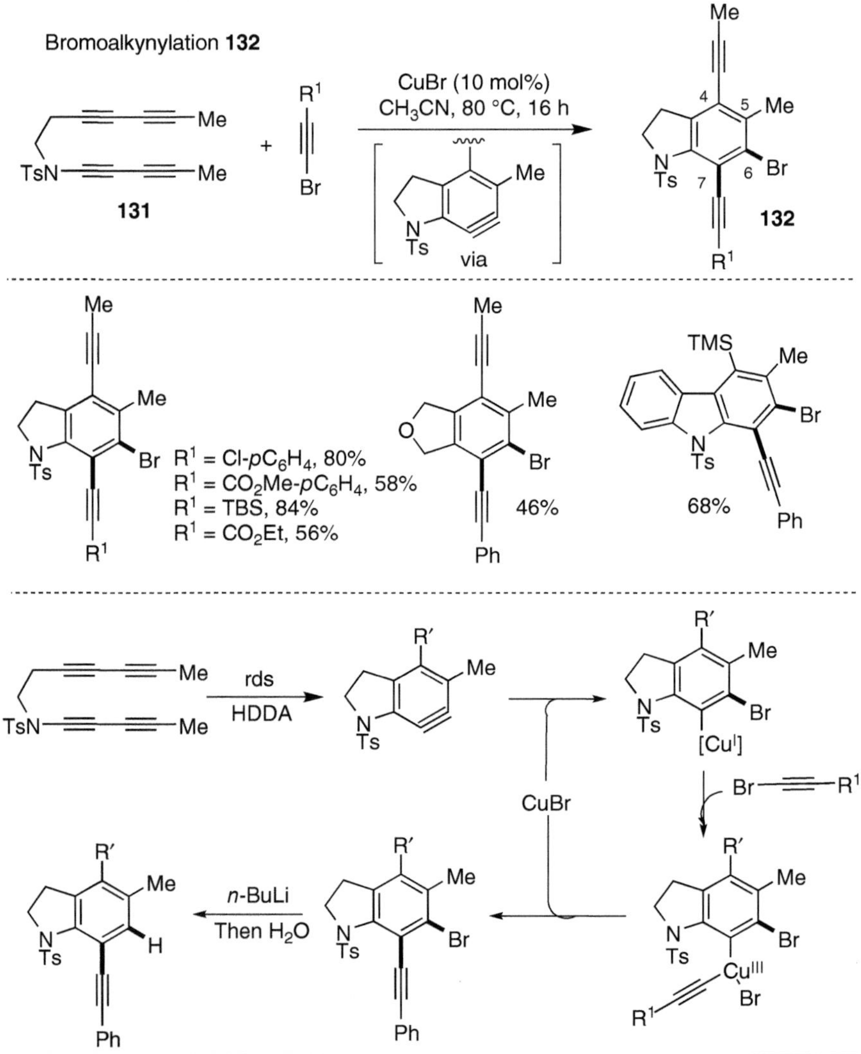

Scheme 6.93 Copper-catalyzed bromoalkynylation and hydroalkynylation from HDDA-generated benzynes.

[95]. This protocol was very effective for the preparation of arylphosphine oxides **136** in good yields (Scheme 6.95). Arylphosphine oxides are key motifs found in materials science and medicinal chemistry. The reaction proceeds via the reaction of CuI with P—H bond of secondary phosphine oxide giving $[R^2{}_2P(O)Cu]$ intermediate. Next, the aryne inserts into the Cu—P bond of active intermediate to give arylcopper intermediate. Further, the protonation of intermediate (**95-I**) with another moiety of secondary phosphine oxide affords a desired product and regenerates the active catalyst for the next catalytic cycle.

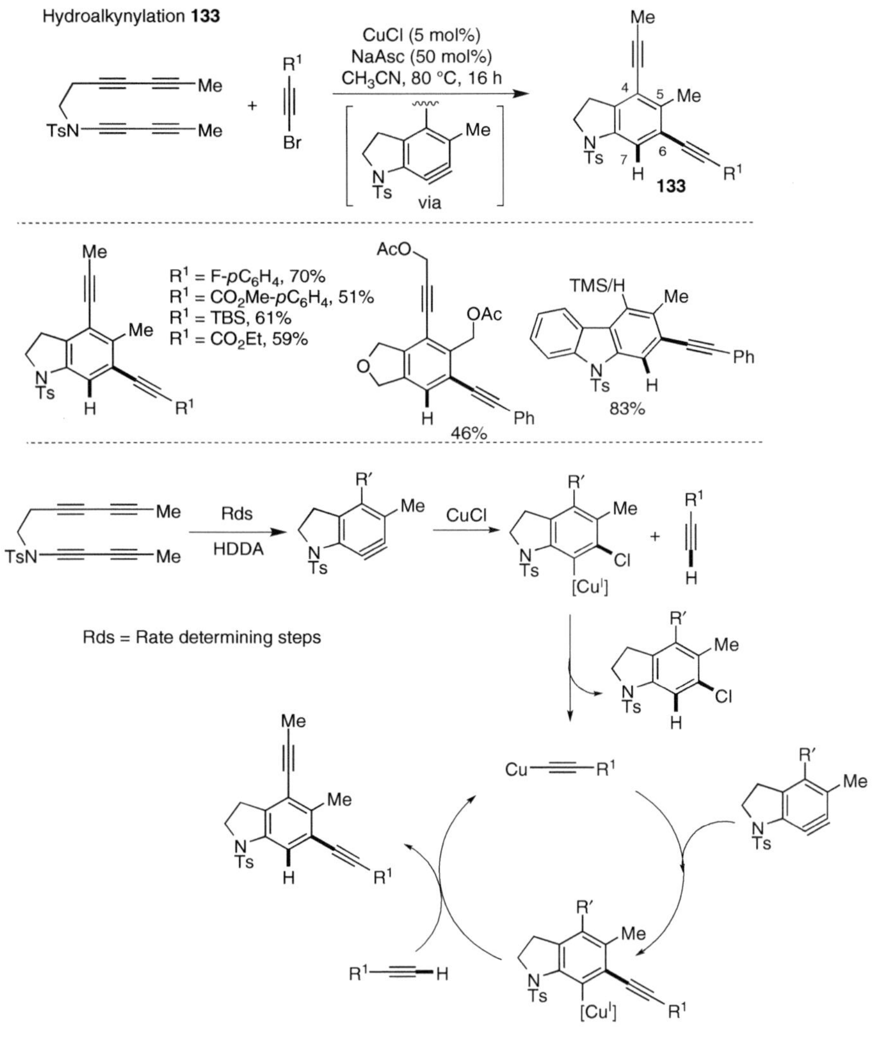

Scheme 6.93 (*Continued*)

6.6 Metal-Catalyzed CO Insertion Reactions of Arynes

6.6.1 Cobalt-, Rhodium-, and Palladium-Catalyzed CO Insertion of Arynes

Transition-metal-catalyzed carbonylation reaction of various organometallic species with π-acceptor property of CO (gas) is a fundamental chemical transformation, which is useful method for the synthesis of wide range of carbonyl-containing compounds. In this context, aryne is also an effective organic species with CO (gas) in the presence of metal-catalyst.

Murai and his coworkers reported cobalt-, rhodium-, and palladium-catalyzed carbonylative cycloaddition of benzynes with CO (Scheme 6.96) [96]. For example,

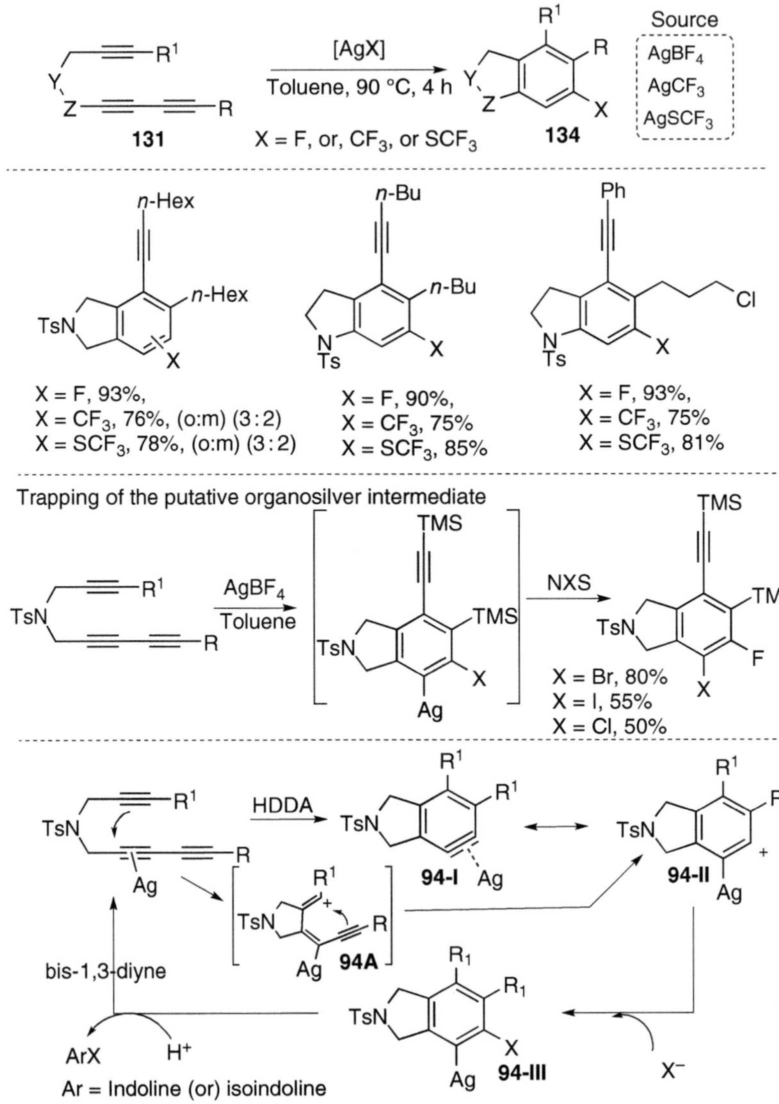

Scheme 6.94 Silver-mediated fluorination, trifluoromethylation, and trifluoromethylthiolation of arynes. Source: Based on Wang et al. [94].

treatment of benzyne precursors **1** with CO (5 atm) in CH_3CN in the presence of $Co_2(CO)_8$ and CsF at 60 °C for 12 hours gave anthraquinones **137** in good-to-excellent yields. The same reaction was carried out in the presence of $[RhCl(cod)]_2$ complex. In the reaction, a mixture of cyclized products **137** in 13% and fluorenone **10** in 30% yields were observed. Similarly, when benzyne precursor **1** was treated with allyl acetate **6** under CO (1 atm) in CH_3CN in the presence of $[(\pi\text{-}C_3H_5)PdCl]_2$, dppe, and CsF at 80 °C for four hours, a 2-methyleneindanone **138** was observed in good-to-excellent yields.

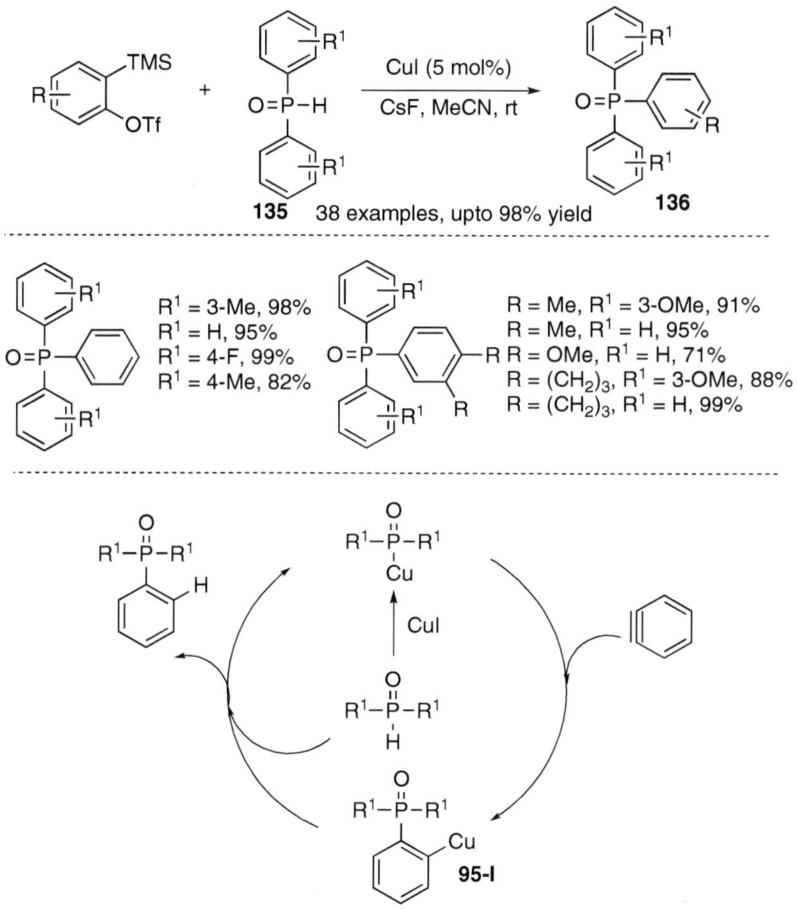

Scheme 6.95 Synthesis of triarylphosphine oxides from diarylphosphine oxides and arynes by copper catalyst.

Interestingly, aryne is selectively inserted into a Pd—C bond of six-membered palladacycle in the presence of CsF, acetonitrile at room temperature giving a stable eight-membered palladacycle **142** (Scheme 6.97) [97]. This insertion reaction allows to synthesize various dimeric eight-membered palladacycles in good-to-excellent yields. The eight-membered palladacycle can also be further stabilized by ligands such as 4-picoline (or) PPh_3. The structure of eight-membered palladacycle **141** was confirmed by a single-crystal X-ray analysis. This eight-membered palladacycle can be converted into various useful organic molecules. For example, 2,2′-functionalized biaryls can be prepared in good yields in the presence of CO/MeOH combination. A similar concept can be extended to synthesize ten-membered heterocycles **143** by the reaction with CO. It is important to note that the ten-membered heterocycles **144** can be formed by the reaction of two molecules of benzynes with the *ortho*-metalated primary phenethylamines (Scheme 6.97).

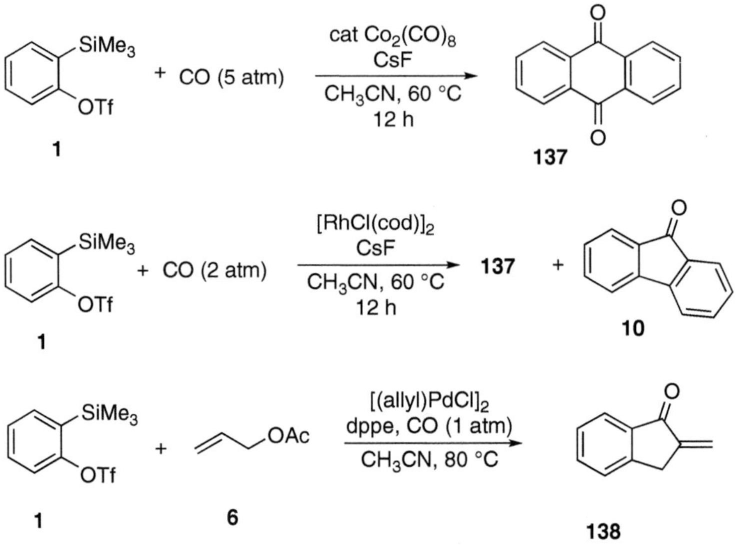

Scheme 6.96 Metal-catalyzed carbonylative coupling of arynes with CO or allylacetate. Source: Based on Chatani et al. [96].

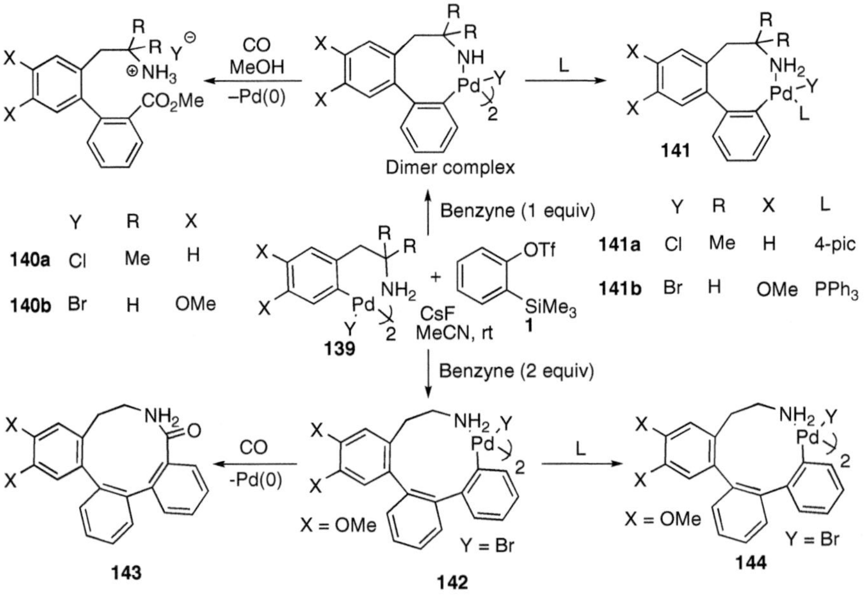

Scheme 6.97 Aryne reactions with six-membered palladacycle. Source: García-López et al. [97a]; Oliva-Madrid et al. [97b]; Oliva-Madrid et al. [97c].

A palladium-catalyzed three-component reaction of arynes with carbon monoxide and 2-iodoanilines **48** was developed to construct phenanthridinone (**47**) and acridone derivatives (**145**) [98]. The product formation is achieved under two distinct reaction conditions (Scheme 6.98). For example, the phenanthridinones are selectively formed under the ligand-free conditions. But, a slow generation of benzyne in the presence of dppm ligand exclusively affords acridones. Further, this method can also be applied to synthesize phenanthridinones and acridones alkaloids. A plausible mechanism was suggested based on the control experiments.

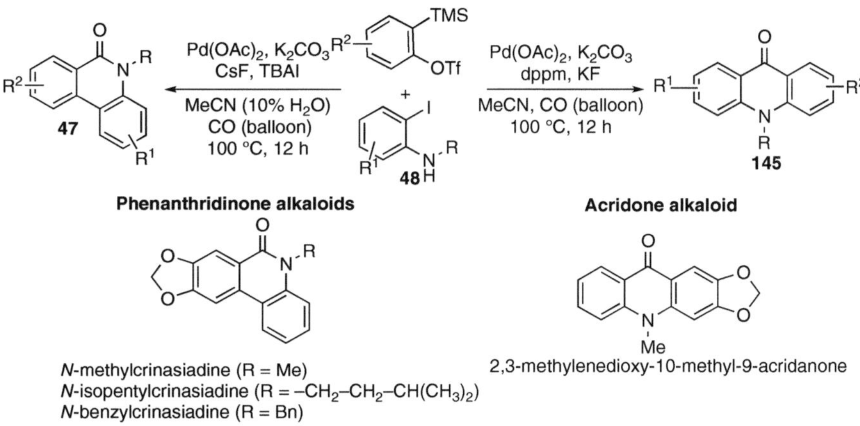

Scheme 6.98 Palladium-catalyzed reaction of aryne, CO, and 2-iodoaniline for the construction of phenanthridinone and acridone alkaloids.

Li and his coworkers observed a ligand-controlled selective synthesis of 2-methylene-3-substituted 2-benzylidene-2,3-dihydro-1*H*-inden-1-ones **138** and 2,3-dihydro-1*H*-inden-1-ones **138a** through a palladium-catalyzed three-component coupling of arynes with allyl carbonates and carbon monoxide (CO) [99]. The reaction proceeds via the formation of π-allylPd(II) intermediate (**98-I** and **98-II**) from allyl carbonates. The π-allyl Pd(II) intermediate undergoes sequential coordinative

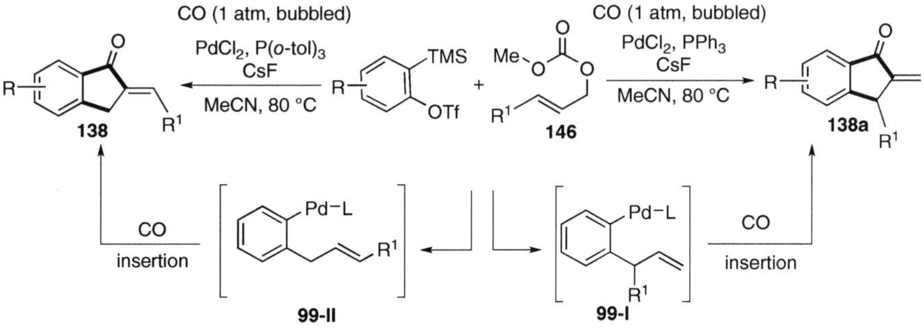

Scheme 6.99 Pd-catalyzed cyclocarbonylation of arynes for the synthesis of 1H-inden-1-ones.

insertion with benzyne and CO followed by subsequent reductive elimination providing the expected product (Scheme 6.99).

6.7 Metal-Catalyzed [3+2] Cycloaddition of Arynes

6.7.1 Silver-Catalyzed [3+2] Cycloaddition of Arynes

In 2018, Guo and his coworkers reported a tandem nucleophilic addition and [3+2] cycloaddition of iminoesters (**147**) with arynes for the synthesis of highly substituted imidazolidine frameworks (**148**) in the presence of a silver catalyst (Scheme 6.100) [100]. This asymmetric transformation provides imidazolidine derivatives with excellent regio-, diastereo-, and enantioselectivities in high yields. Notably, most of the reactions gave only one diasteromer in good yields. As expected, in the reaction of *m*-Me or *p*-Me unsymmetrical benzynes, mixture of two regioisomers (*m*-Me and *p*-Me) was observed in a 1 : 1 ratio.

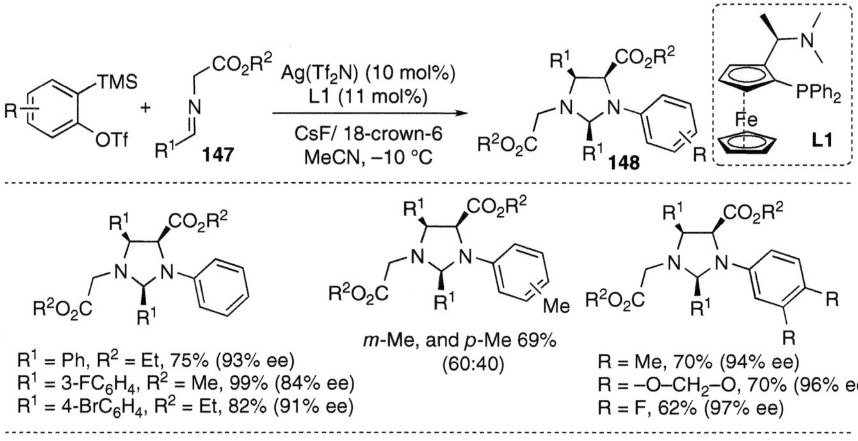

Scheme 6.100 Silver-catalyzed [3+2] cycloaddition reaction of iminoesters with arynes. Source: Jia et al. [100].

In 2011, Wu and his coworkers reported a silver-catalyzed unusual reaction of 2-alkynylbenzaldoxime (**149**) with arynes under mild reaction conditions [101]. This reaction proceeds via 6-*endo*-cyclization, followed by [3+2] cycloaddition, and unusual rearrangement providing 2-oxa-6-azabicylco[3.2.2]nona-6,8-diene derivatives **150** in moderate-to-good yields (Scheme 6.101). Particularly, substituted 2-alkynylbenzaldoxime having electron-withdrawing as well as electron-donating group on the aromatic moieties were well tolerated. Among them, methoxy-substituted arynes gave the expected product in high yield as well as regioselectivity. It indicates that the electronic effect of OMe group on the aromatic moiety played an important role for the high selectivity.

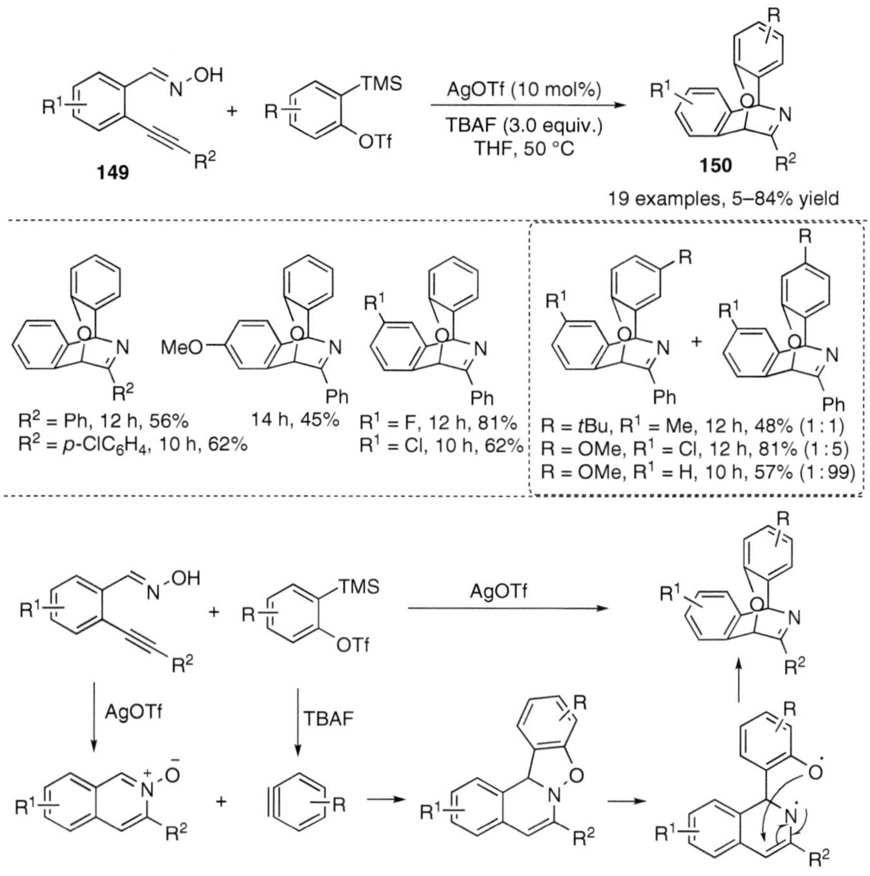

Scheme 6.101 AgOTf-catalyzed reaction of 2-alkynylbenzaldoximes with arynes.

Abbreviations

BINAP	2,2′-Bis(diphenylphosphino)-1,1′-binaphthalene
n-BuLi	*n*-Butyllithium
CsF	cesium fluoride
dba	dibenzylideneacetone
DCE	dichloroethane
DCM	dichloromethane
DIBAL-H	diisobutylaluminum hydride
DDQ	2,3-dichloro-5,6-dicyano-1,4-benzoquinone
DME	dimethyl ether
DMF	dimethylformamide
DMSO	dimethylsulfoxide
DPEphos	bis(2-diphenylphosphinophenyl)ether
dppe	1,2-Bis(diphenylphosphino)ethane
dppf	1,1’-Bis(diphenylphosphino)ferrocene

dppp	1,3-Bis(diphenylphosphino)propane
dppm	1,1-Bis(diphenylphosphino)methane
GC-MS	gas chromatography-mass spectrometry
HMDS	hexamethyldisilazane
HRMS	high-resolution mass spectrometry
KIE	kinetic isotope effect
LED	light emitting diodes
MALDI	matrix-assisted laser desorption/ionization
NHC	*N*-Heterocyclic Carbene
NBS	*N*-Bromosuccinimide
NIS	*N*-Iodosuccinimide
ppm	parts per million
PPh_3	tri-phenylphosphine
TBAB	tetrabutylammonium bromide
TBAI	tetrabutylammonium iodide
TEMPO	2,2,6,6-Tetramethyl-1-piperidinyloxy
THF	tetrahydrofuran
TFP	tri(2-furyl)phosphine
TlOAc	thallous acetate
NMR	nuclear magnetic resonance

References

1 (a) Wenk, H.H., Winkler, M., and Sander, W. (2003). *Angew. Chem. Int. Ed.* 42: 502–528. (b) Sanz, R. (2008). *Org. Prep. Proced. Int.* 40: 215–291. (c) Gandeepan, P., Müller, T., Zell, D. et al. (2019). *Chem. Rev.* 119: 2192–2452. (d) Winkler, M., Wenk, H.H., and Sander, W. (2004). Recent Applications of Aryne Chemistry to Organicsynthesis. A Review (eds. R.A. Moss, M.S. Platz and M.J. Jones), 741–794. Hoboken, NJ: Wiley.

2 Yoshio, H., Takaaki, S., and Hiroshi, K. (1983). *Chem. Lett.* 12: 1211–1214.

3 Ruiz-Castillo, P. and Buchwald, S.L. (2016). *Chem. Rev.* 116: 12564–12649.

4 (a) Peña, D., Escudero, S., Pérez, D. et al. (1998). *Angew. Chem. Int. Ed.* 37: 2659–2661. (b) Peña, D., Pérez, D., Guitián, E., and Castedo, L. (1999). *Org. Lett.* 1: 1555–1557. (c) Peña, D., Cobas, A., Pérez, D. et al. (2000). *Org. Lett.* 2: 1629–1632.

5 (a) Peña, D., Pérez, D., Guitián, E., and Castedo, L. (1999). *J. Am. Chem. Soc.* 121: 5827–5828. (b) Peña, D., Pérez, D., Guitián, E., and Castedo, L. (2000). *J. Org. Chem.* 65: 6944–6950.

6 Alonso, J.M., Díaz-Álvarez, A.E., Criado, A. et al. (2012). *Angew. Chem. Int. Ed.* 51: 173–177.

7 (a) Radhakrishnan, K.V., Yoshikawa, E., and Yamamoto, Y. (1999). *Tetrahedron Lett.* 40: 7533–7535. (b) Yoshikawa, E., Radhakrishnan, K.V., and Yamamoto, Y. (2000). *J. Am. Chem. Soc.* 122: 7280–7286. (c) Yoshikawa, E. and Yamamoto, Y. (2000). *Angew. Chem. Int. Ed.* 39: 173–175.

8 (a) Jayanth, T.T., Jeganmohan, M., and Cheng, C.-H. (2004). *J. Org. Chem.* 69: 8445–8450. (b) Hayes, D.M. and Hoffmann, R. (1972). *J. Phys. Chem.* 76: 656. (c) Epiotis, N.D. (1974). *Angew. Chem. Int. Ed. Engl.* 13: 751.

9 Peña, D., Pérez, D., Guitián, E., and Castedo, L. (2003). *Eur. J. Org. Chem.* 2003: 1238–1243.

10 Sato, Y., Tamura, T., and Mori, M. (2004). *Angew. Chem. Int. Ed.* 43: 2436–2440.

11 Liu, Y.-L., Liang, Y., Pi, S.-F. et al. (2009). *J. Org. Chem.* 74: 3199–3202.

12 Quintana, I., Boersma, A.J., Peña, D. et al. (2006). *Org. Lett.* 8: 3347–3349.

13 Kim, H.S., Gowrisankar, S., Kim, E.S., and Kim, J.N. (2008). *Tetrahedron Lett.* 49: 6569–6572.

14 Cant, A.A., Roberts, L., and Greaney, M.F. (2010). *Chem. Commun.* 46: 8671–8673.

15 (a) García-López, J.-A. and Greaney, M.F. (2014). *Org. Lett.* 16: 2338–2341. (b) Retbøll, M., Edwards, A.J., Rae, A.D. et al. (2002). *J. Am. Chem. Soc.* 124: 8348–8360.

16 Patel, R.M. and Argade, N.P. (2013). *Org. Lett.* 15: 14–17.

17 Lin, J.B., Shah, T.K., Goetz, A.E. et al. (2017). *J. Am. Chem. Soc.* 139: 10447–10455.

18 Hsieh, J.-C. and Cheng, C.-H. (2005). *Chem. Commun.* 2459–2461.

19 Hsieh, J.-C., Rayabarapu, D.K., and Cheng, C.-H. (2004). *Chem. Commun.* 532–533.

20 Hsieh, J.-C. and Cheng, C.-H. (2008). *Chem. Commun.* 2992–2994.

21 Iwayama, T. and Sato, Y. (2010). *Hetrocycles* 80: 917–924.

22 Saito, N., Shiotani, K., Kinbara, A., and Sato, Y. (2009). *Chem. Commun.* 4284–4286.

23 Candito, D.A. and Lautens, M. (2011). *Synlett* 22: 1987–1992.

24 Qiu, Z. and Xie, Z. (2009). *Angew. Chem. Int. Ed.* 48: 5729–5732.

25 Chen, L., Zhang, C., Wen, C. et al. (2015). *Catal. Commun.* 65: 81–84.

26 Asao, N. and Sato, K. (2006). *Org. Lett.* 8: 5361–5363.

27 (a) Liu, Z., Zhang, X., and Larock, R.C. (2005). *J. Am. Chem. Soc.* 127: 15716–15717. (b) Liu, Z. and Larock, R.C. (2007). *J. Org. Chem.* 72: 223–232.

28 Jayanth, T.T. and Cheng, C.-H. (2006). *Chem. Commun.* 894–896.

29 (a) Zhang, X. and Larock, R.C. (2005). *Org. Lett.* 7: 3973–3976. (b) Liu, Z. and Larock, R.C. (2007). *Angew. Chem. Int. Ed.* 46: 2535–2538.

30 Xie, C., Zhang, Y., Huang, Z., and Xu, P. (2007). *J. Org. Chem.* 72: 5431–5434.

31 (a) Bhuvaneswari, S., Jeganmohan, M., and Cheng, C.-H. (2006). *Org. Lett.* 8: 5581–5584. (b) Shih, H.-T., Shih, H.-S., and Cheng, C.-H. (2001). *Org. Lett.* 6: 881. (c) Shih, H.-T., Lin, C.-H., Shih, H.-H., and Cheng, C.-H. (2002). *Adv. Mater.* 14: 1409–1412.

32 (a) Avlasevich, Y., Müller, S., Erk, P., and Müllen, K. (2007). *Chem. Eur. J.* 13: 6555–6561. (b) Lütke Eversloh, C., Li, C., and Müllen, K. (2011). *Org. Lett.* 13: 4148–4150.

33 Li, R.-J., Pi, S.-F., Liang, Y. et al. (2010). *Chem. Commun.* 46: 8183–8185.

34 Parthasarathy, K., Han, H., Prakash, C., and Cheng, C.-H. (2012). *Chem. Commun.* 48: 6580–6582.

35 Lin, Y., Wu, L., and Huang, X. (2011). *Eur. J. Org. Chem.* 2011: 2993–3000.

36 Ni, S., Shu, W., and Ma, S. (2013). *Synlett* 24: 2310–2314.

37 (a) Worlikar, S.A. and Larock, R.C. (2009). *Org. Lett.* 11: 2413–2416. (b) Yoshida, H., Tanino, K., Ohshita, J., and Kunai, A. (2004). *Angew. Chem. Int. Ed.* 43: 5052–5055. (c) Worlikar, S.A. and Larock, R.C. (2009). *J. Org. Chem.* 74: 9132–9139.

38 Huang, X., Sha, F., and Tong, J. (2010). *Adv. Synth. Catal.* 352: 379–385.

39 Yang, Y., Huang, H., Wu, L., and Liang, Y. (2014). *Org. Biomol. Chem.* 12: 5351–5355.

40 Yao, T., Zhang, H., and Zhao, Y. (2016). *Org. Lett.* 18: 2532–2535.

41 Pérez-Gómez, M. and García-López, J.-A. (2016). *Angew. Chem. Int. Ed.* 55: 14389–14393.

42 Yoon, H., Lossouarn, A., Landau, F., and Lautens, M. (2016). *Org. Lett.* 18: 6324–6327.

43 (a) Yao, T. and He, D. (2017). *Org. Lett.* 19: 842–845. (b) Cámpora, J., López, J.A., Palma, P. et al. (1999). *Angew. Chem. Int. Ed.* 38: 147–151.

44 Zuo, Z., Wang, H., Diao, Y. et al. (2018). *ACS Catal.* 8: 11029–11034.

45 Lu, C., Dubrovskiy, A.V., and Larock, R.C. (2012). *J. Org. Chem.* 77: 8648–8656.

46 Lu, C., Markina, N.A., and Larock, R.C. (2012). *J. Org. Chem.* 77: 11153–11160.

47 Gerfaud, T., Neuville, L., and Zhu, J. (2009). *Angew. Chem. Int. Ed.* 48: 572–577.

48 Tang, C.-Y., Wu, X.-Y., Sha, F. et al. (2014). *Tetrahedron Lett.* 55: 1036–1039.

49 Pimparkar, S. and Jeganmohan, M. (2014). *Chem. Commun.* 50: 12116–12119.

50 Peng, X., Wang, W., Jiang, C. et al. (2014). *Org. Lett.* 16: 5354–5357.

51 Wang, W., Peng, X., Qin, X. et al. (2015). *J. Org. Chem.* 80: 2835–2841.

52 Neog, K., Borah, A., and Gogoi, P. (2016). *J. Org. Chem.* 81: 11971–11977.

53 Zhao, J., Li, H., Li, P., and Wang, L. (2019). *J. Org. Chem.* 84: 9007–9016.

54 Asamdi, M., Chauhan, P.M., Patel, J.J., and Chikhalia, K.H. (2019). *Tetrahedron* 75: 3485–3494.

55 Feng, S., Li, S., Li, J., and Wei, J. (2019). *Org. Chem. Front.* 6: 517–522.

56 Meng, Y.-Y., Si, X.-J., Song, Y.-Y. et al. (2019). *Chem. Commun.* 55: 9507–9510.

57 (a) Thorat, V.H., Upadhyay, N.S., Murakami, M., and Cheng, C.-H. (2018). *Adv. Synth. Catal.* 360: 284–289. (b) Cahiez, G., Moyeux, A., Buendia, J., and Duplais, C. (2007). *J. Am. Chem. Soc.* 129: 13788–13789.

58 Zhang, T.-Y., Lin, J.-B., Li, Q.-Z. et al. (2017). *Org. Lett.* 19: 1764–1767.

59 Zhang, T.-Y., Liu, C., Chen, C. et al. (2018). *Org. Lett.* 20: 220–223.

60 Yoshikawa, E., Radhakrishnan, K.V., and Yamamoto, Y. (2000). *Tetrahedron Lett.* 41: 729–731.

61 Jeganmohan, M. and Cheng, C.-H. (2004). *Org. Lett.* 6: 2821–2824.

62 (a) Jeganmohan, M. and Cheng, C.-H. (2005). *Synthesis* 2005: 1693–1697. (b) Jayanth, T.T., Jeganmohan, M., and Cheng, C.-H. (2005). *Org. Lett.* 7: 2921–2924.

63 Bhuvaneswari, S., Jeganmohan, M., Yang, M.-C., and Cheng, C.-H. (2008). *Chem. Commun.* 2158–2160.

64 Jeganmohan, M., Bhuvaneswari, S., and Cheng, C.-H. (2009). *Angew. Chem. Int. Ed.* 48: 391–394.

65 (a) Henderson, J.L., Edwards, A.S., and Greaney, M.F. (2006). *J. Am. Chem. Soc.* 128: 7426–7427. (b) Henderson, J.L., Edwards, A.S., and Greaney, M.F. (2007). *Org. Lett.* 9: 5589–5592.

66 Garve, L.K.B. and Werz, D.B. (2015). *Org. Lett.* 17: 596–599.

67 Li, J., Noyori, S., Nakajima, K., and Nishihara, Y. (2014). *Organometallics* 33: 3500–3507.

68 Jayanth, T.T. and Cheng, C.-H. (2007). *Angew. Chem. Int. Ed.* 46: 5921–5924.

69 Niu, S.-L., Hu, J., He, K. et al. (2019). *Org. Lett.* 21: 4250–4254.

70 Peng, X., Ma, C., Tung, C.-H., and Xu, Z. (2016). *Org. Lett.* 18: 4154–4157.

71 Berti, F., Crotti, P., Cassano, G., and Pineschi, M. (2012). *Synlett* 23: 2463–2468.

72 Xie, C., Liu, L., Zhang, Y., and Xu, P. (2008). *Org. Lett.* 10: 2393–2396.

73 Bhuvaneswari, S., Jeganmohan, M., and Cheng, C.-H. (2008). *Chem. Commun.* 5013–5015.

74 Yoo, W.-J., Nguyen, T.V.Q., and Kobayashi, S. (2014). *Angew. Chem. Int. Ed.* 53: 10213–10217.

75 (a) Yang, X. and Tsui, G.C. (2018). *Org. Lett.* 20: 1179–1182. (b) Chu, H., Sun, S., Yu, J.-T., and Cheng, J. (2015). *Chem. Commun.* 51: 13327–13329. (c) Zeng, Y., Zhang, L., Zhao, Y. et al. (2013). *J. Am. Chem. Soc.* 135: 2955–2958.

76 Cao, W., Niu, S.-L., Shuai, L., and Xiao, Q. (2020). *Chem. Commun.* 56: 972–975.

77 Yang, J., Yu, X., and Wu, J. (2014). *Synthesis* 46: 1362–1366.

78 Yoshida, H., Honda, Y., Shirakawa, E., and Hiyama, T. (2001). *Chem. Commun.* 1880–1881.

79 (a) Yoshida, H., Ikadai, J., Shudo, M. et al. (2003). *J. Am. Chem. Soc.* 125: 6638–6639. (b) Yoshida, H., Tanino, K., Ohshita, J., and Kunai, A. (2005). *Chem. Commun.* 5678–5680.

80 Pawliczek, M., Garve, L.K.B., and Werz, D.B. (2015). *Org. Lett.* 17: 1716–1719.

81 Yuan, W. and Ma, S. (2014). *Org. Lett.* 16: 193–195.

82 Yoshida, H., Okada, K., Kawashima, S. et al. (2010). *Chem. Commun.* 46: 1763–1765.

83 Pareek, M., Fallon, T., and Oestreich, M. (2015). *Org. Lett.* 17: 2082–2085.

84 Yoshida, H., Kawashima, S., Takemoto, Y. et al. (2012). *Angew. Chem. Int. Ed.* 51: 235–238.

85 Tanaka, H., Kuriki, H., Kubo, T. et al. (2019). *Chem. Commun.* 55: 6503–6506.

86 Akubathini, S.K. and Biehl, E. (2009). *Tetrahedron Lett.* 50: 1809–1811.

87 Xie, C., Zhang, Y., and Yang, Y. (2008). *Chem. Commun.* 4810–4812.

88 Yoshida, H., Morishita, T., Nakata, H., and Ohshita, J. (2009). *Org. Lett.* 11: 373–376.

89 Morishita, T., Yoshida, H., and Ohshita, J. (2010). *Chem. Commun.* 46: 640–642.

90 (a) Yang, X. and Tsui, G.C. (2018). *Chem. Sci.* 9: 8871–8875. (b) Okuma, K., Sonoda, S., Koga, Y., and Shioji, K. (1999). *J. Chem. Soc., Perkin Trans. 1* 2997–3000. (c) Yang, X., He, L., and Tsui, G.C. (2017). *Org. Lett.* 19: 2446–2449.

(d) Zhang, B.-S., Gao, L.-Y., Zhang, Z. et al. (2018). *Chem. Commun.* 54: 1185–1188.

91 Uchiyama, M., Naka, H., Matsumoto, Y., and Ohwada, T. (2004). *J. Am. Chem. Soc.* 126: 10526–10527.

92 Zeng, Y. and Hu, J. (2014). *Chem. Eur. J.* 20: 6866–6870.

93 Xiao, X., Wang, T., Xu, F., and Hoye, T.R. (2018). *Angew. Chem. Int. Ed.* 57: 16564–16568.

94 Wang, K.-P., Yun, S.Y., Mamidipalli, P., and Lee, D. (2013). *Chem. Sci.* 4: 3205–3211.

95 Chen, Q., Yan, X., Wen, C. et al. (2016). *J. Org. Chem.* 81: 9476–9482.

96 Chatani, N., Kamitani, A., Oshita, M. et al. (2001). *J. Am. Chem. Soc.* 123: 12686–12687.

97 (a) García-López, J.-A., Oliva-Madrid, M.-J., Saura-Llamas, I. et al. (2012). *Chem. Commun.* 48: 6744–6746. (b) Oliva-Madrid, M.-J., Saura-Llamas, I., Bautista, D., and Vicente, J. (2013). *Chem. Commun.* 49: 7997–7999. (c) Oliva-Madrid, M.-J., García-López, J.-A., Saura-Llamas, I. et al. (2014). *Organometallics* 33: 6420–6430.

98 Feng, M., Tang, B., Wang, N. et al. (2015). *Angew. Chem. Int. Ed.* 54: 14960–14964.

99 Pi, S.-F., Yang, X.-H., Huang, X.-C. et al. (2010). *J. Org. Chem.* 75: 3484–3487.

100 Jia, H., Guo, Z., Liu, H. et al. (2018). *Chem. Commun.* 54: 7050–7053.

101 Ren, H., Luo, Y., Ye, S., and Wu, J. (2011). *Org. Lett.* 13: 2552–2555.

7

Molecular Rearrangements Triggered by Arynes

Lu Han and Shi-Kai Tian

University of Science and Technology of China, Department of Chemistry, 96 Jinzhai Road, Hefei, Anhui 230026, China

7.1 Introduction

Arynes are electronically neutral yet highly electrophilic species due to the presence of a highly strained carbon–carbon triple bond. While arynes can be generated in situ in the reaction mixtures under several sets of conditions, most of which include the use of strong bases, potentially explosive precursors, and toxic metals [1], their preparation from 2-silylaryl triflates via fluoride ion-induced 1,2-elimination, developed by Kobayashi et al. in 1983 [2], remains the most convenient, mild, and efficient method. The Kobayashi method is compatible with a wide variety of functional groups and significantly expands the scope for arynes in various carbon–carbon and carbon–heteroatom bond-forming processes, including molecular rearrangements.

A wide range of carbon and heteroatom nucleophiles, even weak ones such as enolizable ketones, amides, and imides, can add to arynes under mild reaction conditions and the resultant zwitterionic intermediates serve as versatile precursors for a number of rearrangements, delivering structural diverse organic compounds that might otherwise be difficult to be prepared. The aryl moieties of the zwitterionic intermediates originated from the arynes may or may not engage in the following skeletal rearrangements. Exclusion of the original aryne moieties from the skeletal rearrangements leads to the monofunctionalization of arynes. On the other hand, the original aryne moieties are ready to engage in the skeletal rearrangements toward the 1,2-difunctionalization or 1,2,3-trifunctionalization of arynes, often involving the formation of strained cycles as key intermediates. In addition, rearrangement precursors can be formed by cycloaddition between arynes and unsaturated systems, and rearrangements are also involved in the multicomponent reactions with two or more aryne molecules. These aryne-triggered rearrangements enable a variety of synthetically important operations, including skeletal construction, functional group transposition, ring formation, ring expansion, ring contraction, and chirality transfer.

Modern Aryne Chemistry, First Edition. Edited by Akkattu T. Biju.

This book chapter summarizes various molecular rearrangements triggered by arynes in the absence of transition metals [3]. It is organized according to the functionalization modes of arynes forming one, two, or three carbon–carbon and carbon–heteroatom bonds via a variety of reaction pathways.

7.2 Rearrangements Involved in the Monofunctionalization of Arynes

The addition of heteroatom nucleophiles to arynes can significantly weaken the carbon–heteroatom bonds of the heteroatom nucleophiles. Owing to charge acceleration, the resultant zwitterionic intermediates are amenable to engage in rearrangements via carbon–heteroatom bond cleavage under mild reaction conditions. These rearrangements proceed without the participation of the original aryne moieties, resulting in the monofunctionalization of arynes.

7.2.1 Reactions of Arynes with Nitrogen Nucleophiles

As early as 1943, Wittig and Merkle reported that the reaction of triethylamine with benzyne (**2a**) (generated in situ from haloarene **1a** in the presence of phenyllithium) afforded a minor product [4], which was later identified as tertiary amine **3** [5]. This minor product was proposed to be generated via a [1,2]-Stevens rearrangement as depicted in Scheme 7.1. Nucleophilic addition of triethylamine to benzyne (**2a**) leads to the formation of zwitterion **4**, which undergoes proton transfer to form quaternary ammonium ylide **5**. Subsequent [1,2]-migration of the ethyl group affords tertiary amine **3**. A similar rearrangement was observed by Lepley et al. in the addition of *N*,*N*-dialkylanilines to benzyne (**2a**), generated in situ from haloarene **1a** in the presence of *n*-butyllithium [6].

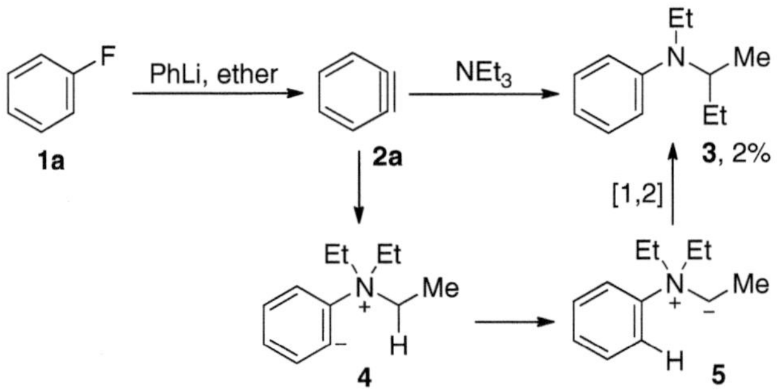

Scheme 7.1 Reaction of benzyne with triethylamine. Source: Based on Wittig and Benz [5].

In the addition of (aminomethyl)silane **6a** to benzyne (**2a**) (generated in situ from haloarene **1b** in the presence of sodium amide), Sato et al. isolated the [1,2]-Stevens rearrangement product **7a** in 60% yield (Scheme 7.2) [7]. The rearrangement reaction is dramatically affected by the substituents on the silicon atom. In sharp contrast, the addition of (aminomethyl)silane **6b** to benzyne (**2a**) afforded the rearrangement products **7b** and **8b** in 5% and 9% yields, respectively. In this case, the [1,2]-Stevens rearrangement competes with a skeletal rearrangement involving the migration of the silyl group to the benzene ring (originated from benzyne (**2a**)) accompanied by the shift of a phenyl group from silicon to the adjacent carbon.

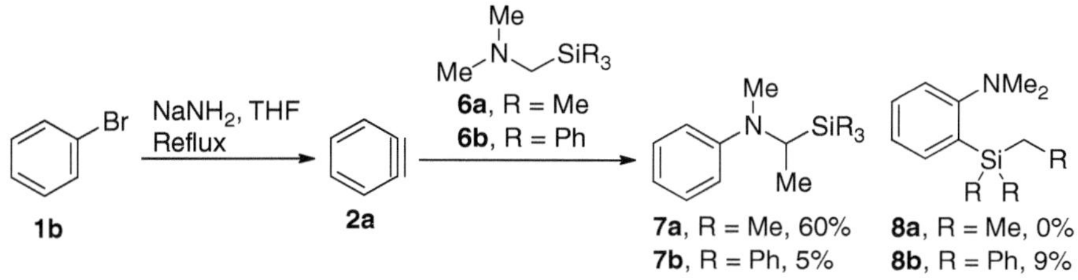

Scheme 7.2 Reaction of benzyne with (aminomethyl)silanes. Source: Based on Sato et al. [7].

The scope of aryne-triggered [1,2]-Stevens rearrangement of tertiary amines has been expanded dramatically through increasing the acidity of amine α-C—H bonds as well as using the Kobayashi method for the generation of arynes under mild reaction conditions [2]. In 2016, Biju and coworkers disclosed the [1,2]-migration of the benzyl group in the reaction of *N*-benzyl glycinate **9** with benzyne (**2a**) (generated in situ from 2-silylaryl triflate **10a** in the presence of KF and 18-crown-6), delivering α-benzyl glycinate **11** in 41% yield (Scheme 7.3) [8]. Pan and Liu extended the aryne-triggered [1,2]-Stevens rearrangement to 1,2,3,4-tetrahydroisoquinolines **12**, resulting in ring expansion to afford 3-aryl-3-benzazepines **13** in moderate-to-good yields (Scheme 7.4) [9]. Arynes **2** are generated from 2-silylaryl triflates **10** using CsF and 18-crown-6, and the addition of potassium bicarbonate minimizes side reactions such as the Hofmann elimination and the C-arylation. Interestingly, lowering the temperature from 80 to 40 °C enables the arylation of the amine α-C—H bonds as the major reaction albeit α-arylated products **14** were obtained in low yields. Mechanistically, both 3-aryl-3-benzazepine **13** and α-arylated product **14** were proposed to be formed via the [1,2]-Stevens rearrangement of quaternary ammonium ylide **16**. Aryne **2** is generated in situ by fluoride ion–induced 1,2-elimination of 2-silylaryl triflate **10**, and then nucleophilic addition of 1,2,3,4-tetrahydroisoquinoline **12** to aryne **2** followed by proton transfer forms quaternary ammonium ylide **16**. [1,2]-migration of the benzyl group of the ammonium ylide occurs at 80 °C to afford 3-aryl-3-benzazepine **13**, and [1,2]-migration of the aryl group occurs at 40 °C to afford α-arylated product **14**.

Although the Sommelet–Hauser rearrangement has not been mentioned in the above reactions of tertiary benzylic amines with arynes (Schemes 7.3 and 7.4), it emerges as a major reaction pathway for the substrates having electron-deficient benzyl groups under milder conditions. In 2019, Biju and coworkers reported the aryne-triggered Sommelet–Hauser rearrangement of tertiary benzylic amines **17** bearing an ester, a cyano, a nitro, or a trifluoromethyl group at the amine

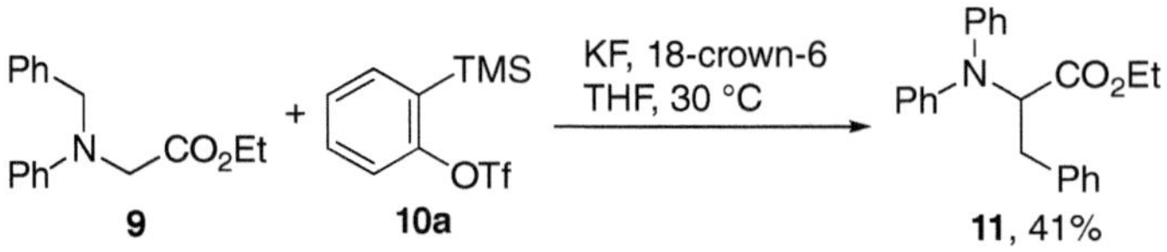

Scheme 7.3 Reaction of benzyne with an *N*-benzyl glycinate. Source: Based on Roy et al. [8].

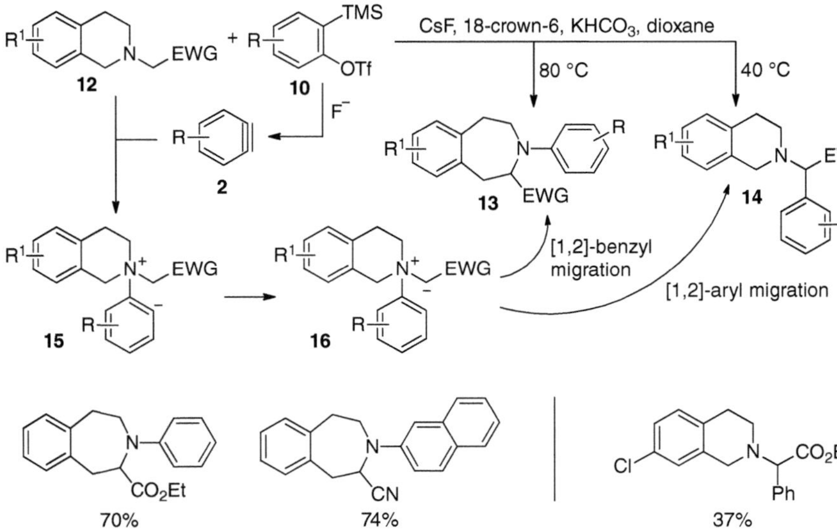

Scheme 7.4 Aryne-triggered [1,2]-Stevens rearrangement of 1,2,3,4-tetrahydroisoquinolines. Source: Based on Pan and Liu [9].

α-C position, taking place at a lower temperature to afford tertiary amines **18** in moderate-to-good yields (Scheme 7.5) [10]. In this case, arynes **2** are generated in situ from 2-silylaryl triflates **10** in the presence of KF and 18-crown-6. Mechanistically, quaternary ammonium ylide **20**, generated from tertiary benzylic amine **17** and aryne **2**, undergoes a [2,3]-sigmatropic rearrangement, involving the aryl ring of benzylic amine **17**, to form dearomatized intermediate **21**, which upon a [1,3]-hydrogen shift results in the formation of tertiary amine **18**. In many cases, the yields are not satisfactory due to the competing [1,2]-Stevens rearrangement. When the reaction was performed on benzylic amine **22** bearing an *N*-cyclohexyl group, tertiary amine **23** was obtained in 40% yield (Scheme 7.6). Hypothetically, the initially formed quaternary ammonium ylide **24** prefers to undergo [1,4]-migration of the cyanomethyl group to generate intermediate **25**, which undergoes aromatization to afford tertiary amine **23**.

The arynes generated from 2-silylaryl triflates via fluoride ion–induced 1,2-elimination can efficiently trigger the [2,3]-Stevens rearrangement of tertiary allylic amines having acidic α-C—H bonds. In 2016, Tian and coworkers reported that the use of 2-silylaryl triflates **10** as aryne precursors in the presence of

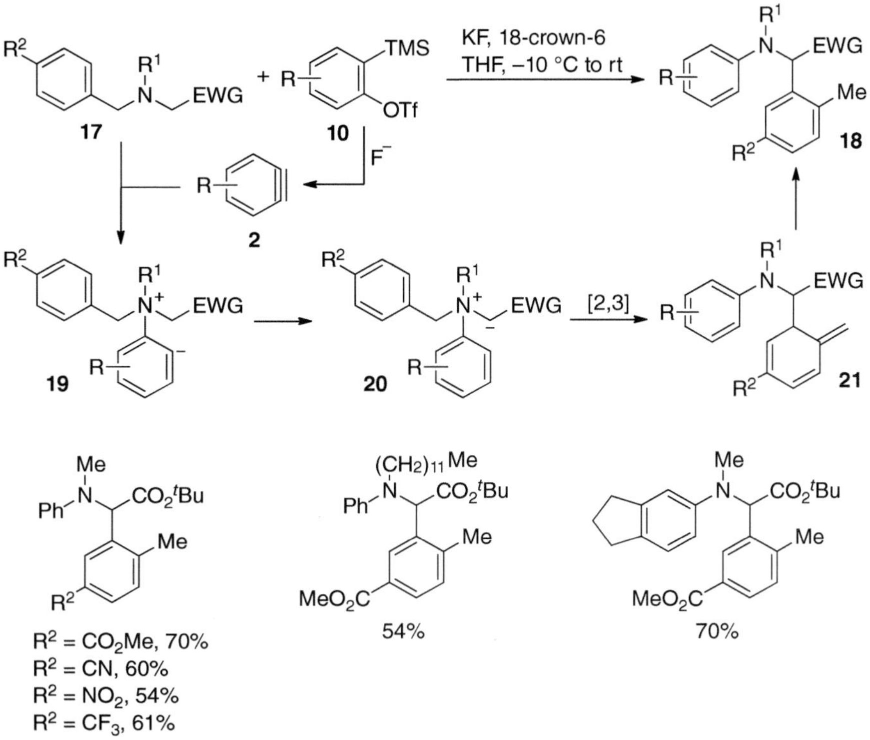

Scheme 7.5 Aryne-triggered Sommelet–Hauser rearrangement of tertiary benzylic amines. Source: Based on Roy et al. [10].

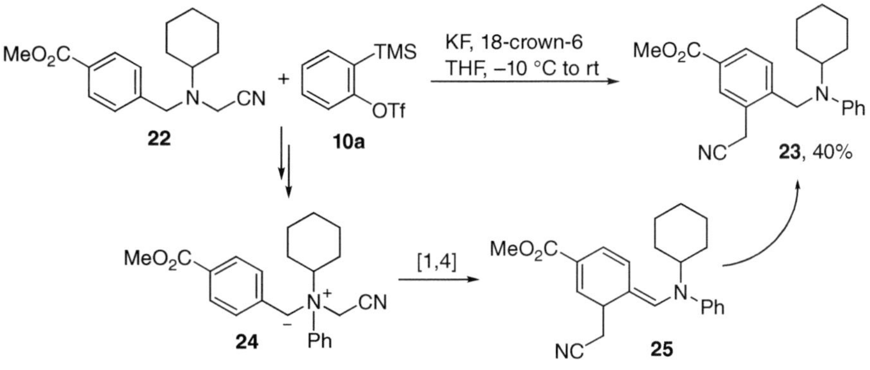

Scheme 7.6 Benzyne-induced [1,4]-rearrangement. Source: Based on Roy et al. [10].

tetrabutylammonium fluoride (TBAF) enables the [2,3]-Stevens rearrangement of tertiary allylic amines **26** bearing an ester, a ketone, an amide, a cyano, a phosphonate, a benzoxazole, or a *p*-nitrophenyl group at the amine α-C position, delivering a range of functionalized homoallylic amines **27** in moderate-to-good yields (Scheme 7.7) [11]. Mechanistically, the reaction proceeds via a [2,3]-sigmatropic rearrangement of quaternary allylic ammonium ylide **29**, generated by nucleophilic addition of tertiary allylic amine **26** to aryne **2** and subsequent proton transfer. Although quaternary allylic ammonium ylide **29** bears an *N*-aryl group, the corresponding aza-Claisen rearrangement product was not observed under the standard reaction conditions.

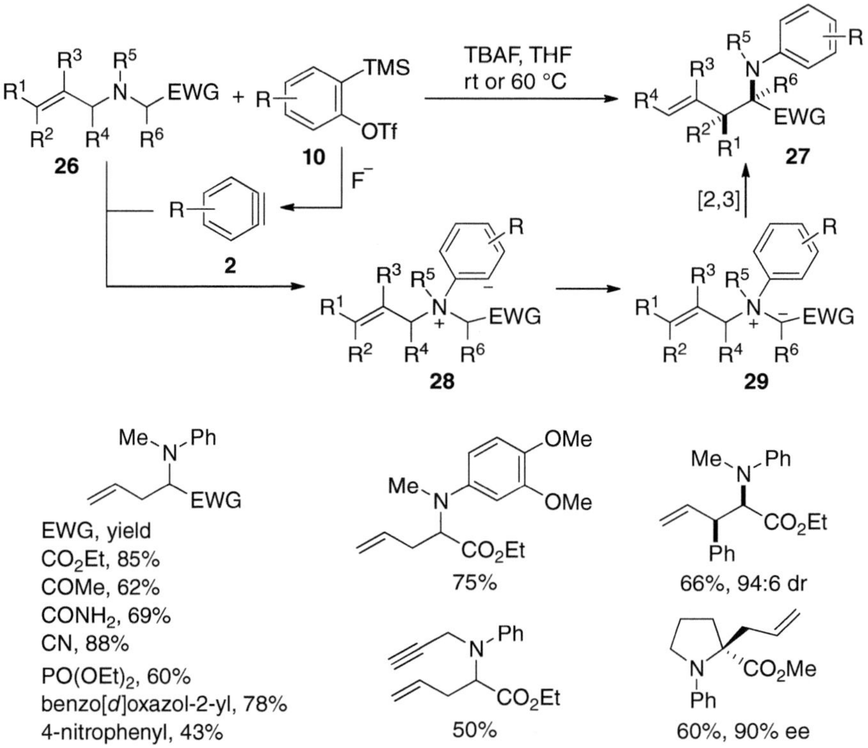

Scheme 7.7 Aryne-triggered [2,3]-Stevens rearrangement of tertiary allylic amines. Source: Based on Zhang et al. [11].

The aryne-triggered [2,3]-Stevens rearrangement has been applied to optically active tertiary allylic amines. The rearrangement reaction of *L*-proline derived tertiary allylic amine provides a nitrogen-substituted quaternary stereocenter with inversion of configuration and excellent retention of enantiopurity (90% ee) (Scheme 7.7) [11]. The high efficiency of chirality transfer, from carbon to nitrogen and then back to carbon, is attributable to the highly diastereoselective attack of the cyclic tertiary amine on benzyne and the concerted nature of subsequent [2,3]-sigmatropic rearrangement. Independent investigation by Biju and coworkers uncovered that higher enantiopurity (96% ee) was achieved in the benzyne-triggered rearrangement of the same *L*-proline-derived tertiary allylic amine by treating the benzyne precursor, 2-silylaryl triflate **10a**, with KF and 18-crown-6 [8].

In addition to chirality transfer, the aryne-triggered [2,3]-Stevens rearrangement of tertiary allylic amines serves as a powerful protocol for ring contraction, ring expansion, and ring reconstruction. As demonstrated by Tian and coworkers, treatment of 1,2,3,6-tetrahydropyridine-2-carboxylate **30**, a cyclic allylic amine, with 2-silylaryl triflate **10a** in the presence of CsF afforded functionalized cyclopropane **31** in 65% yield as a single diastereomer (Scheme 7.8a) [11]. In contrast, the studies by Sweeney and coworkers show that installation of a 2-malonyl group on the nitrogen of the 1,2,3,6-tetrahydropyridine ring resulted in ring opening and subsequent formation of a new nitrogen-containing heterocycle, providing convenient access to various *N*-aryl-2-acylpyrrolidines (Scheme 7.8b) [12]. In 2018, Liu and coworkers reported the ring expansion of 1-vinyl-1,2,3,4-tetrahydroisoquinolines triggered by arynes, leading to the formation of (*E*)-3-aryl-2,3,4,5-tetrahydro-1*H*-3-benzazonines (Scheme 7.8c) [13].

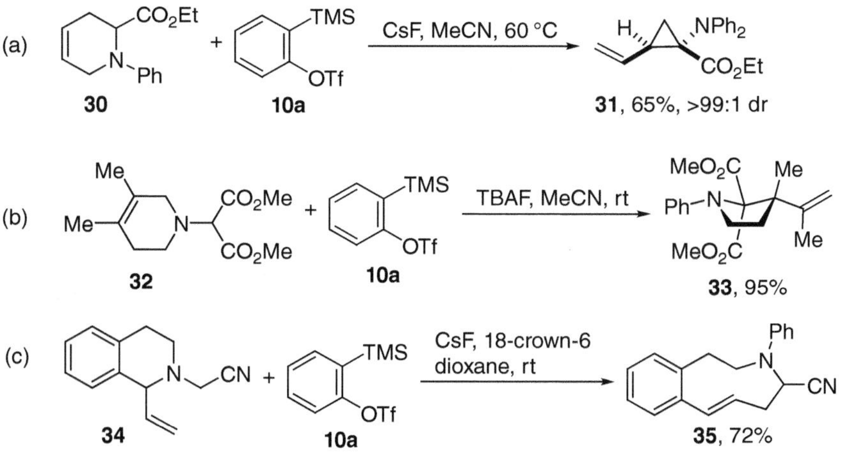

Scheme 7.8 Aryne-triggered [2,3]-Stevens rearrangement of cyclic allylic amines. (a) Ring contraction. Source: Based on Zhang et al. [11]. (b) Ring reconstruction. Source: Based on Moss et al. [12]. (c) Ring expansion. Source: Pan et al. [13].

The aryne-triggered [2,3]-Stevens rearrangement has been extended to tertiary propargylic amines having acidic α-C—H bonds (Scheme 7.9) [14]. Quaternary propargylic ammonium ylide **40** can be generated in a similar reaction pathway from tertiary propargylic amine **36** and aryne **2** (generated in situ from 2-silylaryl triflates **10** in the presence of KF and 18-crown-6), and undergoes a [2,3]-sigmatropic rearrangement to afford 1-(α-aminoalkyl)allene **37**. Under the mild basic reaction conditions, monosubstituted allene **41** isomerizes to 1-amino-1,3-diene **38**. The chemistry was applied to an *L*-pipecolinic acid derived optically active tertiary propargylic amine, and significant erosion of enantiopurity (73% ee) was observed in the construction of a nitrogen-substituted quaternary stereocenter.

A Curtius-type rearrangement has been developed by Tian and coworkers according to the selective addition of *N′*,*N′*-disubstituted acyl hydrazides **42** to benzyne (**2a**), generated in situ from 2-silylaryl triflate **10a** in the presence of CsF (Scheme 7.10) [15]. The resulting zwitterion **44** undergoes proton transfer followed by a Curtius-type rearrangement to afford isocyanate **46**. Trapping the isocyanate

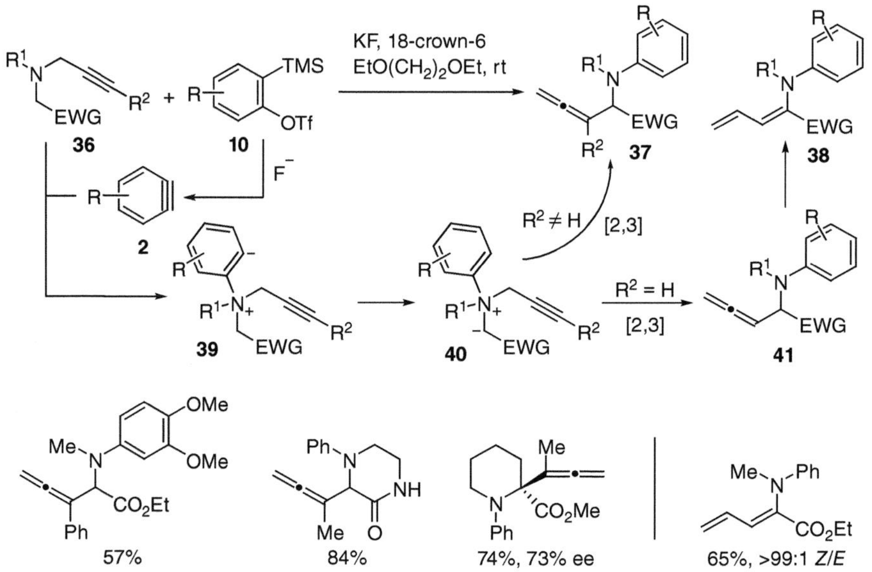

Scheme 7.9 Aryne-triggered [2,3]-Stevens rearrangement of tertiary propargylic amines. Source: Based on Zhou et al. [14].

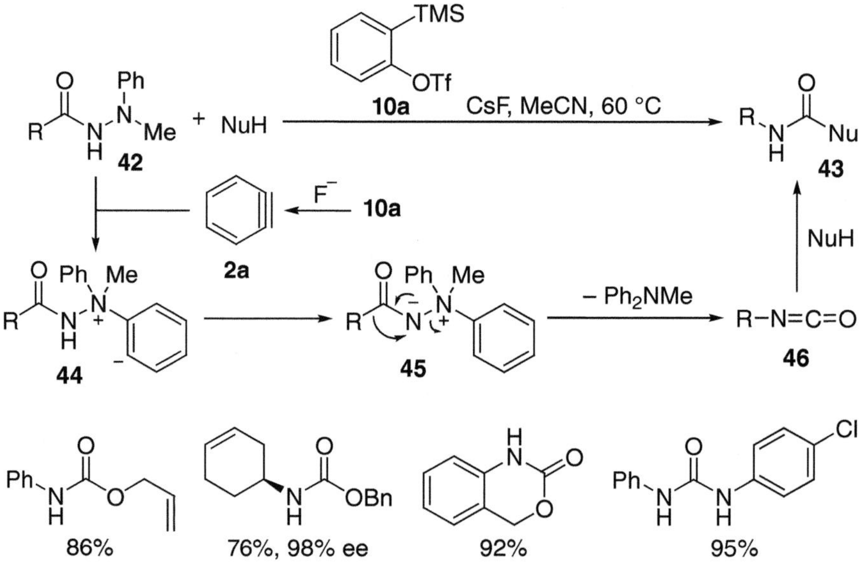

Scheme 7.10 Benzyne-promoted Curtius-type rearrangement of acyl hydrazides. Source: Based on Guo et al. [15].

with an alcohol affords a carbamate product. Complete retention of configuration was observed in the reaction of enantioenriched α-chiral alkanoyl hydrazides. An N'-methyl-N'-phenyl acyl hydrazide tethered with a hydroxyl group proved suitable for the benzyne-promoted Curtius-type rearrangement. Replacing the alcohol with an amine, an N′-unsubstituted acyl hydrazide, or an alkoxyamine in the Curtius-type rearrangement reaction provides convenient access to unsymmetric ureas and analogues.

In addition to tertiary amines and N′,N′-disubstituted acyl hydrazides, pyridines have been identified as suitable nitrogen nucleophiles in the molecular rearrangements through nucleophilic addition to arynes. In 2013, Biju and coworkers disclosed a novel three-component reaction of pyridines **47**, arynes **2** (generated in situ from 2-silylaryl triflates **10** in the presence of KF and 18-crown-6), and isatins **48**, delivering structurally diverse 3,3-disubstituted indolin-2-ones **49** in good yields (Scheme 7.11) [16]. Mechanistically, the reaction is initiated with the nucleophilic addition of pyridine **47** to aryne **2** to generate zwitterion **50**, which undergoes proton transfer to generate pyridylidene **51**. Nucleophilic addition of pyridylidene **51** to isatin **48** forms zwitterion **52**, which undergoes intramolecular nucleophilic aromatic substitution to afford indolin-2-one **49** via spirocycle **53**.

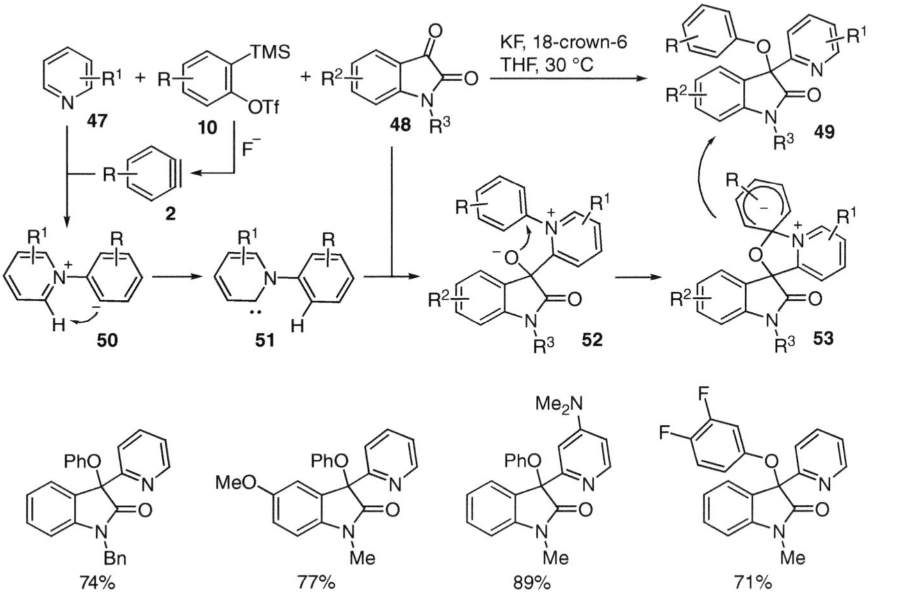

Scheme 7.11 Three-component reaction of pyridines, arynes, and isatins. Source: Based on Bhunia et al. [16].

7.2.2 Reactions of Arynes with Sulfur Nucleophiles

Analogous to tertiary amines, thioethers having acidic α-C—H bonds can be triggered by arynes to participate in [1,2]- and [2,3]-Stevens rearrangements. In 2017, Guo, He, and coworkers reported the aryne-triggered [1,2]-Stevens rearrangement of benzylic thioethers **54** bearing an acyl group at the thioether α-C position, delivering functionalized thioethers in moderate-to-good yields (Scheme 7.12) [17]. In this case, arynes **2** are generated in situ from 2-silylaryl triflates **10** in the presence of KF and 18-crown-6. Mechanistically, the reaction proceeds via the nucleophilic addition of benzylic thioether **54** to aryne **2**. The resulting zwitterion **56** undergoes proton transfer to generate sulfur ylide **57**, which participates in a [1,2]-Stevens rearrangement to afford thioether **55**.

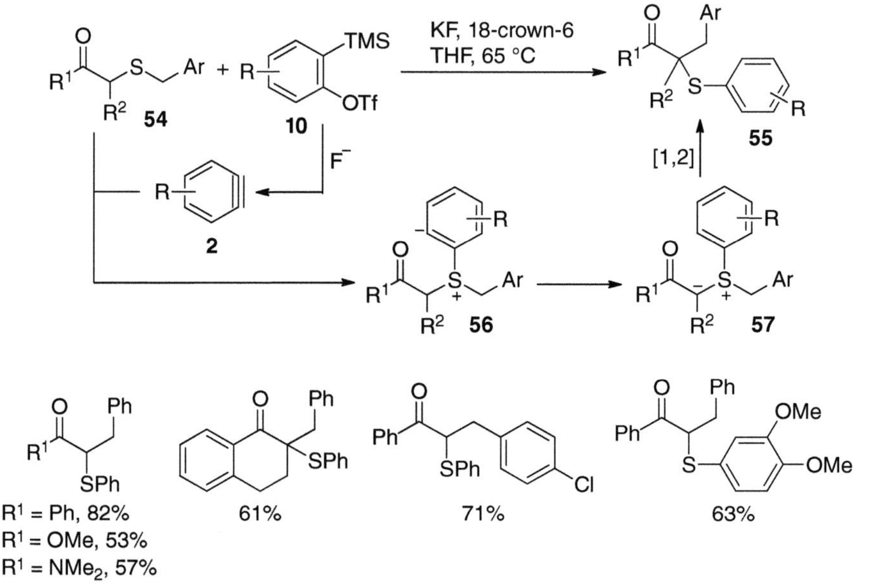

Scheme 7.12 Aryne-triggered [1,2]-Stevens rearrangement of benzylic thioethers. Source: Based on Xu et al. [17].

Guo, He, and coworkers also noted that allylic thioethers **58** bearing an acyl group at the thioether α-C position underwent a [2,3]-Stevens rearrangement in the presence of benzyne (**2a**) (generated in situ from 2-silylaryl triflate **10a** in the presence of KF and 18-crown-6) under the same reaction conditions (Scheme 7.13) [17]. In this case, sulfur ylide **61** is generated from allylic thioether **58** and benzyne (**2a**), and undergoes a [2,3]-Stevens rearrangement to afford thioether **59**. This transformation was also reported independently by two other groups to occur under different reaction conditions. Biju and coworkers employed CsF to promote 2-silylaryl triflates **10** at 25 °C for the generation of arynes **2** and their conditions worked well with cyclic ketone-derived allylic thioethers [18]. Tan, Xu, and coworkers also employed CsF for the generation of arynes **2** but at 70 °C and dramatically expanded the scope of allylic thioethers (Scheme 7.14) [19]. Their reaction conditions are applicable to a variety of functionalized allylic thioethers **62** bearing a vinyl, an acetylenyl, a ketone, an ester, an amide, a cyano, a phosphonate, or a benzoxazole group at the thioether α-C position.

Moreover, Tan, Xu, and coworkers investigated the benzyne-trigger [2,3]-Stevens rearrangement of two propargylic thioethers under similar reaction conditions (Scheme 7.15) [19]. The reaction of benzyne (**2a**) with dipropargyl sulfide (**64**) afforded α-allenyl propargylic thioether **65** in 75% yield. In contrast, the reaction

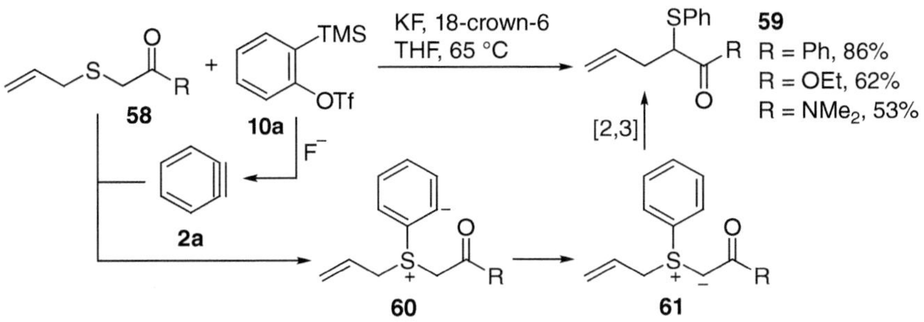

Scheme 7.13 Benzyne-triggered [2,3]-Stevens rearrangement of allylic thioethers. Source: Based on Xu et al. [17].

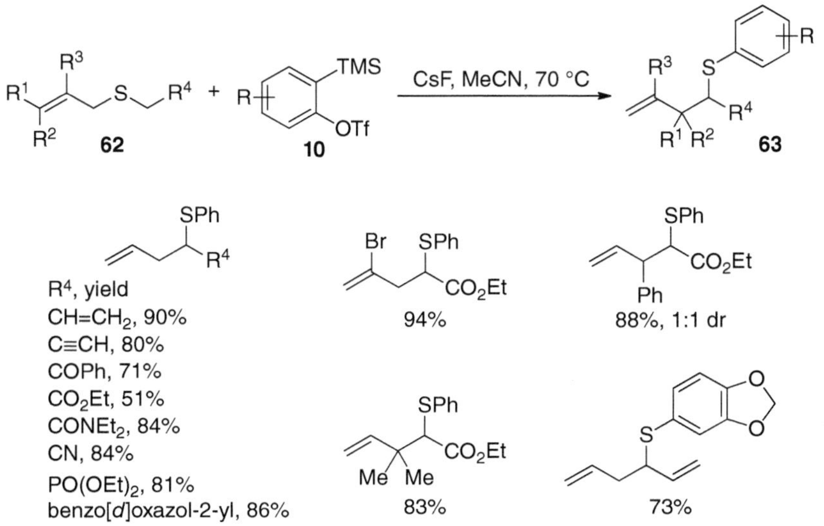

Scheme 7.14 Aryne-trigger [2,3]-Stevens rearrangement of various allylic thioethers. Source: Based on Tan et al. [19].

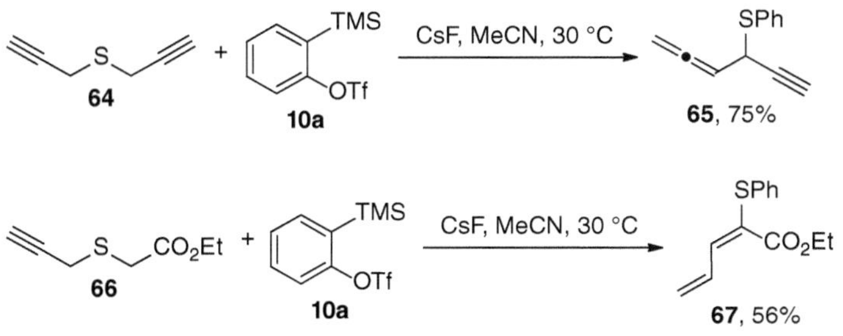

Scheme 7.15 Aryne-trigger [2,3]-Stevens rearrangement of propargylic thioethers. Source: Based on Tan et al. [19].

of propargylic thioether **66**, bearing an ester group at the thioether α-C position, afforded dienyl thioether **67** in 56% yield. In the latter case, the originally formed α-allenyl thioether isomerizes to the dienyl thioether under the weak basic conditions.

7.3 Rearrangements Involved in the 1,2-Difunctionalization of Arynes

A variety of rearrangements occur in the 1,2-difunctionalization of arynes either by formal insertion of arynes into σ-bonds or by other complicated processes for the vicinal formation of two σ-bonds. These rearrangements are very powerful for the synthesis of various functionalized arenes that might otherwise be difficult to access.

7.3.1 Formal Insertion of Arynes into Carbon–Carbon Bonds

Enolizable carbonyl compounds have long been employed in the vicinal difunctionalization of arynes through formal insertion into their carbon–carbon σ-bonds. In earlier studies, carbonyl compounds are preactivated as alkali enolates prior to attack arynes, which are generated from haloarenes in the presence of a strong base. For example, Danheiser and Helgason reported in 1994 a ring-expansive carbon–carbon bond-insertion reaction between cyclic ketone **68** and benzyne (**2a**) under basic conditions, delivering benzocycloheptanone **69** in addition to α-arylated product **70** in a 7 : 3 ratio (Scheme 7.16) [20]. Ketone **68** is converted to enolate **71**

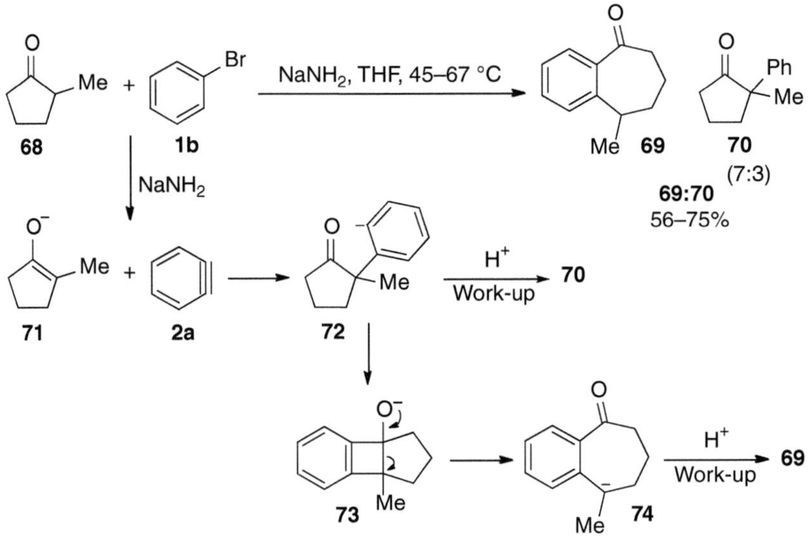

Scheme 7.16 Formal insertion of benzyne into a cyclic ketone. Source: Based on Danheiser and Helgason [20].

with high regioselectivity and benzyne (**2a**) is generated from bromobenzene (**1b**) in the presence of sodium amide. Nucleophilic addition of enolate **71** to benzyne (**2a**) affords phenyl anion **72**, which cyclizes via intramolecular nucleophilic addition to the carbonyl group. This formal [2+2] cycloaddition affords benzocyclobutoxide **73**, which participates in fragmentation followed by protonation (during work-up) to afford benzocycloheptanone **69**.

The generation of arynes from haloarenes in the presence of a strong base is applicable to the formal aryne insertion into some other activated carbonyl compounds and analogues such as an α-lithiated malonate generated from malonate **75** (Scheme 7.17a) [21] and α-lithiated nitrile **78** (Scheme 7.17b) [22]. These reactions proceed via similar tandem formal [2+2] cycloaddition and fragmentation.

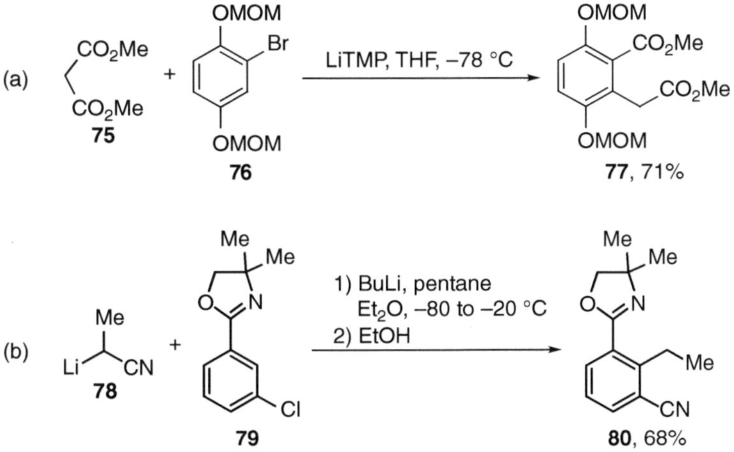

Scheme 7.17 Formal insertion of an aryne into an α-lithiated malonate or an α-lithiated nitrile. (a) Formal insertion of an aryne into an α-lithiated malonate. Source: Based on Shair et al. [21]. (b) Formal insertion of an aryne into an α-lithiated nitrile. Source: Based on Meyers and Pansegrau [22].

The use of 2-silylaryl triflates as aryne precursors permits the formal insertion of arynes into the carbon–carbon σ-bonds of carbonyl compounds under mild conditions, thus dramatically enhancing the selectivity and expanding the substrate scope. In 2005, Tambar and Stoltz reported a mild, direct, and efficient process for the 1,2-acylalkylation of arynes **2** with β-ketoesters **81** (Scheme 7.18) [23]. In this case, arynes **2** are generated in situ from 2-silylaryl triflates **10** in the presence of CsF, and CsF also activates β-ketoester **81** to enolate **83**. Nucleophilic addition of enolate **83** to aryne **2** affords phenyl anion **84**, which cyclizes via intramolecular nucleophilic addition to the carbonyl group to form benzocyclobutoxide **85**. This intermediate undergoes fragmentation followed by protonation to afford 1,2-disubstituted arene **82**. This protocol provides convenient access to a variety of functionalized arenes and benzannulated structures that would otherwise be difficult to obtain. It is noteworthy that cyclic β-ketoesters can be expanded to generate medium-sized carbocycles. In the cases of α-substituted β-ketoesters, α-arylation constitutes a major side reaction.

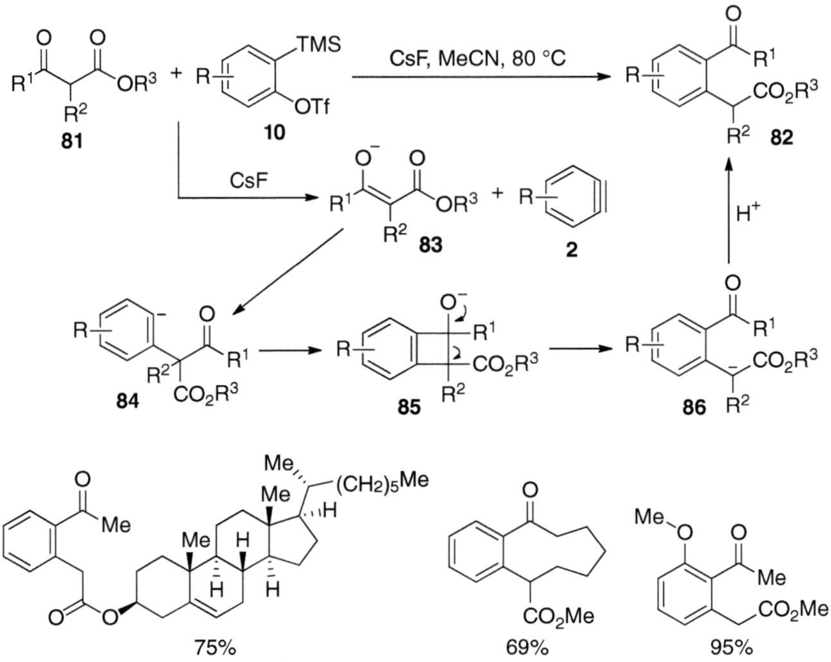

Scheme 7.18 Formal insertion of arynes into β-ketoesters. Source: Based on Tambar and Stoltz [23].

Soon after Stoltz's publication, Yoshida, Kunai, and coworkers reported a similar insertion reaction of arynes (Scheme 7.19) [24]. The carbon–carbon triple bonds of arynes **2** (generated in situ from 2-silylaryl triflates **10** in the presence of KF and 18-crown-6) can be inserted into a range of dialkyl malonates and 1,3-diketones at room temperature. The reaction of arynes with cyclic dialkyl malonates has been demonstrated to be efficient in the construction of large-sized rings. In 2016,

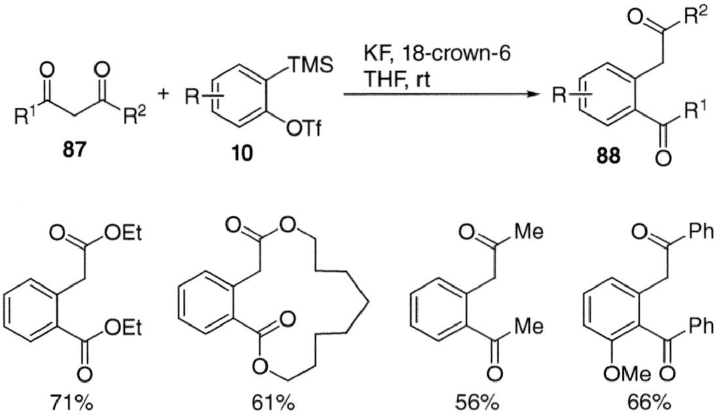

Scheme 7.19 Formal insertion of arynes into malonates and 1,3-diketones. Source: Based on Yoshida et al. [24].

Mehta and coworkers disclosed the synthesis of medium-sized carbocycles from cyclic 1,3-diketones and arynes **2**, generated in situ from 2-silylaryl triflates **10** in the presence of CsF (Scheme 7.20a) [25]. Later, Okuma and coworkers reported the insertion of arynes into trifluoromethylated 1,3-diketones, delivering polysubstituted isocoumarins in moderate-to-good yields (Scheme 7.20b) [26]. In this case, carbanion **93** is generated from 1,3-diketone **91** and benzyne (**2a**) and participates in cyclization to form hemiacetal anion **94**, which extrudes trifluoromethyl anion to afford isocoumarin **92**.

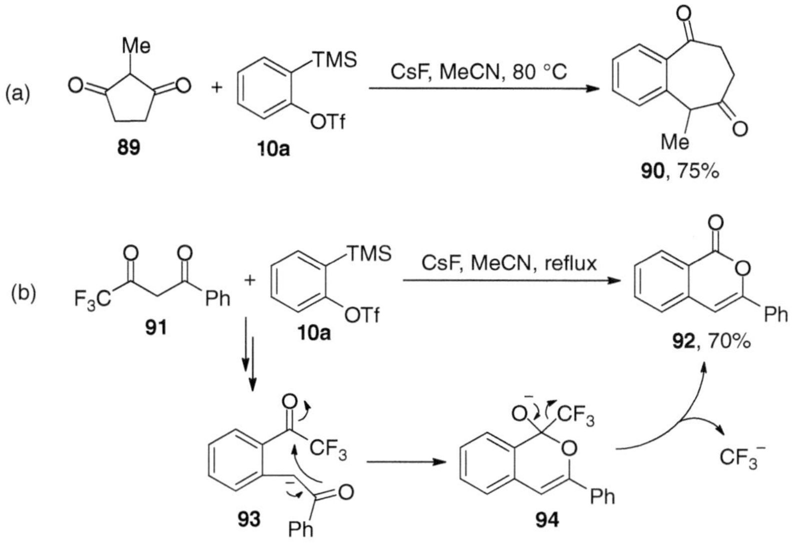

Scheme 7.20 Formal insertion of arynes into 1,3-diketones. (a) Formal insertion of benzyne into a cyclic 1,3-diketone. Source: Based on Samineni et al. [25]. (b) Formal insertion of benzyne into a trifluoromethylated 1,3-diketone. Source: Based on Okuma et al. [26].

With 2-silylaryl triflates as aryne precursors, arynes can be inserted into the carbon–carbon σ-bonds in a number of other carbonyl compounds having at the α-position a cyano (Scheme 7.21a) [27], a nitro (Scheme 7.21b) [28]. phosphonate (Scheme 7.21c) [29], a sulfonyl (Scheme 7.21d) [30], a trifluoromethylthio (Scheme 7.21e) [31], or even an aryl group (Scheme 7.21f) [32]. Moreover, Chandrasekhar and coworkers disclosed the insertion of arynes into *N*-tosylacetimidates or *N*-tosylacetimidamides bearing aryl groups at the α-position (Scheme 7.22) [33]. Hypothetically, these reactions proceed via sequential formal [2+2] cycloaddition and fragmentation.

7.3.2 Formal Insertion of Arynes into Carbon–Heteroatom Bonds

Analogous to the acyl–carbon σ-bonds, a variety of acyl–heteroatom σ-bonds can participate in the 1,2-difunctionalization of arynes through formal insertion involving stepwise [2+2] cycloaddition and fragmentation. In 2005, Liu and Larock

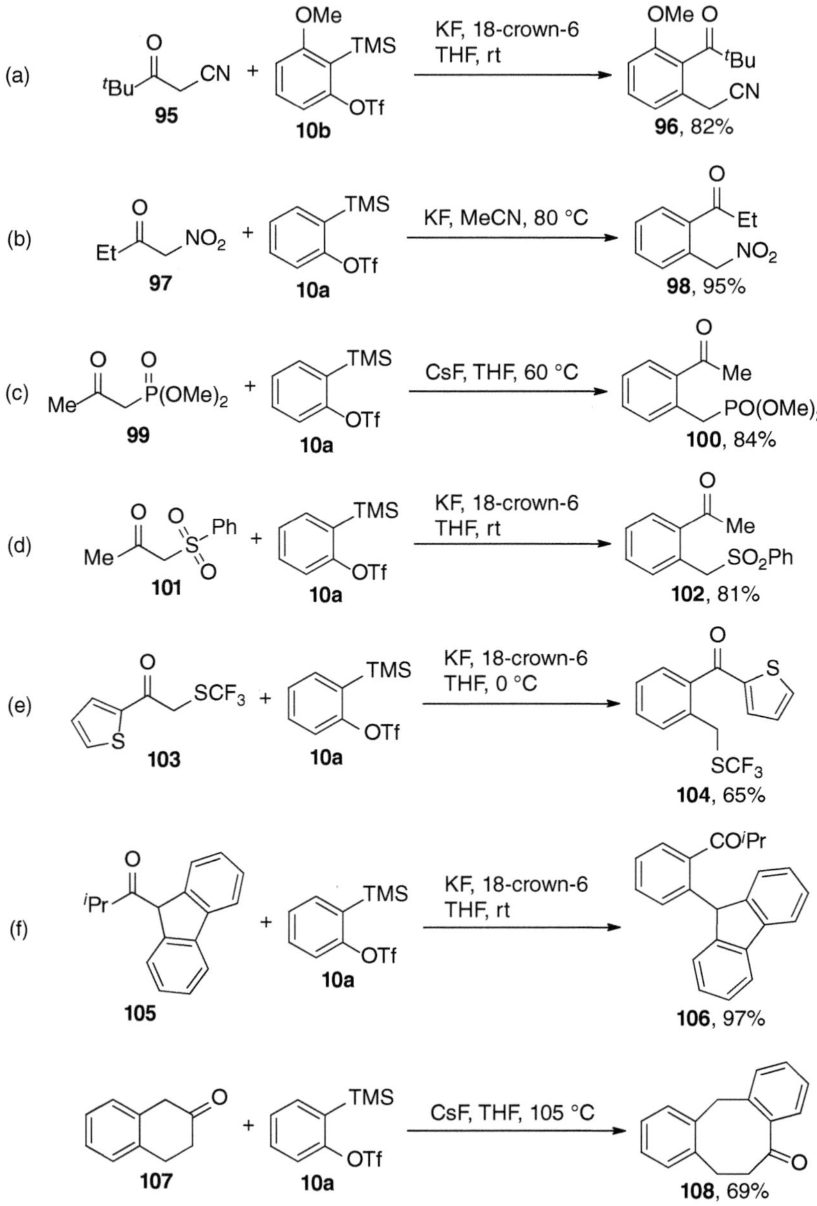

Scheme 7.21 Formal insertion of arynes into α-substituted ketones. (a) Formal insertion of 3-methoxybenzyne into an α-cyano ketone. Source: Based on Yoshida et al. [27]. (b) Formal insertion of benzyne into an α-nitro ketone. Source: Based on Hu and Zheng [28]. (c) Formal insertion of benzyne into dimethyl (2-oxopropyl)phosphonate. Source: Based on Liu et al. [29]. (d) Formal insertion of benzyne into an α-sulfonyl ketone. Source: Based on Zhang et al. [30]. (e) Formal insertion of benzyne into an α-trifluoromethylthio ketone. Source: Based on Ahire et al. [31]. (f) Formal insertion of benzyne into α-aryl ketones. Source: Based on Yoshida et al. [32a]; Yoshida et al. [32b]; Rao et al. [32c]; Rao et al. [32d].

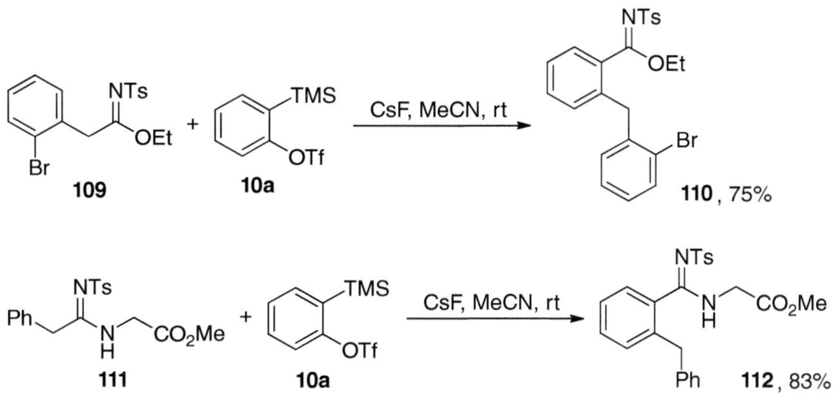

Scheme 7.22 Formal insertion of arynes into *N*-tosylacetimidates or *N*-tosylacetimidamides. Source: Based on Kranthikumar et al. [33].

reported the insertion of arynes into the C—N bond of *N*-aryl trifluoroacetamides using 2-silylaryl triflates as aryne precursors in the presence of CsF [34]. Later, Pintori and Greaney expanded the scope from this specific substrate class to common secondary amides by using tetrabutylammonium triphenyldifluorosilicate (TBAT) for the generation of arynes **2** from 2-silylaryl triflates **10**, delivering 2-aminoaryl ketones in good yields (Scheme 7.23) [35]. Mechanistically, the weakly nucleophilic nitrogen of amide **113** can attack on aryne **2**, and the resulting zwitterion **115** cyclizes via intramolecular nucleophilic addition to the carbonyl group to afford benzazetidinium ion **116**, which participates in fragmentation to afford 2-aminoaryl ketone **114**.

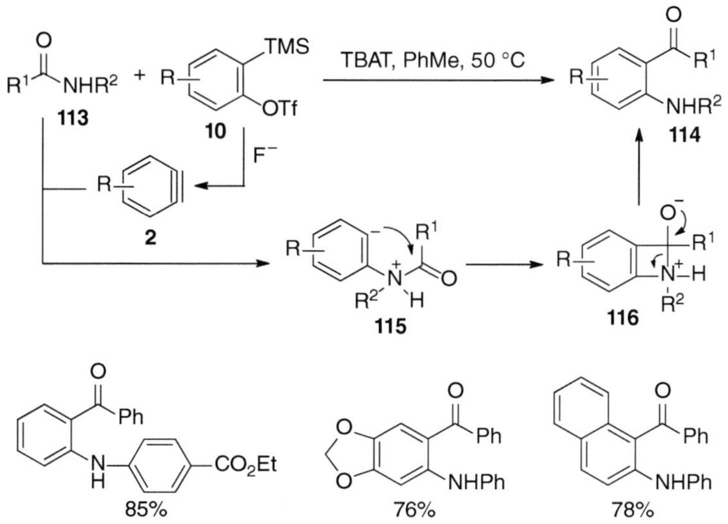

Scheme 7.23 Formal insertion of arynes into secondary amides. Source: Based on Pintori and Greaney [35].

Pintori and Greaney further applied the same reaction conditions to imide **117** and found that the insertion was selective for the amide over the carbamate linkage, delivering Boc-protected amine **118** in 70% yield (Scheme 7.24a) [35]. In 2018, Heretsch, Christmann, and coworkers disclosed a regioselective insertion of arynes into unsymmetric imides (Scheme 7.24b) [36]. The yield of 2-aminoaryl ketone **120** was increased from 25% in batch to 52% when conducted in flow, and a regioselectivity of 2.9 : 1 (**120**:**121**) was observed. The selective insertion of arynes into C—N bonds is also applicable to ureas (Scheme 7.24c) [37] and cyanamides (Scheme 7.24d) [38].

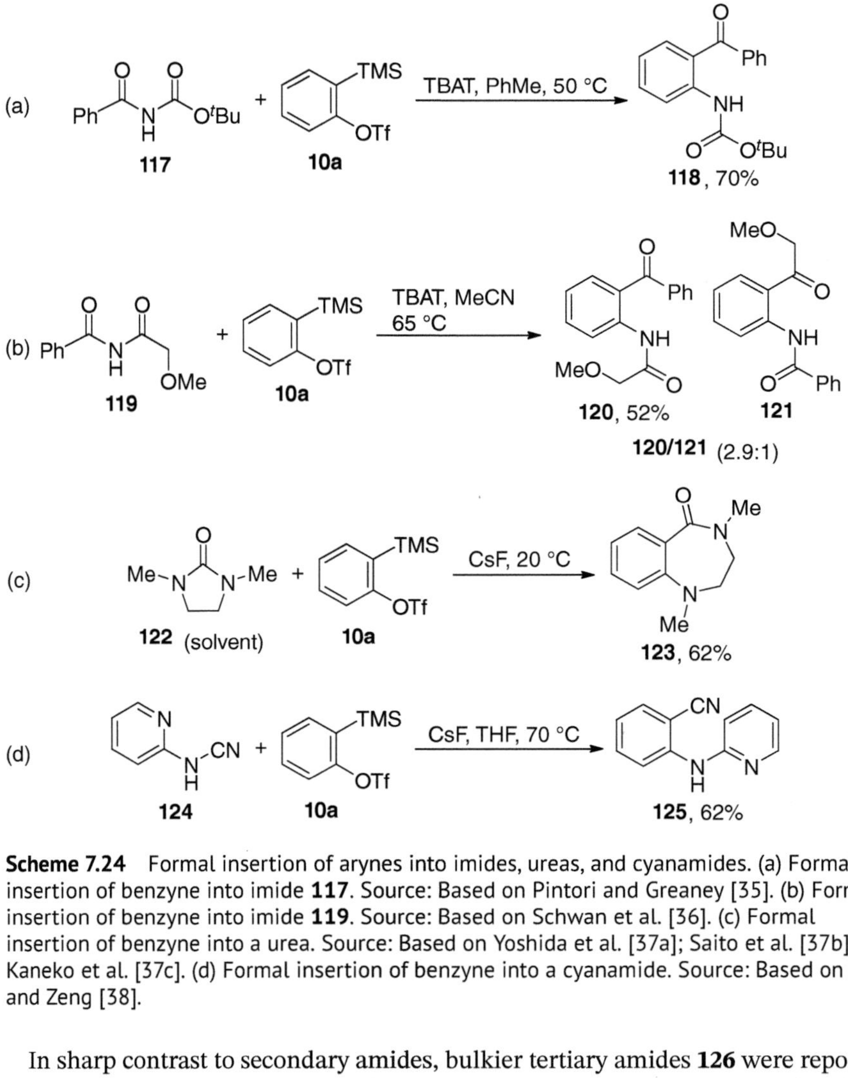

Scheme 7.24 Formal insertion of arynes into imides, ureas, and cyanamides. (a) Formal insertion of benzyne into imide **117**. Source: Based on Pintori and Greaney [35]. (b) Formal insertion of benzyne into imide **119**. Source: Based on Schwan et al. [36]. (c) Formal insertion of benzyne into a urea. Source: Based on Yoshida et al. [37a]; Saito et al. [37b]; Kaneko et al. [37c]. (d) Formal insertion of benzyne into a cyanamide. Source: Based on Rao and Zeng [38].

In sharp contrast to secondary amides, bulkier tertiary amides **126** were reported by Miyabe and coworkers to prefer to attack on arynes as oxygen nucleophiles (Scheme 7.25) [39]. The resulting zwitterion **128** cyclizes via intramolecular nucleophilic addition of the phenyl anion to the iminium ion to afford benzoxetane **129**, which undergoes fragmentation followed by hydrolysis to afford 2-hydroxyaryl ketone **127**. A low yield (34%) was achieved from the reaction of *N*,*N*-dimethylacetamide (**126**, R = Me) due to competitive insertion of the aryne into the C—N bond. The corresponding C—N bond insertion product was obtained in 10% yield.

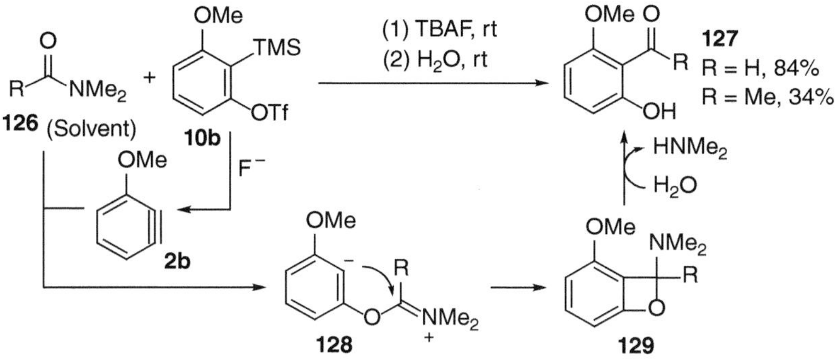

Scheme 7.25 Formal insertion of arynes into tertiary amides. Source: Based on Yoshioka et al. [39].

In addition to acyl–nitrogen σ-bonds, both acyl–oxygen and acyl–halogen σ-bonds can undergo formal insertion of arynes, generated from 2-silylaryl triflates in the presence of a fluoride ion, via stepwise [2+2] cycloaddition and fragmentation (Scheme 7.26). Both carboxylic acids [40] and silyl-protected carboxylic acids [41] are converted to carboxylate ions under the basic reaction conditions prior to the

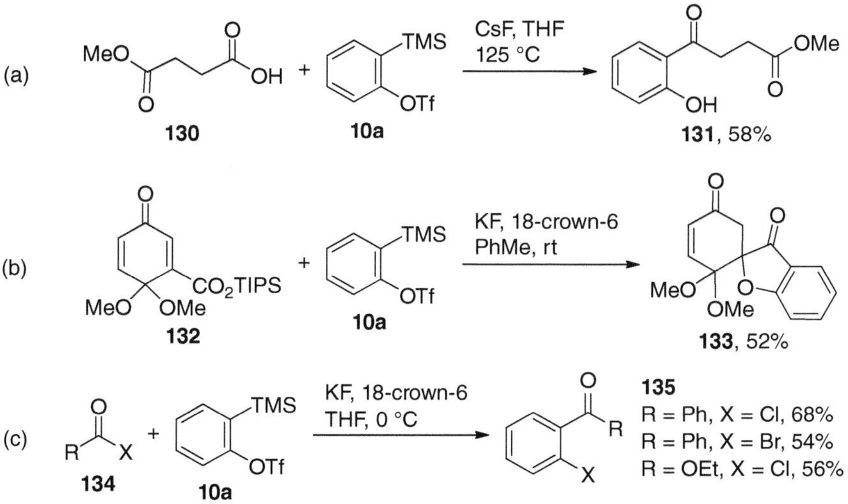

Scheme 7.26 Formal insertion of arynes into acyl–oxygen and acyl–halogen σ-bonds. (a) Formal insertion of benzyne into a carboxylic acid. Source: Based on Dubrovskiy and Larock [40]. (b) Formal insertion of benzyne into a silyl-protected carboxylic acid. Source: Based on Dhokale and Mhaske [41]. (c) Formal insertion of benzyne into an acid chloride, an acid bromide, or a chloroformate. Source: Based on Yoshida et al. [42].

aryne insertion (Scheme 7.26a,b). Arynes can be inserted into the acyl–halogen σ-bonds in acid chlorides, acid bromides, and chloroformates, providing convenient access to various 2-haloaryl ketones (Scheme 7.26c) [42].

Alternatively, the aryne insertion is applicable to the carbon–heteroatom σ-bonds in active methylene compounds bearing heteroatom–oxygen double bonds, which might play roles similar to the carbonyl groups as shown above. In 2005, Yoshida, Kunai, and coworkers reported the 1,2-carbophosphinylation of arynes with cyanomethyldiphenylphosphine oxide (Scheme 7.27a) [43], and in 2017, Xu and coworkers disclosed the 1,2-carbosulfonylation of arynes with aryl trifluoromethyl sulfones (Scheme 7.27b) [44]. Moreover, the C—S bonds of sulfonium salts can also participate in aryne insertion for the synthesis of functionalized aryl thioethers (Scheme 7.27c) [45].

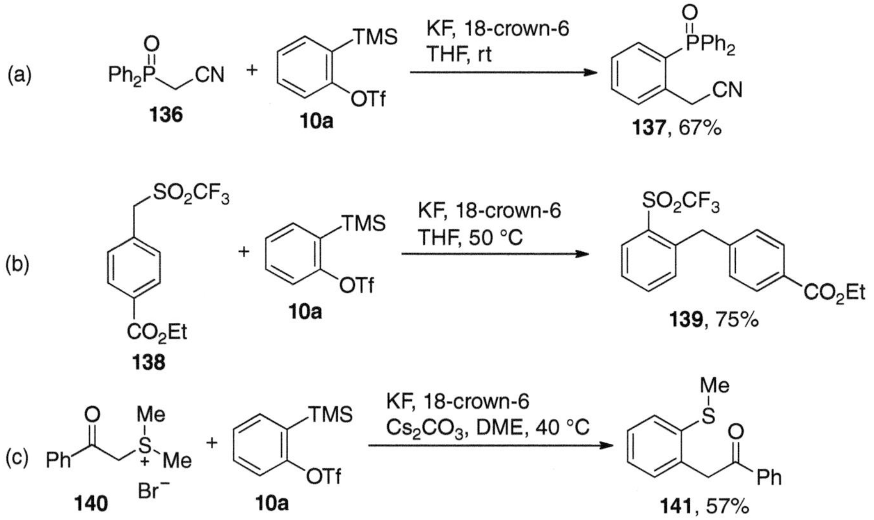

Scheme 7.27 Formal insertion of arynes into C–P and C–S bonds. (a) Formal insertion of benzyne into cyanomethyldiphenylphosphine oxide. Source: Based on Yoshida et al. [43]. (b) Formal insertion of benzyne into a sulfone. Source: Based on Zhao et al. [44]. (c) Formal insertion of benzyne into a sulfonium salt. Source: Based on Ahire et al. [45a]; Li et al. [45b].

In addition to four-membered rings, five-membered rings have also been proposed to be formed as intermediates via stepwise [3+2] cycloaddition in the insertion of arynes into carbon–heteroatom σ-bonds. In 2011, Guitián and coworkers reported a highly chemo- and regioselective insertion of arynes **2** (generated in situ from 2-silylaryl triflates **10** in the presence of CsF) into the C—O bond of ethoxyacetylene (**142**), delivering 2-ethynylaryl ethers **143** in moderate-to-good yields (Scheme 7.28) [46]. According to density functional theory (DFT) calculations, the reaction proceeds via the nucleophilic addition of the triple bond of ethoxyacetylene (**142**) to aryne **2**. The resulting zwitterion **144** cyclizes to afford benzofuran ylide **145**, which evolves through simultaneous [1,2]-hydrogen migration and ring opening via C—O bond cleavage to afford 2-ethynylaryl ether **143**.

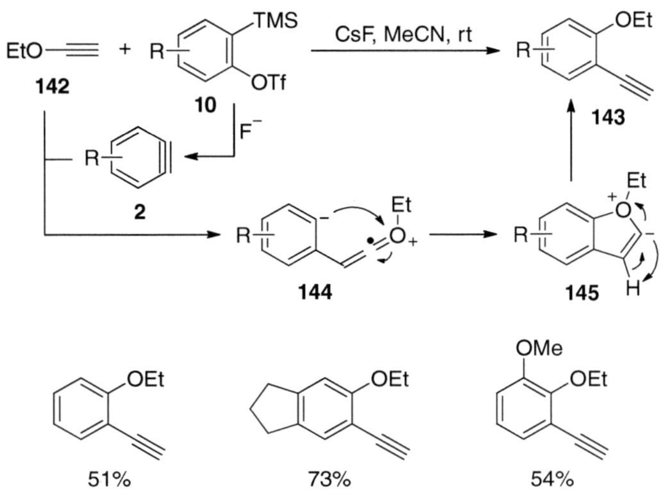

Scheme 7.28 Formal insertion of arynes into ethoxyacetylene. Source: Based on Łaczkowski et al. [46].

The insertion of arynes into carbon–heteroatom σ-bonds may proceed without involving stepwise cycloaddition and fragmentation. In 2019, Jones and coworkers reported the insertion of arynes **2** (generated in situ from 2-silylaryl triflates **10** in the presence of CsF) into the C—N bonds of aminals **146** (Scheme 7.29) [47]. The aryne insertion results in the ring expansion of imidazolidines **146**, leading to the formation of benzodiazepines **147**. The yields are moderate due to competitive formation

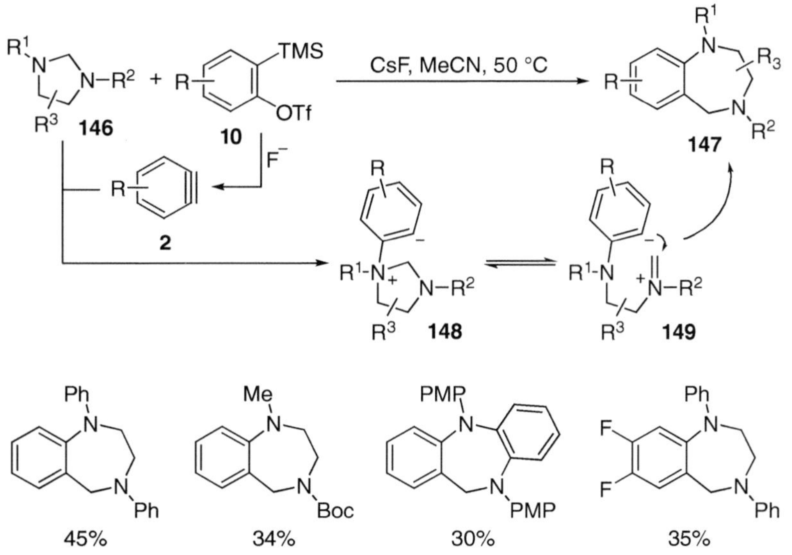

Scheme 7.29 Formal insertion of arynes into imidazolidines. Source: Based on Yang et al. [47].

of linear 1,2-ethylenediamines as byproducts. Mechanistically, the reaction proceeds via initial *N*-arylation of imidazolidine **146**. The resulting zwitterion **148** may be in equilibrium with the ring-opened form **149**, which cyclizes via intramolecular nucleophilic addition of the phenyl anion to the iminium ion to afford benzodiazepine **147**.

7.3.3 Formal Insertion of Arynes into Heteroatom–Heteroatom Bonds

Analogous to the C—N bonds of amides, the P—N bonds of phosphoryl amides **150** participate in the aryne insertion to afford (2-aminoaryl)phosphine oxides **151** in moderate-to-good yields (Scheme 7.30) [48]. In this case, arynes **2** are generated in situ from 2-silylaryl triflates **10** in the presence of KF and 18-crown-6, and the addition of Cs_2CO_3 facilitates the nucleophilic addition of phosphoryl amides **150** to arynes **2**. Mechanistically, *N*-arylation of phosphoryl amide **150** with aryne **2** in the presence of Cs_2CO_3 generates aryl anion **153**, which cyclizes via intramolecular nucleophilic addition to the P=O bond to afford benzannulated four-membered ring **154**. This intermediate undergoes fragmentation followed by protonation to afford (2-aminoaryl)phosphine oxide **151**.

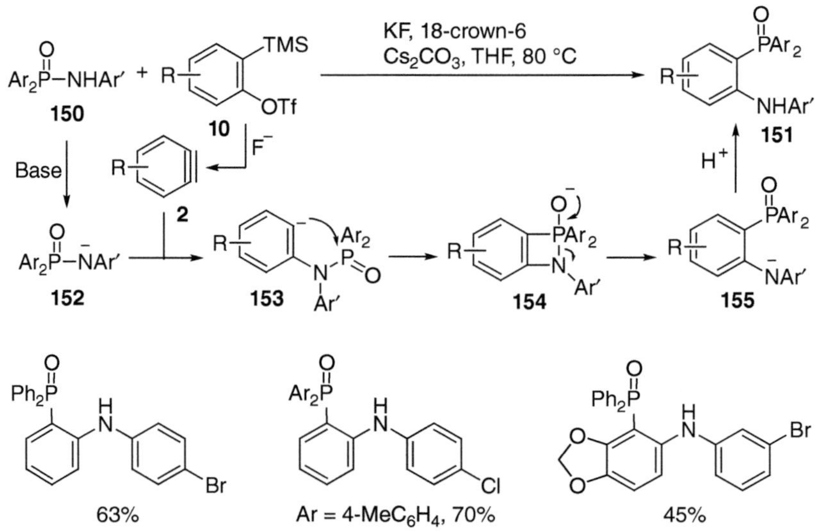

Scheme 7.30 Formal insertion of arynes into phosphoryl amides. Source: Based on Shen et al. [48].

Similar aryne insertion can occur with the P—O bonds of organophosphorus acids **156**, delivering (2-hydroxyaryl)phosphine oxides, (2-hydroxyaryl)phosphinates, and (2-hydroxyaryl)phosphonates in moderate-to-good yields (Scheme 7.31a) [49]. Moreover, the aryne insertion is applicable to the S—N bonds of trifluoromethanesulfinamides bearing S=O bonds (Scheme 7.31b) [34], which might play roles similar to the P=O bonds as shown above.

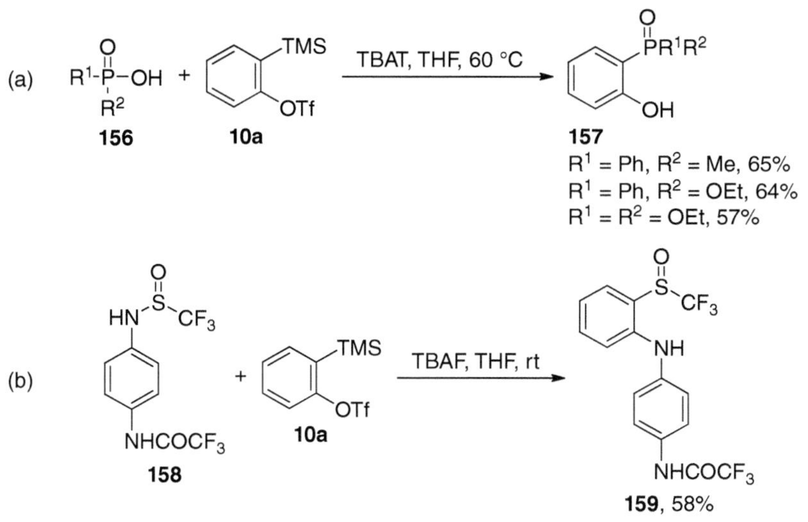

Scheme 7.31 Formal insertion of arynes into organophosphorus acids and sulfinamides. (a) Formal insertion of benzyne into organophosphorus acids. Source: Based on Qi et al. [49]. (b) Formal insertion of benzyne into a sulfinamide. Source: Based on Liu and Larock [34].

Without the aid of heteroatom–oxygen double bonds, certain heteroatom–heteroatom σ-bonds can participate in the aryne insertion reactions. Obviously, a four-membered ring is unlikely to be formed as a rearrangement precursor in such aryne insertion reactions. In 2015, Chen and Wang disclosed the insertion of arynes **2** (generated in situ from 2-silylaryl triflates **10** in the presence of KF and 18-crown-6) into the N—O bonds of *N*-hydroxyindolin-2-ones **160**, delivering sterically hindered 2-aminophenols **161** in moderate-to-good yields (Scheme 7.32) [50]. Mechanistically, the reaction proceeds via the addition of *N*-hydroxyindolin-2-one **160** to aryne **2** followed by a [1,3]-rearrangement, which is possibly facilitated by the introduction of the amide group to enhance the electrophilic nature of nitrogen, in comparison to a hydroxyamine. The structural rigidity of cyclic amides might impede the π–π stacking conformation required for the transition state of a [3,3]-rearrangement.

A number of other heteroatom–heteroatom σ-bonds have been reported to participate in the aryne insertion reactions, probably involving a similar [1,3]-rearrangement (Scheme 7.33). The insertion of arynes into N—Si [51], N—S [52], P—P [53], S—S [54], S—Sn [55], S—Bi [56], Se—Se [57], and F—Sn bonds [58] affords a wide variety of 1,2-difunctionalized arenes, which are difficult to be prepared by alternative methods.

7.3.4 Vicinal Carbon–Carbon/Carbon–Carbon Bond-Forming Reactions of Arynes

In addition to formal insertion of arynes into σ-bonds, there are a number of other methods toward vicinal formation of two σ-bonds for the 1,2-difunctionalization

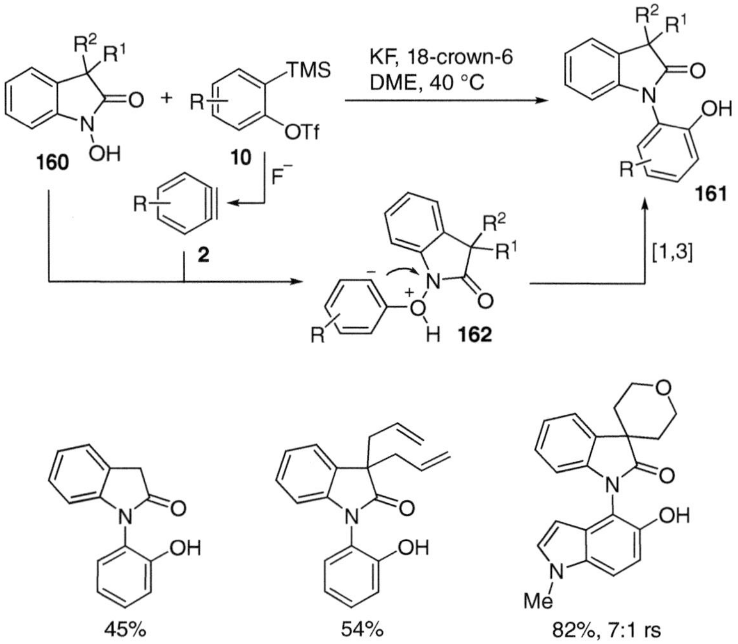

Scheme 7.32 Formal insertion of arynes into N–O bonds. Source: Based on Chen and Wang [50].

of arynes involving molecular rearrangements. In 2018, Li, Chen, and coworkers reported the reaction of arynes **2** (generated in situ from 2-silylaryl triflates **10** in the presence of CsF) with β-(2-isocyanophenyloxy)acrylates **179** (or analogues), delivering 1,2-disubstituted arenes **180** bearing a benzoxazole ring and an alkenyl group in good yields (Scheme 7.34) [59]. Mechanistically, this domino reaction proceeds via the nucleophilic addition of the isocyano group of β-(2-isocyanophenylox)yacrylate **179** (or its analogue) to aryne **2**, affording zwitterion **181**. Intramolecular Michael addition of the aryl anion to the activated carbon–carbon double bond takes place to afford intermediate **182**, which undergoes ring opening via C—O bond cleavage followed by cyclization to afford 1,2-disubstituted arene **180**. An alternative pathway involves first formation of the benzoxazole moiety from zwitterion **181**. Then, the resulting intermediate **184** undergoes [1,4]-migration of the alkenyl group via intermediacy of a five-membered ring to afford the final product **180**.

Taking advantage of the nucleophilic addition of isocyanides to arynes, Biju and coworkers have developed a three-component reaction of isocyanides **186**, arynes **2** (generated in situ from 2-silylaryl triflates **10** in the presence of CsF), and carbon dioxide, delivering *N*-substituted phthalimides **187** in moderate-to-good yields (Scheme 7.35) [60]. Mechanistically, the nucleophilic addition of isocyanide **186** to aryne **2** generates zwitterion **188**, which undergoes a stepwise [3+2] cycloaddition with carbon dioxide leading to the formation of imino isobenzofuranone **190**.

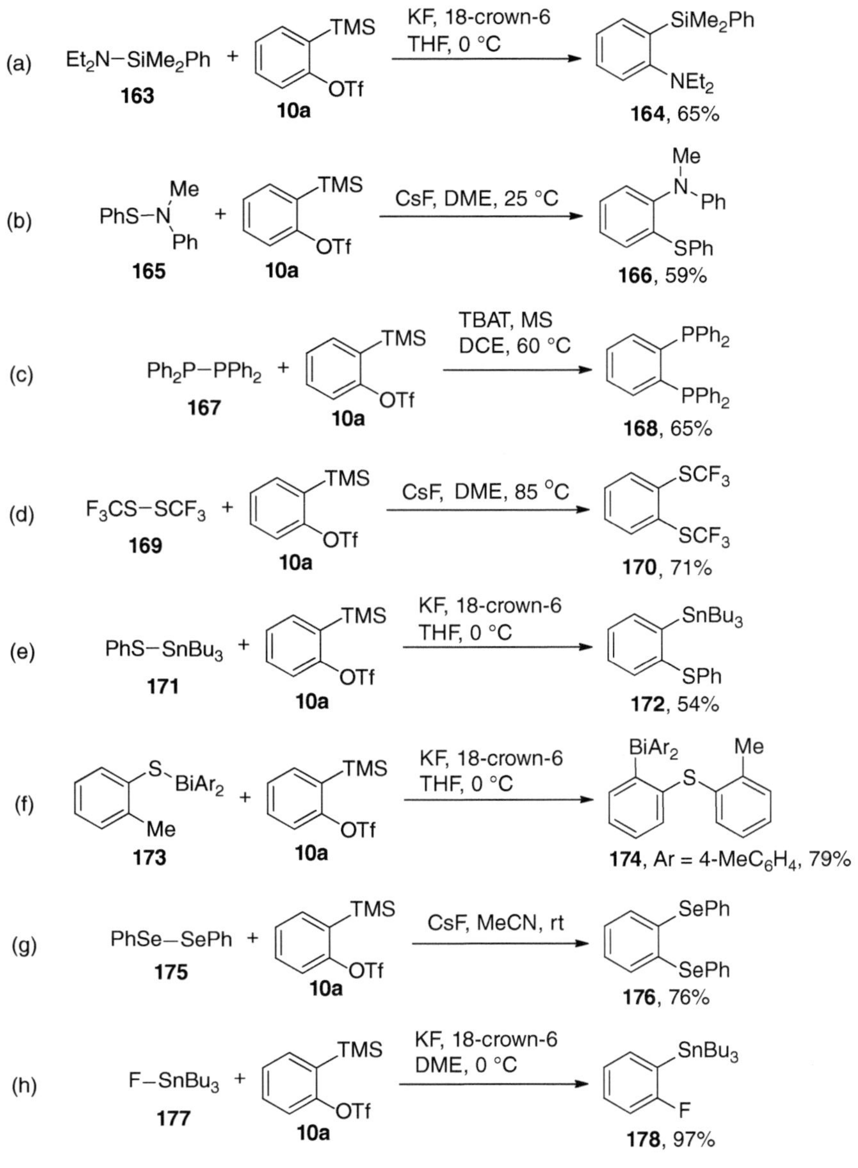

Scheme 7.33 Formal insertion of arynes into various heteroatom–heteroatom bonds. (a) Formal insertion of benzyne into a N–Si bond. Source: Based on Yoshida et al. [51a]; Yoshida et al. [51b]. (b) Formal insertion of benzyne into a N–S bond. Source: Based on Gaykar et al. [52]. (c) Formal insertion of benzyne into a P–P bond. Source: Based on Okugawa et al. [53]. (d) Formal insertion of benzyne into a S–S bond. Source: Based on Mesgar and Daugulis [54]. (e) Formal insertion of benzyne into a S–Sn bond. Source: Based on Yoshida et al. [55]. (f) Formal insertion of benzyne into a S–Bi bond. Source: Based on Chen and Murafuji [56]. (g) Formal insertion of benzyne into a Se–Se bond. Source: Based on Toledo et al. [57]. (h) Formal insertion of benzyne into a F–Sn bond. Source: Based on Yoshida et al. [58].

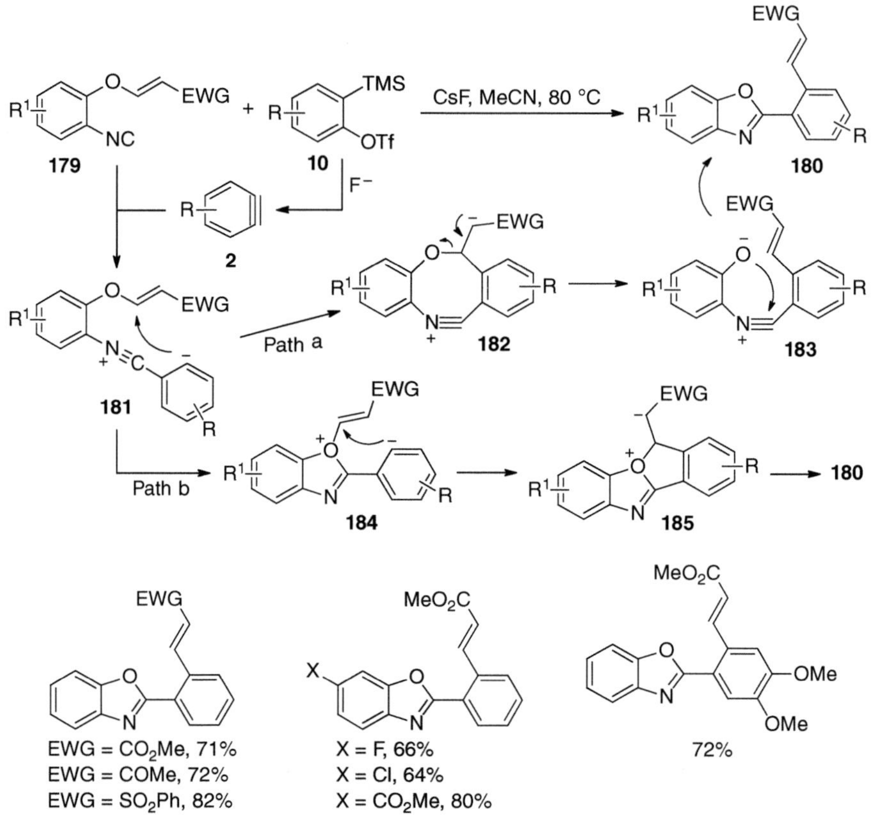

Scheme 7.34 Reaction of arynes with β-(2-isocyanophenyloxy)acrylates (or analogues). Source: Based on Su et al. [59].

A fluoride-induced rearrangement transforms this intermediate to *N*-substituted phthalimide **187** via intermediacy of acylfluoride **191**.

In 2019, Yennam and coworkers disclosed the reaction of 2-arylidene-1,3-indandiones **192** (R = electron-rich or neutral group) with benzyne (**2a**) (generated in situ from 2-silylaryl triflate **10a** in the presence of CsF) under aerobic conditions, delivering dibenzo[*a*,*c*]anthracene-9,14-diones **193** in moderate yields (Scheme 7.36) [61]. Mechanistically, the reaction proceeds via [4+2]-cycloaddition between 2-arylidene-1,3-indandione **192** and benzyne (**2a**), and the resulting spirocycle **194** undergoes intramolecular nucleophilic addition to the carbonyl group to form the cyclopropane ring in strained tricyclic alcohol **195**, which upon retro-aldol reaction and air oxidation affords dibenzo[*a*,*c*]anthracene-9,14-dione **193**.

7.3.5 Vicinal Carbon–Carbon/Carbon–Heteroatom Bond-Forming Reactions of Arynes

The charge-accelerated aza-Claisen rearrangement constitutes one powerful protocol for the introduction of vicinal carbon–carbon and carbon–heteroatom bonds into

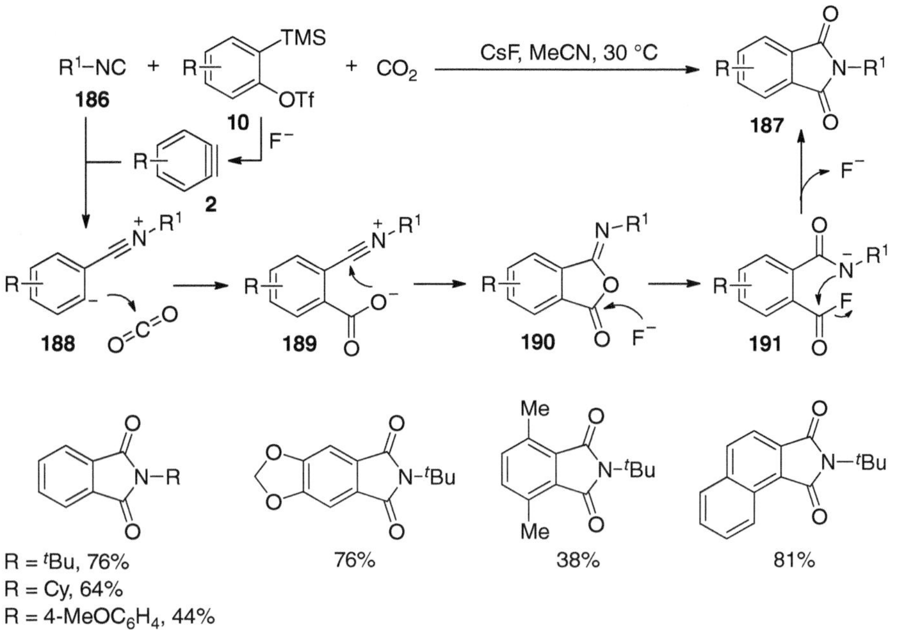

Scheme 7.35 Three-component reaction of arynes, isocyanides, and carbon dioxide. Source: Based on Kaicharla et al. [60].

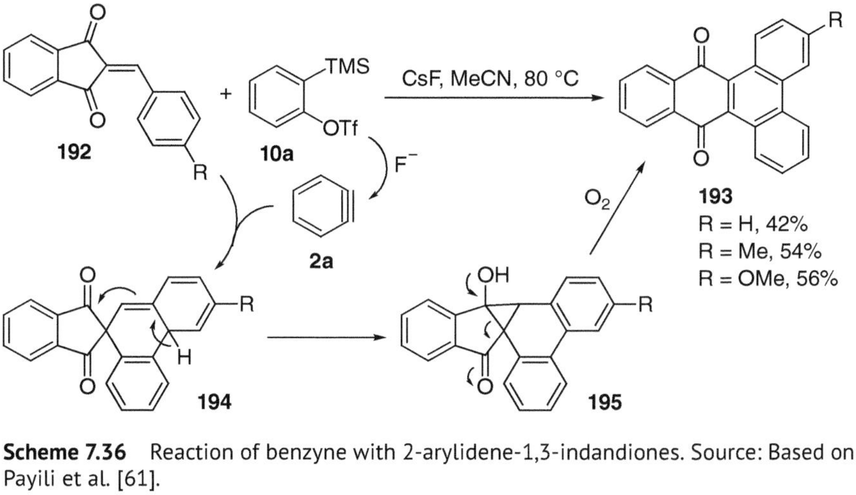

Scheme 7.36 Reaction of benzyne with 2-arylidene-1,3-indandiones. Source: Based on Payili et al. [61].

arynes. In 2009, Greaney and coworkers disclosed the aryne-triggered aza-Claisen rearrangement of tertiary allylic amines **196**, delivering 2-allylanilines **197** in moderate-to-excellent yields (Scheme 7.37) [62]. This reaction avoids high reaction temperature or the use of stoichiometric amounts of Lewis acids as required for conventional aza-Claisen rearrangements. Mechanistically, the reaction is initiated by the nucleophilic addition of tertiary allylic amine **196** to aryne **2** (generated in situ from 2-silylaryl triflate **10** in the presence of CsF), generating zwitterion **198**, which is protonated by the solvent to afford quaternary ammonium salt **199**. This salt undergoes an aza-Claisen rearrangement to afford 2-allylaniline **197**. Notably, the tertiary allylic amines can be prepared in situ from secondary allylic amines and arynes prior to the aryne-triggered aza-Claisen rearrangement [63]. Greaney and coworkers further applied the aryne-triggered aza-Claisen rearrangement of tertiary allylic amines to the construction of unsaturated nine- and ten-membered cyclic amines through ring expansion of five- and six-membered cyclic amines bearing a vinyl group at the amine α-C position [62]. Later, Saito and coworkers extended the ring expansion to 2-vinylazetidines, delivering 1-benzazocine derivatives in moderate-to-good yields [64].

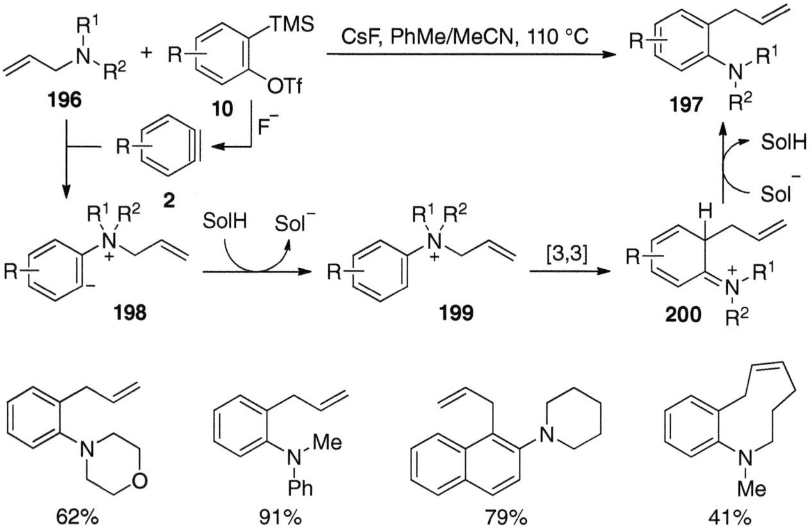

Scheme 7.37 Aryne-triggered aza-Claisen rearrangement of tertiary allylic amines. Source: Based on Cant et al. [62].

The aryne-triggered propargyl Claisen rearrangement was recently reported by Palakodety and coworkers (Scheme 7.38) [65]. 3-Alkynyl-3-hydroxyindolin-2-ones **201** and 2-silylaryl triflates **10** were treated with CsF, 18-crown-6, Cs_2CO_3, and molecular sieves, and the reaction afforded 3-benzofuranylindolin-2-ones **202** in moderate-to-good yields. Mechanistically, deprotonation of propargylic alcohol **201** by the base generates propargyloxide **203**, which undergoes nucleophilic addition to aryne **2** to afford aryl anion **204**. This intermediate participates in a propargyl Claisen rearrangement to afford allene **205**, which undergoes 5-exo-dig cyclization followed by protonation to afford 3-benzofuranylindolin-2-one **202**.

Another type of [3,3]-sigmatropic rearrangement might occur in the reaction of arynes with *N*-hydroxyindoles (Scheme 7.39) [50]. In sharp contrast to the aryne insertion reaction of *N*-hydroxyindolin-2-ones involving a [1,3]-rearrangement (Scheme 7.32) [50], this reaction afforded C3-arylated indoles in moderate yields via a [3,3]-sigmatropic rearrangement.

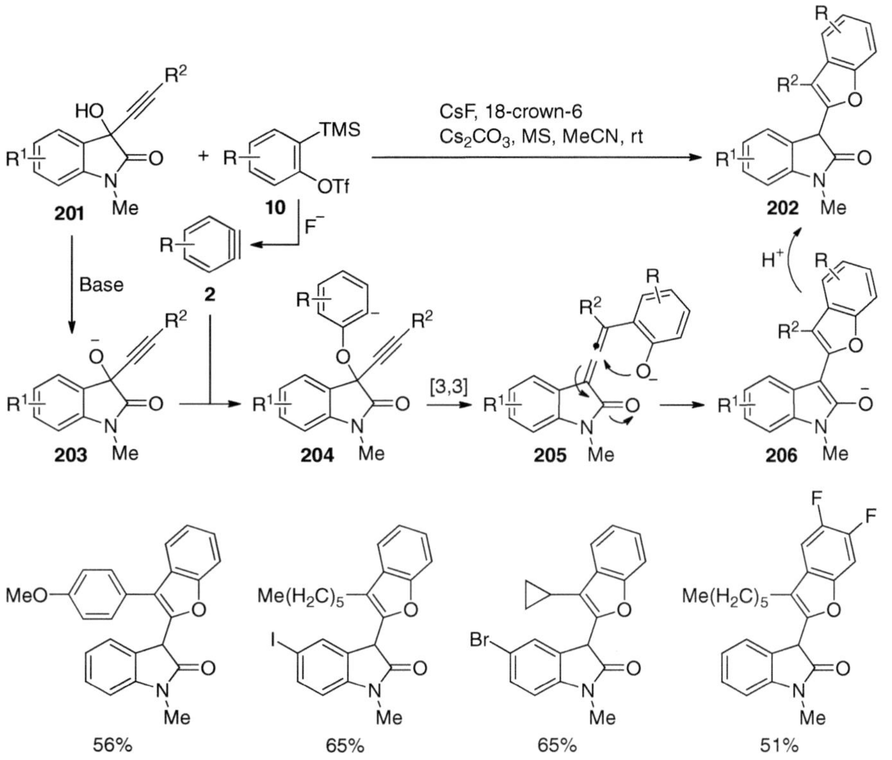

Scheme 7.38 Aryne-triggered propargyl Claisen rearrangement. Source: Based on Kalvacherla et al. [65].

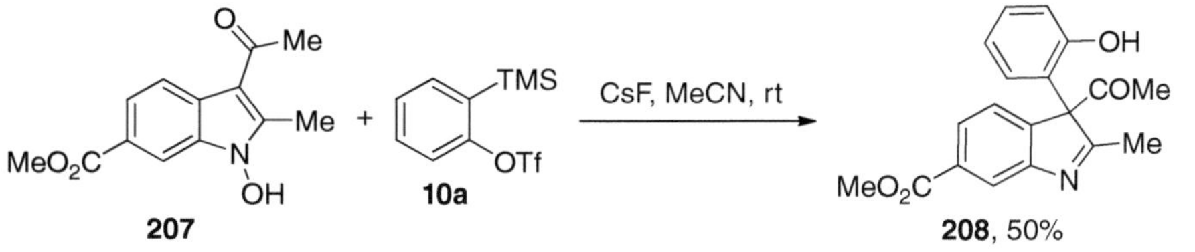

Scheme 7.39 Reaction of arynes with *N*-hydroxyindoles. Source: Based on Chen and Wang [50].

Cycloaddition or formal cycloaddition of arynes followed by rearrangements constitutes another class of protocols for the introduction of vicinal carbon–carbon and carbon–heteroatom bonds into arynes. In 2013, Studer and coworkers reported the reaction of arynes **2** (generated in situ from 2-silylaryl triflates **10** in the presence of CsF) with nitrosoarenes **209**, delivering carbazoles **210** in moderate-to-good yields (Scheme 7.40) [66]. Mechanistically, the reaction proceeds via the [2+2]-cycloaddition between nitrosoarene **209** and aryne **2**, and the resulting benzannulated four-membered ring **211** undergoes ring opening via N—O bond cleavage to afford *o*-quinone derivative **212**. This intermediate cyclizes via intramolecular addition to the carbonyl group to afford intermediate **213**. The C—O bond of intermediate **213** is cleaved by a nucleophile, and subsequent protonation eventually affords carbazole **210**.

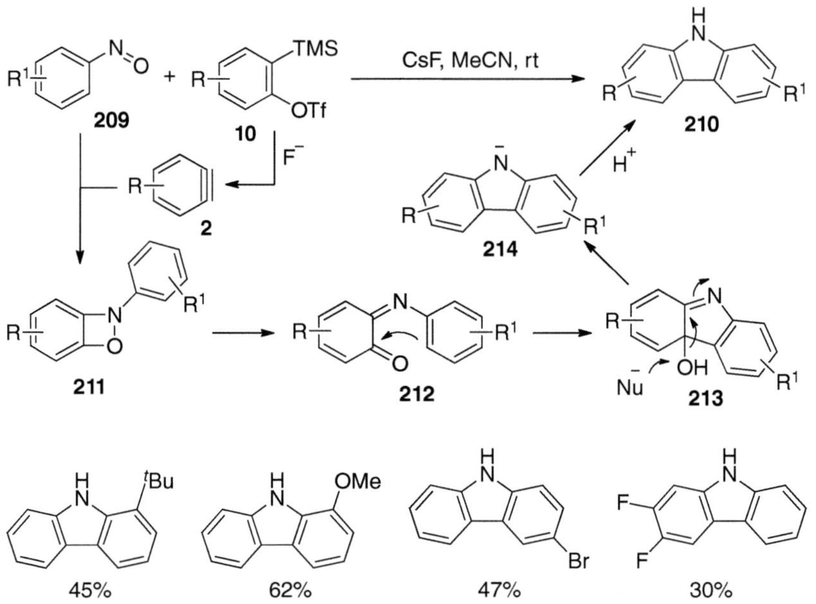

Scheme 7.40 Reaction of arynes with nitrosoarenes. Source: Based on Chakrabarty et al. [66].

In 2016, Hoye and coworkers disclosed the reaction of aromatic thioamides **215** with aryne **218** (generated in situ by the hexadehydro-Diels–Alder cycloisomerization of triyne **216**), delivering structurally diverse dihydrobenzothiazines **217** in a regio- and diastereoselective manner (Scheme 7.41) [67]. It is postulated that the reaction proceeds via the initial [2+2] cycloaddition between thioamide **215** and aryne **218** followed by ring opening of the resulting benzothietene **219** via C—S bond cleavage to form *o*-thiolatoaryliminium **221**. This intermediate undergoes intramolecular [1,3]-hydrogen migration to give iminium zwitterion **222**, which cyclizes via the addition of the thiolate anion to the iminium ion to afford dihydrobenzothiazine **217**.

In addition to [2+2] cycloaddition, [3+2] cycloaddition occurs prior to a variety of rearrangements in the vicinal carbon–carbon and carbon–heteroatom bond-forming reactions of arynes. In 1974, Yamazaki and coworkers disclosed the reaction of benzyne (**2a**) (generated from benzenediazonium-2-carboxylate (**224**)) with diazo compounds **224**, delivering 1-acyl-3-phenylindazole **225** in varied yields (Scheme 7.42) [68]. Mechanistically, the reaction proceeds via initial [3+2] cycloaddition between diazo compound **223** and benzyne (**2a**). The resulting adduct **226** undergoes [1,3]-migration and/or successive [1,2]-migrations of the acyl group to afford indazole **225**.

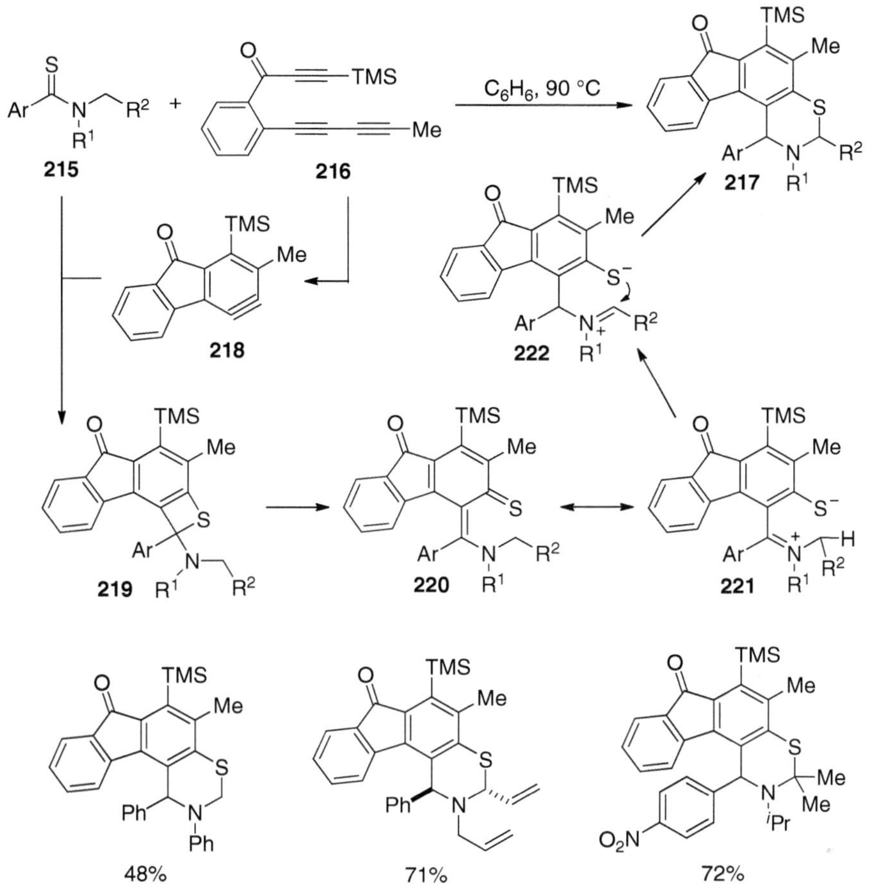

Scheme 7.41 Reaction of arynes with thioamides. Source: Based on Palani et al. [67].

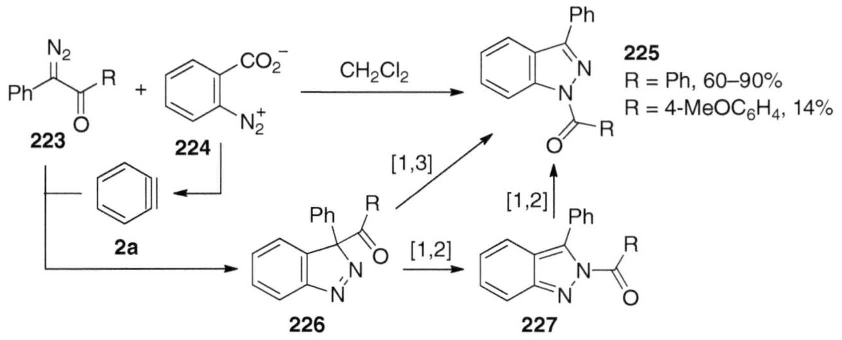

Scheme 7.42 Reaction of benzyne with diazo compounds. Source: Based on Yamazaki et al. [68].

In 2006, Larock and coworkers disclosed the reaction of arynes **2** (generated in situ from 2-silylaryl triflates **10** in the presence of CsF) with pyridine *N*-oxides **228** in a regioselective manner, delivering 3-(2-hydroxyphenyl)pyridines **229** in good yields (Scheme 7.43) [69]. Mechanistically, the reaction proceeds via [3+2] cycloaddition between aryne **2** and pyridine *N*-oxide **228**, and the resulting oxazolopyridine **230** undergoes a simultaneous rearrangement to form the cyclopropane ring of intermediate **231**. This intermediate opens up via C—C bond cleavage to afford 3-(2-hydroxyphenyl)pyridine **229**. 2-Substituted quinoline *N*-oxides and acridine *N*-oxides can undergo a similar rearrangement reaction with arynes to afford 3-(2-hydroxyaryl)quinolines and 4-arylacridines, respectively [70]. The reaction of 3-silylarynes (generated from 2,6-disilylaryl triflates) with pyridine *N*-oxides results in the regioselective incorporation of a hydroxy group on its C2 position, and subsequent one-pot triflation/tosylation allows the preparation of 2,3-aryne precursors [71].

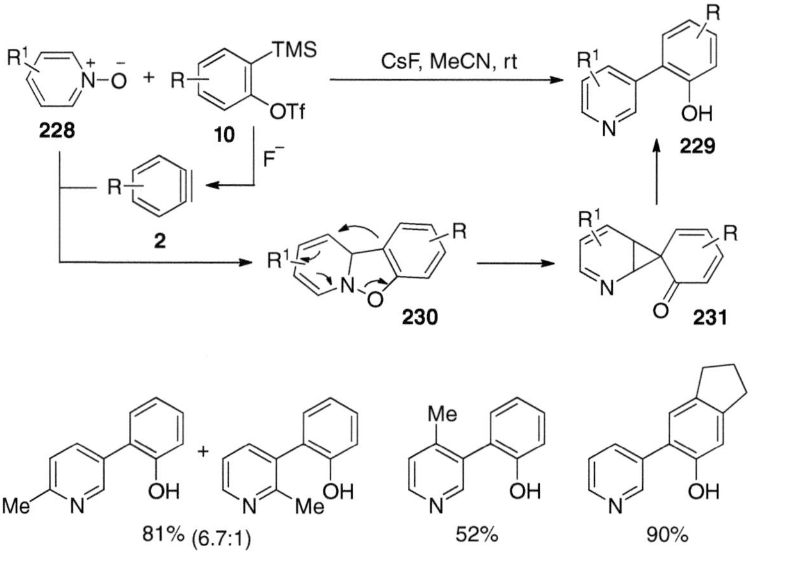

Scheme 7.43 Reaction of arynes with pyridine *N*-oxides. Source: Based on Raminelli et al. [69].

Stepwise formation of five-membered rings as rearrangement precursors also contributes to the vicinal incorporation of carbon–carbon and carbon–heteroatom bonds into arynes. In 2018, Shamsabadi and Chudasama disclosed the reaction of arynes **2** with acyl hydrazides **232**, delivering 2-hydrazobenzophenones **233** in moderate-to-good yields (Scheme 7.44) [72]. Mechanistically, nucleophilic addition of deprotonated acyl hydrazide **234** to aryne **2** affords aryl anion **235**, which cyclizes via intramolecular nucleophilic addition to the carbonyl group to afford 2,3-dihydro-1*H*-indazole **236**. This intermediate participates in fragmentation followed by protonation to afford 2-hydrazobenzophenone **233**.

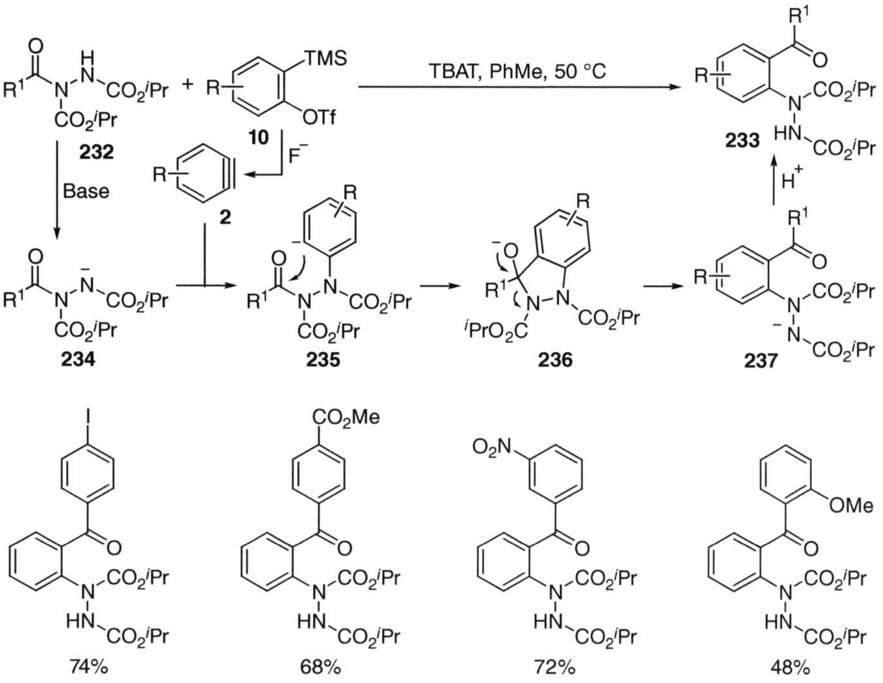

Scheme 7.44 Reaction of arynes with acyl hydrazides. Source: Based on Shamsabadi and Chudasama [72].

In 2016, Voskressensky and coworkers disclosed the reaction of arynes **2** (generated in situ from 2-silylaryl triflates **10** in the presence of CsF) with 1-aroyl-3,4-dihydroisoquinolines **238**, delivering indoloisoquinolinones **239** in good-to-excellent yields (Scheme 7.45) [73]. Mechanistically, nucleophilic addition of dihydroisoquinoline **238** to aryne **2** generates zwitterion **240**, which cyclizes via intramolecular nucleophilic addition of the aryl anion to the carbonyl group to form cyclic iminium **241**. [1,2]-migration of the aryl group to the iminium carbon affords indoloisoquinolinone **239**.

In 2016, Greaney and coworkers disclosed the reaction of arynes **2** (generated in situ from 2-silylaryl triflates **10** in the presence of KF and 18-crown-6) with aryl sulfonamides **242**, delivering sterically hindered 2-amino biaryls **243** in moderate-to-good yields (Scheme 7.46) [74]. Mechanistically, nucleophilic addition of deprotonated aryl sulfonamide **244** to aryne **2** generates aryl anion **245**. This intermediate undergoes sequential Smiles-type ipso substitution via σ-complex **246**, extrusion of sulfur dioxide, and protonation to afford 2-amino biaryl **243**.

The aryne-triggered Smiles rearrangement has also been taken advantage of in multicomponent reactions. In 2015, Biju and coworkers disclosed the three-component reaction of arynes **2** (generated in situ from 2-silylaryl triflates **10** in the presence of KF and 18-crown-6), tertiary aromatic amines **248**, and carbonyl compounds **249**, delivering 2-aminobenzyl aryl ethers **250** in moderate-to-good

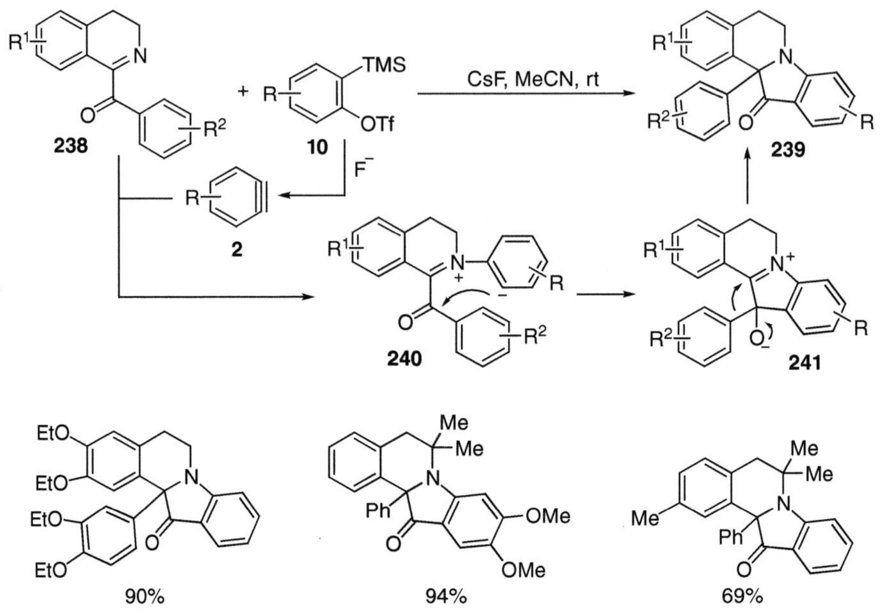

Scheme 7.45 Reaction of arynes with 1-aroyl-3,4-dihydroisoquinolines. Source: Based on Varlamov et al. [73].

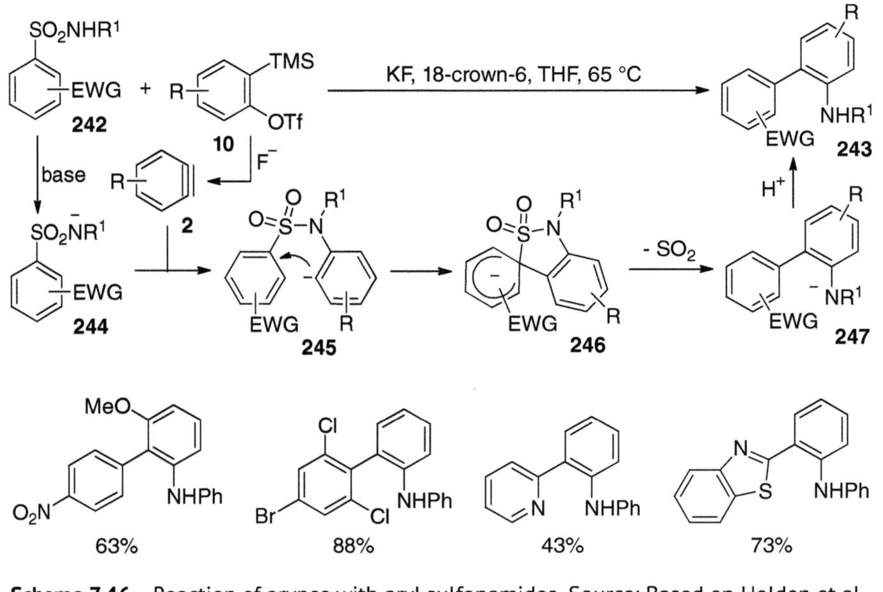

Scheme 7.46 Reaction of arynes with aryl sulfonamides. Source: Based on Holden et al. [74].

yields (Scheme 7.47) [75]. Mechanistically, the reaction proceeds via the nucleophilic addition of tertiary aromatic amine **248** to aryne **2**, and the resulting aryl anion **251** undergoes nucleophilic addition to carbonyl compound **249** to afford zwitterion **252**. An intramolecular nucleophilic aromatic substitution reaction occurs in zwitterion **252** to afford 2-aminobenzyl aryl ether **250** via σ-complex **253**. Later, Okuma and coworkers reported that cyclic tertiary aromatic amines, such as indolines and tetrahydroquinolines, can participate in the three-component reaction with arynes and aldehydes to afford medium-sized dibenzannulated heterocycles such as dibenzo[1,5]oxazonines and dibenzo[1,5]oxazecines [76].

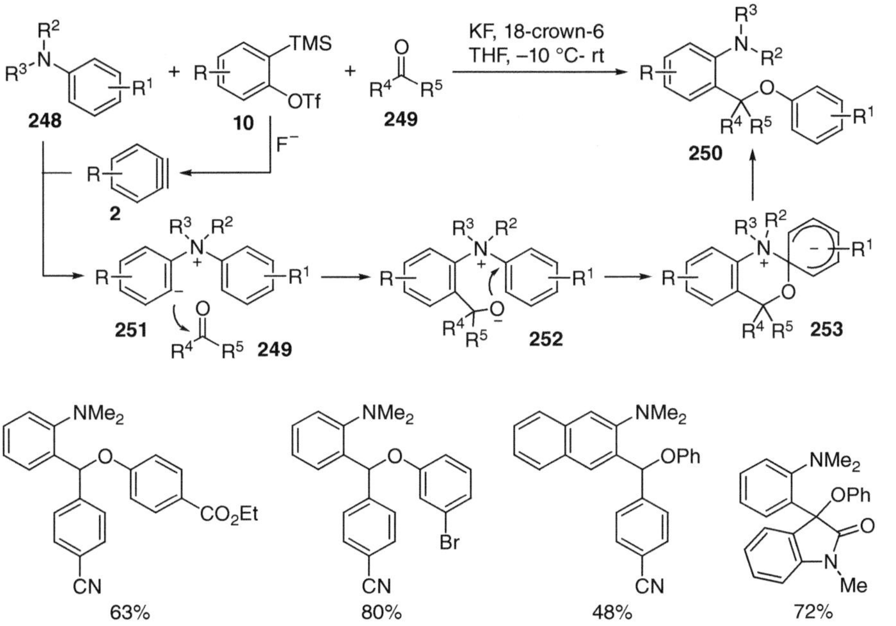

Scheme 7.47 Three-component reaction of arynes, tertiary aromatic amines, and carbonyl compounds. Source: Based on Bhojgude et al. [75].

Biju and coworkers further developed the three-component reaction of arynes **2** (generated in situ from 2-silylaryl triflates **10** in the presence of KF and 18-crown-6), tertiary aromatic amines **248**, and carbon dioxide (Scheme 7.48) [77]. When the tertiary aromatic amine partners possess an electron-withdrawing group on the aromatic ring, the reaction affords aryl 2-aminobenzoates **254** via an aryl migration. Otherwise, the reaction affords alkyl 2-aminobenzoates **255** via an alkyl migration. Mechanistically, the reaction proceeds via the nucleophilic addition of tertiary aromatic amine **248** to aryne **2**, and the resulting aryl anion **251** undergoes nucleophilic addition to carbon dioxide to afford zwitterion **256**. When the R^1 group on the aromatic ring is an electron-withdrawing group, zwitterion **256** undergoes intramolecular nucleophilic aromatic substitution to afford aryl 2-aminobenzoate **254** via σ-complex **257**. Otherwise, zwitterion **256** undergoes *O*-alkylation to afford alkyl 2-aminobenzoate **255**.

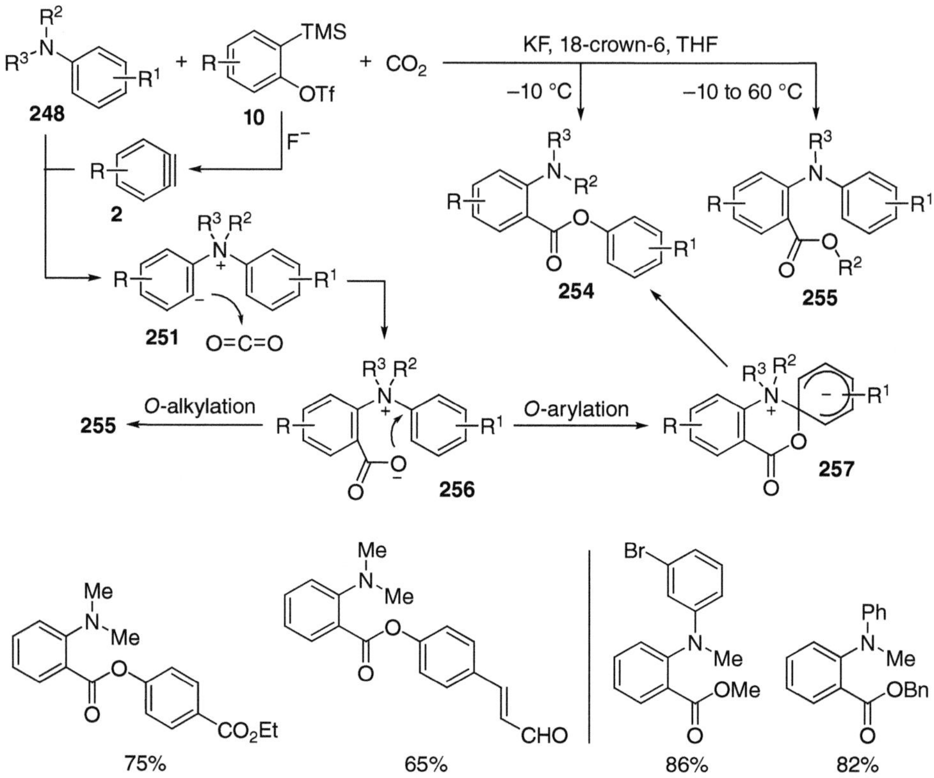

Scheme 7.48 Three-component reaction of arynes, tertiary aromatic amines, and carbon dioxide. Source: Based on Bhojgude et al. [77].

7.3.6 Vicinal Carbon–Heteroatom/Carbon–Heteroatom Bond-Forming Reactions of Arynes

Cycloaddition between arynes and heteroatom–heteroatom bonds followed by rearrangements constitutes an important strategy for the formation of vicinal carbon–heteroatom and carbon–heteroatom bonds with arynes. In 2017, Li and Studer reported the oxythiolation of arynes **2** (generated in situ from 2-silylaryl triflates **10** in the presence of CsF) with aryl vinyl sulfoxides **258**, delivering 2-sulfanylaryl vinyl ethers **259** with complete stereospecificity in moderate-to-good yields (Scheme 7.49) [78]. The use of water as an additive suppresses fluoride-mediated disproportionation of the sulfoxide to the corresponding sulfide and sulfone. Mechanistically, the reaction proceeds via a concerted [2+2] cycloaddition between aryl vinyl sulfoxide **258** and aryne **2**, and the resulting benzannulated four-membered ring **260** undergoes ring opening via S—O bond cleavage to afford zwitterion **261**. The sulfur to oxygen migration of the vinyl group proceeds via intramolecular oxa-Michael addition followed by elimination, affording 2-sulfanylaryl vinyl ether **259**.

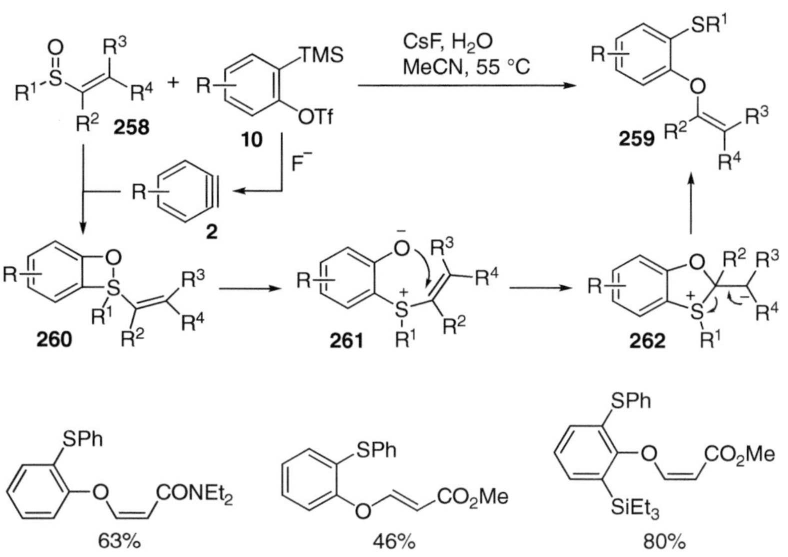

Scheme 7.49 Reaction of arynes with vinyl sulfoxides. Source: Based on Li and Studer [78].

In the absence of a vinyl group, the aryl group of an aryl sulfoxide undergoes sulfur to oxygen migration in a similar reaction pathway when treated with an aryne generated from the corresponding 2-silylaryl triflate under appropriate conditions. In 2017, Hosoya and coworkers disclosed the oxythiolation of arynes with diaryl sulfoxides, delivering 2-sulfanylaryl aryl ethers in moderate-to-good yields (Scheme 7.50a) [79]. Notably, the more electron-deficient aryl group prefers to participate in the sulfur to oxygen migration in the reaction of an unsymmetrical diaryl sulfoxide. Moreover, Hosoya and coworkers observed similar sulfur to nitrogen migration of the aryl group in the thioamination of arynes with sulfilimines, delivering 2-sulfanylanilines in moderate-to-good yields (Scheme 7.50b) [80].

In 2014, Hosoya and coworkers disclosed the reaction of 3-triflyloxybenzyne (**2c**) (generated from 1,3-bis(triflyloxy)-2-iodobenzene (**268**) in the presence of a Grignard reagent) with nucleophiles **267** such as triphenylphosphine and diphenyl sulfide, delivering zwitterionic triflones **269** in good yields (Scheme 7.51) [81]. The reaction proceeds via initial regioselective nucleophilic addition to 3-triflyloxybenzyne (**2c**) at the site distal from the triflyloxy group. The resulting *ortho*-triflyloxyaryl anion **270** undergoes a thia-Fries rearrangement to afford triflone **269**.

7.4 Rearrangements Involved in the 1,2,3-Trifunctionalization of Arynes

1,2,3-Trifunctionalization of arynes involving rearrangements is powerful for the regioselective synthesis of polysubstituted arenes. In 2016, Li and coworkers disclosed the three-component reaction of arynes **2** (generated in situ from

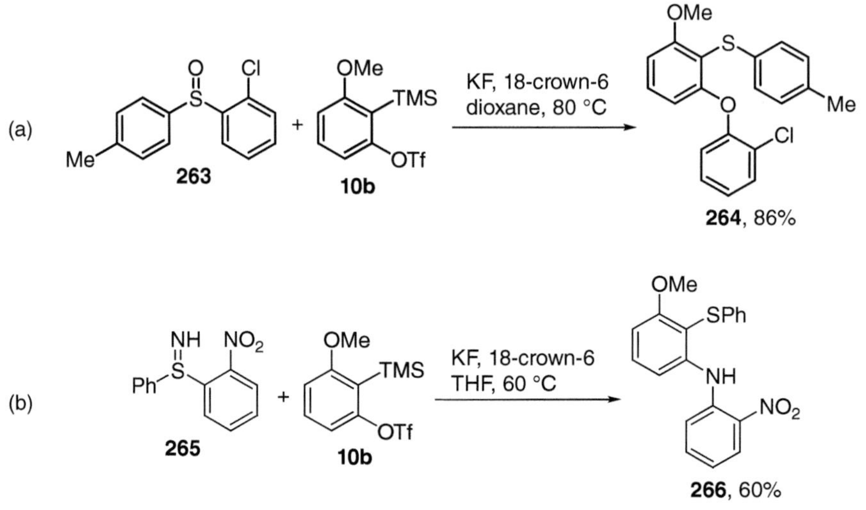

Scheme 7.50 Reaction of arynes with diaryl sulfoxides or sulfilimines. (a) Reaction of 3-methoxybenzyne with a diaryl sulfoxide. Source: Based on Matsuzawa et al. [79]. (b) Reaction of 3-methoxybenzyne with a sulfilimine. Source: Based on Yoshida et al. [80].

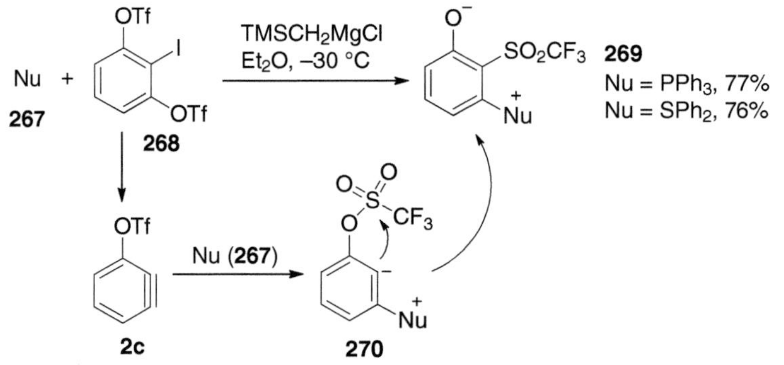

Scheme 7.51 Synthesis of triflones. Source: Based on Yoshida et al. [81].

2-silylaryl triflates **10** in the presence of CsF), allyl aryl sulfoxides **271**, and alkyl bromides **272**, delivering 1,2,3-trifunctionalized arenes **273** in moderate-to-good yields (Scheme 7.52) [82]. The 1,2,3-trifunctionalization of arynes **2** with aryl allyl sulfoxides **271** was achieved through formation of C—S, C—O, and C—C bonds on three consecutive positions of the aromatic ring. Mechanistically, the reaction proceeds via a formal [2+2] cycloaddition between allyl aryl sulfoxide **271** and aryne **2**, and the resulting benzannulated four-membered ring **274** undergoes the sulfur to oxygen migration of the ally group to afford ylide **275**. This intermediate undergoes an oxonium Claisen rearrangement to afford intermediate **276**. Opening of the four-membered ring via S—O bond cleavage affords phenoxide **277**, which is alkylated with alkyl bromide **272** to afford 1,2,3-trifunctionalized arene **273**.

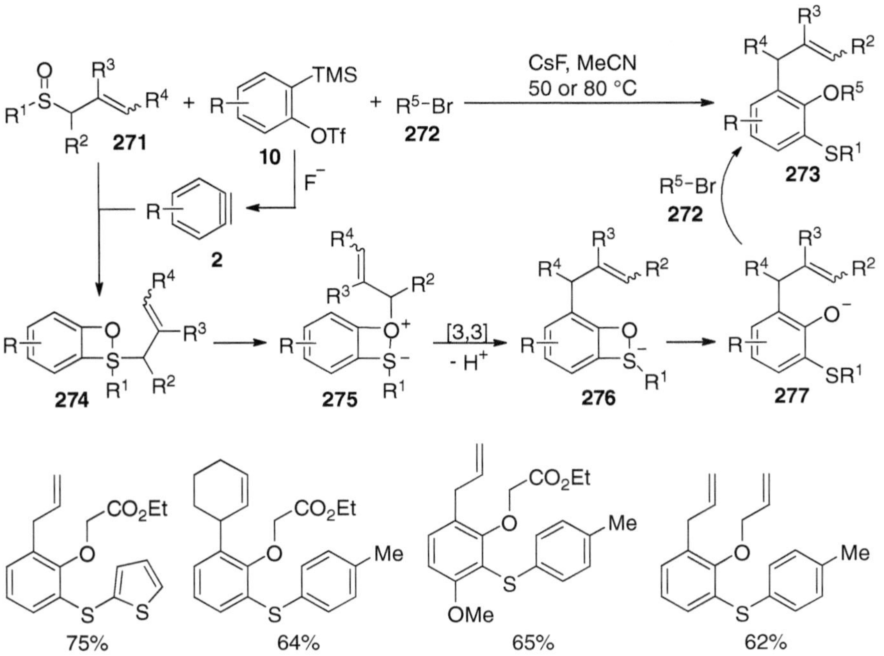

Scheme 7.52 Three-component reaction of arynes, allyl sulfoxides, and alkyl bromides. Source: Based on Li et al. [82].

7.5 Rearrangements Involved in the Multicomponent Reactions with Two or More Aryne Molecules

A variety of active intermediates are generated during the reactions involving aryne-triggered rearrangements, and they might be trapped by the highly electrophilic arynes. To date, two or three aryne molecules have been reported to participate in a number of multicomponent reactions involving rearrangements. These multicomponent reactions with two or three aryne molecules provide convenient access to complex organic molecules.

7.5.1 Three-Component Reactions with Two Aryne Molecules

In the aryne insertion into certain active methylene compounds, the rearrangement intermediates can be trapped by one extra aryne molecule to afford diarylmethanes. In 2007, Kunai and coworkers disclosed the coupling of two molar amounts of arynes **2** (generated in situ from 2-silylaryl triflates **10** in the presence of KF and 18-crown-6) with nitriles **278** bearing an electron-withdrawing group at the α-position, delivering functionalized diarylmethanes **279** in moderate-to-good yields (Scheme 7.53) [83]. Mechanistically, initial nucleophilic addition of deprotonated nitrile **280** to aryne **2** affords aryl anion **281**. Subsequent intramolecular nucleophilic substitution at the cyano moiety generates benzyl anion **282**, which undergoes nucleophilic addition to another molecule of aryne **2** followed by protonation to afford diarylmethane **279**.

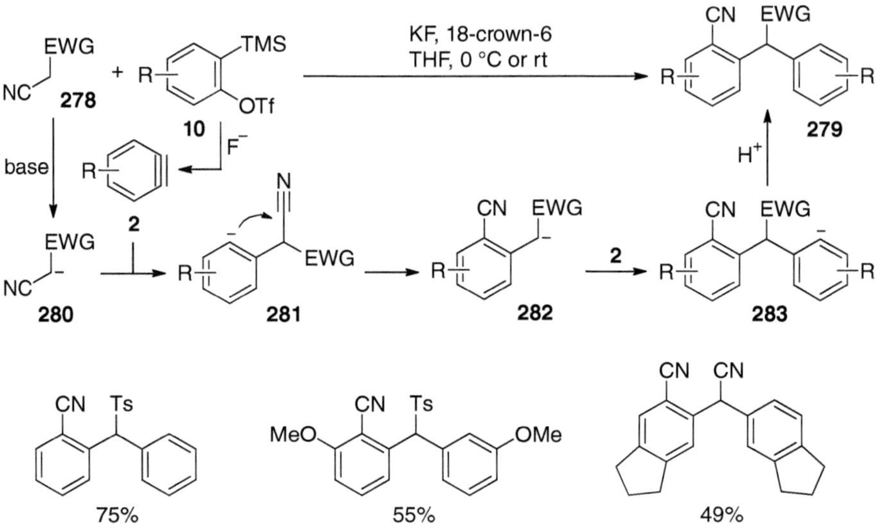

Scheme 7.53 Three-component reaction of nitriles with two aryne molecules. Source: Based on Yoshida et al. [83].

In 2017, Gogoi and coworkers disclosed the three-component reaction of 4-hydroxycoumarins **284** with two aryne molecules (generated in situ from 2-silylaryl triflates **10** in the presence of CsF), delivering 3-substituted isocoumarins **285** in moderate-to-good yields (Scheme 7.54) [84]. Mechanistically, a formal [2+2] cycloaddition between the deprotonated form of 4-hydroxycoumarin **284** and aryne **2** affords benzannulated cyclobutoxide **286**, which undergoes ring opening via C—O bond cleavage and adds to another molecule of aryne **2** to afford benzocyclobutanone **287**. This intermediate opens up via C—C bond cleavage to afford ketene **288**, which undergoes ring closure to afford isocoumarin **285**.

In 2011, Biswas and Greaney disclosed the three-component reaction of thioureas **289** with two aryne molecules (generated in situ from 2-silylaryl triflates **10** in the presence of CsF), involving aryne insertion and *S*-arylation (Scheme 7.55) [85]. The aryne inserts into the thiourea C=S bond, in contrast to C—N bond insertion for related ureas. A range of amidine products **290** were obtained in moderate-to-good yields. Mechanistically, the reaction proceeds via the initial nucleophilic addition of thiourea **289** as a sulfur nucleophile to aryne **2**, and the resulting zwitterion **291** cyclizes via intramolecular nucleophilic addition of the aryl anion to the C=N bond to generate benzannulated four-membered ring **292**. This intermediate collapses via C—S bond cleavage to give thiophenol **293**, which undergoes *S*-arylation via nucleophilic addition to a second molecule of aryne **2** to afford amidine **290**.

In 2017, Yao and coworkers disclosed the three-component reaction of oximes **294** with two aryne molecules (generated in situ from 2-silylaryl triflates **10** in

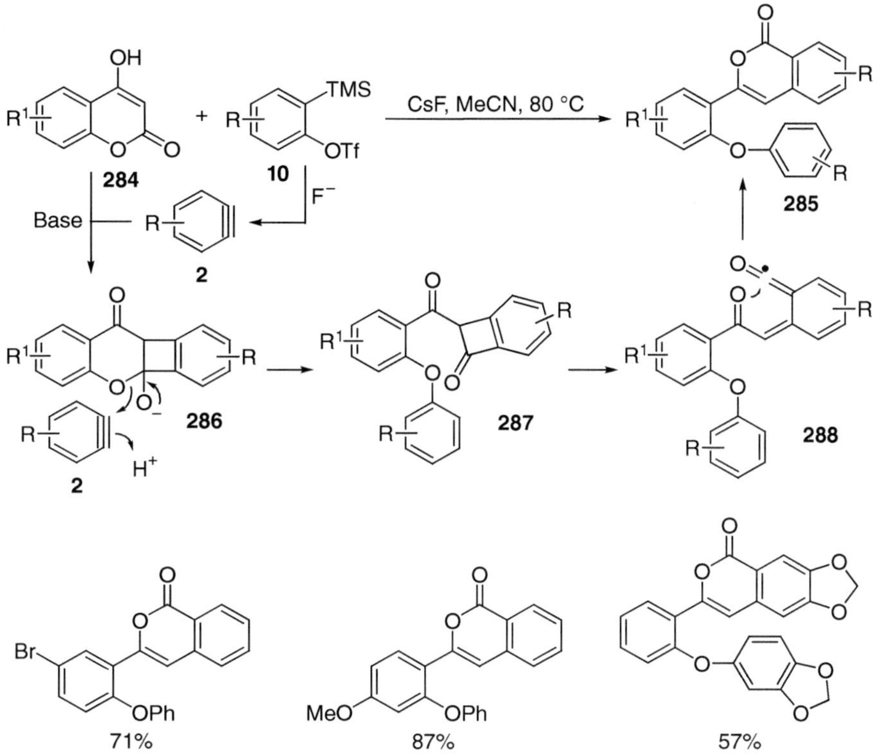

Scheme 7.54 Three-component reaction of 4-hydroxycoumarins with two aryne molecules. Source: Based on Neog et al. [84].

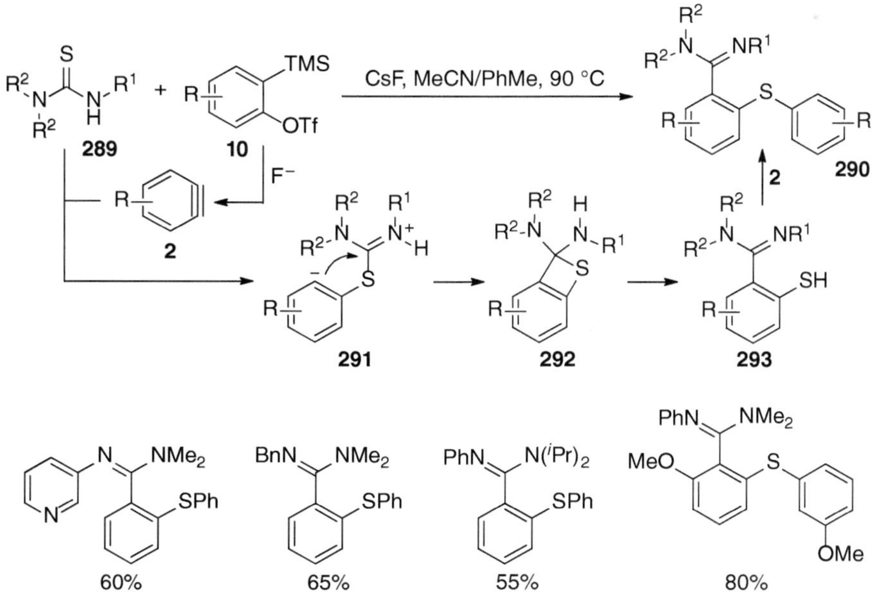

Scheme 7.55 Three-component reaction of thioureas with two aryne molecules. Source: Based on Biswas and Greaney [85].

the presence of CsF), delivering structurally diverse dihydrobenzo[*d*]oxazoles **295** in moderate-to-excellent yields (Scheme 7.56) [86]. Mechanistically, the reaction proceeds via the nucleophilic addition of the nitrogen of oxime **294** to aryne **2**. The aryl anion in the resulting zwitterion **296** abstracts a proton from the hydroxyl group to generate nitrone **297**. [3+2] cycloaddition between nitrone **297** and a second molecule of aryne **2** affords dihydrobenzo[*d*]isoxazole **298**. Thermal N—O bond cleavage, either by a radical fission (Path a) or by an ionic fission (Path b), followed by rebonding affords spiro aziridine **300**. This intermediate undergoes ring opening via C—C bond cleavage to generate iminium **301**, which cyclizes via intramolecular addition of the phenoxide anion to the iminium ion to afford dihydrobenzo[*d*]oxazole **295**.

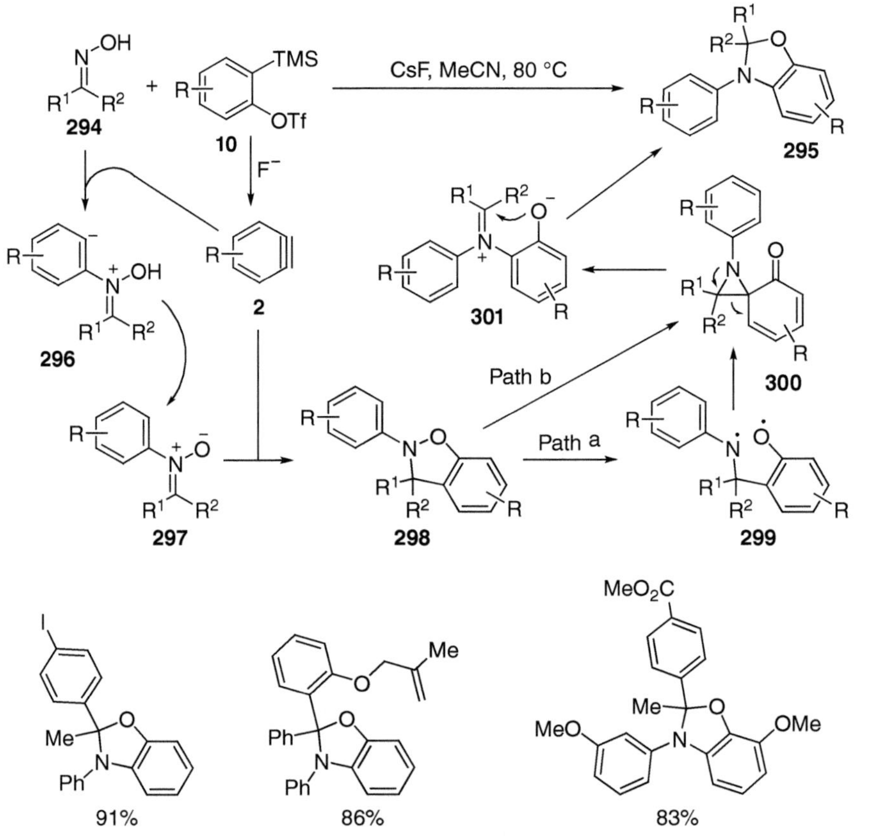

Scheme 7.56 Three-component reaction of oximes with two aryne molecules. Source: Based on Yao et al. [86].

7.5.2 Four-Component Reactions with Three Benzyne Molecules

In 2007, Xie and Zhang disclosed the four-component reaction of *N*-substituted imidazoles **302** with three benzyne molecules (generated in situ from 2-silylaryl triflate **10a** in the presence of CsF), delivering 9-anthracenylamines **303** in moderate-to-good yields (Scheme 7.57) [87]. Mechanistically, the reaction proceeds via multiple cycloadditions and a final nucleophilic addition. Initial Diels–Alder reaction of *N*-substituted imidazole **302** with benzyne (**2a**) leads to the formation of nitrogen-bridged isoquinoline **304**. Elimination of HCN from bridged isoquinoline **304** via a retro Diels–Alder reaction affords isoindole **305**, which undergoes a second Diels–Alder reaction with a second molecule of benzyne (**2a**) to give nitrogen-bridged anthracene **306**. Finally, nucleophilic addition of intermediate **306** to a third molecule of benzyne (**2a**) followed by ring opening via C—N bond cleavage affords 9-anthracenylamine **303**.

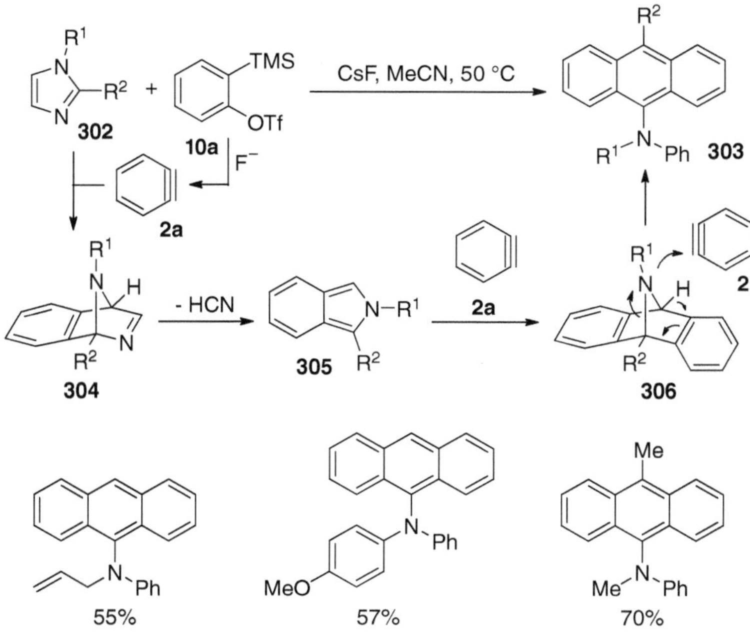

Scheme 7.57 Four-component reaction of imidazoles with three benzyne molecules. Source: Based on Xie and Zhang [87].

7.6 Conclusions

The highly electrophilic nature of arynes enables them to be added by various nucleophiles and unsaturated systems to form active intermediates such as zwitterions and strained cycles, which are ready to serve as versatile precursors for a number of molecular rearrangements such as the Stevens rearrangement, the aza-Claisen rearrangement, and the Smiles rearrangement. The scope for the aryne-triggered rearrangements has recently been expanded dramatically due to the in situ generation of arynes from 2-silylaryl triflates under mild conditions as developed by Kobayashi and coworkers [2]. The aryne-triggered rearrangements proceed under transition-metal-free conditions and enable a variety of synthetically useful operations, including skeletal construction, functional group transposition, ring formation, ring expansion, ring contraction, and chirality transfer. They provide convenient protocols for the formation of one, two, and three carbon–carbon and carbon–heteroatom bonds with arynes, delivering structural diverse organic compounds that might otherwise be difficult to access. Nevertheless, it deserves more efforts to explore the synthetic potentials of arynes in molecular rearrangements. It is highly desirable to develop new convenient methods for the generation of structurally diverse arynes, particularly those functionalized, as well as find new reaction modes of arynes, thereby diversifying the applications of aryne-triggered rearrangements in chemical synthesis. It is reasonable to believe that more fascinating outcomes will be disclosed in future.

References

1 For reviews, see: (a) Idiris, F.I.M. and Jones, C.R. (2017). *Org. Biomol. Chem.* 15: 9044–9056. (b) Goetz, A.E., Shah, T.K., and Garg, N.K. (2015). *Chem. Commun.* 51: 34–45. (c) Tadross, P.M. and Stoltz, B.M. (2012). *Chem. Rev.* 112: 3550–3577. (d) Holden, C. and Greaney, M.F. (2014). *Angew. Chem. Int. Ed.* 126: 5854–5857. *Angew. Chem. Int. Ed.*, 2014, 53, 5746-5749. (e) Diamond, O.J. and Marder, T.B. (2017). *Org. Chem. Front.* 4: 891–910.

2 Himeshima, Y., Sonoda, T., and Kobayashi, H. (1983). *Chem. Lett.*: 1211–1214.

3 For a review, see: Roy, T. and Biju, A.T. (2018). *Chem. Commun.* 54: 2580–2594.

4 Wittig, G. and Merkle, W. (1943). *Ber. Dtsch. Chem. Ges.* 76: 109–120.

5 Wittig, G. and Benz, E. (1959). *Chem. Ber.* 92: 1999–2013.

6 Lepley, R., Giumanini, A.G., Giumanini, A.B., and Khan, W.A. (1966). *J. Org. Chem.* 31: 2051–2055.

7 Sato, Y., Toyo'oka, T., Aoyama, T., and Shirai, H. (1975). *J. Chem. Soc., Chem. Commun.*: 640–641.

8 Roy, T., Thangaraj, M., Kaicharla, T. et al. (2016). *Org. Lett.* 18: 5428–5431.

9 Pan, X. and Liu, Z. (2018). *Org. Chem. Front.* 5: 1798–1810.

10 Roy, T., Gaykar, R.N., Bhattacharjeec, S., and Biju, A.T. (2019). *Chem. Commun.* 55: 3004–3007.

11 Zhang, J., Chen, Z.-X., Du, T. et al. (2016). *Org. Lett.* 18: 4872–4875.

12 Moss, S.G., Pocock, I.A., and Sweeney, J.B. (2017). *Chem. Eur. J.* 23: 101–104.

13 Pan, X., Ma, Y., and Liu, Z. (2018). *Org. Biomol. Chem.* 16: 7393–7399.

14 Zhou, M.-G., Dai, R.-H., and Tian, S.-K. (2018). *Chem. Commun.* 54: 6036–6039.

15 Guo, J.-Y., Zhong, C.-H., He, Z.-Y., and Tian, S.-K. (2018). *Asian J. Org. Chem.* 7: 119–122.

16 Bhunia, A., Roy, T., Pachfule, P. et al. (2013). *Angew. Chem. Int. Ed.* 125: 10224–10227. *Angew. Chem. Int. Ed.*, 2013, 52, 10040-10043.

17 Xu, X.-B., Lin, Z.-H., Liu, Y. et al. (2017). *Org. Biomol. Chem.* 15: 2716–2720.

18 Thangaraj, M., Gaykar, R.N., Roy, T., and Biju, A.T. (2017). *J. Org. Chem.* 82: 4470–4476.

19 Tan, J., Zheng, T., Xu, K., and Liu, C. (2017). *Org. Biomol. Chem.* 15: 4946–4950.

20 Danheiser, R.L. and Helgason, A.L. (1994). *J. Am. Chem. Soc.* 116: 9471–9479.

21 Shair, M.D., Yoon, T., and Danishefsky, S.J. (1995). *Angew. Chem., Int. Ed.* 34: 1721–1723.

22 Meyers, A.I. and Pansegrau, P.D. (1984). *Tetrahedron Lett.* 25: 2941–2944.

23 Tambar, U.K. and Stoltz, B.M. (2005). *J. Am. Chem. Soc.* 127: 5340–5341.

24 Yoshida, H., Watanabe, M., Ohshita, J., and Kunai, A. (2005). *Chem. Commun.*: 3292–3294.

25 Samineni, R., Srihari, P., and Mehta, G. (2016). *Org. Lett.* 18: 2832–2835.

26 Okuma, K., Tanabe, Y., Fukami, T., and Ishibashi, Y. (2018). *Heteroat. Chem* 29: e21444.

27 Yoshida, H., Watanabe, M., Ohshita, J., and Kunai, A. (2005). *Tetrahedron Lett.* 46: 6729–6731.

28 Hu, J.-H. and Zheng, H.-J. (2019). *Synth. Commun.* 49: 558–562.

29 Liu, Y.-L., Liang, Y., Pi, S.-F., and Li, J.-H. (2009). *J. Org. Chem.* 74: 5691–5694.

30 Zhang, T., Huang, X., Xue, J., and Sun, S. (2009). *Tetrahedron Lett.* 50: 1290–1294.

31 Ahire, M.M., Khan, R., and Mhaske, S.B. (2017). *Org. Lett.* 19: 2134–2137.

32 (a) Yoshida, H., Kishida, T., Watanabe, M., and Ohshita, J. (2008). *Chem. Commun.*: 5963–5965. (b) Yoshida, H., Ito, Y., Yoshikawa, Y. et al. (2011). *Chem. Commun.* 47: 8664–8666. (c) Rao, B., Tang, J., Wei, Y., and Zeng, X. (2016). *Chem. Asian J.* 11: 991–995. (d) Rao, B., Tang, J., and Zeng, X. (2016). *Org. Lett.* 18: 1678–1681.

33 Kranthikumar, R., Chegondi, R., and Chandrasekhar, S. (2016). *J. Org. Chem.* 81: 2451–2459.

34 Liu, Z. and Larock, R.C. (2005). *J. Am. Chem. Soc.* 127: 13112–13113.

35 Pintori, D.G. and Greaney, M.F. (2010). *Org. Lett.* 12: 168–171.

36 Schwan, J., Kleoff, M., Hartmayer, B. et al. (2018). *Org. Lett.* 20: 7661–7664.

37 (a) Yoshida, H., Shirakawa, E., Honda, Y., and Hiyama, T. (2002). *Angew. Chem. Int. Ed.* 114: 3381–3383. *Angew. Chem. Int. Ed.*, 2002, 41, 3247-3249. (b) Saito, N., Nakamura, K., Shibano, S. et al. (2013). *Org. Lett.* 15: 386–389. (c) Kaneko, H., Ikaw, T., Yamamoto, Y. et al. (2018). *Synlett* 29: 943–948.

38 Rao, B. and Zeng, X. (2014). *Org. Lett.* 16: 314–317.

39 Yoshioka, E., Kohtani, S., and Miyabe, H. (2010). *Org. Lett.* 12: 1956–1959.

40 Dubrovskiy, A.V. and Larock, R.C. (2013). *Tetrahedron* 69: 2789–2798.

41 Dhokale, R.A. and Mhaske, S.B. (2017). *J. Org. Chem.* 82: 4875–4882.

42 Yoshida, H., Mimura, Y., Ohshita, J., and Kunai, A. (2007). *Chem. Commun.*: 2405–2407.

43 Yoshida, H., Watanabe, M., Ohshita, J., and Kunai, A. (2005). *Chem. Lett.* 34: 1538–1539.

44 Zhao, X., Huang, Y., Qing, F.-L., and Xu, X.-H. (2017). *RSC Adv.* 7: 47–50.

45 (a) Ahire, M.M., Thoke, M.B., and Mhaske, S.B. (2018). *Org. Lett.* 20: 848–851. (b) Li, Z., Jian, H., Wang, W. et al. (2018). *Chin. J. Org. Chem.* 38: 2045–2053.

46 Łaczkowski, K.Z., García, D., Peña, D. et al. (2011). *Org. Lett.* 13: 960–963.

47 Yang, Y., Xu, Y., and Jones, C.R. (2019). *Eur. J. Org. Chem.*: 5196–5200.

48 Shen, C., Yang, G., and Zhang, W. (2013). *Org. Lett.* 15: 5722–5725.
49 Qi, N., Zhang, N., Allu, S.R. et al. (2016). *Org. Lett.* 18: 6204–6207.
50 Chen, Z. and Wang, Q. (2015). *Org. Lett.* 17: 6130–6133.
51 (a) Yoshida, H., Minabe, T., Ohshita, J., and Kunai, A. (2005). *Chem. Commun.*: 3454–3456. (b) Yoshida, S., Nakamura, Y., Uchida, K. et al. (2016). *Org. Lett.* 18: 6212–6215.
52 Gaykar, R.N., Bhattacharjee, S., and Biju, A.T. (2019). *Org. Lett.* 21: 737–740.
53 Okugawa, Y., Hayashi, Y., Kawauchi, S. et al. (2018). *Org. Lett.* 20: 3670–3673.
54 Mesgar, M. and Daugulis, O. (2017). *Org. Lett.* 19: 4247–4250.
55 Yoshida, H., Terayama, T., Ohshita, J., and Kunai, A. (2004). *Chem. Commun.*: 1980–1981.
56 Chen, J. and Murafuji, T. (2011). *Organometallics* 30: 4532–4538.
57 Toledo, F.T., Marques, H., Comasseto, J.V., and Raminelli, C. (2007). *Tetrahedron Lett.* 48: 8125–8127.
58 Yoshida, H., Yoshida, R., and Takaki, K. (2013). *Angew. Chem. Int. Ed.* 125: 8791–8794. *Angew. Chem. Int. Ed.* 2013, 52, 8629–8632.
59 Su, S., Li, J., Sun, M. et al. (2018). *Chem. Commun.* 54: 9611–9614.
60 Kaicharla, T., Thangaraj, M., and Biju, A.T. (2014). *Org. Lett.* 16: 1728–1731.
61 Payili, N., Rekula, S.R., Aitha, A. et al. (2019). *Org. Biomol. Chem.* 17: 9442–9446.
62 Cant, A.A., Bertrand, G.H.V., Henderson, J.L. et al. (2009). *Angew. Chem. Int. Ed.* 121: 5301–5304; *Angew. Chem. Int. Ed.*, 2009, 48, 5199-5202. For corrections, see: Cant, A.A., Bertrand, G.H.V., Henderson, J.L., Roberts, L., and Greaney, M.F. (2012) *Angew. Chem. Int. Ed.*, 124, 10360-10360; *Angew. Chem. Int. Ed.*, 2012, 51, 10214–10214.
63 Mangina, N.S.V.M.R., Guduru, R., and Karunakar, G.V. (2018). *Org. Biomol. Chem.* 16: 2134–2142.
64 Aoki, T., Koya, S., Yamasaki, R., and Saito, S. (2012). *Org. Lett.* 14: 4506–4509.
65 Kalvacherla, B., Batthula, S., Balasubramanian, S., and Palakodety, R.K. (2018). *Org. Lett.* 20: 3824–3828.
66 Chakrabarty, S., Chatterjee, I., Tebben, L., and Studer, A. (2013). *Angew. Chem. Int. Ed.* 125: 3041–3044. *Angew. Chem. Int. Ed.*, 2013, 52, 2968–2971.
67 Palani, V., Chen, J., and Hoye, T.R. (2016). *Org. Lett.* 18: 6312–6315.
68 Yamazaki, T., Baum, G., and Shechter, H. (1974). *Tetrahedron Lett.* 15: 4421–4424.
69 Raminelli, C., Liu, Z., and Larock, R.C. (2006). *J. Org. Chem.* 71: 4689–4691.
70 Dhiman, A.K., Kumar, R., Kumar, R., and Sharma, U. (2017). *J. Org. Chem.* 82: 12307–12317.
71 Lv, C., Wan, C., Liu, S. et al. (2018). *Org. Lett.* 20: 1919–1923.
72 Shamsabadi, A. and Chudasama, V. (2018). *Chem. Commun.* 54: 11180–11183.
73 Varlamov, A.V., Guranova, N.I., Novikov, R.A. et al. (2016). *RSC Adv.* 6: 12642–12646.
74 Holden, C.M., Sohel, S.M.A., and Greaney, M.F. (2016). *Angew. Chem. Int. Ed.* 128: 2496–2499. *Angew. Chem. Int. Ed.*, 2016, 55, 2450–2453.
75 Bhojgude, S.S., Baviskar, D.R., Gonnade, R.G., and Biju, A.T. (2015). *Org. Lett.* 17: 6270–6273.

76 Okuma, K., Kinoshita, H., Nagahora, N., and Shioji, K. (2016). *Eur. J. Org. Chem.*: 2264–2267.
77 Bhojgude, S.S., Roy, T., Gonnade, R.G., and Biju, A.T. (2016). *Org. Lett.* 18: 5424–5427.
78 Li, Y. and Studer, A. (2017). *Org. Lett.* 19: 666–669.
79 Matsuzawa, T., Uchida, K., Yoshida, S., and Hosoya, T. (2017). *Org. Lett.* 19: 5521–5524.
80 Yoshida, S., Yano, T., Misawa, Y. et al. (2015). *J. Am. Chem. Soc.* 137: 14071–14074.
81 Yoshida, S., Uchida, K., Igawa, K. et al. (2014). *Chem. Commun.* 50: 15059–15062.
82 Li, Y., Qiu, D., Gu, R. et al. (2016). *J. Am. Chem. Soc.* 138: 10814–10817.
83 Yoshida, H., Watanabe, M., Morishita, T. et al. (2007). *Chem. Commun.*: 1505–1507.
84 Neog, K., Dutta, D., Das, B., and Gogoi, P. (2017). *Org. Lett.* 19: 730–733.
85 Biswas, K. and Greaney, M.F. (2011). *Org. Lett.* 13: 4946–4949.
86 Yao, T., Ren, B., Wang, B., and Zhao, Y. (2017). *Org. Lett.* 19: 3135–3138.
87 Xie, C. and Zhang, Y. (2007). *Org. Lett.* 9: 781–784.

8

New Strategies in Recent Aryne Chemistry

Yang Li

Chongqing University, School of Chemistry and Chemical Engineering, 174 Shazheng Street, Chongqing 400030, P.R. China

8.1 Introduction

Since the first experimental confirmation of benzyne intermediate by Roberts in 1953 [1], the chemistry of aryne has attracted tremendous attention in the past over 60 years. By the time Hoffman published the first, and also the only, seminal monograph on aryne chemistry in 1967 [2], several characteristic reaction modes of arynes had become well known, i.e. pericyclic reactions and nucleophilic reactions. The development of aryne chemistry, however, slowed down in the next few decades. Until the end of 1990s, along with the growing utilization of mild aryne generation protocols particularly by Kobayashi's method [3] and, recently, by intramolecular hexadehydro-Diels–Alder (HDDA) aryne formation strategy [4–6], both reaction efficiency and the chance to discover new types of transformations have been greatly promoted. Several novel aryne reaction modes, such as transition-metal-catalyzed reactions, insertion reactions, and multicomponent reactions, emerged in the past two decades. Consequently, aryne chemistry stepped into a renaissance era.

This chapter will focus on some of the aspects that, at least in part, can be seen as interesting recent advances in aryne chemistry mainly in the past two decades. The contents of this chapter will include three topics: new aryne generation methods, aryne regioselectivity, and aryne multifunctionalization.

8.2 New Aryne Generation Methods

Along with the advance of aryne chemistry, efforts in searching for mild and efficient aryne generation methods have never stopped. Because arynes possess extremely high activity, they have to be generated and consumed in situ. In this context, conditions for both aryne generation and subsequent reactions should match with each other well. Therefore, only when sufficiently efficient and, particularly, mild aryne generation methods are employed will they promote the rapid

Modern Aryne Chemistry, First Edition. Edited by Akkattu T. Biju.

development of aryne chemistry in terms of enhanced reaction efficiency, better selectivity, more functional group tolerance, and broader synthetic applications. Although Kobayashi aryne generation method has become the most prevailing one within the past two decades, it remains desirable to find alternative protocols that are effective toward some unreachable transformations by current methods. Consequently, new aryne generation approaches emerged in the past two decades.

In general, there are two principal strategies toward generation of aryne intermediates from judiciously designed precursors (Scheme 8.1): (i) strategy I: ortho-deprotonative elimination with strong bases by concomitantly kicking out a leaving group (LG); (ii) strategy II: simultaneous removal of two vicinal substituents on arene ring, one of which serves as an accepting group (AG) and the other LG. A distinct difference between these two strategies is that in strategy I ortho-deprotonation of mono-substituted aryne precursors usually requires strong bases; while the "push–pull" activating mode through a combination of AG and LG in strategy II could provide versatile aryne generation means under much milder conditions. Nevertheless, strategy I remains attractive simply because of atom-economical consideration. Although both strategies have been utilized since the early era of aryne chemistry, in the past two decades, new methods appeared through constant modification on both the structures and activating conditions of aryne precursors. Notably, an outstanding HDDA aryne generation strategy was recently developed by Hoye, which then led to a rapid and intensive investigation on this protocol [4–6]. Because this generation method and its applications will be covered in chapter 10, it will not be discussed in this section.

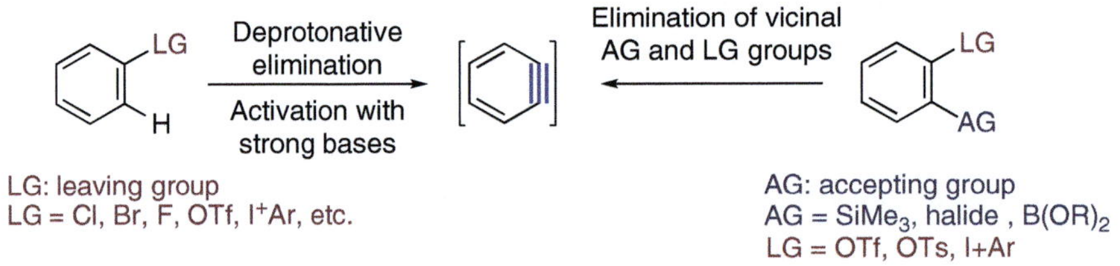

Scheme 8.1 Two general aryne generation strategies.

8.2.1 Revisiting ortho-Deprotonative Elimination Protocols

Although the earliest method to generate benzyne is through dehydrohalogenation of halobenzenes, its application has been hampered by both harsh conditions unavoidable from strongly basic activating reagents and, even worse, undesired nucleophilic attack by those noninnocent bases. As a result, this aryne generation protocol has its inherent drawbacks: limited functional group tolerance and competing side reactions from activating reagents. Despite these unfavorable characters, this deprotonative elimination strategy has not only been attractive with respect to atom-economical consideration, but also owns its merit on ready accessibility of functionalized aryne precursors from either cheap aryl halides or aryl triflates. In this context, people have been motivated to search for alternates of conventional generation conditions in an aim to achieving deprotonative elimination under mild conditions with high tolerance to various functional groups and reaction modes. To this end, modifications by both employing non-nucleophilic bases and enhancing the leaving ability of LGs on aryne precursors have been explored.

In 1973, Olofson and coworkers first disclosed that lithium 2,2,6,6-tetramethylpiperidide (LiTMP) could be utilized as a bulky base to efficiently generate aryne from chlorobenzene, at which moment nucleophilic attack on aryne by TMP anion was prevented due to its bulky size (Scheme 8.2) [7]. In this study, however, only phenylethyn-1-ide and benzenethiolate were used as nucleophiles. In 2011, Daugulis and coworkers revisited this aryne generation protocol and applied it in arylation of heteroaromatic compounds, in which the C—H bonds on heterocycles could be deprotonated by another equivalent of LiTMP (Scheme 8.3a) [8]. Moreover, the employment of both lithium dicyclohexylamide and lithium diisopropylamide (LDA) afforded the arylation product in slightly lower yields. They reasoned that the success of this transformation was ensured by both the bulkiness and low solubility of LiTMP in solvent. Later on, this generation protocol was further extended to the preparation of a broad spectrum of functionalized polyaryls [10]. In 2012, the Daugulis group applied this protocol in ortho-arylation of anilines as well, whereas no *N*-arylation product was detected (Scheme 8.3b) [9]. In marked contrast, anilines have been known to be efficient *N*-nucleophiles in aryne reactions with either Kobayashi's or Suzuki's precursors. Moreover, an efficient preparation of helicenes and 2-arylphenol was demonstrated (Scheme 8.3c) [11]. A divergent synthesis of both types of compounds was discovered by simply varying the base: when LiTMP was used, helicenes were found to be the only products; whereas 2-arylphenols could be obtained in the presence of *t*-BuONa and AgOAc.

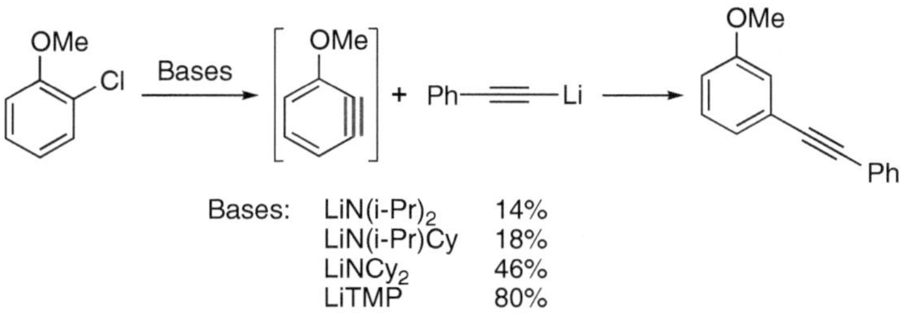

Scheme 8.2 Olofson's early study on PhCl-LiTMP system. Source: Based on Olofson and Dougherty [7].

The strategy by using sufficiently bulky base to prevent undesired side reactions was further examined by Daugulis and coworkers in 2018, where a more hindered amide salt, namely lithium diadamantylamide (LDAM), was prepared and employed as activating reagent to generate aryne intermediates from either aryl halides or aryl triflates (Scheme 8.4) [12]. Unwanted nucleophilic addition by LDAM was suppressed, the base of which was also proven to be superior over LiTMP with higher reaction yields. Furthermore, they exhibited that the aryne intermediates generated

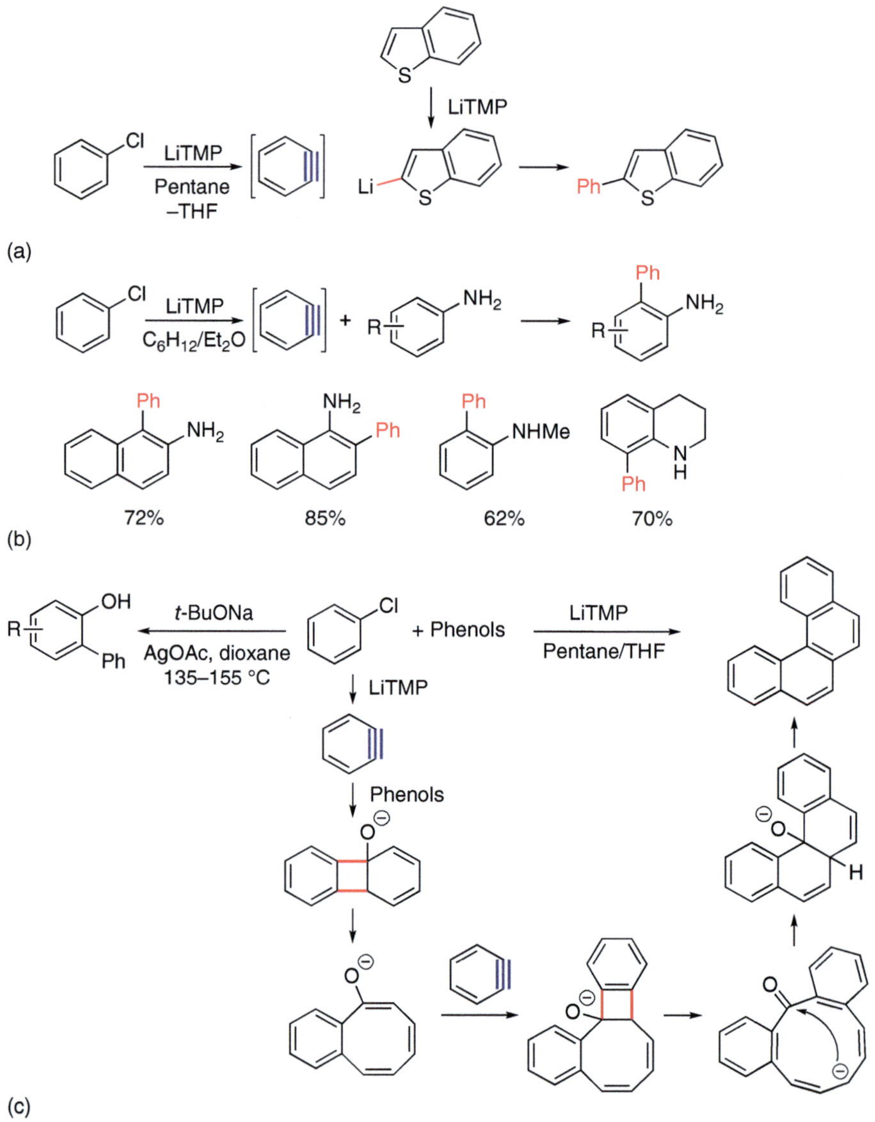

Scheme 8.3 Study on PhCl-LiTMP system. (a) C—H bond arylation of heterocycles. Source: Based on Truong and Daugulis [8], (b) ortho-arylation of anilines. Source: Based on Truong and Daugulis [9], and (c) reactions with phenols.

under these conditions could insert into Si—P, Si—S, Si—N, and C—C σ-bonds with high functional group tolerance. These examples demonstrated that, under certain circumstances, base-promoted dehydrohalogenation protocol could be efficiently utilized in a broad spectrum of transformations as long as undesired side reactions by activating reagents can be prevented.

Beyond the solutions by increasing steric congestion on activating bases to suppress side reactions, Uchiyama and coworkers developed another approach

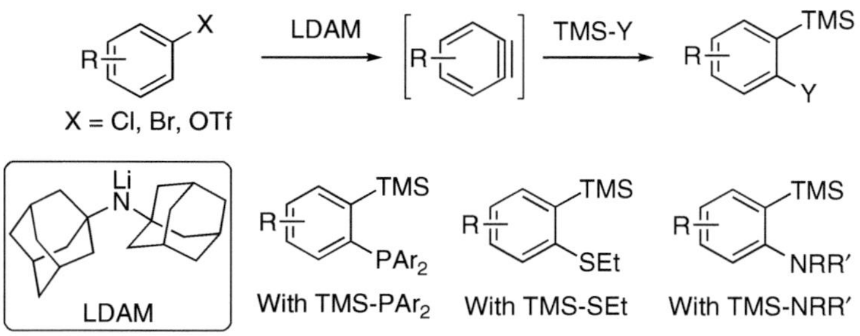

Scheme 8.4 LDAM as bulky base for aryne generation. Source: Based on Mesgar et al. [12].

to circumvent nucleophilic side reactions. In 2002, they found that lithium di-alkyl(2,2,6,6-tetramethylpiperidino)zincate ($R_2Zn(TMP)Li$) could serve as an efficient activating reagent for both halobenzene and phenyl triflate (Scheme 8.5) [13, 14]. Particularly, this generation method avoided the employment of strong bases, making it compatible to a broad scope of functional groups, such as ester, cyano, and amide. Their computational study revealed that the key steps involve a regio- and chemoselective deprotonative zincation and the preferential coordination of dialkylzinc moiety to halogen reduces the activation energy for elimination [14].

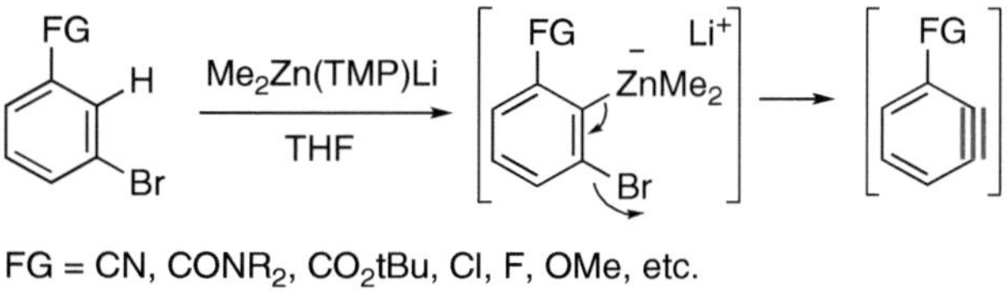

Scheme 8.5 Uchiyama's conditions. Source: Uchiyama et al. [13]; Uchiyama et al. [14].

With a desire to meet the demands on better leaving ability, ready availability, and/or convenient preparation, people have also searched for alternative LGs other than halides. In this context, triflyloxy (OTf) group was chosen to replace halogens, despite the fact that thia-Fries rearrangement might be a competing side reaction after ortho-deprotonation [15]. In 1999, Wickham and Scott first reported a benzyne generation protocol using PhOTf and LDA (Scheme 8.6a) [16]. Unfortunately, this study solely afforded *N*-arylated product on diisopropylamine. This drawback was then solved by Wang and coworkers in 2018, who utilized $LiZnEt_2(TMP)$ as an activating reagent to accomplish an efficient deprotonative zincation-elimination on PhOTf (Scheme 8.6b) [17]. Both benzyne Diels–Alder cycloaddition reaction and insertion into the N—CO σ-bond of urea were examined, affording the corresponding products in high yields. Moreover, in Daugulis' study with LDAM base, they also found that the OTf group has almost the same departure tendency with chloride [12].

Meanwhile, LGs with comparable or even better leaving ability with respect to halogens or OTf were pursued. In a study carried out by Okuyama and Ochiai, they revealed that the leaving ability of an aryliodonio group is million times better

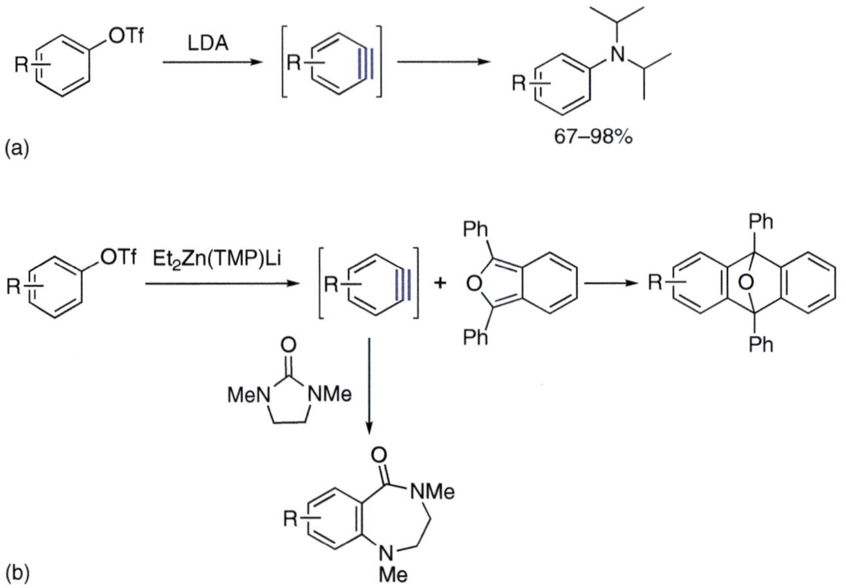

Scheme 8.6 Aryne generation via elimination of HOTf. (a) Elimination of HOTf with LDA by Wickman and Scott. Source: Based on Wickham et al. [16], and (b) mild elimination of HOTf with $E_2Zn(TMP)Li$ by Wang. Source: Based on Cho and Wang [17].

than that of the OTf group [18], making diaryliodonium salts superior candidates as aryne precursors over aryl halides or aryl triflates. Although the idea by using diaryliodonium salts as aryne precursors was first demonstrated by Akiyama and coworkers in 1974, only limited examples were shown in this study with low reaction efficiency [19]. In 2016, Stuart and coworkers re-examined this approach and developed a distinct and efficient aryne generation protocol from readily prepared aryl(mesityl)iodonium tosylate, which could be trapped by furan, benzyl azide, and amine nucleophiles (Scheme 8.7a) [20, 21]. This study unambiguously disclosed the superior leaving ability of hypervalent iodine moiety by showing the inertness of other potential LGs, such as fluoride, chloride, bromide, and triflate on the same molecules. Particularly, a regioselective deprotonation behavior was disclosed in this transformation as well. Using this aryne generation protocol, Han, Wang and coworkers subsequently reported a direct arylation of tertiary amines (Scheme 8.7b) [22].

8.2.2 Arynes from ortho-Difunctionalized Precursors

As the most investigated as well as successfully utilized strategy, aryne generation via elimination of ortho-difunctionalized arenes has received continuous attention. A prominent advantage of this strategy over deprotonative elimination one is that the generation conditions could avoid the utilization of strong bases. Particularly, the "push–pull" activating mode of this strategy would make aryne generation conditions sufficiently mild by concomitantly meeting the requirements for both the

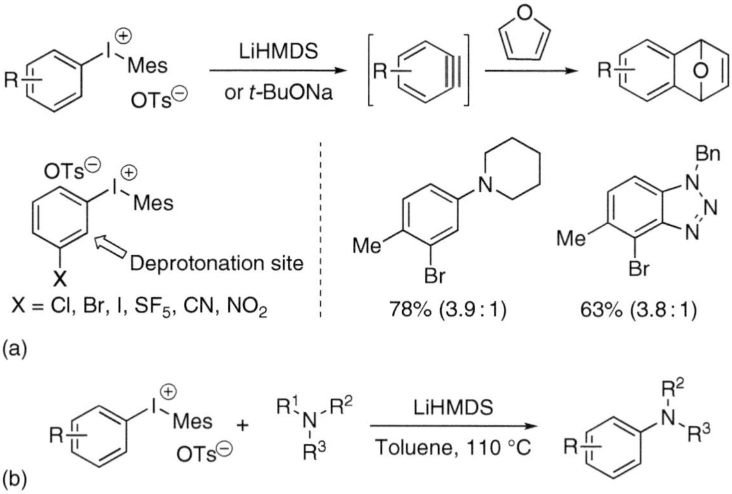

Scheme 8.7 Aryne generation via elimination of aryliodonio group. (a) Staurt's method. Source: Sundalam et al. [20]; Stuart [21], and (b) arylation with teritary amines. Source: Based on Zhang et al. [22].

employment of less harsh, non-nucleophilic bases and high tolerance with respect to various functional groups as well as reaction modes. To this end, an outstanding example is *o*-siylphenyl triflate, commonly known as Kobayashi's benzyne precursor, which employs fluoride-induced generation conditions by concomitant removal of a TMS and an OTf group. The giant success of this generation method in the past two decades stimulated people to further explore variations of aryne generation strategies. Consequently, modifications on either AGs or LGs have been scrutinized in the past decades. As shown in Scheme 8.1, the prevailingly investigated AGs include TMS, halogens (I or Br), and boronic acid/ester, and those frequently utilized LGs are OTf, OTs, bromide, and aryliodonio group. Here in this section, some of the efforts other than Kobayashi's conditions will be summarized.

Beside Kobayashi aryne precursor, presumably the most successfully utilized aryne precursors in the past three decades are *o*-haloaryl sulfonates. In 1991, Suzuki and coworkers reported the first example by using *o*-iodoaryl triflates as aryne precursors (Scheme 8.8a) [23]. *n*-BuLi was used in iodine–lithium exchange and a consequent departure of the OTf group could afford aryne intermediates, those of which were then trapped by furan through Diels–Alder reactions. Similar to Kobayashi's method, this protocol takes the advantage of the excellent leaving ability of the OTf group, which was subsequently proven by Snowden to be a better LG over fluoride, chloride, and OTs [26]. Equipped with this generation method, Suzuki and coworkers discovered a convenient [2+2] cycloaddition of arynes with ketene silyl acetals (KSAs) (Scheme 8.8b) [24, 25]. Distinctively, the presence of alkoxy groups on the 3-position of benzyne promised an excellent regioselective control by discriminating the two carbons on aryne triple bond. Notably, in our previous study, we found that *o*-silylaryl triflates were not as efficient as *o*-iodoaryl triflate in [2+2] cycloaddition with KSAs [27]. Later on, this generation protocol found widespread applications both in different aryne transformations [28–33] and in natural product synthesis [34–36].

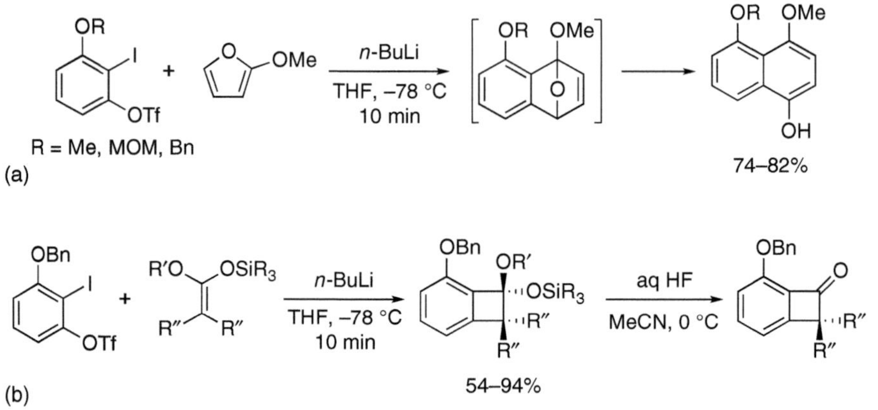

Scheme 8.8 Suzuki's *o*-iodoaryl triflate as aryne precursor. (a) *o*-iodoaryl triflate as benzyne precursar. Source: Matsumoto et al. [23], and (b) [2+2] cycloaddition with *o*-iodoaryl triflate. Source: Hosoya et al. [24]; Hosoya et al. [25].

Encouraged by Suzuki's success on *o*-iodoaryl triflates, variations of this aryne generation protocol were subsequently reported. First, the activating reagents were investigated. In 2015, Hosoya and coworkers found that trimethylsilylmethyl Grignard reagent ($TMSCH_2MgCl$) could serve as a mild activating reagent to generate arynes from *o*-iodoaryl triflates [37]. Distinctively, Suzuki and coworkers also disclosed that alkynyllithium was able to activate *o*-iodophenyl triflate in a catalytic fashion (Scheme 8.15) [38].

Moreover, the OTf group was found to be replaced by other sulfonates with unequal leaving abilities. This modification was based on the expectation that under certain circumstances the diminished departure tendency and/or enhanced stability toward side reactions, such as thia-Fries rearrangement [15], of those LGs could adjust the aryne generation rate and make certain transformations accessible. Indeed, Knochel and coworkers demonstrated that by altering the LGs on *o*-iodoaryl sulfonates, both reaction efficiency and functional group tolerance in Diels–Alder reaction varied significantly (Scheme 8.9) [39–41]. Notably, isopropylmagnesium chloride (*i*-PrMgCl) was found to be able to serve as a mild activating reagent to efficiently produce arynes from various *o*-iodoaryl sulfonates. In marked contrast, *o*-iodoaryl triflate with OTf as the LG was found to be ineffective under this reaction condition (Scheme 8.9).

Within the past two decades, modifications on either silyl-based or OTf-based difunctionalized aryne precursors have also been examined. As shown in Scheme 8.10a, the silyl-based difunctionalized aryne precursors were reported by the groups of Kitamura et al. [42], Akai and coworkers [43], Novák and coworkers [44], Jiang, Wang and coworkers [45], and Daugulis and coworker [46], all of which take the advantage of fluoride-induced mild generation conditions. On the other hand, when OTf group was selected as the LG, the AG side has also been

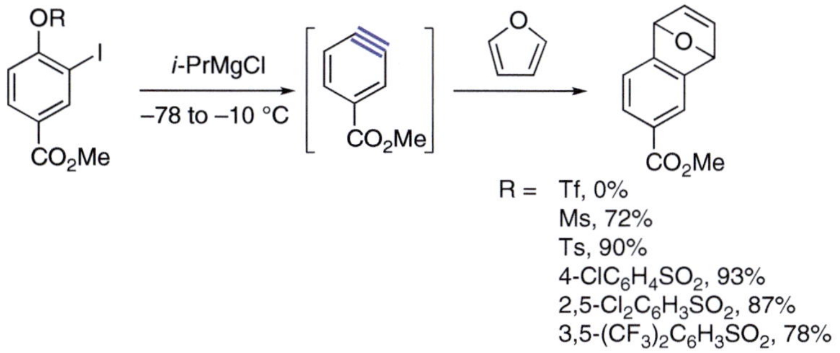

Scheme 8.9 Knochel's aryne precursors. Source: Sapountzis et al. [39]; Lin et al. [40]; Lin et al. [41].

explored by Hosoya and coworkers, in which boronic ester [47], arylsulfoxide [48], and diarylphosphinyl groups [49] were found to be effective alternative AGs (Scheme 8.10b). Harsh activating conditions, however, were usually needed to generate arynes from these precursors. Beside aforementioned achievements, photoinduced aryne generation methods have also been recently discovered independently by the groups of Schnarr and coworkers [50] and Shi and coworkers [51], which would expand aryne chemistry through the utilization of UV-light as activating condition (Scheme 8.10c).

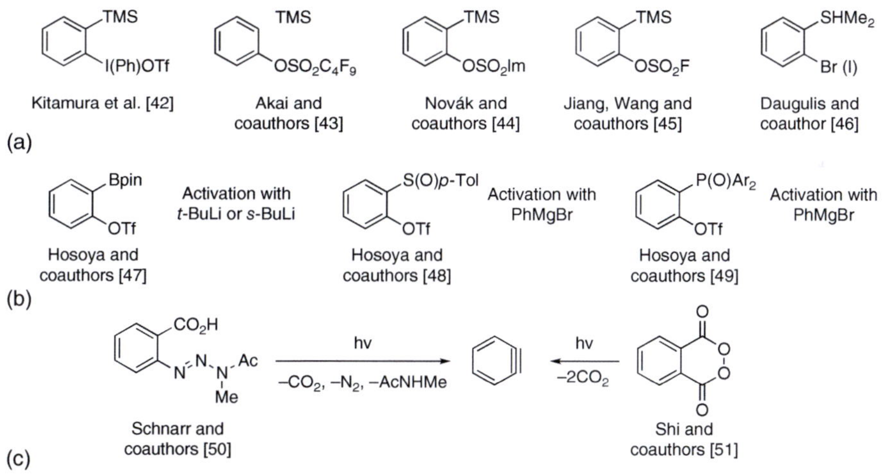

Scheme 8.10 Other ortho-difunctionalized aryne precursors. (a) Silyl-based difunctionalized aryne precursors, (b) OTf-based difunctionalized aryne precursors, and (c) photoinduced generation of benzyne.

8.2.3 Catalytic Aryne Generation Methods

The employment of stoichiometric and frequently excess amount of activating reagents might not be compatible with certain aryne transformations, especially under the circumstances that those activating reagents are not inert to arynes or to substrates. In this context, Hoye's HDDA aryne strategy provides an excellent example, which could avoid both the generation of departed residues from AG and LG and the necessary for activating reagents. Thus, strategies that could produce aryne intermediates under catalytic activation conditions are highly desirable and would be one of the pursuits in aryne chemistry.

In 2006, Hu and coworkers reported a Pd-associated aryne generation strategy from either 1-chloro-2-halobenzenes or 2-haloaryl tosylates with hindered Grignard reagents and revealed a distinct Pd-catalyzed annulative domino process (Scheme 8.11) [52]. In this study, they first disclosed that, in the absence of phosphine or NHC ligands, various palladium catalysts could promote the formation of Pd(II)ClX associated benzyne from 1-chloro-2-halobenzenes via a sequential oxidative addition of Pd(0) and β-chloro elimination. A subsequent carbopalladation with hindered Grignard reagents was followed by intramolecular C—H bond activation and reductive elimination, giving rise to substituted fluorenes (Scheme 8.11). Furthermore, 2-haloaryl tosylates could participate in the same transformation with palladium catalyst as well. Their in-depth mechanistic study revealed that these transformations proceeded through the generation of Pd-aryne species.

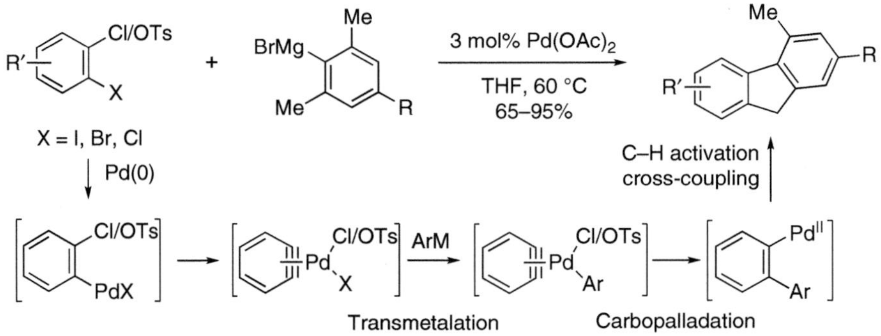

Scheme 8.11 Pd-catalyzed aryne generation by Hu. Source: Based on Dong and Hu [52].

In 2008, Kim and coworkers reported a Pd-mediated aryne generation protocol from methyl 2-bromobenzoates, the generated aryne of which was then trapped in [2+2+2] cyclotrimerization reaction (Scheme 8.12) [53]. A plausible mechanism was proposed: oxidative addition of Pd(0) species with aryl bromide was followed by a δ-carbon elimination-decarboxylation to afford Pd-aryne complex, which then participated in a Pd-catalyzed [2+2+2] cyclotrimerization to furnish triphenylenes.

Greaney and coworkers also investigated different approaches toward catalytic aryne generation. In 2010, they reported a method to produce arynes from benzoic acids via a sequential Pd-catalyzed *ortho* C—H bond activation to form a cyclopalladated complex and a decarboxylation to produce Pd-aryne complex. Subsequent Pd-catalyzed [2+2+2] cyclotrimerization gave rise to triphenylenes as well (Scheme 8.13) [54]. Later on, inspired by Wenger and Bennett's work on aryne generation with stoichiometric amount of either Pd(0) or Ni(0) species [55],

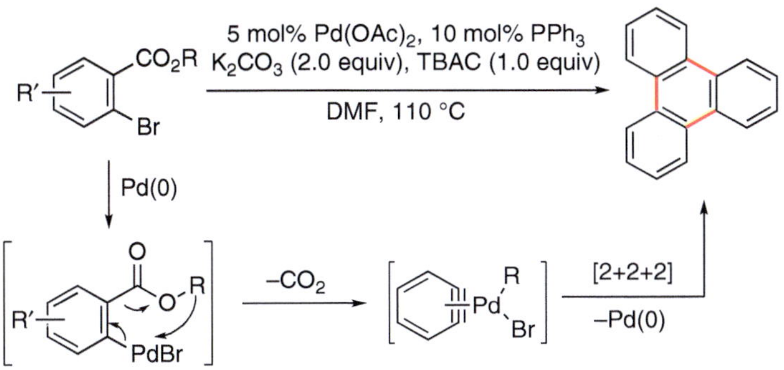

Scheme 8.12 Pd-mediated aryne generation from methyl 2-bromobenzoates. Source: Based on Kim et al. [53].

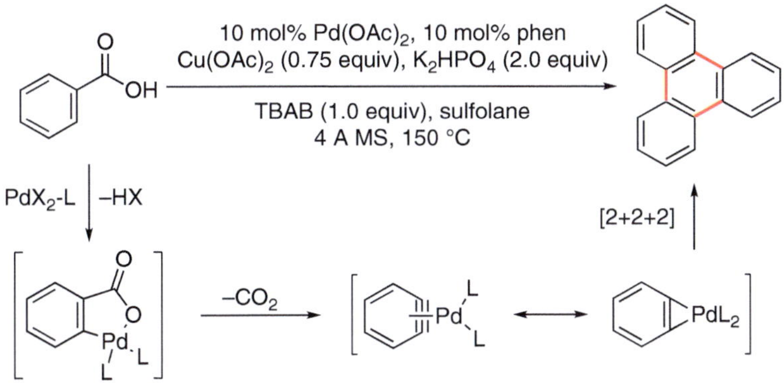

Scheme 8.13 Pd-mediated aryne generation from benzoic acids. Source: Based on Cant et al. [54].

Greaney and coworkers developed a Pd-catalyzed aryne generation protocol from (2-bromoaryl)boronic esters (Scheme 8.14) [56]. *t*-BuOK was found to be an efficient base to activate boronic ester group in this transformation. After careful inspection

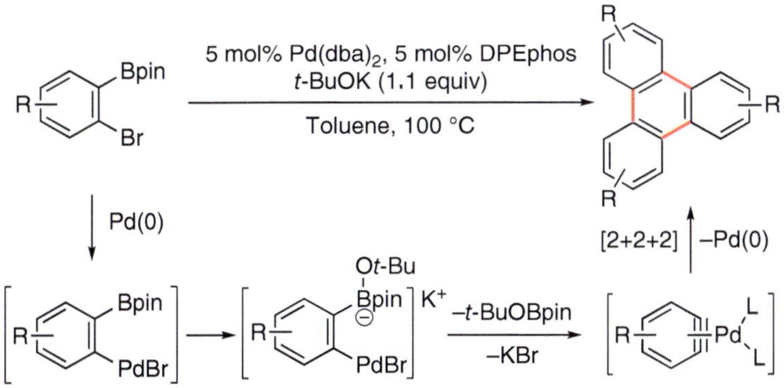

Scheme 8.14 Pd-mediated aryne generation from *o*-bromoaryl boronic esters. Source: Based on García-López and Greaney [56].

on the product ratio by using unsymmetrically substituted aryne precursors, they ruled out the possibility of Suzuki–Miyaura coupling mechanism and suggested that the reaction underwent through an aryne pathway. Moreover, competing experiment with 2,5-dimethylfuran gave no observation of aryne Diels–Alder cycloadduct, indicating that no free aryne was generated in the reaction system.

In 2012, Hamura, Suzuki and coworkers reported an unprecedented catalytic generation of arynes via alkynyllithium catalysis [38]. As shown in Scheme 8.15, this transformation initiated with metal–halogen exchange between *o*-iodophenyl triflate and alkynyllithium. After the generation of benzyne, another alkynyllithium acted as a nucleophile to attack it and produce aryllithium intermediate, which could undergo the second metal–halogen exchange with the previously generated iodoalkyne species to afford iodoarene product by simultaneously releasing alkynyllithium for next catalytic cycle. In this transformation, alkynyllithium served as both nucleophile and aryne-generating catalyst. Particularly, other C-nucleophiles could be also utilized in this transformation in the presence of catalytic amount of alkynyllithium [38].

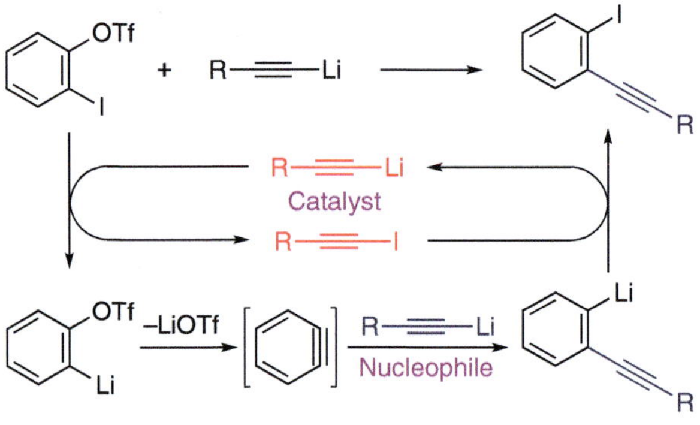

Scheme 8.15 Alkynyllithium-catalyzed aryne generation. Source: Based on Hamura et al. [38].

Although the aforementioned aryne generation protocols might not be a complete coverage in this research area and some of them might find no more future applications as well, we still want to reiterate that, in view of the future exploration on new types of aryne reactions, certain generation protocols might become feasible when those popular ones fail. Therefore, when people plan to study a new aryne transformation, they should keep in mind that consideration on examining different aryne generation methods might be an indispensable part of the project.

8.3 Aryne Regioselectivity

Aryne regioselectivity has long been accompanying with the advances of aryne chemistry. When an unsymmetrically substituted benzyne participates in a reaction, a mixture of regioisomers will be conceived. This lack of regioselectivity situation, however, could be detrimental to an aryne reaction and, in some cases, results in inseparable regioisomers. In this context, both verifying the origin of aryne regioselectivity and paving ways to solve this problem would greatly promote the rapid advance of related fields with respect to developing highly efficient and practical aryne transformations in unambiguous regioselective control.

A conventional approach to access high regioselective aryne reactions was through intramolecular transformation, which has been broadly applied in natural product synthesis [34–36]. This strategy, however, has its limitation, because extra steps are needed for both the preparation of aryne precursors and the removal of the unwanted linkers after aryne reactions. On the other hand, highly efficient regioselective tuning factors in intermolecular aryne transformations have not been systematically explored until recently. Therefore, it is desirable to discuss new strategies on aryne regioselective control as a topic in this chapter, especially in view of the recent achievements in this field. As shown in Scheme 8.16, there are currently three means that could manipulate regioselectivity in an intermolecular aryne transformation: steric effect, electronic effect, and ring strain. In the past decade, new strategies equipped with easily removable or convertible tuning groups were developed by different research groups, making solutions to reach high regioselective control in aryne transformations more and more convenient.

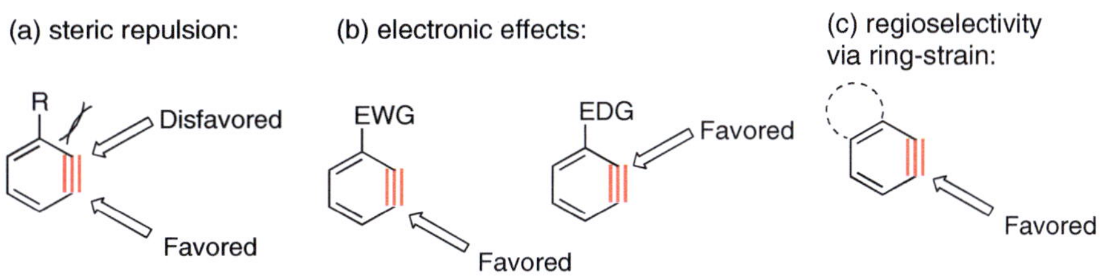

Scheme 8.16 Regioselective controls in intermolecular aryne reactions.

8.3.1 Steric Effect

One of the earliest approaches to realize aryne regioselective control is through steric repulsion, which requires the introduction of a bulky group on the 3-position of a benzyne intermediate (Scheme 8.16). For instance, a 3-*tert*-butylbenzyne usually exhibits high regioselectivity in different reactions and favors less sterically congested products.

In 2011, Kazmaier and coworkers reported an aryne insertion reaction into Sn—H bond [57]. In this work, the product ratio of two substrates was prominently affected by the bulkiness of the substituents on the 3-position of benzyne. As shown in Scheme 8.17a, when aryne intermediate possesses a 3-methyl group, the ratio of regioisomers is about 2 : 1. In contrast, when a 3-*tert*-butylbenzyne was utilized, only meta-selective product was obtained. In a study carried out by Garg and Houk, they examined nucleophilic addition on 3-*tert*-butylbenzyne and found that all three nucleophiles gave rise to solely meta-selective products (Scheme 8.17b) [58]. Unfortunately, this 3-*tert*-butylbenzyne protocol is not a practical solution for aryne regioselectivity, because both the removal and further elaboration on the *tert*-butyl group will be problematic. Apparently, ideal candidates for this steric repulsion-driven regioselective control are those that both possess sufficient bulkiness and could be readily removable or convertible.

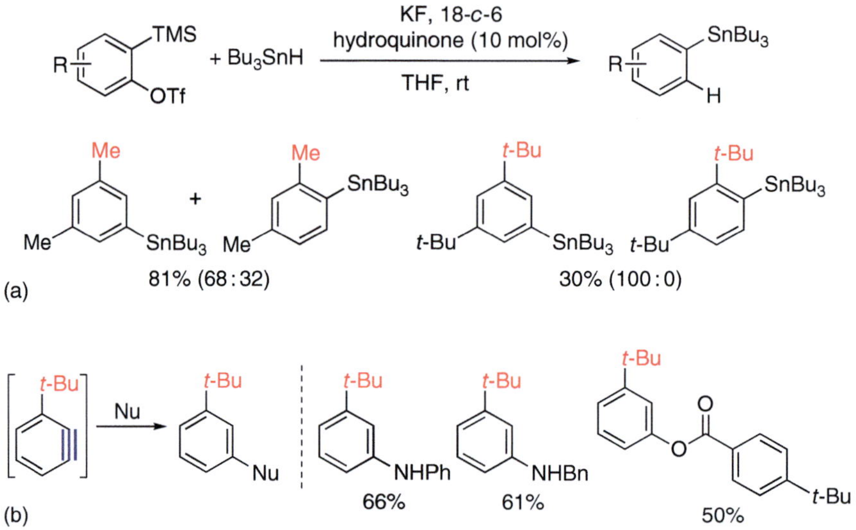

Scheme 8.17 Steric effect by 3-alkyl groups on arynes. (a) Kazmaier's study and (b) Garg–Houk's study. Source: Based on Bronner et al. [58].

In 2005, Schlosser observed that a trimethylsilyl (TMS) group on the 3-position of an aryne intermediate can well tune the regioselectivity in a [4+2] cycloaddition with 2-(trimethylsilyl)furan (Scheme 8.18). In the absence of this TMS group, however, 3-fluorobenzyne gave rise to a 2 : 1 mixture of cycloadducts with an opposite regioselective control [59]. This study disclosed a potential of silyl groups as sterically congested tuning factors in regioselective aryne transformations.

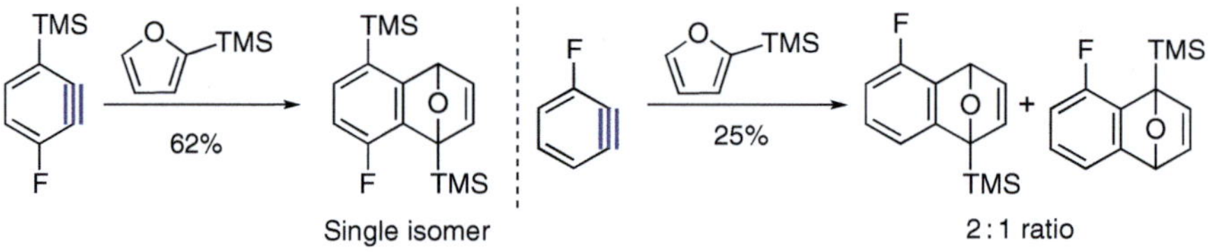

Scheme 8.18 TMS as sterically congested group by Schlosser.

Later on, Akai and coworkers reported a similar property of silyl as sterically congested groups in [4+2] cycloaddition reactions with 2-substituted furans [60]. As shown in Scheme 8.19, the presence of both phenyl and *tert*-butyl groups on the 3-position of benzyne had almost no regioselective control in this Diels–Alder reaction. Interestingly, when TMS group was employed, the isomeric ratio raised to 39 : 1 with preferential formation of anti-adduct. Distinctively, a bulkier *tert*-butyldimethylsilyl (TBDMS) group could further enhance the product ratio up to more than 50 : 1. Moreover, the TBDMS group in cycloadducts was found to be readily converted to aryl groups via palladium-catalyzed Hiyama cross-coupling reactions, exhibiting the versatility of this silyl-based regioselective control protocol in practical synthesis. Subsequently, this silyl-based aryne strategy was successfully utilized by the groups of Akai-Ikawa [43, 61], Du Bois [62], and Hosoya [63].

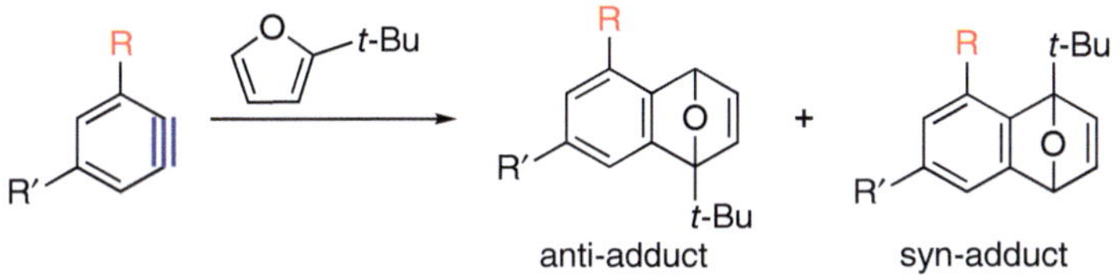

	Yield	anti : syn
R = *t*-Bu, R′ = *t*-Bu	53%	1.7 : 1
R = TMS, R′ = Me	63%	39 : 1
R = TBDMS, R′ = Me	69%	>50 : 1

Scheme 8.19 Silyl as sterically congested group by Akai.

Further exploration on steric effect imposed by 3-silyl group on benzyne revealed that the structure of an arynophile has an indispensable impact on the regioselective outcome as well. As shown in Scheme 8.20, in a study carried out by Garg and Houk on nucleophilic addition with 3-triethylsilylbenzyne, they disclosed that if a nucleophile is not bulky enough, significant amount of ortho-substituted product could be obtained along with meta-substituted product. When a nucleophile possesses sufficient steric bulkiness, meta-position is favored [58].

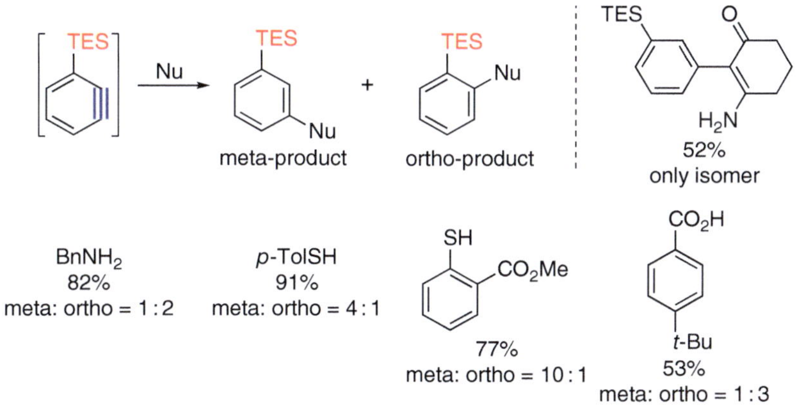

Scheme 8.20 Garg-Houk's study on silyl group as bulky group.

Beside silyl groups as distinct steric congesting groups, boronic esters were found to be efficient steric tuning factors as well. In 2010, Akai and coworkers discovered that a 3-borylbenzyne could reach regioselective [4+2] cycloaddition reactions with 2-substituted furan (Scheme 8.21) [30, 64]. Depending on the size of the 2-substituent on furan ring, ratios of regioisomers varied with preferential formation of less hindered cycloadducts. In view of enormous applications of boronic acids/esters in Suzuki–Miyaura cross-coupling transformations, the products in this approach could be readily converted to diverse molecular architectures.

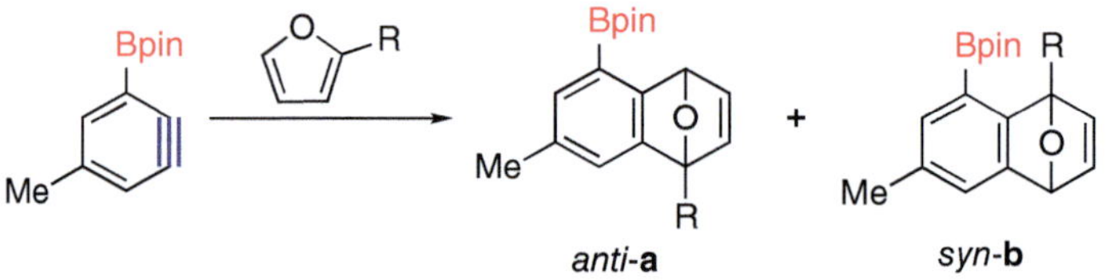

	Yield	anti-a: syn-b
R = Me	85%	88 : 12
R = t-Bu	>99%	94 : 6
R = TMS	91%	>98 : 2
R = CO_2Me	57%	93 : 7

Scheme 8.21 Boronic ester as sterically congested group by Akai. Source: Ikawa et al. [30]; Takagi et al. [64].

8.3.2 Electronic Effect

In comparison with steric factors, electronic effect has been found to be a more versatile way to reach high regioselective control. A variety of approaches have been developed by utilizing both inductively electron-donating (ED) and electron-withdrawing (EW) substituents on aryne ring. Early experimental observations have indicated that a 3-substituted inductively EW group on benzyne ring could lead to preferential meta-selective products in nucleophilic reactions as well as other transformations using polar arynophiles (Scheme 8.16b). Later on, an opposite regioselective outcome was observed when an inductive ED group presents on the 3-position of a benzyne (Scheme 8.16b). Recently, the groups of Garg and Houk established a powerful and practical aryne distortion/interaction model to explain and predict aryne regioselectivity [65–69]. This model divides the activation energy of a bimolecular process into two components: a distortion energy that allows the reactants to reach the transition state geometry; a second energy accounts for the interaction between two distorted species. Moreover, this aryne distortion model could also be well utilized to explain and predict the regioselective outcomes in the chemistry of indolynes, pyridynes, and other heterocyclic arynes (hetarynes) [65, 66, 68–74]. Distinctively, their experimental results were highly consistent with this aryne distortion model.

Scheme 8.22 lists some geometry optimization of commonly substituted benzynes and hetarynes using density functional theory (DFT) calculations DFT methods [58, 67, 75]. Based on this aryne distortion model, the larger the internal angle is on the triple bond carbon, the more electrophilic site is on an aryne species in either nucleophilic or other polar transformations. Moreover, the difference between two internal angles is closely related to the degree of regioselectivity.

When the internal angle difference is larger than 4°, distinct regioisomeric ratios would be obtained [66, 67, 69], which has also been confirmed by many experimental results.

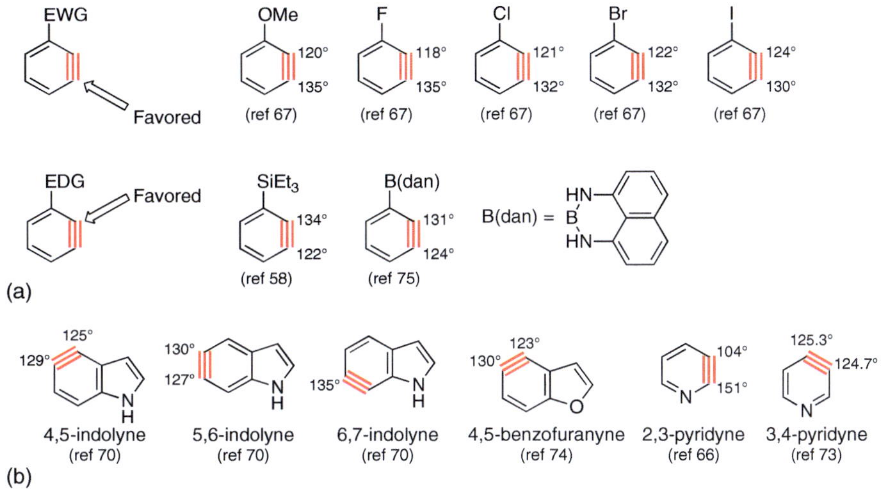

Scheme 8.22 Geometry-optimized substituted benzynes and hetarynes. (a) Substituted arynes and (b) Heterocyclic arynes.

Prior to the establishment of this aryne distortion model, there have been already many experimental examples that could realize decent regioselectivities through electronic effects. For instance, electron-withdrawing groups (EWGs), such as methoxy and halogens, are traditionally used as tuning factors in many transformations. In 1993, Suzuki and coworkers first disclosed that 3-silyl substituents on benzyne exhibited remarkable ED inductive effect (Scheme 8.23) [76]. Thus, in 1,3-dipolar reactions with nitrones, 3-silylbenzynes could afford [3+2] cycloadducts in excellent regioselective control. The aryne generation conditions in this study, however, required *n*-BuLi as the activating reagent. In 2011, Akai and coworkers found that upon activation with Kobayashi's method, 3-trimethylsilylbenzyne could be achieved. Meanwhile, its reaction with primary amines produced products through nucleophilic addition preferentially on the ortho-position of the silyl group (Scheme 8.24a) [77]. Their theoretical study revealed that a silicate-like complex might be formed, which further increases the ED inductive effect, despite

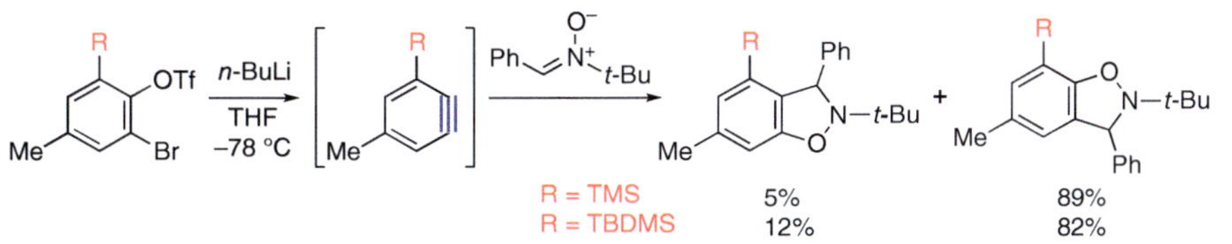

Scheme 8.23 Silyl groups as ED inductive groups by Suzuki. Source: Based on Matsumoto et al. [76].

the fact that the steric demand of this complex might be larger than the corresponding neutral one. In 2013, the same group demonstrated that 3-borylbenzynes could conduct regioselective nucleophilic addition reactions with amines as well (Scheme 8.24b) [75]. In this study, they found that under the conditions of CsF and 18-c-6 in THF/HMPA (1 : 1) at 0 °C TMS group was removed and boronic ester groups remained intact, which could generate 3-borylbenzyne in highly chemoselective manner. Further inspection on the boronic ester groups revealed that 1,8-diaminonaphthalene (dan) protected 3-borylbenzyne could afford the maximum *ortho* to *meta* ratio of the amination products.

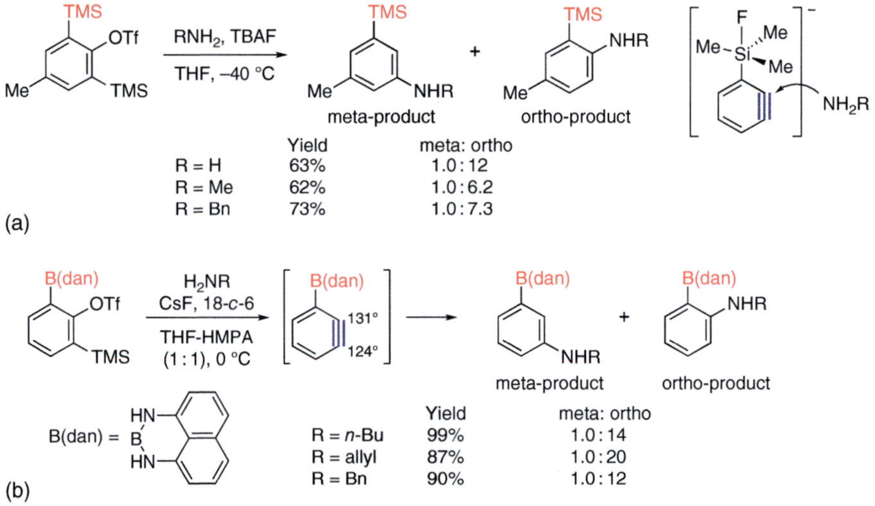

Scheme 8.24 Ikawa-Akai's work on 3-silyl- and 3-borylbenzynes. (a) nucleophilic reaction with 3-trimethylsilylbenzyne. Source: Based on Ikawa et al. [77]; (b) nucleophilic reaction with 3-borylbenzyne. Source: Based on Takagi et al. [75].

It is now known that both 3-boryl and 3-silyl groups are EDGs, both of which also possess certain degree of steric congestion. Interestingly, under certain circumstances, one effect dominates and in other cases the other could become the major one. In a study carried out by Tokiwa, Akai, and coworkers, they reported their observation on regiocomplementary [3+2] cycloaddition reactions by using either 3-boryl- or 3-silylbenzynes [78]. As shown in Scheme 8.25, they found that in aryne [3+2] cycloaddition reactions with 1,3-dipoles, cycloadducts with opposite regioselectivity were obtained from either 3-boryl- or 3-silylbenzynes. When 3-borylbenzyne was employed, the regioisomeric ratios of cycloadducts with azides were predominantly affected by its inductively ED effect [75]. In contrast, the regioselectivity in [3+2] cycloaddition between azides and 3-silylbenzyne was dominated by the steric repulsion of the silyl group [60]. These phenomena were also unambiguously explained by their DFT calculations [78].

In the past decade, the groups of Garg and Houk have systematically investigated a variety of hetarynes, including indolines, pyridynes, and benzofuranyne, which

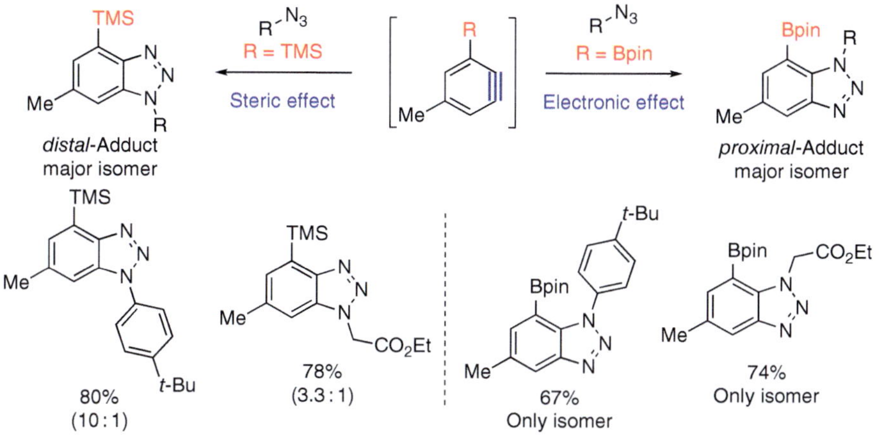

Scheme 8.25 Regiocomplementary [3+2] cycloaddition reactions by Tokiwa-Akai.

led to many outstanding applications in natural product total syntheses. Here, we only want to briefly mention some intriguing aspect related to aryne regioselective control in their study. With the assistance of aryne distortion model, Garg, Houk, and coworkers could easily predict the regioselective preference when different substituents are present on the proximal position of an aryne triple bond. More distinctively, they could intentionally manipulate the regioisomeric ratios on some hetaryne species through the incorporation of inductively EW groups on designated positions.

In their study on 4,5-indolyne, Garg and coworkers demonstrated that by incorporating a bromo group on its 6-position, the original geometry of 4,5-indolyne changed and the regioselective preference upon nucleophilic attack was reversed (Scheme 8.26) [72]. In the absence of 6-bromo group, 4,5-indolyne favored C5 nucleophilic addition product; whereas on 6-bromo-4,5-indolyne various nucleophiles attacked its C4-position preferentially. Encouraged by these experimental observations, they commenced a distinct total synthesis of indolactam V, the key step of which involved a regioselective C4-nucleophilic addition on 6-bromo-4,5-indolyne intermediate.

Subsequently, Garg and coworkers reported another outstanding example by using inductively EW tuning factors to enhance and/or alter the regioselectivity on 3,4-pyridyne [73]. As shown in Scheme 8.27a, their DFT calculation indicated that a simple 3,4-pyridyne has similar internal angles (125.3° and 124.7°) on its triple bond, which means that there is lack of regioselective control upon reaction with arynophiles. Interestingly, by incorporating EWGs on either 5- or 2-positions of 3,4-pyridyne, the internal angles of both aryne carbons changed markedly (Scheme 8.27a). When an EWG was on 5-position, C3-carbon became the more electrophilic site over C4-carbon. In contrast, the C4-position of 2-substituted 3,4-pyridyne is more electrophilic. Next, they examined these DFT calculated structures in various aryne reactions. As shown in Scheme 8.27b, in the absence of substituents on 3,4-pyridyne, there was almost no regioselective control in both

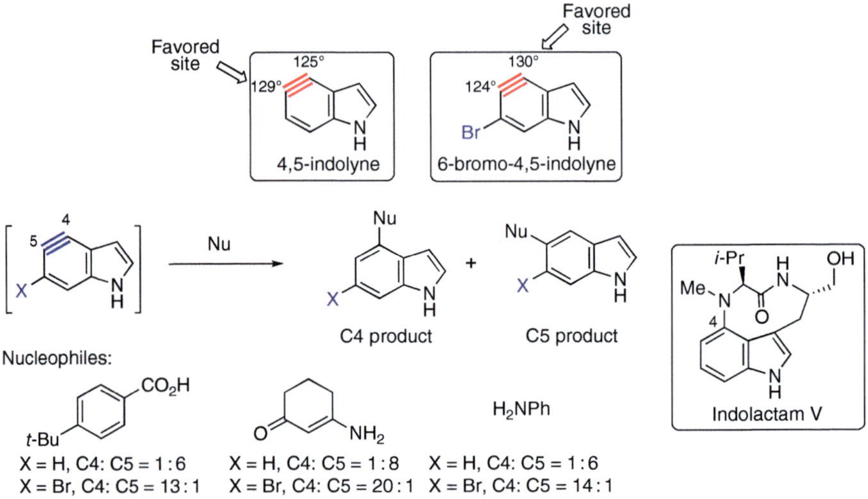

Scheme 8.26 Reversing regioselectivity on 4,5-indolyne. Source: Based on Bronner et al. [72].

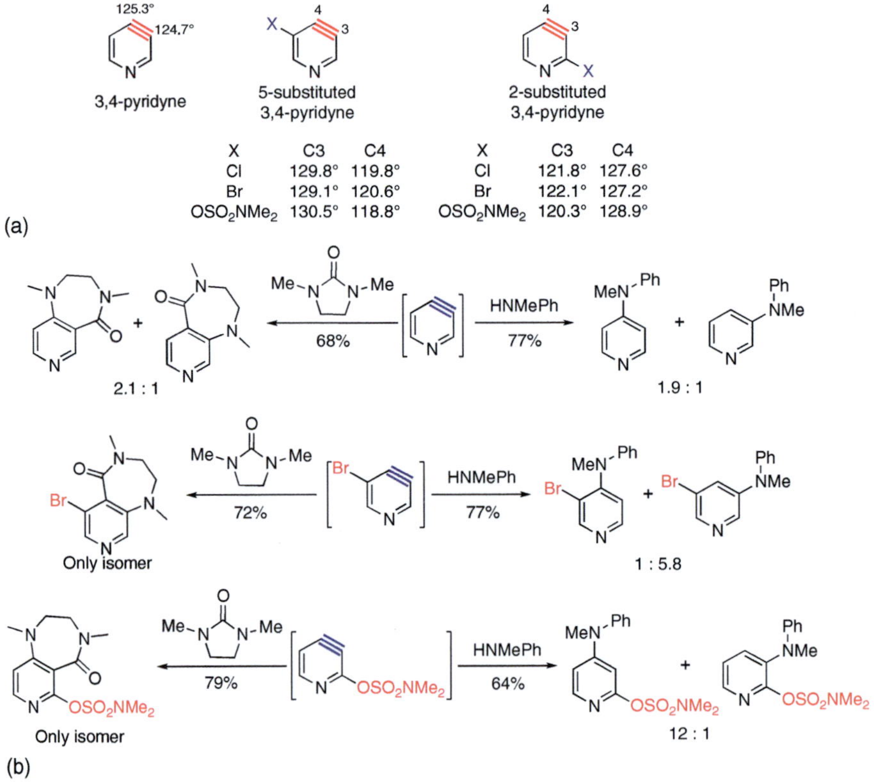

Scheme 8.27 Manipulating regioselectivity on 3,4-pyridyne. (a) internal angels of substituted 3,4-pyridynes; (b) Garg's study on regioselective 3,4-pyridyne chemistry.

nucleophilic addition and C—N σ-bond insertion reaction. Distinctively, both 5- and 2-substituted 3,4-pyridynes could afford products with much higher regioselectivity in the same reactions. Particularly, as shown in their DFT calculation, the regioselective preferences of both 5- and 2-substituted 3,4-pyridynes were opposite to each other.

8.3.3 Regioselectivity on Small Ring-Fused Arynes

Another factor that could tune aryne regioselectivity was through ring strain on small ring-fused arynes, which is uncommon with respect to aryne regioselectivity. This tuning factor was accidentally discovered by Suzuki and coworkers, when they studied the [2+2] cycloaddition of a four-membered ring-fused aryne precursor (Scheme 8.28a) [79]. Two regioisomers were obtained in 31 : 1 ratio. Although they first speculated that steric repulsion might be responsible for this regioselective outcome, it was then ruled out by employing different ring-fused aryne precursors (Scheme 8.28b). Particularly, high regioselectivity was observed when a four-membered ring was fused onto the 3,4-positions of a benzyne intermediate. In marked contrast, both five- and six-membered ring-fused benzyne precursors afforded the products in low regioisomeric ratios (Scheme 8.28b). To elucidate the origin of these regioselective observations, they reasoned that, according to the work by Finnegan [80] and Streitwieser [81], the bridgehead carbon rehybridizes and uses orbital with more p character to bind with the four-membered ring. The consequence of this rehybridization makes the meta-position of the strained ring more electrophilic.

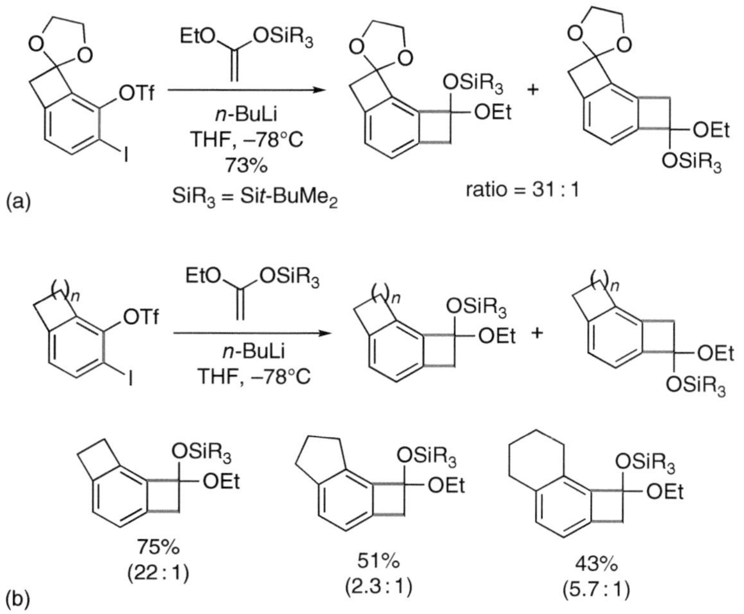

Scheme 8.28 Suzuki's study on small ring-fused arynes. Source: Based on Hamura et al. [79].

Soon after the discovery of regioselective control by fused four-membered ring on benzyne, Suzuki and coworkers quickly utilized this protocol in a series of outstanding syntheses of hexasubstituted benzenes, those of which were otherwise difficult to access via other methods [82–84]. These works will be covered in detail in next section. In view of the growing application of 4,5-indolyne and HDDA aryne chemistry recently developed by Garg-Houk and Hoye, respectively, a question is whether the origin of regioselectivity in some of these aryne intermediates might come from fused ring on arynes as well.

8.4 Recent Advances in Aryne Multifunctionalization

The merit of aryne transformations resides in ready assembly of various vicinal difunctionalized arenes. A massive number of aromatic compounds, however, contain more than two substituents or polycyclic frameworks. Consequently, convenient preparation of polysubstituted arenes through aryne intermediates not only belongs to a natural extension of current difunctionalization strategies, but also occupies an essential portion of aryne chemistry. To break the vicinal difunctionalization restriction on conventional aryne intermediates, Hart and others began their investigation on benzdiyne and benztriyne equivalents in 1980s (Scheme 8.29) [85]. Unfortunately, the synthetic applications of these early studies have been hampered by harsh aryne generation conditions, i.e. *n*-BuLi or LDA, resulting in both low reaction efficiency and limited reaction modes. Along with the growing applications of both Kobayashi aryne precursors and Suzuki's *o*-iodoaryl triflate in the past two decades, breakthroughs in the chemistry of benzdiyne and benztriyne with respect to versatile and intriguing applications were recently unraveled.

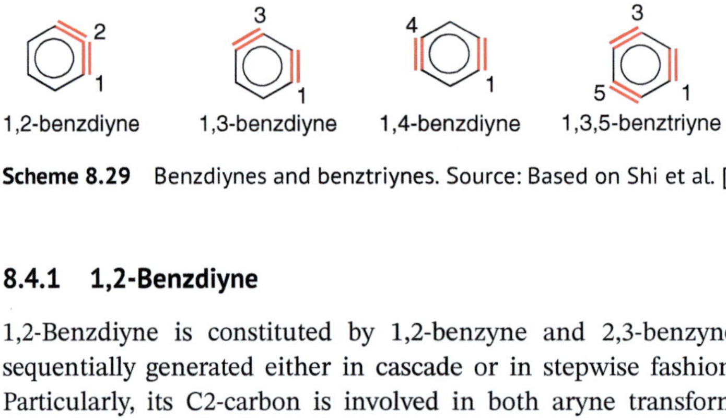

Scheme 8.29 Benzdiynes and benztriynes. Source: Based on Shi et al. [85].

8.4.1 1,2-Benzdiyne

1,2-Benzdiyne is constituted by 1,2-benzyne and 2,3-benzyne, which can be sequentially generated either in cascade or in stepwise fashion (Scheme 8.30a). Particularly, its C2-carbon is involved in both aryne transformations, which is the key factor to allow a successive generation of two aryne intermediates in 1,2-benzdiyne processes. Although this strategy was first demonstrated by Hart in 1980s, 1,2,3-trihalobenzenes were employed at the moment as 1,2-benzdiyne equivalents (Scheme 8.30b) [86, 87]. In addition, only Grignard reagents with either aryl or alkenyl/alkynyl groups could serve as both the activating reagents and double nucleophiles, limiting the application of this early 1,2-benzdiyne investigation. Although efforts have been then tried to find alternative arynophiles, this protocol was still largely restricted by its harsh aryne generation conditions. Recently, new 1,2-benzdiyne reagent by employing Kobayashi's generation method was prepared by Li and coworkers (Scheme 8.30c) [88, 89]. The uniqueness of this reagent is its ability to incorporate three different chemical bonds on a benzene ring in domino processes under mild conditions, both breaking the difunctionalization restriction of conventional aryne chemistry and possessing step-economical advantage. Along with in-depth investigation on this domino 1,2-benzdiyne as well as other 1,2-benzdiyne processes, more and more synthetic applications were recently disclosed.

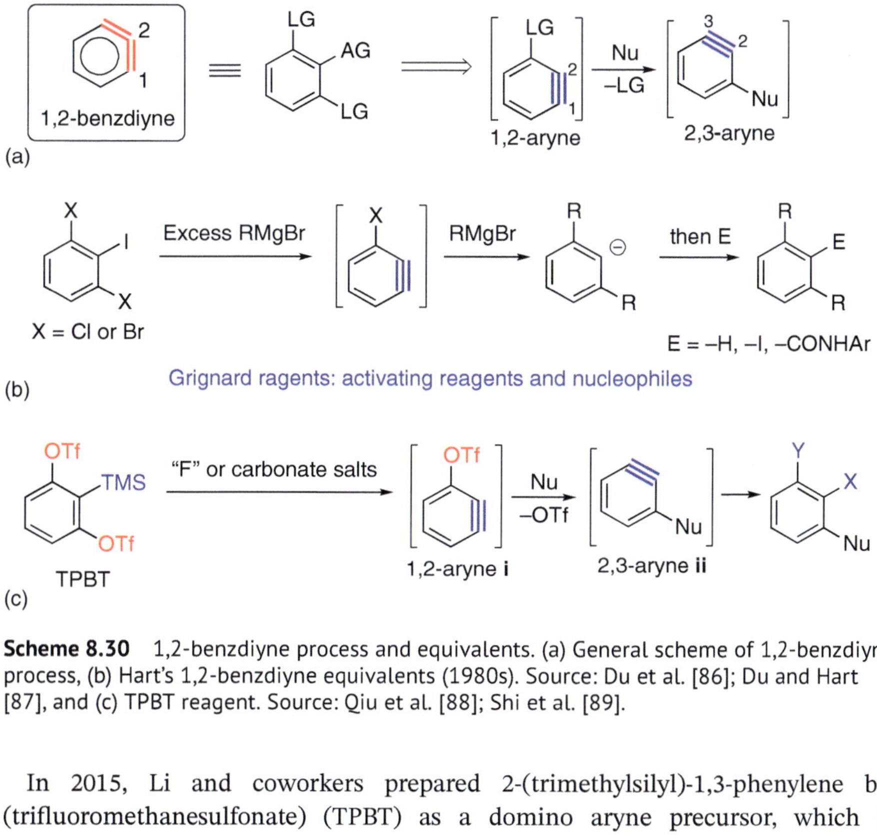

Scheme 8.30 1,2-benzdiyne process and equivalents. (a) General scheme of 1,2-benzdiyne process, (b) Hart's 1,2-benzdiyne equivalents (1980s). Source: Du et al. [86]; Du and Hart [87], and (c) TPBT reagent. Source: Qiu et al. [88]; Shi et al. [89].

In 2015, Li and coworkers prepared 2-(trimethylsilyl)-1,3-phenylene bis (trifluoromethanesulfonate) (TPBT) as a domino aryne precursor, which is constituted by two OTf groups and one TMS group on 1,2,3-positions of an arene ring [88, 89]. As shown in Scheme 8.30c, this modification allows a ready generation of 3-triflyloxybenzyne **i** [90, 91], in which the meta-position of the OTf group is more electrophilic due to the EW inductive effect of the OTf group. At this stage, a nucleophile will preferentially attack this meta-position with concomitant departure of the OTf group to generate 2,3-aryne intermediate **ii**, the process of which is similar to S_N2' mechanism but on aromatic ring. Another interesting property of TPBT reagent as 1,2-benzdiyne equivalent is that no fluoride salt is necessary as the activating reagent; instead, carbonates, such as K_2CO_3 or Cs_2CO_3, were found to be efficient activating reagents.

With this TPBT reagent in hand, Li and coworkers first examined its reaction with protected thiobenzamides (Scheme 8.31a) [89]. 2,4-Disubstituted benzothiazole **2** was readily obtained with a concomitant formation of C—S, C—N, and C—C bonds on three consecutive positions of a benzene ring. As shown in Scheme 8.31b, this domino process involves a first S-nucleophilic addition to 3-triflyloxybenzyne **i** and intramolecular N-nucleophilic attack to the in situ generated 2,3-aryne **iii**. After an intramolecular 1,3-carbonyl group migration, a C—C bond was formed on the C3-position of the benzene ring. Alternatively, when carbonyl groups contain α-proton, 1,5-hydrogen abstraction took place to produce difunctionalized product **1**.

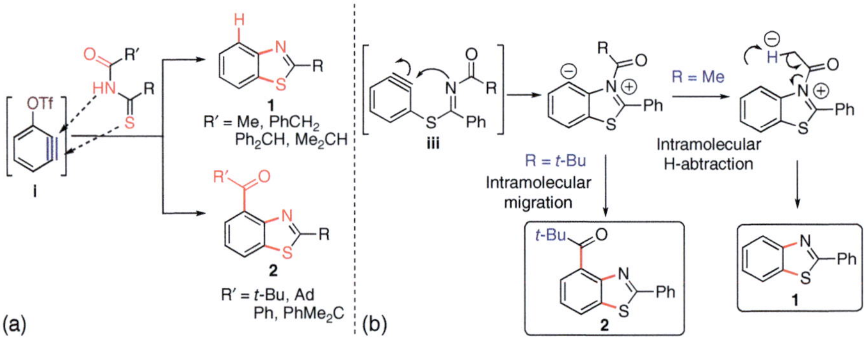

Scheme 8.31 Reaction of TPBT with protected thiobenzamides. (a) reaction between TPBT and protected thiobenzamides; (b) proposed mechanism. Source: Based on Shi et al. [89].

Subsequently, Li and coworkers demonstrated that when sulfamides were employed with SO_2 group as both the bridge and the electron-tuning factor, they could serve as dual nucleophile in domino aryne process. Diverse sulfamides with either *N*-aromatic or *N*-aliphatic substituents could all produce the corresponding vicinal diaminobenzenes (Scheme 8.32a) [92]. Alternatively, when TPBT reagent was treated with sulfonyl protected anilines, the second N-nucleophile attacked the 3-position of 2,3-aryne intermediate, resulting in 1,3-diaminobenzenes (Scheme 8.32b) [93]. Notably, by simply altering the reaction conditions, i.e. using toluene as the solvent and K_2CO_3 as the activating reagent, C2-protonation could be prohibited. Instead, an intramolecular thia-Fries rearrangement of a Tf group took place, affording 1,2,3-trisubstituted 1,3-diaminobenzenes (Scheme 8.32b) [93].

Further investigation by Li and coworkers revealed that this domino aryne process could accommodate two reaction modes, such as nucleophilic addition and ene reaction. They speculated that the strong nucleophilic property of the sulfonamide anion would prefer a nucleophilic attack to 1,2-aryne intermediate; whereas a subsequent intramolecular ene reaction could become more kinetically favored than competing intermolecular nucleophilic reaction after the formation of 2,3-aryne. However, an accurate management on the reactivity difference between *N*-nucleophile and ene arynophile was required for the success of this design.

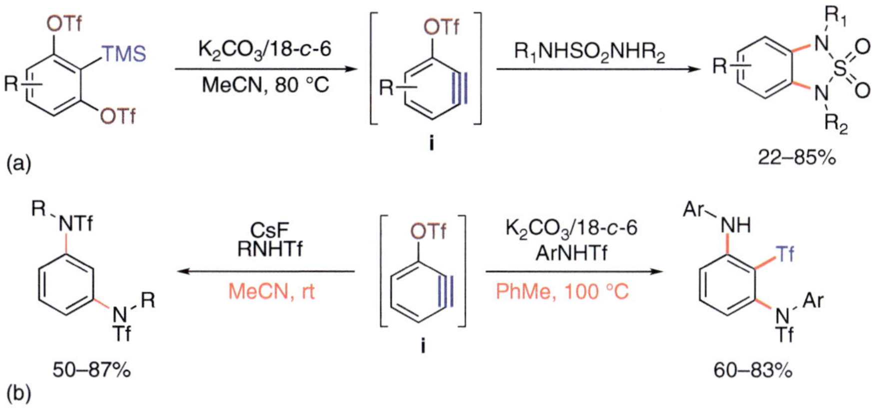

Scheme 8.32 Diamination reactions with TPBT reagent. (a) vicinal diamination with TPBT. Source: Li et al. [92]; (b) reactions with sulfonyl protected anilines; Source: Based on Qiu et al. [93].

To this end, a domino aryne nucleophilic-ene reaction cascade was developed, in which a *N*-nucleophile and an olefin were connected together with a linker (Scheme 8.33) [94]. Notably, a second-generation domino aryne precursor TTPM was developed by simply replacing one of the OTf groups on TPBT reagent with OTs, which could generate 1,2-aryne **iv** preferentially. By using TTPM as domino aryne precursor, a significant enhancement in reaction efficiency was achieved. This result was reasoned by the fact that the retarded departure tendency of the OTs group might be able to better match with the low reactivity in ene reaction, which typically requires higher activation energy than other pericyclic reactions, i.e. Diels–Alder reaction.

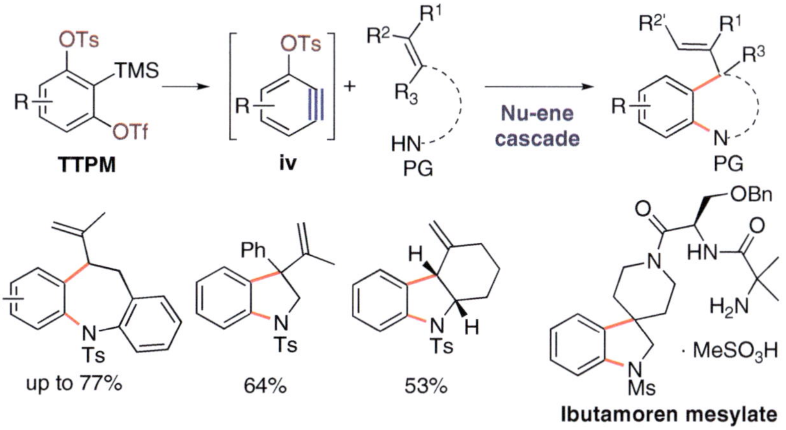

Scheme 8.33 Domino aryne nucleophilic-ene cascade. Source: Based on Xu et al. [94].

Very recently, Hoye and coworkers reported an outstanding process by merging 1,2-benzdiyne chemistry with their HDDA aryne process in a highly efficient cascade (Scheme 8.34) [95]. Upon activation of TTPM, 1,2-aryne **iv** will be formed. A nucleophilic addition with 1,3-diyne substrate, followed by a concomitant departure of the OTs group, could give 2,3-aryne intermediate **v**. An intramolecular HDDA reaction was then took place to afford naphthyne intermediate **vi**. By trapping this intermediate with various arynophiles, either intramolecularly or intermolecularly, diverse polysubstituted naphthalenes could be achieved. A distinct property of this transformation was the combination of three aryne intermediates, namely 1,2-aryne **iv**, 2,3-aryne **v**, and naphthyne **vi**, into a highly efficient and sequential generation pathway.

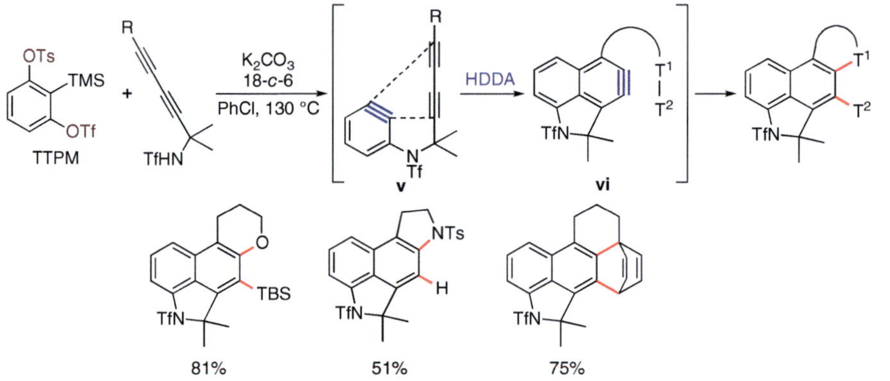

Scheme 8.34 Naphthyne from 1,2-benzdiyne-HDDA cascade. Source: Based on Xiao et al. [95].

To explore other transformation possibilities of 1,2-benzdiyne, Li and coworkers also demonstrated that the four-membered ring of [2+2] cycloadduct **3** of 1,2-aryne **i** with ketene silyl acetyls (KSAs) could be opened in a highly selective manner, in which the Csp2—Csp3 bond on the four-membered ring of compound **3** was cleaved with its Csp3—Csp3 bond intact (Scheme 8.35a) [27]. 2,3-Aryne intermediate **vii** was then generated, and those four chemical bonds between OTf and OTBS groups on benzocyclobutenol scaffold constitute a Grob fragmentation system. Further investigation on this transformation disclosed that sterically congested groups, i.e. *gem*-dimethyl and *gem*-dimethoxy groups, on the benzylic position of benzocyclobutenol could efficiently enhance the regioselectivity of the following transformations. This Grob fragmentation protocol allowed the syntheses of several drug molecules, such as Niraparab (anti-ovarian cancer medicine) and pain relievers Carprofen and Fenoprofen. Later on, Hosoya and coworkers reported a similar aryne generation method via Grob fragmentation process. In this work, they used aryllithium as the nucleophile to attack benzocyclobutenones, followed by a selective ring-opening step to give 2,3-aryne, which was then trapped by arynophiles in high efficiency (Scheme 8.35b) [96].

Other than Kobayashi's method, Hosoya and coworkers demonstrated that 3-triflyloxybenzyne could be generated from iodophenyl 2,6-bis(triflate) using Suzuki's conditions as well, which could in turn insert into the N—Si σ-bond of

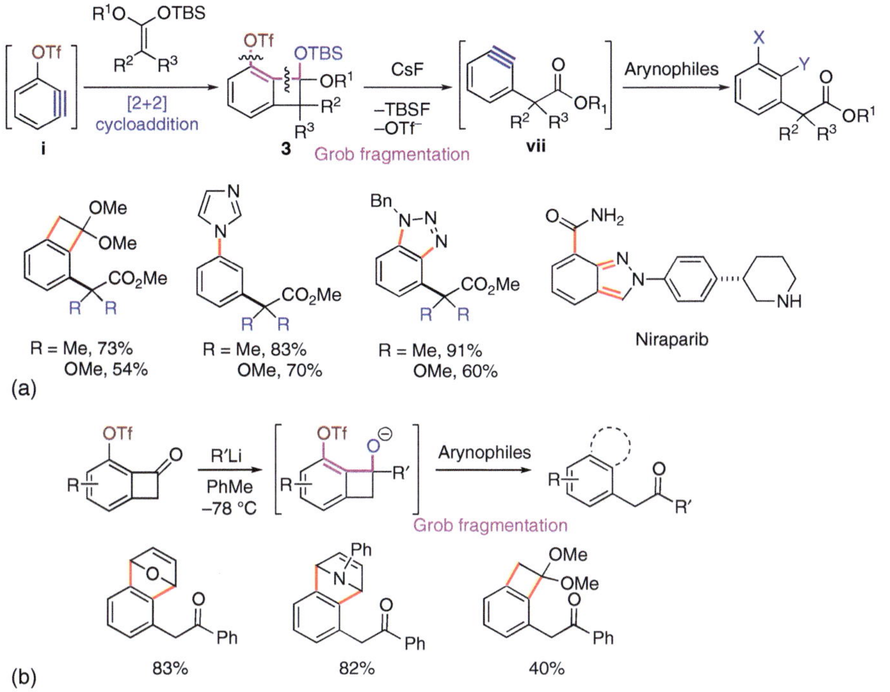

Scheme 8.35 Aryne trifunctionalization through Grob fragmentation. (a) Li's study. Source: Based on Shi et al. [27]; (b) Hosoya's study; Source: Based on Uchida et al. [96].

N-silylamines in excellent regioselectivity (Scheme 8.36a) [97]. Consequently, the isolated products could be activated to produce 2,3-aryne and trapped by various arynophiles in good yields and high regioselectivity. The overall process offered a convenient way to access 1,2,3-trisubstituted arenes containing amino moieties. Very recently, the same group developed a similar two-step protocol. The 1,2-aryne species could insert into S—Si σ-bonds in regiospecific manner (Scheme 8.36b) [98]. After a subsequent generation of 2,3-aryne intermediate, it was trapped by various arynophiles.

In 2018, Li and coworkers demonstrated an uncommon 1,2-benzdiyne process. After redesigning the constitution of 1,2-benzdiyne equivalent, they proposed a composition of two AGs and one LG as 1,2-benzdiyne precursor [99]. As shown in Scheme 8.37, after the formation of 3-(trimethylsilyl)benzyne **viii**, it reacted with pyridine *N*-oxide to generate a C—OH bond and a C-pyridine carbon bond in highly regioselective manner. Due to a strong ED inductive effect of the TMS group, the incoming oxygen selectively attacked its ortho-position. Through a one-pot protection of this OH group with either Tf or Ts group, new sets of Kobayashi precursors **4** were readily prepared with pyridinyl groups on their 3-position. These aryne precursors were then activated and examined with various arynophiles, affording the desired products in both good yields and excellent regioselectivity.

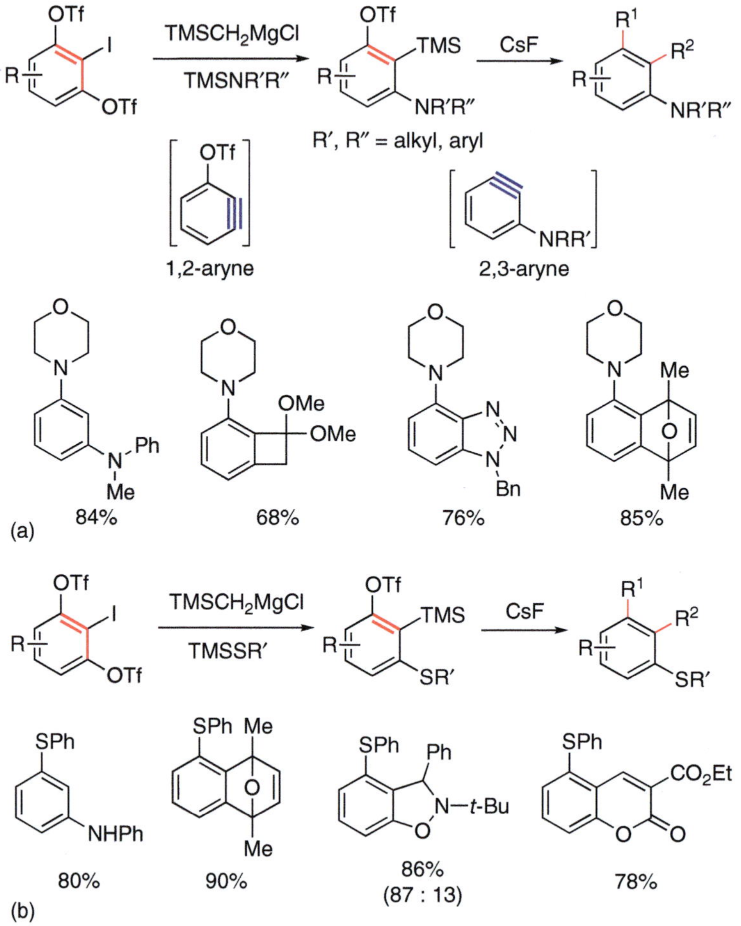

Scheme 8.36 1,2-Benzdiyne processes through σ-bond insertion of N—Si and S—Si bonds. (a) insert into N—Si σ-bond. Source: Based on Yoshida et al. [97]; (b) insert into S—Si σ-bond. Source: Based on Nakamura et al. [98].

8.4.2 1,3-Benzdiyne

1,3-Benzdiynes are composed of two pairs of aryne precursors on 1,2- and 3,4-positions of the same benzene ring, respectively, which could be able to incorporate up to four substituents on a benzene ring. Early study by Hart and coworkers using polyhalobenzenes as 1,3-benzdiyne equivalents received only low reaction efficiency with limited application [100, 101]. Beside the low efficiency caused by harsh generation conditions, there remain other obstacles on 1,3-benzdiyne chemistry. Because two aryne intermediates are independent of each other, both the regioselective control and the reaction modes on these twofold aryne scaffolds are difficult to manipulate. In this context, recent efforts have been tried toward breaking these restrictions. Through properly tuning the generation of two arynes in stepwise manner, a wide range of reaction modes could be accommodated.

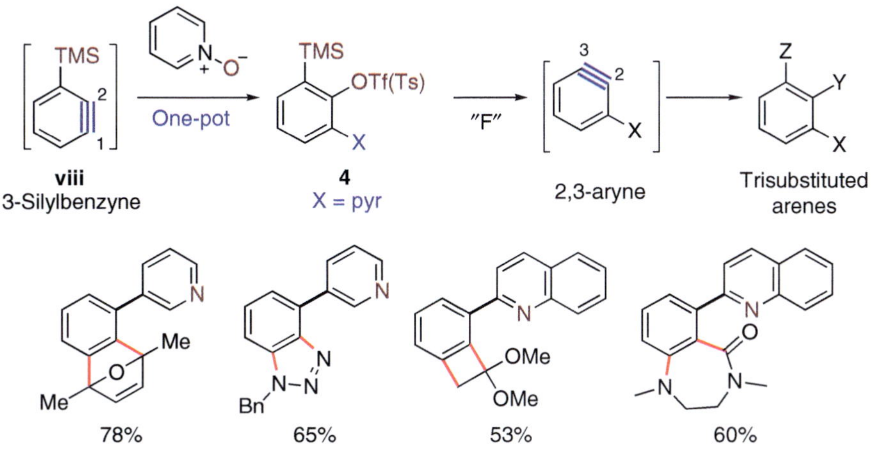

Scheme 8.37 3-Silylbenzyne as 1,2-benzdiyne equivalent.

Moreover, a judiciously designed reaction sequence could help enhance the regioselectivity as well.

In 2003, Suzuki and coworkers found that [2+2] cycloaddition of **5** with KSAs was highly regioselective to produce benzocyclobutenol **6** due to EW inductive effect of oxygen on 3-position (Scheme 8.38) [79]. This benzocyclobutenol was then converted to a 3,4-aryne precursor **7**. Upon activation, 3,4-aryne **ix** was generated and the following reactions were highly regioselective as well due to the presence of fused four-membered ring on the proximal position of 3,4-aryne ix. This work was probably the first example to demonstrate that by properly designing the structure of aryne precursors and reaction modes, 1,3-benzdiyne process could be realized in high efficiency with exclusive regioselective manner.

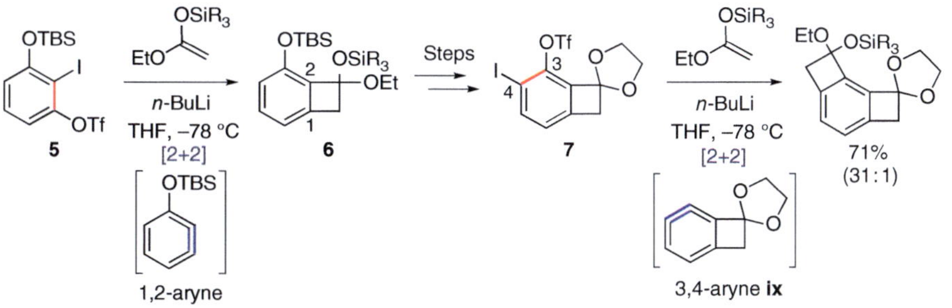

Scheme 8.38 Suzuki's 1,3-benzdiyne strategy. Source: Based on Hamura et al. [79].

In 2016, Ikawa, Akai and coworkers designed a new 1,3-benzdiyne equivalent **8** (Scheme 8.39) [102]. A unique property of this compound is that TMS will be first activated by fluoride over TBS group. Consequently, 1,2-aryne **x** could be generated preferentially. Distinctively, the regioselective control with respect to 1,2-aryne **x** was manipulated by the strong inductively ED effect of the 3-TBS group. After accomplishing highly regioselective cycloaddition reactions on 1,2-aryne **x**, they further realized a time-dependent sequential generation of 1,2- and 3,4-arynes by adding two types of arynophiles at different reaction stages, affording unsymmetrical bisannulated products.

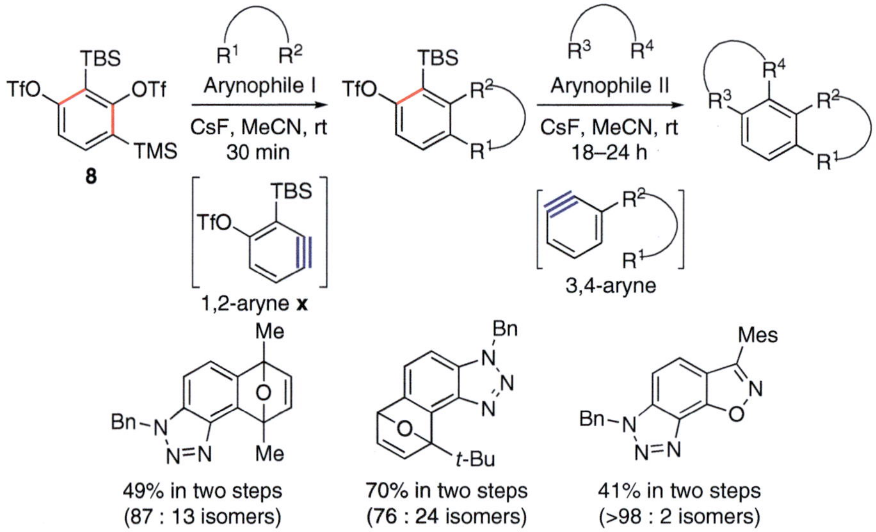

Scheme 8.39 Ikawa-Akai's 1,3-benzdiyne strategy. Source: Based on Ikawa et al. [102].

In 2018, Kitamura and coworkers developed an unprecedented hybrid 1,3-benzdiyne equivalent **9**, in which phenyliodonio and OTf groups were utilized as LGs (Scheme 8.40) [103]. This design was based on the fact that the

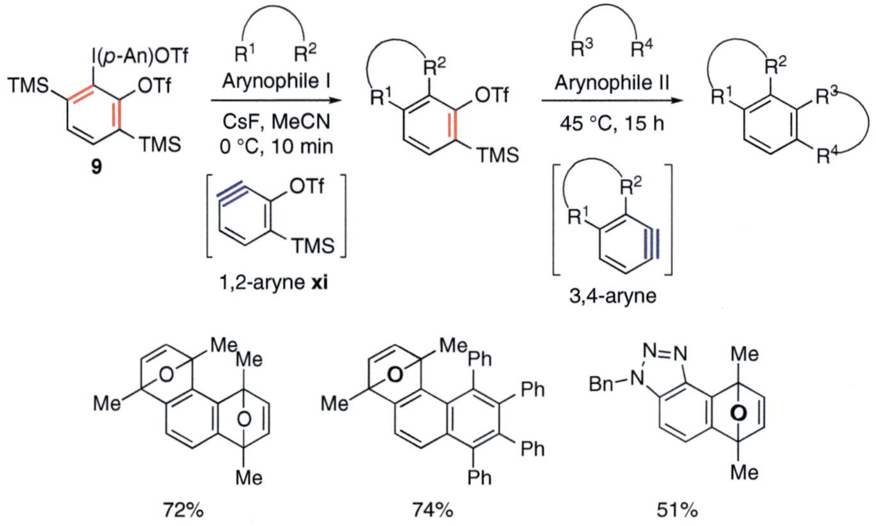

Scheme 8.40 Kitamura's hybrid 1,3-benzdiyne equivalent. Source: Based on Kitamura et al. [103].

phenyliodonio group has a much higher leaving ability than that of the OTf group [18], allowing a chemoselective formation of 1,2-aryne **xi** via preferential extrusion of the phenyliodonio group under mild conditions. With this hybrid 1,3-benzdiyne equivalent in hand, a one-pot double cycloaddition process with different arynophiles was conducted through convenient temperature control, affording angular polycyclic aromatic compounds in high yields.

8.4.3 1,4-Benzdiyne

The utilization of 1,4-benzdiyne equivalents can be traced back to the early stage of benzdiyne chemistry in 1980s, at which moment Hart pioneered a study in this field [104–107]. Unfortunately, unlike both 1,2-benzdiyne and 1,3-benzdiyne, there is almost no way to control the regioselectivity in 1,4-benzdiyne processes, because two aryne intermediates, namely 1,2- and 4,5-arynes, are apart from each other on a benzene ring. However, this drawback does not prevent its synthetic applications. Indeed, not only solutions on regioselective control were discovered in some cases, but also a rich and unprecedented utilization of 1,4-benzdiyne strategy in constructing polycyclic aromatic hydrocarbons (PAHs), i.e. "nanographene" frameworks, has been developed in recent years.

In 2006, Suzuki and coworkers prepared a novel 1,4-benzdiyne equivalent, namely bis(sulfonyloxy)diiodobenzene (**10**), with an OTf and an OTs group as LGs (Scheme 8.41a) [28]. This design could differentiate the generation of two arynes based on the fact that OTf group has a better leaving tendency than OTs group, producing 1,2-aryne **xii** preferentially. In addition, a methoxy group was intentionally positioned on C3-position between two arynes, which could serve as a key factor for regioselective control through its inductively EW effect. With this 1,4-benzdiyne equivalent **10** in hand, they also developed a one-pot process to generate two aryne intermediates in stepwise fashion by simply varying the reaction temperature from −95 °C to above −78 °C. Consequently, different arynophiles were able to react with sequentially generated arynes, namely 1,2- and 4,5-arynes, in highly selective manner, affording various polycyclic products. Later on, they also applied 1,4-benzdiyne equivalent **10** in a temperature-dependent double nucleophilic–iodination reaction with different alkynyllithium in a one-pot fashion, giving rise to asymmetrically bis(alkyn)ylated product **11** (Scheme 8.41b) [38].

Very recently, Suzuki and coworkers employed 1,4-benzdiyne strategy in a total synthesis of actinorhodin, which is a dimeric natural product (Scheme 8.42) [108]. In this synthesis, they first designed and prepared a 1,4-benzdiyne equivalent **12** with *o*-iodoaryl tosylate and *o*-silylaryl triflate as two sets of aryne precursors. Upon activation with *n*-BuLi, the *o*-iodoaryl tosylate side could be first converted to 1,2-aryne intermediate **xiii**, which was then trapped by excess amount of furan. Next, the activation of *o*-silylaryl triflate with KF and 18-crown-6 in dilute THF at room temperature allowed the formation of 4,5-aryne **xiv**, which then participated in an insertion reaction into the C—C σ-bond of methyl acetoacetate to afford **13**. The regioselectivities in both aryne reactions were well tuned by a benzyloxy group on C3-position. After several-step manipulation, monomer **14**

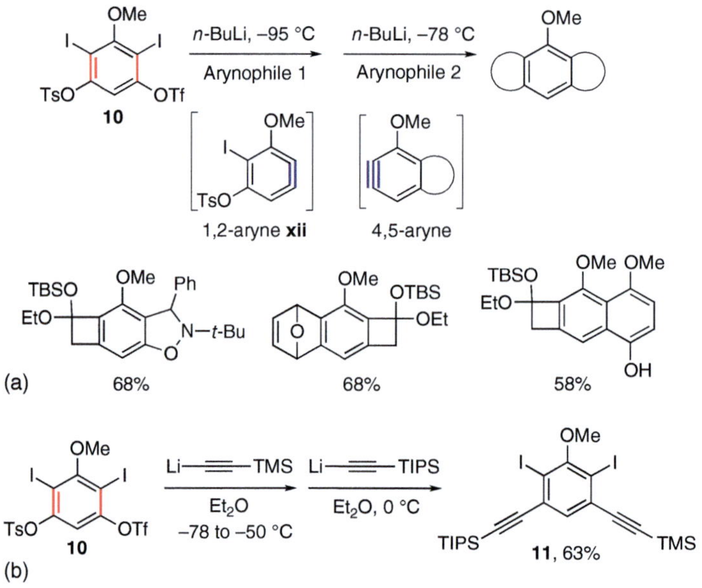

Scheme 8.41 1,4-Benzdiyne from bis(sulfonyloxy)diiodobenzene **10**. Source: Based on Hamura et al. [28]; Hamura et al. [38].

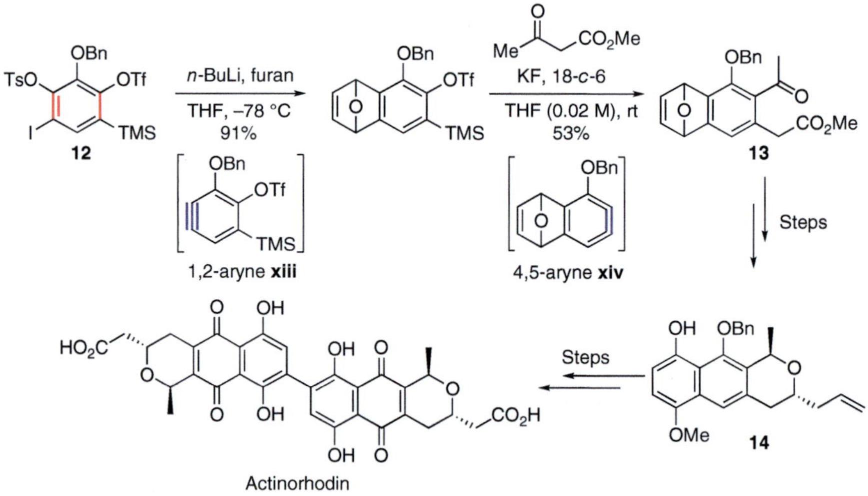

Scheme 8.42 Total synthesis of actinorhodin via 1,4-benzdiyne equivalent **12**. Source: Based on Ninomiya et al. [108].

was prepared. With this monomer in hand, subsequent dimerization, followed by further transformations, completed the synthesis of actinorhodin.

A significant contribution of 1,4-benzdiyne chemistry is its ready assembly of PAHs. In 2003, Wudl and coworkers first prepared 1,4-bis(trimethylsilyl)phenyl 2,5-bis(triflate) (**15**) as 1,4-benzdiyne equivalent, which could be activated through

Kobayashi's method under mild conditions (Scheme 8.43a). A "twistacene" with a seven linear polyacene core was readily prepared from **15** and pyrano-diphenylcyclopentadienone through double [4+2] cycloadditions (Scheme 8.43a) [109]. In 2016, same double [4+2] cycloaddition strategy of compound **15** was employed by Sygula and coworkers to prepare bis-corannulenoanthracene from isocorannulenofuran (Scheme 8.43b) [110]. Recently, Pérez, Martín, and coworkers could use 1,4-benzdiyne equivalent **15** to link either two C_{60} or one C_{60} with few-layered graphene (FLG) unit as well (Scheme 8.43c) [111].

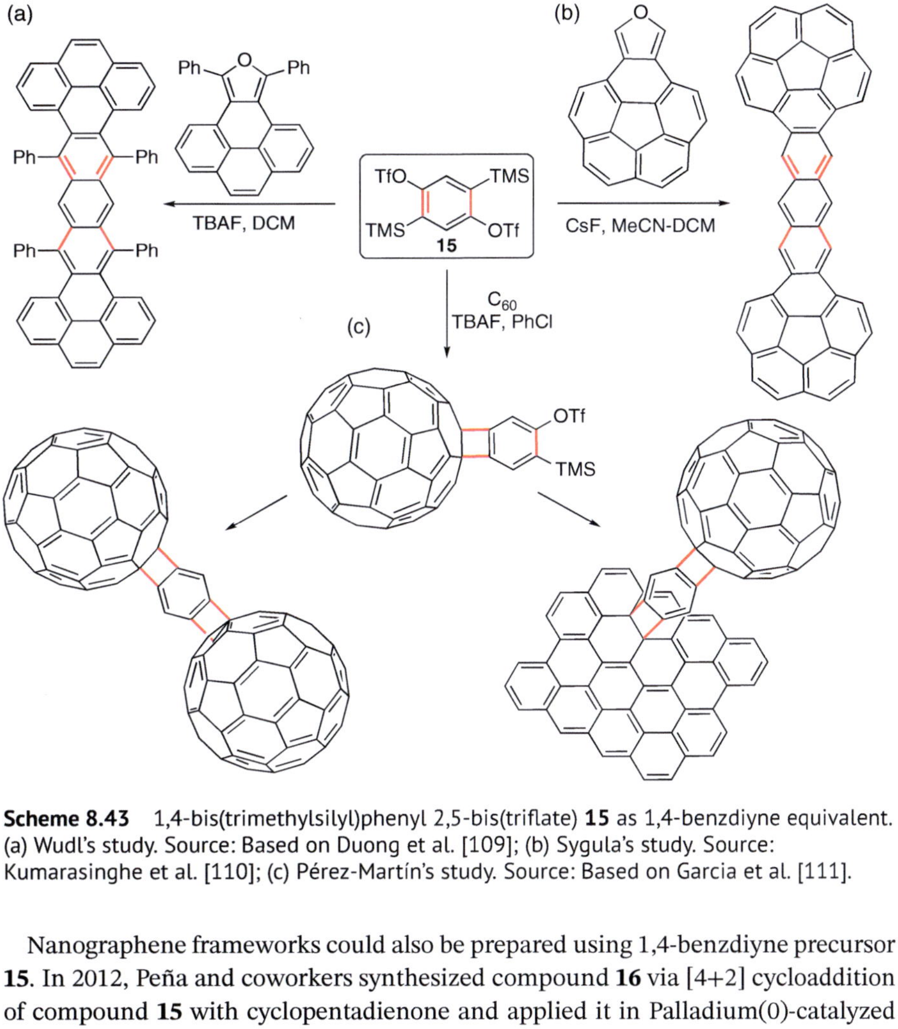

Scheme 8.43 1,4-bis(trimethylsilyl)phenyl 2,5-bis(triflate) **15** as 1,4-benzdiyne equivalent. (a) Wudl's study. Source: Based on Duong et al. [109]; (b) Sygula's study. Source: Kumarasinghe et al. [110]; (c) Pérez-Martín's study. Source: Based on Garcia et al. [111].

Nanographene frameworks could also be prepared using 1,4-benzdiyne precursor **15**. In 2012, Peña and coworkers synthesized compound **16** via [4+2] cycloaddition of compound **15** with cyclopentadienone and applied it in Palladium(0)-catalyzed [2+2+2] cyclotrimerization, affording clover-shaped hexabenzotrianthrycene **17** (Scheme 8.44a) [112]. Subsequently, they could employ the same protocol to prepare nanographene **19** from Pd-catalyzed [2+2+2] cyclotrimerization of aryne precursor **18** with a total of 22 aromatic rings (Scheme 8.44b) [113]. Both studies employed a strategy by preparing 4,5-aryne precursor with polycyclic aromatic skeletons via

[4+2] cycloaddition of 1,2-aryne intermediate, which then underwent Pd-catalyzed [2+2+2] cyclotrimerization to afford nanographene frameworks.

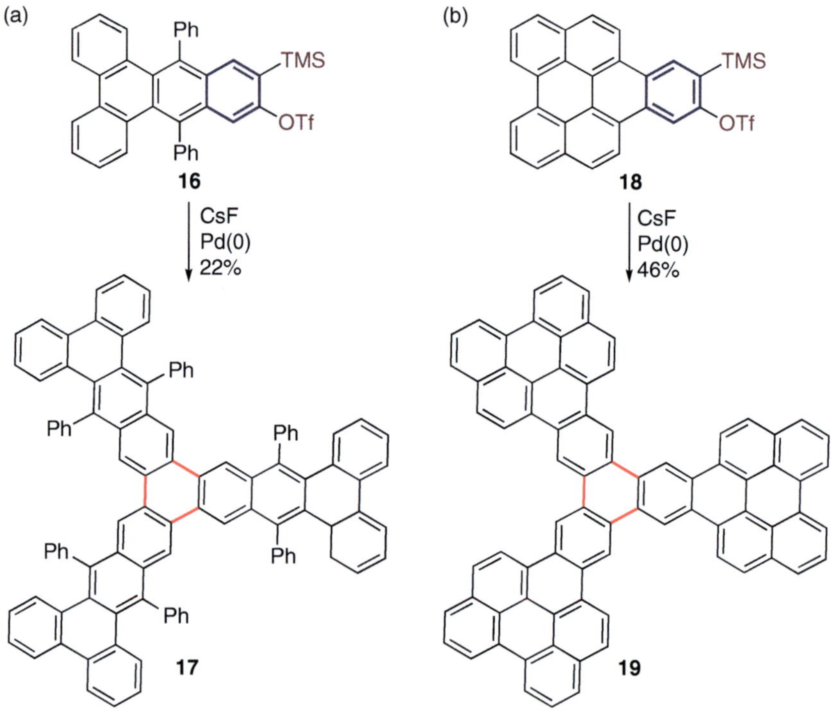

Scheme 8.44 Nanographenes from 1,4-benzdiyne strategy. Source: Based on Alonso et al. [112]; Schuler et al. [113].

Besides, Ikawa, Akai and coworkers investigated the reaction behavior of 1,4-benzdiyne equivalent **15** with different arynophiles, which led to the formation of various unsymmetrical bisannulated products in a one-pot reaction sequence (Scheme 8.45) [102]. Because two arynes are too far away with respect to each other, regioselectivities in this study were generally not satisfied.

Moreover, along with the invention of hybrid 1,3-benzdiyne equivalent **9** using both phenyliodonio and OTf groups as LGs (Scheme 8.40), Kitamura and coworkers also prepared a hybrid 1,4-benzdiyne equivalent **20**, which could generate 1,2-aryne **xv** and 4,5-aryne **xvi** (Scheme 8.46) [103]. By employing temperature-dependent activation strategy on this hybrid 1,4-benzdiyne equivalent, they developed a one-pot double cycloaddition process with different arynophiles as well.

1,4-Benzdiyne strategy has also been utilized in natural product synthesis as well. In 2006, Martin and coworkers reported a convergent synthesis of vineomcinone B_2 methyl ester, in which a double intramolecular Diels–Alder reaction of 1,4-benzdiyne equivalent **21** took place to assemble the core framework **22** (Scheme 8.47) [114]. 1,4-Benzdiyne equivalent **21** was prepared with two removable silicon-containing linkers, which could guarantee regiospecific intramolecular

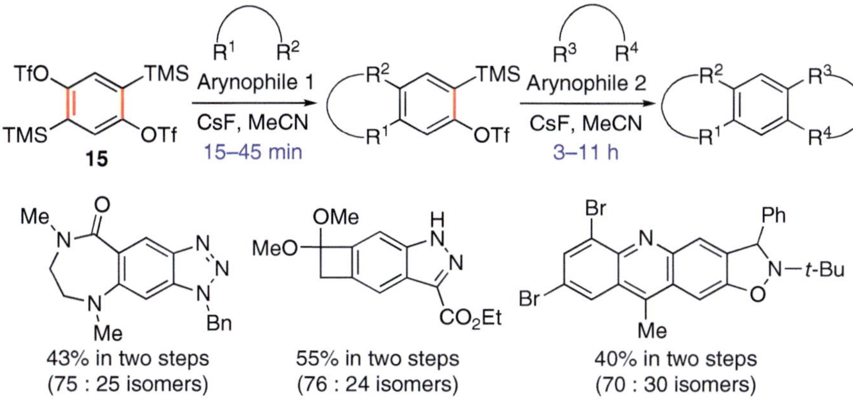

Scheme 8.45 Ikawa-Akai's 1,4-benzdiyne strategy. Source: Based on Ikawa et al. [102].

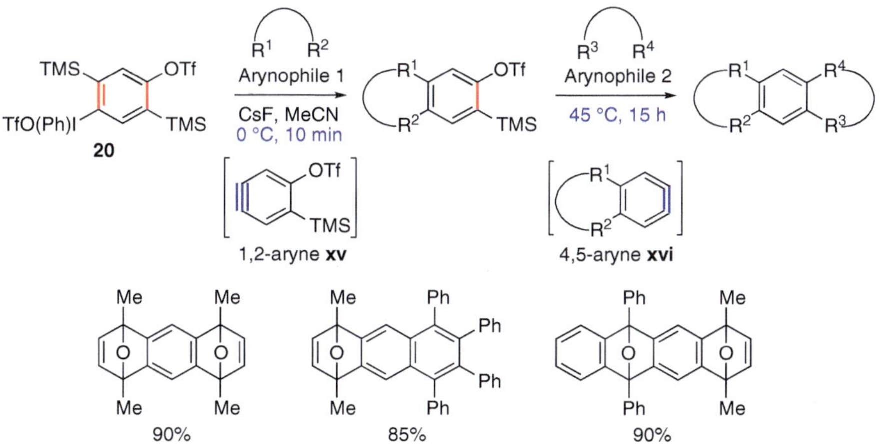

Scheme 8.46 Kitamura's hybrid 1,4-benzdiyne equivalent. Source: Based on Kitamura et al. [103].

double [4+2] cycloaddition reactions. After cleavage of the silicon-bridgehead carbon bonds on cycloadduct **22**, followed by further manipulations, vineomcinone B_2 methyl ester could then be efficiently prepared.

8.4.4 1,3,5-Benztriyne

Presumably one of the ultimate goals of aryne multifunctionalization strategy is to reach hexasubstituted arenes that are otherwise challenging to access via traditional methods. This proposal was first demonstrated in 1980s by Stoddart and coworkers, who employed both hexabromobenzene [115] and 1,2,4,5-tetrabromobenzene [116] as equivalents of 1,3,5-benztriyne. These early efforts on benztriyne hexasubstitution, however, received little attention, presumably due to both unfriendly aryne generation conditions and lack of strategic design on this triple-aryne process. Until recently, 1,3,5-benztriyne chemistry was revisited by Suzuki and coworkers,

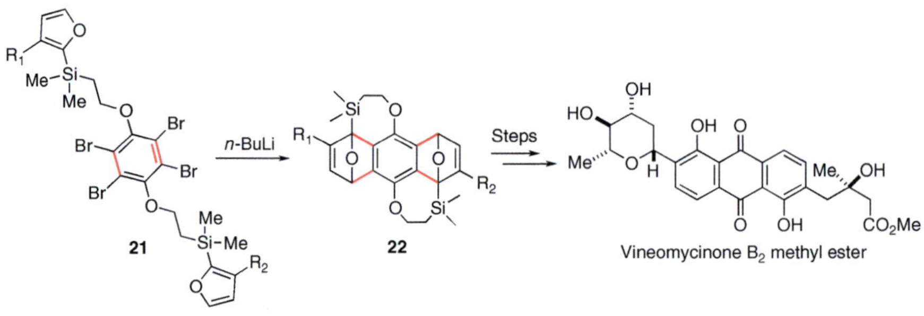

Scheme 8.47 Total synthesis of Vineomcinone B_2 methyl ester by Martin. Source: Based on Chen et al. [114].

where they developed several novel strategies to generate three aryne intermediates, namely 1,2-, 3,4-, and 5,6-arynes, in both stepwise fashion and strictly ordered sequence. In this context, a series of intriguing molecular frameworks could be readily synthesized.

In 2006, Suzuki and coworkers reported a novel preparation of tricyclobutabenzene (TCBB) **29** (Scheme 8.48) [82]. In this synthesis, three sequential [2+2] cycloaddition reactions with ketene silyl acetals (KSAs) were realized in regioselective manner. Starting from aryne precursor **23**, the proximal methoxy group on 1,2-aryne **xvii** ensured a regiospecific [2+2] cycloaddition to afford compound **24**. 3,4-Aryne precursor **25** was then prepared after several-step manipulations. Upon

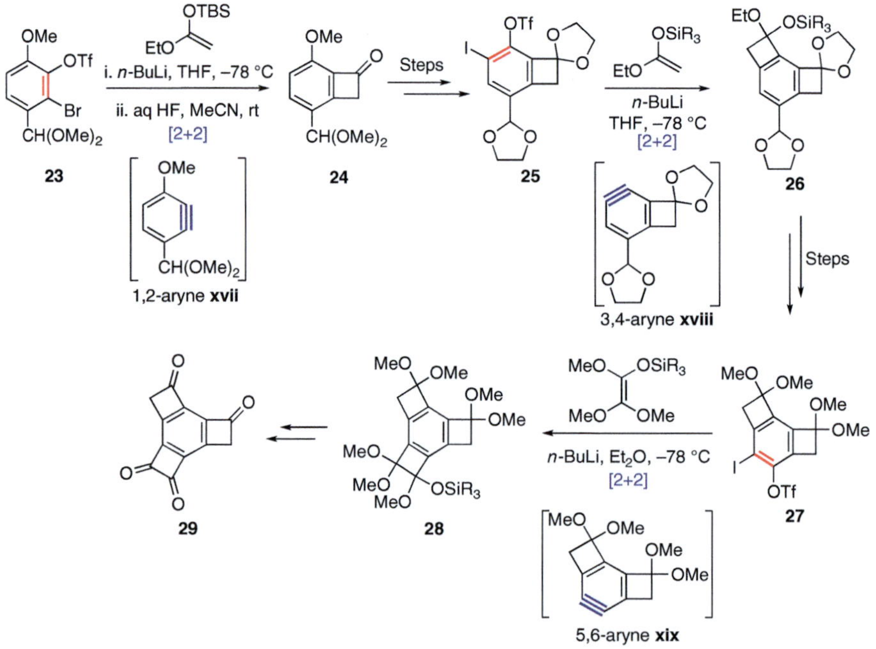

Scheme 8.48 Suzuki's synthesis of tetraketone **29**. Source: Based on Hamura et al. [82].

activation, the second [2+2] cycloaddition took place on 3,4-aryne intermediate **xviii**. At this stage, the regioselective control was realized by proximally fused four-membered ring, which led to a preferential formation of cycloadduct **26** [79]. After the preparation of compound **27**, 5,6-aryne **xix** could be generated and participate in the third [2+2] cycloaddition reaction to give cycloadduct **28**. Hydrolysis of ketene groups on compound **28** prepared compound tetraketone **29**. It is worth noting that the overall route in this synthesis employed three pairs of *o*-iodoaryl triflate as aryne precursors, demonstrating the power of this aryne generation method in the construction of complex molecular framework.

In the same year, Suzuki and coworkers reported the syntheses of TCBBs **33** and **34** (Scheme 8.49) [83]. Once again, this synthetic route was accomplished through three [2+2] cycloaddition reactions as the key steps on sequentially generated 1,2-aryne **xx**, 3,4-aryne **xxi**, and 5,6-aryne **xxii** from their corresponding precursors **30**, **31**, and **32**, respectively. Although both the second and third [2+2] cycloaddition reactions did not give high regioselectivity, after universal conversion of three silyl ethers to methyl ethers, dodecamethoxy-TCBB **33** could be obtained. Upon hydrolysis of ketene groups with concentrated sulfuric acid, hexaoxo-TCBB **34** was achieved.

Later in 2010, Suzuki and coworkers reported an unprecedented synthesis of hexaradialenes **38**, starting from the same compound **30** (Scheme 8.50) [84]. After two regioselective [2+2] cycloadditions with KSAs, compound **35** was obtained, which then underwent the third [2+2] cycloaddition via 5,6-aryne **xxiii** to afford cycloadduct **36**. With further manipulations, compound **37** was prepared. Under elevated temperature, compound **37** could readily dearomatize to afford hexaradialene **38**.

8.4.5 Benzyne Insertion, C–H Functionalization Cascade

The successful employment of both benzdiyne and benztriyne equivalents as the efficient synthons for the preparation of polysubstituted arenes expands the realm of aryne chemistry from traditional vicinal disubstitution to multiple sites on a benzene ring. The expense through these strategies, however, is the utilization of equally substituted aryne equivalents as the starting materials, which requires increasing number of steps in their preparation. Therefore, these strategies fall into lack of functional group economy. In this regard, a better approach toward aryne multifunctionalization is to employ aryne precursors with less substituents. To this end, a possible solution is to combine traditional aryne transformation with sequential C–H functionalization in cascade processes.

In 2016, Li and coworkers reported an unprecedented cascade aryne trifunctionalization protocol by using simple benzyne (Scheme 8.51) [117]. They found that a three-component reaction involving simple benzyne, aryl allyl sulfoxide, and ethyl bromoacetate could afford 1,2,3-trisubstituted arenes with concomitant formation of C—S, C—O, and C—C bonds on the consecutive positions of a benzene ring. This cascade process was proposed to proceed through a benzyne insertion into sulfoxide bond, an intramolecular S to O allyl migration, and an oxonium Claisen rearrangement sequence. Overall, it was highly efficient and can accommodate

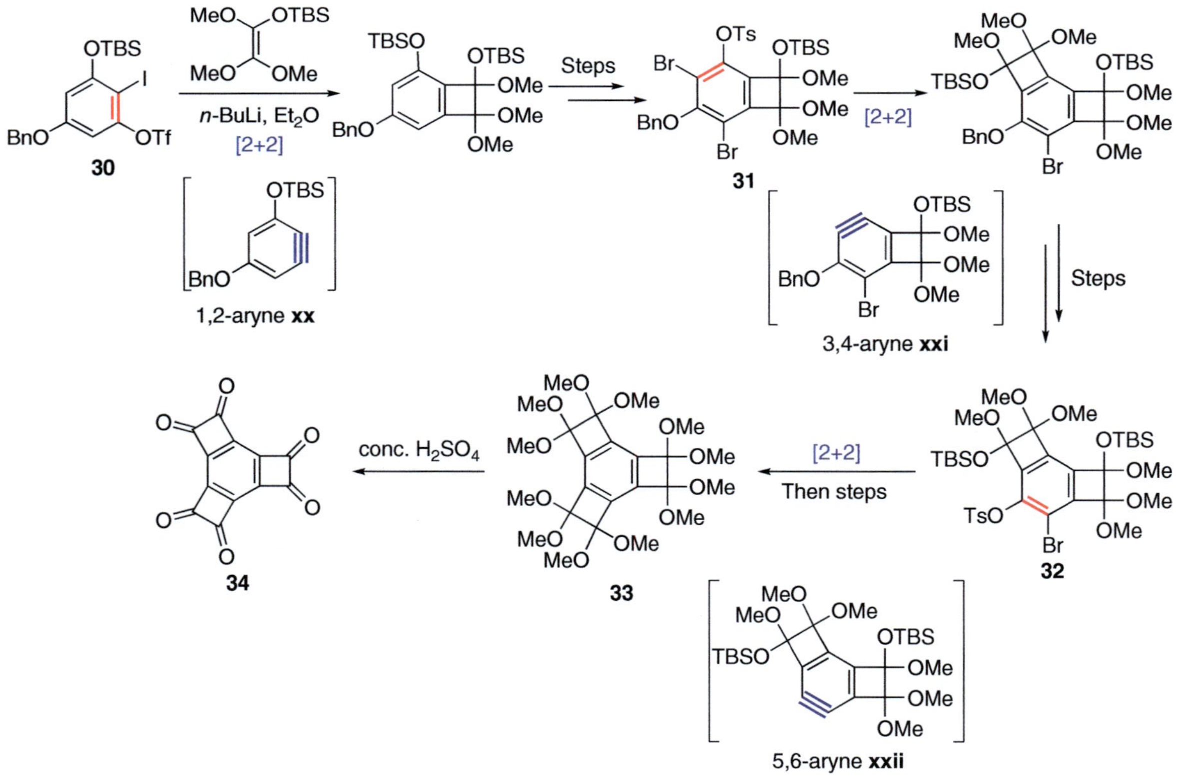

Scheme 8.49 Suzuki's synthesis of TCBBs **33** and **34**. Source: Based on Hamura et al. [83].

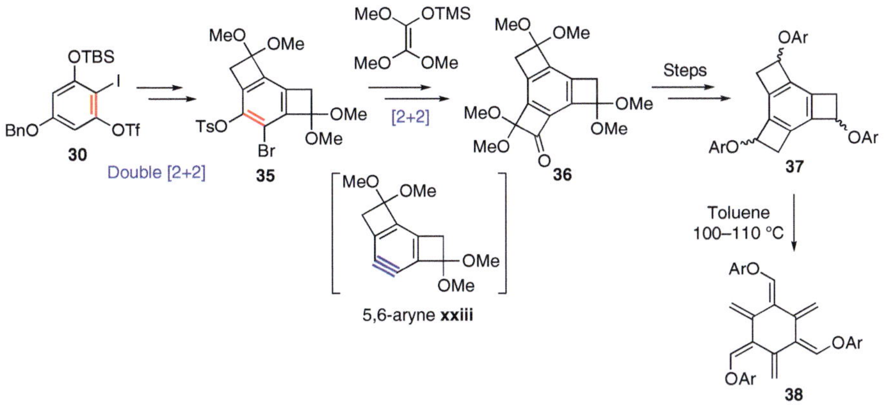

Scheme 8.50 Suzuki's synthesis of hexaredialene **38**. Source: Based on Shinozaki et al. [84].

various substrates on both sulfoxides and aryne rings. Up to five substituted arenes could be readily prepared in high yields and excellent selectivity. This work was also the first example of arene trisubstitution from traditional Kobayashi benzyne precursor, revealing a great potential for further exploration toward atom-economical aryne multifunctionalization solutions.

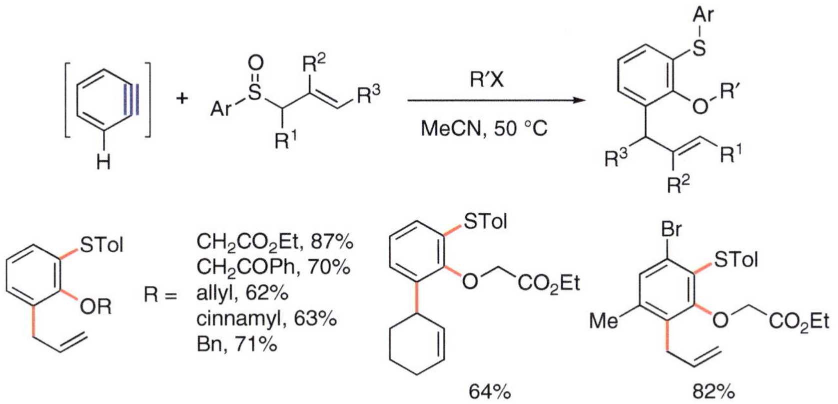

Scheme 8.51 Aryne trifunctionalization from simple benzynes. Source: Based on Li et al. [117].

From these examples, people can easily recognize that the chemistry of benzdiyne and benztriyne possesses two advantages. The first one is to readily prepare multifunctionalized arenes via these polyaryne equivalents, some of which are otherwise challenging through other methods. The other advantage of polyaryne chemistry is the convenient and diverse preparation of various polycyclic aromatic hydrocarbons or nanographenes, making it an interesting expansion in materials science–related new research direction. Consequently, more intriguing and useful transformations should be unraveled using benzdiyne and benztriyne strategies.

8.5 Conclusions

Along with the employment of mild aryne generation conditions, such as Kobayashi's method and HDDA aryne strategy, people have witnessed a rapid advance in aryne chemistry in the past two decades. Many new aryne reaction modes were discovered as well, those of which include transition metal-catalyzed reactions, insertion reaction, and multicomponent reactions. Although certain fields in aryne chemistry remain underdeveloped, they may become new research spotlights in due time. This chapter covers some of the aspects in aryne chemistry that might be underestimated for some reason. Despite the fact that aryne generation strategies have been long accompanying with the development of aryne chemistry, they will continue to play an essential role in the future, because the generation step is indispensable in any aryne transformation. The aryne regioselectivity is normally a "silent" topic, in which there is no systematic summary on it yet. However, regioselectivity stays in an important position in aryne chemistry, not only in reaction efficiency consideration, but also for the sake of more practical synthetic applications. The last one, aryne multifunctionalization, is an expansion of traditional aryne difunctionalization strategies. On one hand, the development of this subfield could provide convenient access to multisubstituted arenes that would be otherwise challenging to reach with respect to atom- and step-economical consideration; on the other hand, recent intensive studies in this subfield unraveled its promising potential on ready preparation of polycyclic aromatic hydrocarbons, which is closely related to materials science. Therefore, advances of these three topics in the past two decades can be seen as new strategies in recent aryne chemistry. Hopefully, along with the continuous exploration on these subfields of benzyne chemistry, more breakthroughs could be unraveled.

References

1 Roberts, J.D., Simmons, H.E. Jr., Carlsmith, L.A., and Vaughan, C.W. (1953). *J. Am. Chem. Soc.* 75: 3290–3291.
2 Hoffmann, R.W. (1967). *Dehydrobenzene and Cycloalkynes*. New York: Academic Press.
3 Himeshima, Y., Sonoda, T., and Kobayashi, H. (1983). *Chem. Lett.*: 1211–1214.
4 Hoye, T.R., Baire, B., Niu, D. et al. (2012). *Nature* 490: 208–212.
5 Niu, D., Willoughby, P.H., Woods, B.P. et al. (2013). *Nature* 501: 531–534.
6 Niu, D. and Hoye, T.R. (2014). *Nat. Chem.* 6: 34–40.
7 Olofson, R.A. and Dougherty, C.M. (1973). *J. Am. Chem. Soc.* 95: 582–584.
8 Truong, T. and Daugulis, O. (2011). *J. Am. Chem. Soc.* 133: 4243–4245.
9 Truong, T. and Daugulis, O. (2012). *Org. Lett.* 14: 5964–5967.
10 Truong, T., Mesgar, M., Le, K.K.A., and Daugulis, O. (2014). *J. Am. Chem. Soc.* 136: 8568–8576.
11 Truong, T. and Daugulis, O. (2013). *Chem. Sci.* 4: 531–535.

12 Mesgar, M., Nguyen-Le, J., and Daugulis, O. (2018). *J. Am. Chem. Soc.* 140: 13703–13710.

13 Uchiyama, M., Miyoshi, T., Kajihara, Y. et al. (2002). *J. Am. Chem. Soc.* 124: 8514–8515.

14 Uchiyama, M., Kobayashi, Y., Furuyama, T. et al. (2008). *J. Am. Chem. Soc.* 130: 472–480.

15 Hall, C., Henderson, J.L., Ernouf, G., and Greaney, M.F. (2013). *Chem. Commun.* 49: 7602–7604.

16 Wickham, P.P., Hazen, K.H., Guo, H. et al. (1991). *J. Org. Chem.* 56: 2045–2050.

17 Cho, S. and Wang, Q. (2018). *Tetrahedron* 74: 3325–3328.

18 Okuyama, T., Takino, T., Sueda, T., and Ochiai, M. (1995). *J. Am. Chem. Soc.* 117: 3360–3367.

19 Akiyama, T., Imasaki, Y., and Kawanisi, M. (1974). *Chem. Lett.* 3: 229–230.

20 Sundalam, S.K., Nilova, A., Seidl, T.L., and Stuart, D.R. (2016). *Angew. Chem. Int. Ed.* 55: 8431–8434.

21 Stuart, D.R. (2017). *Synlett* 28: 275–279.

22 Zhang, Z., Wu, X., Han, J. et al. (2018). *Tetrahedron Lett.* 59: 1737–1741.

23 Matsumoto, T., Hosoya, T., Katsuki, M., and Suzuki, K. (1991). *Tetrahedron Lett.* 32: 6735–6736.

24 Hosoya, T., Hasegawa, T., Kuriyama, Y. et al. (1995). *Synlett* 2: 177–179.

25 Hosoya, T., Hasegawa, T., Kuriyama, Y., and Suzuki, K. (1995). *Tetrahedron Lett.* 36: 3377–3380.

26 Ganta, A. and Snowden, T.S. (2007). *Synlett* 14: 2227–2231.

27 Shi, J., Xu, H., Qiu, D. et al. (2017). *J. Am. Chem. Soc.* 139: 623–626.

28 Hamura, T., Arisawa, T., Matsumoto, T., and Suzuki, K. (2006). *Angew. Chem. Int. Ed.* 45: 6842–6844.

29 Ganta, A. and Snowden, T.S. (2008). *Org. Lett.* 10: 5103–5106.

30 Ikawa, T., Takagi, A., Kurita, Y. et al. (2010). *Angew. Chem. Int. Ed.* 49: 5563–5566.

31 Aikawa, H., Takahira, Y., and Yamaguchi, M. (2011). *Chem. Commun.* 47: 1479–1481.

32 Niu, W.-X., Yang, E.-Q., Shi, Z.-F. et al. (2012). *Org. Chem.* 77: 1422–1434.

33 Li, Y., Chakrabarty, S., Mück-Lichtenfeld, C., and Studer, A. (2016). *Angew. Chem. Int. Ed.* 55: 802–806.

34 Tadross, P.M. and Stoltz, B.M. (2012). *Chem. Rev.* 112: 3550–3577.

35 Gampe, C.M. and Carreira, E.M. (2012). *Angew. Chem. Int. Ed.* 51: 3766–3778.

36 Takikawa, H., Nishii, A., Sakai, T., and Suzuki, K. (2018). *Chem. Soc. Rev.* 47: 8030–8056.

37 Yoshida, S., Uchida, K., and Hosoya, T. (2015). *Chem. Lett.* 44: 691–693.

38 Hamura, T., Chuda, Y., Nakatsuji, Y., and Suzuki, K. (2012). *Angew. Chem. Int. Ed.* 51: 3368–3372.

39 Sapountzis, I., Lin, W., Fischer, M., and Knochel, P. (2004). *Angew. Chem. Int. Ed.* 43: 4364–4366.

40 Lin, W.W., Sapountzis, I., and Knochel, P. (2005). *Angew. Chem. Int. Ed.* 44: 4258–4261.

41 Lin, W., Chen, L., and Knochel, P. (2007). *Tetrahedron* 63: 2787–2797.
42 Kitamura, T., Aoki, Y., Isshiki, S. et al. (2006). *Tetrahedron Lett.* 47: 1709–1712.
43 Ikawa, T., Nishiyama, T., Nosaki, T. et al. (2011). *Org. Lett.* 13: 1730–1733.
44 Kovács, S., Csincsi, Á.I., Nagy, T.Z. et al. (2012). *Org. Lett.* 14: 2022–2025.
45 Chen, Q., Yu, H., Xu, Z. et al. (2015). *J. Org. Chem.* 80: 6890–6896.
46 Mesgar, M. and Daugulis, O. (2016). *Org. Lett.* 18: 3910–3913.
47 Sumida, Y., Kato, T., and Hosoya, T. (2013). *Org. Lett.* 15: 2806–2809.
48 Yoshida, S., Uchida, K., and Hosoya, T. (2014). *Chem. Lett.* 43: 116–118.
49 Nishiyama, Y., Kamada, S., Yoshida, S., and Hosoya, T. (2018). *Chem. Lett.* 47: 1216–1219.
50 Gann, A.W., Amoroso, J.W., Einck, V.J. et al. (2014). *Org. Lett.* 16: 2003–2005.
51 Chang, D., Zhu, D., and Shi, L. (2015). *J. Org. Chem.* 80: 5928–5933.
52 Dong, C.-G. and Hu, Q.-S. (2006). *Org. Lett.* 8: 5057–5060.
53 Kim, H.S., Gowrisankar, S., Kim, E.S., and Kim, J.N. (2008). *Tetrahedron Lett.* 49: 6569–6572.
54 Cant, A.A., Roberts, L., and Greaney, M.F. (2010). *Chem. Commun.* 46: 8671–8673.
55 Retbøll, M., Edwards, A.J., Rae, A.D. et al. (2002). *J. Am. Chem. Soc.* 124: 8348–8360.
56 García-López, J.-A. and Greaney, M.F. (2014). *Org. Lett.* 16: 2338–2341.
57 Lakshmi, B.V., Wefelscheid, U.K., and Kazmaier, U. (2011). *Synlett* 2011: 345–348.
58 Bronner, S.M., Mackey, J.L., Houk, K.N., and Garg, N.K. (2012). *J. Am. Chem. Soc.* 134: 13966–13969.
59 Masson, E. and Schlosser, M. (2005). *Eur. J. Org. Chem.* 2005: 4401–4405.
60 Akai, S., Ikawa, T., Takayanagi, S.-I. et al. (2008). *Angew. Chem. Int. Ed.* 47: 7673–7676.
61 Ikawa, T., Masuda, S., Nakajima, H., and Akai, S. (2017). *J. Org. Chem.* 82: 4242–4253.
62 Devlin, A.S. and Du Bois, J. (2013). *Chem. Sci.* 4: 1059–1063.
63 Yoshida, S. and Hosoya, T. (2013). *Chem. Lett.* 42: 583–585.
64 Takagi, A., Ikawa, T., Kurita, Y. et al. (2013). *Tetrahedron* 69: 4338–4352.
65 Bronner, S.M., Goetz, A.E., and Garg, N.K. (2011). *Synlett* 2011: 2599–2604.
66 Goetz, A.E., Bronner, S.M., Cisneros, J.D. et al. (2012). *Angew. Chem. Int. Ed.* 51: 2758–2762.
67 Medina, J.M., Mackey, J.L., Garg, N.K., and Houk, K.N. (2014). *J. Am. Chem. Soc.* 136: 15798–15805.
68 Goetz, A.E. and Garg, N.K. (2014). *J. Org. Chem.* 79: 846–851.
69 Picazo, E., Houk, K.N., and Garg, N.K. (2015). *Tetrahedron Lett.* 56: 3511–3514.
70 Cheong, P.H.-Y., Paton, R.S., Bronner, S.M. et al. (2010). *J. Am. Chem. Soc.* 132: 1267–1269.
71 Im, G.-Y.J., Bronner, S.M., Goetz, A.E. et al. (2010). *J. Am. Chem. Soc.* 132: 17933–17944.
72 Bronner, S.M., Goetz, A.E., and Garg, N.K. (2011). *J. Am. Chem. Soc.* 133: 3832–3835.

73 Goetz, A.E. and Garg, N.K. (2013). *Nat. Chem.* 5: 54–60.
74 Shah, T.K., Medina, J.M., and Garg, N.K. (2016). *J. Am. Chem. Soc.* 138: 4948–4954.
75 Takagi, A., Ikawa, T., Saito, K. et al. (2013). *Org. Biomol. Chem.* 11: 8145–8150.
76 Matsumoto, T., Sohma, T., Hatazaki, S., and Suzuki, K. (1993). *Synlett* 1993: 843–846.
77 Ikawa, T., Nishiyama, T., Shigeta, T. et al. (2011). *Angew. Chem. Int. Ed.* 50: 5674–5677.
78 Ikawa, T., Takagi, A., Goto, M. et al. (2013). *J. Org. Chem.* 78: 2965–2983.
79 Hamura, T., Ibusuki, Y., Sato, K. et al. (2003). *Org. Lett.* 5: 3551–3554.
80 Finnegan, R.A. (1965). *J. Org. Chem.* 30: 1333–1335.
81 Streitwieser, A. Jr., Ziegler, G.R., Mowery, P.C. et al. (1968). *J. Am. Chem. Soc.* 90: 1357–1358.
82 Hamura, T., Ibusuki, Y., Uekusa, H. et al. (2006). *J. Am. Chem. Soc.* 128: 3534–3535.
83 Hamura, T., Ibusuki, Y., Uekusa, H. et al. (2006). *J. Am. Chem. Soc.* 128: 10032–10033.
84 Shinozaki, S., Hamura, T., Ibusuki, Y. et al. (2010). *Angew. Chem. Int. Ed.* 49: 3026–3029.
85 Shi, J., Li, Y., and Li, Y. (2017). *Chem. Soc. Rev.* 46: 1707–1719.
86 Du, C.-J.F., Hart, H., and Ng, K.-K.D. (1986). *J. Org. Chem.* 51: 3162–3165.
87 Du, C.-J.F. and Hart, H. (1987). *J. Org. Chem.* 52: 4311–4314.
88 Qiu, D., Shi, J., and Li, Y. (2015). *Synlett* 26: 2194–2198.
89 Shi, J., Qiu, D., Wang, J. et al. (2015). *J. Am. Chem. Soc.* 137: 5670–5673.
90 Yoshida, S., Uchida, K., Igawa, K. et al. (2014). *Chem. Commun.* 50: 15059–15062.
91 Ikawa, T., Kaneko, H., Masuda, S. et al. (2015). *Org. Biomol. Chem.* 13: 520–526.
92 Li, L., Qiu, D., Shi, J., and Li, Y. (2016). *Org. Lett.* 18: 3726–3729.
93 Qiu, D., He, J., Yue, X. et al. (2016). *Org. Lett.* 18: 3130–3133.
94 Xu, H., He, J., Shi, J. et al. (2018). *J. Am. Chem. Soc.* 140: 3555–3559.
95 Xiao, X. and Hoye, T.R. (2019). *J. Am. Chem. Soc.* 141: 9813–9818.
96 Uchida, K., Yoshida, S., and Hosoya, T. (2017). *Org. Lett.* 19: 1184–1187.
97 Yoshida, S., Nakamura, Y., Uchida, K. et al. (2016). *Org. Lett.* 18: 6212–6215.
98 Nakamura, Y., Miyata, Y., Uchida, K. et al. (2019). *Org. Lett.* 21: 5252–5258.
99 Lv, C., Wan, C., Liu, S. et al. (2018). *Org. Lett.* 20: 1919–1923.
100 Hart, H. and Shamouilian, S. (1981). *J. Org. Chem.* 46: 4874–4876.
101 Maurin, P., Ibrahim-Ouali, M., and Santelli, M. (2001). *Tetrahedron Lett.* 42: 8147–8149.
102 Ikawa, T., Masuda, S., Takagi, A., and Akai, S. (2016). *Chem. Sci.* 7: 5206–5211.
103 Kitamura, T., Gondo, K., and Oyamada, J. (2017). *J. Am. Chem. Soc.* 139: 8416–8419.
104 Hart, H., Lai, C.-Y., Nwokogu, G. et al. (1980). *J. Am. Chem. Soc.* 102: 6649–6651.
105 Hart, H., Shamouilian, S., and Takehira, Y. (1981). *J. Org. Chem.* 46: 4427–4432.
106 Hart, H., Harada, K., and Du, C.-J.F. (1985). *J. Org. Chem.* 50: 3104–3110.

107 Bashir-Hashemi, A., Hart, H., and Ward, D.L. (1986). *J. Am. Chem. Soc.* 108: 6675–6679.
108 Ninomiya, M., Ando, Y., Kudo, F. et al. (2019). *Angew. Chem. Int. Ed.* 58: 4264–4270.
109 Duong, H.M., Bendikov, M., Steiger, D. et al. (2003). *Org. Lett.* 5: 4433–4436.
110 Kumarasinghe, K.G.U.R., Fronczek, F.R., Valle, H.U., and Sygula, A. (2016). *Org. Lett.* 18: 3054–3057.
111 Garcia, D., Rodríguez-Pérez, L., Herranz, M.A. et al. (2016). *Chem. Commun.* 52: 6677–6680.
112 Alonso, J.M., Díaz-Álvarez, A.E., Criado, A. et al. (2012). *Angew. Chem. Int. Ed.* 51: 173–177.
113 Schuler, B., Collazos, S., Gross, L. et al. (2014). *Angew. Chem. Int. Ed.* 53: 9004–9006.
114 Chen, C.-L., Sparks, S.M., and Martin, S.F. (2006). *J. Am. Chem. Soc.* 128: 13696–13697.
115 Ashton, P.R., Isaacs, N.S., Kohnke, F.H. et al. (1989). *Angew. Chem. Int. Ed.* 28: 1261–1263.
116 Raymo, F., Kohnke, F.H., and Cardullo, F. (1992). *Tetrahedron* 48: 6827–6838.
117 Li, Y., Qiu, D., Gu, R. et al. (2016). *J. Am. Chem. Soc.* 138: 10814–10817.

9

Hetarynes, Cycloalkynes, and Related Intermediates

Avishek Guin, Subrata Bhattacharjee, and Akkattu T. Biju

Indian Institute of Science, Department of Organic Chemistry, Gulmohar Marg, 560012, Bangalore, India

9.1 Introduction to Hetarynes

The term "hetarynes" has been suggested for *ortho*-dehydroaromatic compounds with heteroatom in the ring. The studies of hetarynes are much older than conventional aryne, but the physical evidence for hetarynes is very inadequate. Several challenges are associated with the generation of hetarynes and their use as intermediates for various synthetic targets. The journey of hetarynes started in 1902 invoking the generation of 2,3-benzofuranyne as a reactive intermediate, reported by Stoermer and Kahlert, although there was no direct evidence to prove the formation of aryne intermediate at that time [1]. Hetarynes can be subdivided into two categories depending upon the ring size of the hetaryne ring [2–6]. Five-membered hetarynes are not well explored and difficult to generate due to the inherent strain present in the five-membered ring. Thus, synthetic utility of five-membered hetarynes is very limited. Compared to that, six-membered hetarynes are easy to generate and have been used for complex synthetic targets. Among all the six-membered hetarynes, pyridynes, and indolynes are most well studied. These arynes can be generated in various positions of the ring as well. A discussion on challenges associated with hetarynes, various types of hetarynes, methods of preparation, reactions of hetarynes, applications in synthesis form the context of this hetaryne chapter.

9.2 Challenges in Hetarynes

The studies by Stoermer and Kahlert are already discussed in detail in Chapter 1. The conclusive pieces of evidence for this intermediate were investigated by Robert [7], Huisgen [8], and Wittig [9] to provide symmetry and reactivity of such species. Further evidence from spectroscopic data in the gas phase [10], and argon matrixes [11] has established that the benzyne intermediate exists.

Since hetaryne is an intermediate in a reaction, it cannot be observed freely. Only from the product formation is it understandable whether the hetaryne intermediate is generated or not. But the same product can be generated from other pathways

Modern Aryne Chemistry, First Edition. Edited by Akkattu T. Biju.

also (without the involvement of aryne). Various mechanisms can operate in nucleophilic substitution reactions rather than benzyne mechanism and can give the same product. The elimination–addition (EA) mechanism is one of its kind (so-called benzyne mechanism) (Scheme 9.1) [12–14]. In the EA mechanism, the first deprotonation from **1** takes place from *ortho* to the halogen. Elimination of halide would generate the aryne intermediate **2**. Nucleophilic addition can happen from both sides to give ipso and cine products **3a** and **3b**.

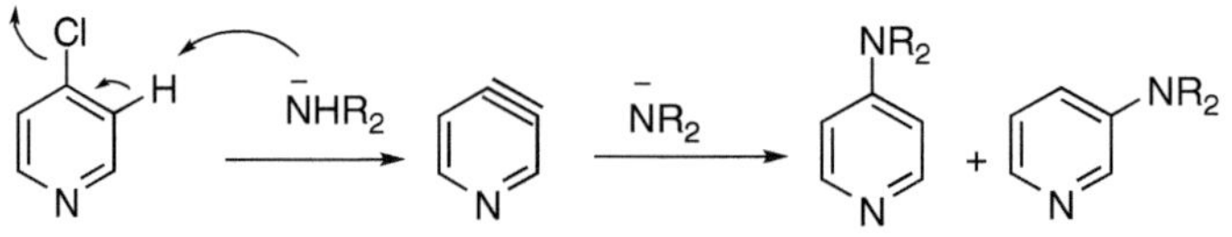

1 **2** **3a** **3b**

Scheme 9.1 Elimination–addition (EA) mechanism. Source: Sauer and Huisgen [12]; Huisgen and Sauer [13]; Kauffmann et al. [14].

The same product can also be obtained via the addition–elimination (AE_n) mechanism [12–14]. In this mechanism, nucleophile/strong base can attack at the C4 position, generating the adduct **4**. Then, the expulsion of the leaving group can give the product **3a** (Scheme 9.2).

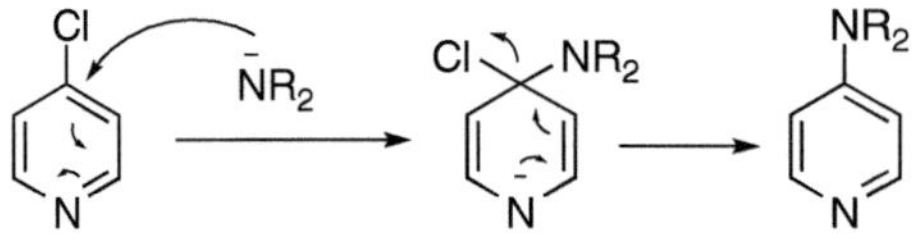

1 **4** **3a**

Scheme 9.2 Addition–elimination (AE_n) mechanism. Source: Sauer and Huisgen [12]; Huisgen and Sauer [13]; Kauffmann et al. [14].

The abnormal addition–elimination (AE_a) mechanism also can operate for nucleophilic substitution reaction, but this mechanism is very substrate sensitive (Scheme 9.3) [14, 15]. In this case, nucleophilic addition can occur from cine substitution to the leaving group on **5** to generate the intermediate **6**. Leaving group elimination from **6** produces the product **7**.

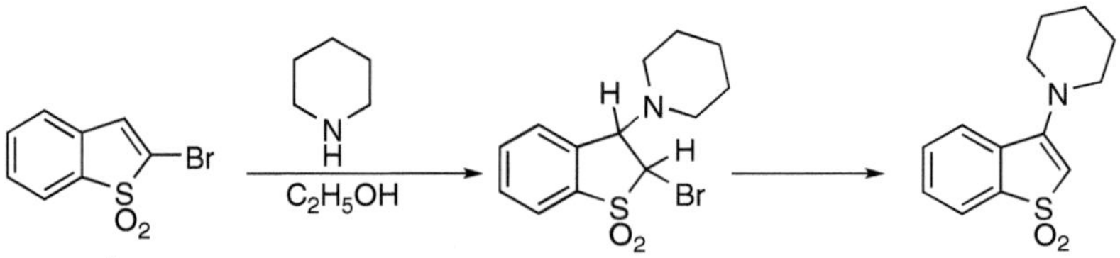

5 **6** **7**

Scheme 9.3 Abnormal addition–elimination (AE_a) mechanism. Source: Kauffmann et al. [14]; Bordwell et al. [15].

Except these three mechanisms, cine-substitution via trans-halogenation (BCHD), cine-substitution via addition–substitution–elimination (ASE), or cine-substitution via addition–ring-opening–elimination–ring-closure (ANRORC cine) also can operate [3]. In another possibility, several mechanisms can operate simultaneously. Thus, predicting the actual mechanism of the reaction is very difficult. However, if the ratio of the isomers in the product obtained is not dependent on the nature of the halogen in the starting precursor, it is valid to conclude that most probably the EA mechanism is followed [2]. Cycloaddition with various conjugated dienes is another indirect proof for the formation of aryne intermediate. But all aryne precursors under definite conditions decompose to cationic, anionic, or radical species, which may also react with a diene to give cycloaddition product [3, 16]. To date, many heterocyclic arynes have been suggested as a reactive intermediate, but in many cases, aryne was not involved as the intermediate [2–6].

9.3 Different Types of Hetarynes

Depending upon the ring size, hetarynes are generally two types. Hetarynes can be generated in five-membered heterocycles. The first-discovered aryne from 3-bromobenzofuran **8** falls in this category (Scheme 9.4). 2,3-indolyne (**9**) and 2,3-benzothiophyne (**10**) are other examples [17]. Computational study by Garg and coworkers suggests that the five-membered hetarynes generation (**11–13**) required high energy, and thus synthetic application for this five-membered hetarynes is very limited [18]. Hetarynes can be generated from thiophene also [19–21]. The detailed studies of these hetarynes will be discussed in Sections 9.4–9.6.

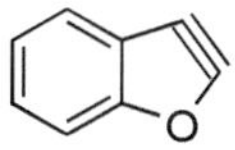

2,3-Benzofuranyne

8

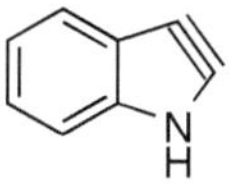

2,3-Indolyne

9

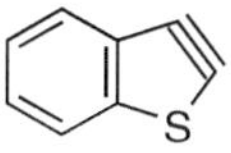

2,3-Benzothiophyne

10

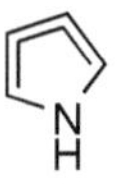

3,4-Pyrrolyne

11

2,3-Thiophyne

12

3,4-Thiophyne

13

Scheme 9.4 5-Membered hetarynes.

Six-membered hetarynes are most well studied. Even pyridyne and indolyne are already used for the synthesis of complex natural products [22–25]. Arynes could be generated at different positions of pyridine and indole ring, thus forming 2,3 or 3,4-pyridyne (**14, 15**) and 4,5/5,6 or 6,7-indolyne (**16–18**) (Scheme 9.5) [6]. Computational study shows that 3,4-pyridyne **15** is more stable than 2,3-pyridyne **14** [26–29]. The idea of generating 1,2-pyridyne was not successful in the solution phase, although the species was found in the gas-phase study [30–32]. Although less explored, quinolynes (**19, 20**) are well known in literature [33, 34]. 3,4-Dehydro-1,5-naphthyridine **20** and 4,5-dehydropyrimidine **21** could be generated in situ and can be trapped via nucleophilic addition or cycloaddition reactions [35, 36]. Martens and Hertog reported the generation of 2,3-dehydropyridine-*N*-oxides **22** [37]. The method of generation of these arynes will be discussed in Section 9.4.

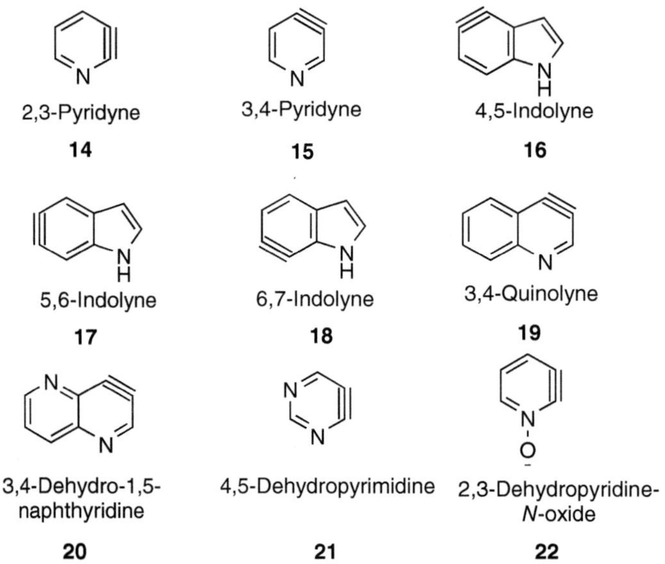

Scheme 9.5 6-Membered hetarynes.

9.4 Methods of Preparation

Due to high reactivity, aryne cannot be isolated, it must be generated in situ. The different groups tried different methodologies for various types of aryne generation. In the early days, many groups claimed to generate aryne, but most claims are lacking conclusive proof. Selected important methods are discussed in this section.

9.4.1 2,3-Benzofuranyne Generation

In 1902, Stoermer and Kahlert first proposed the existence of aryne intermediate to explain the formation of 2-ethoxybenzofuran **24** from 3-bromobenzofuran **23** in the presence of sodium ethoxide (Scheme 9.6) [1]. Notably, a possibility where the starting precursor 3-bromobenzofuran contains some 2-bromobenzofuran as an impurity, which can give the expected product via S_NAr route, was also suggested. Indubitably, this is the starting point for all aryne reactions.

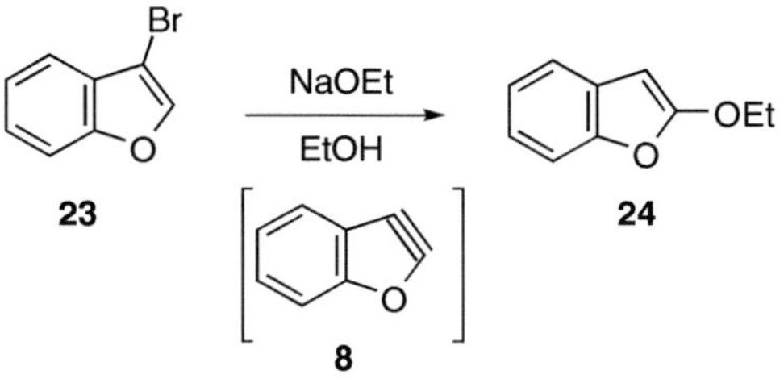

Scheme 9.6 2,3-Benzofuranyne from 3-bromobenzofuran. Source: Based on Stoermer and Kahlert [1].

9.4.2 2,3-Indolyne Generation

Conway and Gribble made several attempts to generate 1-phenylsulfonyl-2,3-indolyne from 2-lithio-3-bromo-1-phenylsulfonylindole **25**, but all attempts resulted in failure and resulted in **26** (Scheme 9.7) [17]. For 2-lithio-3-bromobenzofuran, they found from X-ray crystallography, this species exists as a distorted trans dimer and thus it is not in a proper conformation for LiBr elimination. Garg's computational study also suggested that for the generation of 2,3-indolyne, very high energy is required [18].

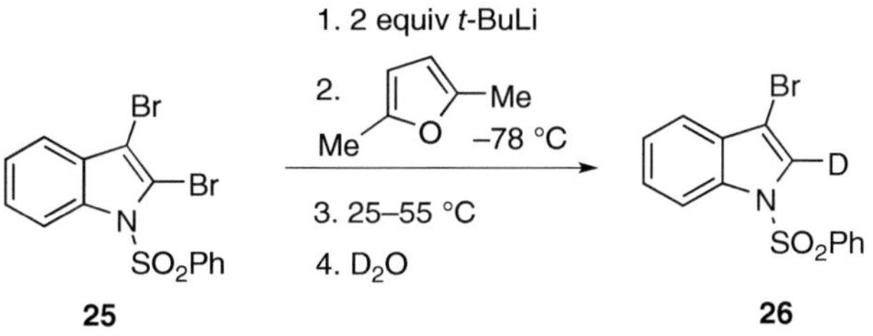

Scheme 9.7 Unsuccessful attempt to generate 2,3-indolyne. Source: Based on Conway and Gribble [17].

9.4.3 2,3-Benzothiophyne Generation

In 1981, Reinecke et al. trapped the intermediate **10** in a cycloaddition reaction with thiophene (Scheme 9.8). In a flash vacuum thermolysis (FVT) experiment of thianaphthene-2,3-dicarboxylic acid anhydride **27** with thiophene produced dibenzothiophene **28b** in 35% yield via elimination of sulfur from **28a** [21].

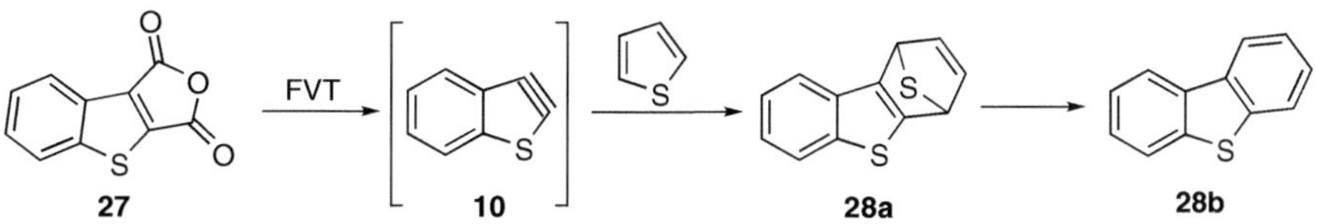

Scheme 9.8 2,3-Benzothiophyne trapping in a Diels–Alder adduct. Source: Based on Reinecke et al. [21].

9.4.4 3,4-Pyrrolyne Generation

In 1999, Wong and coworkers tried to trap 3,4-pyrrolyne **11** generated from the pyrrole derivative **29** via cycloaddition reaction with furan (for the synthesis of **30**) and acrylonitrile (for the synthesis of **31**) (Scheme 9.9) [19]. But the expected adduct was formed only in <5% yield. Thus, it cannot be considered as an exclusive proof for 3,4-pyrrolyne **11** generation.

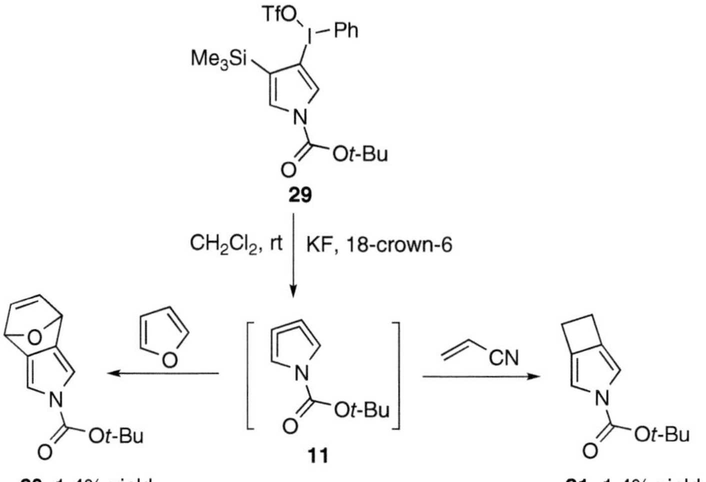

Scheme 9.9 Attempt to trap 3,4-pyrrolyne. Source: Based on Liu et al. [19].

9.4.5 2,3-Thiophyne Generation

Reinecke et al. successfully trapped the 2,3-thiophyne **12** in the same way as 2,3-benzothiophyne **10**. 2,3-Thiophyne **12** was generated from thiophene dicarboxylic acid anhydride **32** reaction, and the hetaryne reacted with thiophene to form benzothiophene **33** in 59% yield (Scheme 9.10). Other dienes were also used in the thiophyne-trapping experiment [21].

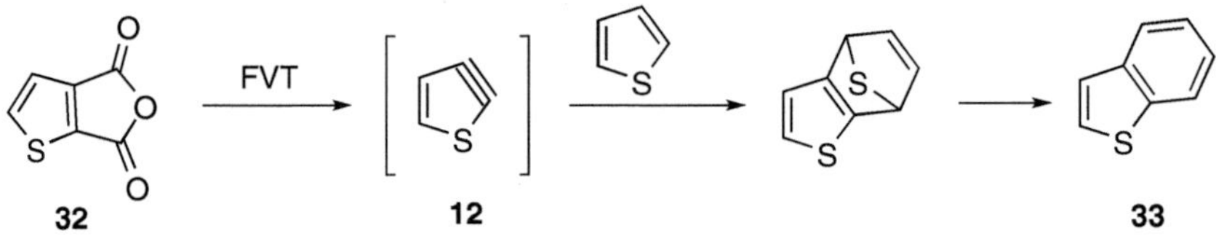

Scheme 9.10 2,3-Thiophyne trapping with thiophene as diene. Source: Based on Reinecke et al. [21].

9.4.6 3,4-Thiophyne Generation

Wong group was successful in generating and trapping 3,4-thiophyne **13**. Thiophyne was generated from the corresponding precursor **34** and this highly reactive intermediate was added to furan and benzene to give the cycloaddition products **35** and **36**, respectively (Scheme 9.11). Other dienes such as anthracene, 2,3-dimethyl-1,3-butadiene, furan derivatives also gave cycloaddition products but in low yield [20]. With butadiene, products **37a** and **37b** are formed. Although aryne generation was difficult for five-membered heterocycles, longer C—S bonds in these molecules help to release the strain and thus aryne generation was successful for these molecules.

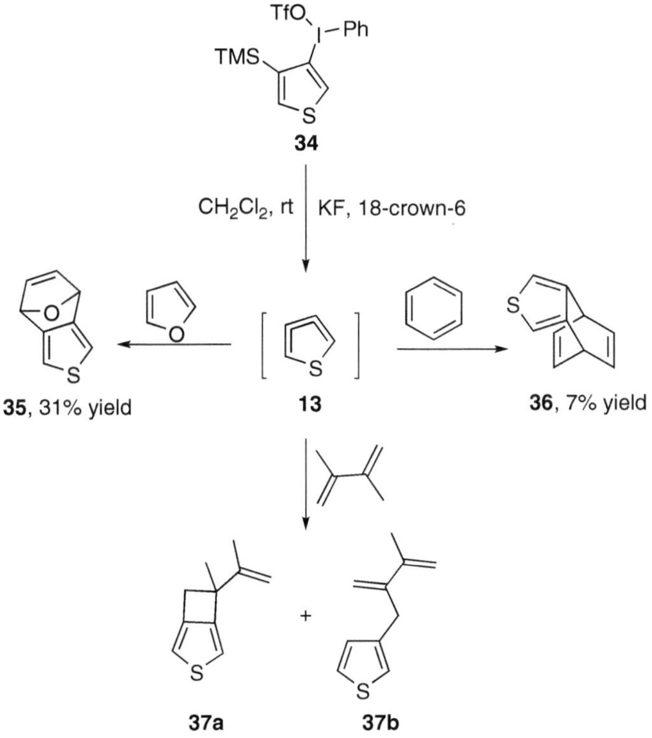

Scheme 9.11 3,4-Thiophyne trapping in a Diels–Alder adduct. Source: Based on Ye et al. [20].

9.4.7 2,3-Pyridyne Generation

Pyridine is an important heterocycle and various drug molecules contain pyridine core. Moreover, in several natural products, pyridine moiety is present. To install pyridine core onto those molecules, pyridyne generation will be one of the important pathways. Thus, the advancement of pyridyne generation has changed with time. Some of the methods are discussed in Sections 9.4.7.1–9.4.7.4.

9.4.7.1 2,3-Pyridyne from 3-Halopyridine

Since the addition of alkali metal amide to 3-halopyridine **38** generates 3,4-pyridynes exclusively, the only way out to form 2,3-pyridynes **14** from 3-halopyridine is to block the 4-position of pyridine. Although the method was not very successful, only 5% product formation was observed as a mixture of 3-amino pyridine (**39a**) 2-amino pyridine (**39b**) (Scheme 9.12) [2, 35]. 2-Halopyridine cannot be used for this purpose since it is difficult to understand whether the product is formed via S_NAr pathway or aryne pathway.

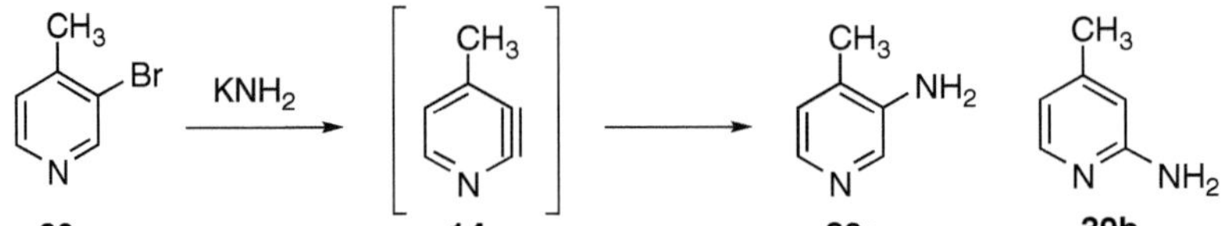

38 14 39a 39b

Scheme 9.12 2,3-Pyridyne generation from metal halide. Source: Kauffmann [2]; Van Der Plas et al. [35].

9.4.7.2 2,3-Pyridyne from Dihalide Precursor

Martens and Hertog demonstrated the use of dihalide pyridine precursor for 2,3-pyridyne **14** generation. Aryne was generated from 3-bromo-2-chloropyridine **40** in presence of lithium amalgam and trapped with furan, which would lead to quinoline **41** as a sole product (Scheme 9.13) [38].

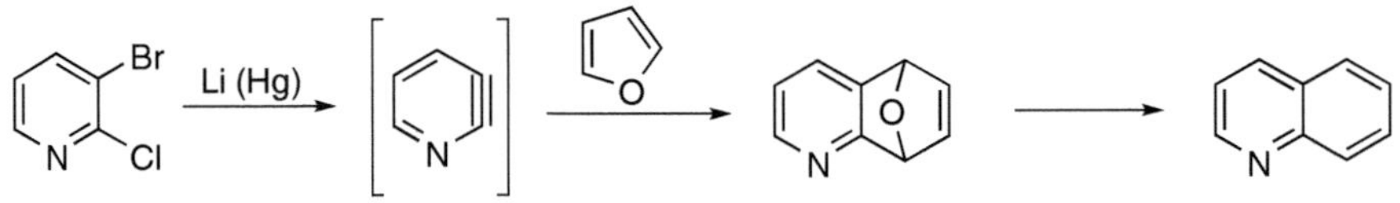

40 14 41

Scheme 9.13 2,3-Pyridyne generation from 3-bromo-2-chloropyridine. Source: Based on Martens and den Hertog [38].

9.4.7.3 From *N*-Aminotriazolo-Pyridine

The idea of generating 2,3-pyridyne **14** was also conceived by the Fleming group. When pyridyne was generated via lead tetraacetate oxidation of *N*-aminotriazolo-pyridine **42**, the expected adduct with tetraphenylcyclopentadienone was formed in only 4% yield (Scheme 9.14) [39].

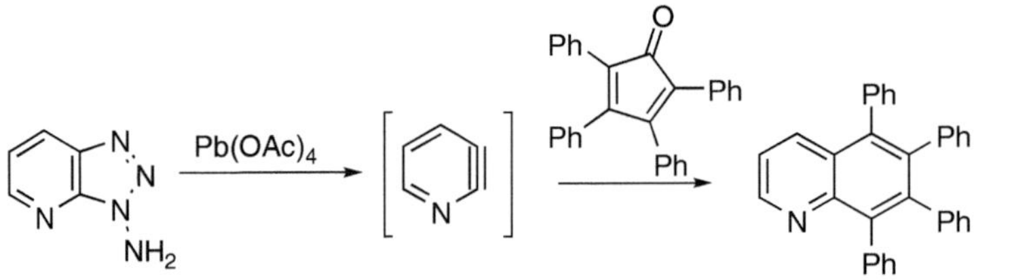

42 14 41a

Scheme 9.14 2,3-Pyridyne generation via oxidation of *N*-aminotriazolo-pyridine. Source: Based on Fleet and Fleming [39].

9.4.7.4 2,3-Pyridyne from 3-(Trimethylsilyl)pyridin-2-yl Trifluoromethanesulfonate

Walters and Shay developed a facile method for the mild generation of 2,3-pyridyne **14**. In presence of CsF, aryne was generated from 3-(trimethylsilyl)pyridin-2-yl trifluoromethanesulfonate **43** and trapped with furan derivative to give the cycloaddition product. Later, the cycloaddition product was converted to quinoline derivative **41b** (Scheme 9.15) [41].

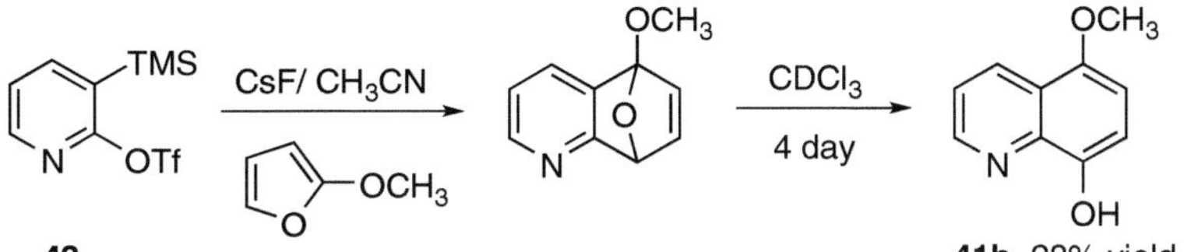

43 **41b**, 28% yield

Scheme 9.15 Mild method for 2,3-pyridyne generation. Source: Based on Nam and Leroi [40].

9.4.8 3,4-Pyridyne Generation

Several methods are available for 3,4-pyridyne **15** generation starting from thermolysis of diazonium carboxylate salt or proto-demetalation of aryl halide to mild aryne generation via fluoride-induced elimination of trimethylsilyl aryl triflates. Although some reactions may not proceed through the aryne mechanism, the isolation of **15** in nitrogen matrix and IR spectroscopy revealed that **15** is a valid and possible intermediate [40]. Each method is discussed in this section.

9.4.8.1 3,4-Pyridyne from Thermolysis of Diazonium Carboxylates

In 1962, Kauffmann and Boettcher demonstrated the generation of 3,4-pyridyne **15** via thermolysis of diazonium carboxylates **43a**. The in situ-generated 3,4-pyridyne **15** undergoes Diels–Alder reaction with furan to produce 5,8-dihydro-5,8-epoxyisoquinoline **44** (Scheme 9.16) [42].

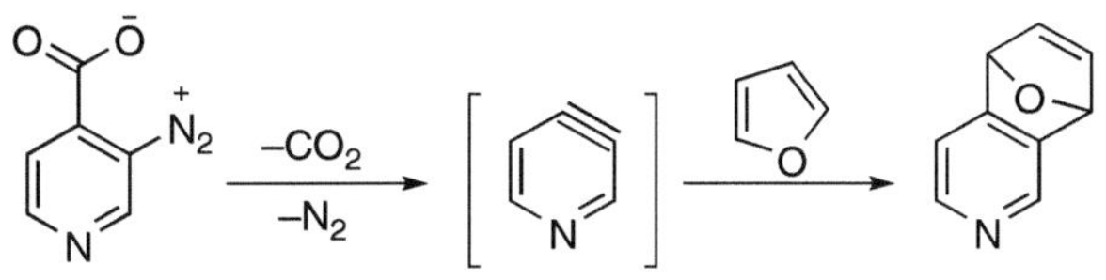

43a **15** **44**

Scheme 9.16 Thermolysis of diazonium carboxylates for 3,4-pyridyne generation. Source: Based on Kauffmann and Boettcher [42].

9.4.8.2 From 3-Halopyridine

Zoltewicz and Smith uncovered the generation of 3,4-pyridyne **15** from 3-halopyridine. Although both pyridynes could be generated, only 3,4-pyridyne **15** was observed under the reaction condition. The formation of 3,4-pyridyne **15** was understood from the partial dehydrohalogenation of 3-chloropyridine-2 or -4*D* in presence of $NaNH_2/NH_3$. For 3-chloropyridine-2-*D*, deuterium content remains same before and after the reaction, which indicated that the aryne was generated from the 3,4-position of pyridine [43].

9.4.8.3 From Oxidation of *N*-Aminotriazolo-pyridine

Although Flaming and Fleet were not very successful with aryne generation at 2,3-position of pyridine, the same idea was helpful for 3,4-pyridyne generation. When *N*-aminotriazolo-pyridine **45** was oxidized in the presence of lead tetraacetate and trapped with diene, the expected isoquinoline derivative **46** was formed in 63% yield (Scheme 9.17) [39].

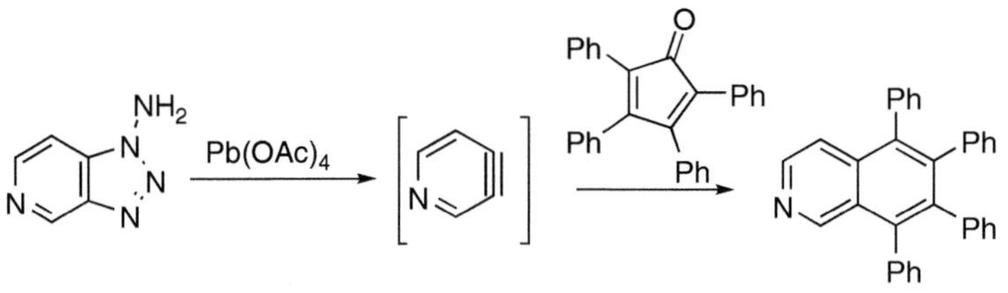

45 **15** **46**

Scheme 9.17 3,4-Pyridyne generation from *N*-aminotriazolo-pyridine. Source: Based on Fleet and Fleming [39].

9.4.8.4 From 3-Bromo-4-(phenylsulfinyl)pyridine

In 1987, Furukawa et al. generated 3,4-pyridyne **15** from 3-bromo-4-(phenylsulfinyl) pyridine **47** via phenyl magnesium bromide addition. The generated aryne was trapped with furan to give the cycloaddition adduct **44** in 42% yield (Scheme 9.18) [44].

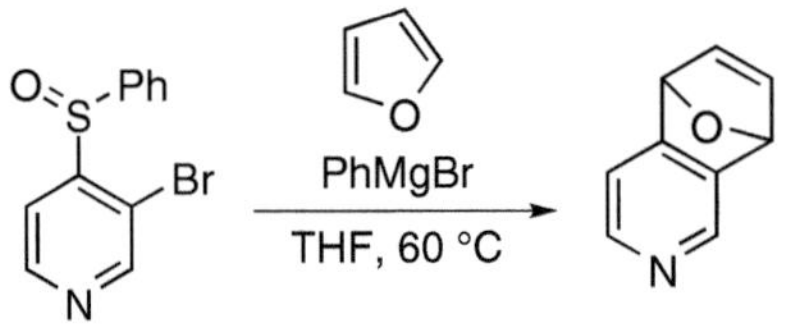

47 **44**

Scheme 9.18 3,4-Pyridyne generation from 3-bromo-4-(phenylsulfinyl)pyridine. Source: Based on Furukawa et al. [44].

9.4.8.5 From *ortho*-Trialkylsilyl Pyridyl Triflates

Tsukazaki and Snieckus synthesized 4-trialkyl silyl-3-pyridyl triflate with the aid of *ortho* metalation chemistry. This starting precursor helped to generate 3,4-pyridyne in a very mild fashion. 4-Trialkylsilyl-3-pyridyl triflate **48** generated pyridyne in presence of fluoride source and can undergo various cycloaddition reactions (Scheme 9.19) [45]. This methodology can tolerate various functional groups and thus is used worldwide for pyridyne generation.

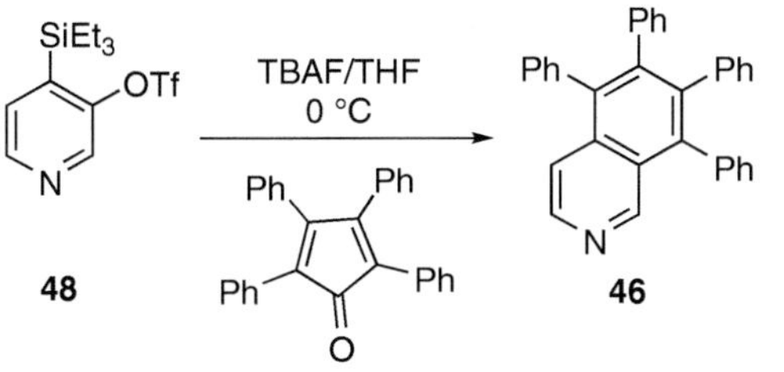

Scheme 9.19 3,4-Pyridyne generation from *ortho*-trialkylsilyl pyridyl triflates. Source: Based on Tsukazaki and Snieckus [45].

9.4.9 4,5-Indolyne Generation

9.4.9.1 From 5-Bromoindole

In 1967, Julia et al. developed the generation of 4,5-indolyne **16** from 5-bromoindole **49**. In the presence of KNH_2, 4,5-indolyne **16** was generated to give 4- and 5-aminoindole (**50a** and **50b**) products as a mixture of regioisomers (Scheme 9.20) [46].

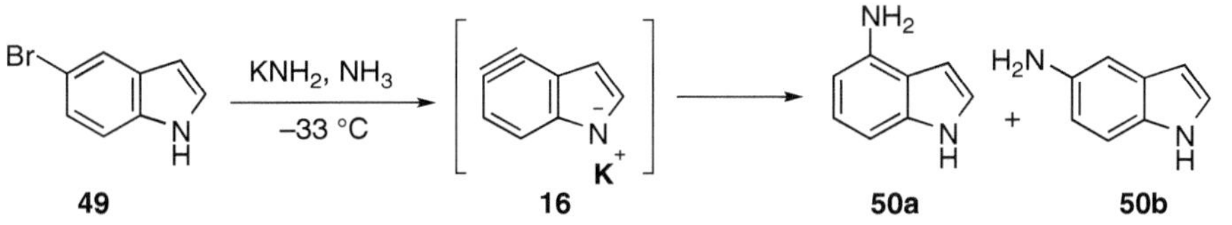

Scheme 9.20 4,5-Indolyne generation from 5-bromoindole. Source: Julia et al. [46].

9.4.9.2 From 4-Chloroindole Derivative

Iwao et al. used the indolyne formation pathway for the total synthesis of makaluvamines. During the synthesis of makaluvamines, the indole derivative **51** needed to cyclize to generate six-membered heterocycle **52** (Scheme 9.21). This reaction is envisioned to proceed via 4,5-indolyne intermediate **16** [47].

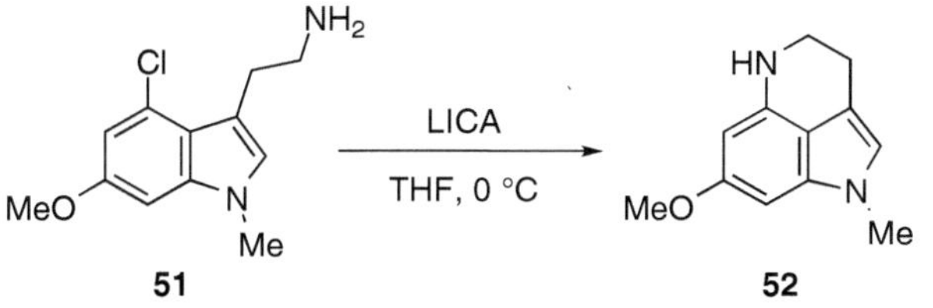

Scheme 9.21 4,5-Indolyne intermediate for the total synthesis of Makaluvamines. Source: Based on Iwao et al. [47].

9.4.9.3 From Dibromoindole

In 2007, Buszek and coworkers successfully trapped 4,5-indolyne **16** intermediate via cycloaddition with furan. The expected cycloadduct **54c** was formed in excellent yield. Not only 4,5-indolyne **16** but all other triple bond–containing benzenoid rings (produced from respective dibromo indole) were also trapped with furan in high yields (Scheme 9.22) [48–51].

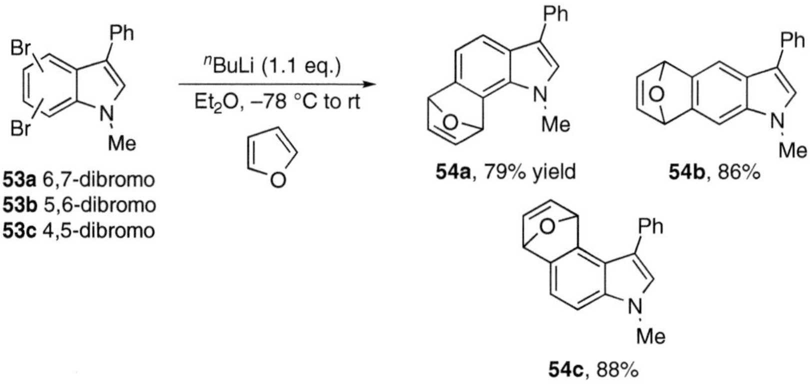

Scheme 9.22 Successful trapping of all three indolyne with furan. Source: Buszek et al. [48]; Brown et al. [49]; Buszek et al. [50].

9.4.9.4 From Silyltriflate Precursor

In 2007, Garg and coworkers demonstrated the 4,5-indolyne **16** generation from 1-methyl-4-(trimethylsilyl)-1*H*-indol-5-yl trifluoromethanesulfonate **55** (Scheme 9.23). The precursor **55** was prepared from 5-(benzyloxy)-1*H*-indole. The in situ–formed indolyne can undergo various reactions such as cycloaddition reaction and nucleophilic addition reaction and can tolerate various functionalities. The same methodology was extended for 5,6-indolyne **17** generations also. The reactivity of these indolynes revealed that nucleophilic addition can happen at the 5-position for both indolynes selectively [52–54].

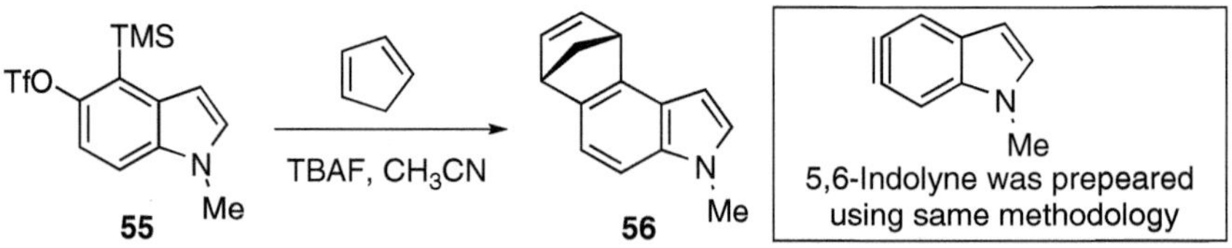

Scheme 9.23 Indolyne generation from silyltriflate precursor. Source: Bronner et al. [52]; Cheong et al. [53]; Im et al. [54].

9.4.10 5,6-Indolyne Generation

As discussed, the expected 5,6-indolyne **17** can also be generated from dibromoindole precursor reported by the Buszek group [48]. Later, in 2007, Garg and coworkers demonstrated the generation of **17**, using the same synthetic manipulation used for the generation of **16** (silyltriflate precursor). Moreover, in 2009, Buszek group showed generation of all the precursors **16, 17, 18** from silyl triflate precursor having a phenyl group present at the C3 position of indole [49].

9.4.11 6,7-Indolyne Generation

9.4.11.1 From Dichloroindole Precursor

Buszek et al. uncovered the generation of 6,7-indolyne **18** from 6,7-dichloro-1-methyl-3-phenyl-1*H*-indole **57** in reaction with *tert*-butyllithium. Intermediate **18** was trapped with furan and further functionalized to fully aromatic compound **58a** (Scheme 9.24) [48].

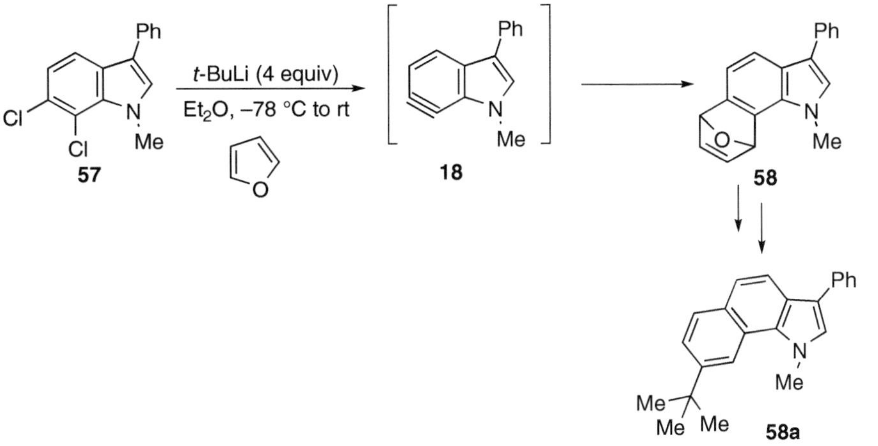

Scheme 9.24 6,7-Indolyne generation from dichloro precursor. Source: Based on Buszek et al. [48].

9.4.11.2 Through Proton–lithium Exchange

From the same group, a different methodology was developed to generate the same intermediate **18**. When they treated 5,6-difluoro-1-methyl-3-phenyl-1*H*-indole **59** with *tert*-butyllithium, **18** was generated through proton–lithium exchange. Again the 6,7-indolyne was trapped with furan to give the cycloadduct **58b** in 80% yield (Scheme 9.25) [48]. It is noted that from the same group the silyl triflate precursor was developed to generate substituted **18**.

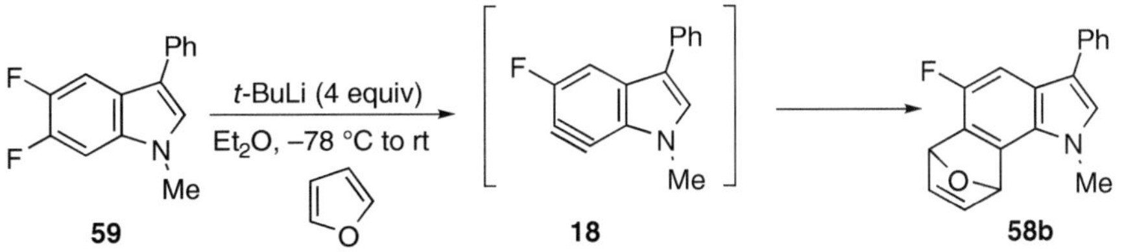

Scheme 9.25 6,7-Indolyne generation through proton–lithium exchange. Source: Based on Buszek et al. [48].

9.4.11.3 From 7-Bromoindole Derivative

Mark Lautens et al. generated indolyne in the presence of LDA using 7-bromoindole derivative **60** as a substrate. The in situ-generated indolyne underwent intramolecular ene reaction with a homoprenyl group attached with nitrogen to afford tricyclic product **61** (Scheme 9.26) [55].

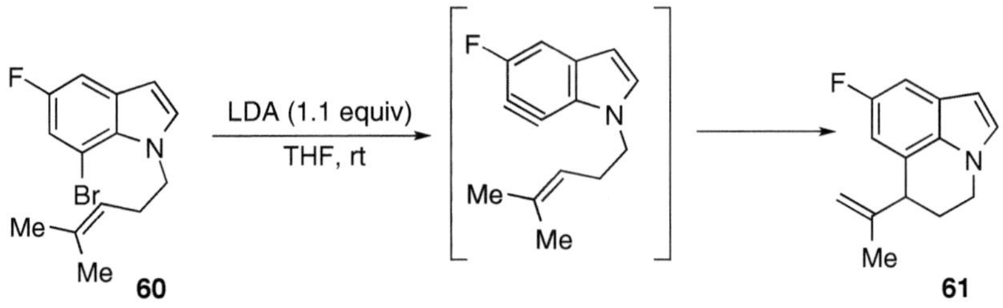

Scheme 9.26 6,7-Indolyne generating and trapping with ene reaction. Source: Based on Candito et al. [55].

9.4.12 Quinolynes Generation

9.4.12.1 3,4-Quinolyne from Halo Derivatives

When 3-Cl/3-Br/3-I quinoline derivatives **62** were reacted in the presence of lithium piperidine/piperidine, almost 1 : 1 regioisomeric mixture of 3- or 4-(piperidin-1-yl) quinoline **63** were produced. These reactions are supposed to proceed via the generation of 3,4-quinolyne **19** (Scheme 9.27) [3]. It is noted that silyltriflate precursor is known for 2,3-quinolyne and it is used for various reactions [56].

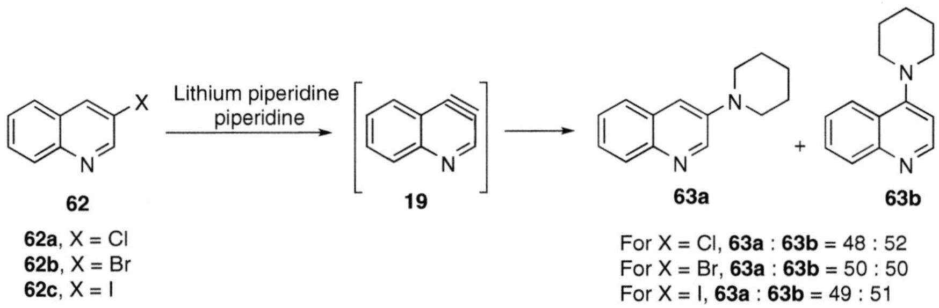

Scheme 9.27 3,4-Quinolyne generation from halo derivative. Source: Based on Reinecke [3].

9.4.12.2 5,6- and 7,8-Quinolynes

Similarly, 5,6- and 7,8-dehydroquinolynes were generated from the corresponding halo derivatives (Cl/Br) in the presence of lithium piperidine/piperidine in boiling ether. In this case also, the products isomeric ratio was independent of the halogen used [57].

9.4.12.3 7,8-Quinolyne from Quinoline 4-Methylbenzenesulfonate Derivatives

Collis and Burrell trapped the 7,8-quinolyne **65** via Diels–Alder reaction with furan. 7,8-Quinolyne **65** was generated via organolithium-induced elimination of 4-methylbenzenesulfonate (Scheme 9.28) [58]. The generated product **66** was further converted to the fully aromatic molecules.

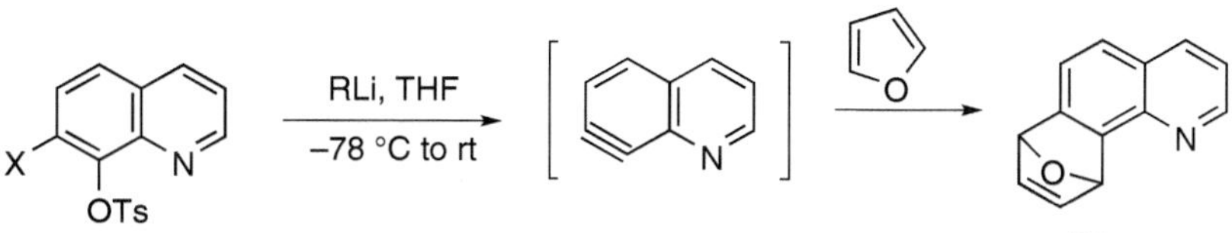

64a, X = H
64b, X = Br

65

66

Scheme 9.28 7,8-Quinolyne generation and trapping. Source: Based on Collis and Burrell [58].

9.4.12.4 3,4-Isoquinolyne Generation

3,4-Isoquinolyne was formed as an intermediate when 4-bromoisoquinoline **67** was heated to 180 °C in presence of anhydrous piperidine. 4-(Piperidin-1-yl)isoquinoline **68a** was formed in major product (35% yield) with 3-(piperidin-1-yl)isoquinoline **68b** (4% yield) (Scheme 9.29). Although the formation of 4-(Piperidin-1-yl)isoquinoline **68a** was major, the possibility of the AE_n mechanism instead of EA cannot be ruled out [3].

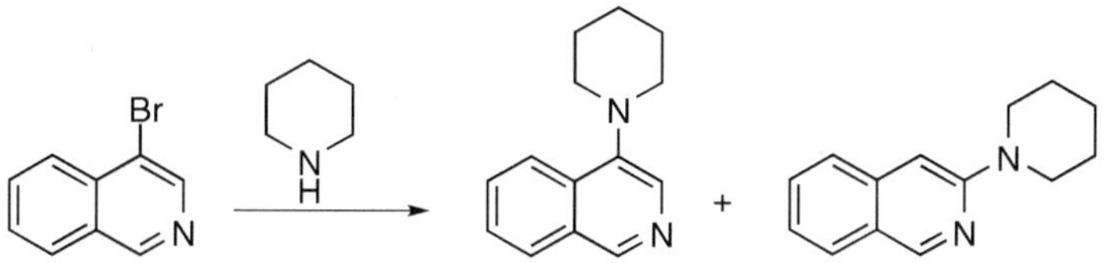

67 **68a**, 35% **68b**, 4%

Scheme 9.29 Possible 3,4-isoquinolyne intermediate generation. Source: Based on Reinecke [3].

9.4.13 3,4-Dehydro-1,5-Naphthyridine

When bromo-1,5-naphthyridines (3- or 4-) **69** were treated with KNH_2 in liquid ammonia at −33 °C, both regioisomers of amino-1,5-naphthyridines **70** were obtained. Most likely, the hetaryne **20** was the common intermediate for the generated products (Scheme 9.30). Depending upon the bromo derivatives used, the product ratio is also changed for amino-1,5-naphthyridines [3, 59].

9.4.14 4,5-Pyrimidyne Generation

Although 4,5-pyrimidyne **21** generation is well known in the literature, the intermediate is not well explored and has only limited synthetic applications. Van der Plas et al. showed the generation of 4-*tert*-butyl-5,6-pyrimidyne from its 2-/3-bromo derivatives. Deuterium-labeling experiment was carried out to confirm that benzyne mechanism was operating [60].

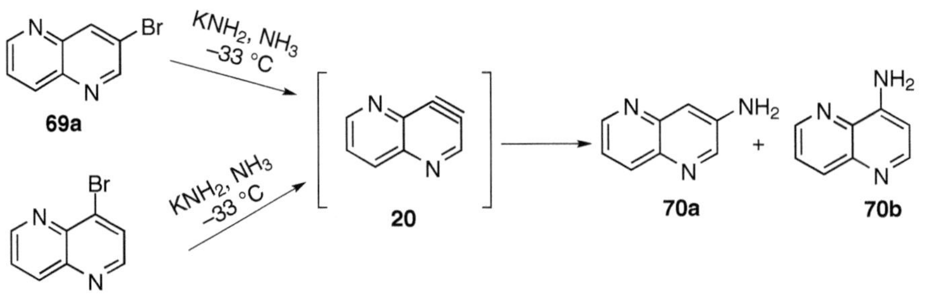

Scheme 9.30 3,4-Dehydro-1,5-naphthyridine generation. Source: Based on Czuba [59].

In 1992, Promel and coworkers reported the occurrence of 2-*tert*-butyl-4,5-didehydropyrimidine intermediate via oxidation of 5-(*tert*-butyl)-3*H*-[1–3]triazolo [4,5-*d*]pyrimidin-3-amine **71**. The generated pyrimidyne was trapped successfully in various Diels–Alder reactions (Scheme 9.31) [61]. Later, in 2016, the Garg group prepared 4-(triethylsilyl)pyrimidin-5-yl trifluoromethanesulfonate to generate aryne via fluoride-induced elimination of silyltriflate, but the method was not successful. Only thia-Fries rearrangement was observed in the presence of CsF [62].

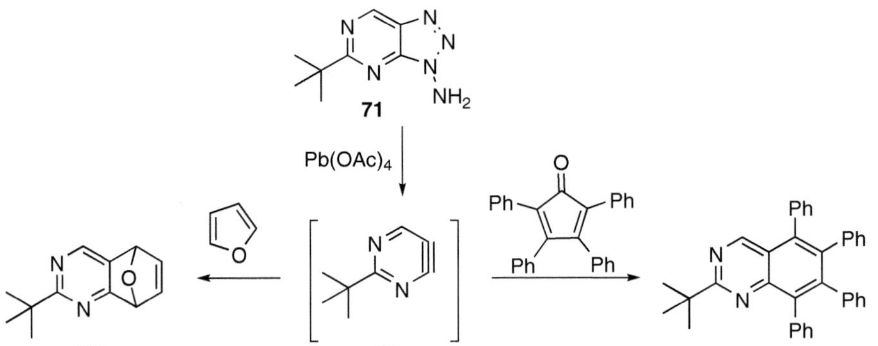

Scheme 9.31 Cycloaddition of 4,5-pyrimidyne. Source: Based on Tielemans et al. [61].

9.4.15 Pyridyne-*N*-oxides Generation

Martenes and Hertog reported the generation of 2,3-pyridyne-*N*-oxide **22**. When 2-chloropyridine-*N*-oxide was treated with KNH_2, 2-amino and 3-amino-pyridine-*N*-oxide (**74a, 74b**) were formed (Scheme 9.32). However, from 3-chloropyridine-*N*-oxide, only 3-aminopyridine-*N*-oxide was formed, which could not confirm that the reaction is proceeding via **22** [37, 63].

Kauffmann identified 3,4-pyridyne as an intermediate when 3-chloropyridine-*N*-oxide was reacted at 100 °C with lithium piperidine to produce 3-piperidino and 4-piperidino compounds [2, 64]. This reaction was envisaged to proceed via 3,4-pyridyne-*N*-oxide intermediate.

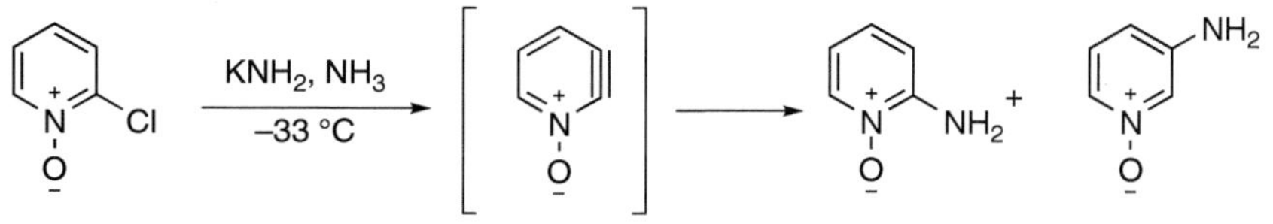

Scheme 9.32 Pyridyne-*N*-oxides generation. Source: Martens and den Hertog [37]; Martens and den Hertog [37, 63].

9.4.16 Indolinyne Generation

Hoye and coworkers established an elegant method for the generation of indolinyne **76** via hexadehydro Diels–Alder reaction (HDDA). In this strategy, the substrate containing three alkynes groups reacts to generate the reactive intermediate **76**, which furnishes the indoline molecule (Scheme 9.33) [65–69]. Using this strategy, one can prepare indoline molecule with multiple substitutions on the benzene ring. It may be noted that Lee group also reported an almost similar aryne generation strategy and explored various reactions of this intermediate [70, 71].

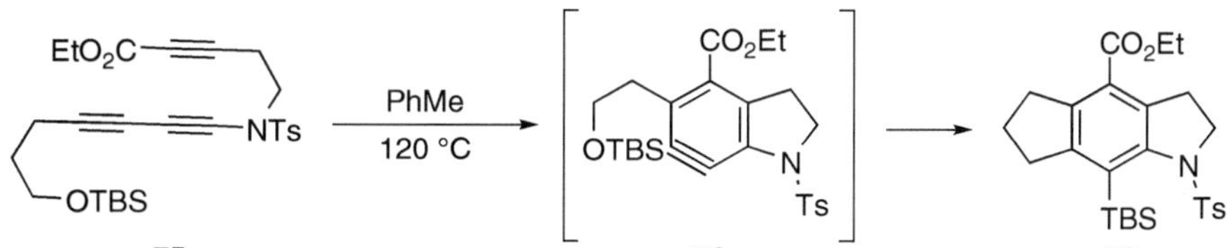

Scheme 9.33 Indolinyne generation via HDDA. Source: Hoye et al. [65]; Niu et al. [66]; Niu and Hoye [67]; Chenet al. [68]; Hoye et al. [69].

9.5 Reactions of Hetarynes

Hetarynes can undergo various types of reactions. Cycloaddition reaction is very popular in hetaryne field. As discussed, cycloaddition product formation is an indirect method to confirm whether aryne is formed or not. Thus, for majority of cases, aryne generation–trapping agents (furan, etc.) were used. So those reactions are not discussed again here. Cycloaddition, where selectivity is controlled, will be discussed because those reactions are synthetically useful. Other than cycloaddition, hetarynes can undergo nucleophilic addition reaction and insertion reaction. When piperidine was used for hetaryne generation, nucleophilic addition of piperidine to hetaryne was observed. Several other nucleophilic addition reactions were known in the literature. Insertion reaction is very well explored in aryne chemistry but not well explored in hetaryne field.

9.5.1 Cycloaddition Reactions

Selectivity is one of the issues for hetaryne cycloaddition reaction. The substitution of different groups has some effect on selectivity. Guitián revealed that halide substitution on the C-2 position of pyridine **77** influences the selectivity of cycloaddition products **78a** and **78b**. Chloro substitution increased the selectivity up to 2.4 : 1, although bromo or fluoro showed no selectivity during cycloaddition (Scheme 9.34) [72, 73].

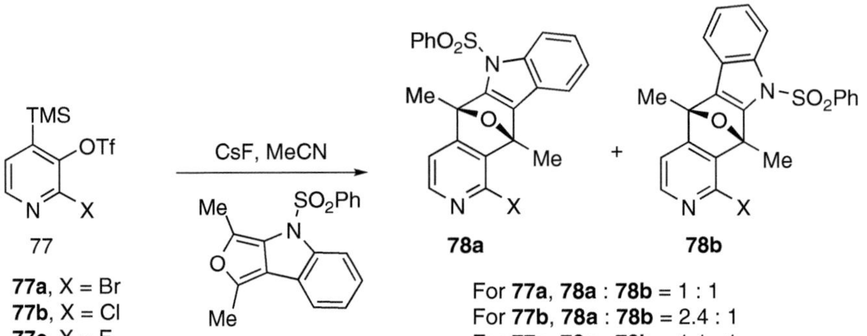

77 | 78a | 78b

77a, X = Br
77b, X = Cl
77c, X = F

For **77a**, **78a** : **78b** = 1 : 1
For **77b**, **78a** : **78b** = 2.4 : 1
For **77c**, **78a** : **78b** = 1.1 : 1

Scheme 9.34 Substitution effect on the selectivity of cycloaddition. Source: Díaz et al. [72]; Díaz et al. [73].

Garg and Goetz showed that selectivity can be fully controlled by the substitution. [3+2] Cycloaddition of 3,4-pyridyne with nitrone has an immense substitution effect. Up to 10.7 : 1 selectivity was observed (**81a** : **81b**) in presence of sulfamoyl group at C-2 position, whereas without any substitution effect selectivity was 1.9 : 1 (**81a** : **81b**). Again, placing bromo group at C-4 position of pyridine reverses the selectivity (**81a** : **81b** = 1 : 3.3). This substrate selective cycloaddition reaction could be useful for various applications (Scheme 9.35) [74]. Iwayama and Sato developed the nickel-catalyzed [2+2+2] cycloaddition with 3,4-pyridyne for synthesis of isoquinoline derivatives [75]

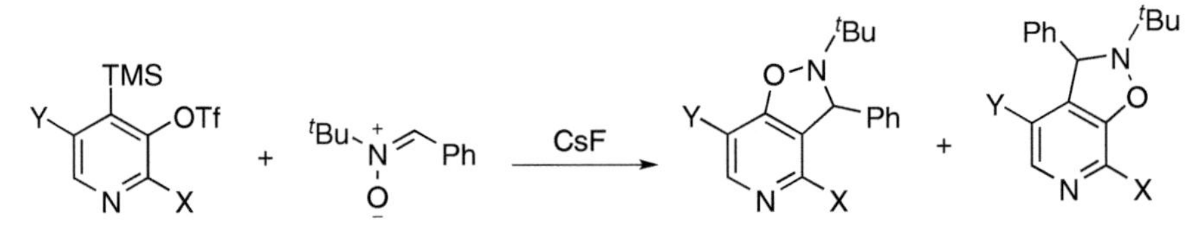

79 | 80 | 81a | 81b

79a, X = H, Y = H
79b, X = OSO_2NMe_2, Y = H
79c, X = H, Y = Br

For **79a**, **81a** : **81b** = 1.9 : 1
For **79b**, **81a** : **81b** = 10.7 : 1
For **79c**, **81a** : **81b** = 1 : 3.3

Scheme 9.35 Selectivity control in cycloaddition reaction. Source: Based on Goetz and Garg [74].

In the case of indole system, 4,5-indolyne showed 2.4 : 1 regioselectivity during cycloaddition with benzyl azide. For 5,6-indolyne selectivity was 1.6 : 1. But in the case of 6,7-indolyne, single regioisomer was obtained (Scheme 9.36) [53]. The expected regioselectivity can be explained by aryne distortion model.

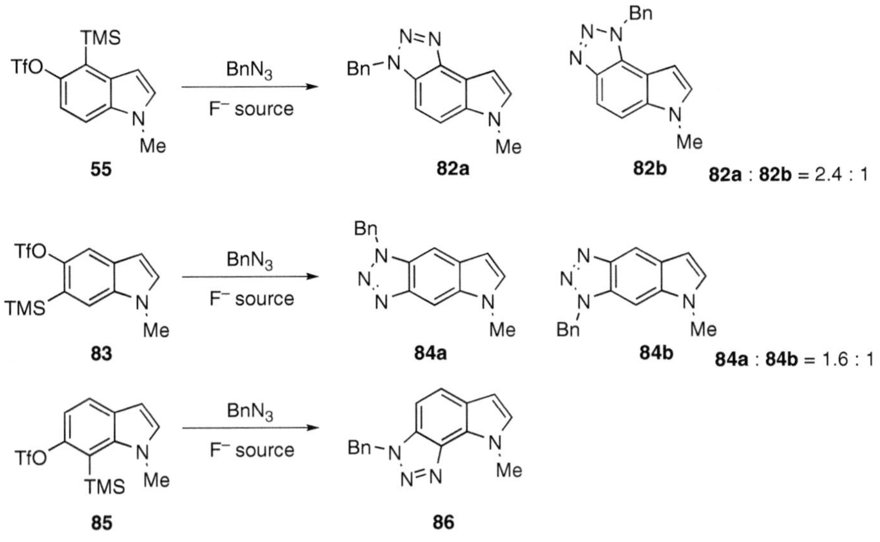

Scheme 9.36 Selectivity of Indolyne for cycloaddition with benzyl azide. Source: Based on Cheong et al. [53].

9.5.2 Nucleophilic Addition Reaction

Various nucleophilic addition reactions are well known for pyridynes and indolynes. Nucleophilic addition reaction on 2,3-pyridyne is selective and nucleophile can add at C-2 position of pyridine exclusively. As Larock and Fang demonstrated, 2,3-pyridyne generated from silyltriflate precursor can undergo nucleophilic addition with *N*-methyl aniline to produce *N*-methyl-*N*-phenylpyridin-2-amine **87** in 65% yield (Scheme 9.37) [56].

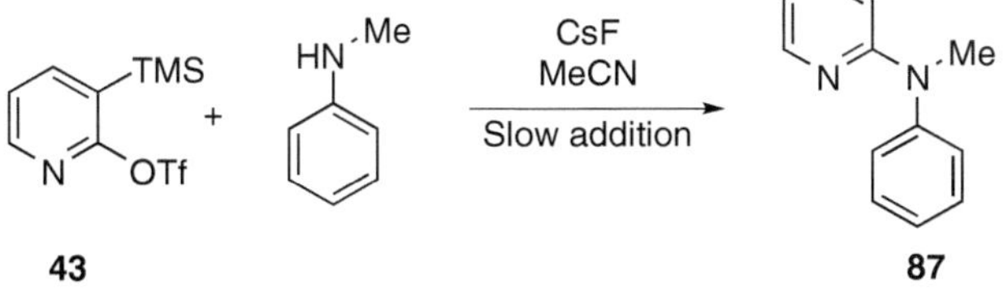

Scheme 9.37 Selective nucleophilic addition on 2,3-pyridyne. Source: Based on Fang and Larock [56].

However, compared to 2,3-pyridine, the addition of 3,4-pyridyne was not selective. Side product formation was the main limitation of previously used aryne generation methodologies. Thus, often strong nucleophile was employed to minimize the side product. As an example, Nisi and Zoltewicz reported the generation of an equal amount of 3- and 4-(methylthio)pyridine **89** using methylmercaptide as nucleophile, where the hetaryne was generated from **88** (Scheme 9.38) [76].

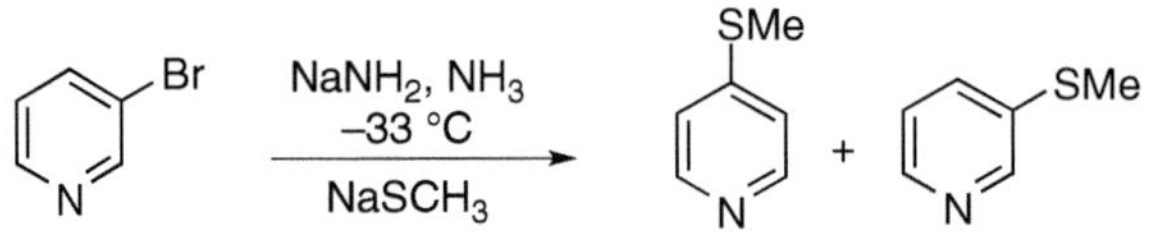

88 **89a** **89b**

89a : **89b** = 1 : 1

Scheme 9.38 Regioselectivity issue in nucleophilic addition reaction for 3,4-pyridyne. Source: Based on Zoltewicz and Nisi [76].

Leake and Levine showed that 3-bromopyridine **88** in presence of acetophenone and $NaNH_2$ produced 4-aminopyridine **90** and 1-phenyl-2-(pyridin-4-yl)ethan-1-one **91**. The product **91** was formed via nucleophilic addition of acetophenone enolate to in situ–generated pyridyne **15**. Although this reaction was free from the regioselectivity issues, both nucleophiles added to highly reactive pyridyne to produce mixture of products (Scheme 9.39) [77]. This was the first report for 3,4-pyridyne **15** generation using 3-bromopyridine **83**.

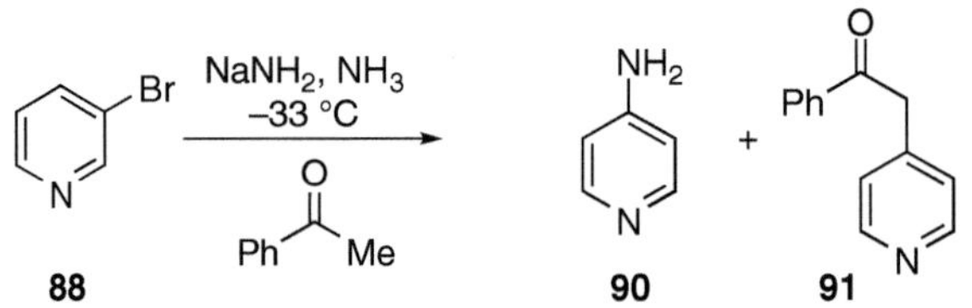

88 **90** **91**

Scheme 9.39 Selectivity in presence of different nucleophiles. Source: Based on Levine and Leake [77].

Silyltriflate precursor of 3,4-pyridyne is also free from side reaction, but selectivity is still not very good. For example, *N*-methylaniline gave only 1.9 : 1 ratio of 3- and 4-regioisomer (**92a** and **92b**). But Garg's methodology was helpful to achieve good selectivity [74]. It showed that sulfamoyl group at the C-2 position forced the upcoming nucleophile to add from C-4 position. However, C-5 bromo substitution restricted the nucleophilic addition at C-4 and **92b** was formed as a major product, where nucleophilic addition happens from C-3 (Scheme 9.40).

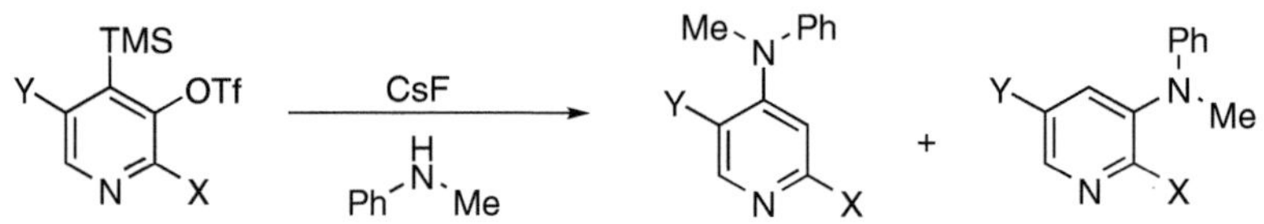

79 **92a** **92b**

79a, X = H, Y = H
79b, X = OSO_2NMe_2, Y = H
79c, X = H, Y = Br

For **79a**, **92a** : **92b** = 1.9 : 1
For **79b**, **92a** : **92b** > 15 : 1
For **79c**, **92a** : **92b** = 1 : 5.8

Scheme 9.40 Selective nucleophilic addition using substitution effect. Source: Based on Goetz and Garg [74].

The nucleophilic addition reaction in indolyne is mostly regioselective. For 4,5- and 5,6-indolyne, nucleophile adds at C-5 position selectively but for 6,7-indolyne, nucleophile adds selectively at C-6 position. The same trend was observed for cycloaddition to indolyne also. Regioselectivity varies depending upon the nucleophile used. Aniline addition to 4,5-indolyne gave 12.5 : 1 regioselectivity, but for 4-methylbenzenethiol addition, selectivity dropped to 2 : 1 [54] For 4,5-indolyne, nucleophilic addition at C-4 is possible if C-6 position is occupied by bromo substitution as demonstrated by Garg and coworkers [22].

9.5.3 Insertion Reaction

The insertion reaction is very much familiar in aryne chemistry. But it is not explored deeply in hetaryne field. Sato et al. developed the insertion of cyclic urea to hetarynes. This methodology is applicable for 2,3-pyridyne, 3,4-pyridyne, as well as 2,3-quinolyne also. For 3,4-pyridyne, regioisomers (**93a** and **93b**) were observed, but for others, exclusively single regioisomer **94** and **96** (for hetaryne generated from **95**) was obtained (Scheme 9.41) [78].

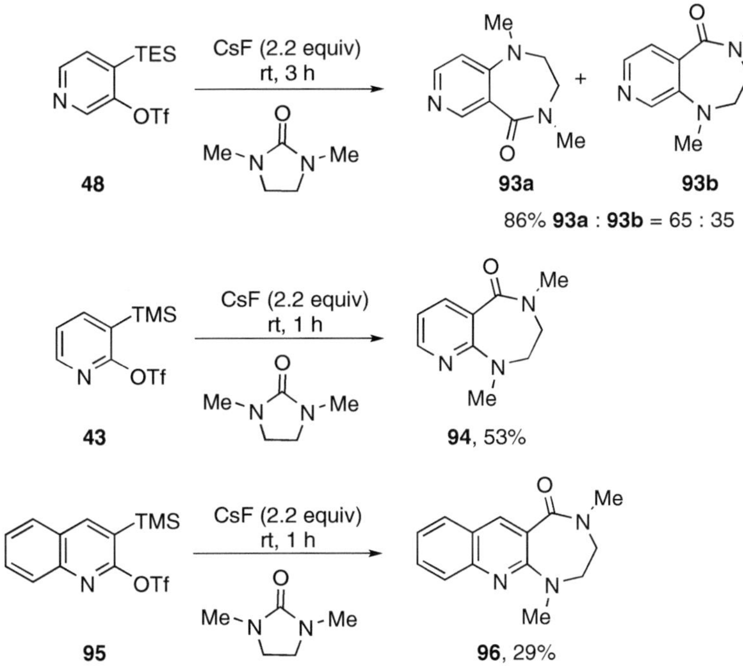

Scheme 9.41 Urea insertion onto hetarynes. Source: Based on Saito et al. [78].

In 2015, Oestreich and coworkers developed the insertion of bis(pinacolato)diboron onto indolynes. The reaction is applicable for all three indolynes (for example generated from **97**) and the insertion product **98** was further functionalized selectively (C-7 substituted) via Suzuki–Miyaura cross-coupling to afford **99**. The remaining boron functionality was also functionalized to produce diaryl indole derivative **100** (Scheme 9.42) [79].

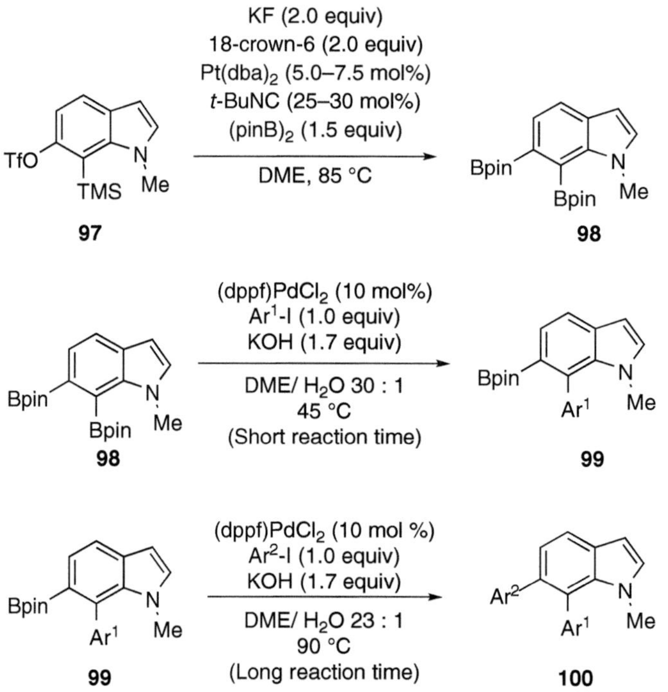

Scheme 9.42 Bis(pinacolato)diboron insertion to indolynes. Source: Based on Pareek et al. [79].

As discussed for 3,4-pyridynes, the regioselectivity was one major issue in insertion reactions. But placing substitution at different positions can improve the selectivity and after the reaction, substitution can also be removed. For insertion to 3,4-pyridyne, Garg's substitution effect was excellent to improve the selectivity. For C-4 substitution, only C-3 addition product (**101**) was observed and for C-2 exclusively C-4 addition product (**101**) was formed. But without any substitution, the products were obtained with 2.1 : 1 regioisomeric ratio (Scheme 9.43) [74].

9.6 Applications in Synthesis

Although earlier methods of hetaryne generation have some limitations, now with the advancement of hetaryne generation methodologies, the hetarynes are widely used for complex synthetic targets (specially pyridyne and indolyne). Using pyridyne as an intermediate ellipticine, macrostomine, eupolauramine, perlolidine can be accessed [24, 25, 72, 73, 80–85]. Indolynes can be used for the synthesis of different makaluvamines, lysergic acid, members of the welwitindolinone,

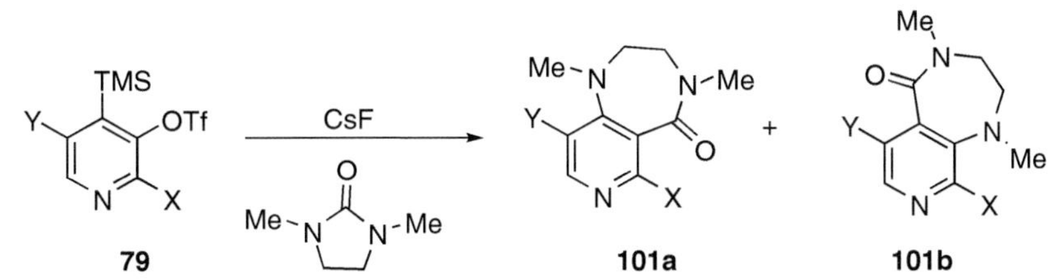

79 **101a** **101b**

79a, X = H, Y = H
79b, X = OSO_2NMe_2, Y = H
79c, X = H, Y = Br

For **79a**, **101a** : 101b = 2.1: 1
For **79b**, **101a** only
For **79c**, **101b** only

Scheme 9.43 Regioselectivity for 3,4-pyridyne insertion. Source: Based on Goetz and Garg [74].

indolactam alkaloid families, and indole diterpenoid tubingensin A, tubingensin B [22, 23, 47, 86–90].

9.6.1 Application of Pyridyne

Since a long time, pyridyne had been used for total synthesis. In 1976, Singh and coworkers synthesized perlolidine **105** using pyridyne as an intermediate. Starting from readily available 2-amino pyridine **102**, after five steps, the derivative **103** was synthesized. Then, treatment of **103** with KNH_2 generated the pyridyne and produced **104**. After successive steps, perlolidine **105** was synthesized from **104** (Scheme 9.44) [84, 85].

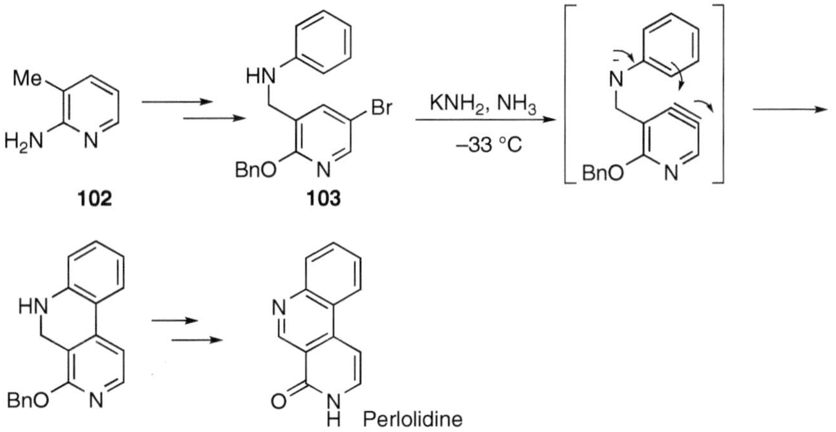

102 **103** **104** **105**

Scheme 9.44 Perlolidine synthesis by Singh and coworkers. Source: Kessar et al. [84]; Kessar and Singh [85].

Moody and May synthesized ellipticine **110** in three steps starting from indole. At the last step, pyridyne was generated from **109** by heating, and the generated pyridyne underwent Diels–Alder reaction with **108** (Scheme 9.45). At the last step, ellipticine **110** and iso ellipticine were formed and were separated via column chromatography [80, 81]. Gribble and Sha also synthesized ellipticine via Diels–Alder reaction with pyridyne [82, 83]. As discussed in Section 9.5.1. (Scheme 9.33), Guitián and coworkers investigated the substitution effect on 3,4-pyridyne with various dienes. This indicated that the chloro substitution can give good selectivity and Guitián also showed that Diels–Alder product **78a** could be elaborated to ellipticine [72, 73].

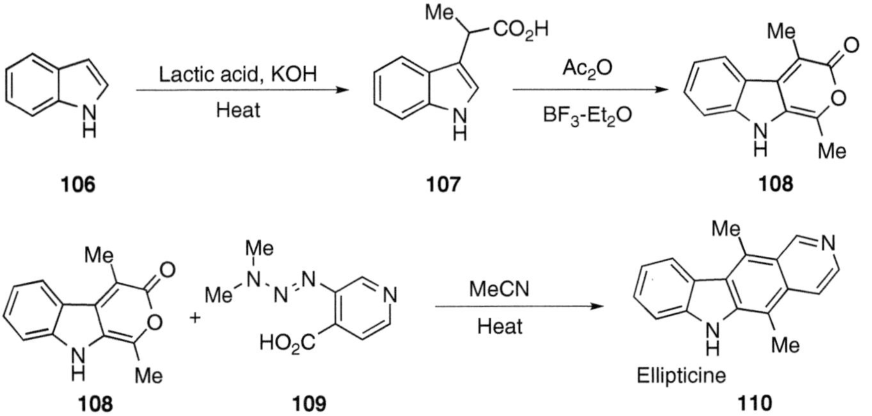

Scheme 9.45 Ellipticine synthesis by Moody and coworkers. Source: May and Moody [80]; May and Moody [81].

Eupolauramine was synthesized by Couture and coworkers using pyridyne as a reactive intermediate. In the early steps of synthesis, pyridyne was generated from 2-chloronicotinic acid derivative **112** formed from **111**, and the intramolecular cyclization afforded the desired products **113a** and **113b**, which underwent sequence of reactions to furnish eupolauramine (Scheme 9.46) [25].

(*S*)-Macrostomine **117** was synthesized from (*S*)-Nicotine by Comins and coworkers. In the intermediate steps of (*S*)-Macrostomine **117** syntheses, pyridyne was generated from **115**, and trapped with 3,4-dimethoxyfuran to afford the cycloadduct **116**. After the cycloaddition, two more steps were performed to achieve the synthesis of **117** (Scheme 9.47) [24].

9.6.2 Application of Indolyne

Like pyridynes, indolynes were also known for total synthesis for a long time. For example, Julia and coworkers reported the synthesis of lysergic acid using indolyne as an intermediate. In their synthesis, indolyne was generated from 5-bromoindoline **118** using $NaNH_2$. $NaNH_2$ also abstracted a proton from **118**, which resulted in an intramolecular cyclization to deliver the tetracycle **120**. The regiochemistry of **120** was controlled by geometric constraints. The cyclized product could be elaborated to lysergic acid **121** (Scheme 9.48) [23].

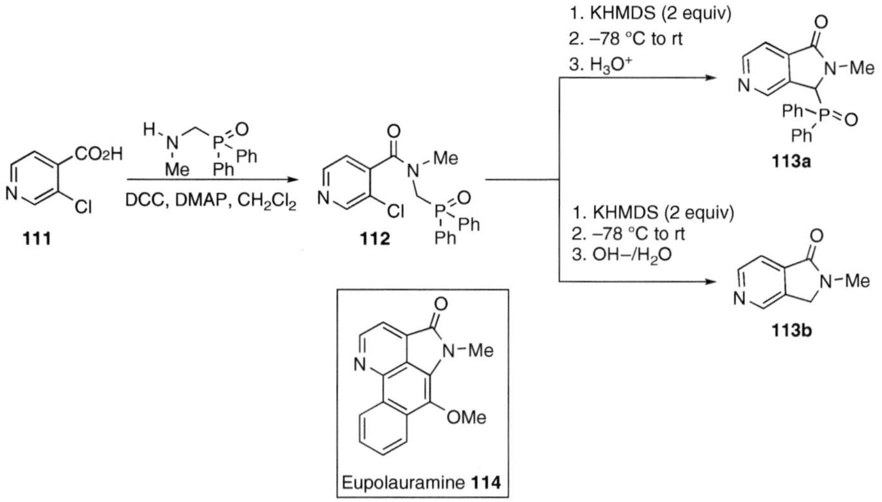

Scheme 9.46 Hetaryne-mediated eupolauramine synthesis by Couture. Source: Based on Hoarau et al. [25].

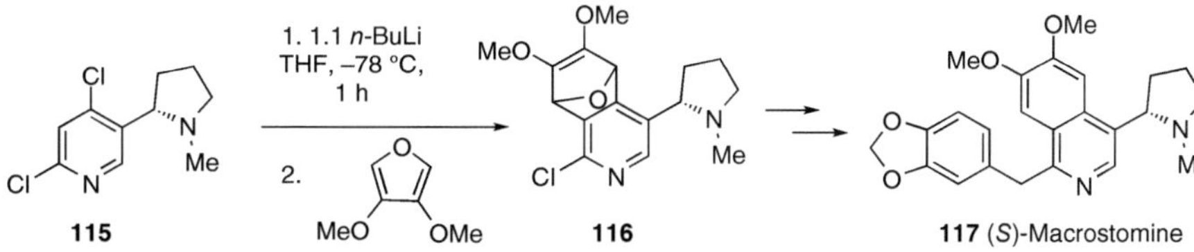

Scheme 9.47 (*S*)-Macrostomine synthesis by Comins. Source: Based on Enamorado et al. [24].

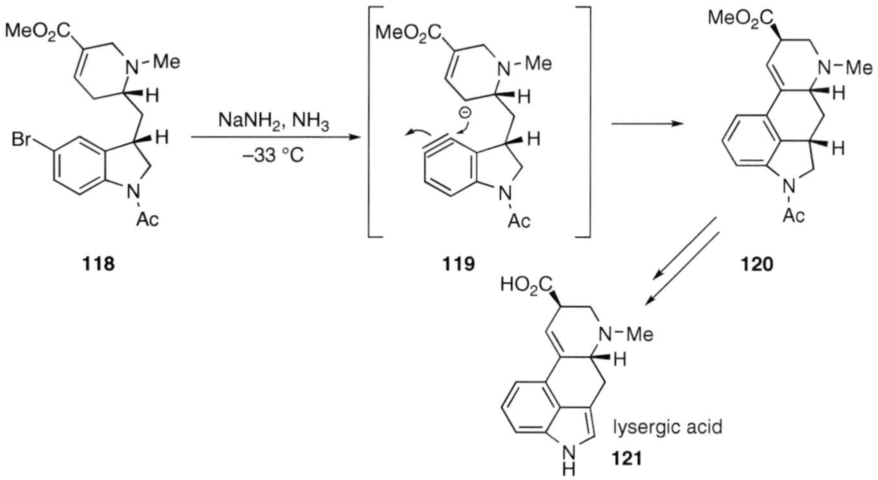

Scheme 9.48 Lysergic acid synthesis by Julia and coworkers. Source: Julia et al. [23].

(±)-*Cis*-trikentrin A **124** and (±)-herbindole A **127** were synthesized by Buszek and coworkers via an intermolecular Diels–Alder reaction using indolyne (generated from **122** and **125**) and cyclopentadiene leading to the formation of **123** and **126** as the key steps. This approach integrated Bartoli indole synthesis for efficient synthesis of the trikentrins and herbindoles (Scheme 9.49) [50, 91].

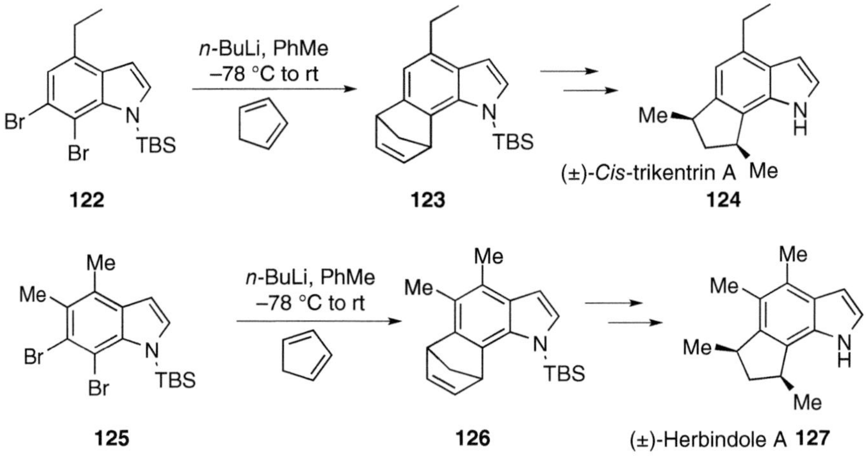

Scheme 9.49 Trikentrin and herbindole synthesis by Buszek group. Source: Buszek et al. [91]; Brown et al. [91].

As discussed, Iwao and coworkers synthesized makaluvamines using 4,5-indolyne formation as a key step (Scheme 9.21) [47]. Members of the welwitindolinone families ((−)-*N*-methylwelwitindolinone C isothiocyanate, oxidized welwitindolinones, (−)-*N*-methylwelwitindolinone C isonitrile, and *N*-methylwelwitindolinone D isonitrile) were synthesized by Garg and coworkers using 4,5-indolyne as an effective intermediate and details of this methodology are discussed in Chapter 11 [86–88].

Garg and coworkers also applied indolyne methodology toward the total synthesis of indolactam alkaloids. The synthesis of the four indolactam alkaloids was performed from one common intermediate. This intermediate was obtained through intermolecular nucleophilic addition at C-4 position of 4,5-indolyne. This method is also discussed in Chapter 11 [22]. Furthermore, the synthesis of tubingensin A and tubingensin B was reported by Garg and coworkers using carbazolyne intermediate and the details are also presented in detail in Chapter 11 [90, 92].

9.7 Introduction to Cycloalkynes

Cyclohexyne or more generally strained cycloalkynes are the aliphatic variants of benzyne intermediate having a strained C—C triple bond in the cyclic aliphatic system. The reactivity and isolability of cycloalkynes depend on the ring size.

Smaller the ring size, higher the reactivity. The smallest possible cycloalkyne, i.e. cyclopropyne, is not even generated or detected spectroscopically due to the enormous angle deformation. Similarly, there is no report on the generation or trapping of cyclobutyne other than the isolation of an osmium complex with cyclobutyne ligand [93]. The five-, six-, and seven-membered cycloalkynes are also very reactive, and these can only be synthesized in situ in presence of a suitable trapping agent. Compared to these three cycloalkyne intermediates, medium-ring cycloalkynes are quite stable and cyclooctyne is the smallest isolable cycloalkyne. These medium-ring cycloalkynes are used extensively in the click chemistry. In the context of this part of the chapter, a brief discussion on the history, methods of preparation, reaction, and application of five-, six-, and seven-membered cycloalkynes and their heteroatom-embedded analogs are presented here. A closely related intermediate strained cyclic allenes also will be discussed in this chapter.

9.8 History of Cycloalkynes

The history of strained cyclic alkynes is closely related to that of aryne intermediate. After the initial assumption for the existence of aryne intermediate by Stoermer and Kahlert in 1902 [1], Favorskii and Boshowskii in 1912 suggested a cyclohexyne intermediate for the formation of dodecahydrotriphenylene **129** from 1,2-dibromocyclohexene **128** (Scheme 9.50) [94]. Similarly, in 1936, a cyclopentyne trimer was prepared by the reaction of 1,2-dibromocyclopentene with sodium [95].

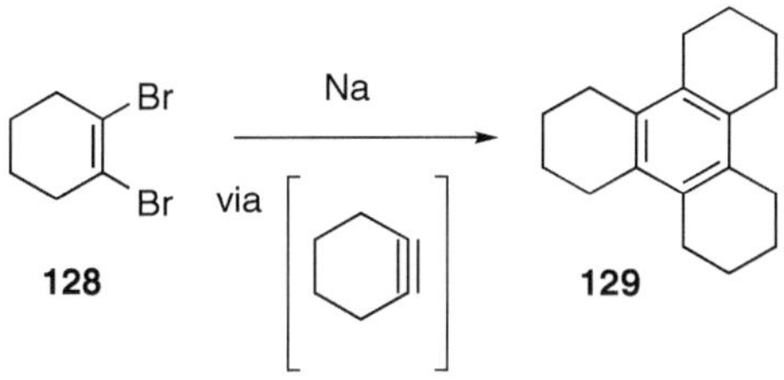

Scheme 9.50 Dodecahydrotriphenylene from 1,2-dibromocyclohexene. Source: Favorskii and Boshowskii [94].

Later in 1944, Wittig and Harboth reported a low-yield method for the synthesis of 1-phenylcyclohexene **131** from the coupling of phenyl lithium with 1-chlorocyclohexene **130** in ether at 100 °C (Scheme 9.51) [96]. The formation of 1-phenylcyclohexene was explained via a direct nucleophilic substitution, and ignored the possibility of cyclohexyne formation as this intermediate was thought to be structurally impossible.

To explore the possibility the formation of 1-phenylcyclohexene **131** from 1-chlorocyclohexene **130** and phenyl lithium via a cyclohexyne intermediate, Scardiglia and Roberts in 1957 devised the reaction with a ^{14}C-labeled

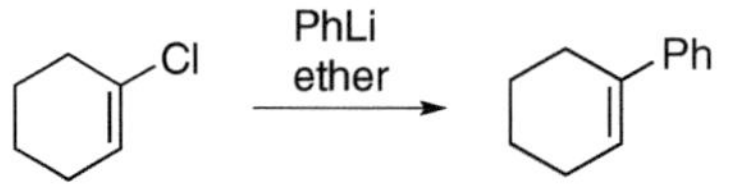

130 131

Scheme 9.51 Coupling of phenyllithium with l-chlorocyclohexene. Source: Based on Wittig and Harborth [96].

1-chlorocyclohexene **132** (Scheme 9.52) [97]. When the equimolecular ratio of ^{14}C-labeled 1-chlorocyclohexene **134** and **135** (prepared from **133**) was heated with phenyl lithium at 150 °C, almost 28% of the 1-phenylcyclohexene was formed, which upon oxidation with sodium permanganate, benzoic acid with 23% ^{14}C-labeling was observed. Thus, 23% of the 1-phenylcyclohexene **141** labeled at 1-position was formed. This is comparable with the theoretical 25% of **141**. This small difference between the theoretical and experimental values can be ignored owing to some nonrearranging reaction. Thus, the coupling reaction of 1-chlorocyclohexene and phenyl lithium was taking place via cyclohexyne intermediate and not via a direct nucleophilic substitution as in the case of direct nucleophilic substitution no 1-phenylcyclohexene **141** labeled at 1-position would have formed.

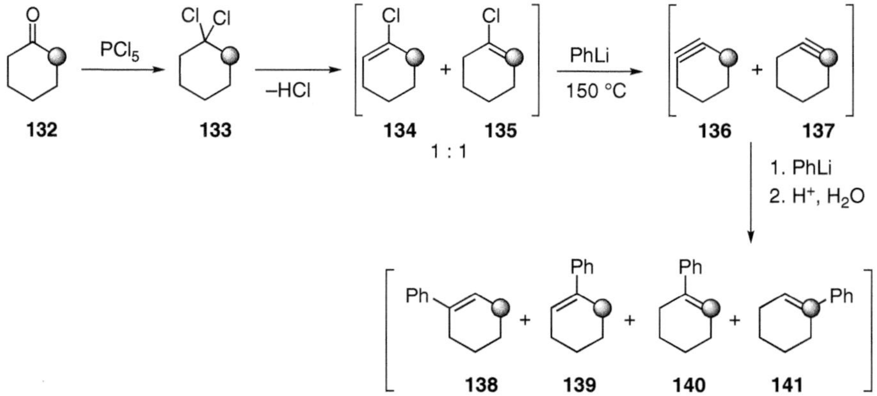

132 133 134 135 136 137 138 139 140 141

Scheme 9.52 ^{14}C-Labelled l-chlorocyclohexene with phenyllithium. Source: Based on Scardiglia and Roberts [97].

1-phenylcyclohexene **131** might have also formed via an allenic intermediate 1,2-cyclohexadiene **142** (Scheme 9.53). But this possibility was eliminated as in this case no 1-phenylcyclohexene **141** labeled at 1-position would have formed. Here, it should be noted that in 1966 Wittig trapped allenic intermediate via a [4+2] cycloaddition reaction.

Finally, in 1960, Wittig successfully trapped cyclopentyne, cyclohexyne, and cycloheptyne intermediate via a [4+2] cycloaddition with 1,3-diphenyl isobenzofuran starting from the dibromide **145** to afford finally the naphthalene derivative **147** via the initially formed cycloadduct **146** (Scheme 9.54) [98].

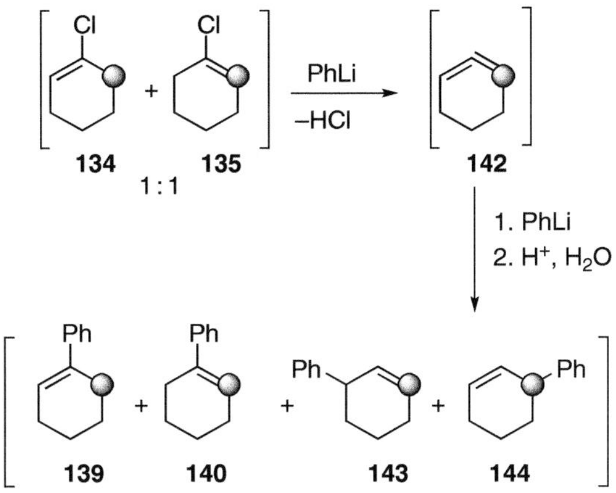

Scheme 9.53 Probability of allenic intermediate. Source: Based on Scardiglia and Roberts [97].

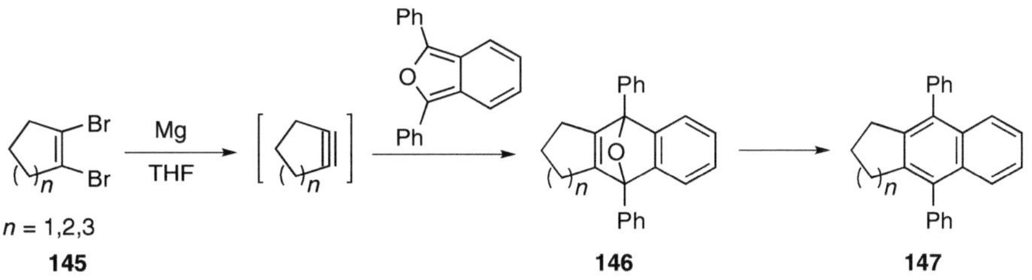

Scheme 9.54 Trapping of cycloalkynes. Source: Based on Wittig et al. [98].

9.9 Different Types of Cycloalkynes

Depending upon the ring size, strained cycloalkynes can be divided broadly into three categories (five-, six-, and seven-membered cycloalkynes), namely cyclopentyne **148**, cyclohexyne **149**, and cycloheptyne **150**. Cycloalkynes, having a heteroatom in the ring, are generally termed as hetero cycloalkynes. If the skeletal ring contains a sulfur atom, it is called thiacycloalkyne, e.g. tetramethylthia cyclopentyne **151**, and if the ring has an oxygen atom, it termed as oxa-cyclohexyne. 3,4-Oxa-cyclohexyne **152** is an example of oxa-cyclohexyne. In the case of nitrogen, the terminology is aza-cyclohexyne, 2,3-piperidyne **153**, and 3,4-piperidyne **154** fall under this category. Similarly, **155** is a cyclohexenynone, a keto group containing cyclohexyne and **156** is norbornyne, a bicyclic cycloalkyne (Scheme 9.55).

9.10 Methods of Cycloalkyne Generation

Strained cycloalkynes are very reactive, so they cannot be isolated and must be generated in situ in the reaction medium. There are various methods available for the

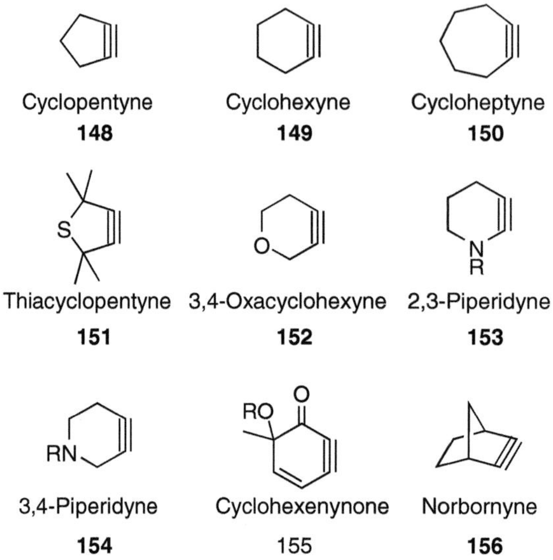

Scheme 9.55 Different types of cycloalkynes.

generation of cycloalkynes, some methods are in principle similar to the generation of aryne intermediates. Selected methods are discussed in Sections 9.10.1–9.10.2.

9.10.1 Traditional Methods of Cycloalkyne Generation

9.10.1.1 Base-Induced 1,2-Elimination

Cycloalkynes can be generated from 1-halocycloalkenes by deprotonation of β-vinyl hydrogen by a strong base. In 1944, Wittig used L-chlorocyclohexene **130** and phenyl lithium for the generation of cyclohexyne **149**, but the yield was only 5% (Scheme 9.56) [96]. The main problem with the base-induced elimination is the competitive generation of 1,2-cycloalkadienes [99].

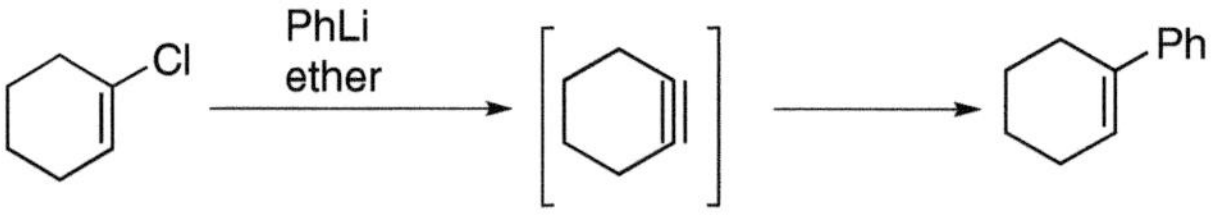

130 149 131

Scheme 9.56 Generation of cyclohexyne from 1-chlorocyclohexene. Source: Based on Wittig and Harborth [96].

Futija and coworkers used iodonium salt **157** for the mild generation of substituted cyclohexyne **158** (Scheme 9.57) [100].

In 2017, Okano and coworkers generated cyclohexynes and cycloheptynes from corresponding enol triflates **159** with a mild bulky base, magnesium bis(2,2,6,6-tetramethylpiperidide) by controlling the reactivity of intermediate anionic species **160** (Scheme 9.58) [101].

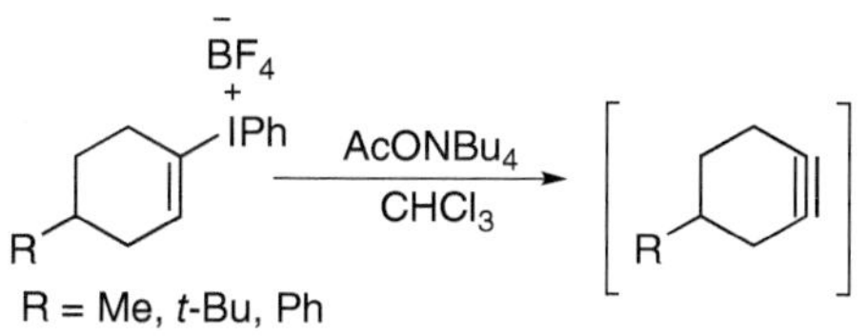

157 **158**

Scheme 9.57 Generation of cyclohexyne from iodonium salt. Source: Based on Fujita et al. [100].

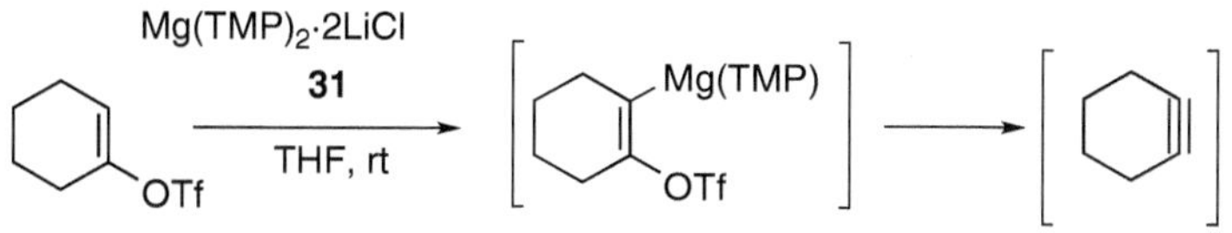

159 **160** **149**

Scheme 9.58 Generation of cycloalkyne from enol triflates. Source: Based on Hioki et al. [101].

9.10.1.2 Metal–Halogen Exchange/Elimination

Another approach involves the metal–halogen exchange/elimination of 1,2-dihalocycloalkenes **145** mediated by metals (Mg or Na) [98]. But one has to use stoichiometric amount of metals for the generation of cycloalkynes. Using this approach, cyclopentyne **148**, cyclohexyne **149**, and cycloheptyne **150** could be generated (Scheme 9.59).

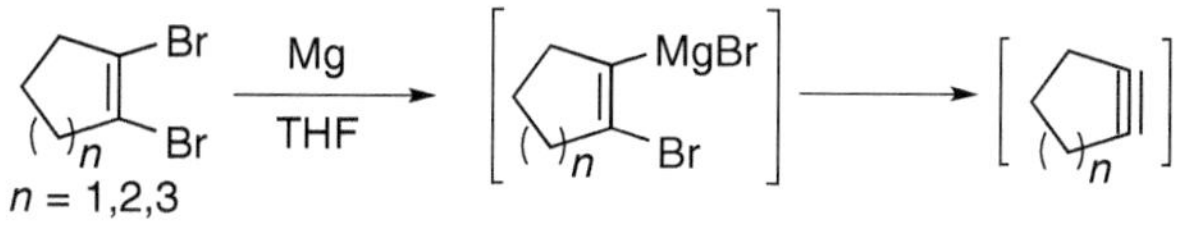

145 **161**

Scheme 9.59 Generation of cycloalkynes from 1,2-dihalocycloalkenes. Source: Based on Wittig et al. [98].

9.10.1.3 Fragmentation of Aminotriazoles

1-Tosylamino-1,2,3-triazole anions **162** upon photolytic or thermal fragmentation produce cycloalkynes with the evolution of nitrogen gas (Scheme 9.60) [102].

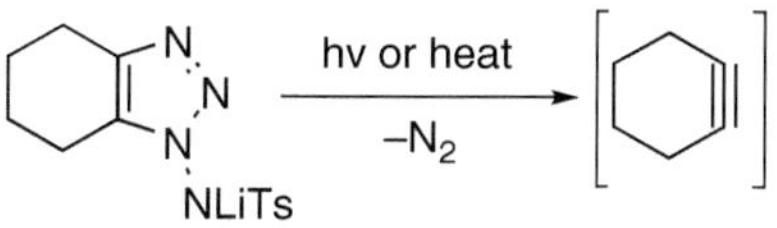

162 **149**

Scheme 9.60 Generation of cycloalkyne from aminotriazole anion. Source: Based on Willey [102].

9.10.1.4 Fragmentation of Diazirine

Thermal or photolytic degradation of isolable bis-diazirine derivatives evolves nitrogen gas and produces cycloalkyne intermediates. This procedure is applicable for the generation of cyclohexyne **149** and cycloheptyne **150** in good yields. Interestingly, norbornyne **156** could be generated by thermolysis of the norbornyl bisdiazirine derivative **163** (Scheme 9.61) [103].

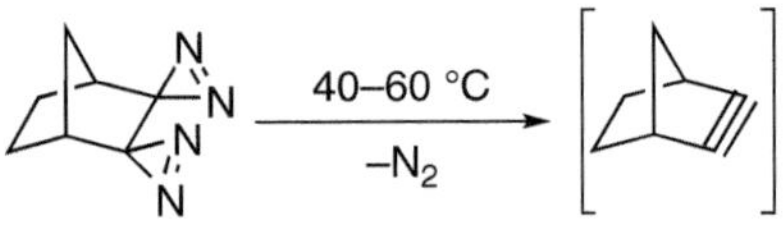

163 **156**

Scheme 9.61 Generation of norbornyne. Source: Based on Al-Omari et al. [103].

9.10.1.5 Oxidation of 1,2-Bis-hydrazones

The Curtius method of alkyne synthesis has been successfully applied for the generation of strained cycloalkynes. Oxidation of 1,2-bis-hydrazonocycloalkanes **164** with strong oxidant (HgO) generates cycloalkynes with the evolution of nitrogen from the initially formed diazo compound **165**. Although the yields are low, this procedure can be used for generation of cyclopentyne **148**, cyclohexyne **149**, cycloheptyne **150**, and tetramethylthiacyclopentyne **151** (Scheme 9.62) [104, 105]. Due to the use of strong oxidant, this method has less functional group tolerance and hence not widely used.

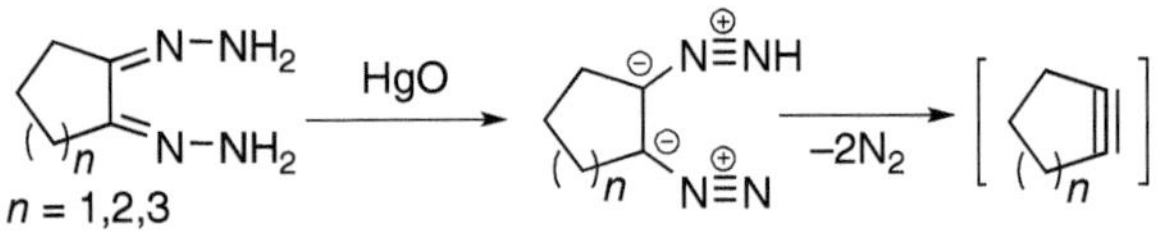

n = 1,2,3

164 **165**

Scheme 9.62 Generation of cycloalkyne from 1,2-bis-hydrazones. Source: Wittig and Krebs [104]; Bolster and Kellogg [105].

9.10.1.6 Rearrangement of Vinylidenecarbenes

When bromomethylene cycloalkanes are heated with a strong base, it generates vinylidene carbene intermediate, which immediately rearranges to cycloalkyne intermediate. For instance, cyclohexyne **149** can be generated by heating bromomethylene cyclopentane **166** with potassium *tert*-butoxide via the rearrangement of the corresponding vinylidene carbene **167**. Similarly, cyclopentyne can be produced from bromomethylene cyclobutane (Scheme 9.63) [106].

9.10.2 Fluoride-Induced Cycloalkyne Generation

In 1983, Kobayashi and coworkers discovered the fluoride-induced 1,2-elimination of 2-(trimethylsilyl)aryl triflates to generate aryne in solution. Following the success

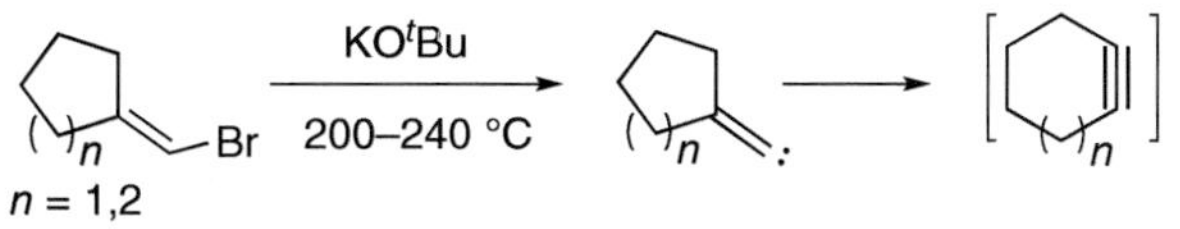

Scheme 9.63 Generation of cycloalkyne from vinylidene carbene. Source: Based on Erickson and Wolinsky [106].

of this mild and base-free method, in 1998 Guitián and coworkers developed the aliphatic variant of Kobayashi's aryne precursor, 2-(trimethylsilyl)cyclohexenyl triflate **169**. This precursor was synthesized from **168** in two steps with 78% yield (Scheme 9.64) [107]. Similarly, 2-(trimethylsilyl)cyclopentenyl triflate was prepared by Guitián and coworkers in 2002 for the mild generation of cyclopentyne [108]. Recently, Garg and coworkers prepared silyl tosylate precursor for milder and controlled generation of cyclohexyne via the fluoride-induced cyclohexyne generation [109]. This fluoride-induced cyclohexyne generation is highly preferred by synthetic chemists over traditional methods as it is a mild, base-free condition, and compatible with various functional groups. Various other silyltriflate precursors to the strained cycloalkynes and heterocycloalkynes have been prepared over the last few years for mild generation of the cycloalkynes.

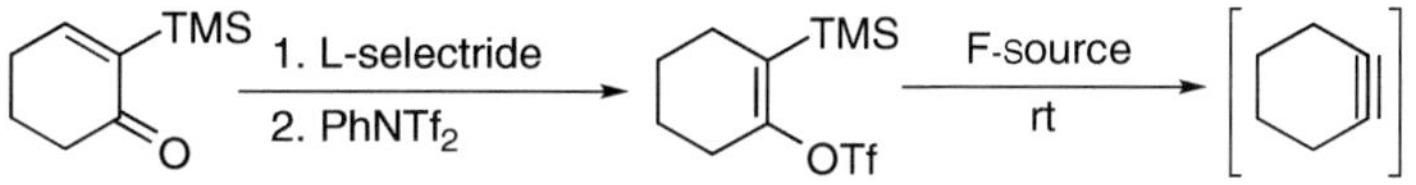

Scheme 9.64 Fluoride-induced cyclohexyne generation. Source: Based on Atanes et al. [107].

9.10.2.1 Generation of 3,4-Oxacyclohexyne

Garg and coworkers reported the generation 3,4-oxacyclohexyne **152** via the fluoride-induced β-elimination of silyl triflate precursor **171**. The precursor was synthesized in four steps starting with commercially available 4-oxotetrahydropyran **170** (Scheme 9.65) [110].

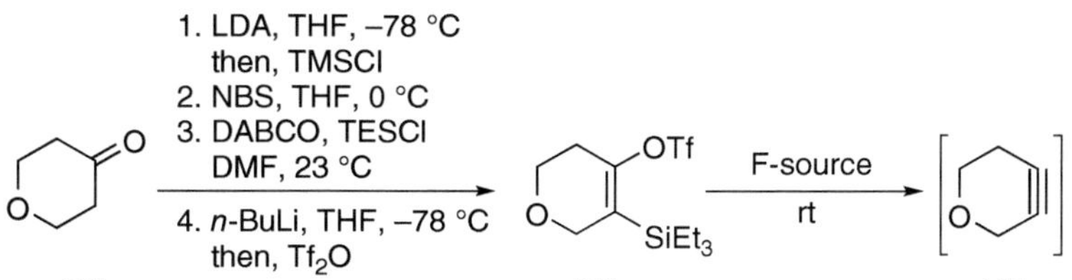

Scheme 9.65 Fluoride-induced generation of 3,4-oxacyclohexyne. Source: Based on Shah et al. [110].

9.10.2.2 Generation of 2,3-Piperidyne

In 1988, Wentrup and coworkers detected the 2,3-piperidyne **153a** at low temperature. This was generated from the pyrolysis of isoxazolone **172** via the rearrangement of the intermediate carbene **173** (Scheme 9.66) [111].

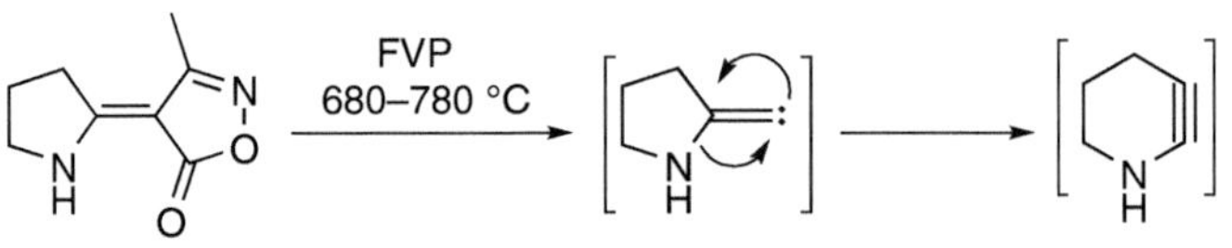

Scheme 9.66 Pyrolysis of isoxazolone. Source: Based on Wentrup et al. [111].

The silyl triflate precursor **175** of the 2,3-piperidyne **153b** was synthesized by Danheiser and coworkers from readily available lactam **174** in a three-step process (Scheme 9.67) [112].

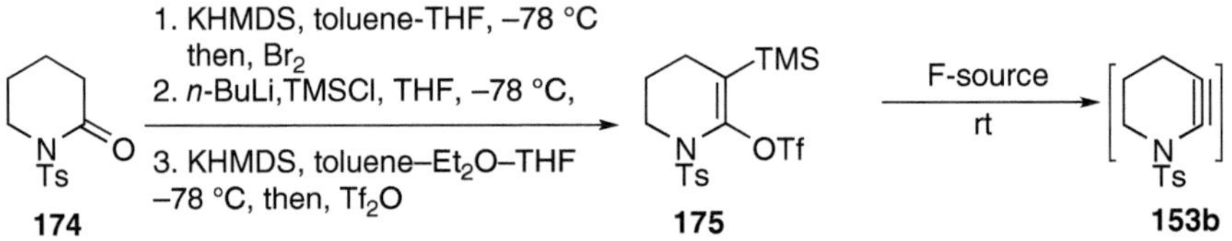

Scheme 9.67 Fluoride-induced generation of 2,3-piperidyne. Source: Based on Tlais and Danheiser [112].

9.10.2.3 Generation of 3,4-Piperidyne

Generation of the 3,4-piperidyne **154a** was described by Garg and coworkers via the fluoride-induced 1,2-elimination of silyl triflate from the corresponding aza-cycloalkyne precursor **177**. It was prepared from the commercially available 4-methoxypyridine **176** by the following synthetic route (Scheme 9.68) [113].

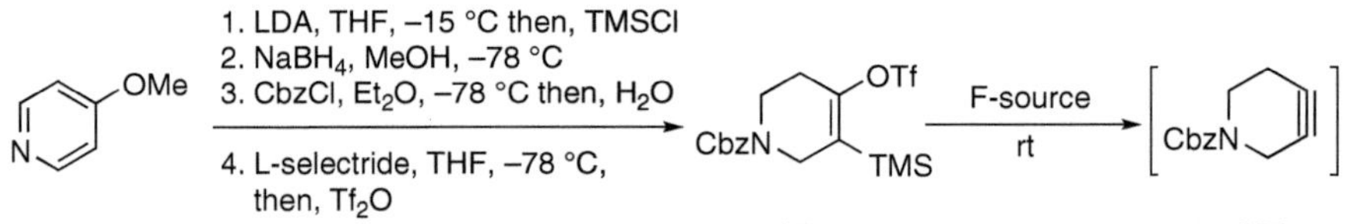

Scheme 9.68 Fluoride-induced generation of 3,4-piperidyne. Source: Based on McMahon et al. [113].

9.10.2.4 Generation of Cyclohexenynone

Recently, Li and coworkers reported the generation of cyclohexenynone **155a** and **182** from the corresponding silyl triflate precursor **179** and **181**. The cyclohexenynone was synthesized from **178** and **180** by oxidative dearomatization (Scheme 9.69) [114].

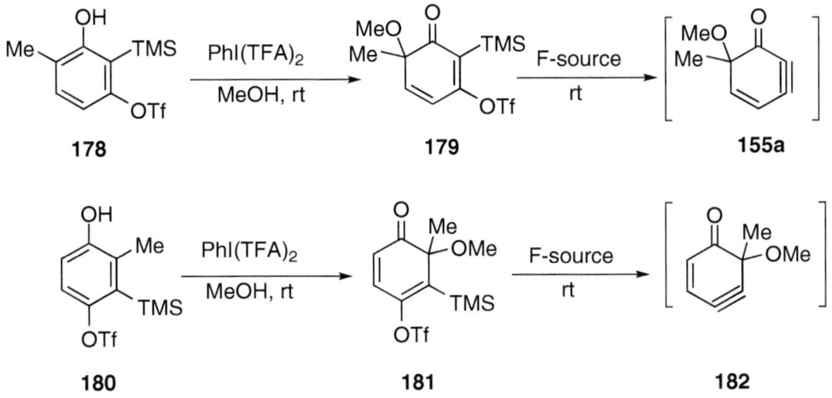

Scheme 9.69 Fluoride-induced generation of cyclohexenynone. Source: Based on Qiu et al. [114].

9.11 Reactions of Cycloalkynes

Strained cycloalkynes and heterocyclohexynes are very reactive, so they can participate in various types of reactions. Cycloalkynes are very good dienophile and it can be trapped by various dienes like furans, 1,3-diphenylisobenzofuran and tetraphenylcyclopentadienone. Cycloalkynes can also participate in arylation reactions and insertion reactions. Multicomponent reactions and molecular rearrangements are still not well explored for cycloalkynes.

9.11.1 Cycloaddition Reactions

Cycloaddition reactions are the most common reaction of cycloalkynes as the generation of cycloalkynes is confirmed via isolation of different cycloadducts. Some of the representative examples with cycloalkynes are presented in Sections 9.11.1.1–9.11.1.3.

9.11.1.1 Diels–Alder Reaction

Cycloalkynes are excellent dienophiles in [4+2] cycloaddition reactions. In 1998, Guitián and coworkers successfully generated cyclohexyne from the silyltriflate precursor, and then trapped the cyclohexyne with pyrones **183a–d** for the synthesis of tetrahydro naphthalenes **184a–d** (Scheme 9.70) [107].

In 2019, Garg and coworkers reported the synthetic route for the generation of heteroatom containing polycyclic aromatic hydrocarbon (PAH) **186** via two successive cycloadditions and elimination of nitrogen and carbon dioxide from **185** (Scheme 9.71) [115].

9.11.1.2 [2+2] Cycloaddition

[2+2] Cycloaddition is not very common in cycloalkynes chemistry. The [2+2] cycloaddition of cyclohexenynone **182** generated from **181** with 1,1-dimethoxyethene generates the cycloadduct **183** in 62% yield (Scheme 9.72) [114].

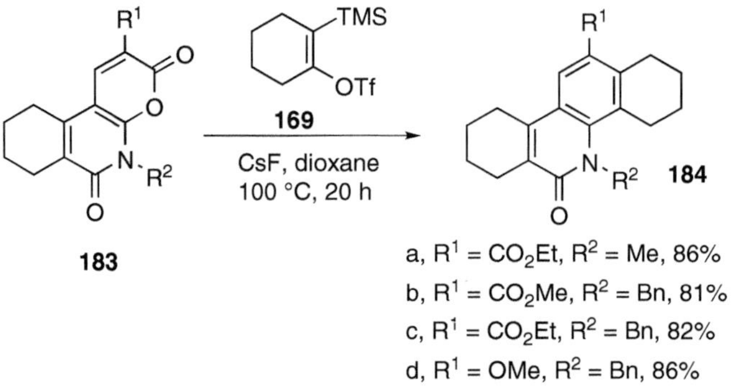

Scheme 9.70 Generation of tetrahydro naphthalenes via Diels–Alder reaction. Source: Based on Atanes et al. [107].

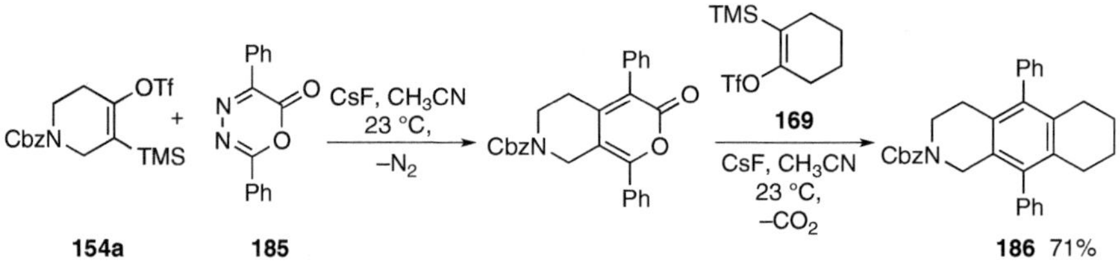

Scheme 9.71 Generation of PAH via Diels–Alder reaction with cyclohexyne. Source: Based on Darzi et al. [115].

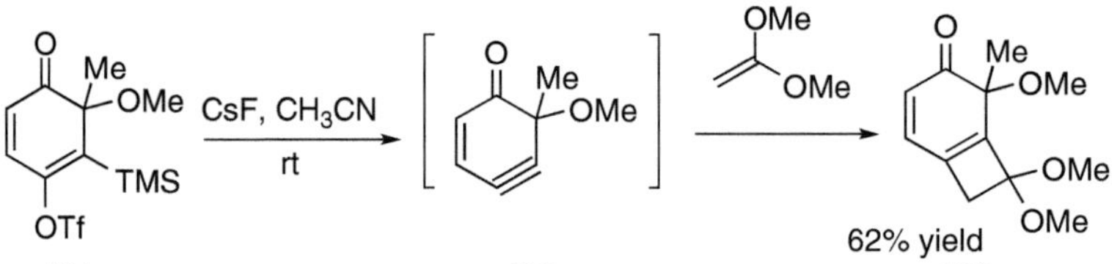

Scheme 9.72 [2+2] Cycloaddition. Source: Based on Qiu et al. [114].

9.11.1.3 1,3-Dipolar Cycloaddition

Cycloalkynes are excellent dipolarophiles, and hence they can add to various 1,3-dipoles such as nitrones, nitrile oxides, nitrile imines, azomethine imines, azides and diazo compounds (Scheme 9.73).

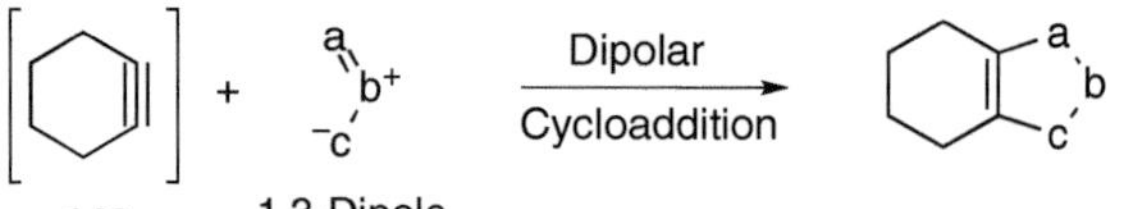

Scheme 9.73 1,3-Dipolar cycloaddition.

For example, Garg and coworkers reported the 1,3-dipolar cycloaddition of cyclohexyne generated from the precursor **169** with benzyl azide **184** leading to the synthesis of tetrahydro benzotriazole **185** (Scheme 9.74) [116]. The use of CsF in THF was best for this transformation.

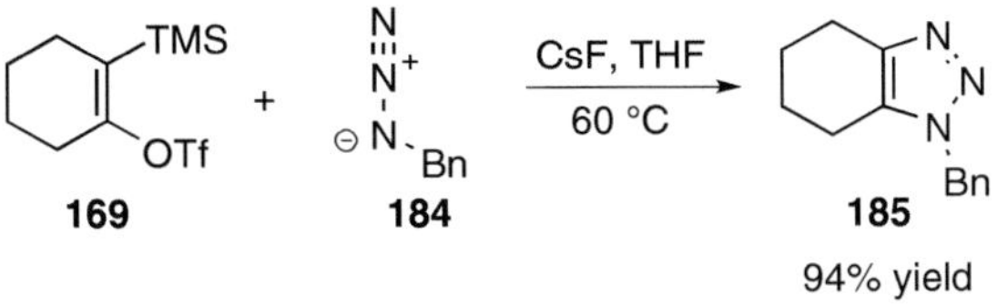

169 184 185

Scheme 9.74 1,3-Dipolar cycloaddition of benzyl azide and cyclohexyne. Source: Based on Medina et al. [116].

9.11.2 Alkenylation Reactions

Similar to arynes, charged or uncharged nucleophiles add to strained cycloalkynes or heterocycloalkynes and upon protonation produce the formal alkenylated product. When 2,3-piperidyne **153b** is generated in presence of thiophenol, it adds to the 2,3-piperidyne **153b** regioselectively at the C-3 position and the generated anion, which was subsequently protonated to furnish a thioalkenylated product **186** (Scheme 9.75) [112].

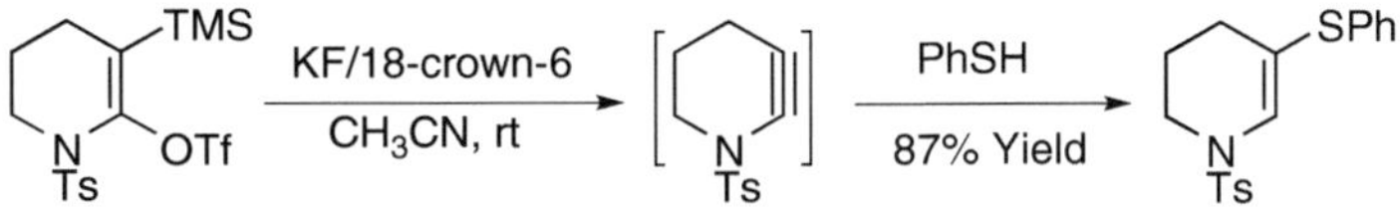

175 153b 186

Scheme 9.75 Alkenylation using thiophenols. Source: Based on Tlais and Danheiser [112].

9.11.3 Insertion Reactions

Insertion reactions are very common in aryne chemistry, but it is not well explored in the strained cycloalkyne field. In 2009, Stoltz and coworkers demonstrated that cyclohexyne can insert into β-keto ester **187** for the formation of C—C σ-bond inserted product **188**. This insertion product was in situ treated with aqueous ammonium hydroxide for the synthesis of tetrahydroisoquinoline derivative **189** in a one-pot procedure (Scheme 9.76) [117].

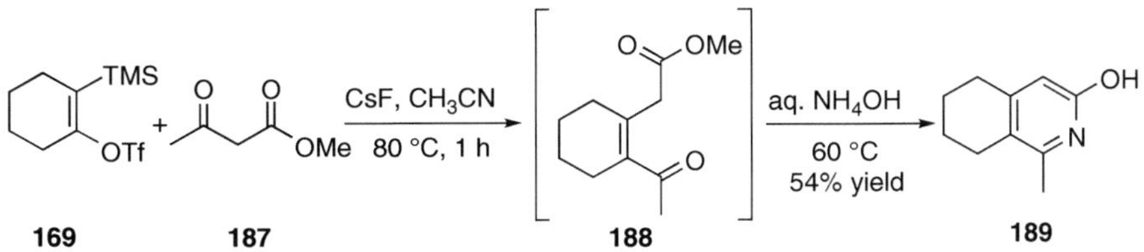

169 187 188 189

Scheme 9.76 Cyclohexyne insertion into β-keto ester. Source: Modified from Allan et al. [117].

Later, in 2010, Carreira and coworkers successfully inserted cyclohexyne generated from **157** into a variety of cyclic ketones **190** for the generation of medium-sized fused rings **191** (Scheme 9.77) [118].

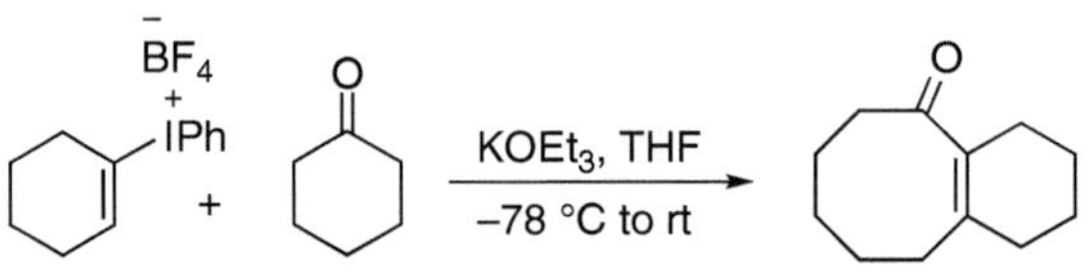

157 **190** **191**

Scheme 9.77 Cyclohexyne insertion into cyclic ketones. Source: Modified from Gampe et al. [118].

9.12 Application in Synthesis

Arynes and heteroarynes are widely used in the natural product synthesis (discussed in detail in Chapter 11), but there are only a few reports where cycloalkynes were used for the synthesis of complex natural products. For example, in 2011, Carreira and coworkers reported the total synthesis of guanacastepenes O and N **197** and **198** via a formal insertion of cyclohexyne **149** into pentalenone **193** as a key step. Treatment of Fujita's cyclohexyne precursor **157** with pentalenone **193** and $KOCEt_3$ furnished cyclobutanol **195**, which under iron-promoted ring opening generated the enone **196**. Finally, this enone **196** was elaborated to guanacastepenes O **197** and guanacastepenes N **198** (Scheme 9.78) [119].

9.13 Strained Cyclic Allenes

Strained cyclic allenes such as 1,2-cycloalkadienes and their hetero analogs are a class of highly strained intermediates related to cyclohexynes. It was reported by Wittig in 1966 [120]. Since then, it was largely ignored by the synthetic chemists, and thus remained relatively underdeveloped than other related intermediates like arynes and cyclohexynes. Depending on the ring size, the strained 1,2-cycloalkadiene can be divided into five-, six-, and seven-membered 1,2-cycloalkadiene. Among them, 1,2-cyclohexadiene **200** is the most studied intermediate. So, in this section, mainly the preparation and reactions of 1,2-cyclohexadiene and their hetero analogues will be discussed along with the 1,2-cyclopentadiene **201** and 1,2-cycloheptadiene **202** (Scheme 9.79).

9.13.1 Generation of 1,2-Cycloalkadienes

Owing to the relatively high angle strain, five-, six-, and seven-membered 1,2-cycloalkadienes are very reactive and thus cannot be isolated and must be generated in situ. 1,2-Cycloalkadienes can be generated by different methods. Selected methods are discussed in this section.

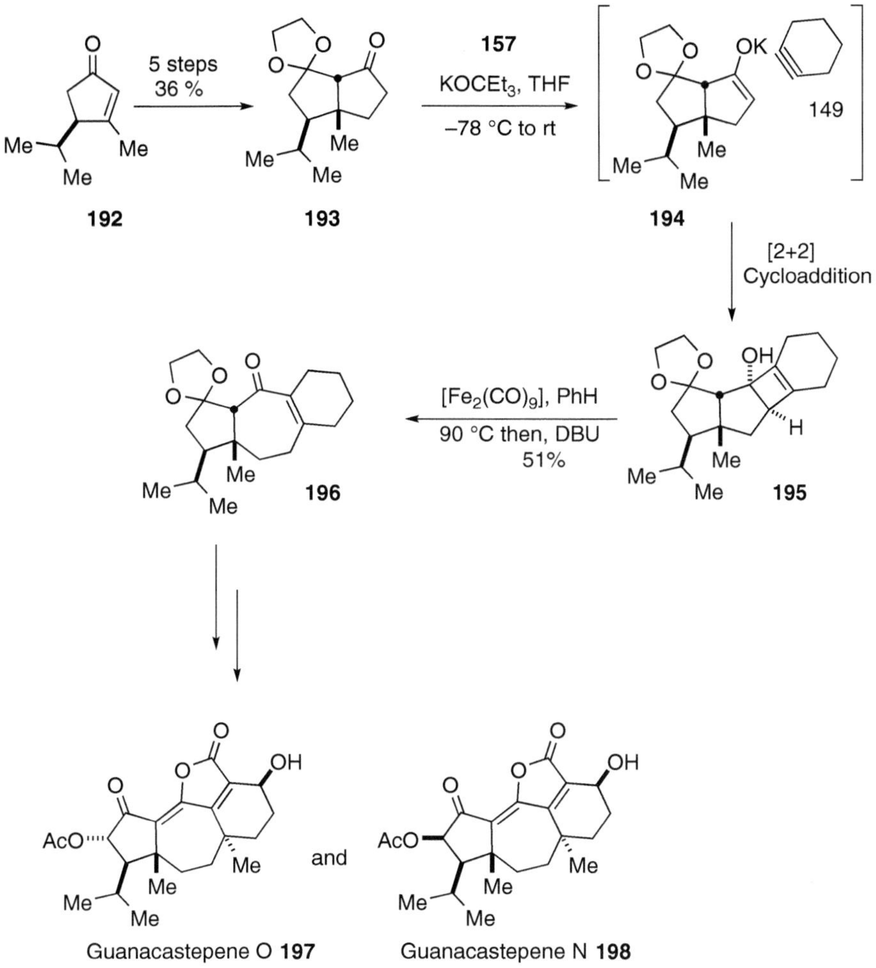

Scheme 9.78 Synthesis of guanacastepenes O and N. Source: Based on Gampe and Carreira [119].

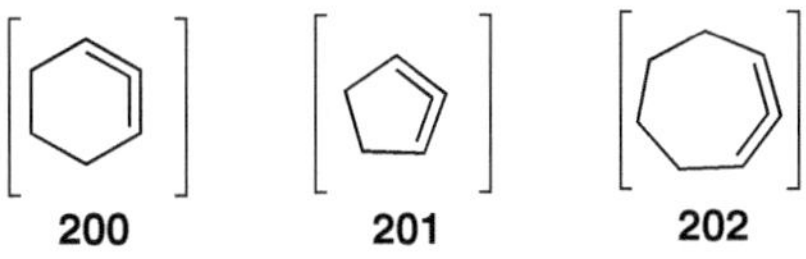

Scheme 9.79 1,2-Cycloalkadienes.

9.13.1.1 Base-Induced 1,6-Elimination

In 1966, Wittig generated 1,2-cyclohexadiene **200** from 1-bromocyclohexene **199** and KO*t*-Bu in DMSO by the 1,6-elimination of HCl. But the competitive 1,2-elimination for the generation of cyclohexyne is the main drawback of this type of base-induced process (Scheme 9.80) [120]. The generated 1,2-cyclohexadiene **200** could undergo a cycloaddition to afford the [4+2] adduct **203**.

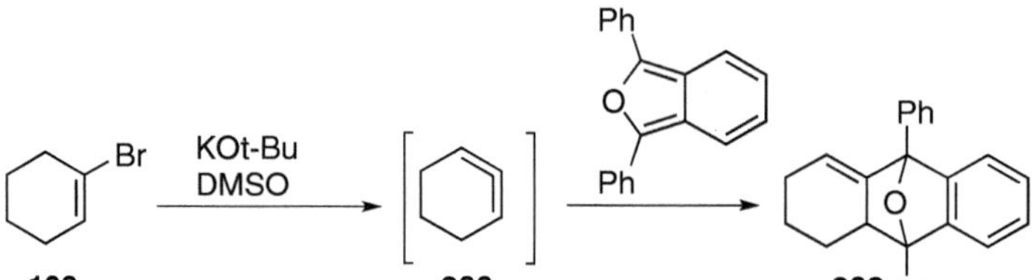

199 200 203

Scheme 9.80 Generation of 1,2-cyclohexadiene from 1-bromocyclohexene. Source: Based on Wittig and Fritze [120].

9.13.1.2 Rearrangement of Cyclopropylidenes

This approach involves the rearrangement of the generated cyclopropylidenes to the cyclic allene. When 6,6-dibromo bicyclo[3.1.0]hexane **204** was treated with methyl lithium at low temperature, the 1,2-cyclohexadiene **200** was generated by the rearrangement of the initially formed cyclopropylidene **205** (Scheme 9.81) [121]. Similarly, 1,2-cyclopentadiene **201** [122] and 1,2-cycloheptadiene **202** [123] were also generated using analogous procedure.

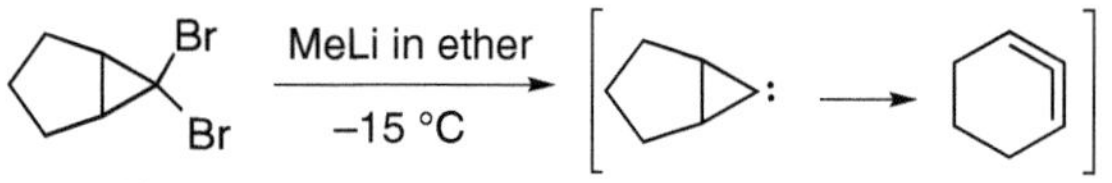

204 205 200

Scheme 9.81 Generation of 1,2-cyclohexadiene via rearrangement of cyclopropylidene. Source: Based on Moore and Moser [121].

This cyclopropylidene **205** can also be generated from bicyclo[3.1.0]hexane-carbonyl chloride **206** by vacuum pyrolysis at 800 °C. The carbonyl chloride **206** was first converted to the ketene **207**, which loses a molecule of CO to generate the carbene **205**, which, in turn, rearranges to the 1,2-cyclohexadiene **200** (Scheme 9.82) [124].

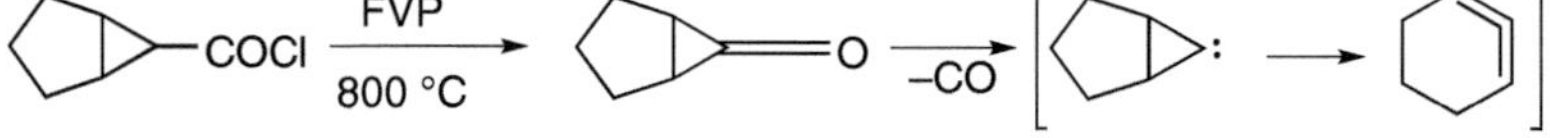

206 207 205 200

Scheme 9.82 Generation of 1,2-cyclohexadiene via pyrolysis of ketene. Source: Modified from Wentrup et al. [124].

9.13.1.3 Fluoride-Induced Elimination

In 1990, Shakespeare and Johnson reported the fluoride-induced generation of 1,2-cyclohexadiene **200** by the 1,6-elimination of silyl bromide precursor **208** (Scheme 9.83) [125]. Similarly, 1,2-cycloheptadiene **202** can also be prepared by this method [126].

Later in 2009 Guitián and coworkers prepared the silyl triflate precursor **210** for the mild generation of 1,2-cyclohexadiene **200**. The precursor **210** can be

Scheme 9.83 Fluoride-induced generation of 1,2-cyclohexadiene. Source: Based on Shakespeare and Johnson [125].

prepared easily from **209** in a two-step procedure (Scheme 9.84) [127]. Recently, Garg and coworkers reported the silyl tosylate precursor for milder generation of 1,2-cyclohexadiene **200** and 1,2-cycloheptadiene **202** [109].

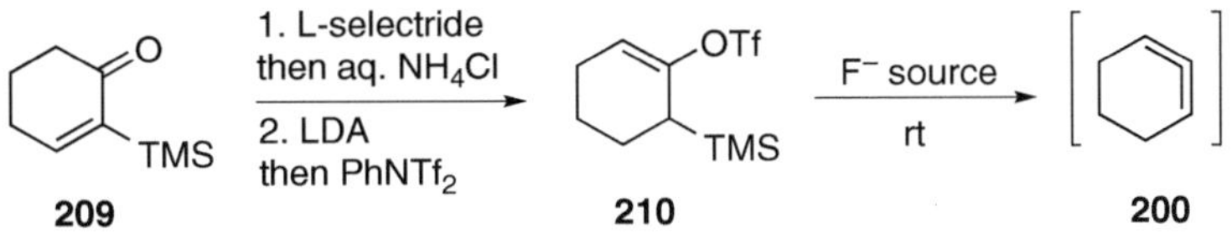

Scheme 9.84 Preparation of the silyl triflate precursor of 1,2-cyclohexadiene. Source: Based on Quintana et al. [127].

The fluoride-induced method for 1,2-cycloalkadienes is mild and base-free, and thus compatible with various functional groups. A variety of strained cyclic allene and heterocyclic allenes are prepared by this procedure; some examples are given below.

9.13.1.3.1 Generation of Azacyclic Allene

In 2018, Garg and coworkers demonstrated the fluoride-induced generation of azacyclic allene **214** from the silyl triflate precursor **213**. The precursor **213** was synthesized from commercially available 4-methoxy pyridine **211**, which was elaborated to silyl derivative **212** in few steps, followed by a 1,4-reduction and triflation (Scheme 9.85) [128].

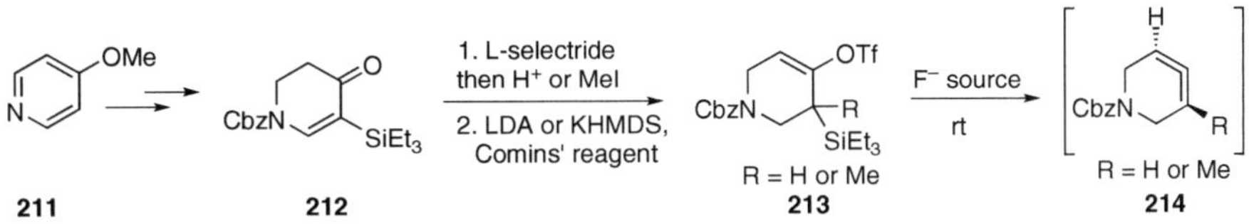

Scheme 9.85 Preparation of azacyclic allene precursor. Source: Modified from Barber et al. [128].

9.13.1.3.2 Generation of Oxacyclic Allene

The silyl triflate precursor **216** to the oxacyclic allene **217** was prepared by Garg and coworkers in 2019. It was prepared from the easily available bromo-ketone **215** in good yields by a four-step synthetic strategy outlined in Scheme 9.86 [129].

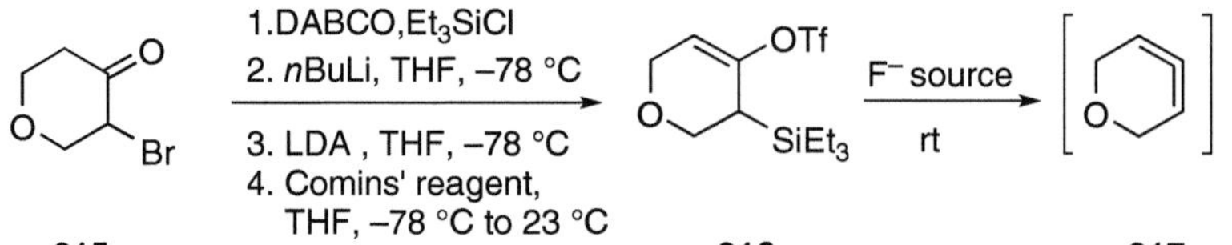

215 216 217

Scheme 9.86 Preparation of oxacyclic allene precursor. Source: Modified from Yamano et al. [129].

9.13.1.3.3 *Generation of Cycloalkatriene*

In 1990, Shakespeare and Johnson generated the 1,2,3-cyclohexatriene intermediate **219** via the fluoride-induced degradation of the cyclohexadiene precursor **218** and successfully trapped the intermediate **219** with 1,3-diphenylisobenzofuran as a Diels–Alder cycloadduct **220** (Scheme 9.87) [126].

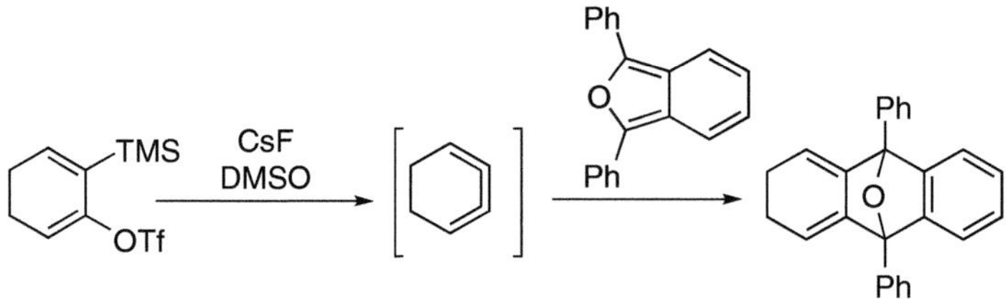

218 219 220

Scheme 9.87 Generation of 1,2,3-cyclohexatriene. Source: Based on Almehmadi [126].

9.13.2 Reaction of 1,2-Cycloalkadienes

Reactions of 1,2-cycloalkadienes are not well explored as in cyclohexyne or aryne field. These strained cyclic allenes are very good dienophiles; thus, they can be trapped by various dienes via Diels–Alder [4+2] cycloaddition reaction. [2+2] and [3+2] cycloadditions are also common in 1,2-cycloalkadienes. But still, this strained intermediate has not been utilized for other types of reactions common for similar strained intermediates like insertion reaction, multicomponent coupling, and molecular rearrangements.

9.13.2.1 Diels–Alder Addition

The Diels–Alder reaction of 1,2-cycloalkadienes began in 1966 with the successful trapping of the 1,2-cyclohexadiene intermediate **200** with 1,3-diphenylisobenzofuran via a [4+2] cycloaddition reaction. Since then, a variety of dienes like furan and cyclopentadiene have been used to trap the cyclic allenes (Scheme 9.88) [120].

In 2019, Garg and coworkers demonstrated the Diels–Alder reaction of enantioenriched cyclic allene intermediate **222** generated from the corresponding cyclic allene precursor **221** with 2,5-dimethylfuran without any loss of enantiopurity in the Diels–Alder cycloadduct **223** (Scheme 9.89) [126]. Here, the point chirality of

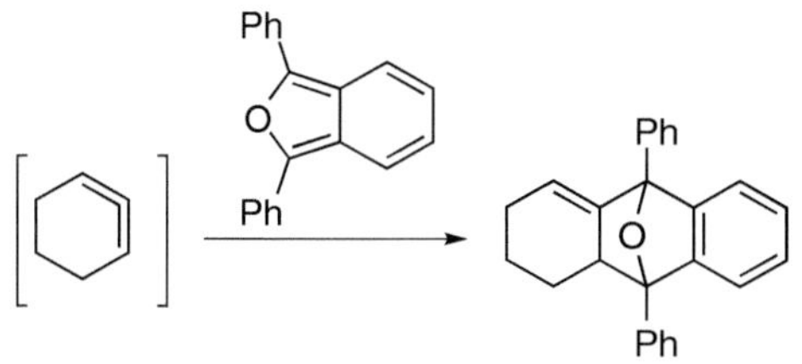

200 203

Scheme 9.88 Diels–Alder reaction of 1,2-cycloalkadiene. Source: Based on Wittig and Fritze [120].

the cyclic allene precursor **221** was transferred to the intermediate **222** as axial chirality, which further relayed to the cycloadduct as point chirality.

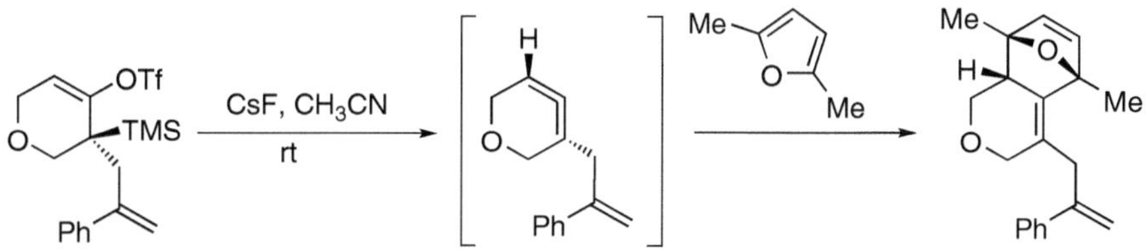

221
81% ee
222
223
81% ee

Scheme 9.89 Diels–Alder reaction of enantioenriched cyclic allene. Source: Based on Almehmadi and West [126].

9.13.2.2 [2+2] Cycloaddition

In the absence of any trapping agents, cyclic allenes **200** dimerize via [2+2] cycloaddition reaction. Additionally, in 1970, Moore and Moser successfully trapped 1,2-cyclohexadiene with styrene in a [2+2] annulation to form the cyclobutane derivative **224** (Scheme 9.90) [121].

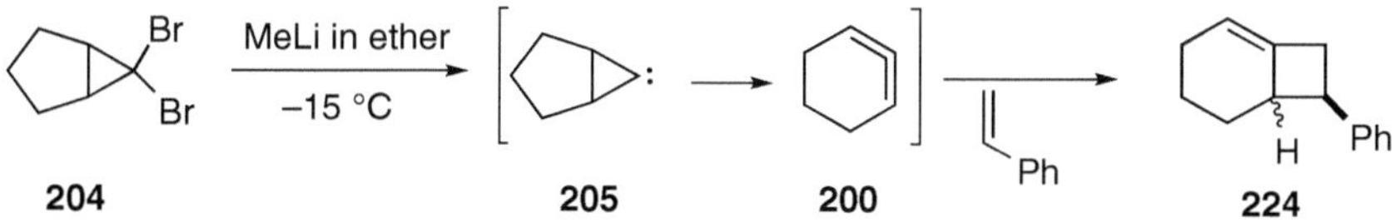

204 205 200 224

Scheme 9.90 [2+2] cycloaddition with styrene. Source: Modified from Moore and Moser [121].

9.13.2.3 1,3-Dipolar Cycloaddition

Until recently, cycloaddition reactions of cyclic allenes were restricted to [4+2] and [2+2] cycloaddition only. In 2016, Garg and coworkers reported the first 1,3-dipolar cycloaddition of 1,2-cyclohexadiene **200** with nitrones **105** leading to the synthesis of adduct **226** (Scheme 9.91) [130]. West and coworkers in 2020 demonstrated the 1,3-dipolar cycloaddition of 1,2-cycloheptadiene with different nitrones, nitrile oxides, and azomethine imines [126].

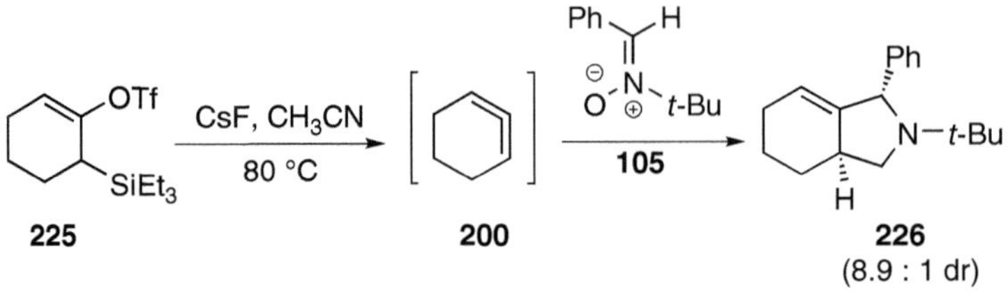

Scheme 9.91 1,3-Dipolar cycloaddition with nitrones. Source: Based on Almehmadi and West [126].

9.14 Conclusions

With the advent of new strategies for the mild generation of hetarynes and cycloalkynes, these intermediates are now much more accessible. Although cycloaddition and nucleophilic addition reactions are well known in these fields, multicomponent reactions are underdeveloped with hetarynes or cycloalkynes, unlike aryne chemistry. Many natural products were synthesized using these in situ-generated intermediates and more total syntheses are expected in the future employing the hetaryne generation. It is reasonable to believe that easy generation, compatible reaction conditions, and broad applications of these intermediate will inspire organic chemists to find more possibilities and stimulating applications in this field.

References

1 Stoermer, R. and Kahlert, B. (1902). *Ber. Dtsch. Chem. Ges.* 35: 1633–1640.
2 Kauffmann, T. (1965). *Angew. Chem. Int. Ed.* 4: 543–557.
3 Reinecke, M.G. (1982). *Tetrahedron* 38: 427–498.
4 Bronner, S.M., Goetz, A.E., and Garg, N.K. (2011). *Synlett*: 2599–2604.
5 Goetz, A.E. and Garg, N.K. (2014). *J. Org. Chem.* 79: 846–851.
6 Goetz, A.E., Shah, T.K., and Garg, N.K. (2015). *Chem. Commun.* 51: 34–45.
7 Roberts, J.D., Simmons, H.E., Carlsmith, L.A., and Vaughan, C.W. (1953). *J. Am. Chem. Soc.* 75: 3290–3591.
8 Wittig, G. and Pohmer, L. (1956). *Chem. Ber.* 89: 1334–1351.
9 Huisgen, R. and Rist, H. (1954). *Naturwiss.* 41: 358–359.
10 Berry, R.S., Spokes, G.N., and Stiles, M. (1962). *J. Am. Chem. Soc.* 84: 3570–3577.
11 Chapman, O.L., Mattes, K., McIntosh, C.L. et al. (1973). *J. Am. Chem. Soc.* 95: 6134–6135.
12 Sauer, J. and Huisgen, R. (1960). *Angew. Chem. Int. Ed.* 72: 294–315.
13 Huisgen, R. and Sauer, J. (1960). *Angew. Chem. Int. Ed.* 72: 91–108.
14 Kauffmann, T., Risberg, A., Schulz, U.J., and Weber, R. (1964). *Tetrahedron Lett.*: 3563.

15 Bordwell, F.G., Lampert, B.B., and McKellin, W.H. (1949). *J. Am. Chem. Soc.* 71: 1702–1705.

16 Reinecke, M.G. (1982). Five-membered hetarynes. In: *Reactive Intermediates*, vol. 2 (ed. R.A. Abramovitch), 367–510. New York, Chapter 5: Plenum.

17 Conway, S.C. and Gribble, G.W. (1992). *Heterocycl.* 34: 2095–2108.

18 Goetz, A.E., Bronner, S.M., Cisneros, J.D. et al. (2012). *Angew. Chem. Int. Ed.* 124: 2812–2816.

19 Liu, J.-H., Chan, H.-W., Xue, F. et al. (1999). *J. Org. Chem.* 64: 1630–1634.

20 Ye, X.-S., Li, W.-K., and Wong, H.N.C. (1996). *J. Am. Chem. Soc.* 118: 2511–2512.

21 Reinecke, M.G., Newsom, J.G., and Chen, L.-J. (1981). *J. Am. Chem. Soc.* 103: 2760–2769.

22 Bronner, S.M., Goetz, A.E., and Garg, N.K. (2011). *J. Am. Chem. Soc.* 133: 3832–3835.

23 Julia, M., Le Goffic, F., Igolen, J., and Baillarge, M. (1969). *Tetrahedron Lett.* 10: 1569–1572.

24 Enamorado, M.F., Ondachi, P.W., and Comins, D.L. (2010). *Org. Lett.* 12: 4513–4515.

25 Hoarau, C., Couture, A., Cornet, H. et al. (2001). *J. Org. Chem.* 66: 8064–8069.

26 SánchezSanz, G., Alkorta, I., Trujillo, C., and Elguero, J. (2012). *Tetrahedron* 68: 6548–6556.

27 Nam, H.H., Leroi, G.E., and Harrison, J.F. (1991). *J. Phys. Chem.* 95: 6514–6519.

28 Rau, N.J. and Wenthold, P.G. (2011). *J. Phys. Chem. A* 115: 10353–10362.

29 Debbert, S.L. and Cramer, C.J. (2000). *Int. J. Mass spectrom.* 201: 1–15.

30 Bunnett, J.F. and Singh, P. (1981). *J. Org. Chem.* 46: 4567–4569.

31 Sparrapan, R., Mendes, M.A., Carvalho, M., and Eberlin, M.N. (2000). *Chem. Eur. J.* 6: 321–326.

32 Gozzo, F.C. and Eberlin, M.N. (1999). *J. Org. Chem.* 64: 2188–2193.

33 Kauffmann, T., Boettcher, F., and Hansen, J. (1961). *Angew. Chem. Int. Ed.* 73: 341.

34 Kauffmann, T. (1962). *Ann. Chem.* 659: 102.

35 Van Der Plas, H.C., Woźniak, M., and Van Den Haak, H.J.W. (1983). Reactivity of naphthyridines toward nitrogen nucleophiles. In: *Advances in Heterocyclic Chemistry*, vol. 33 (ed. A.R. Katritzky), 95–146. Elsevier.

36 Kauffmann, T., Hansen, J., Udluft, K., and Wirthwein, R. (1964). *Angew. Chem. Int. Ed.* 76: 590.

37 Martens, R.J. and den Hertog, H.J. (1964). *Recl. Trav. Chim. Pays-Bas* 83: 621–630.

38 Martens, R.J. and den Hertog, H.J. (1962). *Tetrahedron Lett.* 3: 643–645.

39 Fleet, G.W.J. and Fleming, I. (1969). *J. Chem. Soc. C*: 1758–1763.

40 Nam, H.-H. and Leroi, G.E. (1988). *J. Am. Chem. Soc.* 110: 4096–4097.

41 Walters, M.A. and Shay, J.J. (1997). *Synth. Commun.* 27: 3573–3579.

42 Kauffmann, T. and Boettcher, F.-P. (1962). *Chem. Ber.* 95: 949–955.

43 Zoltewicz, J.A. and Smith, C.L. (1969). *Tetrahedron* 25: 4331–4337.

44 Furukawa, N., Shibutani, T., and Fujihara, H. (1987). *Tetrahedron Lett.* 28: 2727–2730.
45 Tsukazaki, M. and Snieckus, V. (1992). *Heterocycl.* 33: 533–536.
46 Julia, M., Huang, Y., and Igolen, J. (1967). *C.R. Acad. Sci., Ser. C* 265: 110–112.
47 Iwao, M., Motoi, O., Fukuda, T., and Ishibashi, F. (1998). *Tetrahedron* 54: 8999–9010.
48 Buszek, K.R., Luo, D., Kondrashov, M. et al. (2007). *Org. Lett.* 9: 4135–4137.
49 Brown, N., Luo, D., VanderVelde, D. et al. (2009). *Tetrahedron Lett.* 50: 63–65.
50 Buszek, K.R., Brown, N., and Luo, D. (2009). *Org. Lett.* 11: 201–204.
51 Garr, A.N., Luo, D., Brown, N. et al. (2010). *Org. Lett.* 12: 96–99.
52 Bronner, S.M., Bahnck, K.B., and Garg, N.K. (2009). *Org. Lett.* 11: 1007–1010.
53 Cheong, P.H.-Y., Paton, R.S., Bronner, S.M. et al. (2010). *J. Am. Chem. Soc.* 132: 1267–1269.
54 Im, G.-Y.J., Bronner, S.M., Goetz, A.E. et al. (2010). *J. Am. Chem. Soc.* 132: 17933–17944.
55 Candito, D.A., Dobrovolsky, D., and Lautens, M. (2012). *J. Am. Chem. Soc.* 134: 15572–15580.
56 Fang, Y. and Larock, R.C. (2012). *Tetrahedron* 68: 2819–2826.
57 Kauffmann, T., Boettcher, F.-P., and Hansen, J. (1962). *Liebigs Ann. Chem.* 659: 102–109.
58 Collis, G.E. and Burrell, A.K. (2005). *Tetrahedron Lett.* 46: 3653–3656.
59 Czuba, W. (1963). *Recl. Trav. Chim. Pays-Bas* 82: 997.
60 Van der Plas, H.C., Smit, P., and Koudijs, A. (1968). *Tetrahedron Lett.* 9: 9–13.
61 Tielemans, M., Areschka, V., Colomer, J., and Promel, R. (1992). *Tetrahedron* 48: 10575–10586.
62 Medina, J.M., Jackl, M.K., Susick, R.B., and Garg, N.K. (2016). *Tetrahedron* 72: 3629–3634.
63 Martens, R.J. and den Hertog, H.J. (1967). *Recl. Trav. Chim. Pays-Bas* 86: 655–669.
64 Kato, T. and Niitsuma, T. (1973). *Heterocycl.* 1: 233–236.
65 Hoye, T.R., Baire, B., Niu, D. et al. (2012). *Nature* 490: 208–212.
66 Niu, D., Willoughby, P.H., Woods, B.P. et al. (201). *Nature* 501: 531–534.
67 Niu, D. and Hoye, T.R. (2014). *Nat. Chem.* 6: 34–40.
68 Chen, J., Baire, B., and Hoye, T.R. (2014). *Heterocycl.* 88: 1191–1200.
69 Hoye, T.R., Baire, B., and Wang, T. (2014). *Chem. Sci.* 5: 545–550.
70 Yun, S.Y., Wang, K.-P., Lee, N.-K. et al. (2013). *J. Am. Chem. Soc.* 135: 4668–4671.
71 Wang, K.-P., Yun, S.Y., Mamidipalli, P., and Lee, D. (2013). *Chem. Sci.* 4: 3205–3211.
72 Dıaz, M.T., Cobas, A., Guitián, E., and Castedo, L. (1998). *Synlett*: 157–158.
73 Dıaz, M.T., Cobas, A., Guitián, E., and Castedo, L. (2001). *Eur. J. Org. Chem.*: 4543–4549.
74 Goetz, A.E. and Garg, N.K. (2013). *Nat. Chem.* 5: 54–60.
75 Iwayama, T. and Sato, Y. (2009). *Chem. Commun.*: 5245–5247.
76 Zoltewicz, J.A. and Nisi, C. (1969). *J. Org. Chem.* 34: 765–766.

77 Levine, R. and Leake, W.W. (1955). *Science* 121: 780.
78 Saito, N., Nakamura, K., Shibano, S. et al. (2013). *Org. Lett.* 15: 386–389.
79 Pareek, M., Fallon, T., and Oestreich, M. (2015). *Org. Lett.* 17: 2082–2085.
80 May, C. and Moody, C.J. (1984). *J. Chem. Soc., Chem. Commun.*: 926–927.
81 May, C. and Moody, C.J. (1988). *J. Chem. Soc., Perkin Trans.* 1: 247–250.
82 Gribble, G.W., Saulnier, M.G., Sibi, M.P., and Obaza-Nutaitis, J.A. (1984). *J. Org. Chem.* 49: 4518–4523.
83 Sha, C.-K. and Yang, J.-F. (1992). *Tetrahedron* 48: 10645–10654.
84 Kessar, S.V., Gupta, Y.P., Pahwa, P.S., and Singh, P. (1976). *Tetrahedron Lett.* 17: 3207–3208.
85 Kessar, S.V. and Singh, P. (2001). *Indian J. Chem., Sect. B: Org. Chem. Incl. Med. Chem.* 40: 1129–1131.
86 Huters, A.D., Quasdorf, K.W., Styduhar, E.D., and Garg, N.K. (2011). *J. Am. Chem. Soc.* 133: 15797–15799.
87 Quasdorf, K.W., Huters, A.D., Lodewyk, M.W. et al. (2012). *J. Am. Chem. Soc.* 134: 1396–1399.
88 Styduhar, E.D., Huters, A.D., Weires, N.A., and Garg, N.K. (2013). *Angew. Chem. Int. Ed.* 52: 12422–12425.
89 Fine Nathel, N.F., Shah, T.K., Bronner, S.M., and Garg, N.K. (2014). *Chem. Sci.* 5: 2184–2190.
90 Goetz, A.E., Silberstein, A.L., Corsello, M.A., and Garg, N.K. (2014). *J. Am. Chem. Soc.* 136: 3036–3039.
91 Brown, N., Luo, D., Decapo, J.A., and Buszek, K.R. (2009). *Tetrahedron Lett.* 50: 7113–7115.
92 Corsello, M.A., Kim, J., and Garg, N.K. (2017). *Nat. Chem.* 9: 944–949.
93 Adams, R.D., Chen, G., Qu, X. et al. (1992). *J. Am. Chem. Soc.* 114: 10977–10978.
94 Favorskii, A.E. and Boshowskii, V. (1912). *Ann.* 390: 122.
95 Favorskii, A.E. (1936). *J. Gen. Chem. USSR* 6: 720.
96 Wittig, G. and Harborth, G. (1944). *Ber.* 77: 306.
97 Scardiglia, F. and Roberts, J.D. (1957). *Tetrahedron* 1: 343–344.
98 Wittig, G., Krebs, A., and Pohlke, R. (1960). *Angew. Chem. Int. Ed.* 73: 324.
99 Bottini, A.T., Corson, F.P., Fitzgerald, R., and Frost, K.A. II, (1972). *Tetrahedron* 28: 4883–4904.
100 Fujita, M., Kim, W.H., Sakanishi, Y. et al. (2004). *J. Am. Chem. Soc.* 126: 7548–7558.
101 Hioki, Y., Okano, K., and Mori, A. (2017). *Chem. Commun.* 53: 2614–2617.
102 Willey, F. G. (1964). *Angew. Chem. Int. Ed.* 3: 138.
103 Al-Omari, M., Banert, K., and Hagedorn, M. (2006). *Angew. Chem. Int. Ed.* 45: 309–311.
104 Wittig, G. and Krebs, A. (1961). *Chem. Ber.* 94: 3260–3275.
105 Bolster, J.M. and Kellogg, R.M. (1981). *J. Am. Chem. Soc.* 103: 2868–2869.
106 Erickson, K.L. and Wolinsky, J. (1965). *J. Am. Chem. Soc.* 87: 1142–1143.
107 Atanes, N., Escudero, S., Perez, D. et al. (1998). *Tetrahedron Lett.* 39: 3039–3040.
108 Iglesias, B., Peña, D., Perez, D. et al. (2002). *Synlett*: 486–488.

109 McVeigh, M.S., Kelleghan, A.V., Yamano, M.M. et al. (2020). *Org. Lett.* 22: 4500–4505.

110 Shah, T.K., Medina, J.M., and Garg, N.K. (2016). *J. Am. Chem. Soc.* 138: 4948–4954.

111 Wentrup, C., Blanch, R., Briehl, H., and Gross, G. (1988). *J. Am. Chem. Soc.* 110: 1874–1880.

112 Tlais, S.F. and Danheiser, R.L. (2014). *J. Am. Chem. Soc.* 136: 15489–15492.

113 McMahon, T.C., Medina, J.M., Yang, Y.-F. et al. (2015). *J. Am. Chem. Soc.* 137: 4082–4085.

114 Qiu, D., Shi, J., Guo, Q. et al. (2018). *J. Am. Chem. Soc.* 140: 13214–13218.

115 Darzi, E.R., Barber, J.S., and Garg, N.K. (2019). *Angew. Chem. Int. Ed.* 58: 9419–9424.

116 Medina, J.M., McMahon, T.C., Jimenez-Ose's, G. et al. (2014). *J. Am. Chem. Soc.* 136: 14706–14709.

117 Allan, K.M., Hong, B.D., and Stoltz, B.M. (2009). *Org. Biomol. Chem.* 7: 4960–4964.

118 Gampe, C.M., Boulos, S., and Carreira, E.M. (2010). *Angew. Chem. Int. Ed.* 49: 4092–4095.

119 Gampe, C.M. and Carreira, E.M. (2011). *Angew. Chem. Int. Ed.* 50: 2962–2965.

120 Wittig, G. and Fritze, P. (1966). *Angew. Chem. Int. Ed.* 5: 846.

121 Moore, W.R. and Moser, W.R. (1970). *J. Org. Chem.* 35: 908–912.

122 Algi, F., Özen, R., and Balci, M. (2002). *Tetrahedron* 43: 3129–3131.

123 Taylor, K.G., Hobbs, W.E., Clark, M.S., and Chancy, J. (1972). *J. Org. Chem.* 37: 2436–2443.

124 Wentrup, C., Gross, G., Maquestiau, A., and Flammang, R. (1983). *Angew. Chem. Int. Ed.* 22: 542–543.

125 Shakespeare, W.C. and Johnson, R.P. (1990). *J. Am. Chem. Soc.* 112: 8578–8579.

126 Almehmadi, Y.A. and West, F.G. (2020). *Org. Lett.* 22: 6091–6095.

127 Quintana, I., Peña, D., Pérez, D., and Guitián, E. (2009). *Eur. J. Org. Chem.*: 5519–5524.

128 Barber, J.S., Yamano, M.M., Ramirez, M. et al. (2018). *Nat. Chem.* 10: 953–960.

129 Yamano, M.M., Knapp, R.R., Ngamnithiporn, A. et al. (2019). *Angew. Chem. Int. Ed.* 58: 5653–5657.

130 Barber, J.S., Styduhar, E.D., Pham, H.V. et al. (2016). *J. Am. Chem. Soc.* 138: 2512–2515.

10

Hexadehydro Diels–Alder (HDDA) Route to Arynes and Related Chemistry

Rachel N. Voss and Thomas R. Hoye

University of Minnesota, Department of Chemistry, 207 Pleasant St SE, Minneapolis, MN 55455, USA

10.1 Introduction

The hexadehydro Diels–Alder (HDDA) reaction involves the thermal generation of benzynes by cycloisomerization of precursors having, minimally, a 1,3-diyne tethered to an additional alkyne, the latter the diynophile. In situ trapping agents then capture the reactive benzyne. Since 2012, there has been an explosion of reports that, collectively, demonstrate considerable generality of this process. This chapter represents an effort to highlight some of the key features of these developments. It is not intended to be all encompassing in its scope, a task beyond the purview of this *Handbook*, nor are the examples presented in chronological order. However, one of the authors here (TRH) has recently coauthored a comprehensive review where citations to (and examples from) all reports of experimental results of HDDA reactions (through mid-2020) are included [1]. That document is organized quite differently from this chapter; there, reactions are categorized according to the types of new bonds that are formed during each known mode of trapping process. Interested readers are directed to that resource for details beyond those captured here. Finally, the topics presented in this chapter are somewhat randomly organized, so readers might want to scan the headers and/or figure graphics to identify those items of greatest interest.

10.2 History

10.2.1 Overview of the Family of Dehydro-Diels–Alder Reactions

The Diels–Alder (DA) reaction is one of the most venerable of all chemical transformations. The DA transformation is ubiquitous, being the subject of myriad synthetic and mechanistic studies since being first reported [2]; for example, SciFinder® records over 16 600 publications having "Diels" + "Alder" in their *titles*. In the example introduced in most textbooks of elementary organic chemistry, the conjugated diene 1,3-butadiene engages the dienophile ethylene to produce the

Modern Aryne Chemistry, First Edition. Edited by Akkattu T. Biju.

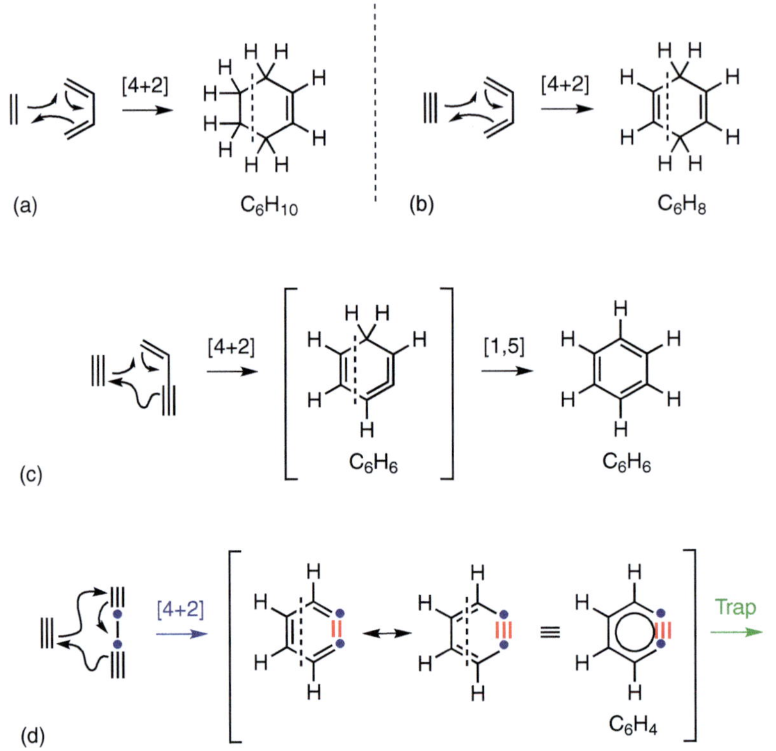

Figure 10.1 Diels–Alder reactions of progressively less saturated pairs of 4π- and 2π-components. (a) Diels–Alder, (b) didehydro-Diels–Alder, (c) tetradehydro-Diels–Alder, and (d) hexadehydro-Diels–Alder.

cyclic six-membered hydrocarbon cyclohexene (Figure 10.1a). Pairwise removal of dihydrogen (i.e., formal oxidation) from these prototypical reactants leads to cyclic products having progressively higher oxidation states (i.e., higher degrees of unsaturation). Collectively, these are recognized as dehydro-Diels–Alder reactions [3]. The most highly oxidized variant (Figure 10.1d) corresponds to the reaction between 1,3-butadiyne and, as a diynophile, ethyne. But this transformation is unique compared with the others of intermediate levels of oxidation, because it would produce the hydrocarbon *o*-benzyne (C_6H_4), a well-known (and highly versatile) reactive intermediate that can only be observed under conditions where it does not encounter other reaction partners (e.g., in the gas-phase [4], in inert frozen matrices [5], or inside a protective cavitand [6]).

10.2.2 First Example of a Tetradehydro-Diels–Alder (TDDA) Reaction

The first example of what today can be recognized as a dehydro-Diels–Alder reaction was reported in 1898 (Figure 10.2) [2a]. Phenylpropiolic acid (**1**), when refluxed in acetic anhydride, gave the naphthalene derivative **3**, as recognized today

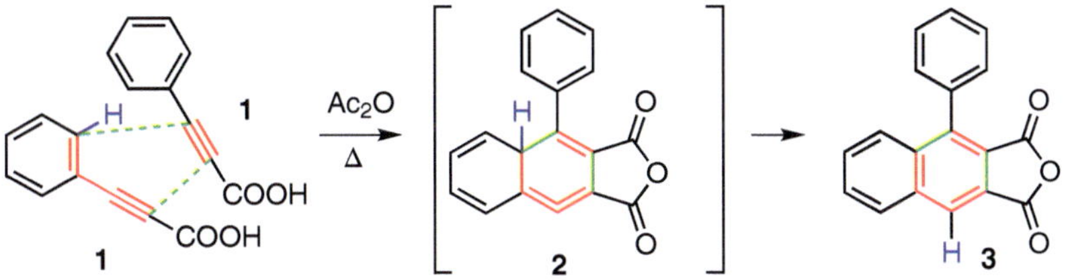

Figure 10.2 First example, from 1898 [2a], of what today can be termed a tetradehydro-Diels–Alder (TDDA) reaction. Source: Michael et al. [2a].

by way of the intermediate strained allene **2**. Interestingly, this work predates the initial report of Diels and Alder of reactions between dienes and dienophiles by three decades! [2b] Many dehydro-Diels–Alder reactions have been described in the intervening century, and this body of work is captured in the comprehensive review by Müller in 2008 [3].

10.2.3 Earliest Triyne to Benzyne Cycloisomerization (i.e., HDDA) Reactions

The earliest experimental evidence that clearly established the viability of the transformation of a conjugated diyne with a third alkyne [7] to produce a benzyne intermediate was reported more or less simultaneously in 1997 in two independent studies carried out in the laboratories of Johnson at New Hampshire and Ueda at Osaka [8]. These are summarized in Figures 10.3a,b, respectively. Johnson and coworkers carried out the flash vacuum pyrolysis of 1,3,8-nonatriyne (**4**) [8a], the prototypical substrate for an intramolecular cyclization of the type introduced in Figure 10.1d. This resulted in a highly efficient transformation to indene (**6**), indane (**7**), and "some soot." Indene is at the same oxidation state as the triyne substrate **4** and indane is reduced by one unsaturation unit. On the basis of both computations of the parent butadiyne plus ethyne reaction and a deuterium-labeling study, the researchers demonstrated that this transformation in all likelihood proceeded through the indanyne **5**. In the initial Ueda report, the tetrayne **8** gave rise, at room temperature, to the pair of anthracene-trapped products **11** and **12**, which were proposed to arise by capture of the "1,2-didehydrobenzene diradicals" **9** and **10**, respectively [8b]. The Ueda team proceeded to investigate a number of related systems over the next decade, nearly always with an eye toward studying the DNA-cleaving potential of these reactive intermediates as analogs of the Bergman 1,4-diradical species (and the enediyne family of natural products) [9].

Subsequently, the Sterenberg group at Regina in 2004 reported the interesting, unusual transformation of the dinuclear, mixed metal complex **13** (Figure 10.3c) [10]. At ambient temperature in a solution of tetrahydrofuran (THF), this gave rise to the bridged benzene derivative **15**. THF was identified as the source of the hydrogen atoms delivered to the intermediate **14** by using THF-d_8.

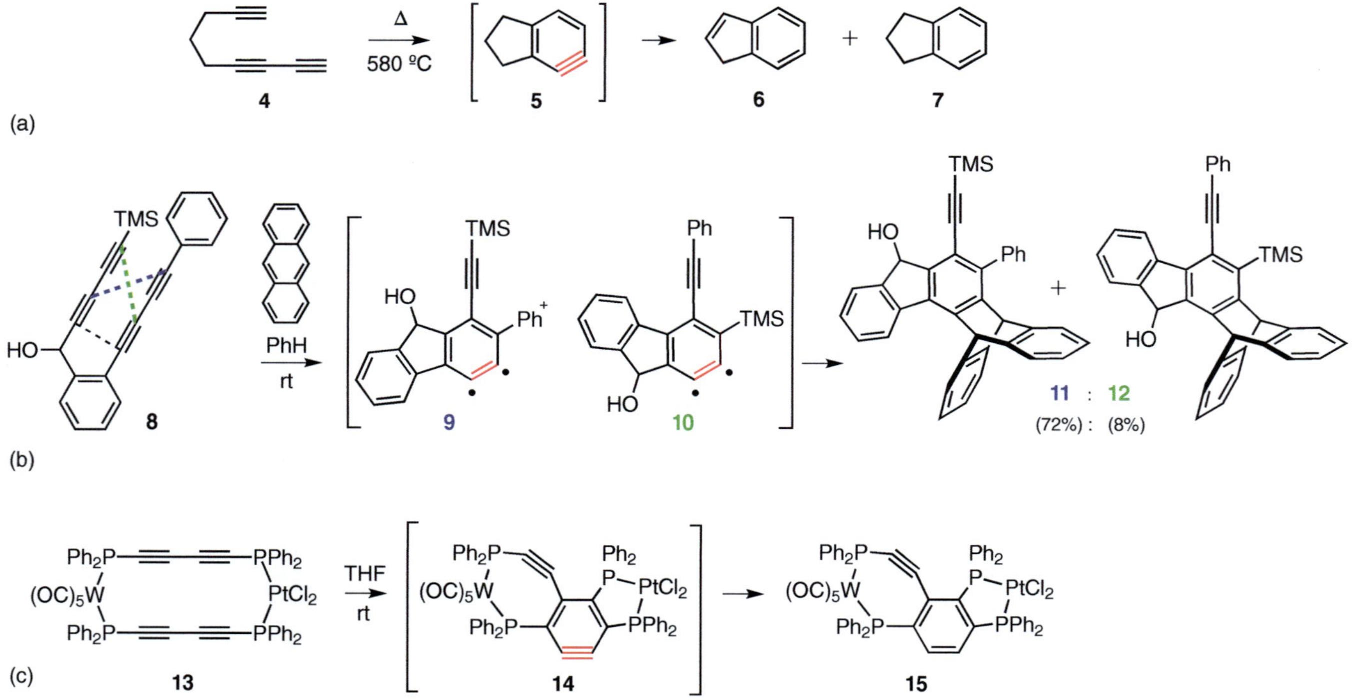

Figure 10.3 Earliest examples of triyne cycloisomerizations to benzynes. (a) Johnson flash vacuum prolysis of 1,3,8-nonatriyne, (b) Ueda trapping of "1,2-didehydrobenzene diradical" with anthracene, and (c) Sterenberg metal-templated [4+2] cycloaddition.

10.2.4 First Minnesota Examples (and the Naming) of the "HDDA" Reaction

Our initial foray at Minnesota into this class of reaction arose unintentionally [11]. In 2011, a researcher here attempted to prepare the enynone **17** (by oxidation of the alcohol precursor **16**) but observed that the unexpected, isomeric tricyclic compound **19** had formed instead. We surmised that this ambient-temperature cycloisomerization reaction, occurring in the setting of the fortuitous presence of a silyl ether substituent, had proceeded by way of the benzyne **18**, an isomer of **17**. This transformation initiated a focused effort to explore additional aspects of this reaction. Our first designed substrate, the enone **20**, smoothly and spontaneously converted into the isomeric indenone derivative **22** (via **21**) as the only observed product. This reaction was measured to have a half-life of c. seven hours at ambient temperature. Even in this first example, a new type of aryne-trapping reaction, captured by a nucleophilic silyl ether [12], had been revealed. This was the first of many previously unknown types of reaction whose discovery has been enabled by the reagent-free, thermal reaction conditions used for generation of HDDA-benzynes.

This cycloisomerization was quickly established to have considerable generality [13]. In the initial 2012 report [11], the nomenclature of the broader "dehydro-Diels–Alder" terminology was parsed into subcategories that reflect the specific degree of oxidation of the alkenes/alkynes participating in the net [4+2] cycloaddition leading to six-membered carbocycles. Thus, didehydro-Diels–Alder, tetradehydro-Diels–Alder (TDDA), and HDDA reactions, all variants on the classic textbook Diels–Alder (hereafter, DA) cycloaddition, reflect the progressively increased degrees of unsaturation of the reactants and products (cf. Figure 10.1).

Various examples of our first HDDA reactions (Figure 10.5) demonstrated that the 1,3-diyne and diynophile partners needed to be tethered, nearly always, by a

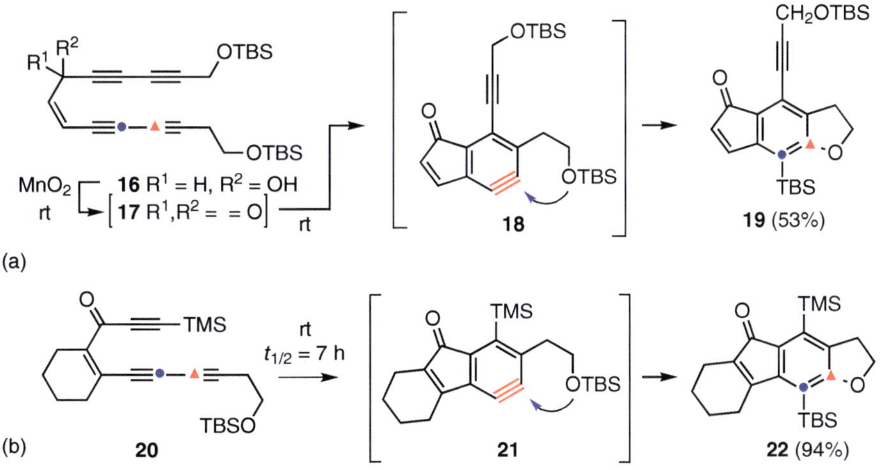

Figure 10.4 (a) The unanticipated transformation leading to the recognition that the HDDA reaction was general and (b) the first (at Minnesota) designed substrate launching that eventuality.

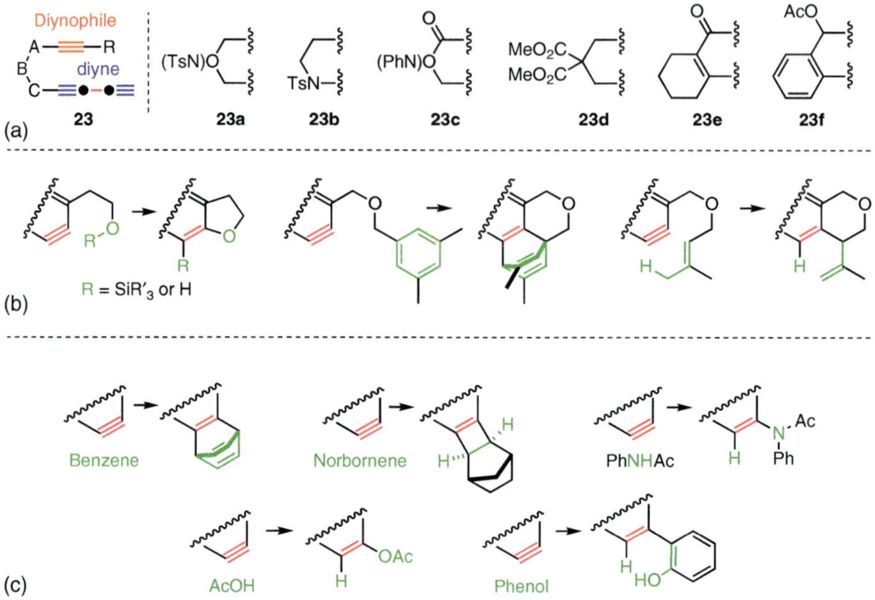

Figure 10.5 (a) Examples of three-atom tethers (ABC) that enable the HDDA cycloisomerization; (b) examples of *intra*molecular trapping reactions; (c) examples of *inter*molecular trapping reactions. Source: Hoye et al. [11a]; Baire et al. [11b].

three-atom linker (cf., ABC in **23**, Figure 10.5a; one substrate had a five-atom linker). Benzyne formation is always the rate-limiting event. The subsequent trapping of the reactive benzyne intermediate, whether intramolecular (Figure 10.5b) or intermolecular (Figure 10.5c) in nature, is always a fast, subsequent step. Substrates **23** vary in the ease with which they isomerize to the initial benzyne intermediate. Reactions were typically performed between 85 and 120 °C for one to two days to reach >95% consumption of the triyne, although it should be noted that the rate of the HDDA reaction is also dependent on the nature of the substituent R on the terminus of the diynophile. Most dramatic in that regard, when R is an alkynyl group in tetrayne substrates with the linkers shown in **23a**, the cyclization temperature was reduced from 110 to 65 °C to achieve a comparable rate of conversion [14].

10.3 Early Demonstration of New Modes of Aryne-Trapping Reactivity: Ag- and B-Promoted Carbene Chemistry

In another early study in this emerging field, researchers in the laboratory of Daesung Lee at Illinois, Chicago, demonstrated novel carbene-like reactivity in HDDA benzynes [15]. For example, when the symmetrical tetrayne **24** was heated in the presence of various silver(I) salts, the cyclopentanoisoindoline derivative **26** was efficiently generated. This outcome was rationalized by invoking an intramolecular

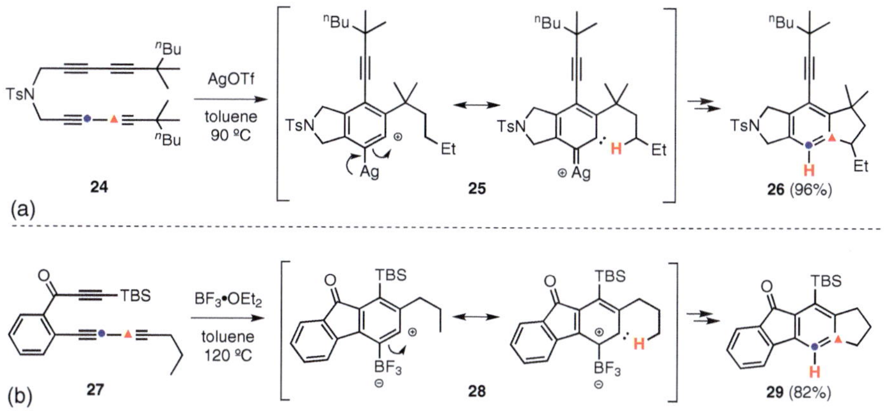

Figure 10.6 (a) Ag(I) promotion of carbene reactivity (C–H insertion) within an HDDA-benzyne. (b) The first instance of boron activation of a vicinal dicarbene.

CH-insertion within the silver carbenoid **25**. In a related overall transformation, Shen et al. recently showed that the triyne **27** produced the tetracyclic product **29**, but this process was effected by the action of boron trifluoride etherate [16]. Lewis acidic boron reagents were not previously known to induce carbene-like behavior; thus, this represents another example in which benzynes produced thermally by the HDDA reaction have allowed the discovery of new modes of reactivity. Computational support for boron-benzyne adducts like **28** was also provided (Figure 10.6).

10.4 De novo Construction of Arenes: A New Paradigm for Synthesis of Highly Substituted Benzenoid Natural Products

The de novo nature of arene construction via HDDA reactions provides a strategically unique approach for the chemical synthesis of certain benzenoid natural products. This new paradigm is exemplified by the key event used to construct the arene in the target structures in Figure 10.7. In the first, herbindole B (**35**, Figure 10.7a), the HDDA reaction of **30** gives the net HBr addition product **31** with methylene bromide serving as the source of those atoms. Notable about the synthesis is the manner in which the various sites on this initial HDDA product are manipulated to reveal (i) the ethyl and methyl groups (**31** to **32**), the atoms required for the cyclopentannulation reaction (**32** to **34** via **33**), and the ultimate oxidation of indoline to indole in herbindole B (**35**) [17].

The fluorene core of selaginpulvin C (**39**) [18] was efficiently created by the facile cyclization of the tetrayne **37**, prepared in situ by the MnO_2 oxidation of alcohol **36** (Figure 10.7b). The fluorenone **38** was readily elaborated to the quaternized fluorene in **39**.

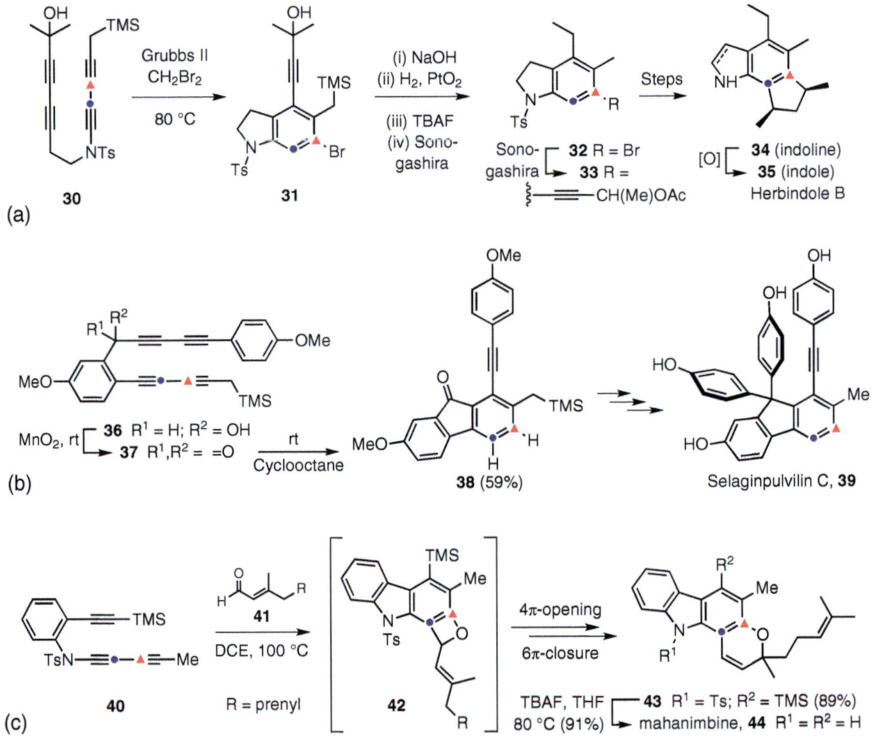

Figure 10.7 Strategic advantage of de novo arene ring construction demonstrated in syntheses of (a) herbindole B, (b) selaginpulvin C, and (c) mahanimbine.

A third example is seen in the synthesis of the carbazole-containing alkaloid mahanimbine (**44**) [19]. The diynamide **40**, when heated in the presence of citral (**41**), gave rise to a carbazolyne that was trapped in a net [2+2] reaction with the carbonyl group to produce the benzoxetene intermediate **42**. Under the reaction conditions, this proceeded through consecutive electrocyclic ring-opening (4π) and -closing (6π) events [20] to produce the pyranocarbazole **43** in a notably efficient fashion (89%).

10.5 Diradical Mechanism of the HDDA Cycloisomerization of Triyne to Benzyne

The mechanism of the HDDA cycloisomerization of three alkynes to a benzyne has been studied computationally as well as experimentally. Both concerted and stepwise pathways, the latter by way of a (delocalized) diradical intermediate, are worthy for consideration. In fact, computations suggest that, depending on the exact pair of reacting species, these can have quite similar energies of activation [8a, 21]. The computed activation barriers typically range from c. 30–40 $kcal\ mol^{-1}$.

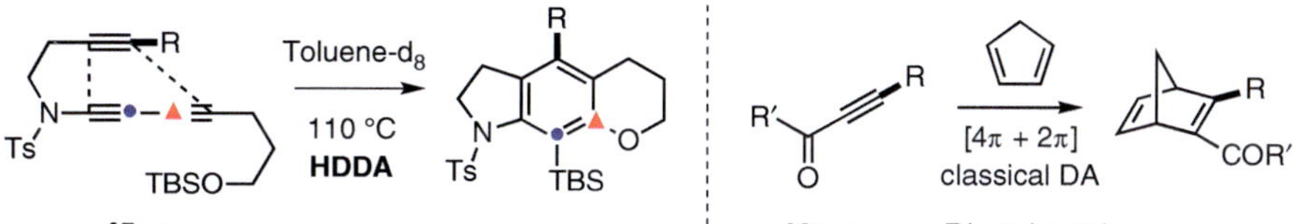

45a-g | **46a-g** R′ = *n*-heptyl

R	~k_{rel}		RSE
CF_3	0.67	Slow	+1.9
H	1		0
$CONEt_2$	4.7		−4.9
CO_2Me	43		−4.9
COEt	80		−6.7
CHO	490		−7.7
C≡CMe	1,500	Fast	−12.1

R	~k_{rel}		σ_p
C≡CMe	0.1	Slow	0.03
H	1		0
$CONEt_2$	5		0.26
CO_2Me	84		0.34
COEt	210		0.34
CF_3	1,680		0.54
CHO	4,700	Fast	0.42

Figure 10.8 Impact of substituents on rates of HDDA reactions (of **45**) vs. classical DA reactions (of **46**) of alkynyl diynophiles and dienophiles, respectively. (RSE = radical stabilizing energy). Source: Modified from Wang et al. [22].

Whether the diradical or concerted mechanism is energetically favored depends on the nature of the linker group, ABC. To the extent it has been explored, the exact computational method used appears to have a relatively minor effect on that conclusion. Importantly, and somewhat surprisingly at first glance, the overall free energy change that accompanies the conversion of triyne to benzyne is downhill c. 40–50 kcal mol^{-1} – that is, the HDDA cyclization is a rare example of a highly *exergonic* transformation that gives rise to a highly reactive intermediate! The nature of the substituent at the terminus of the diynophile also significantly impacts the ease of the reaction. A bystander alkyne, the presence of which is, perforce, the case for tetraynyl HDDA substrates (tethered bis-1,3-diynes), is computed to be a particularly effective activator [21f, j].

To distinguish between a diradical vs. a concerted cycloisomerization pathway experimentally, Wang et al. designed a study in which the impact of various substituents at the terminus of the diynophile was examined (Figure 10.8) [22]. The relative rates of reaction for these triynes (**45a–g**) were found to correlate with resonance stabilization energies (RSE) significantly better than for the Hammett σ_p values (i.e., cation stabilization) for each substituent. Conversely, the rates of classical [4+2] Diels–Alder cycloaddition reactions of **46a–g**, in which the dienophile was substituted with the same set of substituents as in the triyne substrates **45**, showed an expected response to the electron-withdrawing strength of the substituents. In the two most extreme examples, the CF_3 and 1-propynyl groups showed a dramatic rate-enhancing and -diminishing effect, respectively, on the HDDA reactions of **45**, just the opposite of their impact on the DA reactions of **46**. This is strong evidence in support of a stepwise cycloisomerization pathway in which radical stabilization is much more important than simple electronic perturbation and highest-occupied molecular orbital (HOMO)/lowest-unoccupied molecular orbital (LUMO) interaction, as is true of concerted DA reactions.

10.6 Additional Contributions from the Lee Group (University of Illinois, Chicago (UIC))

Researchers at UIC have made many additional (cf. Figures 10.6a and 10.7a,b, above) important contributions to the body of HDDA chemistries. Some of the most prominent are summarized in Figure 10.9. In each instance, a key feature of the trapping event is indicated in brackets. They demonstrated in 2013 that not only would Ag(I) promote C–H insertion reactions with appropriately tethered alkanes (Figure 10.6a), but that alkenes would efficiently participate in Alder-ene processes (e.g., **47** to **48**, Figure 10.9a) [23a–c] and that fluorination (and trifluoromethylation) of the benzyne could also be achieved (e.g., **49** to **50**, Figure 10.9b) [23d]. They established that various nucleophiles (e.g., amines [**49** to **51**] in Figure 10.9c) will trap the thermally generated benzynes [23e, f]. They also showed that acetonitrile would engage the electrophilic benzyne in Ritter-type reactions, leading to benzamide derivatives (e.g., **49** to **52**, Figure 10.9d) [23g]. In another recent advance, they reported a remarkable dearomatization reaction that used the high potential energy of the aryne to fuel the otherwise energetically unlikely transformation shown by the example of **53** to **54** in Figure 10.9e [23h]. Most recent of all, isocyanides, in the presence of various weak acids, were shown to trap the benzynes derived from a variety of tetraynes related to **49** (cf. **49′**) to give **55** [23h].

10.7 Additional Notable Modes of Aryne Reactivity

10.7.1 HDDA Benzynes as Dienophiles in Diels–Alder [4π+2π] Cycloaddition Reactions with Aromatic Dienes

Ever since the pioneering studies of Wittig (trapping of *o*-benzyne by furan) [24], arynes have long been recognized for their effectiveness in engaging a wide variety of dienes; especially notable are those cases where the thermodynamics of the cycloaddition might otherwise be unfavorable, as is often the case with DA cycloadditions of furan or simple benzene derivatives. HDDA reactions culminating with such reactions are known. In addition to Ueda and coworkers trapping with anthracene (**11** and **12**, Figure 10.3b [in benzene solution, incidentally]), they demonstrated that benzene itself, in the absence of another sufficiently reactive trapping agent, would engage the benzyne **57** (**56** to **58**, Figure 10.10a) [25]. Bimolecular reactions with substituted furans (**59** to **60**, Figure 10.10b) [26] as well as intramolecular trapping of arenes, including naphthalenes (**61** to **62**, Figure 10.10c) [27], are also possible. *N*-Methyimidazole gives rise to [4+2] adducts such as **63** that undergo further fragmentation and oxidation to benzomaleimides (**64**, Figure 10.10d) [28].

10.7.2 Trapping of Natural Products: Phenolics

Some remarkable levels of selectivity were revealed by a study in which HDDA-benzynes were generated in the presence of multifunctional natural products

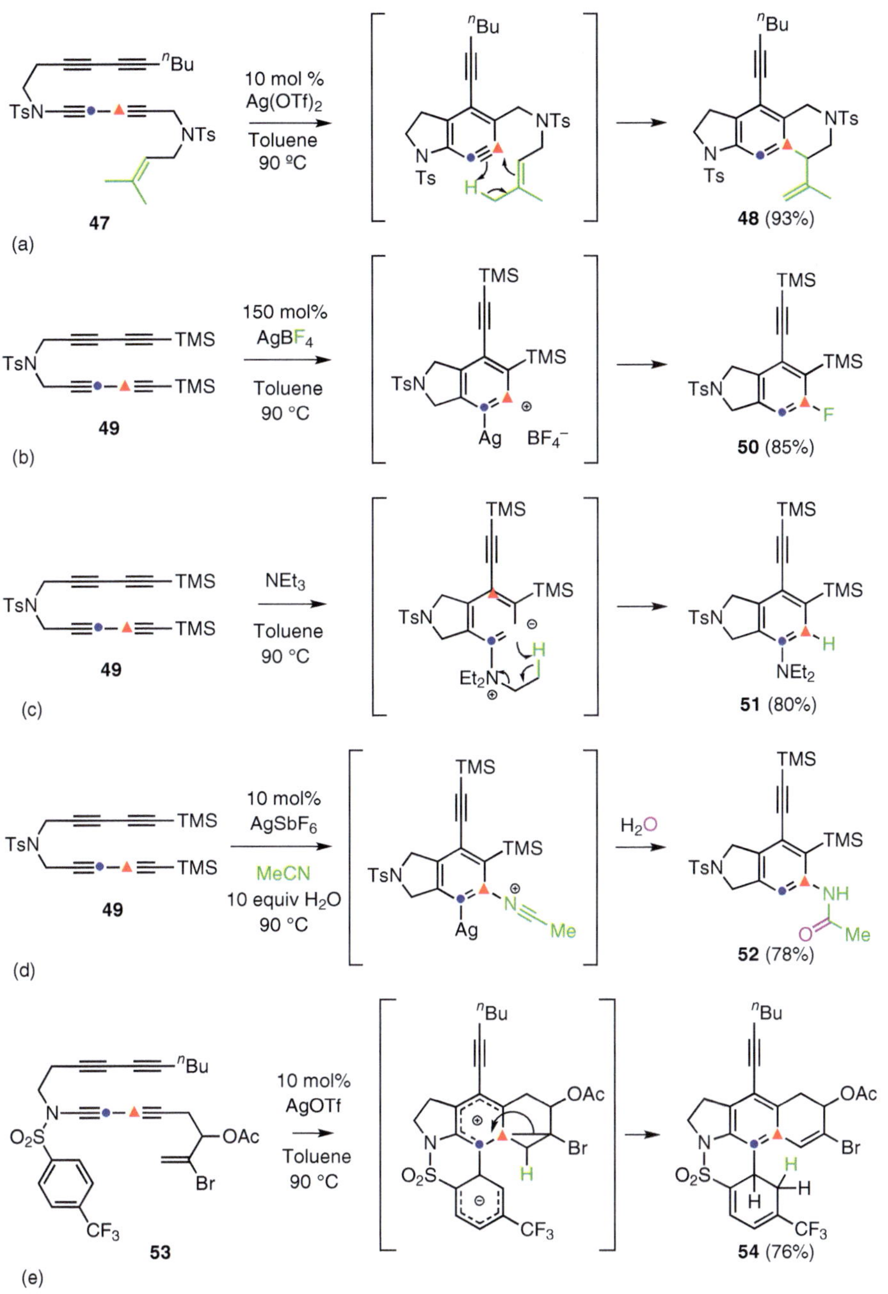

Figure 10.9 Highlights of contributions from the Lee group (University of Illinois, Chicago). (a) Alder-ene. Source: Karmakar et al. [23a]; Gupta et al. [23b]; Gupta et al. [23c], (b) silver-mediated fluorination. Source: Modified from Wang et al. [23d], (c) amine (nucleophile) trapping, (d) Ritter-type transformations. Source: Modified from Ghorai and Lee [23g], (e) arene dearomatization by HDDA benzynes. Source: Modified from Karmakar et al. [23h], and (f) isonitrile trapping gives products complementary to those of the Ritter-type (d, above).

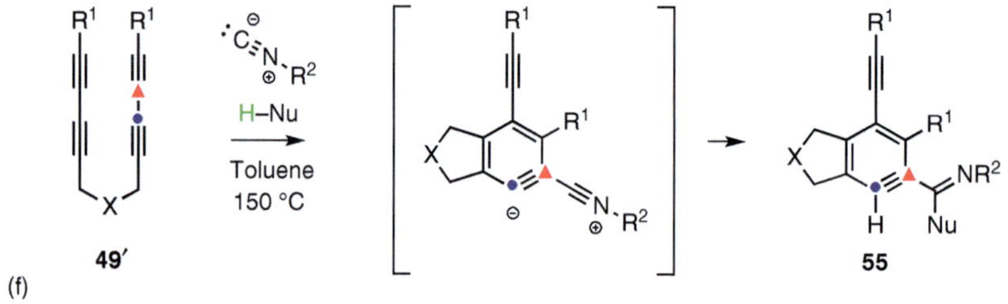

Figure 10.9 (*Continued*)

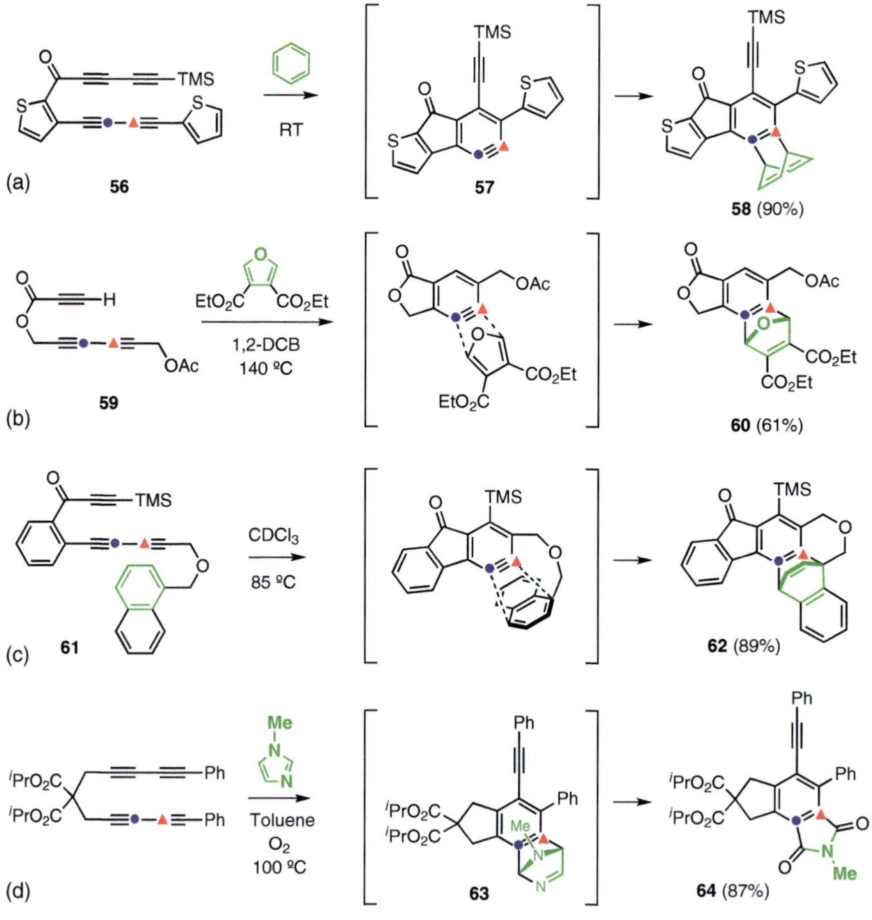

Figure 10.10 Examples of Diels–Alder trapping of often inert, aromatic "dienes" by HDDA benzynes. (a) Benzene DA reaction, a rare event. Source: Modified from Kawano et al. [25], (b) substituted furan trapping. Source: Based on Chen et al. [26], (c) intramolecular capture by a naphthalene. Source: Modified from Pogula et al. [27], and (d) imidazole capture, fragmentation, and oxidation. Source: Based on Hu et al. [28].

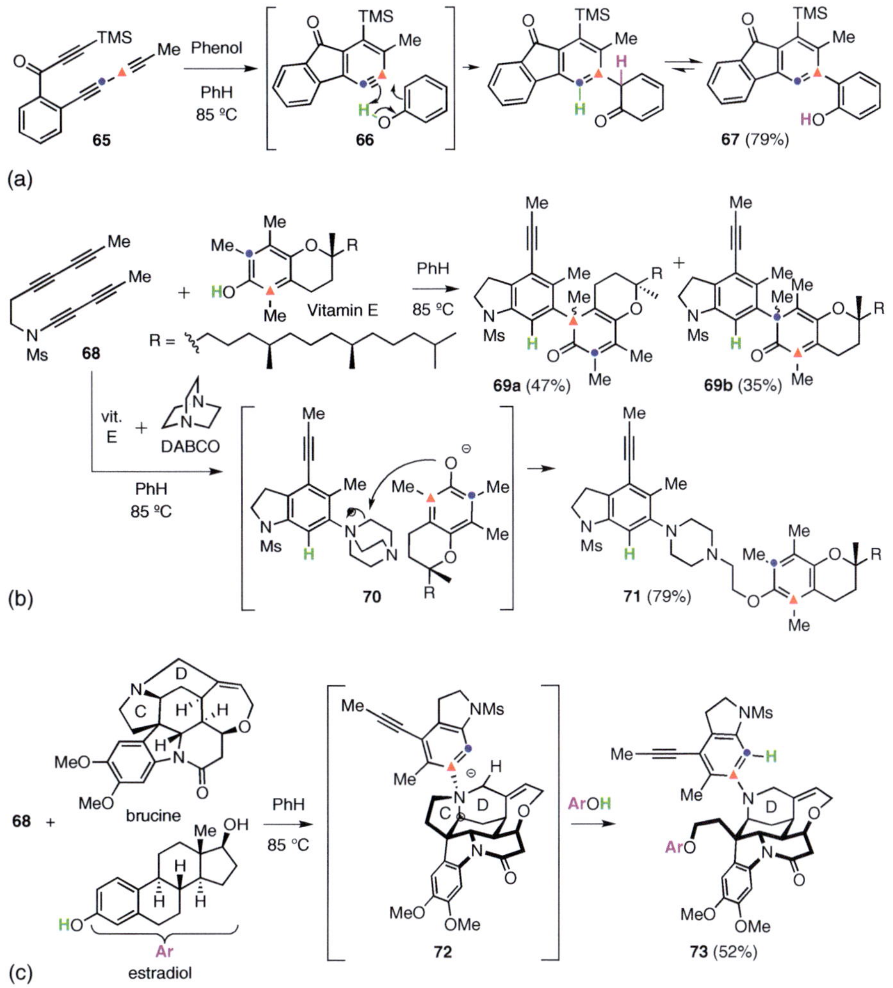

Figure 10.11 Examples of (notably chemoselective) reactions of HDDA benzynes with multifunctional, phenolic natural products. (a) Phenol-ene-like reaction. Source: Modified from Zhang et al. [30], (b) Vitamin E, and (c) three-component reaction with brucine and estradiol

possessing numerous potential sites for reaction [29]. Phenol itself has been shown to react with HDDA benzynes by way of an ene-type process portrayed in **66** (**65** to **67**, Figure 10.11a) [30]. Even the fully substituted phenolic ring in vitamin E is efficiently captured by the benzyne to provide the dearomatized cyclohexadienones **69** from **68** in excellent combined yield (Figure 10.11b). However, if a trapping agent more reactive than phenol, 1,4-diazabicyclo[2.2.2]octane (DABCO) in the case of formation of **71**, is also present, then a staged three-component reaction intervenes. That is, the role of the phenol in vitamin E is completely diverted to produce the dabconium phenoxide ion pair **70**, which then proceeds to **71**. In a reaction of yet greater complexity, the two natural products brucine and estradiol were trapped, in sequence, via the 1,3-zwitterion **72** from capture of the benzyne by the alkaloid's aliphatic nitrogen atom, only then to abstract the more acidic phenolic proton in estradiol en route to the product **73** (Figure 10.11c).

10.7.3 Trapping of Natural Products: Colchicine and Quinine

Colchicine traps the benzyne from **68** in an unprecedented fashion, engaging the carbonyl oxygen in a net (3+2) cyclization to the tropylium ion-like zwitterion **74** that proceeds to product **75** (Figure 10.12a). In perhaps the most dramatic example of a selective cyclization, quinine, in which 11 different potential sites of reactivity can be envisioned, proceeds by largely one pathway to **77**, presumably via the zwitterion **76** (Figure 10.12b). Collectively, these results underscore an often-overlooked important adage – namely that highly reactive is not a synonym for unselective [31]. Thus, even though the cycloisomerization of triyne to benzyne is the rate-limiting step in HDDA reactions, there still is considerable selectivity expressed in the product-determining manifold of potential competing pathways.

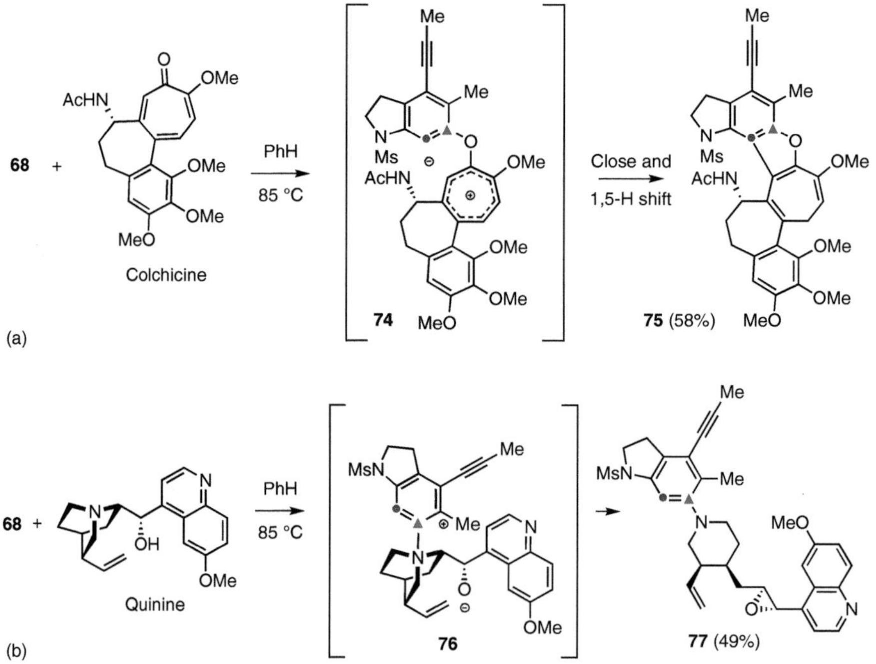

Figure 10.12 Additional examples of (notably chemoselective) reactions of HDDA benzynes with multifunctional natural products. (a) Reactions of HDDA benzynes with colchicine and (b) reactions of HDDA benzynes with quinine.

10.8 New Reaction Modes and New Mechanistic Understanding

10.8.1 Three-Component Reactions

Trapping of arynes with a pair of complementary and suitably coordinated reactants, processes amounting to three-component reactions, is well established [32]. Not surprisingly, similar reactions can be identified for the thermally produced HDDA benzynes, although the types of compatible coreactants that can be orchestrated to perform can often be different from, and complementary to, those that succeed with classically produced benzynes. Cyclic aliphatic amines generate zwitterionic species upon their initial addition to the benzyne (cf. the DABCO in Figure 10.11b). In the presence of protic acids, these can be protonated and ring-opened (e.g., **78** to **80** via **79**, Figure 10.13a) [33]. By a similar process, heteroaromatic amines of the pyridine family generate pyridinium-like zwitterions (cf. **82**) that can then be

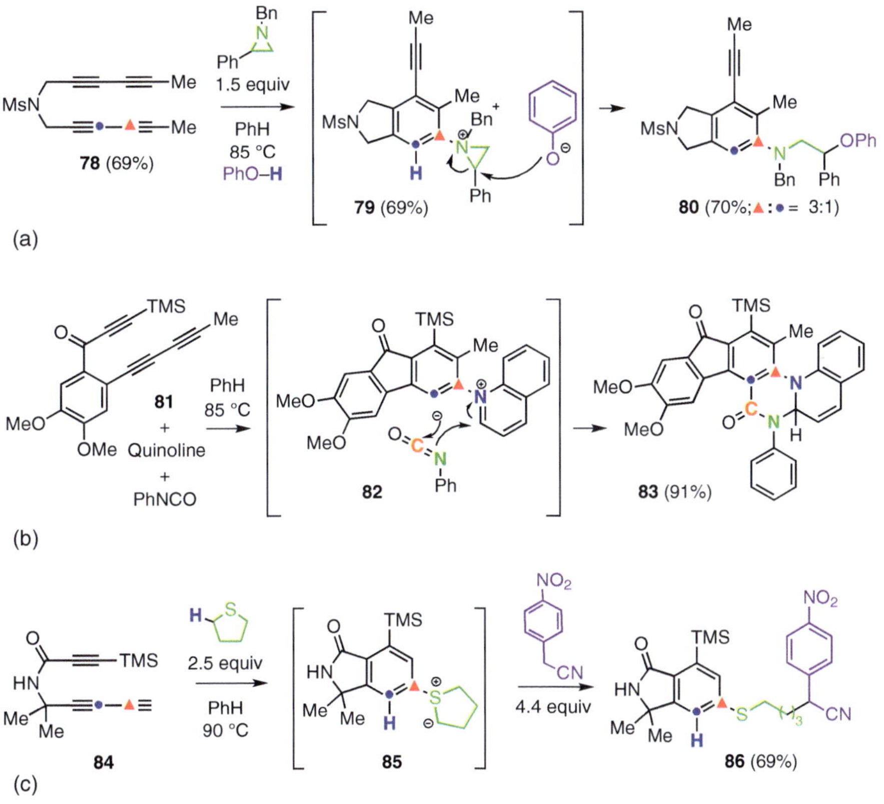

Figure 10.13 Examples of three-component reactions of HDDA benzynes. (a) Aliphatic cyclic amines (here, aziridine) and protic nucleophiles (here, phenol). Source: Modifiied from Ross and Hoye [33], (b) pyridinyl aromatics (here, quionline) and electrophiles (here, phenylisocynate). Source: Modified from Arora et al. [34], and (c) sulfide (here, tetrahydothiophene) and carbon nucleophile (here, *p*-nitrophenylacetonitrile). Source: Modified from Chen et al. [35].

captured by various electrophiles, including ketones and aldehydes, electron-poor alknyes and alkenes, and isocyanates (e.g., **81** to **83** via **82**, Figure 10.13b) [34]. In a third example, cyclic sulfides generate zwitterions that undergo intramolecular proton transfer to produce transient sulfonium ylides, which then can engage, for example, suitable acidic carbon-based protic acids (e.g., **84** to **86** via **85**, Figure 10.13c) [35].

10.8.2 Dihydrogen Transfer Reactions

HDDA-benzynes allowed the discovery of an unprecedented, concerted removal of two hydrogen atoms from vicinal H–CX–H arrays (X = C or O). Thus, dihydrogen transfer to benzynes **87**, **90**, or **93** was seen when each was generated in the presence of cyclic hydrocarbons (e.g., **65** to **88**, Figure 10.14a) [36], cyclic ethers (cf. Figure 10.3b [10] and, e.g., **89** to **91**, Figure 10.14b [36]), or primary or secondary alcohols (e.g., **92** to **94**, Figure 10.14c) [37]. Both experimental and computational studies indicate that these redox processes proceed through previously unrecognized, concerted transition-state structures. For example, when **89** was heated in a solution of equimolar amounts of THF-h_8 and THF-d_8, only nondeuterated and

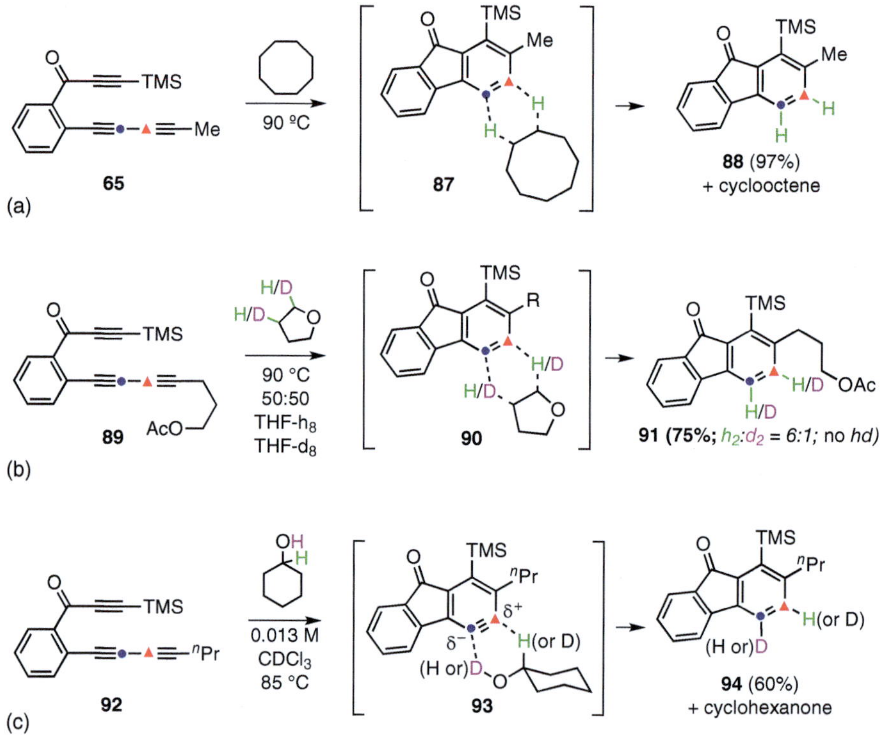

Figure 10.14 Various dihydrogen-transfer reagents reduce the benzyne by concerted processes. (a) H_2 transfer from hydrocarbons. Source: Modified from Niu et al. [36], (b) H_2 transfer from THF. Source: Modified from Niu et al. [36], and (c) H_2 transfer from alcohols. Source: Modified from Willoughby et al. [37].

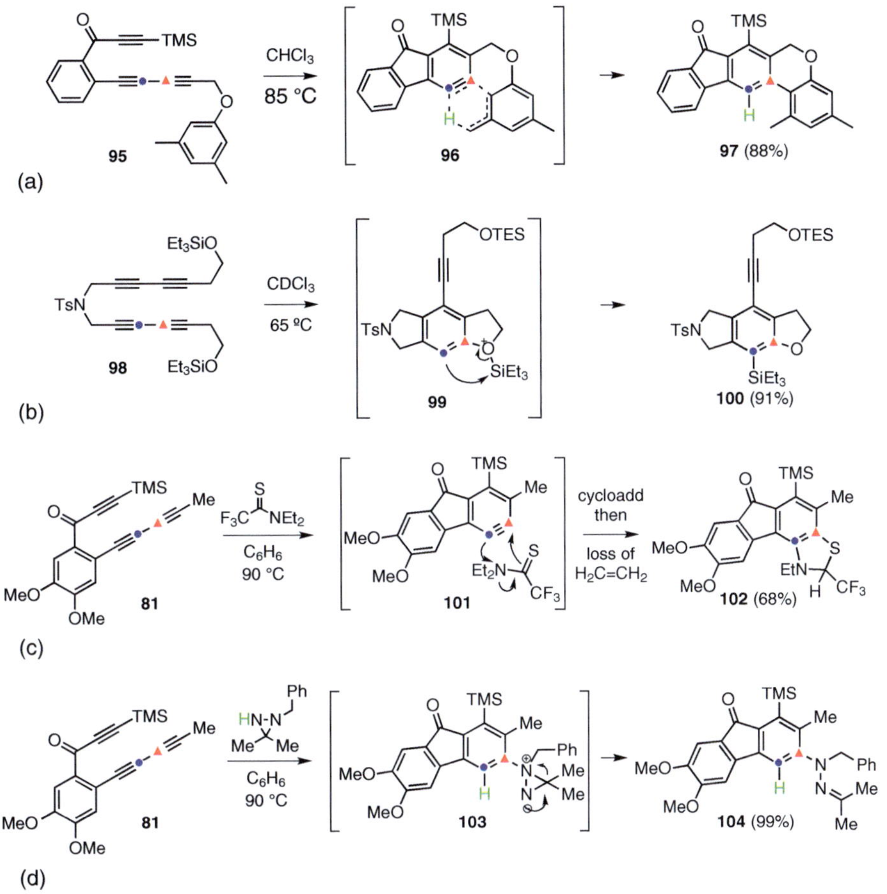

Figure 10.15 Additional examples of new classes of aryne-trapping reactions. (a) The aromatic ene reaction. Source: Modified from Niu and Hoye [39], (b) trapping by silyl ethers, (c) atypical [3+2]-cycloaddition, (d) diaziridine trap. Source: Modified from Arora et al. [41].

dideuterated forms of **91** were observed, consistent with the simultaneous transfer shown in **90** rather than a stepwise net H_2-transfer that would be involved if hydrogen atom abstraction were the first step. The alcohol-mediated redox reactions are effective at low concentrations of the alcohol; at higher concentrations, addition of the RO–H to produce ArOR ethers occurs via a purported concerted addition of an alcohol dimer to the benzyne [37]. As depicted in **93**, the monomeric alcohol reacts in a regioselective manner, as established by complementary deuterium-labeling studies; the C–H can be viewed to add in hydride-like fashion to the more electrophilic benzyne carbon [38] and the O–H as a lagging proton-like donation to the more electron-rich carbon [37].

10.8.3 Aromatic ene, Silyl Ether, Thioamide, and Diaziridine Reactions

Other examples of new types of reaction or of new mechanistic insights are shown in Figure 10.15. First, a properly poised benzylic hydrogen atom, as in the triyne substrate **95**, is able to be transferred in a rare type of dearomatizing ene reaction. The TS **96** capitalizes on that arrangement to produce, following rearomatization, the pentacyclic **97** (Figure 10.15a) [39]. The initial ene product, an intermediate isotoluene, is sufficiently long lived that it can also be captured if Alder enophiles are present in the reaction medium (not shown). The length of the tether connecting the pendant arene to diyne is critical, as demonstrated by those having less ideally situated arenes, which undergo competitive [4+2] cycloadditions with the pendant arene itself (cf. Figure 10.10b). Second, in the serendipitous observation of HDDA cycloisomerization of **17**, a silyl ether was fortuitously poised to intercept the benzyne, producing an aryl silane. This formal O—Si bond insertion reaction (cf. **18** to **19** and **21** to **22**, Figure 10.4) was previously unknown for arynes. Density functional theory (DFT) analysis suggests that the energetics for a stepwise process via a zwitterion such as **99** (Figure 10.15b, in the conversion of **98** to **100**) vs. a concerted, single-step insertion of the benzyne directly into the silyl ether proceed with comparable activation energies [12]. Third, an unprecedented, pseudo-1,3-dipolar cycloaddition process (cf. **101**) was discovered when the benzyne was generated in the presence of thioamides (e.g., **81** to **102**, Figure 10.15c). Fourth, diaziridines, long known to react with simple electron-deficient alkynes [40], captured an HDDA benzyne to produce ring-cleaved hydrazones in a mechanistically similar fashion (e.g., **81** to **104**, Figure 10.15d) [41]. Interestingly, the more substituted nitrogen atom of the diaziridine selectively engages the electrophilic benzyne sp-carbon (cf. **103**). One can think of the large potential energy of the strained formal triple bond in the aryne as the thermodynamic fuel that drives these atypical types of reaction.

10.9 New Routes to Polycylic, Highly Fused Aromatic Products

10.9.1 Naphthynes via Double-HDDA, Intramolecular-HDDA, and Highly Functionalized Naphthalenes

The HDDA reaction can serve as a powerful platform for accessing structurally complex polycyclic aromatic frameworks, often with high efficiency and, sometimes, through novel transformations. This was first demonstrated in early examples from the Ueda group, in which a sequential cycloisomerization provided a naphthyne intermediate (e.g., **105** to **106**), revealed by the formation of trapped products such as **107** (Figure 10.16a) [42]. In a rare example of an intramolecular HDDA reaction (cf. Figure 10.3c), the cyclic triyne **108** was cyclized to the furan-trapped adduct **110** (via **109**, Figure 10.16b) [43]. Symmetrical tetraynes having the general structure of **111** have been efficiently trapped by furan derivatives (as well as pyrroles, thiophenes, and cyclopentadienes) to produce adducts **112** [44], which can, in principle [45], be deoxygenated to reveal an additional arene (cf. **113**, Figure 10.16c). Tetraynes like **111** have also been cyclized and trapped by tetraphenylcyclopentadienone (TPCPD) to produce adducts such as **114**. These were demonstrated to be efficient blue emitters in prototypical OLED devices [46].

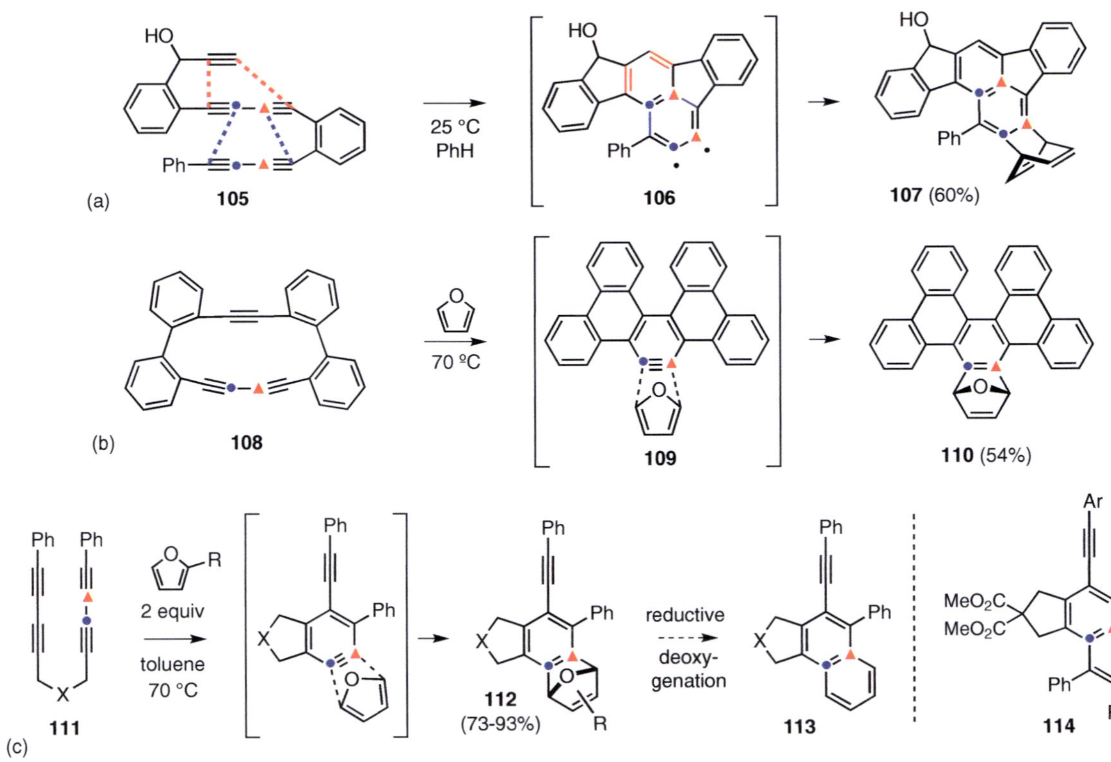

Figure 10.16 Examples of readily accessed, fused-ring, polycyclic aromatic motifs by HDDA reaction strategies. (a) Naphthyne generation via double HDDA. Source: Based on Miyawaki et al. [42], (b) a rare intramolecular HDDA. Source: Based on Nobusue et al. [43], and (c) furan and TPCPD trapping en route to new arenes.

10.9.2 Trapping with Perylene, Domino HDDA, and Tandem HDDA/TDDA

The polycyclic aromatic agent perylene can be used to capture an HDDA benzyne in its bay region (cf. **116**) to give, following spontaneous ejection of dihydrogen, adducts such as **117** in a single step (e.g., **115** to **117**, Figure 10.17a) [47]. In a domino-HDDA process, the polyyne **118** proceeded, consecutively, via naphthyne and anthracyne intermediates to the tetracyne **119**, which could be trapped by the cyclopentadienone fused-ring derivative **120** en route to the hexacene derivative **121** (Figure 10.17b) [48]. Very recently, an interesting sequential HDDA-then-TDDA double cyclization of the symmetrical tetrayne **122** was reported (Figure 10.17c) [49]. It gave rise to, initially, **123** in which the benzyne became the enynophile in a TDDA reaction en route to **124**. This result is more intriguing in light of the fact that the corresponding disulfide (i.e., where the Me2Si moiety is replaced by a sulfur atom) substrate does, instead, a double-TDDA cyclization, highlighting a redirecting effect of the silicon atom.

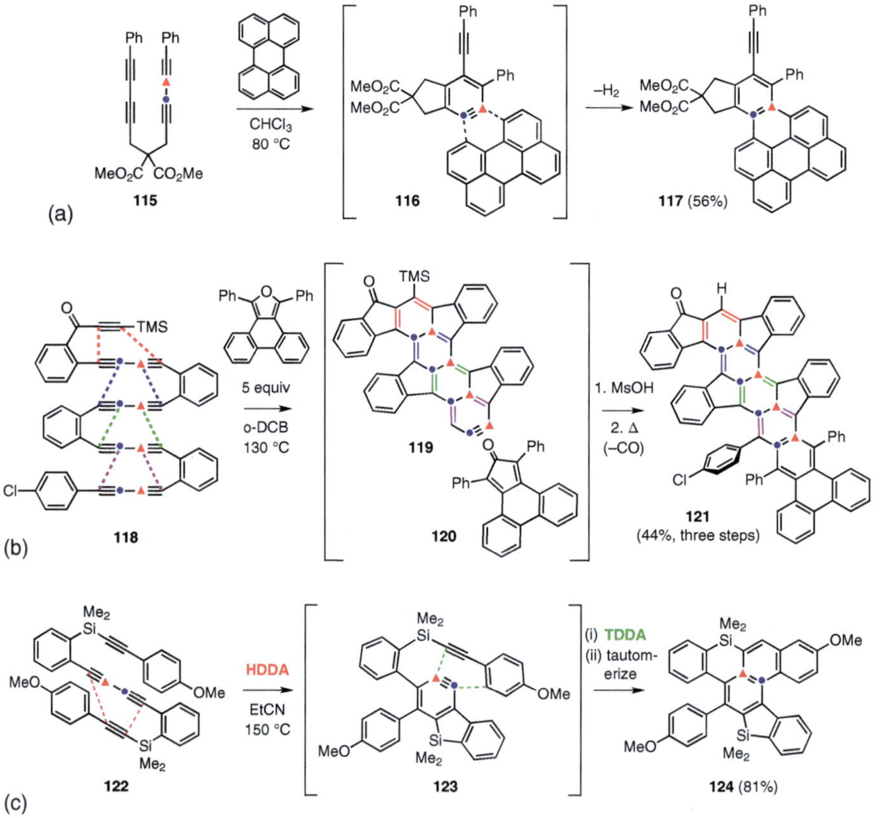

Figure 10.17 More examples of readily accessed fused-ring, polycyclic aromatic motifs by HDDA reaction. (a) Perylene trap. Source: Modified from Xu et al. [47], (b) domino HDDA via a tetracyne. Source: Modified from Xiao and Hoye [48], and (c) domino HDDA via a naphthacyne. Source: Mitake et al. [49].

10.10 One-Offs

10.10.1 Enal, Formamide, Diselenide, and (N-heterocyclic carbene) NHC-Borane Trapping

A group of otherwise unrelated but interesting examples of transformations is gathered in Figure 10.18. As with conventional benzynes [50], engagement with conjugated enals leads to benzopyrans via oxetanes (cf. Figure 10.7d) and *o*-quinonemethides such as **126** (e.g., **125** to **127**, Figure 10.18a) [51]. In a similar fashion, when **128** is heated in 1:1 DMF:H_2O, the carbonyl group of the amide

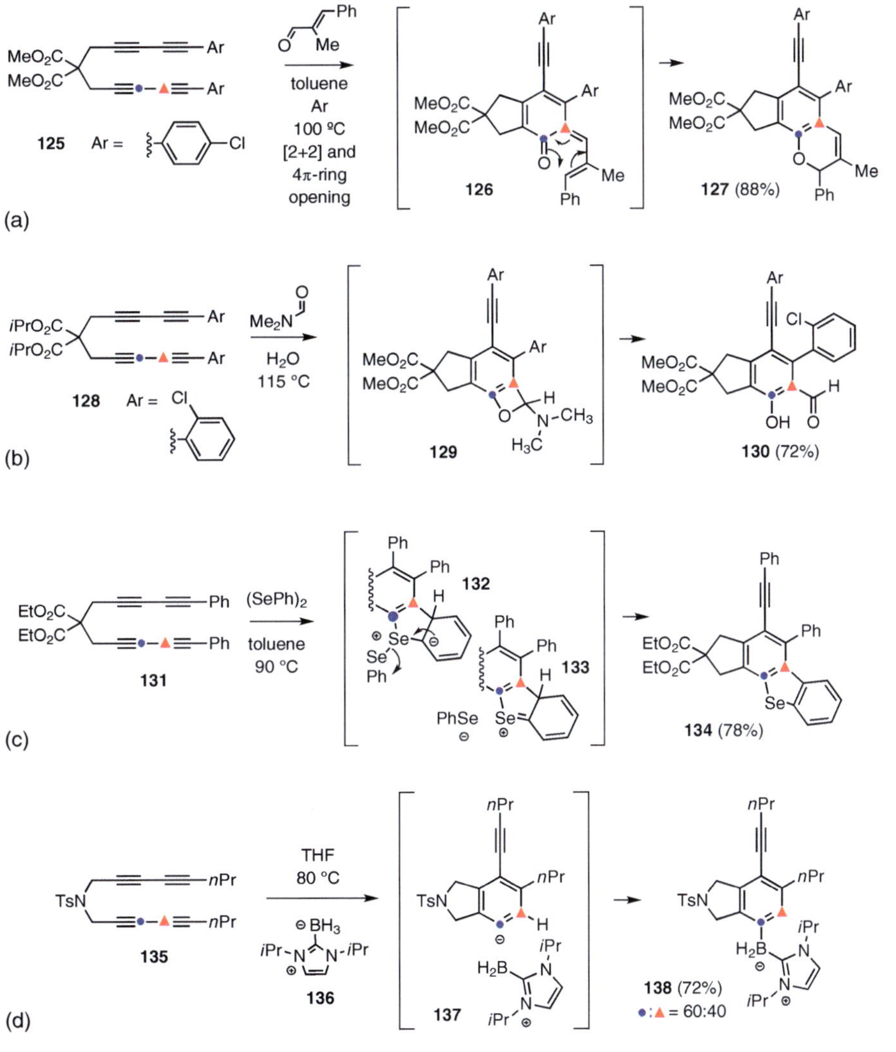

Figure 10.18 Miscellaneous examples, demonstrating novel modes of trapping. (a) Enal trapping. Source: Meng et al. [51a]; Wang et al. [51b], (b) formamide trapping. Source: Based on Hu et al. [52], (c) diselenide trapping. Source: Based on Hu et al. [53], and (d) net hydroboration. Source: Modified from Watanabe et al. [54].

engages the benzyne to produce the oxetane derivative **129**, which undergoes electrocyclic ringopening and hydrolysis to generate the salicylaldehyde derivative **130** (Figure 10.18b) [52]. Diphenyldiselenide traps the HDDA benzyne from **131** in an unexpected fashion, providing the selenophene **134**, we might suggest via species such as **132** and **133** (Figure 10.18c) [53]. In another unusual transformation, when used to capture the benzyne from **135**, the borane·NHC complex **136** gives rise to the net hydroboration product **138**, likely by way of the transient zwitterion **137** (Figure 10.18d) [54].

10.10.2 Cu(I)-Catalyzed Hydroalkynylation, Ether vs. Alcohol Competition, Photo-HDDA, and a Kobayashi Benzyne as an HDDA Diynophile

Cu(I)-Catalyzed hydro- and halocupration of HDDA benzynes is feasible [55]. The chemoselectivity observed for the reactions of but-3-yn-1-ol is instructive (Figure 10.19a). In the absence of Cu, simple alcohol addition to the benzyne from **139** ensues (to give the aryl ether **140**), but the chemoselectivity of reaction is entirely reversed when substoichiometric CuCl is present, leading to the arylalkyne **142**, likely via a 1,2-alkynylaryl copper species such as **141**. Reactions of glycidol derivatives (and other hydroxyalkyl-substituted cyclic ethers) show a different mode of reaction with HDDA benzynes than they do in the reactions with benzynes made using conventional fluoride ion generation. The latter predominantly gives, at low temperature, simple O–H addition products (e.g., **143**, Figure 10.19b) [56]. In contrast, the HDDA benzyne from **144** gives products resulting from attack, instead, by the epoxide oxygen to give, e.g., **145**, which then undergoes rearrangement and ring fragmentation en route to the ketone **146** [57]. A photochemical variant of the HDDA (the hv-HDDA) reaction is possible, at least for substrates containing a bisaryl-substituted tetrayne such as that present in **147** (Figure 10.19c). This proceeds, even at subambient temperature, through an intermediate benzyne that shows all the hallmarks (e.g., it gives the same distribution of minor byproducts) of the same reactive species that is generated by heating **147** ($t_{1/2}$ = 12 hours at 75 °C). Capture by the internal methoxy group produces **148** and demethylation (by traces of water, Cl_3C^-, or Cl^-) results in the formation of the dibenzofuran **149** [58]. The multiaryne cascade (benzyne-to-benzyne-to-naphthyne) reaction between the diynyl triflamide **150** and the 1,2-bisbenzyne precursor **151** [59] integrates a key HDDA reaction as one component (**153** to **154**, Figure 10.19d). In this example, the naphthyne is trapped by a tethered silyl ether, leading to **155**. This demonstrates how new strategic developments incorporating an HDDA event can be conceived and reduced to practice [60].

10.11 Outgrowths from HDDA Chemistry

10.11.1 Processes that Outcompete Aryne Formation in Potential HDDA Substrates

A final pair of topics (Figure 10.20) demonstrates how the HDDA platform can be expanded in atypical fashion, again in each of these instances in unanticipated

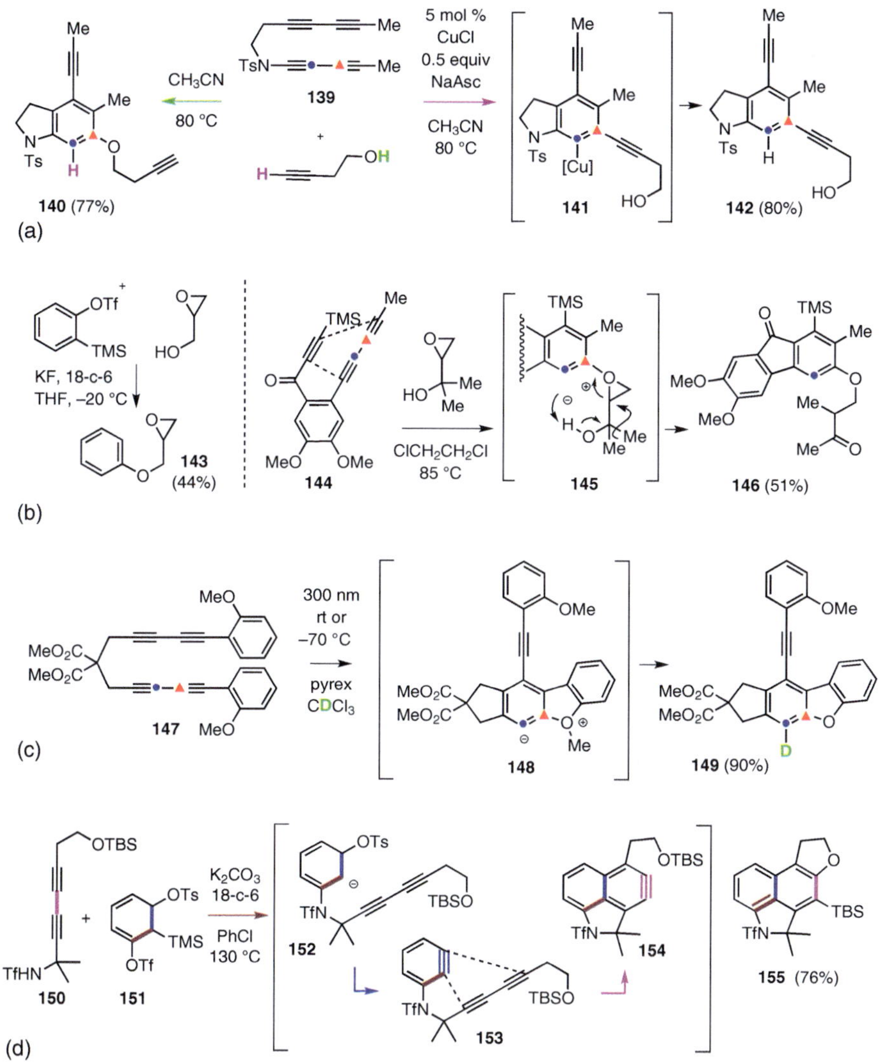

Figure 10.19 Miscellaneous examples, demonstrating novel modes of trapping. (a) Cu(I)-catalyzed hydroalkynylation, (b) ether > hydroxyl competition. Source: Based on Thangaraj et al. [56], (c) the photochemical HDDA reaction, and (d) benzyne to benzyne to naphthyne.

ways. These outgrowths also result from the thermodynamic potential that the multialkynes endow into the substrates. When treated with a weak base, the cycloisomerization of the diyne nitrile **156** takes a different course resulting in formation of **159** (Figure 10.20a). This new process was named a "pentadehydro"-Diels–Alder reaction, not because of any mechanistic relationship to DA or HDDA reactions, but because the intermediate allene **157** has five sp-hybridized atoms (not six, as with HDDA cycloisomerizations) that become incorporated in the strained, α,3-dehydroazatoluene intermediate **158** [61]. This is then trapped by piperidine to

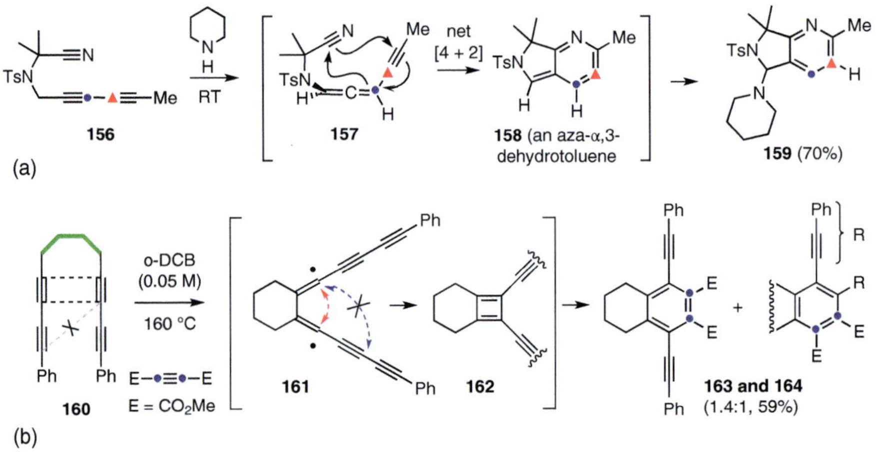

Figure 10.20 Intervention of other, faster processes starting from potential HDDA substrates. (a) The pentadehydro Diels–Alder reaction and (b) Tether length matters: cyclobutadienes instead of benzynes.

give the pyridine derivative **159**. This transformation was also significant because it represented the first time that a nitrile, the aza-surrogate of an alkyne, had participated in a cycloisomerization of a substrate with six degrees of unsaturation (cf. Figure 10.21). In a second unexpected outcome, the tetrayne **160**, in which the tether comprises four (sp^3-C) atoms rather than the three-atom linkage essentially required for HDDA cyclizations (cf. Figure 10.5a), produced the four-membered cyclobutadiene **162** (red arrow) rather than the six-membered (blue arrow) benzyne. This presumably reflects a kinetic preference for the cyclization of the diradical **161**; the ring fused to the cyclobutadiene is now six-membered, and the bicyclic[4.2.0]substructure is less strained than in the cases where the fused ring is five-membered [62]. The intermediacy of **162** was only recognized once the reaction was performed in the presence of various electron-deficient traps, here dimethyl acetylenedicarboxylate to give **163** and **164** (Figure 10.20b), because no tractable adducts were formed when typical, electron-rich aryne capture reagents were used. Clearly, subtle ring-strain effects arising from the size of the eventual fused ring have a strong influence on how the diradical cyclizes.

10.11.2 Aza-HDDA Reaction

Replacement of a terminal carbon atom in a HDDA triyne substrate with a nitrogen atom presents the opportunity to effect a cycloisomerization to form a pyridyne rather than benzyne intermediate. This would constitute an aza-HDDA reaction (Figure 10.21) [63]. Based on the location of the nitrogen atom, either a 3,4-pyridyne such as **166** (Figure 10.21a) or a 2,3-pyridyne (Figure 10.21b) could be formed. DFT studies suggest that these have considerably different levels of geometric distortion [38], which is consistent with the differences in selectivity upon reaction with nucleophilic trapping agents. For example, the nitrile–diyne substrate **165** gives

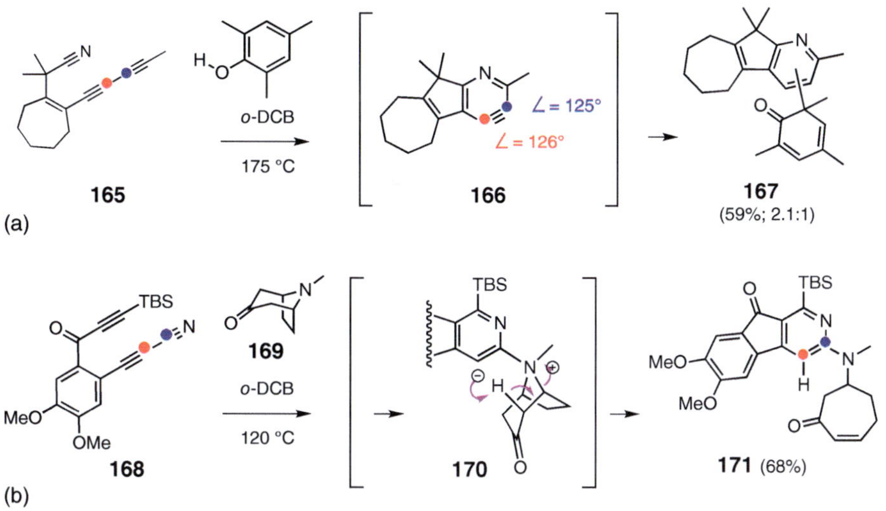

Figure 10.21 The aza-HDDA reaction gives 3,4-pyridynes (cf. **166**) or 2,3-pyridynes (precursor to **170**). (a) Aza-HDDA reactions involving nitriles as diynophiles and (b) aza-HDDA reactions involving yne-nitriles as 1-aza-1,3-diynes. Source: Modified from Thompson and Hoye [63].

the dienone **167** as a mixture of substantial amounts of the regioisomers arising from each of the pyridyne carbons in **166** by 2,4,6-trimethylphenol. In contrast, the aza-diyne-containing substrate **168** reacts with, for example, tropinone (**169**) to give the amine-trapped product **171**, via zwitterion **170**, as the sole regioisomer.

The aza-HDDA variant has several limitations. The reaction is slower; some substrates require reaction temperature greater than 200 °C. This greater activation barrier stems from the fact that a C—N triple bond is considerably stronger than a C—C triple bond. Product yields for many of the reported aza-HDDA reactions are below 50%, and the synthesis of substrates is not as easy as for analogous HDDA triynes. Even with this more limited scope, the aza-HDDA reaction still represents a useful process for accessing certain highly substituted pyridine derivatives.

10.12 *Guidelines and Practical Issues*: Strategic Considerations

10.12.1 Complementarity of Classical vs. HDDA Benzyne Chemistries

From a strategic perspective, the HDDA reaction is quite complementary to classical methods for aryne generation and utilization. The latter constitutes a net substitution reaction of a precursor arene (Figure 10.22a). That is, a benzenoid aromatic substrate **172** is activated by a reagent(s) to produce a benzyne **173**, which is then trapped to produce a product **174** that contains the same six-membered ring that was present in the starting substrate. By contrast, HDDA reactions (Figure 10.22b) create new benzenoid rings in de novo fashion (cf. **23** to **176** via **175**). Because of the

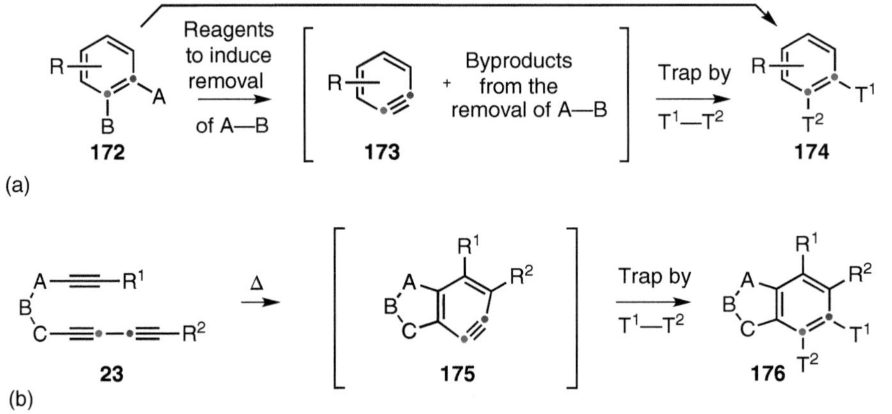

Figure 10.22 Strategic difference between (a) classical aryne net substitution chemistry and (b) de novo construction of the benzyne and its trapped products.

requirement for the triyne to be tethered in the precursor to render the cycloisomerization intramolecular, HDDA benzynes (and the derived final trapped products) are of greater structural complexity compared to those involved in the majority of classical benzyne reactions. For these primary reasons, the HDDA reaction should be viewed as orthogonally complementary to classical approaches to making and using arynes. Neither strategy is superior in all settings – far from it.

10.12.2 Regioselectivity Issues and the Nature of the Nucleophilic Trapping Agent

Because HDDA benzynes are, of necessity, unsymmetrically substituted (cf. **175**, Figure 10.22b), even when the precursor is a symmetrical tetrayne, regiochemical issues are a consideration when the trapping reagent is also not symmetrical. The great majority of benzyne-trapping reactions can and should be viewed as an in-plane attack on the electrophilic benzyne π-bond by a nucleophile. The distortion analysis [38] mentioned earlier in the context of the aza-HDDA reaction (cf. Figure 10.21) is also quite effective at rationalizing/predicting the preferred site of nucleophilic addition. More specifically, the benzyne carbon atom computed to have the larger internal bond angle is more electrophilic because of the greater p-character imparted to its in-plane component of the strained benzyne π-bond [38].

For symmetrical trapping agents regioselectivity is often not an issue, as seen from many examples above, include furan, H_2-transfer agents, benzene, dihalogenation (e.g., with $CuCl_2$) [64], and bis-sulfide formation (via disulfide [disulfane] trapping) [65]. Because of the electrophilic nature of the benzyne, when two traps of the same class of nucleophile simultaneously present in a reaction mixture, the one with greater electron density typically will react faster. For example, anisole, present in <1 vol% in a reaction solvent of benzene will far outcompete the trapping of the benzyne by the benzene in a [4+2] cycloaddition process. The exceptionally high degree of polarizability of the aryne π-bond is responsible for the polar nature of its reactions and to its remarkable versatility. "Said differently, an aryne represents one of the softest and most malleable of all carbon-based electrophiles." [37] Consequently, soft, polarizable nucleophiles can be expected to add in preference to harder nucleophiles of higher electron density. For example, small amounts of sulfides effectively trap even in alcohol or aqueous solutions. As a corollary to the reality of this property of polar reactivity, radical additions to the strained π-bond are rare.

10.12.3 Limitations Imposed by Trapping Agents

In planning a potential HDDA reaction strategy, it is worth keeping two limiting conditions in mind. If the trapping reagent reacts too quickly with the polyyne substrate (i.e., the benzyne precursor), it will prematurely intervene prior to the requisite cycloisomerization event. Examples of this incompatibility include the reaction of primary or secondary amines with ynone or ynoate substrates and the excessive reactivity of typical hydroboration reagents and dihalogens. If, on the other extreme, the trapping event is too slow, the benzyne will often turn back and cannibalize more of its polyyne precursor, often in myriad manners leading to intractable, highly conjugated, brown–black material. Fortunately, the "window of opportunity" for successful HDDA reactions defined by those limiting extremes is quite large, as the myriad examples described in this chapter attest. Rare instances of selective reactions between an HDDA benzyne and its precursor triyne are shown in Figure 10.23 (e.g., **177** to **178** and **179** to **180**), rationalized by the intermediates in brackets [62]. All of the successful substrates **177** shared the property of having R^1 and R^2 be either a bulky substituent (quaternized alkyl or trialkylsilyl) or a hydrogen atom.

10.12.4 Formal Equivalent of the Elusive Bimolecular HDDA Reaction

The need for the HDDA substrates to include a three-atom tether between the 1,3-diyne and the diynophile has been evident throughout the examples described in this chapter. The temperature requirements to achieve convenient reaction times

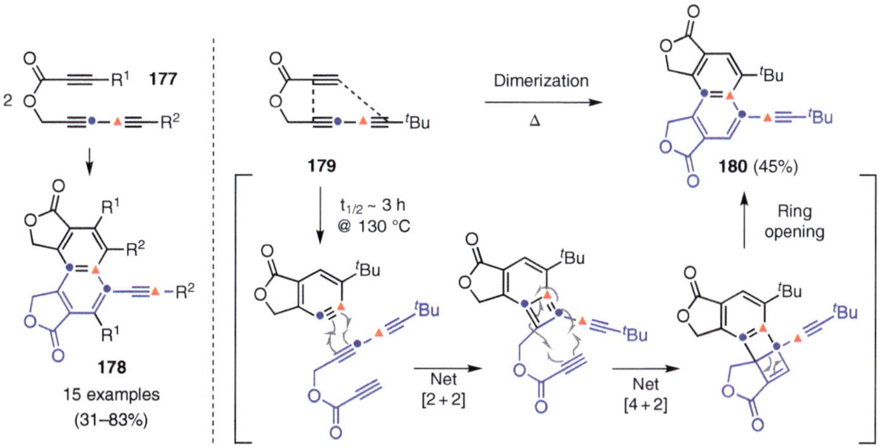

Figure 10.23 Selective self-dimerization reactions of a class of triyne substrates **177**.

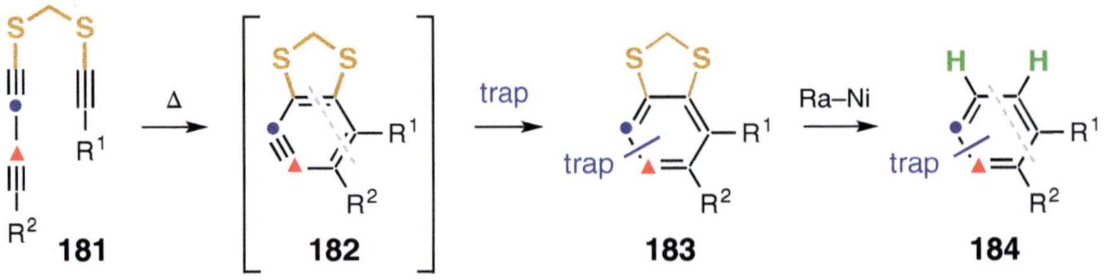

Figure 10.24 Traceless tethers, demonstrated here by reductive cleavage of a dithia-acetal linker, establish, in effect, an intermolecular HDDA reaction.

vary widely, depending largely upon the exact nature of the tether structure. However, in one study, it was shown that the tether can be deleted after having played its role of bringing the reactive diyne/diynophile pair into proximity. This is achieved by the use of sulfur-containing substrates such as **181** (Figure 10.24). The benzyne **182** was efficiently trapped to provide **183**. Removal of the tether by, in this case, Raney nickel reduction provided **184** [66]. This tactic establishes the formal equivalent of an *inter*- or *bi*-molecular HDDA reaction, a process otherwise yet to be realized.

10.12.5 Aspects of Substrate Design

The energy of activation of the initial, rate-limiting cycloisomerization events varies widely, largely dictated by two factors. *First*, the exact nature of the three-atom linker plays a significant role. Second- vs. third-row atoms impact the intervening bond lengths. The presence of an embedded ring or *Z*-alkene reduces the degrees of freedom in the linker, thereby increasing the fraction of polyynes in a reactive conformation or geometry at any instant in time. More sp^2-hydridized atoms tend to increase rates by shortening the distance between the proximal carbon atoms of the diyne and diynophile; partial bond formation between those two atoms is a component of the transition structure of the rate-determining step regardless of whether the mechanism is stepwise-diradical (cf. Figure 10.8) or concerted in nature. *Second*, the nature of the substituents on the terminus of the diyne and, especially, the diynophile impacts the ease of cyclization considerably [21g]. These substituent effects do not follow the classical Diels–Alder dienophile activation paradigm in which frontier molecular orbital mixing energies dictate reactivity (for example and most typically, lowering of the dienophile LUMO by the incorporation of more strongly electron-withdrawing substituents). Instead and to the extent it has been systematically studied, the trend is that substituents more capable of stabilizing the radical character of the putative diradical intermediate (i.e., having greater radical stabilizing energies [RSEs]) formed as the first elementary step in the rate-determining cycloisomerization enhance the rate of reaction (cf. Figure 10.8) [22].

10.12.6 Limitations Imposed by Substituents on the Diynophile

Finally, it should be noted that triyne substrates in which the terminal substituent on a mono-alkynyl diynophile (i.e., R^1 in substrates such as **23** [Figure 10.22b], **177** [Figure 10.23], or **181** [Figure 10.24]) is an arene, alkene, or alkyl group bearing a propargylic C—H bond are generally not suitable benzyne precursors. Arene and alkene groups often enter into TDDA reactions faster than HDDA and alkyl groups are susceptible to faster or at least competitive propargylic ene reactions.

10.13 *Guidelines and Practical Issues*: Experimental Considerations

10.13.1 Pristine Reaction Conditions (and Solvent Choices)

As pointed out earlier, the thermal-only nature of the HDDA cycloisomerization stands in sharp contrast, and is quite complementary, to the classical methods that are used to generate arynes [67]. Those nearly always employ reagents such as strong bases, nucleophiles, oxidants, or reducing agents to activate the benzyne precursor, and they produce byproducts that accompany benzyne generation. As also mentioned earlier, because of this pristine reaction-free environment, previously unobserved or incompatible types of benzyne-trapping reaction and/or new fundamental aspects of the mechanisms of many trapping reactions are able to be observed or learned from these "heat-only" reactions.

Solvent compatibility with the HDDA-benzyne is an issue for consideration. Some solvents are quite reactive and will interfere with an intended trapping reaction. For example, THF, cyclopentane, and cyclooctane are known to donate two hydrogen atoms to the benzyne (cf. Figure 10.14); cyclohexane, decalins [66], and acyclic hydrocarbons also do so, but at a slower rate. Some solvents are only marginally reactive – benzene, toluene, chlorobenzene, and 1,2-dichlorobenzene are examples. These will eventually trap many benzynes in a [4+2] DA reaction, but in many instances trapping reactions with other added trapping agents are faster. Acetonitrile, ethyl acetate, and chloroform are often good choices, although for certain trapping processes, carbanionic character can emerge on the benzyne carbon adjacent to the site of nucleophilic attack and protonation by these weak Brønsted acids is observed, sometimes with productive outcomes. 1,2-Dichloroethane is a particularly inert candidate, rendering it often a good choice as a reaction medium. Because they react readily as nucleophiles, dimethyl sulfoxide (DMSO), dimethyl formamide (DMF), amines, and alcohol solvents are not good choices for HDDA reactions. In this vein, it should be remembered that reagent bottles of chloroform quite often contain substantial amounts of ethanol (0.75% is typical) as an inhibitor (of autoxidation and subsequent phosgene and HCl formation), which can interfere with other desired trapping processes. Finally, because the deuterated variants of several of the solvents just mentioned are conveniently available (i.e., $CDCl_3$, C_6D_6, and CD_3CN), they are well suited for use in direct nuclear magnetic resonance (NMR) monitoring of initial scouting reactions or for monitoring reaction rate, progress, selectivity, and overall reaction cleanliness by direct inspection (e.g., [27]) of the reaction mixture.

10.13.2 Reaction Conditions: Tolerance for Water and Oxygen

Two features of HDDA reactions lend favorably to the relative ease and convenience of performing these transformations. *First*, the rate of reaction of HDDA-generated benzynes with water, a relatively hard nucleophile, is slow. That means that experiments can be carried out in normal laboratory solvents without the need to predry them exceptionally carefully. As mentioned above, the high degree of polarizable character of bent, strained aryne π-bonds are a major contributor to their preference for reaction with soft (i.e., polarizable) nucleophilic trapping reagents. *Second,* because it is rare that any (long-lived) radical intermediates are present on the potential energy surface of the overall reaction, the HDDA cycloisomerization and subsequent trapping events are not affected in any appreciable way by the presence of oxygen. Hence, careful deoxygenation of the reaction environment prior to heating the polyyne to promote reaction is not crucial and components of trapping agents that might otherwise be susceptible to radical polymerization are tolerated. In other words, because of the tolerance of water and oxygen, special care of the reaction conditions is typically not needed.

10.13.3 Reaction Conditions: Temperature, Pressure, and Alkyne Stability

The temperature of the reaction medium required to achieve a convenient rate of reaction is another consideration that factors into the choice of solvent. Because reaction temperatures required to achieve a convenient rate of reaction are often higher than those of the atmospheric boiling points of solvents that are otherwise convenient choices for the reaction medium, it is often advantageous to perform the reaction in a sealed vessel at elevated pressure. The Antoine equation, an empirical correlation between a solvent's vapor pressure and its temperature, is a very useful tool for judging the internal pressure that will be reached. We use an Excel spreadsheet to inform many of our experiments [68]. As one example, we often use chloroform at temperatures above its ambient pressure boiling point (61.75 °C at 760 mm Hg^{-1}). The Antoine equation shows the following P/T relationships: 30 psi@85 °C, 60 psi@112 °C, and 90 psi@130 °C. Incidentally, the vapor pressure of decalin at 300 °C is estimated to be 115 psi. Pyrex glass pressure ratings are dependent on the diameter and wall thickness of the object [69]. As an added precaution, we prefer to use threaded culture tubes with rounded bottoms, likely stronger than the squared-off, flat bottoms of glass vials. Use of screw-caps with an internal Teflon lining, backed by a soft rubberized layer, creates an excellent and inert seal to the lip of the threaded culture tube. Solvent loss is typically unobservable. We once held a capped culture tube, c. 1 cm in diameter and containing c. 2 mL of dichloromethane, at 200 °C (!) for 24 hours. There was no loss of solvent or failure of the glass culture tube; the Antoine equation suggests that the internal pressure was c. 550 psi. Of course, this experiment was performed in a hood and behind a safety blast shield, and we recommend that every experiment done in a sealed vessel, where >1 atm of pressure is anticipated, be done behind a safety shield.

While on the topic of safety, it is appropriate to comment on the thermodynamic stability of alkynes themselves. Because of the inherently low bond dissociation energy of a carbon–carbon triple bond, alkynes carry an inherently high degree of potential energy. This is the very source of the energy that fuels triyne conversion to the highly reactive intermediate benzyne (computed exergonicities of, typically, $\Delta G^{\circ} = -40$ to $-50\,\text{kcal mol}^{-1}$(!)) [21]. Unsubstituted alkynes are kinetically prone to rapid decomposition – for example, ethyne (acetylene) detonates under pressure and 1,3-butadiyne [70] and 1,3,5-hexatriyne (a natural product! [71]) decompose violently as condensed materials. When handling low-molecular-weight, terminal alkynes with additional weak bonds (e.g., 1-bromopropyne), it is advisable to prepare, analyze, and then react them in solution rather than to purify them as neat substances. Finally, differential scanning calorimetry (DSC) analysis of multiyne HDDA substrates of relative low molecular weight, especially if one or both termini are unsubstituted, is recommended for initial assessment of inherent stability properties [72, 73].

10.13.4 Reaction Conditions: Substrate Concentration

Another point to be considered is the substrate concentration of the reaction solution. Because of the ability of the HDDA-benzynes to react further in, necessarily, bimolecular events with functionality in the substrate polyyne (or in some instances the trapped products) the initial substrate concentration can be an issue. In cases where the trapping reaction is quite fast, this is rarely a concern, but some otherwise-desired trapping agents have an inherently slow rate constant for capturing the reactive benzyne. This can sometimes be countered by use of a larger excess of that coreactant, but if the steady-state concentration of the benzyne rises too high because of its slow consumption by the external agent, then undesired reactions can intervene. These are instances where performing the experiment at higher dilution (i.e., lower $[\text{polyyne}]_{t=0}$) can improve the selectivity for the desired trapping pathway. At the other extreme, one study using the highest possible concentration of substrate (i.e., neat) and trapping agent (i.e., intramolecular) showed that the reaction could be performed in the absence of solvent and still provide remarkably efficient conversion to monomeric, trapped product [72]. Additionally, it was observed that the onset temperatures of the exothermic events measured by DSC of the neat samples, correlated, at least qualitatively, with the rates of their HDDA cyclization in solution.

10.13.5 The Value of Half-Life Measurements

Finally, there have been several reports of HDDA reactions being catalyzed by various transition metal complexes [74]. In such studies, it is incumbent to clearly demonstrate that the presence of the metal is actually enhancing the rate of the rate-determining cycloisomerization reaction. We suggest that this be done by measuring the approximate *half-life* for the uncatalyzed, thermal conversion rather than drawing inferences from, e.g., isolated yields of products at time points corresponding to many $t_{1/2}$s. One convenient protocol we have used for measuring the half-life of the rate-determining, unimolecular, HDDA cycloisomerization involved heating a substrate in the high-boiling (and nearly inert) solvent 1,2-dichlorobenzene, removing an aliquot of the reaction solution at various time points (including t_0), diluting *directly* into $CDCl_3$, and integrating the diminution over time of substrate resonances vs. the ^{13}C-coupled satellite resonances of the aromatic protons in the solvent (see Supporting Information of Ref. [14]). Regardless of the analytical method used, it is always helpful to establish the approximate half-life of the HDDA cyclization for any new substrate that is examined. This information adds to the fundamental body of knowledge and understanding associated with the ever-growing compendium [1] of HDDA chemistry.

References

1 Fluegel, L.L. and Hoye, T.R. The hexadehydro-Diels–Alder reaction: Benzyne generation via cycloisomerization of tethered triynes. *Chem. Rev.* 121: 2413–2444.

2 (a) Michael, A. and Bucher, J.E. (1898). Über die Einwirkung von Eissigsäureanhydrid auf Phenylpropiolsäure. *Chem. Zentrblt.*: 731–733. (b) Diels, O. and Alder, K. (1928). Syntheses in the hydroaromatic series. *Justus Liebigs Ann. Chem.* 460: 98–122.

3 Wessig, P. and Müller, G. (2008). The dehydro-Diels–Alder reaction. *Chem. Rev.* 108: 2051–2063.

4 Diau, E.W.-G., Casanova, J., Roberts, J.D., and Zewail, A.H. (2000). Femtosecond observation of benzyne intermediates in a molecular beam: Bergman rearrangement in the isolated molecule. *Proc. Natl. Acad. Sci. U.S.A.* 97: 1376–1379.

5 (a) Chapman, O.L., Mattes, K., McIntosh, C.L. et al. (1973). Photochemical transformations. LII. Benzyne. *J. Am. Chem. Soc.* 95: 6134–6135. (b) Münzel, N. and Schweig, A. (1988). UV/VIS absorption spectrum, geometry and electronic structure of transient *o*-benzyne. *Chem. Phys. Lett.* 147: 192–194.

6 Warmuth, R. (1997). *o*-Benzyne: strained alkyne or cumulene?—NMR characterization in a molecular container. *Angew. Chem. Int. Ed.* 36: 1347–1350.

7 Fields, E.K. and Meyerson, S. (1966). Arynes by pyrolysis of acid anhydrides. *J. Org. Chem.* 31: 3307–3309.

8 (a) Bradley, A.Z. and Johnson, R.P. (1997). Thermolysis of 1,3,8-nonatriyne: evidence for intramolecular [2+4] cycloaromatization to a benzyne intermediate. *J. Am. Chem. Soc.* 119: 9917–9918. (b) Miyawaki, K., Suzuki, R., Kawano, T., and Ueda, I. (1997). Cycloaromatization of a non-conjugated polyenyne system: synthesis of 5*H*-benzo[*d*]fluoreno[3,2-*b*]pyrans via diradicals generated from 1-(2-(4-(2-alkoxyethylphenyl)butan-1,3-diynyl))phenylpetan-2,4-diyn-1-ols and trapping evidence for the 1,2-didehydrobenzene diradical. *Tetrahedron Lett.* 38: 3943–3946.

9 (a) Ueda, I., Sakurai, Y., Kawano, T. et al. (1999). An unprecedented arylcarbene formation in thermal reaction of non-conjugated aromatic enetetraynes and DNA strand cleavage. *Tetrahedron Lett.* 40: 319–322. (b) Torikai, K., Otsuka, Y., Nishimura, M. et al. (2008). Synthesis and DNA cleavage activity of water-soluble non-conjugated thienyl tetraynes. *Bioorg. Med. Chem.* 16:

5441–5451. (c) Bergman, R.G. (1973). Reactive 1,4-dehydroaromatics. *Acc. Chem. Res.* 6: 25–31.

10 Tsui, J.A. and Sterenberg, B.T. (2009). A metal-templated 4+2 cycloaddition reaction of an alkyne and a diyne to form a 1,2-aryne. *Organometallics* 28: 4906–4908.

11 (a) Hoye, T.R., Baire, B., Niu, D. et al. (2012). The hexadehydro-Diels–Alder reaction. *Nature* 490: 208–212. (b) Baire, B., Niu, D., Willoughby, P.H. et al. (2013). Synthesis of complex benzenoids via the intermediate generation of o-benzynes through the hexadehydro-Diels–Alder reaction. *Nat. Protoc.* 8: 501–508.

12 Hoye, T.R., Baire, B., and Wang, T. (2014). Tactics for probing aryne reactivity: mechanistic studies of silicon–oxygen bond cleavage during the trapping of (HDDA-generated) benzynes by silyl ethers. *Chem. Sci.* 5: 545–550.

13 (a) Diamond, O.J. and Marder, T.B. (2017). Methodology and applications of the hexadehydro-Diels–Alder (HDDA) reaction. *Org. Chem. Front.* 4: 891–910. (b) Holden, C. and Greaney, M.F. (2014). The hexadehydro-Diels–Alder reaction: a new chapter in aryne chemistry. *Angew. Chem. Int. Ed.* 53: 5746–5749.

14 Woods, B.P., Baire, B., and Hoye, T.R. (2014). Rates of hexadehydro-Diels–Alder (HDDA) cyclizations: impact of the linker structure. *Org. Lett.* 16: 4578–4581.

15 Yun, S.Y., Wang, K., Lee, N. et al. (2013). Alkane C–H insertion by aryne intermediates with a silver catalyst. *J. Am. Chem. Soc.* 135: 4668–4671.

16 Shen, H., Xiao, X., Haj, M. et al. (2018). BF_3-promoted, carbene-like, C–H insertion reactions of benzynes. *J. Am. Chem. Soc.* 140: 15616–15620.

17 Karmakar, R., Wang, K.-P., Yun, S.Y. et al. (2016). Hydrohalogenative aromatization of multiynes promoted by ruthenium alkylidene complexes. *Org. Biomol. Chem.* 14: 4782–4788.

18 Karmakar, R. and Lee, D. (2016). Total synthesis of selaginpulvilin C and D relying on in situ formation of arynes and their hydrogenation. *Org. Lett.* 18: 6105–6107.

19 Wang, T. and Hoye, T.R. (2016). Hexadehydro-Diels–Alder (HDDA)-enabled carbazolyne chemistry: single step, de novo construction of the pyranocarbazole core of alkaloids of the *Murraya koenigii* (curry tree) family. *J. Am. Chem. Soc.* 138: 13870–13873.

20 (a) Heaney, H. and Jablonski, J.M. (1968). Reactions of arynes in the synthesis of 2*H*-chromens. *Chem. Commun.* 1139–1139. (b) Wang, T., Oswood, C.J., and Hoye, T.R. (2017). Trapping of hexadehydro-Diels–Alder benzynes with exocyclic, conjugated enals as a route to fused spirocyclic benzopyran motifs. *Synlett* 28: 2933–2935.

21 For reports that include computational studies of both the forward (i.e., triynes to benzyne) and back (i.e., benzyne pyrolysis back to triynes) HDDA reactions, see: (a) Deng, W.-Q., Han, K.-L., Zhan, J.-P., and He, G.-Z. (1998). Ab initio and RRKM calculations of *o*-benzyne pyrolysis. *Chem. Phys. Lett.* 288: 33–36. (b) Moskaleva, L.V., Madden, L.K., and Lin, M.C. (1999). Unimolecular isomerization/decomposition of ortho-benzyne: ab initio MO/statistical theory study. *Phys. Chem. Chem. Phys.* 1: 3967–3972. (c) Diau, E.W., Casanova, J., Roberts, J.D., and Zewail, A.H. (2000). Femtosecond observation of benzyne intermediates in a molecular beam: Bergman rearrangement in the isolated molecule. *Proc. Natl. Acad. Sci. U.S.A* 97: 1376–1379. (d) Zhang, X., Maccarone, A.T., Nimlos, M.R. et al. (2007). Unimolecular thermal fragmentation of ortho-benzyne. *J. Chem. Phys.* 126: 044312(-1)–044312(-20). (e) Cahill, K.J., Ajaz, A., and Johnson, R.P. (2010). New thermal routes to ortho-benzyne. *Aust. J. Chem.* 63: 1007–1012. (f) Ajaz, A., Bradley, A.Z., Burrell, R.C. et al. (2011). Concerted vs stepwise mechanisms in dehydro-Diels–Alder reactions. *J. Org. Chem.* 76: 9320–9328. (g) Liang, Y., Hong, X., Yu, P., and Houk, K.N. (2014). Why alkynyl substituents dramatically accelerate hexadehydro-Diels–Alder (HDDA) reactions: stepwise mechanisms of HDDA cycloadditions. *Org. Lett.* 16: 5702–5705. (h) Ghigo, G., Maranzana, A., and Tonachini, G. (2014). *o*-Benzyne fragmentation and isomerization pathways: a CASPT2 study. *Phys. Chem. Chem. Phys.* 16: 23944–23951. (i) Marell, D.J., Furan, L.R., Woods, B.P. et al. (2015). Mechanism of the intramolecular hexadehydro-Diels–Alder reaction. *J. Org. Chem.* 80: 11744–11754. (j) Skraba-Joiner, S.L., Johnson, R.P., and Agarwal, J. (2015). Dehydropericyclic reactions: symmetry-controlled routes to strained reactive intermediates. *J. Org. Chem.* 80: 11779–11787. (k) Chen, M., He, C.Q., and Houk, K.N. (2019). Mechanism and regioselectivity of an unsymmetrical hexadehydro-Diels–Alder (HDDA) reaction. *J. Org. Chem.* 84: 1959–1963.

22 Wang, T., Niu, D., and Hoye, T.R. (2016). The hexadehydro-Diels–Alder (HDDA) cycloisomerization reaction proceeds by a stepwise mechanism. *J. Am. Chem. Soc.* 138: 7832–7835.

23 (a) Karmakar, R., Mamidipalli, P., Yun, S.Y., and Lee, D. (2013). Alder-ene reactions of arynes. *Org. Lett.* 15: 1938–1941. (b) Gupta, S., Lin, Y., Xia, Y. et al. (2019). Alder-ene reactions driven by high steric strain and bond angle distortion to form benzocyclobutenes. *Chem. Sci.* 10: 2212–2217. (c) Gupta, S., Xie, P., Xia, Y., and Lee, D. (2018). Reactivity of arynes toward functionalized alkenes: Intermolecular Alder-ene vs. addition reactions. *Org. Chem. Front.* 5: 2208–2213. (d) Wang, K., Yun, S.Y., Mamidipalli, P., and Lee, D. (2013). Silver-mediated fluorination, trifluoromethylation, and trifluoromethylthiolation of arynes. *Chem. Sci.* 4: 3205–3211. (e) Karmakar, R., Yun, S.Y., Wang, K., and Lee, D. (2014). Regioselectivity in the nucleophile trapping of arynes: the electronic and steric effects of nucleophiles and substituents. *Org. Lett.* 16: 6–9. (f) Karmakar, R., Yun, S.Y., Chen, J. et al. (2015). Benzannulation of triynes to generate functionalized arenes by spontaneous incorporation of nucleophiles. *Angew. Chem. Int. Ed.* 54: 6582–6586. (g) Ghorai, S. and Lee, D. (2017). Aryne formation via the hexadehydro Diels–Alder reaction and their Ritter-type transformations catalyzed by a cationic silver complex. *Tetrahedron* 73: 4062–4069. (h) Karmakar, R., Le, A., Xie, P. et al. (2018). Aryne-mediated dearomatization of arylsulfonyl group. *Org. Lett.* 20: 4168–4172. (i) Ghorai, S. and Lee, D. (2019). Synthesis of imides, imidates, amidines, and amides by intercepting the aryne–isocyanide adduct with weak nucleophiles. *Org. Lett.* 21: 7390–7393.

24 Wittig, G. and Pohmer, L. (1955). Intermediäre Bildung von dehydrobenzol (cyclohexadienin). *Angew. Chem. Int. Ed.* 67: 348–348.

25 Kawano, T., Inai, H., Miyawaki, K., and Ueda, I. (2005). Synthesis of indenothiophenone derivatives by cycloaromatization of non-conjugated thienyl tetraynes. *Tetrahedron Lett.* 46: 1233–1236.

26 Chen, J., Baire, B., and Hoye, T.R. (2014). Cycloaddition reactions of azide, furan, and pyrrole units with benzynes generated by the hexadehydro-Diels–Alder (HDDA) reaction. *Heterocycles* 88: 1191–1200.

27 Pogula, V.D., Wang, T., and Hoye, T.R. (2015). Intramolecular [4+2] trapping of a hexadehydro-Diels–Alder (HDDA) benzyne by tethered arenes. *Org. Lett.* 17: 856–859.

28 Hu, Q., Li, L., Yin, F. et al. (2017). Fused multifunctionalized isoindole-1,3-diones via the coupled oxidation of imidazoles and tetraynes. *RSC Adv.* 7: 49810–49816.

29 Ross, S.P. and Hoye, T.R. (2017). Reactions of hexadehydro-Diels–Alder benzynes with structurally complex multifunctional natural products. *Nat. Chem.* 9: 523–530.

30 Zhang, J., Niu, D., Brinker, V.A., and Hoye, T.R. (2016). The phenol-ene reaction: biaryl synthesis via trapping reactions between HDDA-generated benzynes and phenolics. *Org. Lett.* 18: 5596–5599.

31 Mayr, H. and Ofial, A.R. (2006). The reactivity-selectivity principle: an imperishable myth in organic chemistry. *Angew. Chem. Int. Ed.* 45: 1844–1854.

32 Bhunia, A., Yetra, S.R., and Biju, A.T. (2012). Recent advances in transition-metal-free carbon–carbon and carbon–heteroatom bond-forming reactions using arynes. *Chem. Soc. Rev.* 41: 3140–3152.

33 Ross, S.P. and Hoye, T.R. (2018). Multiheterocyclic motifs via three-component reactions of benzynes, cyclic amines, and protic nucleophiles. *Org. Lett.* 20: 100–103.

34 Arora, S., Zhang, J., Pogula, V., and Hoye, T.R. (2019). Reactions of thermally generated benzynes with six-membered *N*-heteroaromatics: Pathway and product diversity. *Chem. Sci.* 10: 9069–9076.

35 Chen, J., Palani, V., and Hoye, T.R. (2016). Reactions of HDDA-derived benzynes with sulfides: mechanism, modes, and three-component reactions. *J. Am. Chem. Soc.* 138: 4318–4321.

36 Niu, D., Willoughby, P.H., Baire, B. et al. (2013). Alkane desaturation by concerted double hydrogen atom transfer to benzyne. *Nature* 501: 531–534.

37 Willoughby, P.H., Niu, D., Wang, T. et al. (2014). Mechanism of the reactions of alcohols with *o*-benzynes. *J. Am. Chem. Soc.* 136: 13657–13665.

38 (a) Hamura, T., Ibusuki, Y., Sato, K. et al. (2003). Strain-induced regioselectivities in reactions of benzyne possessing a fused four-membered ring. *Org. Lett.* 5: 3551–3554. (b) Cheong, P.H.Y., Paton, R.S., Bronner, S.M. et al. (2010). Indolyne and aryne distortions and nucleophilic regioselectivities. *J. Am. Chem. Soc.* 132: 1267–1269. (c) Garr, A.N., Luo, D., Brown, N. et al. (2010). Experimental and theoretical investigations into the unusual regioselectivity of 4,5-, 5,6-, and 6,7-indole aryne cycloadditions. *Org. Lett.* 12: 96–99. (d) Bickelhaupt, F.M. and Houk, K.N. (2017). Analyzing reaction rates with the distortion/interaction-activation strain model. *Angew. Chem. Int. Ed.* 56: 10070–10086.

39 Niu, D. and Hoye, T.R. (2014). The aromatic ene reaction. *Nat. Chem.* 6: 34–40.

40 Heine, H.W., Hoye, T.R., Williard, P.G., and Hoye, R.C. (1973). Diaziridines II. The addition of diaziridines to electrophilic acetylenes. *J. Org. Chem.* 38: 2984–2988.

41 Arora, S., Palani, V., and Hoye, T.R. (2018). Reactions of diaziridines with benzynes give N-arylhydrazones. *Org. Lett.* 20: 8082–8085.

42 Miyawaki, K., Kawano, T., and Ueda, I. (1998). Multiple cycloaromatization of novel aromatic enediynes bearing a triggering device on the terminal acetylene carbon. *Tetrahedron Lett.* 39: 6923–6926.

43 Nobusue, S., Yamane, H., Miyoshi, H., and Tobe, Y. (2014). [4.2](2,2′)(2,2′)Biphenylophanetriyne: a twisted biphenylophane with a highly distorted diacetylene bridge. *Org. Lett.* 16: 1940–1943.

44 Zhang, M., Shan, W., Chen, Z. et al. (2015). Diels-Alder reactions of arynes in situ generated from DA reaction between bis-1,3-diynes and alkynes. *Tetrahedron Lett.* 56: 6833–6838.

45 Jung, K. and Koreeda, M. (1989). Synthesis of 1,4-, 2,4-, and 3,4-dimethylphenanthrenes: a novel deoxygenation of arene 1,4-endoxides with trimethylsilyl iodide. *J. Org. Chem.* 54: 5667–5675.

46 Xu, F., Hershey, K.W., Holmes, R.J., and Hoye, T.R. (2016). Blue-emitting arylalkynyl naphthalene derivatives via a hexadehydro-Diels–Alder (HDDA) cascade reaction. *J. Am. Chem. Soc.* 138: 12739–12742.

47 Xu, F., Xiao, X., and Hoye, T.R. (2016). Reactions of HDDA-derived benzynes with perylenes: rapid construction of novel polycyclic aromatic compounds. *Org. Lett.* 18: 5636–5639.

48 Xiao, X. and Hoye, T.R. (2018). The domino hexadehydro-Diels–Alder reaction transforms polyynes to benzynes to naphthynes to anthracynes to tetracynes (and beyond?). *Nat. Chem.* 10: 838–844.

49 Mitake, A., Nagai, R., Sekine, A. et al. (2019). Consecutive HDDA and TDDA reactions of silicon- tethered tetraynes for the synthesis of dibenzosilole-fused polycyclic compounds and their unique reactivity. *Chem. Sci.* 10: 6715–6720.

50 Heaney, H. and Jablonski, J.M. (1968). Reactions of arynes in the synthesis of 2*H*-chromens. *Chem. Commun.* 1139–1139.

51 (a) Meng, X., Lv, S., Cheng, D. et al. (2017). Fused multifunctionalized chromenes from tetraynes and α,β-unsaturated aldehydes. *Chem. Eur. J.* 23: 6264–6271. (b) Wang, T., Oswood, C.J., and Hoye, T.R. (2017). Trapping of hexadehydro-Diels–Alder benzynes with exocyclic, conjugated enals as a route to fused spirocyclic benzopyran motifs. *Synlett* 28: 2933–2935.

52 Hu, Y., Hu, Y., Hu, Q. et al. (2017). Direct access to fused salicylaldehydes and salicylketones from tetraynes. *Chem. Eur. J.* 23: 4065–4072.

53 Hu, Y., Ma, J., Li, L. et al. (2017). Fused multifunctionalized dibenzoselenophenes from tetraynes. *Chem. Commun.* 53: 1542–1545.

54 Watanabe, T., Curran, D.P., and Taniguchi, T. (2015). Hydroboration of arynes formed by hexadehydro-Diels–Alder cyclizations with *N*-heterocyclic carbene boranes. *Org. Lett.* 17: 3450–3453.

55 Xiao, X., Wang, T., Xu, F., and Hoye, T.R. (2018). Cu(I)-Mediated bromoalkynylation and hydroalkynylation reactions of unsymmetrical benzynes: complementary modes of addition. *Angew. Chem. Int. Ed.* 57: 16564–16568.

56 Thangaraj, M., Bhojgude, S.S., Maneb, M.V., and Biju, A.T. (2016). From insertion to multicomponent coupling: temperature dependent reactions of arynes with aliphatic alcohols. *Chem. Commun.* 52: 1665–1668.

57 Zhang, J. and Hoye, T. (2019). R. Divergent reactivity during the trapping of benzynes by glycidol analogs: ring cleavage via pinacol-like rearrangements vs. oxirane fragmentations. *Org. Lett.* 21: 2615–2619.

58 Xu, F., Xiao, X., and Hoye, T.R. (2017). Photochemical hexadehydro-Diels–Alder (hν-HDDA) reaction. *J. Am. Chem. Soc.* 139: 8400–8403.

59 Shi, J., Li, Y., and Li, Y. (2017). Aryne multifunctionalization with benzdiyne and benztriyne equivalents. *Chem. Soc. Rev.* 46: 1707–1719.

60 Xiao, X. and Hoye, T.R. (2019). A one-pot, three-aryne cascade strategy for naphthalene formation from 1,3-diynes and 1,2-benzdiyne equivalents. *J. Am. Chem. Soc.* 141: 9813–9818.

61 Wang, T., Naredla, R.R., Thompson, S.K., and Hoye, T.R. (2016). The pentadehydro-Diels–Alder reaction. *Nature* 532: 484–488.

62 Xiao, X., Woods, B.P., Xiu, W., and Hoye, T.R. (2018). Benzocyclobutadienes: an unusual mode of access reveals unusual modes of reactivity. *Angew. Chem. Int. Ed.* 57: 9901–9905.

63 Thompson, S.K. and Hoye, T.R. (2019). The aza-hexadehydro-Diels–Alder reaction. *J. Am. Chem. Soc.* 141: 19575–19580.

64 Niu, D., Wang, T., Woods, B.P., and Hoye, T.R. (2014). Dichlorination of (hexadehydro-Diels–Alder generated) benzynes and a protocol for interrogating the kinetic order of bimolecular aryne trapping reactions. *Org. Lett.* 16: 254–257.

65 Xu, F., Xiao, X., and Hoye, T.R. (2017). Photochemical hexadehydro-Diels–Alder reaction. *J. Am. Chem. Soc.* 139: 8400–8403.

66 Pierson Smela, M. and Hoye, T.R. (2018). A traceless tether strategy for achieving formal intermolecular hexadehydro-Diels–Alder reactions. *Org. Lett.* 20: 5502–5505.

67 For a summary of these methods, see, for example: (a) Tadross, P.M. and Stoltz, B.M. (2012). A comprehensive history of arynes in natural product total synthesis. *Chem. Rev.* 112: 3550–3577. (b) Gampe, C.M. and Carreira, E.M. (2012). Arynes and cyclohexyne in natural product synthesis. *Angew. Chem. Int. Ed.* 51: 3766–3778. (c) Kitamura, T. (2010). Synthetic methods for the generation and preparative application of benzyne. *Aust. J. Chem.* 63: 987–1001.

68 See “Antoine Equation Excel file” at http://hoye.chem.umn.edu/teaching (accessed 3 January, 2021).

69 http://www.public.asu.edu/~aomdw/MAIP/index.html (accessed 3 January 2021).

70 Maretina, I.A. and Trofimov, B.A. (2000). Diacetylene: a candidate for industrially important reactions. *Russ. Chem. Rev.* 69: 591–608.

71 Glen, A.T., Hutchinson, S.A., and McCorkindale, N.J. (1966). Hexa-1, 3, 5-triyne – a metabolite of fomes annosus. *Tetrahedron Lett.* 4223–4225.

72 Woods, B.P. and Hoye, T.R. (2014). Differential scanning calorimetry (DSC) as a tool for probing the reactivity of polyynes relevant to hexadehydro-Diels-Alder (HDDA) cascades. *Org. Lett.* 16: 6370–6373.

73 https://www.yumpu.com/en/document/read/3619825/dsc-differential-scanning-c-l-iaormetry (accessed 3 January 2021).

74 (a) Vandavasi, J.K., Hu, W.-P.P., Hsiao, C.-T.T. et al. (2014). A new approach for fused isoindolines via hexadehydro-Diels–Alder reaction (HDDA) by Fe(0) catalysis. *RSC Adv.* 4: 57547–57552. (b) Ghorai, S. and Lee, D. (2018). Silver-catalyzed cycloaddition reactions. In: *Silver Catalysis in Organic Synthesis* (eds. C.-J. Li and X. Bi), 33–83. New York: Wiley-VCH. (c) Karmakar, R., Wang, K.-P.P., Yun, S.Y. et al. (2016). Hydrohalogenative aromatization of multiynes promoted by ruthenium alkylidene complexes. *Org. Biomol. Chem.* 14: 4782–4788. (d) Sieck, C., Tay, M.G., Thibault, M.-H. et al. (2016). Reductive coupling of diynes at rhodium gives fluorescent rhodacyclopentadienes or phosphorescent rhodium 2,2′-biphenyl complexes. *Chem. Eur. J.* 22: 10523–10532.

11

Applications of Benzynes in Natural Product Synthesis

Hiroshi Takikawa[1] *and Keisuke Suzuki*[2]

[1] *Kyoto University, Graduate School of Pharmaceutical Sciences, Yoshidashimoadachi-cho, Sakyo-ku, Kyoto 606-8501, Japan*
[2] *Tokyo Institute of Technology, Department of Chemistry, 2-12-1, O-okayama, Meguro-ku, Tokyo 152-8551, Japan*

11.1 Introduction

Complex molecular architectures embedded in bioactive natural products give us various synthetic issues, requiring innovative approaches for (i) expeditious skeletal construction, (ii) introduction/management of sensitive functionalities, and (iii) stereochemical control. Complex polycyclic natural products belong to such challenging targets, which provide high levels of synthetic difficulties. As one of the tactical bases to cope with such problems, potential utility of benzynes [1] has been recognized from the early days, expecting for their inherently high reactivity and unique characteristics. In recent years, we have witnessed significant advances along these lines, as reviewed in some recent articles [2].

This chapter is an overview of benzyne chemistry in natural product synthesis [3]. Examples are classified by the reaction patterns, i.e. additions of nucleophiles (Section 11.3), addition–fragmentation reactions (Section 11.4), [4+2] cycloadditions (Section 11.5), [2+2] cycloadditions (Section 11.6), and ene reactions (Section 11.7). Recent advances, including the multiple use of benzyne, transition metals in benzyne reactions, and a de novo generation of benzynes from triyne precursors, are also outlined (Section 11.8).

11.2 General Reactivities of Benzynes

The frontier orbital of benzyne is the unusually low-lying LUMO, leading to the high electrophilicity (Figure 11.1) [1, 2]. Nonanionic nucleophiles, such as tertiary amines and even ethers, may add to benzynes (Figure 11.1, (1)) [5]. The resulting aryl anions can be protonated, giving simple adducts, but trapping with other electrophiles (E^+) leads to vicinal difunctionalization [6], reminiscent of the 1,4-addition

Modern Aryne Chemistry, First Edition. Edited by Akkattu T. Biju.

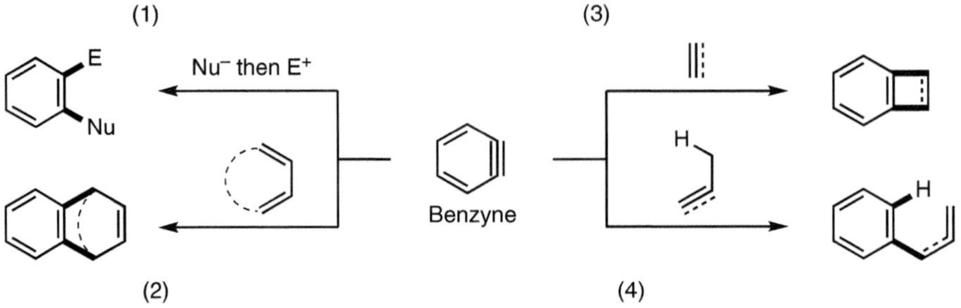

Figure 11.1 Four typical reactivities of benzynes: (1) nucleophilic addition; (2) [4+2] cycloaddition; (3) [2+2] cycloaddition; and (4) ene reaction. Source: Based on Kametani and Ogasawara [4].

to α,β-enones and enolate trapping. In addition, the formal triple bond of a benzyne has the ability to undergo pericyclic reactions, including [4+2] cycloadditions, [2+2] cycloadditions, and ene reactions (Figure 11.1, (2), (3), (4)). Interestingly, benzynes undergo *thermal* [2+2] cycloaddition reaction with alkenes or alkynes [7], normally a symmetry-forbidden process, which can be explained by the frontier molecular orbital theory [8].

For the benzyne reactions, the choice of the benzyne precursor is critical. Depending on the reactivity of the reaction partner, one needs to choose a suitable precursor and associated conditions. Table 11.1 lists four representative methods for benzyne generation, showing the precursors and the typical reaction conditions used in natural product synthesis. Aryl halides are the most conventionally used precursor, which are treated with bases to generate benzyne via deprotonation followed by β-elimination (no. 1). The halogen–metal exchange reactions are also exploited, using aryl halides possessing an ortho-leaving group (no. 2). Another means is the fluoride-induced desilylation of arylsilanes with an ortho-leaving group (no. 3). Thermolysis of benzenediazonium-2-carboxylates was frequently used in the early days (no. 4), although special precautions are required owing to the highly explosive properties.

11.3 Strategies Based on Nucleophilic Additions to Benzynes

Among the various reactivities of benzynes, additions of nucleophiles have been most extensively utilized for natural product synthesis. This section will describe related topics in three parts categorized by the difference in the nucleophilic atoms, nitrogen nucleophiles (Section 11.3.1), oxygen nucleophiles (Section 11.3.2), carbanions (Section 11.3.3.1), and enamines or enolates (Section 11.3.3.2). Section 11.4 will deal with the addition–fragmentation reaction of benzynes, a sequential process initiated by the nucleophilic addition.

Table 11.1 Benzyne precursors used in natural product synthesis.

No.	Benzyne precursor	Reagent	Applications in natural product syntheses
1	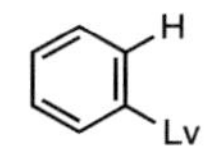 Lv = F, Cl, Br, I	Amide base RLi	Nucleophilic addition [4+2] Cycloaddition [2+2] Cycloaddition Ene reaction
2	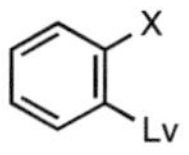 X = Br, I; Lv = Br, OSO_2R	RLi, RMgX	[4+2] Cycloaddition [2+2] Cycloaddition
3	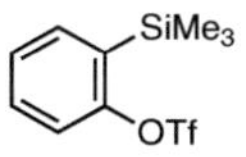	Fluoride ion	Nucleophilic addition [4+2] Cycloaddition [2+2] Cycloaddition Transition-metal-catalyzed reaction Ene reaction
4	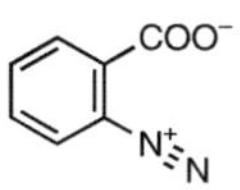	(Heat)	Nucleophilic addition [4+2] Cycloaddition

11.3.1 Additions of Nitrogen Nucleophiles

Addition of nitrogen nucleophiles to benzynes is one of key methods for constructing aza-cycles in alkaloid scaffolds. An early example is the seminal synthesis of cryptowoline by Kametani and Ogasawara in 1967, featuring an intramolecular benzyne addition of a nitrogen nucleophile for constructing the indoline skeleton of the dibenzopyrrocoline alkaloids (Scheme 11.1) [4]. By the action of sodium amide in liquid ammonia, aryl chloride **1** undergoes 1,2-elimination to generate benzyne **2**. The amino group is concurrently deprotonated to give an amido anion, which adds to the internal benzyne to give indoline **3**.

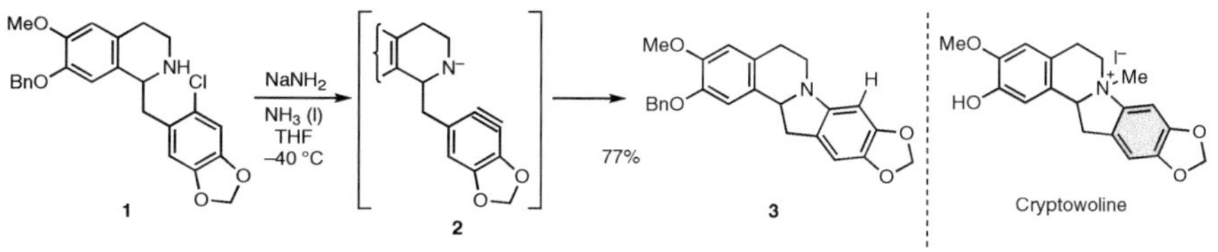

Scheme 11.1 Syntheses of dibenzopyrrocoline alkaloids [4].

In 1922, Meyers and coworker reported the asymmetric total synthesis of (+)-cryptaustoline (Scheme 11.2) [9]. The key transformation is a diastereoselective alkylation of isoquinoline **4** with benzyl bromide **5** by the assistance of a chiral amidine based auxiliary, which is followed by intramolecular amination of benzyne **6**.

It is notable that even a nonanionic nitrogen nucleophile within an amidine can attack the benzyne species.

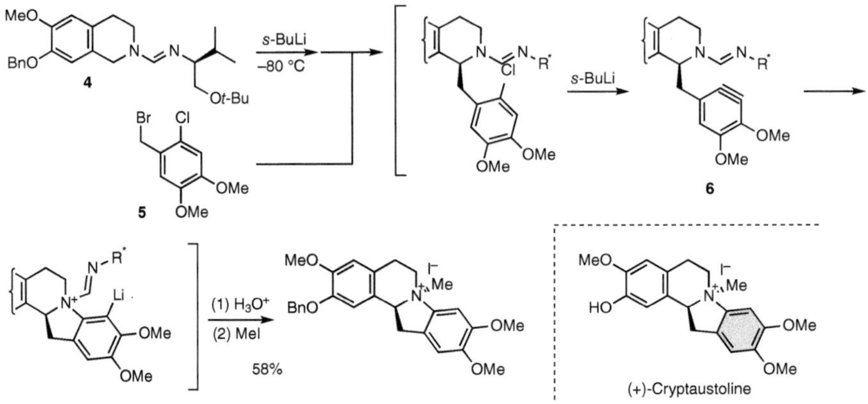

Scheme 11.2 Asymmetric synthesis of cryptaustoline. Source: Based on Sielecki and Meyers [9].

Another example of an intramolecular benzyne amination is the Iwao synthesis of the makaluvamines, a class of the pyrroloiminoquinone alkaloid (Scheme 11.3a) [10]. Construction of the pyrroloquinoline skeleton is effected by using 4-chloroindole **7** as the benzyne precursor and LiN*i*-PrCy as the base. A similar approach is employed by Tokuyama and coworkers [11], using 4-iodoindoline **8** with an *N*-Boc-aminoethyl group is treated with LiTMP at −78 °C (Scheme 11.3b). Importantly, the in situ-generated aryl anion can be trapped with an electrophile. Indolyne **9** undergoes an intramolecular attack of the carbamate anion and an in situ chlorination by treatment with $Cl(CCl_2)_2Cl$, giving the corresponding chloride **10**, allowing access to halo congeners, such as isobatzelline C.

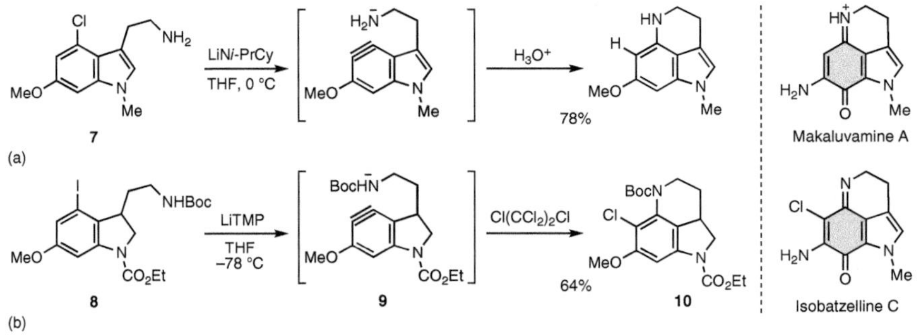

Scheme 11.3 (a) Syntheses of makaluvamine A. Source: Based on Iwao et al. [10] and (b) isobatzelline C. Source: Based on Oshiyama et al. [11].

Tokuyama and coworkers expanded such in situ-trapping protocol to the cross-coupling reactions, as illustrated by the total synthesis of dictyodendrin A, a telomerase-inhibitory marine natural product (Scheme 11.4) [12]. Treatment of aryl dibromide **11** with $Mg(TMP)_2 \cdot 2LiBr$ allows the formation of arylmagnesium intermediate **12** via addition of the *N*-Boc-amido anion to the generated benzyne. After the Mg → Cu transmetalation, Pd-catalyzed coupling with iodo-*p*-methoxybenzene gives arylindoline **13**. Other bases such as LiTMP and $[Me_2Zn(TMP)]Li$ are less effective, and $Mg(TMP)_2 \cdot 2LiBr$ proved to be a suitable base.

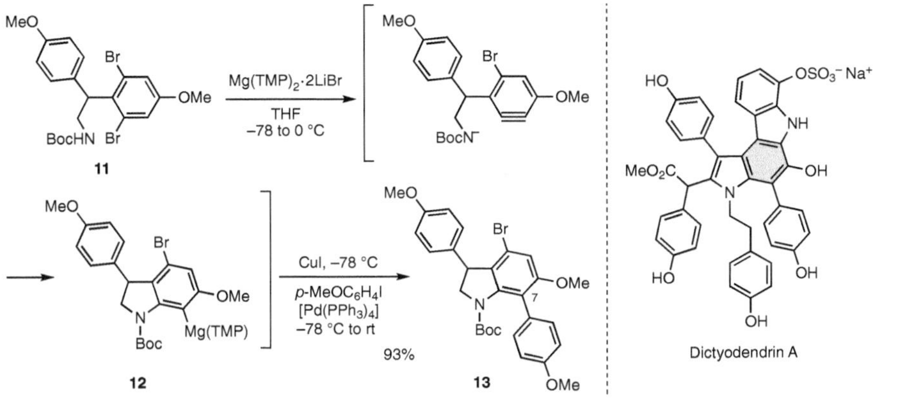

Scheme 11.4 Syntheses of dictyodendrin A. Source: Based on Okano et al. [12].

The intermolecular versions of the benzyne reactions have also been used in alkaloid syntheses, as demonstrated by the Watanabe synthesis of acronycine, an antitumor acridone alkaloid (Scheme 11.5a) [13]. The annulation of aminoester **14** with bromochromene **15** is achieved via addition of amido anion **16** to the benzyne **17** followed by cyclocondensation, giving acronycine. Recently, Argade and coworker reported the related reaction of quinozalinone **18** and silylaryl triflate **19** in the syntheses of quinazolinone alkaloids (Scheme 11.5b). Anion **20**, derived from **18**, undergoes an intermolecular addition to benzyne **21**, and subsequent cyclocondensation gives tryptanthrin, which is further converted to cruciferane [14].

An intermolecular addition of alkylamines to benzyne is less common. In 2011, Garg and coworkers reported the total synthesis of indolactam V, an alkaloid with an ansa bridge (Scheme 11.6) [15]. Aminoindole **25** is synthesized by the regioselective intermolecular addition of dipeptide **22** to indolyne **24**, generated from silylaryl triflate **23** and CsF. The enhanced electrophilicity of the C-4 position in **24** is rationalized by the geometrical distortion effect by the bromo group [16].

A recent example of an intermolecular reaction of alkylamines can be found in the Zhu synthesis of hinckdentine A via the benzyne annulation of an α-aminoimide (Scheme 11.7) [17]. Treatment of silylaryl triflate **19** with CsF generates benzyne, which undergoes nucleophilic addition of cyclic α-amino imide **26**, followed by cyclocondensation to give indolinone **27**. The reaction of the corresponding α-amino ester was not fruitful.

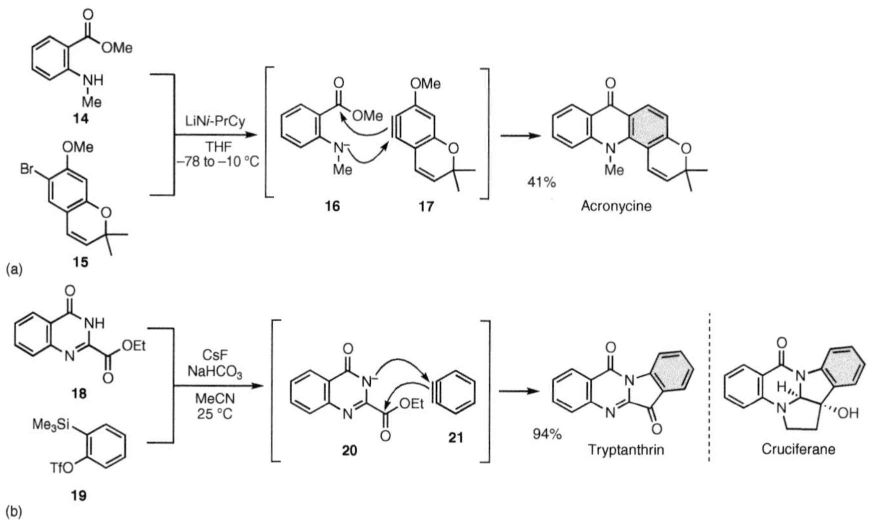

Scheme 11.5 (a) Syntheses of acronycine. Source: Based on Watanabe et al. [13] and (b) quinazolinone alkaloids. Source: Based on Vaidya and Argade [14].

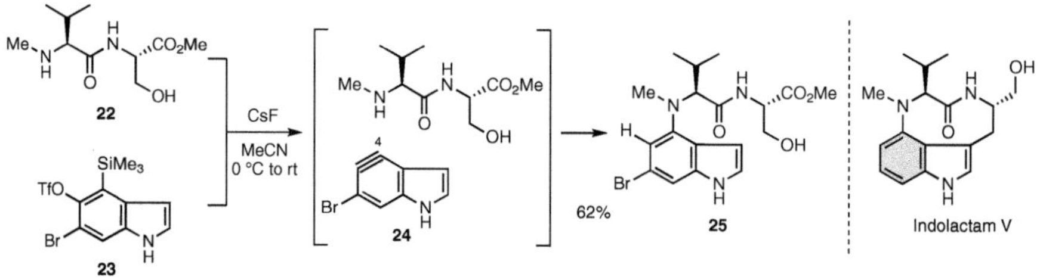

Scheme 11.6 Synthesis of indolactam V. Source: Based on Bronner et al. [15].

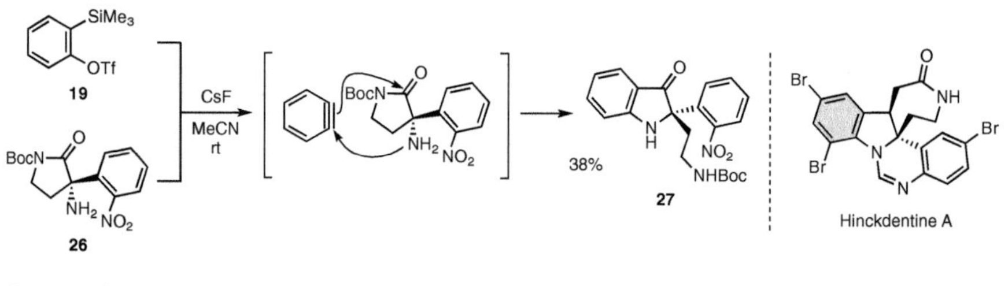

Scheme 11.7 Synthesis of hinckdentine A. Source: Based on Torres-Ochoa et al. [17].

11.3.2 Additions of Oxygen Nucleophiles

This part describes the synthetic strategies based on the addition of oxygen nucleophiles to benzynes. As benzynes are soft electrophiles, addition of hard nucleophiles is relatively difficult. Indeed, oxygen nucleophiles have been little applied to natural product syntheses, which stand in contrast to nitrogen nucleophiles.

One of the few examples is Stoltz's asymmetric synthesis of liphagal, a meroterpenoid (Scheme 11.8) [18]. Benzyne **29**, generated from aryl bromide **28**, undergoes nucleophilic addition of an internal alkoxide, giving dihydrobenzofuran **30**. Note that the high reactivity of the benzyne enables the bond formation at a congested position. By contrast, related attempts via a Cu-catalyzed cyclization failed [19].

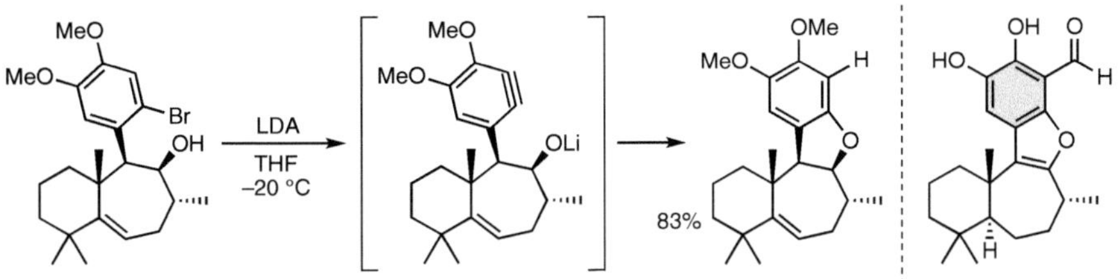

28 29 30

Scheme 11.8 Synthesis of liphagal. Source: Based on Day et al. [18].

Larock and coworker reported the construction of xanthone structures via the benzyne reactions with phenolates [20], which is exploited by Moody and coworkers for the total synthesis of toxyloxanthone B (Scheme 11.9) [21]. Benzyne **33**, generated from silylaryl triflate **31**, reacts with the phenolate **32** in a regioselective manner, giving xanthone **34**, after an intramolecular condensation reaction. Use of NaH suppresses formation of the byproduct derived from simple protonation after the addition stage.

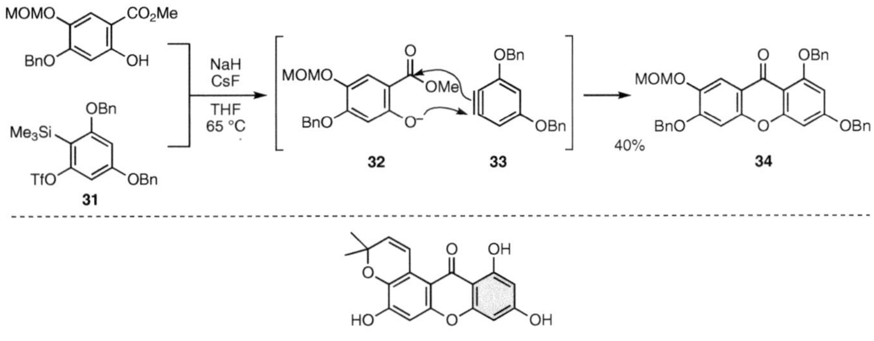

31 32 33 34

Scheme 11.9 Synthesis of toxyloxanthone B. Source: Based on Giallombardo et al. [21].

Knight et al. reported a formal synthesis of vitamin E (Scheme 11.10) [22]. Aminotriazole **35** allows the reaction under nearly neutral conditions, demonstrating the addition of a nonionized alcohol. After Boc-deprotection in **35**, exposure to NIS under neutral conditions generates benzyne **36**, inducing a nucleophilic addition of an internal hydroxy group. The resulting aryl anion is trapped by NIS, giving aryl iodide **37**.

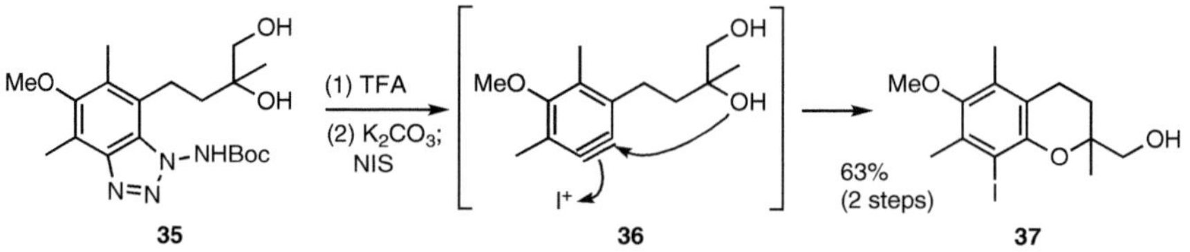

Scheme 11.10 Synthesis of vitamin E core. Source: Knight and Qing [22a]; Knight and Xu [22b].

11.3.3 Addition of Carbon Nucleophiles

Benzyne addition of carbon nucleophiles provides opportunities for constructing densely substituted polycyclic natural products. The carbon nucleophiles are categorized into (a) carbanions, and (b) π-nucleophiles, such as enamines and enolates.

11.3.3.1 Carbanions

An early example is Oppolzer's synthesis of chelidonine, featuring a benzocyclobutene formation via the benzyne addition of an internal carbanion (Scheme 11.11) [23]. Upon treatment of bromobenzene **38** with KNH_2, the α-cyano carbanion attacks the internal benzyne, generated in situ, giving benzocyclobutene **39**. After assembly of the D ring unit to afford alkyne **40**, heating of **40** triggers pericyclic ring opening to generate quinodimethane **41** that undergoes an intramolecular [4+2] cycloaddition to give hexacyclic compound **42**. Similar reactions were extensively used for various steroid syntheses by Kametani et al. [24].

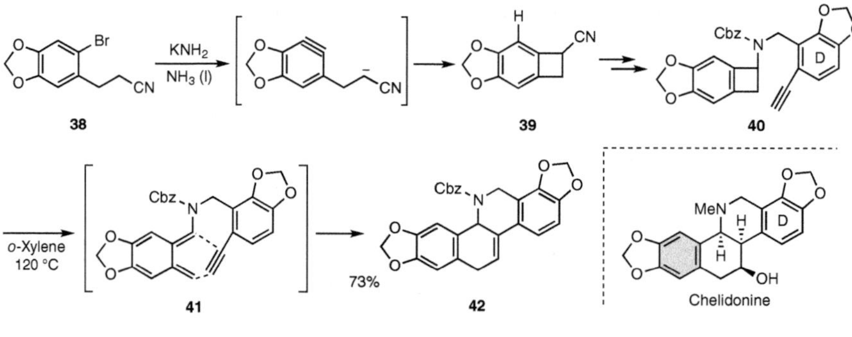

Scheme 11.11 Synthesis of chelidonine. Source: Based on Oppolzer and Keller [23].

In 2006, Barrett and coworkers achieved the asymmetric total synthesis of *ent*-clavilactone B by using a benzyne-based three-component coupling (Scheme 11.12) [25]. The benzyne, generated from fluorobenzene **43**, reacts with Grignard reagent **44** to give aryl magnesium **45**, which is trapped with epoxyaldehyde **46**, giving epoxyalcohol **47**. The synthesis proved the unknown absolute stereochemistry of the natural product. The approach was expanded to a four-component variant, and exploited in the total synthesis of dehydroaltenuene B, an antibacterial marine natural product [26].

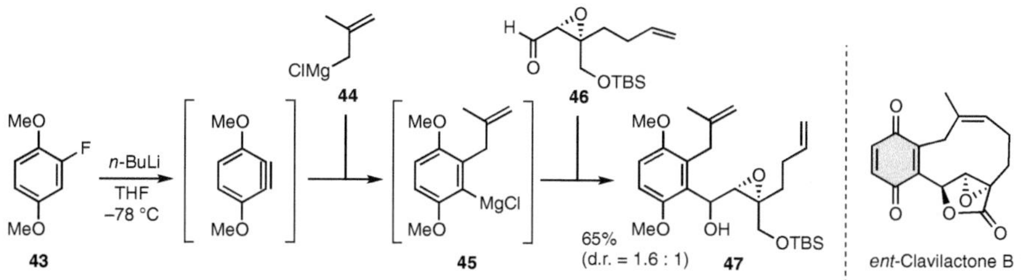

Scheme 11.12 Synthesis of *ent*-clavilactone B. Source: Based on Larrosa et al. [25].

11.3.3.2 π-Nucleophiles: Enamines and Enolates

Enamines are suitable for alkaloid syntheses since the nitrogen atom becomes a part of the target. Kametani et al. reported in 1977 the total synthesis of xylopinine, a protoberberine alkaloid (Scheme 11.13) [27]. Treatment of bromobenzene **48** with $NaNH_2$ generates benzyne **49**, which undergoes nucleophilic attack of the internal enamine, giving tetracycle **50**. A similar scheme is exploited by Kibayashi and coworkers in the total synthesis of an *Amaryllidaceae* alkaloid [28].

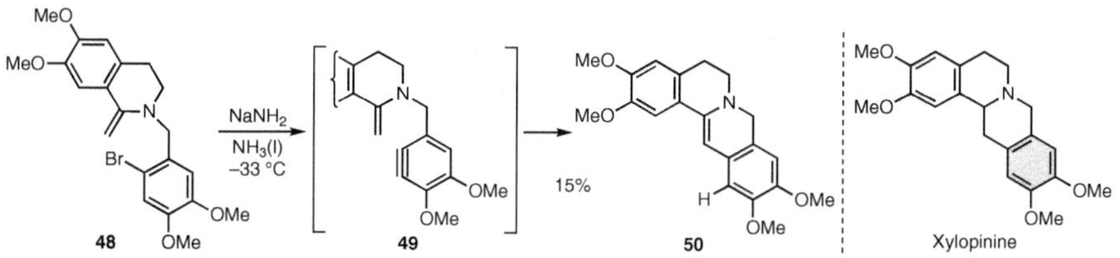

Scheme 11.13 Intramolecular addition in the synthesis of xylopinine. Source: Based on Kametani et al. [27].

In 2008, Stoltz and coworkers reported the construction of the isoquinoline skeleton of papaverine, a biosynthetic precursor of the pavine-class opiate natural products by an annulation of benzynes with *N*-acyl enamines (Scheme 11.14) [30]. Benzyne **53**, generated from silylaryl triflate **51**, undergoes nucleophilic addition of enamide **52**, and the subsequent cyclocondensation gives isoquinoline **54**. The strategy was further exploited for the asymmetric total synthesis of (−)-quinocarcin, a tetrahydroisoquinoline antitumor antibiotic [29]. Note that orthogonal reactivities

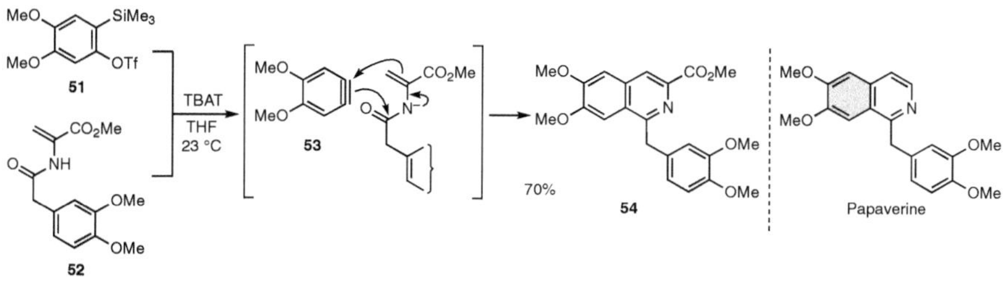

Scheme 11.14 Synthesis of papaverine. Source: Based on Allan and Stoltz [29].

are observed by differentially substituted enamine derivatives, in that reaction of *N*-Boc enamines with benzyne gives the corresponding indolines via a formal [3+2] cycloaddition.

Recently, Guo, He, and coworkers reported the total syntheses of the dictyodendrins, potential compounds for treatment of Alzheimer disease (Scheme 11.15) [31]. The indole skeleton is constructed by a formal [3+2] cycloaddition of 2-aminoquinone methide **56** and the benzyne intermediate generated from silylaryl triflate **55** upon treatment with TBAT. Initially, regioselective addition of the enamine moiety in **56** as a π-nucleophile to the benzyne occurs to give zwitterionic intermediate **57**, whose aryl anion moiety adds to the iminium nitrogen, activated by the 14-carbonyl group, followed by dehydrogenation to give the desired annulated product **58**, along with the regioisomer **59**.

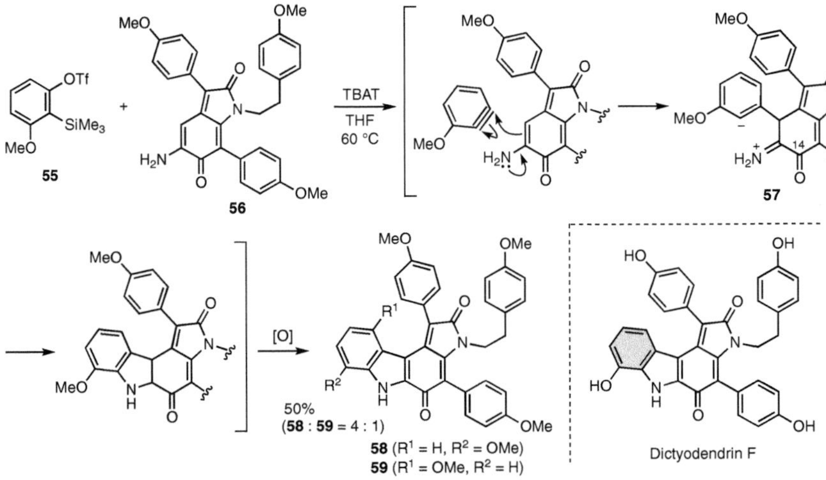

Scheme 11.15 Annulation of a benzyne in the synthesis of dictyodendrins. Source: Guo et al. [31a]; Banne et al. [31b].

In the total synthesis of welwitindolinone alkaloid, a marine natural product, Garg and coworkers used an intramolecular indolyne addition of an enolate, constructing the bicyclo[4.3.1]decenone skeleton (Scheme 11.16a) [32]. Upon treatment of bromobenzene **60** with $NaNH_2$, benzyne **61** undergoes addition of an internal enolate, giving tetracycle **62**. As side product, **63** is formed by an *O*-addition. A similar cyclization was used for the total synthesis of tubingensin A, constructing the characteristic fused ring system with vicinal quaternary carbon centers (Scheme 11.16b) [33].

11.4 Addition–Fragmentation Reactions

The addition–fragmentation reaction of benzyne is a useful process, which is initiated by the nucleophilic addition to benzyne followed by reaction of the

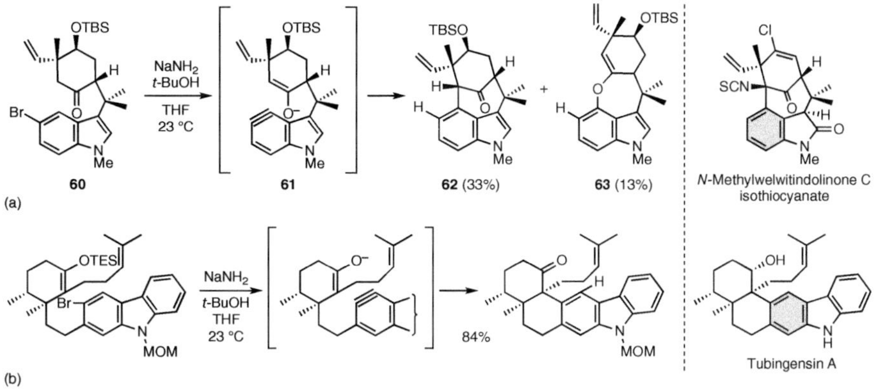

Scheme 11.16 Synthesis of (a) welwitindolinone alkaloid. Source: Based on Huters et al. [32] and (b) tubingensin A. Source: Based on Goetz et al. [33].

resulting aryl anion to an internal carbonyl and bond cleavage. In 1967, Caubère reported a prototype of this sequential process, where simple ketone enolates were the nucleophiles (Scheme 11.17a) [34]. Later, Guyot and Molho used 1,3-dicarbonyl compounds in related reactions (Scheme 11.17b) [35]. Benzyne **64**, generated from *o*-bromoanisole by reaction with $NaNH_2$, undergoes an attack of malonate anion **65**, concomitantly generated from diethyl malonate. Migration of the ethoxycarbonyl group followed by hydrolysis gives ester **66**, which is converted to the natural product, melleine [36].

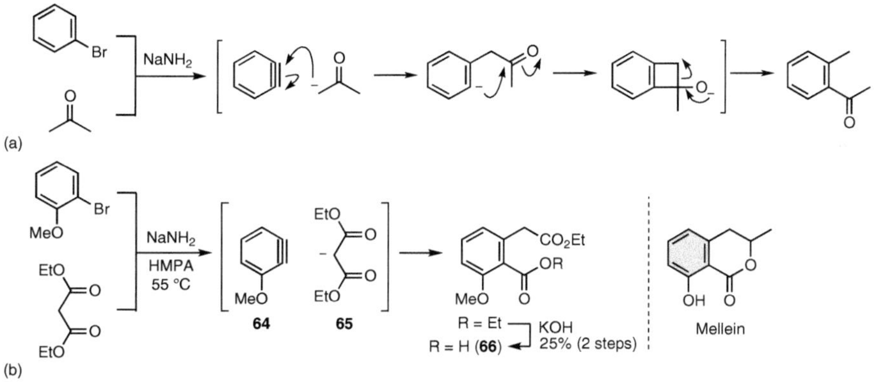

Scheme 11.17 Addition–fragmentation reactions of benzyne with (a) acetone or (b) diethyl malonate. Source: Caubère [34]; Guyot and Molho [35].

The Danishefsky synthesis of dynemicin A in 1995 [37] used the addition–fragmentation reaction for annulation to the E-ring. Benzyne **68**, generated from bromobenzene **67**, reacts with an enolate of dimethyl malonate to give ester **69** (Scheme 11.18). A suitable set of reaction conditions is established (LiTMP, THF, −78 °C), giving higher yield compared to the conditions using $NaNH_2$. More recently, Lin, Shia, and coworkers exploited this reaction to the synthesis of fasamycin A [38], where the optimized conditions involve the use of sodium malonate as a nucleophile and LDA as a base [39].

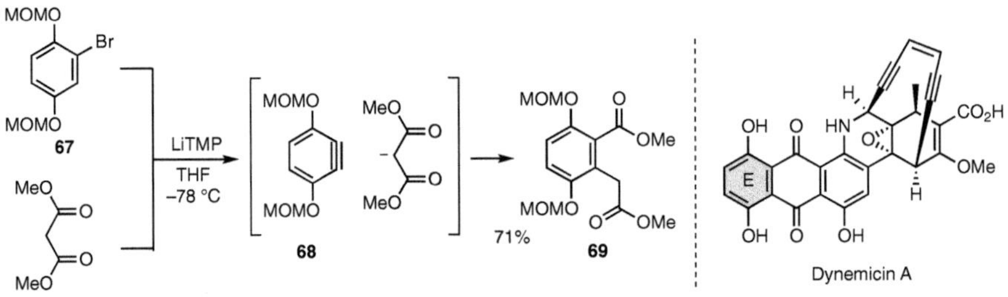

Scheme 11.18 Synthesis of dynemicin A. Source: Based on Shair et al. [37a].

The addition–fragmentation strategy is involved as one of the key steps in Kita's synthesis of fredericamycin A (Scheme 11.19) [40]. The benzyne, generated from aryl bromide **70**, reacts with the enolate derived from dimethyl malonate to give regioisomeric esters **71** and **72** (3 : 2 ratio), which are used for the syntheses of both enantiomers of the natural product, clarifying the hitherto-unknown absolute configuration.

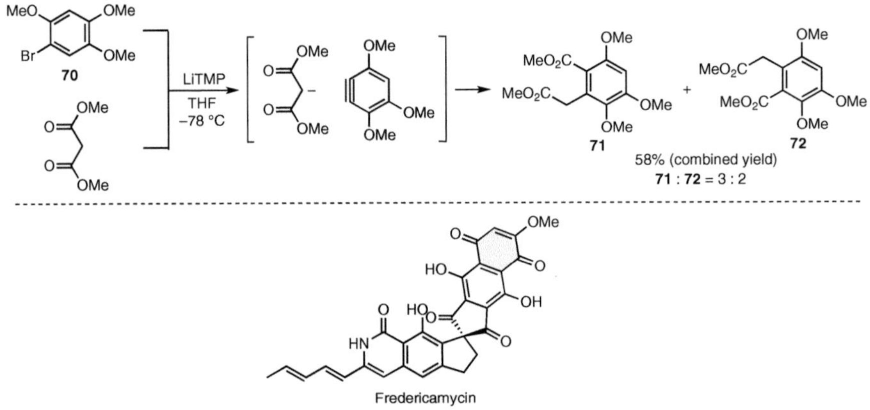

Scheme 11.19 Synthesis of fredericamycin A.Source: Based on Kita et al. [40a].

Application of the addition–fragmentation reaction to cyclic substrates provides the corresponding ring-expanded products, as initially demonstrated by Danheiser and Helgason in the synthesis of salvilenone, a phenalenone diterpene [36c]. Silylaryl triflates have been extensively exploited as effective benzyne precursors (Scheme 11.20) [44]. Stoltz and coworkers achieved the total synthesis of curvularin, a natural product having a benzannulated macrolactone structure (Scheme 11.20a) [41]. By treating silylaryl triflate **31** and cyclic ketoester **73** with CsF, a regioselective addition proceeds, and subsequent expansion gives macrocyclic lactone **74**. In 2016, Iwabuchi and coworkers reported the synthetic study of turkiyenine

(Scheme 11.26) [42]. Treatment of silylaryl triflate **75** and β-ketolactam **76** with CsF induces a facile addition–fragmentation reaction, giving benzazepineone **77**, which is reduced to give the originally proposed structure of turkiyenine. Recently, Sarpong and coworkers demonstrated the utility of this reaction in the total synthesis of cossonidine (Scheme 11.20c) [43]. In the presence of β-ketoester **79**, silylaryl triflate **78** is treated with CsF, the ketoester anion derived from **79** adds to benzyne **80** regioselectively to give tricyclic compound **81**.

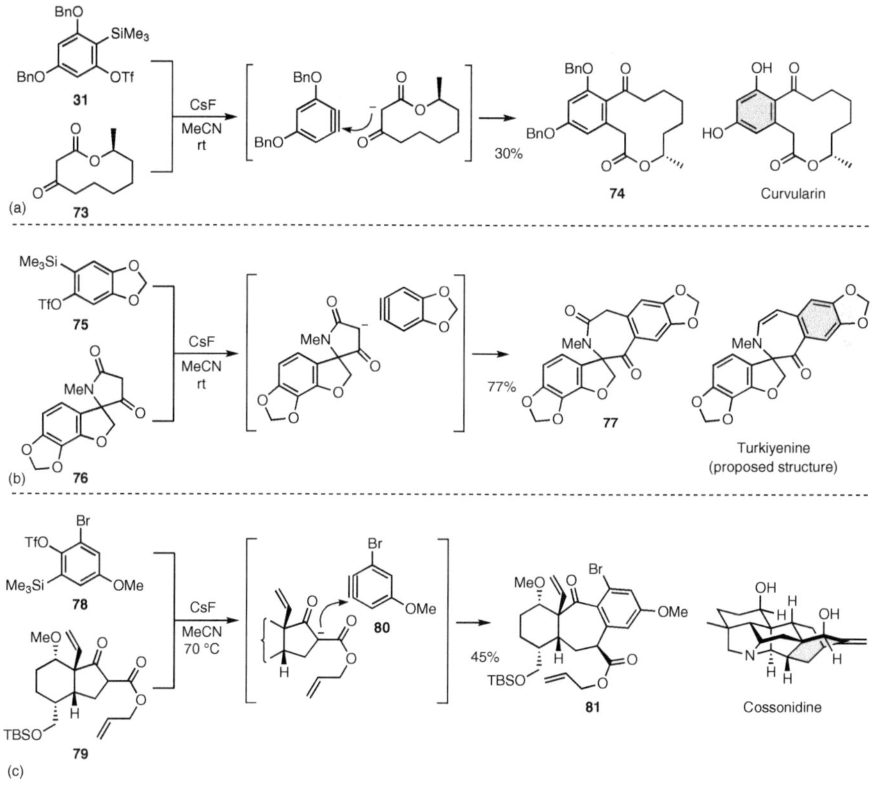

Scheme 11.20 Syntheses of (a) curvularin. Source: Based on Tadross et al. [41], (b) turkiyenine. Source: Based on Kobayashi et al. [42], and (c) cossonidine. Source: Based on Kou et al. [43] .

11.5 Strategies Based on [4+2] Cycloadditions

This section describes the [4+2] cycloadditions of benzynes,[1] which provide effective tools for constructing various fused six-membered ring systems. One of the major issues is the periselectivity, in that the [2+2] and the ene reactions may compete,

1 The mechanism of the [4+2] and [2+2] cycloadditions has been the subject of controversy, i.e. concerted vs. stepwise. Since the reaction pathway usually depends on arynophiles, it is not straightforward to discriminate between these mechanisms. In this review, the categorization follows the original articles.

although the [4+2] cycloaddition is normally the predominant pathway [47]. The regioselectivity could be controlled by the substituent effect of the benzynes and the arynophiles [16].

Among the heterocyclic dienes used as arynophiles, furans are the most commonly used. After the benzyne [4+2] cycloadditions, the cycloadducts are easily aromatized, allowing facile access to fused aromatic ring systems. Early examples include the syntheses of averufin by Townsend et al. [48], ellipticine by Gribble et al. [49], and morindaparvin A by Biehl and coworker [50]. A pioneering report on the intramolecular variant in the syntheses of naphthoquinone natural products by Best and Wege is worth noting [51].

Suzuki and coworkers reported in 1992 the total synthesis of *ent*-gilvocarcin M, an aryl *C*-glycoside antibiotic (Scheme 11.21) [52, 53]. Iodoaryl triflate **82** having a sugar moiety is used as the benzyne precursor, which is treated with *n*-BuLi in the presence of methoxyfuran **83** (THF, −78 °C), inducing an intermolecular [4+2] cycloaddition and subsequent aromatization gives naphthol **84** in 88% yield (regioisomer: 7% yield). The regioselectivity is attributed to the distortion of the benzyne [16] and the polarization of the furan, both of which are due to the alkoxy groups. A similar reaction was subsequently applied to the total synthesis of C104, an aryl *C*-glycosyl angucyclinone [54]. In 2014, Hosoya, Sumida, and coworkers reported the synthesis of gilvocarcin aglycon [55], using aryl boronic acid ester with an *ortho*-triflate as a leaving group [56].

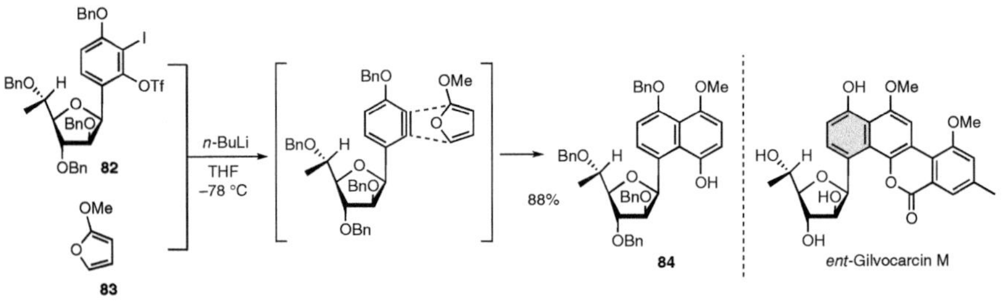

Scheme 11.21 Synthesis of *ent*-gilvocarcin M. Source: Based on Matsumoto et al. [52a].

Brückner and coworker recently reported the asymmetric total syntheses of arizonin B1 using the [4+2] cycloaddition of benzyne and siloxyfuran (Scheme 11.22) [57]. Treatment of a mixture of bromoaryl tosylate **85** and siloxyfuran **86** with *n*-BuLi gives a mixture of regioisomeric cycloadducts **87** and **88** (84 : 16), which are converged into the naphthalene **89** by treatment with TBAF. More than 10 naphthoquino-γ-lactone analogues were synthesized by this approach [58].

In 2016, MacMillan, DeBrabander, and coworkers reported the isolation and total synthesis of rifsaliniketal, a biosynthetic congener of rifamycin (Scheme 11.23) [59]. Based on Kinoshita's report [60], the central naphthoquinone is constructed by a benzyne [4+2] cycloaddition. The benzyne intermediate, generated from symmetrical dibromide **90**, reacts with furan, giving cycloadduct **91**. The total synthesis is completed by an introduction of the side chain after construction of the naphthoquinone skeleton from **91**.

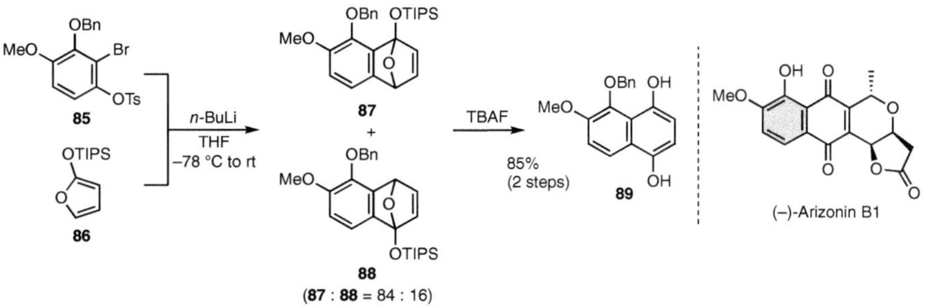

Scheme 11.22 Synthesis of arizonins B1. Source: Based on Neumeyer and Brückner [57].

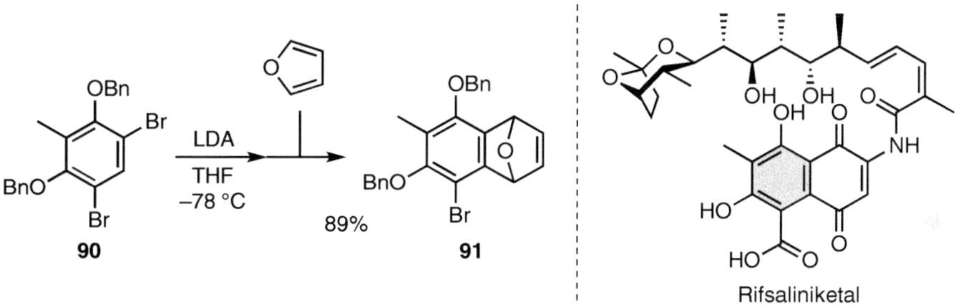

Scheme 11.23 Synthesis of rifsaliniketal. Source: Based on Feng et al. [59].

In 2017, Suzuki and coworkers used a benzyne–furan [4+2] cycloaddition in the enantioselective total synthesis of a marine antibiotic, spiroxin C, (Scheme 11.24) [61]. Upon treatment of a mixture of aryl bromide **92** and furan with LDA at −78 °C, a [4+2] cycloaddition proceeds to give cycloadduct **93**, which is converted to naphthalene **94** by a ring opening by acid treatment.

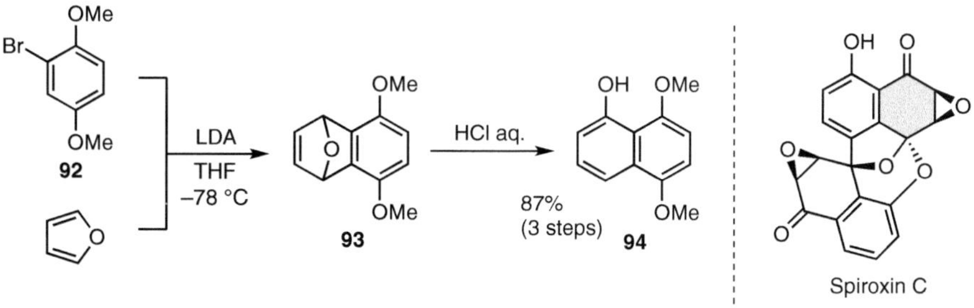

Scheme 11.24 Synthesis of spiroxin C. Source: Based on Ando et al. [61].

In 2017, Odagi, Nagasawa, and coworkers reported the total synthesis of rishirilide B using a benzyne–furan [4+2] cycloaddition for constructing the tricyclic framework (Scheme 11.25) [62]. Note that in the reaction of the benzyne from bromoaryl triflate **95** with furan, two unprotected tertiary alcohols are tolerated. Rishirilide B is synthesized by a regioselective ring opening followed by deprotection.

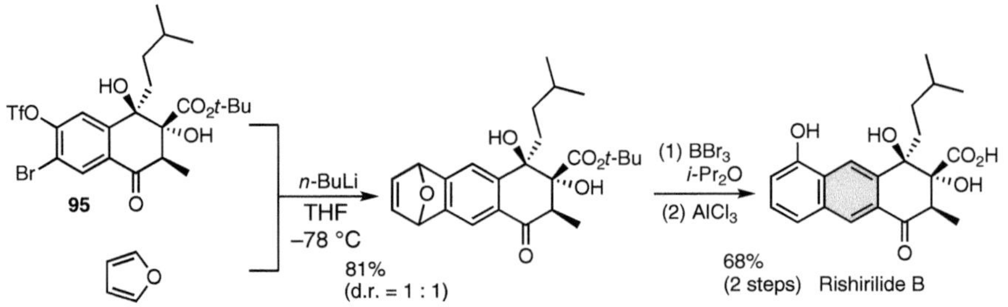

Scheme 11.25 Synthesis of rishirilide B. Source: Based on Odagi et al. [62].

The use of elaborated furans as an arynophile is attractive for rapid construction of highly functionalized polycyclic systems. Koert and coworkers reported a synthetic study of granaticin A, a pyranonaphthoquinone antibiotic, using a benzyne–furan [4+2] cycloaddition (Scheme 11.26a) [45]. The benzyne, generated from aryl bromide **92**, reacts with tricyclic furan **96** having a pyranyl moiety, giving cycloadduct **97** quantitatively. Lewis and coworkers reported the total syntheses of arnottins I and II (Scheme 11.26b) [46]. Silylaryl triflate **75** is treated with CsF in the presence of furan having a functionalized aryl group **98**, giving cycloadduct **99**. Acid treatment of **99** induces a ring opening and lactonization to give arnottin I. Use of a *t*-butyl ester is essential; the yield is low with the corresponding methyl ester.

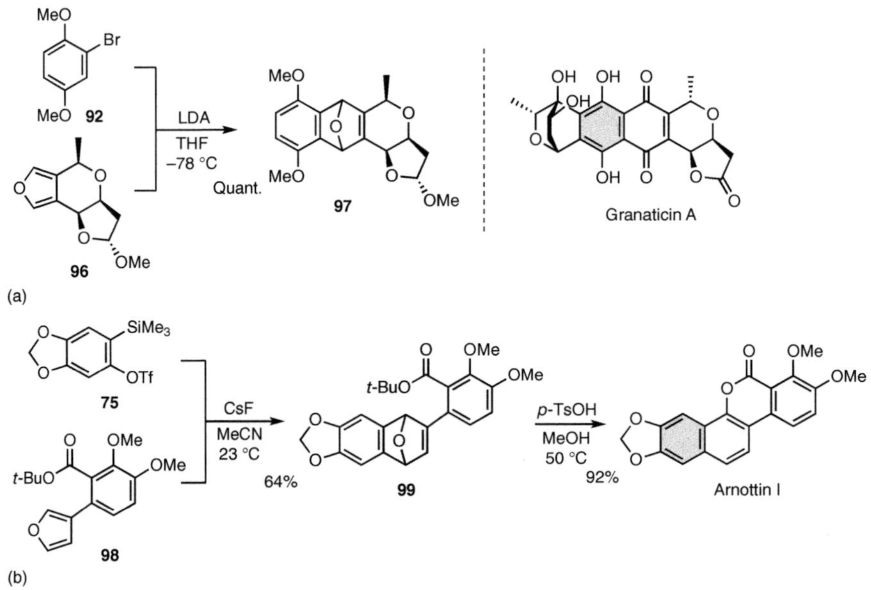

Scheme 11.26 Syntheses of (a) granaticin A. Source: Based on Bartholomäus et al. [45], and (b) arnottin I. Source: Based on Moschitto et al. [46].

Martin and coworkers exploited an intramolecular benzyne–furan [4+2] cycloaddition for the total synthesis of isokidamycin (Scheme 11.27). A disposable silicon tether is used to connect a benzyne precursor and a furanyl arynophile [63]. Dibromide **100**, flanked by a glycosyl furan, is treated with *n*-BuLi, to generate a benzyne species that undergoes a [4+2] cycloaddition to give cycloadduct **101**. A four-atom linker is identified as an appropriate tether length [64]. The tether is cleaved by treatment of TBAF·3H_2O to give, after methylation, naphthalene **102**.

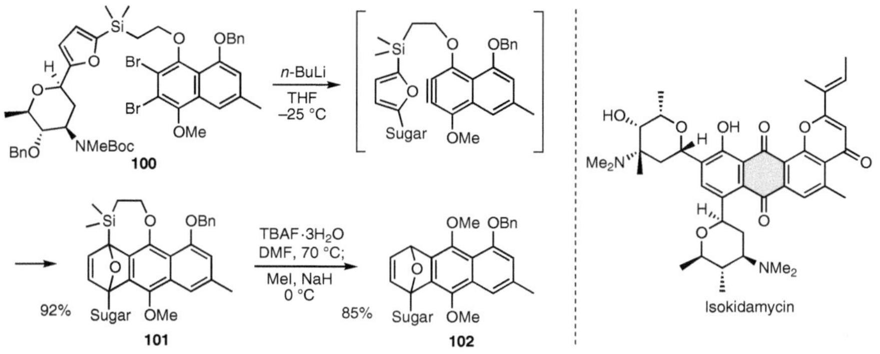

Scheme 11.27 Synthesis of isokidamycin. Source: Based on O'Keefe et al. [63a].

Among various dienes other than furans in benzyne [4+2] cycloaddition, pyrroles are useful for the construction of benzo-fused aza-bicyclic skeletons. Based on the early work of Lautens and coworkers on the asymmetric total synthesis of (+)-homochelidonine, a phenanthridine alkaloid [65], Coquerel and coworkers used a benzyne–pyrrole [4+2] cycloaddition in the total synthesis of nomitidine (Scheme 11.28) [66]. The Lautens intermediate **104** is obtained by the [4+2] cycloaddition of the benzyne, generated from silylaryl triflate **75**, and *N*-Boc-pyrrole **103**. After several steps, the isoquinoline skeleton is constructed by another [4+2] cycloaddition of aza-diene **105** and benzyne **53**. The use of the electron-rich diene **105** is essential; otherwise, the [2+2] cycloaddition of the benzyne with the imine moiety in **105** would compete. The dimethylamino group also serves as a leaving group for the aromatization.

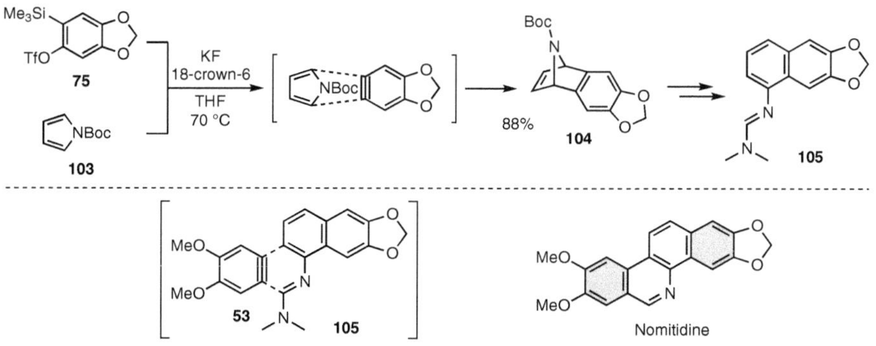

Scheme 11.28 Synthesis of nomitidine. Source: Based on Castillo et al. [66].

α-Pyrones are often exploited in the benzyne [4+2] cycloaddition. After the cycloaddition, extrusion of CO_2 ensures smooth aromatization to afford the naphthalene core. In an early report on the syntheses of pyridocarbazole alkaloids by Moody and coworker, the benzyne [4+2] cycloaddition of a pyranoindolone is exploited as an arynophile for constructing the isoquinoline skeleton [67]. Cho and coworkers's synthesis of ningalin D highlights this reaction in alkaloid syntheses (Scheme 11.29) [68]. The benzyne–pyrone [4+2] cycloaddition followed by spontaneous aromatization via CO_2 extrusion gives naphthalene **106**, which is exploited for the construction of the characteristic biphenylene quinonemethide framework.

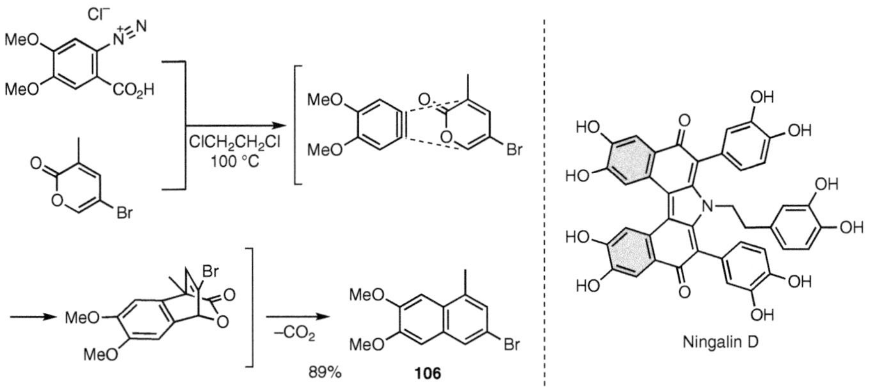

Scheme 11.29 Syntheses of ningalins D and G. Source: Based on Kim et al. [68].

As a limited example of using cyclopentadiene as an arynophile, Buszek and coworkers reported the total syntheses of *cis*-trikentrin B and other related indole alkaloids (Scheme 11.30) [69]. The characteristic fused-ring structure is constructed by the [4+2] cycloaddition of indolyne **108**, generated from tribromoindole **107**, and cyclopentadiene. It should be noted that the Br–Li exchange only occurred at the C-7 position, leading to the regioselective formation of **109**.

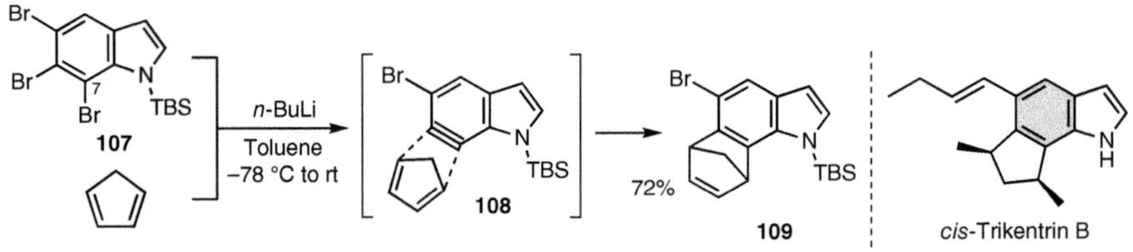

Scheme 11.30 Synthesis of trikentrins. Source: Based on Chandrasoma et al. [69a].

Use of cyclohexadienes in benzyne cycloaddition has been rare in natural product syntheses, due to the periselectivity problem: the [4+2] reactions of carbocyclic dienes or acyclic dienes are often annoyed by competing [2+2] and ene reactions [47]. However, the situation is different in the intramolecular variant. In 1995,

Buszek and Bixby reported an intramolecular benzyne [4+2] cycloaddition of cyclohexadienes in the total synthesis of pseudopterosin aglycon (Scheme 11.31) [70]. Aryl bromide **110** having a cyclohexadienyl moiety is treated with LDA to generate benzyne **111** that undergoes an intramolecular [4+2] cycloaddition to give cycloadduct **112** as a mixture of diastereomers. One of the diastereomers is used for the synthesis of the aglycon via oxidative cleavage of the double bond.

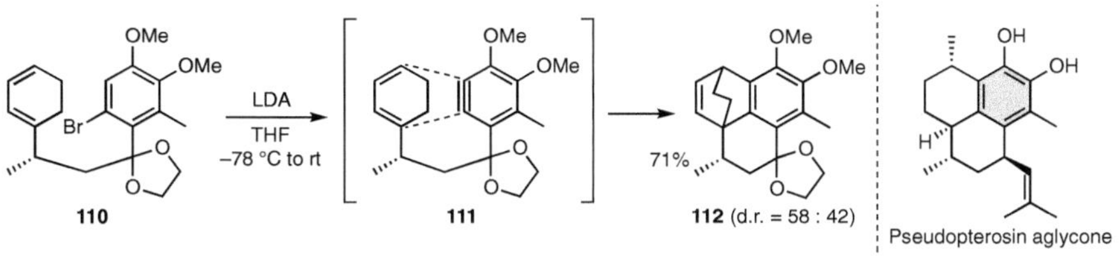

Scheme 11.31 Synthesis of pseudopterosin aglycon. Source: Based on Buszek and Bixby [70].

Suzuki reported an intramolecular benzyne–diene [4+2] cycloaddition with broad substrate scope by using a cleavable silicon tether, which allows an access to various polycyclic structures (Scheme 11.32) [71]. 2-Bromo-6-(chlorodiisopropylsilyl)phenyl tosylate **113** serves as an efficient platform for (i) facile attachment of various arynophiles to the benzyne precursor via a Si—O bond and (ii) facile generation of benzyne via halogen–metal exchange. For instance, **113** is combined with alcohol **114** having a cyclohexadienyl moiety to give silyl ether **115**. Precursor **115**, thus obtained, is treated with Ph_3MgLi, where the desired cycloadduct **116** is obtained. This strategy is proved to be applicable to various other arynophiles, including furans, pyrroles, acyclic dienes, and naphthalenes, and is expected to be applied to complex polycyclic natural product syntheses.

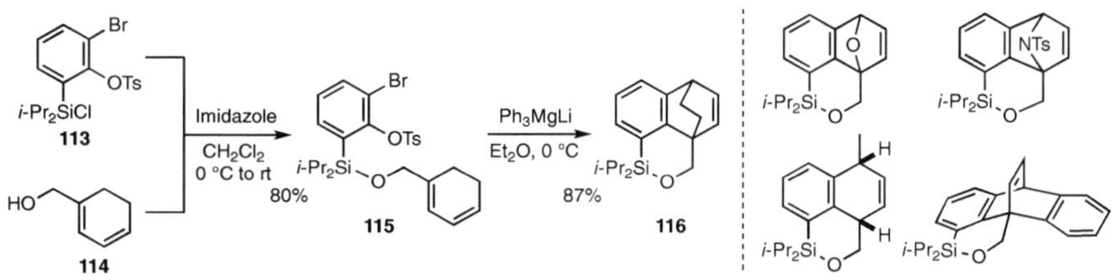

Scheme 11.32 Intramolecular Benzyne–Diene [4+2] cycloaddition. Source: Nishii et al. [71].

The benzyne [4+2] cycloaddition is used also in the syntheses of the aporphine-class alkaloids. Castedo et al. [72] reported the synthesis of an oxoaporphine alkaloid PO-3 (Scheme 11.33) [72c]. Upon refluxing the mixture of methylene tetrahydroisoquinoline **117** and diazonium carboxylate **118**, a regioselective [4+2] cycloaddition proceeds to give tetracycle **119** after air oxidation. Inspired by these syntheses, Raminelli and coworkers reported syntheses of nornuciferine

(Scheme 11.33) [73]. When using a silylaryl triflate as a benzyne precursor, no air oxidation occurs during the benzyne cycloaddition with methylene tetrahydroisoquinoline **117**, allowing a rapid access to the characteristic dihydrophenanthrene skeleton.

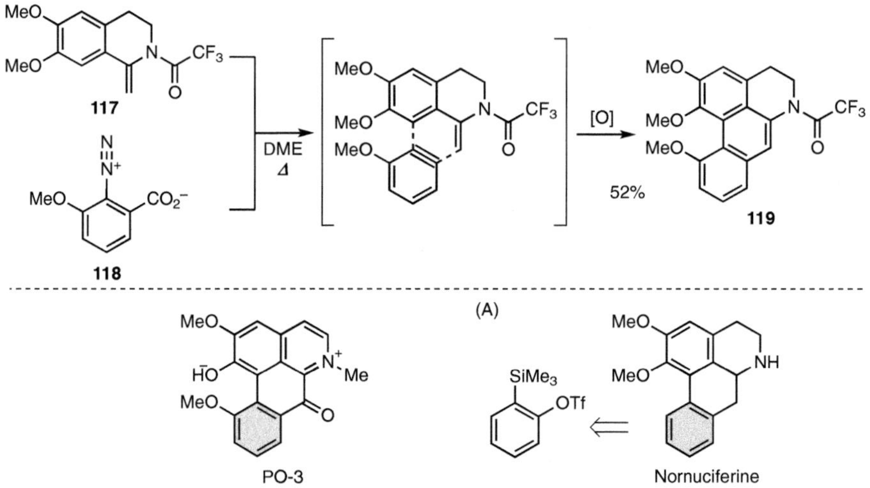

Scheme 11.33 Syntheses of aporphine alkaloids. Source: Based on Perecim et al. [73a].

In 2017, Jeganmohan and coworker reported the total syntheses of the aristolactam alkaloids such as cepharanone B [74]. The key phenanthrene skeleton is constructed by the parent benzyne [4+2] cycloaddition with methylene isoindolinone **120** (Scheme 11.34). Note that the introduction of a benzenesulfonyl group at the methylene moiety in **120** is crucial for securing the periselectivity. The corresponding isoindolinone without the sulfonyl group gives lower yields of the product, due to the competing undesired [2+2] cycloaddition [72a, b]. The sulfonyl group also serves as a leaving group to effect the aromatization.

11.6 Strategies Based on [2+2] Cycloadditions

The benzyne–olefin [2+2] cycloaddition[1] constitutes a simple method for the construction of benzocyclobutenes. Since the final targets seldom contain the so-generated four-membered rings, the cycloadducts are rather used as synthetic intermediates, which are useful for constructing polycyclic ring systems via the pericyclic ring opening – due to inherent strain [75]. In this context, benzocyclobutenones, readily accessible from benzyne [2+2] cycloadditions with a ketene acetals, are widely used for the synthesis of polyketides as shown in Scheme 11.35, among which some synthetic uses will be described below [76–82].

In the early days, electron-deficient olefins were employed as arynophiles, and the [2+2] cycloaddition suffered from low yields and poor regioselectivity [83]. However,

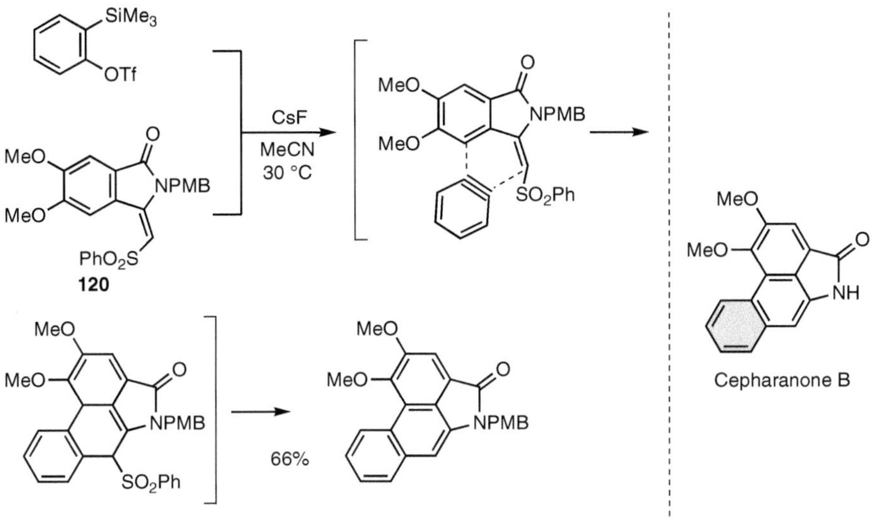

Scheme 11.34 Syntheses of aristolactam alkaloids. Source: Based on Reddy and Jeganmohan [74].

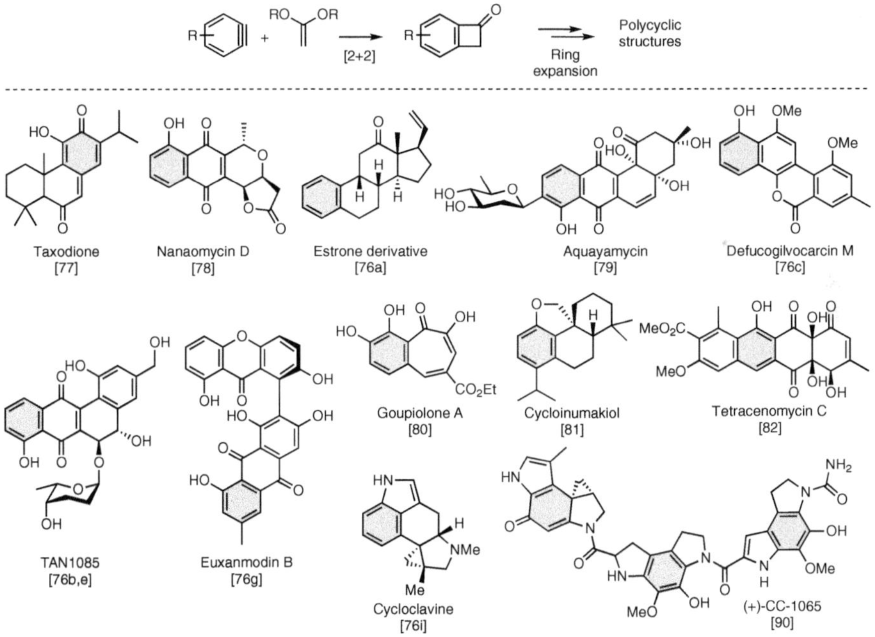

Scheme 11.35 The [2+2] cycloaddition strategy.

the situation is different with the use of electron-rich olefins. In 1982, Stevens and Bisacchi reported that the [2+2] cycloaddition of alkoxybenzynes and ketene acetals proceeds in high yield and high regioselectivity (Scheme 11.36) [77]. In the presence of $NaNH_2$, aryl bromide **121** and ketene dimethyl acetal **122** react in THF upon heating and following acidic workup gave benzocyclobutenone **123**. The regioselectivity is rationalized by the benzyne distortion due to the adjacent benzyloxy group [16]. The addition of anion **124** to **123** gives alcohol **125**. Treatment of **125** with base opens the four-membered ring, giving ketone **126**, an intermediate *en route to* taxodione.

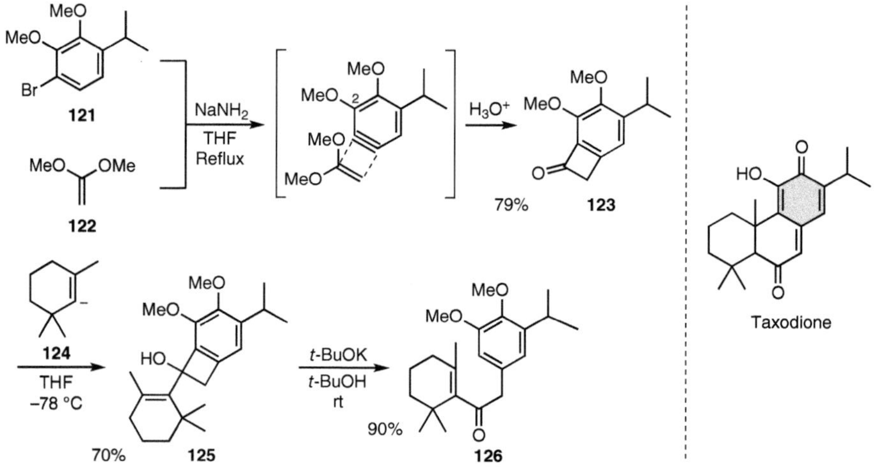

Scheme 11.36 Synthesis of taxodione. Source: Based on Stevens and Bisacchi [77].

In 1994, Moore and coworkers achieved a total synthesis of nanaomycin D using benzocyclobutenone **127**, prepared by Stevens' method (Scheme 11.37) [78]. After converting **127** into the corresponding α-siloxyketone **128** with a dihydropyranyl moiety, heating followed by acid hydrolysis and oxidation gives pyranonaphthoquinone **130** via quinodimethane intermediate **129**. Removal of the methyl group with $AlCl_3$ gave the target compound.

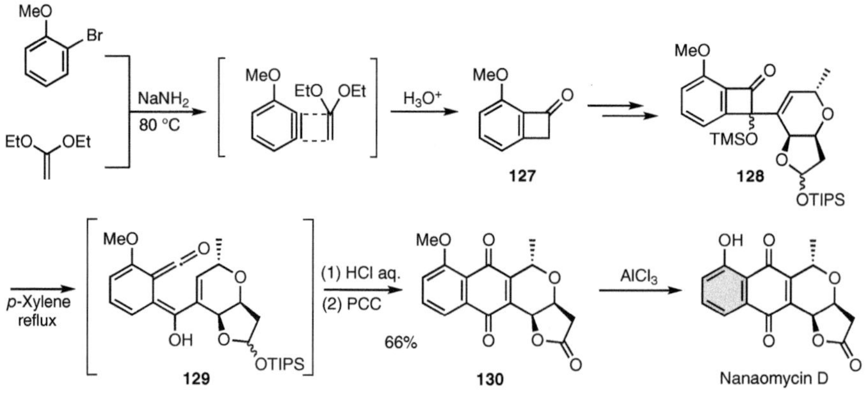

Scheme 11.37 Synthesis of nanaomycin D. Source: Based on Winters et al. [78].

In 1995, Suzuki, Matsumoto, Hosoya, and coworkers identified the ketene silyl acetal (KSA) as an especially reactive arynophile [84]. The reaction with an alkoxybenzyne, generated at −78 °C from an iodoaryl triflate by the action of *n*-BuLi, allows regioselective [2+2] cycloaddition. In 2000, the protocol was applied for the first total synthesis of aquayamycin, an aryl *C*-glycoside antibiotic (Scheme 11.38) [79]. Benzocyclobutenone **133** is obtained by a [2+2] cycloaddition of KSA **132** and the benzyne, generated from *C*-glycosyl substrate **131**. Several steps, including a regioselective Baeyer–Villiger oxidation, converts **133** to sulfonylphthalide **134**, which is subjected to a Hauser annulation with enone **135** followed by methylation to give dimethyl ether **136**.

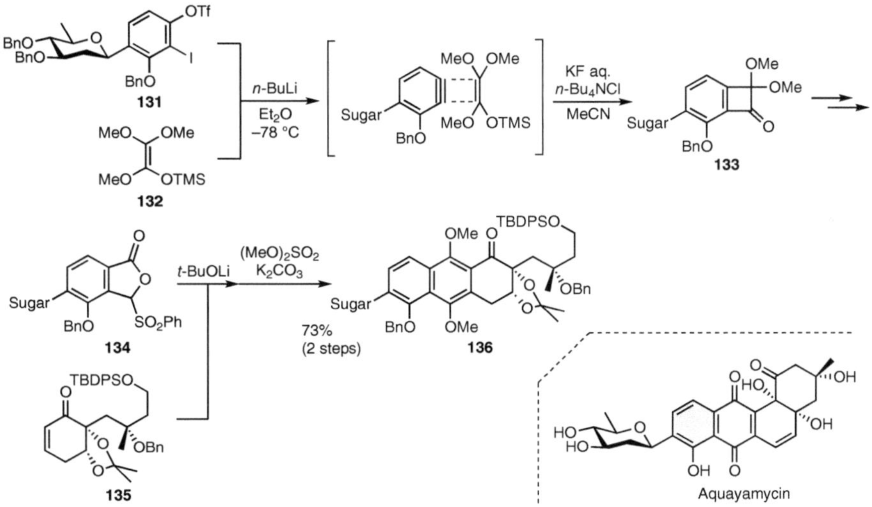

Scheme 11.38 Synthesis of aquayamycin. Source: Based on Matsumoto et al. [79a].

In 2012, Suzuki and coworkers applied this regioselective benzyne–KSA [2+2] cycloaddition also to the total synthesis of goupiolone A, a benzotropolone natural product (Scheme 11.39) [80]. The benzyne, generated from iodoaryl triflate **137**, reacts with the coexisting KSA **138**, and hydrolysis gives benzocyclobutenone **139**. After conversion of **139** to benzocyclobutene **140** bearing a cyclopropane moiety, heating in refluxing *p*-xylene induces a tandem electrocyclic ring opening–sigmatropic rearrangement to give benzocycloheptadiene **141**.

In 2014, Dong and coworkers reported the total synthesis of the proposed structure of cycloinumakiol (Scheme 11.40) [81]. Upon treatment of a mixture of aryl bromide **141** and enolate **142**, generated from THF with *n*-BuLi [85], with LiTMP, a [2+2] cycloaddition proceeds to cyclobutenol **143**. After elaboration into ketone **144** followed by union with allyl alcohol **145**, the resulting benzocyclobutenone **146** is subjected to a Rh-catalyzed carboacylation of the internal olefin to give ketone **147**.

In connection with the synthetic studies on the pyranonaphthoquinone-class antibiotics such as β-naphthocyclinone by Suzuki and coworkers, a two-step procedure for the preparation of *ortho*-methylbenzaldehyde derivatives is devised

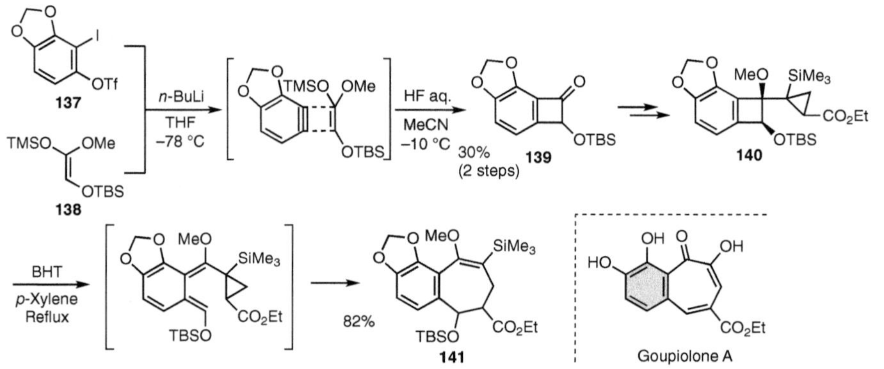

Scheme 11.39 Synthesis of goupiolone A. Source: Based on Fukui et al. [80].

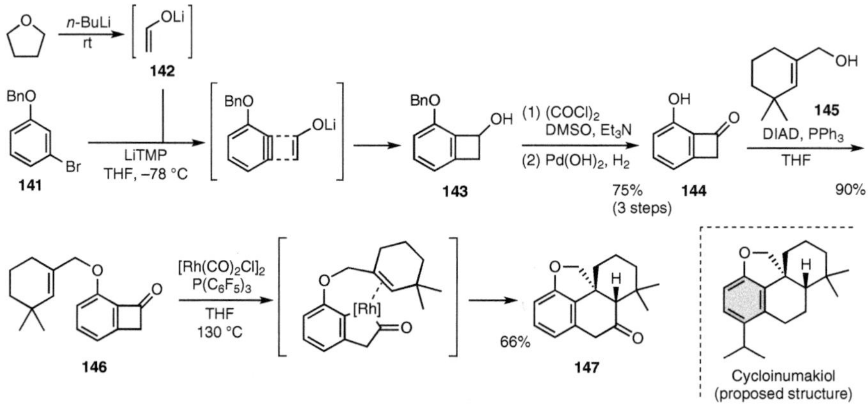

Scheme 11.40 Synthesis of cycloinumakiol. Source: Based on Chen et al. [81a].

(Scheme 11.41) [86]. Benzocyclobutenol **148**, prepared via a benzyne [2+2] cycloaddition with the acetaldehyde enolate **142**, is used as a key intermediate, which undergoes a ring opening [87] by treatment with sodium methoxide to give benzaldehyde **149**.

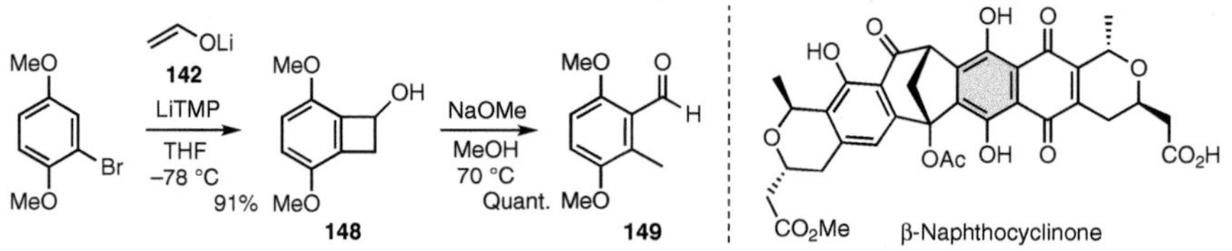

Scheme 11.41 Synthesis of *ortho*-methylbenzaldehydes. Source: Based on Maturi et al. [86].

Hsung and coworkers reported the total synthesis of chelidonine by exploiting the [2+2] cycloaddition and a ring expansion (Scheme 11.42) [88]. Treatment of a mixture of silylaryl triflate **75** and enamide **150** with CsF gives benzocyclobutene **151**, which is heated at 120 °C inducing electrocyclic ring opening to give the corresponding quinodimethane **152**, which undergoes an intramolecular cycloaddition. The silyl protection of the alkyne is necessary, as the free terminal alkyne undergoes a nucleophilic addition to the benzyne.

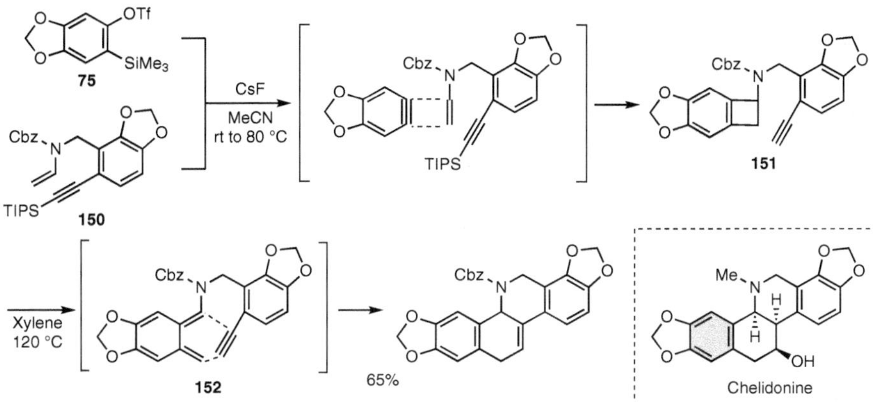

Scheme 11.42 Synthesis of chelidonine. Source: Based on Ma et al. [88].

In 2017, Garg and coworkers reported the total synthesis of tubingensin B, an indole diterpenoid of fungal origin, where the characteristic seven-membered ring is constructed via a carbazolyne (Scheme 11.43) [89]. Carbazolyne **154**, generated by treatment of aryl bromide **153** with $NaNH_2$, undergoes intramolecular [2+2] cycloaddition with the internal silyl enol ether, giving benzocyclobutene **154**. Of particular note is the reactivity difference between the sizes of the ring to be formed. While the six-membered ring formation gives the addition–protonation product (vide supra, Scheme 11.16b), the formation of [2+2] cycloadduct **155** is predominant in the seven-membered ring construction. Rhodium-catalyzed ring opening via cleavage of the desired C—C bond leads to ketone **156**.

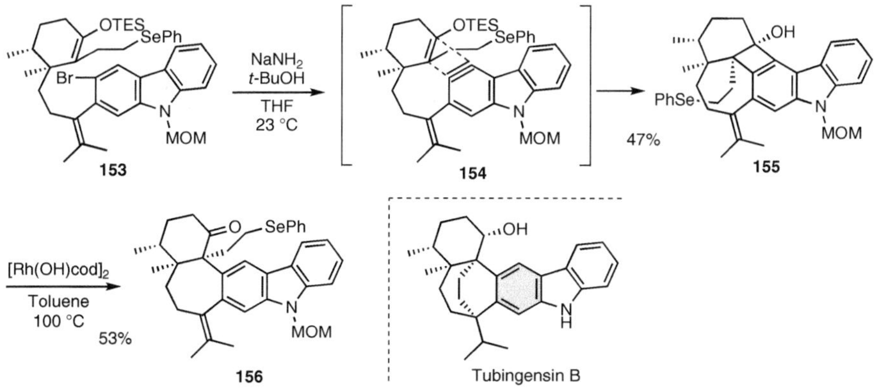

Scheme 11.43 Synthesis of tubingensin B. Source: Based on Corsello et al. [89].

Tokuyama and coworkers reported a convergent total synthesis of CC-1065 (Scheme 11.44) [90]. The left segment, i.e. the cyclopropa[*c*]-pyrrolo[3,2-*e*]indol-4(5*H*)-one unit, is constructed from pyrrolotetrahydroquinoline **160** via ring contraction [91]. The bottom line is a $NaBH_4$-mediated ring expansion of benzocyclobutenone oxime sulfonate **159**, prepared from arylbromide **157** via regioselective benzyne [2+2] cycloaddition with KSA **158**, to give indole **160**. Another key point is the construction of two pyrroroindoline units through a two-directional ring expansion of bis-oxime derivative **161** (vide infra).

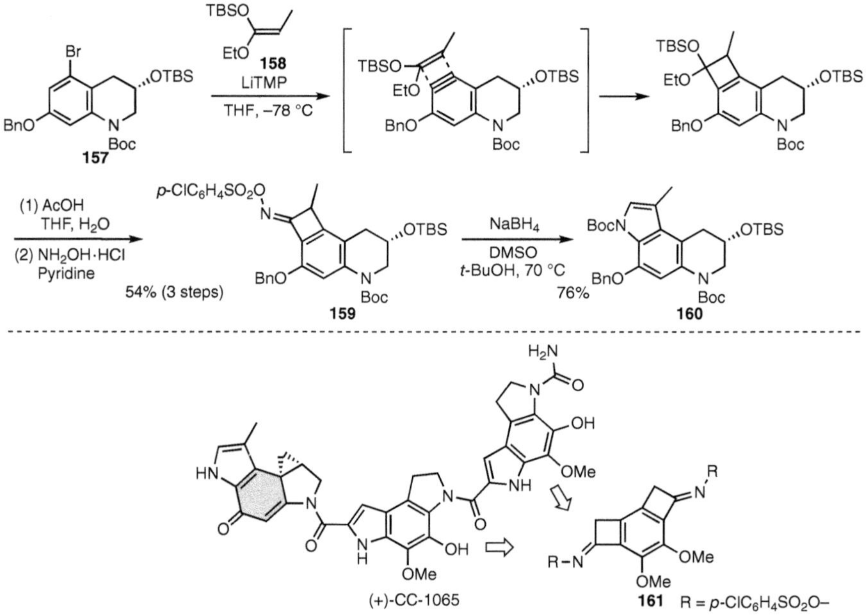

Scheme 11.44 Synthesis of (+)-CC-1065. Source: Based on Imaizumi et al. [90].

11.7 Strategies Based on Benzyne–Ene Reactions

While the *intermolecular* benzyne–ene reaction is often problematic due to the poor reactivity and periselectivity, two recent alkaloid syntheses circumvented these problems by an intramolecular approach.

Lautens and coworkers reported a formal total synthesis of crinine, an *Amaryllidaceae* alkaloid (Scheme 11.45) [92], in which the tetrahydroisoquinoline skeleton is constructed by an intramolecular benzyne–ene reaction. Treatment of aryl bromide **162** with LDA allows formation of benzyne **163**, which undergoes intramolecular ene reaction to afford tetracyclic compound **164**. Note that two alkoxy groups on the benzene ring are essential for the smooth cyclization. Another recent example of this reaction, reported by Li and coworkers, involves an in situ organization of an alkene moiety by nucleophilic addition to benzyne followed by a subsequent benzyne generation from 1,2-benzdiyne precursors (vide infra, Scheme 11.51) [93].

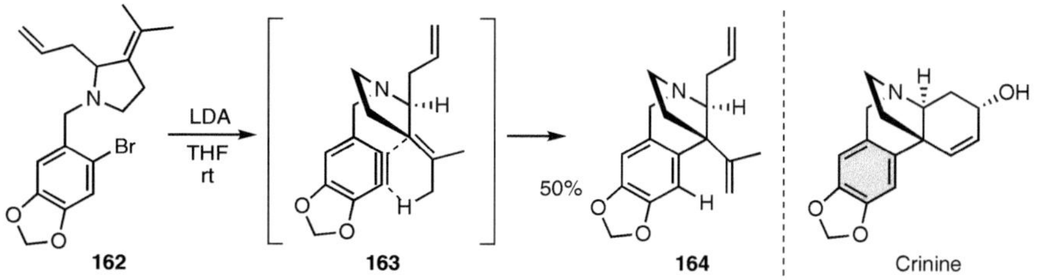

Scheme 11.45 Synthesis of crinine. Source: Based on Candito et al. [92].

11.8 Recent Advances

11.8.1 Strategies Based on Multiple Use of Benzyne

Multiple use of benzyne allows for the expeditious assembly of polycycles with diverse functionalities. In this context, many equivalents of a benzdiyne, a six-membered carbon ring consisting of two formal C≡C bonds and one C=C bond, have been widely investigated. Among three benzdiyne equivalents, i.e. 1,4-benzdiyne, 1,3-benzdiyne, and 1,2-benzdiyne equivalents, 1,4-bendiyne has been well investigated from the dawn of benzyne chemistry as in the pioneering contribution by Wittig and Härle [94].

1,4-Benzdiyne equivalents started to be exploited in natural product syntheses from the mid-2000s, as exemplified in an intermolecular dual benzyne–furan [4+2] cycloaddition in the synthetic study of *ent*-Sch 47555 by Barrett and coworker (Scheme 11.46) [95]. Difluorobenzene **165** can be regarded as a formal equivalent to bis-benzyne. The first cycloaddition is carried out with the benzyne **166**, generated by treatment with *n*-BuLi in the presence of furan, giving mono-cycloadduct **167** in 80% yield. After glycosylation under the Suzuki conditions [53, 96], aryl fluoride **168** is treated with *n*-BuLi for the second cycloaddition with furan to give cycloadduct **169**.

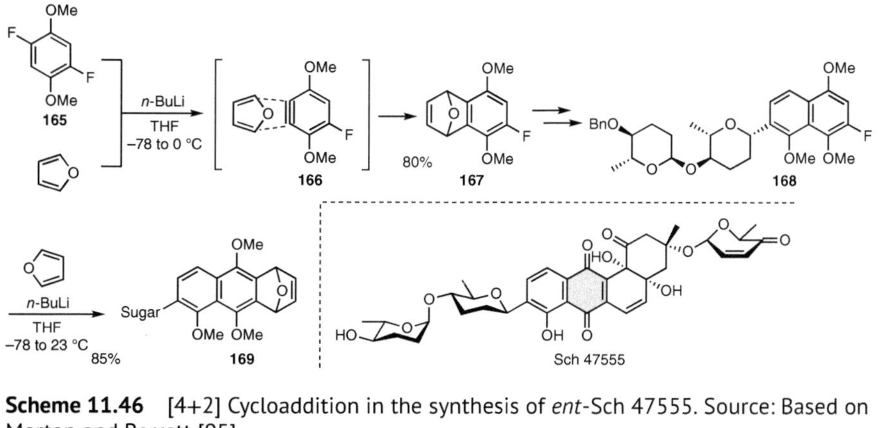

Scheme 11.46 [4+2] Cycloaddition in the synthesis of *ent*-Sch 47555. Source: Based on Morton and Barrett [95].

In 2006, Martin and coworkers reported the synthesis of vineomycinone B2 methyl ester, in which iterative intramolecular benzyne–furan [4+2] cycloaddition is demonstrated (Scheme 11.47) [97]. Upon treatment of tetrabromoarene **170** bearing two furanyl side chains connected by silicon tethers with *n*-BuLi at −20 °C, two benzynes are successively generated in one pot, which undergo dual cycloadditions with two internal furans to give bis-cycloadduct **171**.

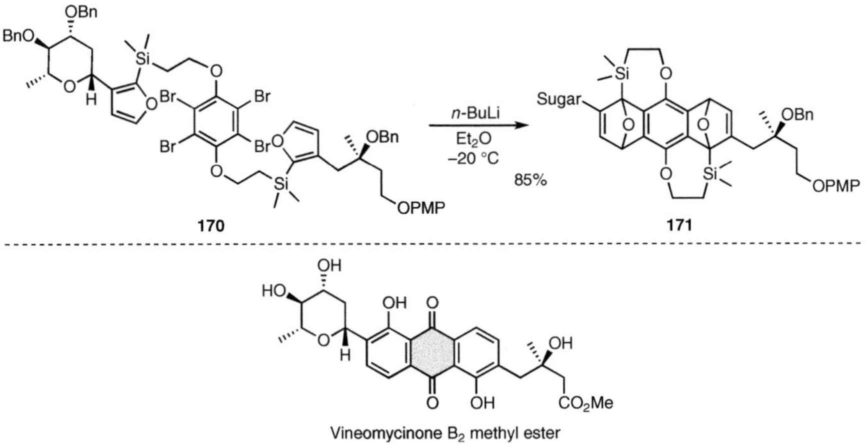

Scheme 11.47 Synthesis of vineomycinone B_2 methyl ester. Source: Based on Chen et al. [97a].

Strategic use of 1,4-benzdiyne to assemble poylicyclic scaffolds was made by Suzuki in the asymmetric total syntheses of tetracenomycins C and X (Scheme 11.48) [82]. The strategy utilized a highly selective dual benzyne cycloaddition sequence comprising a [2+2] and a [4+2] cycloaddition. Bis-tosylate **172** serves as a 1,4-benzdiyne equivalent, allowing sequential generation of two benzyne species, which are regioselectively trapped with KSA **173** and furan **175**, affording, via benzocyclobutene **174**, cycloadduct **176**. After converting **176** to oxime **177**, treatment with NCS and Et_3N, in the presence of MeOH, induces an oxidative ring opening to afford naphthonitrile oxide **178** [98], which serves as a useful platform to allow the total syntheses.

As a synthetic equivalent to allow stepwise generation of two benzyne species in the Suzuki synthesis of actinorhodin, two distinct benzyne precursors are installed in the starting material **179** (Scheme 11.49) [99]. The first run exploits the facile iodine–lithium exchange at low temperature, and the reaction in the presence of furan gives cycloadduct **180**. The second run uses F^--promoted benzyne generation, and the anion of methyl acetoacetate, generated under the reaction conditions, undergoes addition–fragmentation sequence to give advanced intermediate **181**. Several elaborations, including an enantioselective reduction, Lewis-acid-mediated allylation and concomitant ring opening in a regioselective manner, give nanaomycin-related compound **182**. Dimerization allows the first total synthesis of actinorhodin.

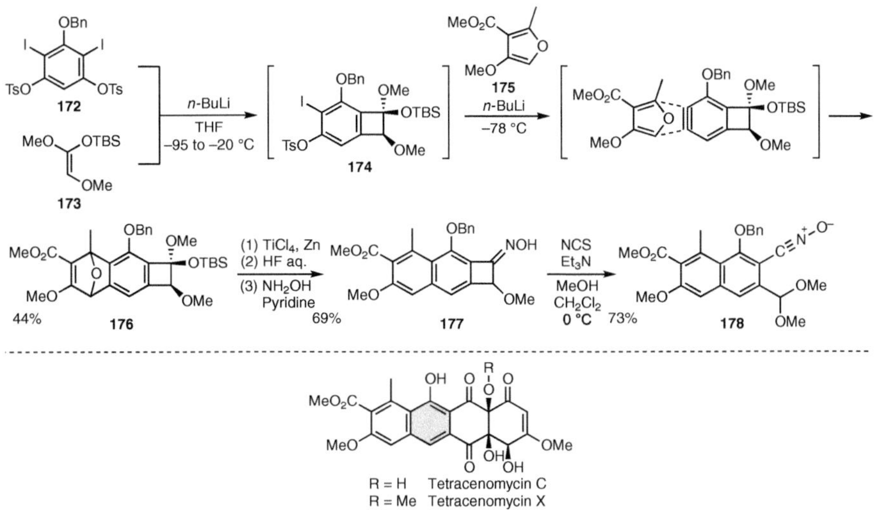

Scheme 11.48 Synthesis of tetracenomycins C and X. Source: Based on Sato et al. [82].

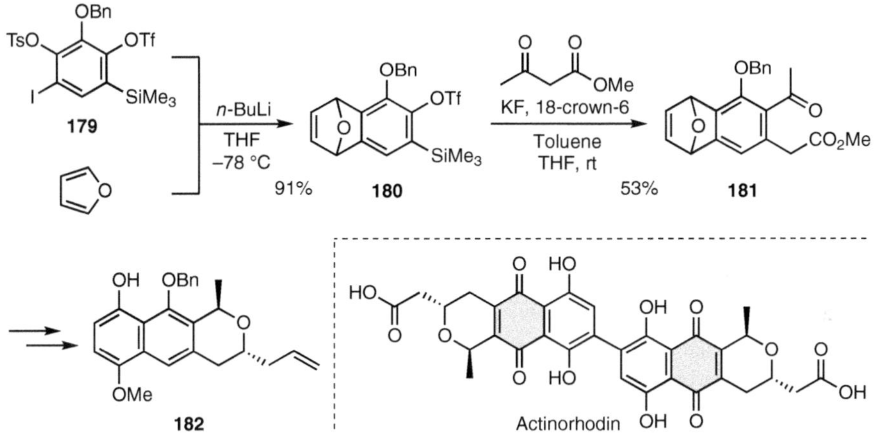

Scheme 11.49 Synthesis of actinorhodin . Source: Based on Ninomiya et al. [99].

Although less investigated than 1,4-benzdiyene equivalents, 1,3-benzdiyne equivalents have also been applied to natural product synthesis. In the Tokuyama synthesis of CC-1065 (vide supra, Scheme 11.44) [90], two pyrroloindoline units were synthesized through a two-directional ring expansion of bis-cyclobutenone oxime derivative **186** (Scheme 11.50). 1,2-Dibromo-4,5-dimethoxybenzene (**183**) is employed as a 1,3-benzdiyne precursor to access to bis-benzocyclobutene **185** via the double benzyne [2+2] cycloaddition with ketene acetal **184**. After conversion into bis-oxime sulfonate **186**, the indole and 2-cyanoindole moieties are constructed via sequential ring expansion triggered by $NaBH_4$ and KCN, respectively. It should be noted that the order of two-ring expansions is crucial, in that initial treatment of **186** with KCN affords only a trace amount of the corresponding indole.

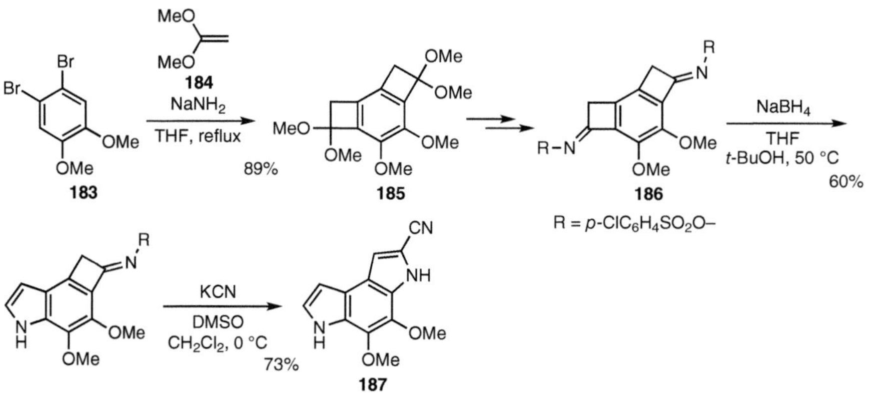

Scheme 11.50 Synthesis of pyrroloindole **187.** Source: Based on Imaizumi et al. [90].

1,2-Benzdiyne is a particular class of benzdiynes, where an addition of nucleophiles to a first-generated benzyne triggers a second benzyne generation. Hart and coworkers first reported a synthetic method for *m*-terphenyls by the reaction of a 1,2,3-trihalobenzene as 1,2-benzdiyne equivalent with an arylmagnesium reagent [100]. Initiated by a halogen–metal exchange, a cascade reaction occurs involving first benzyne generation and nucleophilic addition, second benzyne gene.

Li and coworkers recently reported a related cascade reaction, consisting of a nucleophilic addition and an intramolecular ene reaction (Scheme 11.51) [93]. Benzyne **190**, generated from silylaryl triflate **189** using Cs_2CO_3 and crown ether, undergoes a nucleophilic addition of the sulfonamido anion **191** derived from amide **188** followed by 1,2-elimination. The second benzyne **192**, thus generated, undergoes an ene reaction to form a spiro structure, embedded in ibutamoren mesylate, an orphan drug for growth hormone deficiency.

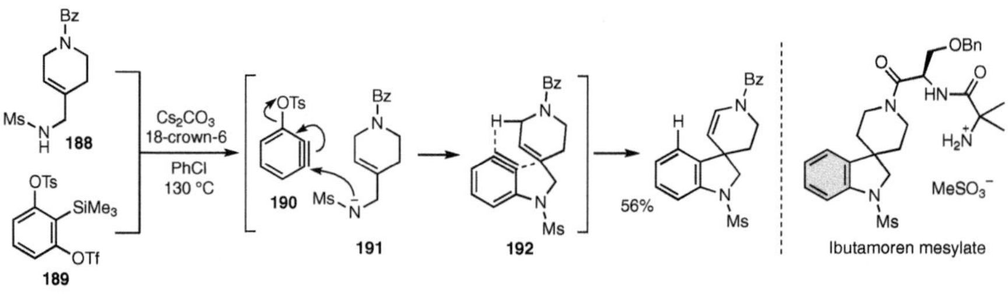

Scheme 11.51 Benzyne cascade reaction and the synthesis of ibutamoren mesylate. Source: Based on Xu et al. [93].

11.8.2 Strategies Based on Transition-Metal-Catalyzed Reactions

Transition-metal catalysis is gaining increasing importance in benzyne chemistry, allowing rapid assembly of benzo-fused polycycles [2p, q]. Silylaryl triflates have been mostly used as the benzyne precursor, due to the mild reaction conditions compatible with transition-metal catalysis. Recent applications in natural product synthesis are discussed below.

A pioneering contribution is in Mori's total syntheses of taiwanins using a Pd-catalyzed [2+2+2] cyclization involving a benzyne (Scheme 11.52) [101]. In the presence of a palladium catalyst, diyne **193** provides the corresponding palladacycle intermediate **194**, which reacts with the benzyne formed from silylaryl triflate **75** by the action of CsF, giving naphthalene **195**.

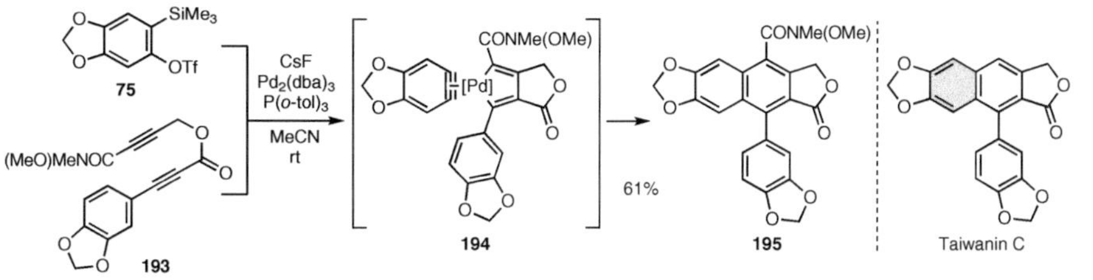

Scheme 11.52 Syntheses of taiwanins. Source: Based on Sato et al. [101].

In 2013, Argade and coworker reported the syntheses of justicidin B and retrojusticidin B [102]. The key naphthalene **197** is synthesized using a Pd-catalyzed [2+2+2] cyclization of a benzyne and diene **196** (Scheme 11.53). Careful choice of the catalyst/ligand is essential.

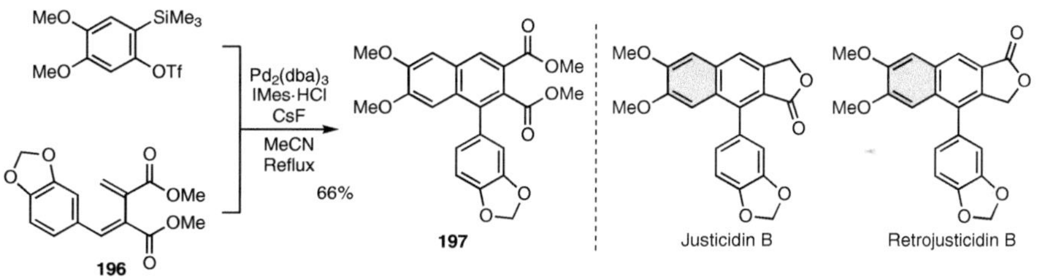

Scheme 11.53 Syntheses of retrojusticidin B and justicidin B. Source: Based on Patel and Argade [102].

Among various methods to construct a phenanthridinone skeleton [103], a Pd-catalyzed annulation of benzynes with *o*-halobenzamides was developed by Larock and coworkers in 2012 [104]. Simple benzamides without an *ortho* halogen substituent can be used for this Pd-catalyzed C–H activation reaction. In 2014, Jeganmohan and coworker reported the synthesis of *N*-methylcrinasiadine. The benzyne, generated from silylaryl triflate **19**, reacts with benzamide **198** to give the target compound (Scheme 11.54) [105].

Xu, Jiang, and coworkers also reported the synthesis of the phenanthridinone alkaloids, including *N*-methylcrinasiadine, by using a Pd-catalyzed three-component reaction of iodoaniline **199**, the benzyne, generated from silylaryl triflate **75**, and carbon monoxide [106]. The same group reported a similar reaction involving a C–H activation of aniline **200** [107]. A slow generation of the benzyne is crucial,

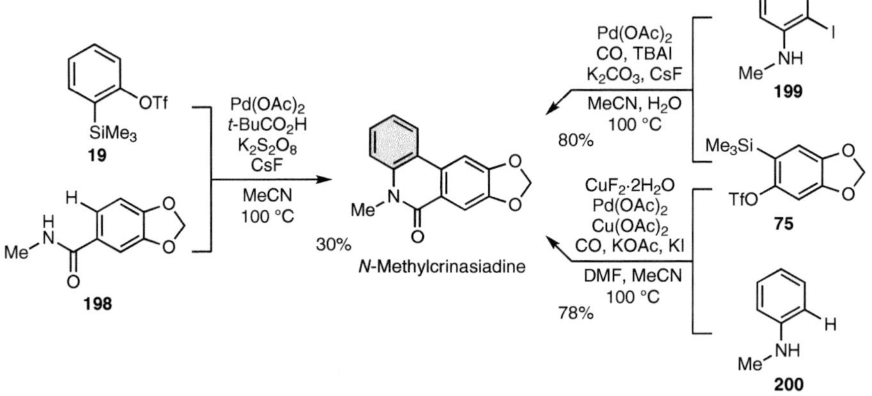

Scheme 11.54 Pd-catalyzed benzyne reactions toward *N*-methylcrinasiadine [105–107]

since the C–H activation step is rate determining $CuF_2 \cdot 2H_2O$ is the most effective fluoride source.

Recently, similar annulations have been exploited, such as the Zhang synthesis of isocryptolepine (Scheme 11.55) [108]. The key skeleton in tetracycle **202** is constructed by the Cu-catalyzed cycloaddition of the benzyne, generated from silylaryl triflate **19**, and indole-3-carboxamide **201**, having a directing group (MQ: 5-methoxy-8-quinolyl) for C–H activation.

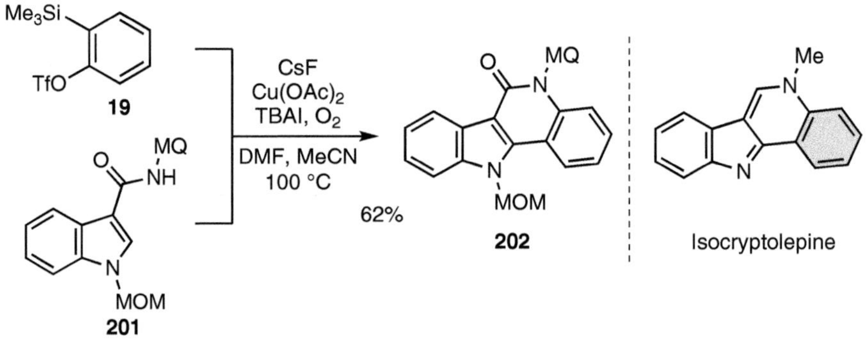

Scheme 11.55 [4+2] Cycloaddition in the synthesis of isocryptolepine. Source: Based on Zhang et al. [108].

Larock and coworker reported the syntheses of 9-fluorenylidenes and 9-phenanthrenes by a Pd-catalyzed benzyne annulation via Heck-type reaction [109]. Functional-group compatibility is demonstrated by the Yao–Zhang synthesis of tylophorine (Scheme 11.56) [110].

Gogoi and coworkers reported the total synthesis of flemichapparin C, an analogue of coumestans, using a Pd-catalyzed oxidative annulation (Scheme 11.57) [111]. The target compound is directly synthesized by the migratory insertion of the

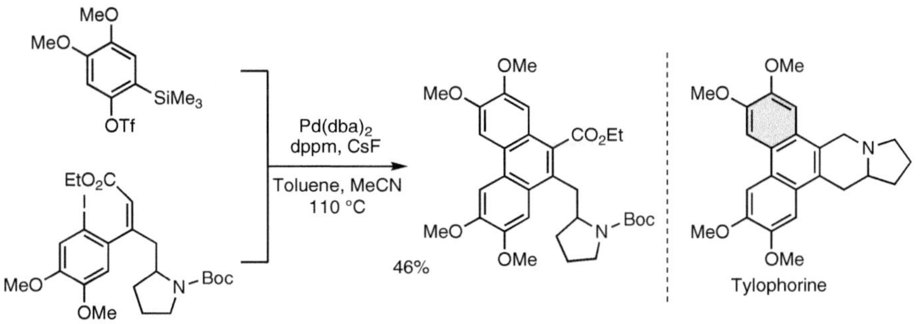

Scheme 11.56 Synthesis of tylophorine. Source: Based on Yao et al. [110].

palladium enolate **204**, generated from coumarin **203**, and the benzyne, generated from silylaryl triflate **75**, followed by the intramolecular C—O bond formation.

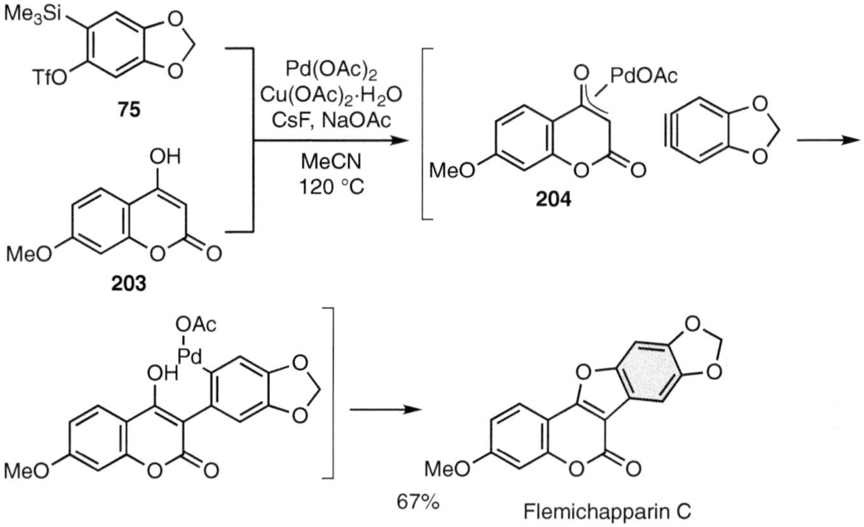

Scheme 11.57 Palladium-catalyzed reaction toward flemichapparin C. Source: Based on Neog et al. [111].

11.8.3 Benzyne Generation via Hexadehydro-Diels–Alder Reaction

Conventional methods of benzyne generation involve a release of two adjacent substituents from a benzenoid, while a conceptually distinct approach has emerged in late 1990s, involving an intramolecular hexadehydro-Diels–Alder reaction of triyne precursors (Scheme 11.58). After the early report by Ueda, Johnson, and coworkers [113], the potential of this unique method remained underscored until Hoye et al. [114] and Lee and coworkers [115] expanded the scope and utility. This section describes two natural product syntheses, exploiting this method of benzyne generation.

In 2016, Hoye and coworker reported the synthesis of koenidine using hexadehydro-Diels–Alder cascade (Scheme 11.58) [112]. Carbazolyne **207**, generated from triyne **205**, reacts with aldehyde **206** to form oxa-cyclobutene **208**. Ring opening followed by 6π-electrocyclization gives carbazole **209**.

Lee and coworker applied this approach to the total synthesis of selaginpulvilin C, a phosphodiesterase-4 inhibitor (Scheme 11.59) [116]. Upon oxidation of tetrayne

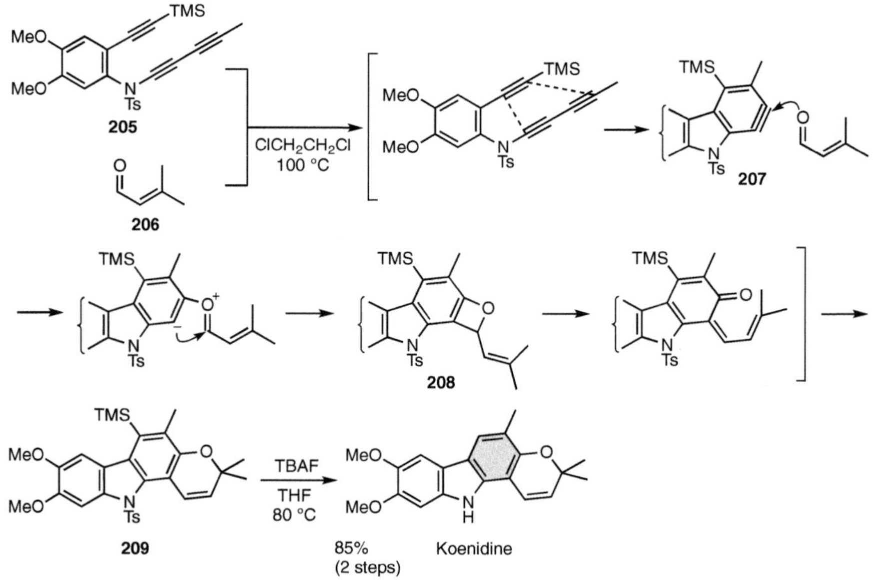

Scheme 11.58 Synthesis of koenidine. Source: Based on Wang and Hoye [112].

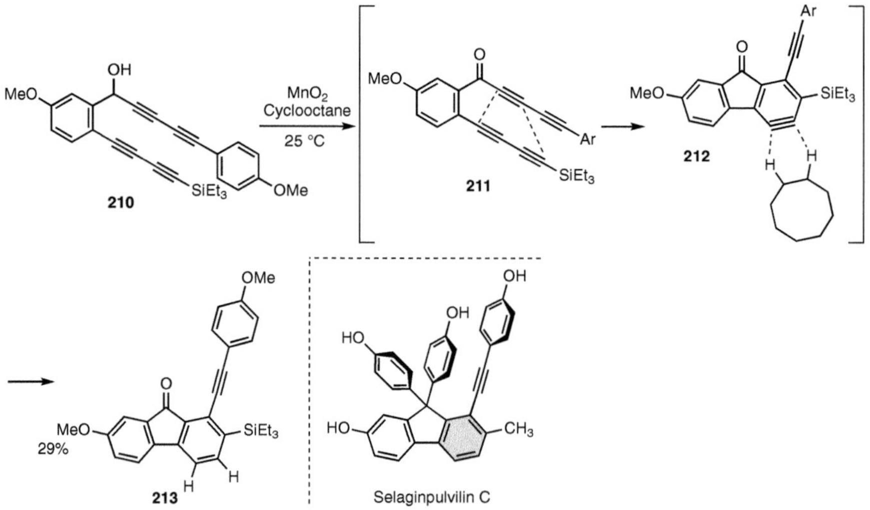

Scheme 11.59 Syntheses of selaginpulvilins. Source: Based on Karmakar and Lee [116].

210 with MnO_2, the resulting ketone **211** allowed facile generation of benzyne **212** that undergoes an unusual hydrogen transfer from cycloalkanes [117] to give the fluorenone **213**. A similar approach was applied to the formal synthesis of selaginpulvilin D [118].

References

1 Hoffmann, R.W. (1967). *Dehydrobenzene and Cycloalkynes*. New York: Academic Press.

2 (a) Pellissier, H. and Santelli, M. (2003). *Tetrahedron* 59: 701–730. (b) Wenk, H.H., Winkler, M., and Sander, W. (2003). *Angew. Chem. Int. Ed.* 42: 502–528. (c) Sanz, R. (2008). *Org. Prep. Proced. Int.* 40: 215–291. (d) Kitamura, T. (2010). *Aust. J. Chem.* 63: 987–1001. (e) Wentrup, C. (2010). *Aust. J. Chem.* 63: 979–986. (f) Yoshida, H., Ohshita, J., and Kunai, A. (2010). *Bull. Chem. Soc. Jpn.* 83: 199–219. (g) Bhunia, A., Yetra, S.R., and Biju, A.T. (2012). *Chem. Soc. Rev.* 41: 3140–3152. (h) Wu, C. and Shi, F. (2013). *Asian J. Org. Chem.* 2: 116–125. (i) Dubrovskiy, A.V., Markina, N.A., and Larock, R.C. (2013). *Org. Biomol. Chem.* 11: 191–218. (j) Pérez, D., Peña, D., and Guitián, E. (2013). *Eur. J. Org. Chem.*: 5981–6013. (k) Miyabe, H. (2015). *Molecules* 20: 12558–12575. (l) Yoshida, S. and Hosoya, T. (2015). *Chem. Lett.* 44: 1450–1460. (m) García-López, J.-A. and Greaney, M.F. (2016). *Chem. Soc. Rev.* 45: 6766–6798. (n) Karmakar, R. and Lee, D. (2016). *Chem. Soc. Rev.* 45: 4459–4470. (o) Idiris, F.I.M. and Jones, C.R. (2017). *Org. Biomol. Chem.* 15: 9044–9056. (p) Shi, J., Li, Y., and Li, Y. (2017). *Chem. Soc. Rev.* 46: 1707–1719. (q) Feng, M. and Jiang, X. (2017). *Synthesis* 49: 4414–4433. (r) Dhokale, R.A. and Mhaske, S.B. (2018). *Synthesis* 50: 1–16. s Roy, T. and Biju, A.T. (2018). *Chem. Commun.* 54: 2580–2594.

3 (a) Tadross, P.M. and Stoltz, B.M. (2012). *Chem. Rev.* 112: 3550–3577. (b) Gampe, C.M. and Carreira, E.M. (2012). *Angew. Chem. Int. Ed.* 51: 3766–3778. (c) Goetz, A.E., Shah, T.K., and Garg, N.K. (2015). *Chem. Commun.* 51: 34–45. (d) Takikawa, H., Nishii, A., Sakai, T., and Suzuki, K. (2018). *Chem. Soc. Rev.* 47: 8030–8056.

4 Kametani, T. and Ogasawara, K. (1967). *J. Chem. Soc. C*: 2208–2212.

5 (a) Wolthuis, E., Bouma, B., Modderman, J., and Sytsma, L. (1970). *Tetrahedron Lett.* 11: 407–408. (b) Richmond, G.O. and Spendel, W. (1973). *Tetrahedron Lett.* 14: 4557–4560.

6 Chapdelaine, M.J. and Hulce, M. (1990). *Org. React.* 38: 225–653.

7 For explanation by invoking pseudo-excitation effect, see;(a)Hayes, D.M. and Hoffmann, R. (1972). *J. Phys. Chem.* 76: 656–663. (b) Epiotis, N.D. (1974). *Angew. Chem. Int. Ed. Engl.* 13: 751–828.

8 (a) Woodward, R.B. and Hoffmann, R. (1965). *J. Am. Chem. Soc.* 87: 395–397. (b) Woodward, R.B. and Hoffmann, R. (1969). *Angew. Chem. Int. Ed. Engl.* 8: 781–932.

9 Sielecki, T.M. and Meyers, A.I. (1992). *J. Org. Chem.* 57: 3673–3676.

10 Iwao, M., Motoi, O., Fukuda, T., and Ishibashi, F. (1998). *Tetrahedron* 54: 8999–9010.

11 (a) Oshiyama, T., Satoh, T., Okano, K., and Tokuyama, H. (2012). *Tetrahedron* 68: 9376–9383. (b) Oshiyama, T., Satoh, T., Okano, K., and Tokuyama, H. (2012). *RSC Adv.* 2: 5147–5149.

12 Okano, K., Fujiwara, H., Noji, T. et al. (2010). *Angew. Chem. Int. Ed.* 49: 5925–5929.

13 Watanabe, M., Kurosaki, A., and Furukawa, S. (1984). *Chem. Pharm. Bull.* 32: 1264–1267.

14 Vaidya, S.D. and Argade, N.P. (2013). *Org. Lett.* 15: 4006–4009.

15 Bronner, S.M., Goetz, A.E., and Garg, N.K. (2011). *J. Am. Chem. Soc.* 133: 3832–3835.

16 (a) Cheong, P.H.-Y., Paton, R.S., Bronner, S.M. et al. (2010). *J. Am. Chem. Soc.* 132: 1267–1269. (b) Medina, J.M., Mackey, J.L., Garg, N.K., and Houk, K.N. (2014). *J. Am. Chem. Soc.* 136: 15798–15805, and references therein.

17 Torres-Ochoa, R.O., Buyck, T., Wang, Q., and Zhu, J. (2018). *Angew. Chem. Int. Ed.* 57: 5679–5683.

18 Day, J.J., McFadden, R.M., Virgil, S.C. et al. (2011). *Angew. Chem. Int. Ed.* 50: 6814–6818.

19 Carril, M., SanMartin, R., Tellitu, I., and Domínguez, E. (2006). *Org. Lett.* 8: 1467–1470.

20 Zhao, J. and Larock, R.C. (2005). *Org. Lett.* 7: 4273–4275.

21 Giallombardo, D., Nevin, A.C., Lewis, W. et al. (2014). *Tetrahedron* 70: 1283–1288.

22 (a) Knight, D.W. and Qing, X. (2009). *Tetrahedron Lett.* 50: 3534–3537. (b) Knight, D.W. and Xu, Q. (2016). *Heterocycles* 93: 647–672.

23 Oppolzer, W. and Keller, K. (1971). *J. Am. Chem. Soc.* 93: 3836–3837.

24 (a) Kametani, T., Nemoto, H., Ishikawa, H. et al. (1977). *J. Am. Chem. Soc.* 99: 3461–3466. (b) Kametani, T., Matsumoto, H., Nemoto, H., and Fukumoto, K. (1978). *J. Am. Chem. Soc.* 100: 6218–6220. (c) Kametani, T., Suzuki, K., and Nemoto, H. (1979). *J. Chem. Soc., Chem. Commun.*: 1127–1128. (d) Kametani, T., Suzuki, K., and Nemoto, H. (1981). *J. Am. Chem. Soc.* 103: 2890–2891. (e) Kametani, T., Suzuki, K., and Nemoto, H. (1982). *J. Org. Chem.* 47: 2331–2342. (f) Nemoto, H., Nagai, M., Fukumoto, K., and Kametani, T. (1985). *Tetrahedron Lett.* 26: 4613–4616. (g) Nemoto, H., Nagai, M., and Fukumoto, K. (1986). *J. Chem. Soc., Perkin Trans. 1*: 1621–1625. (h) Nemoto, H., Nagai, M., Moizumi, M. et al. (1988). *Tetrahedron Lett.* 29: 4959–4962. (i) Nemoto, H., Nagai, M., Moizumi, M. et al. (1989). *J. Chem. Soc., Perkin Trans. 1*: 1639–1645.

25 Larrosa, I., Da Silva, M.I., Gómez, P.M. et al. (2006). *J. Am. Chem. Soc.* 128: 14042–14043.

26 Soorukram, D., Qu, T., and Barrett, A.G.M. (2008). *Org. Lett.* 10: 3833–3835.

27 Kametani, T., Sugai, T., Shoji, Y. et al. (1977). *J. Chem. Soc., Perkin Trans. 1*: 1151–1155.

28 Iida, H., Yuasa, Y., and Kibayashi, C. (1979). *J. Org. Chem.* 44: 1074–1080.

29 Allan, K.M. and Stoltz, B.M. (2008). *J. Am. Chem. Soc.* 130: 17270–17271.

30 Gilmore, C.D., Allan, K.M., and Stoltz, B.M. (2008). *J. Am. Chem. Soc.* 130: 1558–1559.

31 (a) Guo, J., Kiran, I.N.C., Gao, J. et al. (2016). *Tetrahedron Lett.* 57: 3481–3484. (b) Banne, S., Reddy, D.P., Li, W. et al. (2017). *Org. Lett.* 19: 4996–4999.

32 Huters, A.D., Quasdorf, K.W., Styduhar, E.D., and Garg, N.K. (2011). *J. Am. Chem. Soc.* 133: 15797–15799.

33 Goetz, A.E., Silberstein, A.L., Corsello, M.A., and Garg, N.K. (2014). *J. Am. Chem. Soc.* 136: 3036–3039.

34 Caubère, P. (1967). *Bull. Soc. Chim. Fr.*: 3451–3457.

35 Guyot, M. and Molho, D. (1973). *Tetrahedron Lett.* 14: 3433–3436.

36 (a) Caubère, P., Derozier, N., and Loubinoux, B. (1971). *Bull. Soc. Chim. Fr.*: 302–307. (b) Caubère, P., Guillaumet, G., and Mourad, M.S. (1971). *Tetrahedron Lett.* 12: 4673–4676. (c) Danheiser, R.L. and Helgason, A.L. (1994). *J. Am. Chem. Soc.* 116: 9471–9479.

37 (a) Shair, M.D., Yoon, T.Y., and Danishefsky, S.J. (1995). *Angew. Chem. Int. Ed. Engl.* 34: 1721–1723. (b) Shair, M.D., Yoon, T.Y., Mosny, K.K. et al. (1996). *J. Am. Chem. Soc.* 118: 9509–9525.

38 Huang, J.-K., Lauderdale, T.-L.Y., Lin, C.-C., and Shia, K.-S. (2018). *J. Org. Chem.* 83: 6508–6523.

39 Bauta, W.E., Lovett, D.P., Cantrell, W.R. Jr., and Burke, B.D. (2003). *J. Org. Chem.* 68: 5967–5973.

40 (a) Kita, Y., Higuchi, K., Yoshida, Y. et al. (1999). *Angew. Chem. Int. Ed.* 38: 683–686. (b) Kita, Y., Higuchi, K., Yoshida, Y. et al. (2001). *J. Am. Chem. Soc.* 123: 3214–3222.

41 Tadross, P.M., Virgil, S.C., and Stoltz, B.M. (2010). *Org. Lett.* 12: 1612–1614.

42 Kobayashi, H., Sasano, Y., Kanoh, N. et al. (2016). *Eur. J. Org. Chem.*: 270–273.

43 Kou, K.G.M., Pflueger, J.J., Kiho, T. et al. (2018). *J. Am. Chem. Soc.* 140: 8105–8109.

44 (a) Tambar, U.K. and Stoltz, B.M. (2005). *J. Am. Chem. Soc.* 127: 5340–5341. (b) Yoshida, H., Watanabe, M., Ohshita, J., and Kunai, A. (2005). *Chem. Commun.*: 3292–3294. (c) Ebner, D.C., Tambar, U.K., and Stoltz, B.M. (2009). *Org. Synth.* 86: 161–171.

45 Bartholomäus, R., Bachmann, J., Mang, C. et al. (2013). *Eur. J. Org. Chem.*: 180–190.

46 Moschitto, M.J., Anthony, D.R., and Lewis, C.A. (2015). *J. Org. Chem.* 80: 3339–3342.

47 (a) Huisgen, R. and Knorr, R. (1963). *Tetrahedron Lett.* 4: 1017–1021. (b) Wittig, G. and Dürr, H. (1964). *Justus Liebigs Ann. Chem.* 672: 55–62. (c) Crews, P. and Beard, J. (1973). *J. Org. Chem.* 38: 522–528. (d) Waali, E.E. (1975). *J. Org. Chem.* 40: 1355–1356.

48 Townsend, C.A., Davis, S.G., Christensen, S.B. et al. (1981). *J. Am. Chem. Soc.* 103: 6885–6888.

49 Gribble, G.W., Saulnier, M.G., Sibi, M.P., and Obaza-Nutaitis, J.A. (1984). *J. Org. Chem.* 49: 4518–4523.

50 Khanapure, S.P. and Biehl, E.R. (1989). *J. Nat. Prod.* 52: 1357–1359.

51 Best, W.M. and Wege, D. (1981). *Tetrahedron Lett.* 22: 4877–4880.

52 (a) Matsumoto, T., Hosoya, T., and Suzuki, K. (1992). *J. Am. Chem. Soc.* 114: 3568–3570. (b) Hosoya, T., Takashiro, E., Matsumoto, T., and Suzuki, K. (1994). *J. Am. Chem. Soc.* 116: 1004–1015.

53 For a recent review on total syntheses of aryl *C*-glycoside natural products, see:Kitamura, K., Ando, Y., Matsumoto, T., and Suzuki, K. (2018). *Chem. Rev.* 118: 1495–1598.

54 (a) Matsumoto, T., Sohma, T., Yamaguchi, H. et al. (1995). *Synlett*: 263–266. (b) Matsumoto, T., Sohma, T., Yamaguchi, H. et al. (1995). *Tetrahedron* 51: 7347–7360.

55 Sumida, Y., Harada, R., Kato-Sumida, T. et al. (2014). *Org. Lett.* 16: 6240–6243.

56 Sumida, Y., Kato, T., and Hosoya, T. (2013). *Org. Lett.* 15: 2806–2809.

57 Neumeyer, M. and Brückner, R. (2017). *Eur. J. Org. Chem.*: 2512–2539.

58 Neumeyer, M., Kopp, J., and Brückner, R. (2017). *Eur. J. Org. Chem.*: 2883–2915.

59 Feng, Y., Liu, J., Carrasco, Y.P. et al. (2016). *J. Am. Chem. Soc.* 138: 7130–7142.

60 (a) Nakata, M., Kinoshita, M., Ohba, S., and Saito, Y. (1984). *Tetrahedron Lett.* 25: 1373–1376. (b) Nakata, M., Wada, S., Tatsuta, K., and Kinoshita, M. (1985). *Bull. Chem. Soc. Jpn.* 58: 1801–1806.

61 (a) Ando, Y., Matsumoto, T., and Suzuki, K. (2017). *Synlett* 28: 1040–1045. (b) Ando, Y., Hanaki, A., Sasaki, R. et al. (2017). *Angew. Chem. Int. Ed.* 56: 11460–11465.

62 Odagi, M., Furukori, K., Takayama, K. et al. (2017). *Angew. Chem. Int. Ed.* 56: 6609–6612.

63 (a) O'Keefe, B.M., Mans, D.M., Kaelin, D.E. Jr., and Martin, S.F. (2010). *J. Am. Chem. Soc.* 132: 15528–15530. (b) O'Keefe, B.M., Mans, D.M., Kaelin, D.E. Jr., and Martin, S.F. (2011). *Tetrahedron* 67: 6524–6538.

64 Kaelin, D.E. Jr., Sparks, S.M., Plake, H.R., and Martin, S.F. (2003). *J. Am. Chem. Soc.* 125: 12994–12995.

65 McManus, H.A., Fleming, M.J., and Lautens, M. (2007). *Angew. Chem. Int. Ed.* 46: 433–436.

66 Castillo, J.-C., Quiroga, J., Abonia, R. et al. (2015). *Org. Lett.* 17: 3374–3377.

67 May, C. and Moody, C.J. (1984). *J. Chem. Soc., Chem. Commun.*: 926–927.

68 Kim, J.-Y., Kim, D.-H., Jeon, T.-H. et al. (2017). *Org. Lett.* 19: 4688–4691.

69 (a) Chandrasoma, N., Brown, N., Brassfield, A. et al. (2013). *Tetrahedron Lett.* 54: 913–917; For related studies of the [4+2] cycloaddition of indolyne and furan, see:(b)Buszek, K.R., Luo, D., Kondrashov, M. et al. (2007). *Org. Lett.* 9: 4135–4137; (c) Buszek, K.R., Brown, N., and Luo, D. (2009). *Org. Lett.* 11: 201–204. (d) Chandrasoma, N., Pathmanathan, S., and Buszek, K.R. (2015). *Tetrahedron Lett.* 56: 3507–3510.

70 Buszek, K.R. and Bixby, D.L. (1995). *Tetrahedron Lett.* 36: 9129–9132.

71 Nishii, A., Takikawa, H., and Suzuki, K. (2019). *Chem. Sci.* 10: 3840–3845.

72 (a) Castedo, L., Guitián, E., Saá, J.M., and Suau, R. (1982). *Tetrahedron Lett.* 23: 457–458. (b) Atanes, N., Castedo, L., Guitián, E. et al. (1991). *J. Org. Chem.* 56:

2984–2988. (c) Saá, C., Guitián, E., Castedo, L., and Saá, J.M. (1985). *Tetrahedron Lett.* 26: 4559–4560.
73 (a) Perecim, G.P., Rodrigues, A., and Raminelli, C. (2015). *Tetrahedron Lett.* 56: 6848–6851; (b) Rossini, A.F.C., Muraca, A.C.A., Casagrande, G.A., and Raminelli, C. (2015). *J. Org. Chem.* 80: 10033–10040.
74 Reddy, M.C. and Jeganmohan, M. (2017). *Chem. Sci.* 8: 4130–4135.
75 (a) Klundt, I.L. (1970). *Chem. Rev.* 70: 471–487. (b) Oppolzer, W. (1978). *Synthesis*: 793–802. (c) Thumme, R.P. (1980). *Acc. Chem. Res.* 13: 70–76. (d) Mehta, G. and Kotha, S. (2001). *Tetrahedron* 57: 625–659.
76 (a) Pellissier, H. and Santelli, M. (1996). *Tetrahedron* 52: 9093–9100. (b) Ohmori, K., Mori, K., Ishikawa, Y. et al. (2004). *Angew. Chem. Int. Ed.* 43: 3167–3171. (c) Takemura, I., Imura, K., Matsumoto, T., and Suzuki, K. (2004). *Org. Lett.* 6: 2503–2505. (d) Mori, K., Tanaka, Y., Ohmori, K., and Suzuki, K. (2008). *Chem. Lett.* 37: 470–471. (e) Mori, K., Ohmori, K., and Suzuki, K. (2009). *Angew. Chem. Int. Ed.* 48: 5633–5637. (f) Ben, A., Hsu, D.-S., Matsumoto, T., and Suzuki, K. (2011). *Tetrahedron* 67: 6460–6468. (g) Takahashi, N., Kanayama, T., Okuyama, K. et al. (2011). *Chem. Asian J.* 6: 1752–1756. (h) Yamaguchi, S., Takahashi, N., Yuyama, D. et al. (2016). *Synlett* 27: 1262–1268. (i) Deng, L., Chen, M., and Dong, G. (2018). *J. Am. Chem. Soc.* 140: 9652–9658.
77 (a) Stevens, R.V. and Bisacchi, G.S. (1982). *J. Org. Chem.* 47: 2393–2396. (b) Stevens, R.V. and Bisacchi, G.S. (1982). *J. Org. Chem.* 47: 2396–2399.
78 Winters, M.P., Stranberg, M., and Moore, H.W. (1994). *J. Org. Chem.* 59: 7572–7574.
79 (a) Matsumoto, T., Yamaguchi, H., Hamura, T. et al. (2000). *Tetrahedron Lett.* 41: 8383–8387. (b) Yamaguchi, H., Konegawa, T., Tanabe, M. et al. (2000). *Tetrahedron Lett.* 41: 8389–8392. (c) Matsumoto, T., Yamaguchi, H., Tanabe, M. et al. (2000). *Tetrahedron Lett.* 41: 8393–8396.
80 Fukui, N., Ohmori, K., and Suzuki, K. (2012). *Helv. Chim. Acta* 95: 2194–2217.
81 (a) Chen, P.-H., Savage, N.A., and Dong, G. (2014). *Tetrahedron* 70: 4135–4146. (b) Xu, T. and Dong, G. (2014). *Angew. Chem. Int. Ed.* 53: 10733–10736.
82 Sato, S., Sakata, K., Hashimoto, Y. et al. (2017). *Angew. Chem. Int. Ed.* 56: 12608–12613.
83 South, M.S. and Liebeskind, L.S. (1982). *J. Org. Chem.* 47: 3815–3821.
84 (a) Hosoya, T., Hasegawa, T., Kuriyama, Y. et al. (1995). *Synlett*: 177–179. (b) Hosoya, T., Hasegawa, T., Kuriyama, Y., and Suzuki, K. (1995). *Tetrahedron Lett.* 36: 3377–3380. (c) Hosoya, T., Hamura, T., Kuriyama, Y. et al. (2000). *Synlett*: 520–522.
85 Fleming, I. and Mah, T. (1975). *J. Chem. Soc., Perkin Trans. 1*: 964–965.
86 Maturi, M.M., Ohmori, K., and Suzuki, K. (2018). *Chimia* 72: 870–873.
87 Choy, W. and Yang, H. (1988). *J. Org. Chem.* 53: 5796.
88 Ma, Z.-X., Feltenberger, J.B., and Hsung, R.P. (2012). *Org. Lett.* 14: 2742–2745.
89 Corsello, M.A., Kim, J., and Garg, N.K. (2017). *Nat. Chem.* 9: 944–949.
90 Imaizumi, T., Yamashita, Y., Nakazawa, Y. et al. (2019). *Org. Lett.* 21: 6185–6189.

91 Okano, K., Tokuyama, H., and Fukuyama, T. (2006). *J. Am. Chem. Soc.* 128: 7136–7137.

92 (a) Candito, D.A., Panteleev, J., and Lautens, M. (2011). *J. Am. Chem. Soc.* 133: 14200–14203. (b) Candito, D.A., Dobrovolsky, D., and Lautens, M. (2012). *J. Am. Chem. Soc.* 134: 15572–15580.

93 (a) Shi, J., Qiu, D., Wang, J. et al. (2015). *J. Am. Chem. Soc.* 137: 5670–5673. (b) Xu, H., He, J., Shi, J. et al. (2018). *J. Am. Chem. Soc.* 140: 3555–3559.

94 Wittig, G. and Härle, H. (1959). *Liebigs Ann. Chem.* 623: 17–34.

95 (a) Morton, G.E. and Barrett, A.G.M. (2005). *J. Org. Chem.* 70: 3525–3529. (b) Morton, G.E. and Barrett, A.G.M. (2006). *Org. Lett.* 8: 2859–2861.

96 (a) Matsumoto, T., Katsuki, M., and Suzuki, K. (1988). *Tetrahedron Lett.* 29: 6935–6938. (b) Ben, A., Yamauchi, T., Matsumoto, T., and Suzuki, K. (2004). *Synlett*: 225–230.

97 (a) Chen, C.-L., Sparks, S.M., and Martin, S.F. (2006). *J. Am. Chem. Soc.* 128: 13696–13697. (b) Sparks, S.M., Chen, C.-L., and Martin, S.F. (2007). *Tetrahedron* 63: 8619–8635.

98 Takikawa, H., Sato, S., Seki, R., and Suzuki, K. (2017). *Chem. Lett.* 46: 998–1000.

99 Ninomiya, M., Ando, Y., Kudo, F. et al. (2019). *Angew. Chem. Int. Ed.* 58: 4264–4270.

100 (a) Du, C.J.F., Hart, H., and Ng, K.K.D. (1986). *J. Org. Chem.* 51: 3162–3165. (b) Du, C.J.F. and Hart, H. (1987). *J. Org. Chem.* 52: 4311–4314.

101 Sato, Y., Tamura, T., and Mori, M. (2004). *Angew. Chem. Int. Ed.* 43: 2436–2440.

102 Patel, R.M. and Argade, N.P. (2013). *Org. Lett.* 15: 14–17.

103 (a) Mohanakrishnan, A.K. and Srinivasan, P.C. (1995). *J. Org. Chem.* 60: 1939–1946. (b) Kleppinger, R., Lillya, C.P., and Yang, C. (1997). *J. Am. Chem. Soc.* 119: 4097–4102. (c) Padwa, A., Dimitroff, M., Waterson, A.G., and Wu, T. (1998). *J. Org. Chem.* 63: 3986–3997. (d) Furuta, T., Kitamura, Y., Hashimoto, A. et al. (2007). *Org. Lett.* 9: 183–186. (e) Bernardo, P.H., Fitriyanto, W., and Chai, C.L.L. (2007). *Synthesis*: 1935–1939. (f) Rajeshkumar, V., Lee, T.-H., and Chuang, S.-C. (2013). *Org. Lett.* 15: 1468–1471.

104 Lu, C., Dubrovskly, A.V., and Larock, R.C. (2012). *J. Org. Chem.* 77: 8648–8656.

105 Pimparkar, S. and Jeganmohan, M. (2014). *Chem. Commun.* 50: 12116–12119.

106 Feng, M., Tang, B., Wang, N. et al. (2015). *Angew. Chem. Int. Ed.* 54: 14960–14964.

107 Feng, M., Tang, B., Xu, H.-X., and Jiang, X. (2016). *Org. Lett.* 18: 4352–4355.

108 Zhang, T.-Y., Liu, C., Chen, C. et al. (2018). *Org. Lett.* 20: 220–223.

109 (a) Worlikar, S.A. and Larock, R.C. (2009). *Org. Lett.* 11: 2413–2416. (b) Worlikar, S.A. and Larock, R.C. (2009). *J. Org. Chem.* 74: 9132–9139.

110 Yao, T., Zhang, H., and Zhao, Y. (2016). *Org. Lett.* 18: 2532–2535.

111 Neog, K., Borah, A., and Gogoi, P. (2016). *J. Org. Chem.* 81: 11971–11977.

112 Wang, T. and Hoye, T.R. (2016). *J. Am. Chem. Soc.* 138: 13870–13873.

113 (a) Miyawaki, K., Suzuki, R., Kawano, T., and Ueda, I. (1997). *Tetrahedron Lett.* 38: 3943–3946. (b) Bradley, A.Z. and Johnson, R.P. (1997). *J. Am. Chem. Soc.*

119: 9917–9918; For a review, see: (c) Holden, C. and Greaney, M.F. (2014). *Angew. Chem. Int. Ed.* 53: 5746–5749.

114 (a) Hoye, T.R., Baire, B., Niu, D. et al. (2012). *Nature* 490: 208–211. (b) Hoffmann, R.W. and Suzuki, K. (2013). *Angew. Chem. Int. Ed.* 52: 2655–2656.

115 Yun, S.Y., Wang, K.-P., Lee, N.-K. et al. (2013). *J. Am. Chem. Soc.* 135: 4668–4671.

116 Karmakar, R. and Lee, D. (2016). *Org. Lett.* 18: 6105–6107.

117 Niu, D., Willoughby, P.H., Woods, B.P. et al. (2013). *Nature* 501: 531–534.

118 Shankar, B. and Baire, B. (2017). *Org. Biomol. Chem.* 15: 5908–5911.

Index

Modern Aryne Chemistry, First Edition. Edited by Akkattu T. Biju.

i

O

t

u

v